The Freshwater Mussels of Ohio

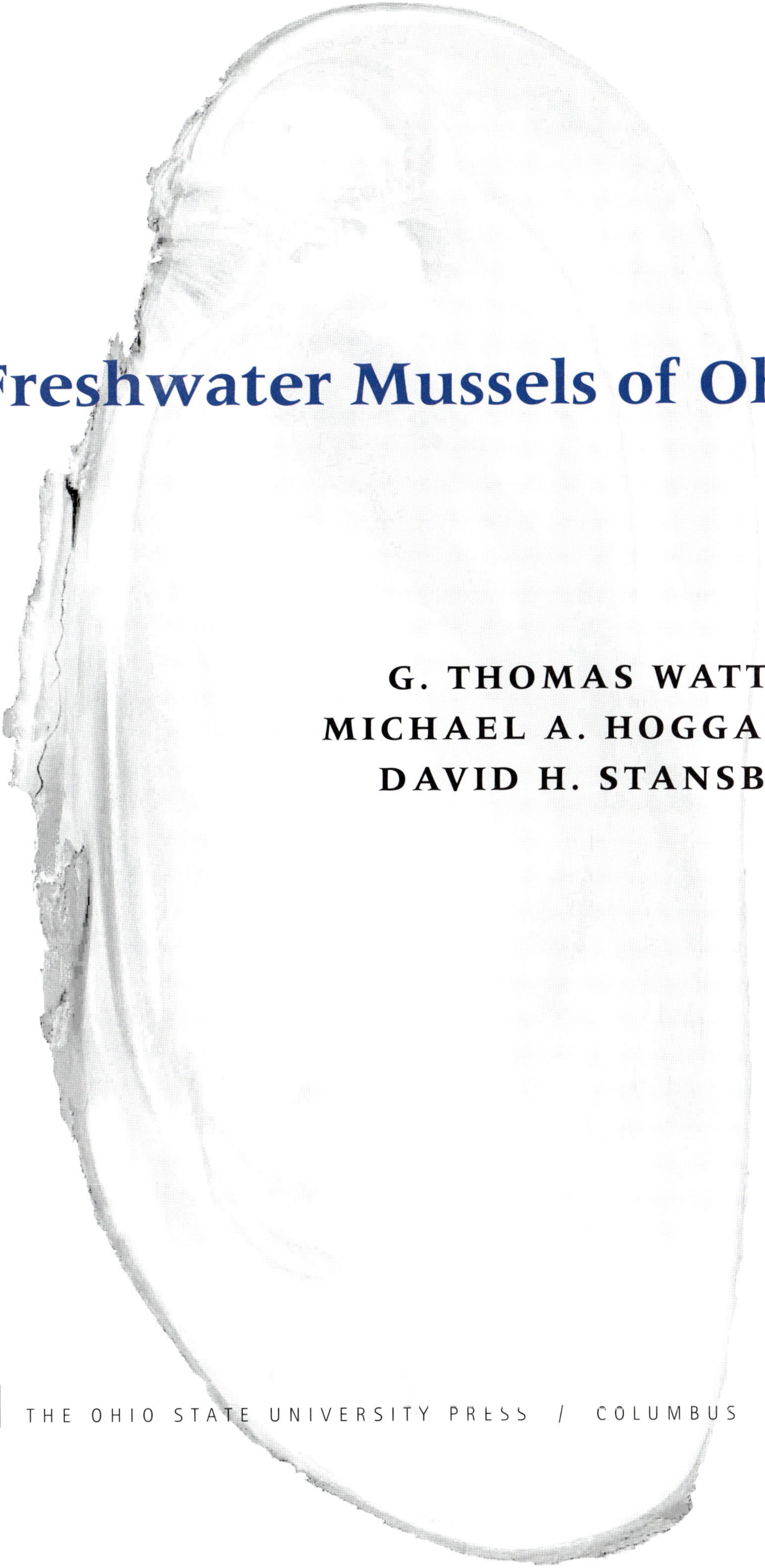

The Freshwater Mussels of Ohio

G. THOMAS WATTERS
MICHAEL A. HOGGARTH
DAVID H. STANSBERY

THE OHIO STATE UNIVERSITY PRESS / COLUMBUS

Library of Congress Cataloging-in-Publication Data
Watters, G. Thomas.
The freshwater mussels of Ohio / G. Thomas Watters, Michael A. Hoggarth, [and] David H. Stansbery.
p. cm.
Includes bibliographical references and index.
ISBN 978-0-8142-1105-2 (cloth : alk. paper)
1. Freshwater mussels—Ohio—Identification. 2. Margaritiferidae—Ohio—Identification. 3. Unionidae—Ohio—Identification. I. Hoggarth, Michael A. (Michael Alan), 1955– II. Stansbery, David H. (David Honor), 1926– III. Title.
QL430.6.W38 2009
594.'409771—dc22
2009021038

Cover design by Janna Thompson-Chordas
Text design by Juliet Williams
Type set in Adobe Meridien
Printed by Sheridan Books, Inc.

♾ The paper used in this publication meets the minimum requirements of the American National Standard for Information Sciences—Permanence of Paper for Printed Library Materials. ANSI Z39.48–1992.

9 8 7 6 5 4 3 2 1

The true mystery of the world is the visible, not the invisible.
—Oscar Wilde

Contents

Preface

The reasons for gathering together the material for a book of this kind are found in human needs. These needs may be current or anticipated, or they may rest in the conviction that all living organisms, including the human species, have relationships with and effects upon each other, regardless of how small or how large, how obvious or how obscure, and that these relationships are well worth knowing for reasons both pure and applied. We have, looking at our history, excellent reason to believe that more benefits to humankind remain to be discovered than we have anticipated in our wildest speculations. It is apparent that we have a growing need for knowledge and for the efficiencies necessary for its ease of ready access. This volume is one such accumulation of information and is arranged so as to be readily available for use.

There are those among us who have the conviction that all species of life have not only the right to exist but also the privilege to be understood and valued by humankind. Our subject is the unionoid molluscs of our freshwater habitats, especially streams. Some species may be found in ponds, lakes, and impoundments, but by far the highest diversity of this life-form is in our flowing freshwater systems. While some future uses of these bivalves must necessarily remain speculative, an examination of our history (literature) reveals a number of ways these molluscs have been used in times past.

Native Americans found these animals not only easy to gather but also quite edible, with shells that could be fashioned into tools of several sorts. The occasional natural pearl was treasured as much then as in more recent times. Pearl buttons fashioned from discs cut from mother-of-pearl shell seem not to have appeared in the Ohio country until introduced about 1880 by immigrants along the Ohio River. By the early 1900's, this art became an industry, and shortly the most common source of button material was mother-of-pearl from freshwater shell. Those collecting the raw material for the button factories were careful to search out the valuable natural pearls, which were then sold to itinerant buyers who resold them in Eastern and foreign markets for impressive sums.

The development of a technology to produce molded-glass and plastic buttons led to the replacement of pearl for all but the most expensive garments. The production of cultured pearls using pearl spheres brought strands of these near-natural gems within the price range of those other than the very wealthy.

And so since before the time of recorded history, these larger riverine bivalves, commonly called clams or mussels, less commonly naiads or unionids, have been used by humans in various ways. Over the same time span, the human population has been becoming ever more numerous and through its carelessness has been instrumental in the extirpation of many populations and the recent extinction of too many species. These extinctions and extirpations have come to pass largely through habitat degradation, although over-harvesting in some streams may have played a role. Major factors responsible for this relatively rapid decline have been the construction of impoundments, the dredging of streams for navigation and flood control, the addition to our streams of the sediments of agricultural and construction activities, and the addition of a wide spectrum of solid, semisolid, and liquid waste materials from the industries of a rapidly growing human population.

More recent observations by field malacologists have revealed that these animals are, with rare exception, sensitive to many of these pollutants. They are considered, in view of their rapidly decreasing numbers, to be the most endangered group of organisms in North America by the U.S. Fish and Wildlife Service. This very sensitivity may well lead to what could be their most valuable service yet to human society: they may indeed become our most efficient and effective indicators of pollution of riverine habitats that have been damaged in hazardous ways not visible to casual examination. Even now mussels are being used as the "miners' canaries" of our waterways.

Ohio's streams are a valuable resource that we must protect and restore. As we move in the direction of wiser use of these resources, we should not forget the importance of protecting and restoring the faunas that have made these resources so rich. Just as many species of unionids have been eliminated from streams throughout the state, or their populations have been reduced, so too we must begin to work toward their restoration. Some species, given their remarkable means of distribution (as larvae on a more motile host), will become restored naturally because they will have access once again to improved streams. Others will need our assistance.

It would seem that the need for a book such as this one is eminently due, if not overdue. This volume identifies the various species, describes their geographic distribution, and relates the history these animals have had in the state. The authors sincerely hope and trust that the volume will help us focus our conservation efforts into the future.

— David H. Stansbery

Acknowledgments

We wish to thank Dr. Arthur Bogan, North Carolina Museum of Natural History, Raleigh, NC, for discussions of unionacean nomenclature; Kevin Cumming, Illinois Natural History Survey, Champaign, IL, for information on type depositories; and Richard Petit, South Myrtle Beach, SC, and Dr. Henk K. Mienis, Curator, National Mollusc Collection, Hebrew University of Jerusalem, Israel, for information on obscure literature.

The authors wish to particularly thank Mr. Randy Sanders, Mr. Joe Mion, and Mr. John Navarro of the Ohio Department of Natural Resources Division of Wildlife for keeping the project not only on track but also alive. The Ohio Division of Wildlife contributed all funds toward this book, and clearly it would not exist in any form without their support.

Dr. Watters thanks his immediate and extended family for support, badgering, and the occasional kick in the seat to complete this work. I thank my parents and my wife for indulging me in such an odd pursuit, and my grandfather for introducing me to the natural history of Ohio. I hope that mussels will still be around for my grandchildren (or even just my children!) to enjoy.

Dr. Hoggarth wishes to thank his parents, Robert Hoggarth and Helen Hoggarth. You helped me to see beauty and wonder in nature and encouraged the pursuit of my dream of becoming a biologist. My teachers helped me to put that dream into action. Four teachers I would especially like to acknowledge are Mr. Charles Randall, Upper Arlington High School; Dr. Ross F. Shaw, Seattle Pacific University; Dr. Paul Yokley, Jr., the University of North Alabama; and Dr. David H. Stansbery, The Ohio State University. The knowledge and enthusiasm each of you shared with me still seems unlimited. You have helped me fashion my own teaching and learning after your example. Many of my students at Otterbein College and Franz Theodore Stone Laboratory are to some small degree a product of your teaching. Students of mine who have made a special contribution to the knowledge of the mussels of Ohio are Diana (Lee) Dickson, Tara (Chinn) Grove, Jim Gates, Melissa Haltuch, Bill Rodman, Scott Ross, Erin (Stewart) Miller, Megan (Hysell) Michael, Leah Gillig, and Wayne Rossiter. Daniel L. Rice accompanied my students and me on many collecting trips, and his help and friendship are gratefully acknowledged. I am most grateful to my wife, Karen Cook-Hoggarth, and my sons, Mark Hoggarth and Kyle Hoggarth. You have helped me collect and process shells, and you have shared my excitement with each new species and each new discovery. Your continued support and encouragement made this volume possible.

Ms. Carol Myers Flaute prepared the original maps and labored long and hard to prepare the supporting ArcView files. Ms. Sarah Hazzard and Mr. Kody Kuehnl were invaluable in proofreading the final version and assisting with last-minute details; Mr. Kuehnl also was instrumental in filling in the sections on genetics and phylogeny. Dr. Robert Warren, Illinois State Museum, provided invaluable information on the use of mussels by prehistoric man. Mr. Fred Obhof organized the final specimens for photography. Photography for this book was the tortuous, accumulated efforts of Ms. Janna Thompson-Chordas, Mr. Tim Daniel, and G. T. Watters. Ms. Cynthia Bishop, then at the Ohio Division of Wildlife, made the pen-and-ink drawings in the Glossary.

Although the great majority of records used in this book were deposited at The Ohio State University Museum of Biological Diversity, we endeavored to include as many additional museums as possible to fill in gaps. To this end we are indebted to Dr. Tim Pearce, Carnegie Museum of Natural History; Dr. Diarmaid Ó Foighil and Dr. Jack Burch, University of Michigan Museum of Zoology; Dr. Robert Kennedy and Ms. Paula Work, Museum of Natural History and Science, Cincinnati; Dr. James Bissell, Cleveland Museum of Natural History; and Gary Coovert, Boonshoft Museum of Discovery, Dayton.

We thank Dr. George Davis for permission to use previously published photos from Hoggarth, *Malacologia* (1999), and Roland Houart for permission to use previously published photos from Watters, *Novapex* (2008). And without the gracious attention to detail of Jennifer Shoffey Forsythe, Juliet Williams, and Maggie Diehl, The Ohio State University Press, this book would never have become a reality.

Finally, we thank everyone who has ever deposited a mussel shell into the The Ohio State University Museum of Biological Diversity or has surveyed Ohio's streams—a list of people numbering in the thousands. A book like this is the result of so many individual efforts by so many people that it would be impossible to thank everyone. So to anyone who has deposited a shell in the collection of the museum or has collected a shell in the state and deposited it in other collections, we say thank you. Many of the dots on the maps in this book are yours.

Abbreviations Used in Species Accounts

ANSP	Academy of Natural Sciences, Philadelphia, PA
BMNH	Natural History Museum, London
CMNH	Carnegie Museum of Natural History, Pittsburgh, PA
MAH	Collection of Michael A. Hoggarth
ICZN	International Code of Zoological Nomenclature
MCZ	Museum of Comparative Zoology, Harvard University, Cambridge, MA
MNHN	Muséum national d'Histoire naturelle, Paris, France
OSUM	The Ohio State University Museum of Biological Diversity, Columbus, OH
SMF	Forschungsinstitut und Naturmuseum Senckenberg, Frankfurt am Main, Germany
UMMZ	University of Michigan Museum of Zoology, Ann Arbor, MI
USNM	Smithsonian National Museum of Natural History, Washington, DC

Introduction

Ohio has an especially rich heritage of freshwater mussels, from both a biological and a historical perspective. Eighty species have been reported from the state. This number is 27% of all mussel species known to be from North America. More than 60 mussel taxa were described from specimens collected from Ohio, although many of them are now recognized as synonyms of other species.

This bounty of Ohio taxa is a result of the efforts of some of Ohio's early naturalists. Isaac Lea (1792–1886), the undaunted Philadelphia describer of more than 800 mussel species, was sent specimens from his brother Thomas, a resident of the Cincinnati area. Their efforts alone accounted for numerous Ohio type localities. Samuel Hildreth (1783–1863), a Marietta physician, collected in the mussel-rich Muskingum River, documenting species that sadly no longer occur there today. Benjamin Tappan (1773–1857), Ohio senator, co-founder of the Smithsonian Institution, and founder of Ravenna, Ohio, was a patron of natural history and aided some of the best malacologists of his day, including Isaac Lea and Charles Adams. Jared Kirtland (1793–1877), legislator, physician, and naturalist, collected in eastern Ohio, while Victor Sterki (1846–1933), physician and malacologist, and Arnold Ortmann (1863–1927), Curator of Mollusks at the Carnegie Museum of Natural History in Pittsburgh, spent a great deal of time collecting mussels from the northeastern portions of the state. In particular, their collections from the Tuscarawas River at New Philadelphia represent one of the most comprehensive pictures of what occurred in a single Ohio stream prior to the mass extirpations that were soon to follow.

Today, Ohio is one of the best-surveyed states, thanks to the work of many dedicated collectors, the support of state and federal agencies, environmental regulations, the interest of conservation groups, and curious citizens. However, there are still discoveries to be made, as evidenced by the remarkable finding of the Purple Catspaw, *Epioblasma obliquata obliquata,* in Killbuck Creek (Hoggarth *et al.*, 1995). It is because of the place Ohio has in the study of mussels and the interest that so many Ohioans have shown and continue to show in these animals—and because we want to see that interest continue—that we have written this book.

The Higher Classification of North American Freshwater Mussels

The recognition of higher taxa in North American mussels has been at best contentious. What constitutes a family or subfamily rank characteristic? To date, we still do not have a set of unifying characteristics that allows us to unite taxa at the family or subfamily level. Historically these differences were first based on shell characteristics, then anatomical characteristics, and most recently molecular DNA differences. Unfortunately, there has been a tendency to use one, and only one, of these types of data at a time while ignoring the larger picture. For example, no study has even considered the extensive fossil record of this group.

Linnaeus (1758) originally considered mussels as members of *Mya,* now reduced to a very small marine genus. The first division was made by Lamarck (1799) based solely on the presence or absence of hinge teeth—those without teeth were placed in *Anodonta.* Philipsson (1788) created *Unio* for those with teeth, thus removing all freshwater mussels from *Mya.* Lea added a further refinement by recognizing those with a wing as *Symphynota.* Further splitting based on shell characteristics continued with abandon.

Ortmann (1919) recognized the importance of anatomy and organized mussels into the Unionidae and Margaritiferidae based on gill morphology. His Unionidae was further divided into the Unioninae, Anodontinae, and Lampsilinae, again based on gill morphology, particularly the structure of the marsupium. Heard & Guckert (1971) further refined the divisions but proposed three separate families: Margaritiferidae, Amblemidae, and Unionidae. However, few subsequent workers have used this scheme. Davis & Fuller (1981) reexamined the anatomical evidence as well as new immunoelectrophoretic data and essentially agreed with Ortmann except for making Margaritiferidae a subfamily of the Unionidae. Smith & Wall (1984) reinstated this group as a family. Hoeh *et al.* (1998), based on Cytochrome C Oxidase Subunit I DNA (COI) sequences, found evidence for separation of the Margaritiferidae from the Ambleminae and Anodontinae—suggesting either three families or three subfamilies.

Lydeard *et al.* (1996) used 16S rRNA DNA sequences and morphological data to construct a phylogenetic tree of unionoid relationships. This approach, the most comprehensive to date, maintained the separation of the Unionidae and Margaritiferidae. In their study, Lydeard *et al.* (1996) identified two subfamilies within the Unionidae: the Anodontinae and Ambleminae. Within the Ambleminae groups representing *Quadrula, Megalonaias, Elliptio,* and *Amblema* showed support, as did the Lampsilinae of other studies.

Graf & Ó Foighl (2000) performed a phylogenetic analysis using brooding and COI characteristics. Their results supported separation of Margaritiferidae, Unioninae, Anodontinae, Pleurobeminae, and Lampsilinae. *Amblema* was associated not with the pleurobemines, its usual place, but with the lampsilines.

The study of Graf (2000), using larval and adult characteristics, suggested that the Unionidae are paraphyletic and that all groups except for the Unionidae and Margaritiferidae represent a separate clade, the Etherioidea. We would like to see this controversial theory tested using molecular DNA data.

While the phylogeny of a group of organisms can never truly be known, our current level of understanding of the relationships between and among freshwater mussels, particularly those relationships dealing with the Unionidae, is elementary at best. The relationships that have been identified have not been thoroughly tested using multiple independent data sets and therefore may be inaccurate. Until such studies are undertaken, our current understanding of these relationships will be limited. However, based on the information currently available for unionoids, the consensus is that the Unionidae and Margaritiferidae are separate entities. Several DNA studies also suggest that the Anodontinae may be at least as distinct from other Unionidae as the Margaritiferidae (e.g., Hoeh *et al.,* 1998). There also appears to be support for a lampsiline and a pleurobemine/amblemine group. For the most current review of historical relationships among unionoids see Roe & Hoeh, 2003.

Recognition of lower level taxa including species and genera among freshwater mussels (Margaritiferidae and Unionidae) has been no less contentious among malacologists. As with higher level taxa, species and genera are delineated principally by shell characteristics, even though it has been extremely difficult to delineate mussel species based on these characteristics alone. This undoubtedly has led to the description and redescription of numerous taxa of freshwater mussels, while in the process complicating our ability to recognize accurate species, which in turn has hindered our conservation efforts in this imperiled group. However, the advent of molecular methods, particularly DNA sequencing, has provided a new way to assess species boundaries as well as address hypotheses of relationships between and among different mussel taxa. These studies have proved invaluable in identifying unique or cryptic species that require systematic changes as well as those requiring various conservation activities. Such studies have also helped identify morphological characters that set apart monophyletic groups of mussels from similar species (e.g., genus *Hamiota;* Roe *et al.,* 2001) or have shown that certain morphological characters which we thought were good characters for uniting certain taxa actually may not be (e.g., genus *Potamilus;* Roe & Lydeard, 1998).

However, the majority of the studies utilizing molecular data to test current species and generic appellations have showed para- and polyphyletic relationships. To date, numerous molecular studies have provided important insights for the delineation of species as well as the relationships between species of freshwater mussels. Lydeard *et al.* (2000) used mtDNA sequences to show that several of the genera endemic to the Gulf Coast region were polyphyletic and not representative of natural groups. Other studies that found para- or polyphyletic genera include the genus *Quadrula* (Serb *et al.,* 2003) and *Lampsilis, Obovaria, Fusconaia, Pleurobema,* and *Quincuncina,* among others (Campbell *et al.,* 2005).

Still other studies utilizing molecular data have found that while our current understanding of what constitutes a single species may be accurate (i.e., molecular analyses identify all individuals sampled to comprise a monophyletic group), many of the populations that make up that species can and often do contain unique genetic variation (e.g., Mulvey *et al.,* 1996; Jones *et al.,* 2006; Serb, 2006; Burdick & White, 2007; Elderkin *et al.,* 2007). In a few cases, morphological variability supporting unique species is not supported by molecular data (e.g., Mulvey *et al.,* 1996; Buhay *et al.,* 2002). Regardless the result, molecular data have proven invaluable in assisting researchers in generating a more realistic taxonomic scheme and assisting in generating meaningful and appropriate conservation activities (e.g., captive propagation, population augmentation, and reintroductions).

Nomenclatorial and Systematic Concepts

Every species that is described in this book has a unique, binomial scientific name. That name is composed of genus and species names (both in italics), the author of the description, and the year in which the name was first applied to that species. In some instances, the species name is followed by a subspecies name that is also printed in italics. The author and date are surrounded by parentheses if the species name has been moved from the genus in which it was first placed by the author of the species; author and date are not enclosed in parentheses if the species remains in its original genus. Genera are organized into families, families into orders, orders into classes, classes into phyla, and phyla into kingdoms. North American freshwater mussels are members of the Kingdom Animalia, Phylum Mollusca, Class Bivalvia, Order Unionoida, and Families Unionidae and Margaritiferidae.

Five families of freshwater bivalve molluscs occur in Ohio: Margaritiferidae, Unionidae, Sphaeriidae, Corbiculidae, and Dreissenidae. Only the first three are native members of the Ohio molluscan fauna. The Margaritiferidae and Unionidae are commonly known as the pearly freshwater mussels (the subject of this book) while the Sphaeriidae are the fingernail or pea clams. The other two families are represented by *Corbicula fluminea* (Müller, 1774) and *Dreissena polymorpha* (Pallas, 1771) and *Dreissena bugensis* Andrusov, 1897, all exotics.

The specimens from which a species or subspecies were described are referred to as type specimens (holotype if so designated by the author of the name at the time of description or by monotypy). Other type specimens commonly encountered in the literature include paratype (all other specimens in the type series other than the holotype), lectotype (designated by a subsequent author from the type series when no holotype has been designated), paralectotype (all other specimens in the type series other than the lectotype), and syntypes (for which neither a holotype nor a lectotype has been designated). Topotypes are specimens that are not included in the type series but have been collected from the type locality. The type locality is the site from which the nominal species or subspecies was collected and described. Cotype, a term no longer used by the ICZN, may mean either a primary type (usually a syntype) or simply a specimen, not of the type lot, later identified by the author as a specimen of that species. In the synonymies given here, cotypes are listed because future study may reveal them to be primary types. Allotypes refer to specimens of a gender opposite that of the holotype.

Each name listed in the Species Accounts is followed by a list of synonyms in chronological order. Species names are listed in the synonymy if the specimens (or description or illustration in some cases where the type specimens have been lost) are clearly identical to another named species and that other named species has priority. Priority is established by date of publication and availability of the name. Some earlier names located in the synonymy for each species are unavailable for a number of reasons, the most common being that the name had already been used for another species (whether a valid species or not). Each species account following will therefore include the genus name, the species name, and the subspecies name (if appropriate); the author and date of publication (in parentheses if the species is no longer considered to be in the originally described genus); a citation to the original description; the type locality (if known); the location of the type material (if known); and a list of synonyms. The list of synonyms includes only novel combinations; it does not include every citation of a species (which would have doubled the length of this book). Vernacular names are also given. Those recognized by the American Fisheries Society (Turgeon *et al.*, 1998) are shown first and in italics.

There is a tendency among today's biologists, including some systematists, to be overly critical of the early descriptive authors. Convoluted synonymies, lack of type material, and vague type localities all seem to be the result of carelessness on the part of these scientists. This is an unfair assessment. We must remember that the naturalists of the 1700's and 1800's often worked in a near-vacuum. With Europe in a nearly constant state of war or revolution, workers in one country were often unaware of what their counterparts in another country had published. As a result, the same species was often described at least twice. Today we are spoiled by the ease with which we can access information from any point on Earth from our desks. We tend to forget the parochial nature of early scientific journals and books and the difficulty that even a simple literature search must have presented in those days. The lack of types arose not from carelessness but from the fact that the "type concept" as used today simply did not exist in those times. Our nomenclatorial rules have slowly evolved, and it is unfair to hold a worker in the 1700's accountable for concepts that would not emerge for over one hundred years. These were years of great exploration, with scientists often relying on sailors or soldiers to bring back specimens from foreign lands. Given this source it is not surprising that some species were described with little or no locality data (or even with spurious data).

On the other hand, some early systematists clearly seemed to operate beyond the norm. Rafinesque is often vilified for his crude drawings and brief descriptions. Isaac Lea described species based on only the slightest of differences. He was an American unknowingly adherent to the contemporary French group of systematists known as the Nouvelle École (New School). Championed by Jules René Bourguignat (1828–92), they believed that if a mollusc specimen differed from any others by three characters or more, it was a different species. The characters could be quite minor, resulting in a "plethora of ill-conceived new species" (Dance, 1966). In more recent times De Gregorio (1914) has redescribed numerous Ohio species, apparently ignoring the work of previous workers.

Origin of the Ohio Fauna

Descendants of marine bivalves have invaded the freshwater environment several times during the past 400 million years. These were independent invasions, occurring at different times and comprising diverse groups. The result is a freshwater bivalve fauna composed of unrelated groups now found living side by side: zebra mussels, fingernail clams, freshwater mussels, Asian clams, marsh clams, etc. Several other freshwater groups, such as the Anthracosiacea and Archanodontacea, which flourished between 200 and 400 million years ago, resembled unionoids but seem to have been unrelated; these groups are now extinct.

Unionoid freshwater mussels, the subject of this book, may be traced back to at least the Triassic Period, 250 million years ago. At that time, all of the present-day continents were assembled into a single supercontinent, Pangaea. Freshwater mussel groups that today are found only on other continents apparently enjoyed a much wider distribution in Pangaea. Fossils very similar to those of the Hyriidae, a mussel group now occurring only in South America, Australia, and New Zealand, were found in the Triassic of Pennsylvania, Texas, and Arizona (Simpson, 1896; Richards, 1948). Mycetopodidae-like species, now found only in South America, were found in the Cretaceous of Colorado (White, 1878). An Iridinidae-like species, a group now found only in Africa, was found in the Cretaceous of Montana (Morris & Williamson, 1988). By the close of the Mesozoic Era, 65 million years ago, a diverse mussel fauna comparable to that found today occurred in North America. Ironically, this fauna is best known from western North America, an area that today is very depauperate in mussels. Climatic and orogenic changes drove most of these freshwater mussels to extinction. The extent of unionoid diversity in eastern North America is unknown due to the paucity of fossil material. The great radiation of species in the Cenozoic Era, culminating in our present fauna, may have been derived from the few survivors of the western fauna, migrations from Asia across the Bering land bridge and from Europe across Greenland, and whatever eastern fauna may have previously existed. The evolution of the North American unionoids is described in greater detail in Watters (2001).

No pre-Pleistocene mussel species are known from Ohio, but this does not necessarily mean that none occurred here. Ohio has no outcrops from the Mesozoic Era, that time during which fossil mussels reached their zenith. The Paleozoic Era outcrops that cover Ohio are too early for freshwater mussels to occur and, in any event, are predominately marine.

A series of glaciers scoured most of Ohio in the Pleistocene such that any earlier mussel faunas must have been eliminated. Most of Ohio's mussels are derived from unglaciated rivers to the south, including the Ohio and Wabash rivers. Relatively few Ohio mussels owe their origin to the Lawrentian system, which includes the Great Lakes. Indeed, most of the Lake Erie fauna is of Wabash River origin. The Maumee River, now a Lake Erie tributary, was once part of the Wabash River. As the last glacier retreated, the river level fell below a divide at Fort Wayne, Indiana, and the Maumee River reversed direction and flowed not to the Wabash River but to Lake Erie. The river took its mussels with it, so that today western Lake Erie is populated by Ohio River mussels, not Lawrentian mussels.

Human Utilization of Freshwater Mussels

From at least the Archaic Period (ca. 8000–1500 B.C.), Native Americans employed freshwater mussels for a variety of uses: ornamental, ceremonial, and utilitarian, and as a food item. Having an iridescent nacre, or mother-of-pearl lining, mussels were valued as jewelry. Shells were carefully formed into beads and discs and were drilled for stringing. Especially cherished were pearls, which were occasionally placed in ceremonial mounds. Kunz (1898) reported a Hopewell mound in Ross County having more than 100,000 pearls. As tools, mussel shells were used as knives, tweezers, hoes, scrapers, and bowls. Some shells used for tools appear to have been brought from over 100 miles away (Theler, 1991). Mussel shells also were charred and crushed and then added to pottery clay to increase durability.

Some Native Americans consumed great quantities of mussels as food, as evidenced by the extensive shell middens still found along some rivers. The Aztalan Site in Wisconsin (ca. A.D. 1000–1150) had a midden composed of 8,500 shells comprising 23 species (Theler, 1991). Middens along the Green, Cumberland, and Tennessee rivers may cover up to 6 hectares each (Stansbery, 1966b). Some middens show few modified shells, suggesting that the mussels were gathered solely for food, and not for tools or ornamentation (Warren, 1995). In many middens these shells are charred, indicating that they were cooked before being eaten.

The importance of mussels in Native American diets, both as a source of calories and as a favored food, is the subject of debate. Mussel meat is relatively nutrient-poor. Theler (1991) showed that a one-hour harvest in a good-quality stream would produce one-half the daily caloric requirement of an individual, but required the ingestion of 200 mussels. Other studies suggest even less food value in mussels (Warren, pers. comm.), and thus mussels probably were harvested seasonally to supplement other food sources. Harvested areas included the deep channels of large rivers and muddy lakes, as well as the more accessible smaller streams (Warren, 1991). All sizes and species were harvested, indicating that Native Americans did not selectively gather the larger species for their greater food value but rather collected all mussels they encountered (Peacock, 2000). The impact of Native Americans on mussels through harvesting may have been localized and negligible, but this has not been adequately determined. Only people living close to the mussel beds utilized them for food—inland groups did not travel to gather them (Warren, 1983). However, there is some evidence that as Native American society became more agrarian, runoff began to impact mussels, just as it does today on a much larger scale (Peacock *et al.*, 2005).

The arrival of Europeans eventually dispossessed Native Americans of their lands, including access to harvestable mussel beds. Initially Europeans made only limited use of mussels, primarily as a source of mother-of-pearl for inlay work and the serendipitous pearl. Europeans did not consume mussels on anywhere near the scale that had Native Americans, and mussels did not become a true commercial commodity until the last half of the 19th century. Beginning then, the commercial use of mussels occurred in three stages: the Pearl Rush; the Pearl Button Industry; and the Cultured Pearl Industry, the latter of which continues to this day.

The Pearl Rush began in 1857 when a pearl was found in a mussel from a New Jersey stream. Valued at $25,000, the discovery launched a whirlwind of mussel collecting that eventually swept all of the United States west to the Rocky Mountains and persisted until the early 1900's. Although gem-quality pearls were quite rare, thousands of men, women, and children shucked every mussel they could find in hopes of easy riches (Kunz, 1908). How many millions of mussels were killed will never be known. To little avail, the U.S. Fish Commission called for legislation to curb the "reckless prodigality and waste of such resources by man" (Kunz, 1898). Eventually, a glutted market and over-harvested beds brought an end to the Pearl Rush, but the mussels had little respite before the next wave of overexploitation arrived.

In 1887, Johanne Boepple, a German immigrant, brought his knowledge of pearl inlay artistry to America. He quickly realized the untapped potential of American mussels—as buttons. By a process of "punching" discs out of shells, mother-of-pearl could be tooled into buttons (fig. 1). In 1891, Boepple established his first factory in Muscatine, Iowa. The production of mother-of-pearl buttons was to become a multi-million-dollar business, with factories appearing over much of the eastern United States, employing tens of thousands of people. In Ohio, factories were established in Cincinnati, Marietta, and Manchester, where punched shells may occasionally be found to this day. Through the use of rakes, hooks, and particularly brails, deepwater beds previously untouched by the Pearl Rush were accessible (Claassen, 1994).

Fig. 1. A mussel "punched" for buttons.

As with the Pearl Rush, the U.S. Fish Commission realized that this unregulated harvest was quickly leading to an economic disaster in the industry. But clammers refused to consider closed seasons: they wasted most of the shells they harvested; they harvested species they did not need; and they collected juveniles, too small to be of use (Smith, 1899; Coker, 1914). Failure to contain the commercial industry led to a bold measure by the Commission—the attempt to artificially grow and stock mussels, with the aim to rejuvenate the depleted resource. Established at the Fisheries Biological Station in Fairport, Iowa, a team of researchers engaged in the most in-depth study of mussels ever undertaken, the results of which still form the basis of our knowledge of these animals. Experiments were carried out to determine (a) if mussels could be raised in hatcheries and impoundments (Howard, 1914b; Isely, 1914; Howard, 1916; Corwin, 1920, 1921; Howard, 1922); (b) what mussels ate (Churchill & Lewis, 1924); and (c) how mussels infected hosts and what hosts were needed (Lefevre & Curtis, 1910a, b; Young, 1911; Surber, 1913; Howard, 1914a, 1914c, 1915). In addition, they made observations on morphology (Coker & Surber, 1911; Surber, 1912) and investigated the ingenious possibility of rearing glochidia in artificial media (Lefevre & Curtis, 1912a). From a practical point of view, the Fairport Station was a failure—no means were found to supply mussels to meet the demands of a wasteful and self-destructive industry. From a scientific point of view, the Station was a well-spring of knowledge about mussels. In 1917, the Station burned to the ground, taking many of the researchers' invaluable notes with it.

By the turn of the century, the button industry was suffering. Over-harvesting had depleted most beds beyond the point of being profitable. The final blow came with the production of the plastic button. Cheaper and more durable, it signaled the end of the Pearl Button Industry founded some 70 years, and untold millions of mussels, ago.

The most recent wave of commercial use is by the Cultured Pearl Industry, centered largely in Japan, Australia, China, and other Pacific Rim countries. Cultured pearls are formed by inserting beads, or nuclei, cut from North American mussel shells into marine pearl oysters (*Pinctada* sp., family Pteriidae, the wing-oysters). The oyster lays down a veneer of nacre over the bead, forming a pearl. Whereas a non-cultured, natural pearl requires decades to reach a gem size, cultured pearls can be made in less than three years. Because the pearl is formed from a man-made implant, nearly any size and shape of pearl can be produced. Cultured pearls are significantly cheaper than natural pearls, allowing the general public to acquire them. Not surprisingly, the culture of pearls has become a multi-billion-dollar industry.

Beads originate from mussels in North America, particularly those in the Ohio and Mississippi river drainages. Shells are delivered to buyers, who then steam out the animals. Shells may be cut and tumbled into beads before shipment or shipped whole for bead production overseas. As of 1990, more than 26,000 tons of shells were harvested annually from North America. Buyers typically pay several dollars per pound of raw shells. Shells are turned into beads costing thousands of dollars per pound. In turn, the beads become gem-quality pearls worth millions of dollars. Kentucky, Alabama, Tennessee, and a few other states have an active industry supporting shellers, buyers, and others. Commercial clamming is licensed and regulated in these states. Other states, including Ohio, Indiana, Pennsylvania, West Virginia, and Michigan, are closed to all commercial mussel harvest. Recently, mussel poaching has become a chronic problem in some of these closed states.

As with the Pearl Rush and the Pearl Button Industry, there is concern that the Cultured Pearl Industry will over-exploit the mussel resource. Mussels are slow growing, have little recruitment, and are already assailed by habitat destruction, pollution, and zebra mussels. At a 26,000-ton annual harvest, is this resource renewable? As evidence that it is not, some people believe that poaching is increasing because the legally harvested beds have been depleted. Others point to the fact that clammers may harvest, unwittingly or not, endangered species. Because of the difficulties of identifying some species, even by an expert under the best conditions, it is highly unlikely that clammers can distinguish all endangered species in the field. As zebra mussels eradicate the commercial beds, and as new methods and materials are devised to form pearls, the future of mussel harvesting in North America is far from clear. It remains to be seen if the mussel resource can continue to support this industry.

Anatomy, Life History, and Reproduction

Unionoid bivalves, the subject of this book, are commonly called freshwater mussels, naiads, clams, or even oysters, but they are only distantly related to those predominately marine groups. However, those names are entrenched in our vernacular (and those of other languages as well), and it would be useless to attempt to change them. Still, the reader should be aware that these animals are a group very distinct from true mussels (Mytilidae), clams (Veneridae), or oysters (Ostreidae). Their closest living relatives may be an obscure marine group (the Trigonoideans), now found only off the coast of Australia, although this group had a long and diverse fossil history.

Freshwater mussels have two shells, or valves, arranged left and right. The earliest part of the shell is called the beak or umbo. The shell expands along the margins as the animal grows. Most freshwater mussels have a dorsal area called the hinge, which has interdigitating projections called teeth. These teeth serve to keep the shells aligned and prevent shearing during burrowing. The anterior-most teeth are called the cardinal (or pseudocardinal) teeth, whereas the posterior teeth are the lateral (or pseudolateral) teeth. Some mussels lack teeth altogether. The shells are held together in life by two adductor muscles which close the shells. These muscles counteract the ligament, a non-living proteinaceous structure which acts as a spring to open the shells. The muscular foot protrudes from the anterior half of the shells; the siphons, the openings through which water enters and exits the shells, are located posteriorly. On the inner surface of the shells are scars, sites of attachment for various muscles, including the adductors and the pallial line—the linear scar where the mantle tissue is anchored to the shell. The major landmarks of the freshwater mussel shell are illustrated in figures 2 and 3.

Freshwater mussels live by filter-feeding food from the surrounding water with their gills, or ctenidia. Because of their food-gathering function, these gills are much larger than is needed for respiration. North American species lack true siphons, or tubes for water intake and release, such that many species are confined to burrowing only to the posterior edge of the shell during much of the year. This renders them susceptible to predators, desiccation, and temperature and other environmental extremes. Never-

Fig. 2. Morphology of freshwater mussel shells. Also see the Glossary.

theless, many species live for 20–30 years, and some up to 200 years or more (Bauer & Wächtler, 2001). Other species, such as *Pleurobema clava*, may spend much of their life buried several centimeters beneath the surface, relying on water to percolate between the substrate particles for food and oxygen.

Unionoid food continues to be the subject of debate. Allen (1914, 1921), Churchill & Lewis (1924), and Fikes (1972) found the gut to contain mostly diatoms and other algae, although the diatoms passed through the digestive system intact. However, Imlay & Paige (1972) believed that mussels fed on bacteria and protozoans. Bisbee (1984) found different proportions of algal species in the guts of two mussel species, suggesting that not all unionids fed upon the same food. Recently, Nichols & Garling (1998) demonstrated that mussels were omnivores, feeding on detritus and zooplankton, as well as algae and bacteria. Newly metamorphosed juveniles do not filter-feed with their gills, but may feed on interstitial nutrients using cilia on their foot, gills, and mantle. This stage may last several years before changing to a filter-feeding mode (Tankersley *et al.*, 1997). Yeager *et al.* (1993) believed that food for juveniles consisted of interstitial bacteria, yet an algal mix including silt was suggested as food by Humphrey & Simpson (1985) and Gatenby *et al.* (1993). Small amounts of silt have been found to enhance survivorship in cultured mussels, both adults and juveniles (Hudson & Isom, 1984; Humphrey, 1987b; Hove & Neves, 1991), probably by introducing bacteria and zooplankton.

Gametogenesis, the formation of eggs and sperm, is initiated by changes in water temperature and/or light levels. There appear to be threshold temperatures or light levels that cue reproductive events. For those species relying on some upper temperature threshold, constant low water temperatures, such as are found below some dams, may prevent reproduction from ever taking place. In such conditions, populations of adult mussels may live out their normal lives and die without ever producing offspring. For most Ohio mussels there is only one breeding season a year, although the extirpated *Cumberlandia monodonta* may breed twice a year (Howard, 1915; Gordon & Smith, 1990). Ortmann (1919) believed that *Lampsilis fasciola* and *Lampsilis cardium* had two broods per year and that *Potamilus alatus* and *Utterbackia imbecillis* bred year round.

Typically, sexes are separate, although small numbers of hermaphrodites have been found in many species (Poupart, 1706; Fischerstrom, 1761; van der Schalie, 1966, 1970; Heard, 1979). Some species, such as *Toxolasma parvum*, *Lasmigona compressa*, and *Utterbackia imbecillis*, are believed to be wholly hermaphroditic (Ortmann, 1912a, 1916; Utterback, 1916). The relative proportion of hermaphrodites may increase under low population densities, perhaps to bolster declining population numbers (Kat, 1983; Bauer, 1987c). Males liberate sperm into the water, sometimes as spherical (Utterback, 1931; Lynn, 1987; Barnhart & Roberts, 1997) or disc-shaped aggregates termed spermatozeugmata. Females downstream take up the sperm with incoming water. Fertilization success may be related to population density, with a threshold density required for any reproductive success to occur (Downing *et al.*, 1993). For at least *Elliptio complanata*, there is evidence that mussels of both sexes may horizontally migrate to form aggregates during spawning, thus increasing fertilization success (Amyot & Downing, 1998). Eggs are fertilized in the suprabranchial chambers of the gills and then apparently are moved to the marsupia. The marsupia are regions of the gill that act as brood chambers for the glochidia. Their placement and structure vary from genus to genus and have been used as key taxonomic characteristics. The marsupium may change in shape and structure during the breeding season, particularly in lampsiline mussels (Smith, 1979; Kays *et al.*, 1990; Richard *et al.*, 1991). During this time, the marsupium either does not function as a site of respiration (Richard *et al.*, 1991) or operates at greatly reduced efficiency (Allen, 1921; Tankersley & Dimock, 1992). This region may remain non-respiratory during the non-breeding season as well (Richard *et al.*, 1991).

The developing embryos are physiologically isolated in the marsupium from the outside water (Kays *et al.*, 1990). Muscles associated with the water tubes may be responsible for maintaining this isolation (Gardiner *et al.*, 1991). Larval shells are formed indirectly from mineral deposits in the gills that act as a source of calcium carbonate (Silverman *et al.*, 1985). These minute bivalved larvae, or glochidia, develop over a period of days to months, depending on water temperature and species. Glochidia are a type of specialized veliger larvae. Freshwater mussels, and freshwater molluscs in general, lack the trochophore larvae so common to marine molluscs. When first discovered by Leeuwenhoek in 1697, glochidia were considered by some (but not by Leeuwenhoek) to be parasites living in the mussel's gills, and they were given the scientific name *Glochidium parasiticum* by Rathke (1797). For nearly three-quarters of a century, a lively debate ensued as to whether these "agglomerations of animicules," as Prévost called them (1826), were mussel parasites or mussel larvae. Houghton (1862) appears to have been the first to identify glochidia on fishes, and Forel (1866) experimentally demonstrated their parasitic role and their true identity as larval mussels.

North American freshwater mussels are historically divided into two behavioral groups based upon the duration that glochidia are held in the marsupia. Tachytictic or short-term breeders spawn in the spring or summer and release their glochidia later the same year, usually by July

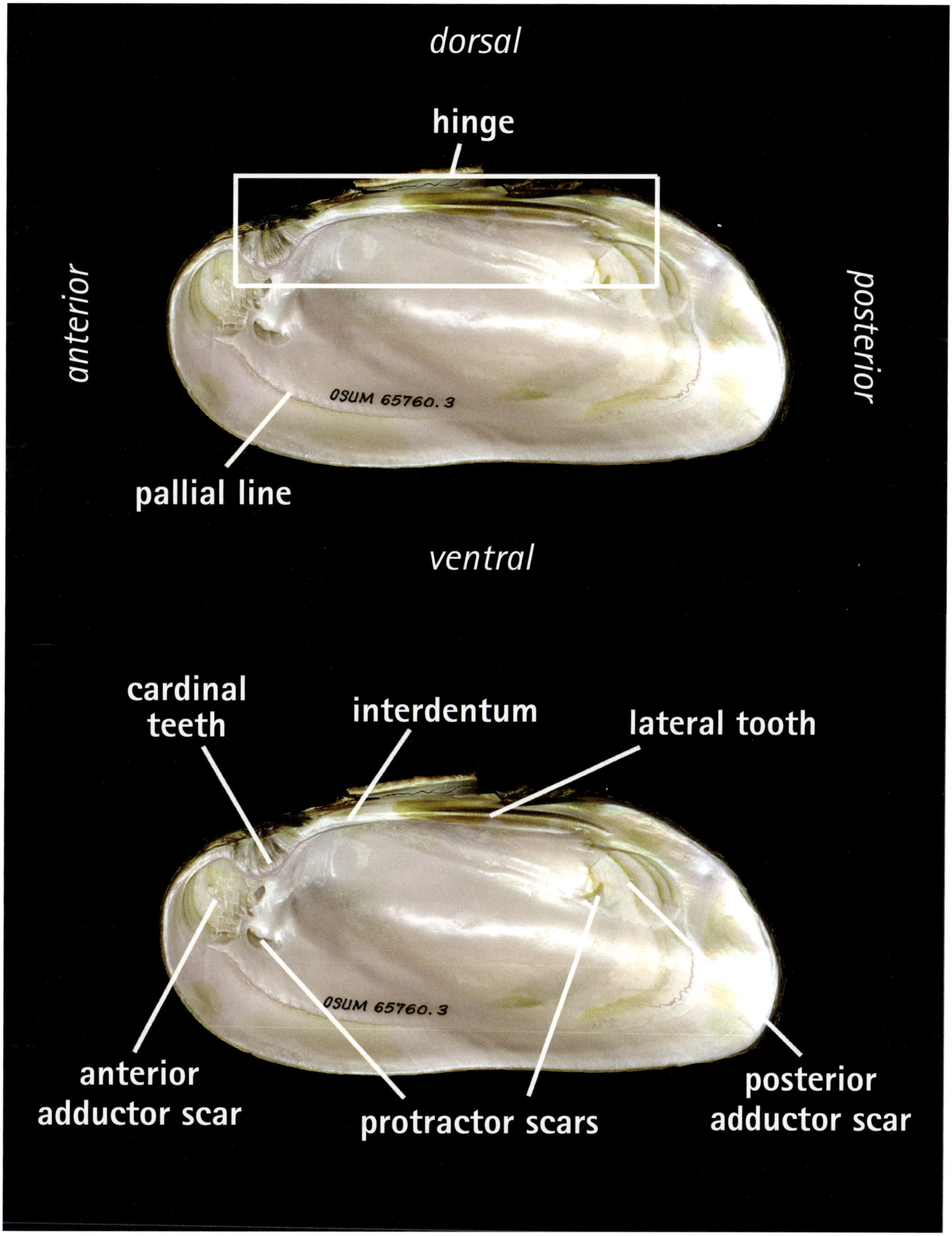

Fig. 3. Morphology of freshwater mussel shells (cont.). Also see the Glossary.

or August. Bradytictic or long-term breeders spawn in the summer or early autumn, form glochidia, and typically hold these larvae in the marsupium until the following spring or summer. There is evidence that glochidia that overwinter till spring in the marsupium experience less mortality once on the host than those glochidia of the same species that are released in the autumn and metamorphose that same year (Corwin, 1921; Higgins, 1930; Tedla & Fernando, 1969; Zale & Neves, 1982a). But some otherwise "bradytictic" individuals release glochidia in autumn or winter to overwinter on their hosts. They remain dormant on their host until a threshold temperature is reached the following spring, at which time they metamorphose and excyst (Watters & O'Dee, 1999). Some species, such as *Pyganodon grandis* and *Leptodea fragilis,* apparently utilize host-overwintering glochidia as part of their normal life history. The ability to overwinter on hosts is related to the duration the glochidia spend attached to the host, with overwintering success declining with time (Watters & O'Dee, 1999). Overwintering of glochidia on hosts may aid in dispersal of the species and confer greater fitness on those glochidia capable of remaining attached till spring. Whether these glochidia overwinter in the marsupia or on their hosts, this prolonged developmental stage may be necessary for bradytictic species.

Nearly all freshwater mussels are obligate vertebrate parasites as larvae. Of the approximately 300 species in North America, only one, the Salamander Mussel, *Simpsonaias ambigua,* is believed to use exclusively a non-fish host, the Mudpuppy, *Necturus maculosus* (Howard, 1915, 1951). The remainder are believed to use fishes but may be capable of parasitizing amphibians as well (Watters, 1997a; Watters & O'Dee, 1998a). Although claims have been made that several species may complete their metamorphosis without a host (Lefevre & Curtis, 1911; Howard, 1914b), evidence suggests that most North American mussels must parasitize a host to complete their life cycle. Two North American species, *Lasmigona subviridis* (Barfield & Watters, 1998; Lellis & King, 1998) and *Utterbackia imbecillis* (Dickinson & Sietman, 2008); an African irinid species (Kondo, 1990); and several South American mussels (Mansur & Campos-Velho, 1990) are known to bypass the parasitic stage completely.

Hosts are infested with glochidia when they come into contact with them in the water, or on the substrate, or by attempting to ingest them (fig. 4). Different mussel species have different methods of releasing these larvae. Some simply expel the glochidia along with water and waste products. Hosts take in suspended glochidia and pass them over their gills, where they attach (fig. 5), or they contact larvae on the substrate, where the parasites attach to the fins or skin (fig. 6). Some mussels, such as *Anodonta suborbiculata, Anodontoides ferussacianus,* and *Utterbackia imbecillis,* release "webs" or "trot lines" of mucus

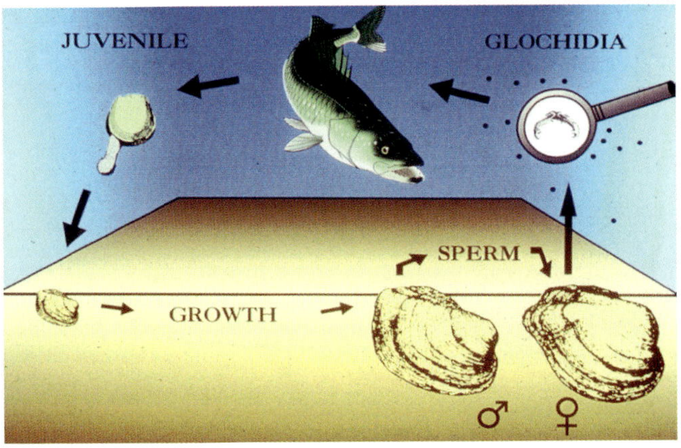

Fig. 4. Life cycle of a freshwater mussel.

containing glochidia that ensnare passing fishes (Hove *et al.,* 1995a; Watters & O'Dee, 1997f). Other mussels bind glochidia into matrices called conglutinates (Chamberlain, 1934; Fuller, 1971). Several types of conglutinates have been identified (Watters, 2008). The simplest are formed entirely of glochidia, each enclosed in a fragile envelope. A more complex type is formed from eggs and glochidia. In these, the majority of the mass is composed of sterile eggs. In some species up to 85% of the total number of eggs have this structural, rather than reproductive, function and apparently are destined never to be fertilized (Barnhart, 1997). The eggs function to give shape and color to the entire mass, which has functional glochidia embedded in it or attached to it. Another type of conglutinate, found in *Strophitus undulatus,* is composed of a mucus rod containing the glochidia. When released by the female, the glochidia squeeze through pores in the conglutinate wall and move to the outside (Watters, 2002). Other conglutinates are even more complex, such as the fry lures of *Ptychobranchus fasciolaris,* which are multi-layer constructions complete with discrete pigmented areas and adhesive pads (Watters, 1999). Conglutinates also may resemble worms, insect larvae, or other food items (Barnhart & Roberts, 1997), or they may simply be fragile structures that rapidly disassociate into free glochidia. The latter probably act only to facilitate the movement of the glochidia out of the marsupia. Some mussels, such as *Lampsilis,* apparently can release either conglutinates or free glochidia (Watters, unpubl.), perhaps dependent on the time of year or the presence or absence of hosts. Glochidia or conglutinates may be moved to the suprabranchial chamber and expelled through the excurrent opening, but in many lampsilines, the distal portion of the marsupia rupture and conglutinates are expelled directly to the outside. Recently, "superconglutinates" were discovered in several southern species (Haag *et al.,* 1995; Hartfield & Butler, 1997). Superconglutinates are groups of individual conglutinates formed into a single lure. The lure is played out in the water on a strand of mucus up to

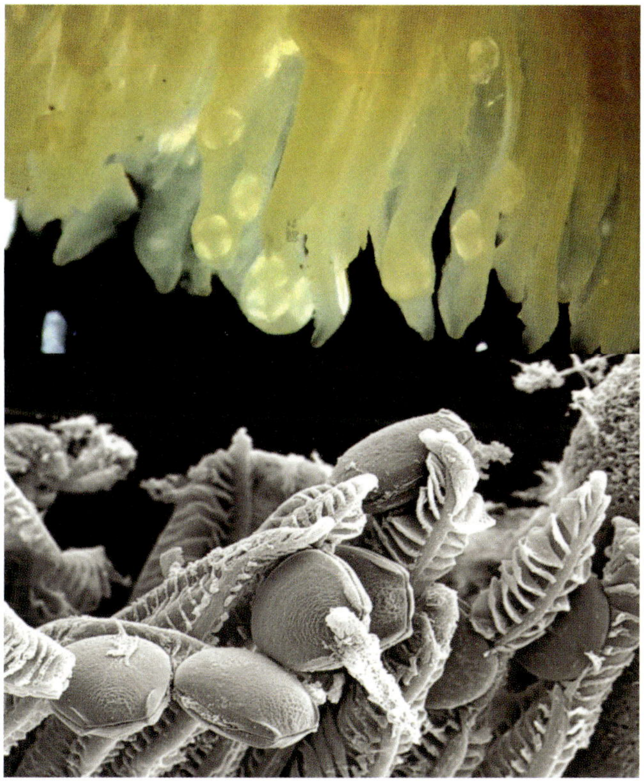

Fig. 5. Top—Glochidia encapsulated on gill filaments. Bottom—Newly attached, unencapsulated glochidia on gill filaments.

Fig. 6. Top—Glochidia encapsulated on the fin of a darter. Bottom—The same fin under scanning electron microscopy.

2.5 m long. No Ohio species has been found with superconglutinates.

Some lampsiline species have modified the mantle into lures resembling fishes, insects, and other food items (fig. 7). These structures may undulate in swimming-like motions (Kraemer, 1970). Often the marsupial portion of the gills is protruded out of the shell between these flaps. When struck by a predatory fish, females snap their shells closed on the protruded marsupia, causing the marsupia to rupture and release clouds of glochidia which attach to the would-be predator. Radically different mantle lures have been found in the same "species," such as *Lampsilis fasciola,* suggesting either the presence of sibling species or a startling polymorphism in mantle pigmentation and morphology within single species, even in a single population. There is some evidence that mussels and glochidia can detect nearby hosts and modify their behavior to increase their chances of attracting or attaching to them (Henley & Neves, 1997). Mantle tissue is sensitive to changes in light intensity and may react to the shadows of passing fishes (Hove & Anderson, 1997). In addition, it has been suggested that mantle tissue may release chemical attractants for hosts (Pepi & Hove, 1997).

Once shed by the female, glochidia must acquire a suitable host or die, usually within 24–48 hours. "Free" glochidia are preyed upon by numerous predators: insects, bottom-feeding fishes, flatworms, hydras, etc. Infective glochidia may travel some distance downstream in currents (Clark & Stein, 1921; Neves & Widlak, 1988). The estimated chances of a glochidium surviving to metamorphose and excyst range from 0.0001% (Jansen & Hanson, 1991) to 0.000001% (Young & Williams, 1984). Although some species may compensate for this with high fecundity over many years (Bauer, 1987b), other species have been shown to develop late and then reach early senescence (Downing *et al.,* 1993). Species having lures and conglutinates generally produce fewer glochidia but are not able to parasitize as wide a range of hosts as those that broadcast large numbers of glochidia (Watters, 1997b, 2007). Because of the manner in which hosts acquire glochidia (lures, conglutinates, webs, etc.), it is not surprising to find that glochidia are over-dispersed, that is, most hosts either are unparasitized or carry only a few glochidia, whereas a very small number of hosts bear most of the parasite burden (Bangham, 1940, 1955; Weir, 1977; Dartnall & Walkey, 1979; Neves & Widlak, 1988). However, even heavily infected hosts show few ill effects, with few exceptions. Recently parasitized fishes may become lethargic and secretive but usually recover within 24 hours. Large, hooked glochidia, such as occur in *Strophitus undulatus,* may cause great agitation to hosts when attaching.

Glochidia clamp down on the host tissue, causing cells to rupture. This cellular fluid forms part of the food for the developing parasite (Arey, 1924b, 1932b; Blystad, 1924).

Fig. 7. A female *Lampsilis ovata* with mantle flaps mimicking a fish. Note eyespots and "lateral lines."

A host wound reaction forms a capsule by the movement of host cells over the glochidium (Faussek, 1895; Arey, 1921, 1932a). During growth, the larva will resorb much of its own tissue, including the adductor muscle and much of the mantle, as well as feeding on the host (Young, 1911; Blystad, 1924). The glochidia of most mussel species, except the margaritiferids and the axe-head-shaped glochidia of *Potamilus,* do not appreciably change size during encystment. After a certain amount of time (from days to months, depending on water temperature and mussel species), the glochidium metamorphoses into a juvenile and excysts (Schierholz, 1889; Howard & Anson, 1923). Threshold temperatures necessary for completion of metamorphosis were identified for *Lampsilis cardium* and undoubtedly exist for all species (Watters & O'Dee, 1999). At temperatures below these thresholds, metamorphosis may be delayed indefinitely. The duration of metamorphosis decreases with increasing temperature until an upper threshold is reached. At this point, glochidia excyst, fail to metamorphose, and die (Dudgeon & Morton, 1984). Barnhart & Roberts (1997) found that although higher water temperatures greatly decreased the time to complete metamorphosis, metamorphosis occurred on more hosts species at lower temperatures.

Glochidia metamorphose into juvenile mussels. Whereas the glochidium lacks a foot and gills, and has a single adductor muscle and a larval thread, the juvenile has a functional foot, gill "buds," and two adductor muscles. In many species (except anodontines), the glochidial shell is white but turns tan or brown after metamorphosis. Anodontione glochidia typically are tan. The glochidial shell is incorporated into the adult shell and may be visible in exceptionally well preserved specimens (Hoggarth, 1987).

Once metamorphosed, the juvenile drops from the host and burrows into the substrate (Bauer, 1986; Clarke, 1986; Buddensiek *et al.,* 1993) or attaches to a larger object with a byssal thread. This buried juvenile stage may last several years and has been referred to as the mesoconch (Clarke, 1986). The thread is lost in most adults but persists for a year or more in some species (e.g., *Truncilla donaciformis*). The thread, which is formed by a gland within the foot, is very similar to the byssal threads of marine bivalves, such as *Mytilus* (Smith, 2000b).

Potential hosts may possess one of two types of immunity to attached glochidia. Natural immunity occurs in unsuitable hosts, which have tissue responses against the glochidia (Howard, 1914a; Bauer & Vogel, 1987). Acquired immunity occurs when a suitable host has been previously parasitized and has built up a temporary immunity. The number of exposures needed to achieve acquired immunity depends on the degree of prior infestations and the duration between them (Lefevre & Curtis, 1910a; Surber, 1913; Reuling, 1919; Arey, 1924a; Bauer, 1987a). Although acquired immunity may be demonstrated in the laboratory, acquired immunity in wild-caught fishes has been observed only once, and its overall prevalence in wild fishes is unknown (Watters & O'Dee, 1996). In both natural and acquired immunity, encysted glochidia are killed by the host and are either sloughed off or absorbed (Arey, 1932c; Fustish & Millemann, 1978; Zale & Neves, 1982b; Waller & Mitchell, 1989).

Host specificity varies greatly among mussels. Some species, usually widespread and abundant, are able to parasitize many species of fishes, including exotics (Trdan & Hoeh, 1982; Watters & O'Dee, 1997a; Watters & O'Dee, 1998a). Other mussel species, usually localized and rare, may use only a few fish species. Because this symbiosis is obligatory, plans to conserve and manage rare and endangered mussels must include management of the host as well. It is suspected that some mussel populations are not recruiting because their hosts no longer occur with them.

This host-parasite relationship apparently arose as a means of dispersal for the unionoids. Lacking internal fertilization and great motility as adults, mussel populations would be doomed to be carried back to the sea over many generations. By attaching themselves to a highly motile host, such as a fish, they are dispersed within and between drainages. Records of dispersal on the feet of waterfowl largely are apocryphal (Rees, 1965), and the idea that glochidia swim by clapping their shells, once the preferred theory, has been abandoned.

Natural predators of metamorphosed mussels consist of fishes, birds, muskrats, and raccoons (fig. 8). In Europe, the hooded crow has been shown to drop mussels from the air to crack them open (Berrow, 1991), and North American birds may have a similar behavior. Baker (1918) listed Freshwater Drum, Lake Sturgeon, Spotted Sucker, Common Redhorse, Pumpkinseed, and others as predators on freshwater mussels. Hanson *et al.* (1989) and Convey *et al.* (1989) reported muskrats eating up to 37,000 mussels a year in a Canadian lake. Muskrats are selective in the spe-

Fig. 8. A mussel bearing claw or tooth marks from a muskrat.

cies and sizes of mussels they consume, preferring smaller species and individuals (Watters, 1995b), and they may be an important source of predation on endangered species (Neves & Odom, 1989). Trapping of muskrats has been suggested in some areas as a means of protecting rare mussels. Juvenile mussels and small species are consumed by all of these predators, but large species are probably nearly immune to predation. Once a mussel has become an adult, most natural mortality arises as the result of catastrophes to habitat: scouring; droughts; beaver impoundments, etc.; or disease.

Unionids often are parasitized by unionicolid mites (Mitchell & Wilson, 1965; Davids, 1973; Vidrine, 1989) and monogenic trematodes (see review of Hendrix *et al.*, 1985), which feed on gill and mantle tissue. Chironomid larvae may consume up to 50% of the mussel gill (Gordon *et al.*, 1978), interfering with respiration and reproduction. Leeches also may infest unionids.

Mussel-Host Associations

Within the Species Accounts are lists of presumed mussel-host relationships. Each listing is followed by an abbreviation indicating how the relationship was determined: NS—not stated; NI—natural infestation; LI—laboratory infestation; NT—natural transformation; LT—laboratory transformation. These distinctions were first employed by Hoggarth (1992).

Natural infestations (NI) are based on wild-caught fishes parasitized with glochidia. Most studies reporting natural infestations were not continued to determine whether the glochidia metamorphose. Because glochidia will attach to almost any fish with which they come into contact, including unsuitable hosts, these associations must be viewed with caution.

Laboratory infestations (LI) are similar to natural infestations in that fishes were parasitized (but by artificial methods), but the studies were never carried to completion (i.e., metamorphosis). Again, these associations may be incorrect.

Natural transformations (NT) are the least common evidence of a mussel-host association. In these studies, wild-caught fishes bearing glochidia were kept in the laboratory until the glochidia metamorphosed. Because glochidia may be difficult to identify to species, the determination of the juveniles is often inferred from what species co-occurred with the host. These studies may suffer from misidentified glochidia.

Laboratory transformations (LT) are the most common type of mussel-host association studies. Hosts are infested with glochidia and kept in captivity until metamorphosis occurs. This procedure identifies *potential* hosts. It suffers from identifying associations that may occur in an experimental setting but never occur in nature.

Effects of Habitat on Shell Morphology

Freshwater mussels are unusually plastic in their shell morphology—specimens of a species taken from different parts of its range may look completely unrelated. This great variation has led to the description of numerous superfluous names for some species, as evidenced by their tortuous synonymies. More so than nearly any other group of molluscs, freshwater mussel shells are the product of their immediate surroundings. Mussel shells vary in their degree of lateral compression or inflatedness, in the strength of their sculpture, and in their color patterns. By and large, these variations are not random.

Early on it was recognized that some mussel shells changed in a predictable way within a given river reach—the Law of Stream Distribution (Ortmann, 1920). Headwater forms often are compressed and thin-shelled, with low umbos, whereas big river forms are usually inflated and thick-shelled, with prominent umbos. This gradient in characteristics occurs across taxonomic lines but is most obvious in amblemines. Watters (1994) interpreted this as an adaptation to life in flashy headwaters, where streamlining is at a premium.

Sculpture also varies in strength but can usually be associated with habitat. Savazzi & Peiyi (1992) and Watters (1994) experimentally demonstrated that unionoid shell sculpture was used for anchoring and antiscouring. Thus, for any sculptured mussel species we should find the greatest sculpture in fast-moving water and the least in slow-moving water. This is born out by *Amblema plicata*, whose stream individuals are coarsely sculptured but lake specimens are nearly smooth.

Color patterns probably represent the sequestering of metabolic byproducts in the periostracum. Freshwater mussels differ from the great majority of molluscs in that the outward shell color and pattern are not present in the shell itself but reside only in the periostracum. Stripped of periostracum, mussel shells are uniformly white. The colors and patterns in freshwater mussel shells cannot be construed as camouflage, as most mussels live their lives buried in the substrate. What portion of the shell is exposed is often covered with algae, larval insect cases, and marl. The colors are probably non-adaptive.

For the same reason, nacre color cannot be considered to have any apparent value to the mussel. Variations in nacre color often occur within specific populations. For example, *Obovaria subrotunda* nearly always has white nacre, yet the population in Middle Fork Salt Creek of the Scioto River system often has purple nacre. *Elliptio dilatata* is usually purple nacred but may have purple, white, or salmon nacre in individuals living side by side in some streams. *Pleurobema sintoxia* usually has white nacre except for those in the Maumee River system, where nearly half have pink nacre. Furthermore, whereas *Pleurobema sintoxia* may have white or pink nacre, its close cognate *Pleurobema cordatum* very rarely has pink nacre. It is difficult to ascribe an environmental cause that would allow three different nacre colors in some populations and only one in others or that would make some individuals strongly rayed and others unrayed. Clearly, color in mussels is the byproduct of genetics. Variations in patterns and colors probably represent polymorphisms and linked genes rather than environmental effects.

Effects of Habitat Modifications and Pollutants

IMPOUNDMENTS

Rivers have been dammed by humans for millennia, for many purposes: to run mills and hydroelectric turbines; to irrigate otherwise inarable land; to control floods; to allow navigation of waterways; and to create bodies of water for recreation. But impoundments are not the same as naturally occurring pools in a river: impoundments have a hydrology different from those of natural pools, with different flow patterns, topographies, and temperatures. These differences often result in a change in the aquatic fauna, including mussels. For example, 15 mussel species were found in Lake Cooper, a man-made impoundment on the Mississippi River, but in adjacent Lake Pepin, a naturally formed pool, 30 species were encountered (van der Schalie, 1938).

The loss of these mussels is a result of habitat modifications caused by impoundment. A free-flowing river has riffles, runs, pools, shoals, water-willow stands, and meanders (fig. 9). But impoundment dramatically reduces

Fig. 9. The Elk River in West Virginia supports a high diversity of mussels and fishes because of numerous habitat types: riffles, runs, pools, meanders, canopy, gravel bars, water willow, etc.

this habitat heterogeneity (fig. 10) (Bates, 1962; Blalock & Sickel, 1996). The original channel remains but is often buried under deeper water and fine sediments. But most mussels do not occur in deep water, and those that do are often stunted (Haukioja & Hakala, 1974; Lewandowski & Stanczykowska, 1975). This channel begins to accumulate sediments, smothering most mussels. Ellis (1936) showed that a silt accumulation of ¼-inch to 1-inch depth resulted in mortality approaching 90% in the mussels he tested.

If not smothered outright by accumulating sediments, mussels may be affected adversely in other ways. Reproduction may be reduced or stopped entirely, and growth may be slowed. Bates (1962) found no evidence of mussel reproduction in the channel of Kentucky Lake of the Tennessee River, even though adult mussels were present. Rivers having high levels of suspended solids result in slower mussel growth (Semenova *et al.*, 1992). Juvenile mussels, which live buried for several years, may be killed by overlaying sediments (Bauer *et al.*, 1980; Buddensiek *et al.*, 1993). Silt results in the loss of light penetration, reducing algal food available to mussels, and leads to oxygen depletion. The bottom waters of the channel may become cold enough to stunt the growth of mussels (Harman, 1974; Ghent *et al.*, 1978; Semenova *et al.*, 1992) and interfere with reproduction (Hruska, 1992; Heinricher & Layzer, 1999).

After impoundment, the original mussel fauna may be eliminated or greatly reduced (Holland-Bartels, 1990) or may be changed in favor of silt-tolerant species, such as anodontines and species of *Leptodea* and *Potamilus* (Clark & Gillette, 1911; Ellis, 1931; Bates, 1962; Isom, 1969; Klippel & Parmalee, 1979; Parmalee & Hughes, 1993; Blalock & Sickel, 1996). These species often occur in the now permanently inundated floodplains. Thus, while some impoundments may contain numerous mussels, they invariably are invading, soft-substrate-adapted species that have replaced the original fauna.

Dams may represent physical barriers to mussel hosts.

Fig. 10. The Ohio River at Willow Island Lock and Dam. Impoundments are often diversity-poor because many of the original habitats have been lost. Compare with the Elk River, left.

Some mussel populations may no longer be recruiting juveniles because their hosts are absent (Suloway *et al.*, 1981; Jones, 1991; Burkhead *et al.*, 1992). Watters (1996c) showed that dams as low as 1 m in height affected host movements and restricted the range of some mussels. Tailwaters tend to accumulate mussels, another source of evidence that the dam is acting as a barrier.

Tailwaters are affected as well (Ligon *et al.*, 1995). Fluctuating water levels may strand or otherwise expose mussels to extreme temperatures, resulting in mass mortalities (Riggs & Webb, 1956; Neck & Howells, 1994). Low temperatures associated with some tailwaters may adversely affect mussel reproduction and growth. High-volume water discharges and abrupt stoppages cause instability of substrate below dams (Miller & Payne, 1992; Nagel, 1992; Yeager, 1993). Sedimentation in tailwaters reduces fish diversity and abundance by altering necessary habitats, thereby removing necessary mussel hosts (Petts, 1984; Holland & Huston, 1985; Nelson *et al.*, 1987). Mussels may be absent for miles below some dams because of these problems.

That impoundments are detrimental to aquatic life in general, and to most mussels in particular, has been demonstrated by many case studies. Perhaps several dozen mussel species in North America, and numerous more freshwater snails, were driven to extinction by the effects of impoundments (Stansbery, 1973a; Layzer *et al.*, 1993; Lydeard & Mayden, 1995). Although mussel faunas change through time, with or without dams, and species become extinct as a matter of course, the interference of humanity has dramatically accelerated this process. For example, the Tennessee River's mussel diversity decreased from 100 species to 44, owing mainly to impoundment (Isom, 1969). In the Fort Loudoun Reservoir alone on the Tennessee River, Isom (1971) found only four mussel species, while Ortmann (1918) reported 64 species from the same general area before impoundment. The area of the Chickamauga Reservoir of the Tennessee River supported 46 species for perhaps 2,000 years prior to impoundment (Parmalee *et al.*, 1982). After impoundment, 28 species were extirpated, and several are now extinct. Similar reductions in diversity after impoundment were documented for the Cumberland River (Layzer *et al.*, 1993; Schmidt, 1986), where the construction of the Center Hill hydroelectric dam resulted in the loss of 78% of the original mussel species. In the Little Tennessee River, only 6 of the original 50 mussel species at Tellico Lake remained after impoundment (Parmalee & Hughes, 1993). The mussels of other rivers, such as the Kaskaskia River (Suloway *et al.*, 1981) and the Tombigbee River (Williams *et al.*, 1992), have met a similar fate.

LAND USE PRACTICES

Land use practices (fig. 11), such as logging, mining, construction, farming, livestock, etc., often impact mussel populations by releasing runoff of silt, salt, pesticides, fertilizers, and other pollutants. Proximity of streams to roads may increase the amounts of salt, heavy metals, and other pollutants that enter a stream (Van Hassel *et al.*, 1980). Mussels are smothered or poisoned, and contaminants may remain in the sediments for years, precluding recolonization. Runoff also causes changes in fish composition, perhaps removing a necessary host from a mussel population. But detrimental effects are not confined to the physical destruction of the riparian corridor. Alterations in land cover and canopy are important as well. Morris and Corkum (1996) found greater temperature fluctuations, and higher ammonia and nitrogen concentrations, in a stream bordered by a grassy site than one bordered by a wooded one. The two sites had the same mussel diversity, but the grassy site was dominated by an anodontine.

CHANNELIZATION, DREDGING, AND SNAGGING

Channelization, dredging, and snagging reduce available host and mussel habitat (Nelson, 1993) and alter circulation patterns and substrate composition (Loar *et al.*, 1980). Dredging and channelization affect mussels in many ways. Mussels caught in the dredge path are destroyed, and sediments churned into the water column may travel downstream and affect mussels outside the construction area. Mussels may be smothered by or exposed to resuspended contaminants (Engler, 1979). Valovirta (1990) showed that simply removing large rocks from a stream caused the substrate to become unstable, killing mussels. Channelization on a greater scale is catastrophic to mussels and has been implicated in the decline of some populations (Yokley & Gooch, 1976; Grace & Buchanan, 1981a; Schuster

Fig. 11. Land use in Ohio's watersheds.

et al., 1989; Valovirta, 1990; Hartfield, 1993). Dredge spoil may reenter the river through upland runoff, and contaminants may reenter through groundwater. Although soft-substrate-adapted mussels may invade dredged areas, most affected reaches are not recolonized by mussels of any kind.

The snagging of fallen trees and debris is a common practice to ostensibly prevent a river from inundating its natural floodplains. Like channelization, snagging reduces the available habitat (Marzolf, 1978). It also increases bank erosion, creates unstable substrates as the stream recovers, and generally reduces aquatic diversity. Channelization and snagging actually may increase flood heights (Belt, 1975), creating additional runoff and the need for additional remediation. Thus the effects of channelization, dredging, and snagging may become chronic, affecting aquatic organisms long after the actual "improvement" has been completed.

POLLUTANTS

Havlik & Marking (1987) and Farris & Van Hassel (2007) gave reviews of the effects of contaminants on mussels. Readers are urged to consult these sources for more information on specific contaminants.

Although effects of pesticides are species-specific, in general, sub-lethal levels of PCBs, DDT, Malathion, Rotenone, and other compounds inhibit respiratory efficiency and accumulate in the tissues. Mussels are more sensitive to pesticides than many other animals tested.

Mussels are particularly sensitive to heavy metals (Keller & Zam, 1991), and responses may be species-specific (see the example for copper in Jacobson *et al.*, 1993). Adult mussels may be able to survive short-term exposure through behavioral responses (Keller, 1993), but chronic exposure at lower levels may impact mussel populations. For example, low levels of metals may interfere with the ability of glochidia to attach to the host (Huebner & Pynnönen, 1992).

Glochidia are very sensitive to ammonia from wastewater treatment plants (Goudraeu *et al.*, 1993). At sub-lethal exposures adult mussels exhibit decreased respiratory efficiency (Anderson *et al.*, 1978). Ellis (1931) discovered that mussels found below sewage outfalls had dead glochidia in the marsupia contaminated with bacteria and fungi.

Acidic water from mine runoff and sandy soils may eliminate mussels and preclude recolonization (Simmons & Reed, 1973; Humphrey, 1987a). Mussels may be able to survive several weeks of exposure to relatively low pH because of buffering in the blood (Mäkelä & Oikari, 1992), but chronic exposures are lethal. Low pH also interferes with the glochidia's ability to close its shells on a host (Huebner & Pynnönen, 1992). Houselet & Layzer (1997) found that mussels living below a coal strip mine were significantly smaller than those of the same age occurring above the influence of the mine.

Liqouri & Insler (1985) and Anders & Wiese (1993) gave circumstantial evidence that salinity was lethal to some glochidia. This may be a problem in runoff from salt used for clearing roads in winter. Salinity also is a concern near oil and gas production areas.

Effects of Introduced Bivalves

ZEBRA AND QUAGGA MUSSELS

Zebra and quagga mussels are not true mussels (Mytilidae), but belong to the family Dreissenidae (fig. 12). Although there are native members of this group in North America, they are mainly estuarine species. The zebra mussel, *Dreissena polymorpha* (Pallas, 1771), is native to the Caspian and Black Sea region, and the Volga and Ural rivers. However, as canals were built westward across Europe in the 1700's, zebra mussels followed, eventually colonizing most of Europe and even reaching the British Isles in the 1800's. Zebra mussels probably reached North America in 1985 or 1986 in the ballast water of a cargo ship (Hebert *et al.*, 1989). That ship had taken on freshwater in Europe for the transoceanic voyage that ended in Lake St. Clair. There it discharged its ballast water to take on cargo, releasing either larvae or juveniles of the mussels.

In 1991, a second species of exotic dreissenid mussel was found in North America and named the quagga mussel. It now occurs in Lake Erie, Lake Ontario, parts of Lake Huron, and the St. Lawrence Seaway. Eventually it was identified as *Dreissena bugensis* Andrusov, 1897, another eastern European species (Rosenberg & Ludyanskiy, 1994). Quagga mussels tolerate deeper water (to 107 m) and muddier substrates than do zebra mussels (Snyder *et al.*, 1990). Despite these slightly different environmental tolerances, the two species are often found together.

Zebra mussels become sexually mature within their first year. Spawning takes place when spring water temperatures warm to 12°C and continues until water temperatures fall below this threshold. Depending on the weather in any given year, spawning may begin as early as May and last until October or December, but typically spawning is most concentrated in July and August. Females may produce 1 million eggs per year, which hatch into free-swimming veliger larvae. Unlike our native mussels, dreissenids do not have a parasitic larval stage. Their larvae may float in the water column for 3–4 weeks before transforming into a juvenile and settling to the bottom. Adults may live for 6 years (Snyder *et al.*, 1990).

Zebra and quagga mussels produce fibers from the byssal gland in the base of the foot that are used to anchor

Fig. 12. Top—*Dreissena polymorpha* (Pallas, 1771), the zebra mussel. Bottom—*Dreissena bugensis* Andrusov, 1897, the quagga mussel.

the mussels to the substrate. These fibers, called byssal threads, may be released by the mussel to allow movement. They are extremely strong. Native unionoid mussels also produce a single byssal thread as a juvenile, but it is usually lost in the first year's growth.

Zebra and quagga mussels are highly detrimental to native mussels, sphaeriid clams, and snails (Strayer, 1999). They congregate on the shell around the native unionoid mussel siphons, removing both food and oxygen before these necessities can become available to the native species. Gametes released by male native mussels may be ingested by zebra mussels, thereby interfering with reproduction. The extremely strong byssal threads of the zebra mussels may "sew" the shells of native mussels shut. Clusters of zebra mussels attached to the ends of native mussels may create drag, pulling the native mussels out of the substrate where they are swept ashore to die. At sub-lethal levels of infestation, native mussels experience lowered glycogen levels and increased stress, resulting in decreased fitness (Haag *et al.*, 1993).

Up to 10,000 zebra mussels have been found on a single native mussel. Lewandowski (1976) found that 85% of the native mussels in a Polish lake were infested with zebra mussels but the degree of infestation varied with season and year. He observed that zebra mussel spat preferred other zebra mussel shells and living unionids over dead shells or stones. In the Mississippi River, colonization rates of zebras on natives increased from 27% to 99.7% within a year (Tucker, 1994). Zebra mussels formed a pavement on the gravel substrate such that dislodged unionids were not able to rebury. Tucker also found that thick-shelled, sculptured species were more heavily infested than smooth, thin-shelled species. Conversely, Haag *et al.* (1993) reported that the heavy-shelled, sculptured *Amblema* was not as susceptible to infestation as the smooth-shelled species of anodontines and lampsilines.

In Ohio, native mussels have been nearly extirpated from the western basin of Lake Erie (ESI, 2000), although a few refugia may still exist. Zebra mussels are present in the Ohio River all along Ohio's shore and have invaded the lower Hocking and Muskingum rivers. They have colonized numerous reservoirs as the result of human introductions. In some cases, creeks below the outfalls of these reservoirs have been contaminated as well, sometimes for several miles (Tymochtee Creek, for example).

In retrospect this invasion seems to have been inevitable. The reason it did not occur earlier was likely the long sailing time between Europe and North America. But new technology, faster ships, and the opening of the St. Lawrence Seaway finally allowed these creatures to make the journey successfully. The U.S. EPA required ships to retain their ballast and to exchange old freshwater ballast for new freshwater ballast. This change, designed to protect inland waters of the United States, allowed for the introduction of zebra mussels and other non-native species. At least seven species of *Dreissena* are known from Europe and Asia, and it is possible that they may make their way to our shores as well.

CORBICULA, THE ASIAN CLAM

In North America, the veneroid Asian clam *Corbicula* (Corbiculidae) (fig. 13) first appeared in the Columbia River in Washington in 1938, probably the result of intentional introductions by Asian immigrants to propagate an exotic foodstuff (Mills *et al.*, 1993). By 1958, it had reached Arizona; in 1959 it was found in the Tennessee River; and by 1963 it had appeared in the Ohio River at Cincinnati (Sinclair & Isom, 1963). Today, few water bodies are free of this exotic clam. Similar introductions of other corbiculid species have taken place in other countries (Darrigan & Pastorino, 1993).

The systematics of the genus *Corbicula* are not well-known. It is not clear which species, or how many species, currently occur in North America. Most recent literature refers to a single species, *Corbicula fluminea* (Müller, 1774), but recent studies suggest at least two lineages of *Corbicula* in North America (Siripattrawan *et al.*, 2000). Older literature also lists species such as *C. manilensis* (Philippi, 1844) and *C. leana* (Prime, 1864).

Larvae are brooded as trochophores and early veligers in the gills and released as late veligers to the plankton.

Fig. 13. *Corbicula fluminea* (Müller, 1774), the Asian clam.

There is no parasitic stage in their life cycle. Upon settling, they may attach to objects with a byssal thread, but this thread is soon lost. Unlike zebra mussels, adult Asian clams do not attach themselves to objects with byssal threads as adults. But like zebra mussels, the Asian clam may be present in enormous numbers, clogging water pipes, drains, and other structures.

At one time it was believed that Asian clams would eradicate native mussels wherever the two co-occurred. This has not happened, at least not to any great degree, and there is little evidence to support the idea that Asian clams are detrimental to native mussels (Strayer, 1999). It is possible that the two compete for food, and Asian clams may ingest the gametes of unionoids. But, superimposed on a general decline in native mussel populations due to habitat alteration, pollution, commercial use, and zebra mussels, the effects of the Asian clam are difficult to gauge.

Conservation of Species

The Unionidae are the most endangered of our aquatic animal species (Neves, 1993). Over one-half (54%) of Ohio's native freshwater mussel species are now endangered, extirpated, or extinct. Whether as a result of habitat degradation, competition with non-native species, pollution, or any other of a long list of negative environmental impacts (reviewed above), the unionoids have steadily lost the ecological battle with abundance and distribution. It is because of these losses that conservation of freshwater mussels is even an issue. However, as others have pointed out, the battles these species are waging with survival should send us a message. All life is dependent on all other life, and so in a sense our survival depends on the survival of these often overlooked animals. This message might direct us to begin to understand the problem but must lead to habitat restoration, water quality improvements, biotic reconstruction, and conservation.

Conservation broadly defined is the act of setting aside for the future. It is not the identification of existing populations of mussels, the preservation of those populations, or the reintroduction of organisms now extirpated, although conservation efforts might include any one of these activities. Instead, conservation is acknowledging the importance of a resource to society and then working to protect and restore that resource. The following sections of this Introduction will discuss the activities that are currently being applied to identify the condition of freshwater mussels in Ohio and the methods that are being used to protect and restore this resource. However, none of these discussions is important if we do not value the resource to begin with.

THE VALUE OF A MUSSEL

Value is subjective. What one person finds valuable, another may not. Still, if we look closely at any resource, we should be able to identify the attributes that make it valuable, all of which may be important to one member of society, but only a few of which may be significant to someone else. The accumulated list of value statements describes the ultimate perceived value of a resource. However, any single item on that list, if important to any single member of society, brings value to that resource. In this case, it is the least of all of these that establishes the conceptual mandate to protect, restore, and conserve. So what are the values of freshwater mussels? They can be divided into ecological, economic, cultural, aesthetic, and intrinsic value categories.

Freshwater mussels play important roles in the ecology of the streams and lakes where they live. Not only are they among the largest of our freshwater invertebrates; they are also among the longest-lived. During their lifetimes, some of which may be as long as 40 to 50 years (for Ohio species), they serve as substrate for other organisms, remove silt and other pollutants from the water through their filtering activities, serve as food for other wildlife species, and stabilize the substrate (Vaughn & Hakenkamp, 2001). They contribute to the biodiversity of aquatic habitats by being members of that community and by the symbiotic relationships (commensal, trophic, parasitic, etc.) in which they engage. Every species adds to the complexity of an ecosystem, and that complexity results in what is often referred to as ecosystem service to that community and to us (Reaka-Kudla *et al.*, 1997).

Generally when one speaks of the economic value of freshwater mussels, one speaks of the pearl industry or, historically, the Pearl Button Industry. These industries were (buttons), and continue to be (cultured pearl), dependent on the shells of North American freshwater mussels (Coker, 1919; DeVillez, 1991; Fassler, 1991; McGregor & Gordon, 1992). The North American mussel supply was valued at $50 million in the 1980's (Fassler, 1991). Other economic values associated with freshwater mussels include their use as water quality indicators, their use as food by humans (much more important historically than now), and their use in the shell trade industry. Naturally produced pearls have been a commodity bought and sold as well (Fassler, 1991).

Many scientists who work with these animals share a sense of fascination with their shells. Shell collecting may be where scientists first became interested in freshwater mussels, which ultimately led to questions about the ecology, life history, anatomy, and systematics of the group. Freshwater mussels are exciting to study. They are mostly big, long-lived, and fairly complex animals living their entire lives in freshwater habitats. The way they live their relatively mundane lives and the associations they make with other living things provide ample opportunity to study ecology and evolution. One might examine the behavior of these animals, or one might become fascinated by some aspect of their anatomy. Any aspect of biology that can be studied can be studied using freshwater mussels.

Still, for other people it is enough to know that these mussels are embedded in the substrate of a nearby stream or lake. For them the fact that mussels exist is enough to bring them a sense of joy or wonder. It is not easy to promote freshwater mussels as the poster organism for the conservation of freshwater habitats—perhaps because of their sedentary existence and the fact that many people have never seen one as a living animal. Yet many would say that these animals are worth preserving because they are native, they are natural, and they are a part of this world that is both very old (in a geologic sense) and unique (in a biological sense).

CONSERVATION ACTIONS

Just as there are threats to the continued existence of freshwater mussels in Ohio, so too are there actions we can take to protect, restore, and conserve this fauna. The simplest and most effective method is to become aware of this group of organisms. This book, then, is a means of freshwater mussel conservation. Other methods include the investigation of freshwater mussel biology, the preservation of water and riparian resources, and the control and/or elimination of threats to these animals.

The investigation of the biology of freshwater mussels has taken many forms. Some scientists are examining the distributions of species to attempt to reconstruct past distributions to determine, among other benefits, where else to look for rare and endangered species. Hoggarth *et al.* (1995) reported finding a species of mussel in Ohio long thought to be extirpated from the state. Other scientists are examining the host-parasite relationships between fish and mussel because we realize that it will be impossible to protect mussels without protecting their fish hosts. Still other scientists are involved in captive propagation programs and the storage of genetic material to be used someday in the artificial propagation of freshwater mussels. The examination of relocation techniques that may someday play an important role in mussel conservation, once threats such as water pollution or competition with non-native species has been eliminated, is ongoing as well (Cope & Waller, 1995).

Habitat protection and restoration play equal roles in freshwater mussel conservation. One effective way of setting aside mussels for future generations is by setting aside land near streams. Protecting wooded stream banks and wetlands helps to preserve streams by stabilizing their banks, removing sediment and other pollutants, and controlling fluctuations in stream flow. These wooded and wet streamside habitats provide food to the stream in the form of leaves and other vegetation; they moderate the temperature of the water; and they often provide for the subtle differences in habitat necessary to allow for greater biodiversity in an area. As much as possible, habitat restoration activities should attempt to mimic as closely as possible the natural habitat conditions found in a stream before it was altered. The elimination of dams (see Joseph, 1998), the natural recovery of a channelized stream, and the return of wooded habitats adjacent to a stream all contribute to habitat restoration.

The elimination and/or control of non-native species is a more difficult problem. Currently there are no effective means of eliminating a nuisance species such as zebra mussels other than protecting waterways from them to begin with. Ohio Sea Grant (1995) has compiled a booklet that will assist in understanding and controlling this threat, and the Ohio Department of Natural Resources Division of Wildlife has prepared a State Management Plan for Aquatic Nuisance Species (1997).

The conservation of mussels, then, boils down to just these three concepts: realizing that mussels have value, identifying the threats to their continued existence in the state, and working to protect and restore their populations in Ohio. If there is a take-home message here, it is that even with all of the declining species, even with all of the threats, even with all of the unknowns regarding the basic biology and distribution of these animals in the state—even with all of these—it is not too late. It is not too late to protect what dwindling mussel resources we currently have, and it is not too late to restore some of the rich bounty Ohio has known in the past.

Distribution of Mussels in Ohio

Much of Ohio's mussel diversity is due to Ohio's physiography (fig. 14). To the south lies the Ohio River and to the north Lake Erie. Three physiographic provinces extend into Ohio: the Central Lowland (Huron-Erie Lake Plains and Till Plains sections), the Appalachian Plateaus (Glaciated Allegheny Plateaus and Allegheny Plateaus sections), and the Interior Low Plateau (Bluegrass Section). While the great majority of Ohio's mussels are derived from the Ohio River fauna, a few species, such as the Eastern Pondmussel, *Ligumia nasuta,* are part of the Lawrentian fauna of the Eastern Seaboard. Most of our mussels occur in the glaciated portion of the state, within the Central Lowland and Interior Low Plateau provinces and within the Glaciated Allegheny Plateaus Section of the Appalachian Plateaus Province. There are notable exceptions, such as the Muskingum River system, most of which flows through the unglaciated portion of the state but which possesses a streambed composed of glacial outwash high in carbonates and high in mussel diversity. Ohio's major watersheds are shown in figure 15.

Ohio is composed of two large regions, differing in their geological history, soil types, and vegetation. The southeast third of the state was never glaciated and consists of portions of the Appalachian Mountains Foothills. This area stretches roughly from Wheeling in the northeast to Brown County in the southwest and includes the Hocking Hills region and Wayne National Forest. The bedrock there is composed of sandstone and shale, largely of Permian and Pennsylvanian Period origin, with mostly aluminum- and iron-enriched soils (alfisol) and a mixture of very old, nutrient-depleted soils (ultisol) and very young, nutrient-poor soils (entisols and inseptisols). The original vegetation consisted mainly of mixed oak forests. Some of Ohio's largest rivers cut through this region: the Muskingum, Hocking, and Scioto rivers. Streams leading to these rivers and to the Ohio River beyond often flow over exposed bedrock and have high gradients. They may be relatively poor in calcium—a mineral that mussels need to construct their shells. As a consequence, these streams often have a depauperate mussel fauna.

Glaciated Ohio was overlain to varying degrees by the Kansan, Illinoisan, and Wisconsin glaciers, beginning approximately 600,000 years ago and ending (for now) about 10,000 years ago. Mussels were obviously absent in this region during the last glaciation. Mussels therefore did not colonize this area of Ohio until at least 10,000 years ago—the same time Ohio's earliest humans, the Paleoindians, colonized the area. As the glacier retreated, meltwater flowed mainly south to the newly formed Ohio River, and the corridor for most of Ohio's mussels was established.

Mussel distributions also have been influenced by the construction of canals (fig. 16). These structures linked disparate drainages and allowed mussels and their hosts to migrate to places outside their natural range. Two great canal systems were built in Ohio linking Lake Erie to the Ohio River. The Ohio and Erie Canal was completed in 1832, linking Cleveland and Portsmouth. The Miami and Erie Canal was completed in 1845, linking Toledo and Cincinnati. Spurs of the Ohio and Erie Canal connected to the Muskingum and Hocking rivers, and other spurs reached as far as Pennsylvania. A spur of the Miami and Erie Canal connected to the Wabash River. Mussels were apparently abundant in at least some canals. Of the Columbus feeder canal of the Ohio and Erie Canal, Higgins (1858) noted: "Many species have traversed the whole length of the canal, and many species there thrive and become abundant which are quite rare in the adjacent rivers." Species living in the canals were not just soft-substrate taxa but also included riverine species, as evidenced by records at the Carnegie Museum of Natural History, Pittsburgh, of now-endangered species living in these canals. Several species were even described from canals. But Ohio's canal system was out-competed by the railroad, and its demise was complete in 1913 when floods demolished canals state-wide. How these canals changed the distributions of mussels in Ohio is unknown, but we are certain that some genetic mixing must have occurred.

Regulation of Mussels in Ohio

No one denies that Ohio's mussel fauna has experienced dramatic declines. Much effort and money has been spent to ensure that whatever remains will be protected for posterity. Federally endangered species are under the jurisdiction of the U.S. Fish and Wildlife Service and the Ohio Department of Natural Resources (ODNR) Division of Wildlife. All other species, regardless of rarity, are under the jurisdiction of the ODNR Division of Wildlife. No other state, local, or private agency has statutory authority over these animals.

Several laws are of importance to anyone wishing to work with mussels in Ohio. Early on, the ODNR Division of Wildlife recognized that the mussels that were being commercially harvested, mostly in the lower Muskingum River, were not a renewable resource given the current rate of harvest. All commercial collecting was stopped statewide in 1975. However, a limited number of mussels could still be used for bait purposes. Realizing that endangered species were occasionally being shucked for bait prompted the Division to end that practice as well, and currently it is illegal to collect any live mussel or dead shell, including Asian clams, zebra mussels, and quagga mussels, regardless of rarity, without a Scientific Collecting Permit.

The state endangered status of Ohio mussels is based strictly upon their occurrence within the state's borders. A species may be abundant just across the state line, but that does not affect its Ohio status. For this reason, some Ohio species considered endangered may be quite common globally. Some of these species occur in the lower Muskingum River, where they are considered endangered. Endangered status does not necessarily mean that a species exists in few numbers. *Pleurobema cordatum*, an Ohio endangered species, is abundant in the lower Muskingum River. It is considered endangered because that is one of the *only* places it occurs in the state, and, abundant or not, it could be extirpated by a single ecological catastrophe.

Records Used in This Book

The distributional maps figured in this book are based on 35,000+ records at The Ohio State University Museum of Biological Diversity (OSUM). Additional records were included from the Carnegie Museum of Natural History, Pittsburgh; the University of Michigan Museum of Zoology, Ann Arbor; the Cincinnati Museum of Natural History and Science; the Cleveland Museum of Natural History; and the Boonshoft Museum of Discovery (formerly the Dayton Museum of Natural History). Collection sites are shown in figure 17.

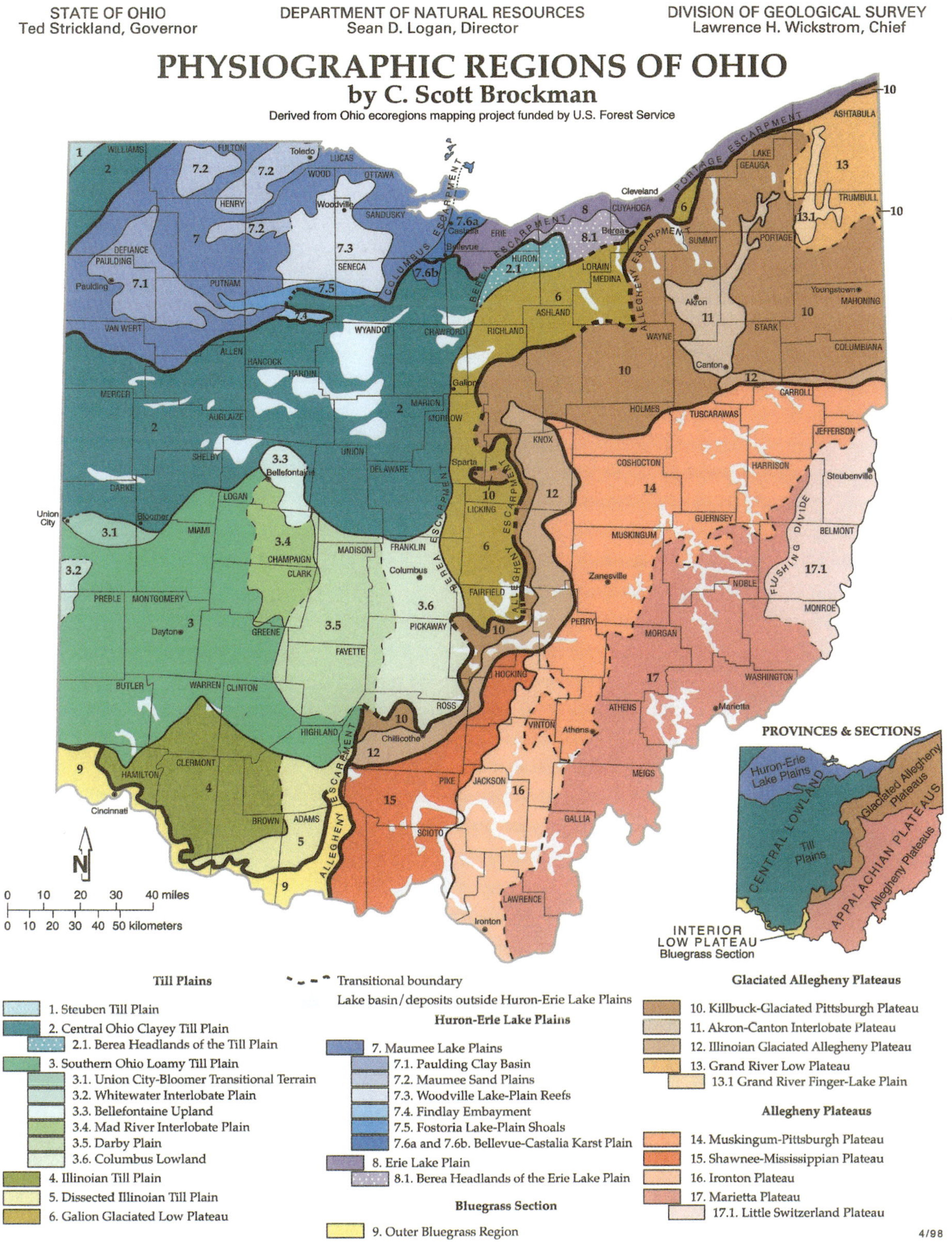

Fig. 14. Ohio's physiographic regions.

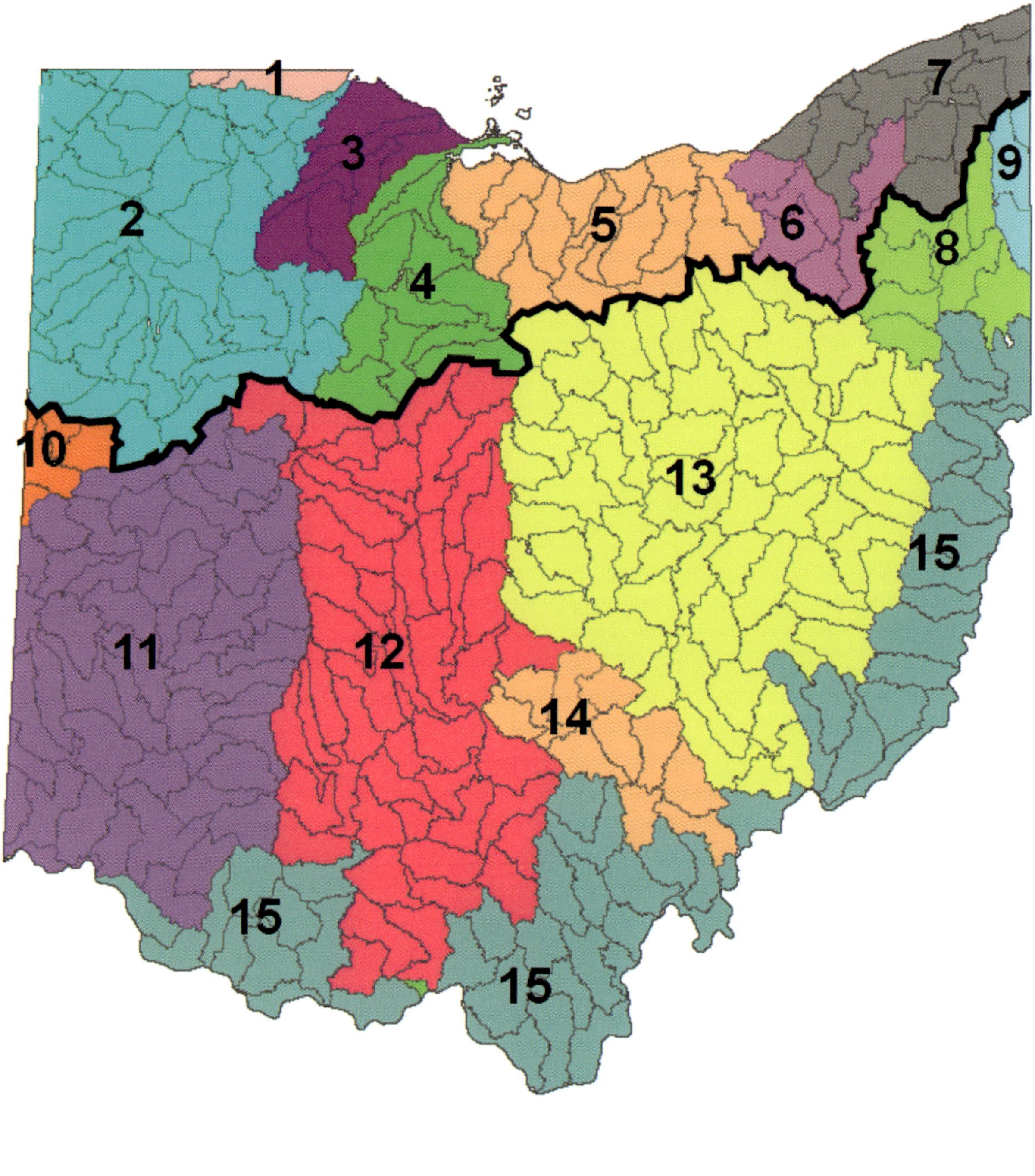

Fig. 15. Ohio's principal drainages.
1—Tenmile Creek. 2—Maumee River. 3—Portage River. 4—Sandusky River. 5—Huron, Vermilion, and Black rivers. 6—Cuyahoga River. 7—Grand and Ashtabula rivers. 8—Mahoning River. 9—Pymatuning Creek. 10—Wabash River. 11—Miami River. 12—Scioto River. 13—Muskingum River. 14—Hocking River. 15—minor Ohio River creeks. Black line—divide between Lake Erie and Ohio River drainages. Gray lines—eleven-digit hydrologic units.

Fig. 16. Ohio's canal system: Miami and Erie Canal (left) and Ohio and Erie Canal (right).

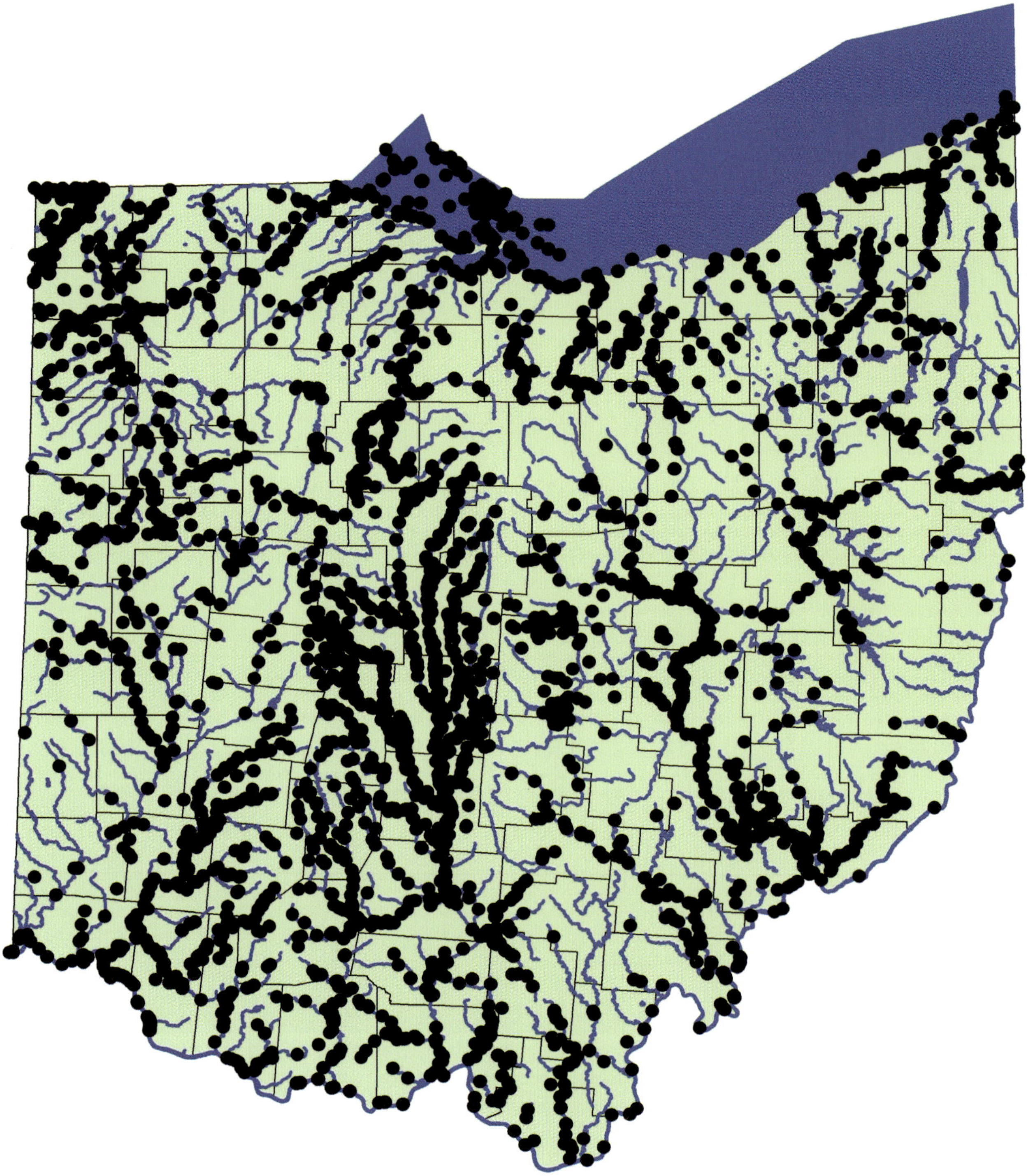

Fig. 17. Collection sites used in this book.

Species Accounts

Family Margaritiferidae Henderson, 1929

Cumberlandia Ortmann, 1912

Etymology. From the Cumberland region of North America. "Discovered" by Dr. Thomas Walker in 1748 and named after the then Prime Minister of England, the Duke of Cumberland.

Original description

Cumberlandia Ortmann, 1912b: 13–14.

Type species. *Unio monodonta* Say, 1829, by original designation.

Geographic range. This monotypic genus is found only within the Mississippi River basin. It still lives in some parts of the main stem above St. Louis as well as in less damaged parts of its former range in both eastern and western tributary systems.

Comments. There is not complete agreement that the species of *Margaritifera* Schumacher, 1816, are sufficiently distinct from *C. monodonta* to justify the removal of the latter to a separate monotypic genus. Haas (1969: 9–14) proposed the use of subgenera to solve this and several similar problems within the Margaritiferidae, but Heard & Guckert (1971) raised *Cumberlandia* to subfamily status. Vaught (1989) synonymized *Cumberlandia* with *Margaritifera* without justification. Smith (2001) placed *C. monodonta* in *Margaritanopsis*, a change that has not been adopted by most workers.

Cumberlandia monodonta (Say, 1829)

Figures 18, 19

Etymology. From G. *monos*, one, single + G. *odontos*, tooth. Say (1829: 293) was of the opinion that each valve of this species developed only one cardinal tooth. While many specimens do fit this description, others do not.

Vernacular name(s). *Spectaclecase*, Cumberland Pearl Shell, Mule's Ear.

Original description

Unio monodonta Say, 1829: 293, no plates or figures [redescribed in Say, 1830a: 2 pp., no pagination, pl. 6. The redescription is essentially a reprinting of the original description accompanied by the information that "Ms. Lesueur has presented me with a specimen found in the Wabash [River]" and a plate (VI) whereupon " . . . two views present the inner and exterior pages of the shell."].

Type locality. In 1829 Say gave the type locality as "the Falls of the Ohio [River]" (p. 293). The Falls of the Ohio are at Louisville, Kentucky. Say's (1830a) subsequent locality of "Wabash [River]" is not to be construed as a correction of the type locality.

Type material. A search of the North American museums having major unionoid collections has not revealed primary type material of this species. Following the death of Thomas Say in 1834 his widow, Lucy W. Say, stated (1840: 3) in a republication of some species described by her late husband: "If any specimens here described are not contained in the Cabinet of the Academy of Natural Sciences, they will probably be found in my own private collection, which has not been opened since it was put up five years since. If this should be the case, it will afford me much pleasure to place them in that collection for future reference." Johnson & Baker (1973: 146) noted, with reference to the presence of Say's other unionid types at the Philadelphia Academy of Natural Sciences, that ". . . there are none there in spite of the statement of Mrs. Say. . . ." It appears that the primary types of *C. monodonta* are lost.

Synonymy

Unio monodonta Say, 1829: 293.

Unio soleniformis Lea, 1831: 87–89, pl. 10, fig. 17 [Type locality: "Ohio," probably Ohio River at Cincinnati. Type not located].

Unio monodontus Say, 1829. Say, 1834: unpaginated.

Alasmodonta monodonta (Say, 1829). Férussac, 1835: 26.

Margarita (*Unio*) *monodontus* (Say, 1829). Lea, 1836: 40.

Margaron (*Unio*) *monodonta* (Say, 1829). Lea, 1852b: 39.

Margaritana monodonta (Say, 1829). Conrad, 1853: 262.

Margaritana soleniformis (Lea, 1831). Paetel, 1890: 173.

Margaritana monodonta (Say, 1829). Ortmann, 1912a: 233.

Cumberlandia monodonta (Say, 1829). Ortmann, 1912b: 13.

Margaritifera (*Cumberlandia*) *monodonta* (Say, 1829). Haas, 1969: 14.

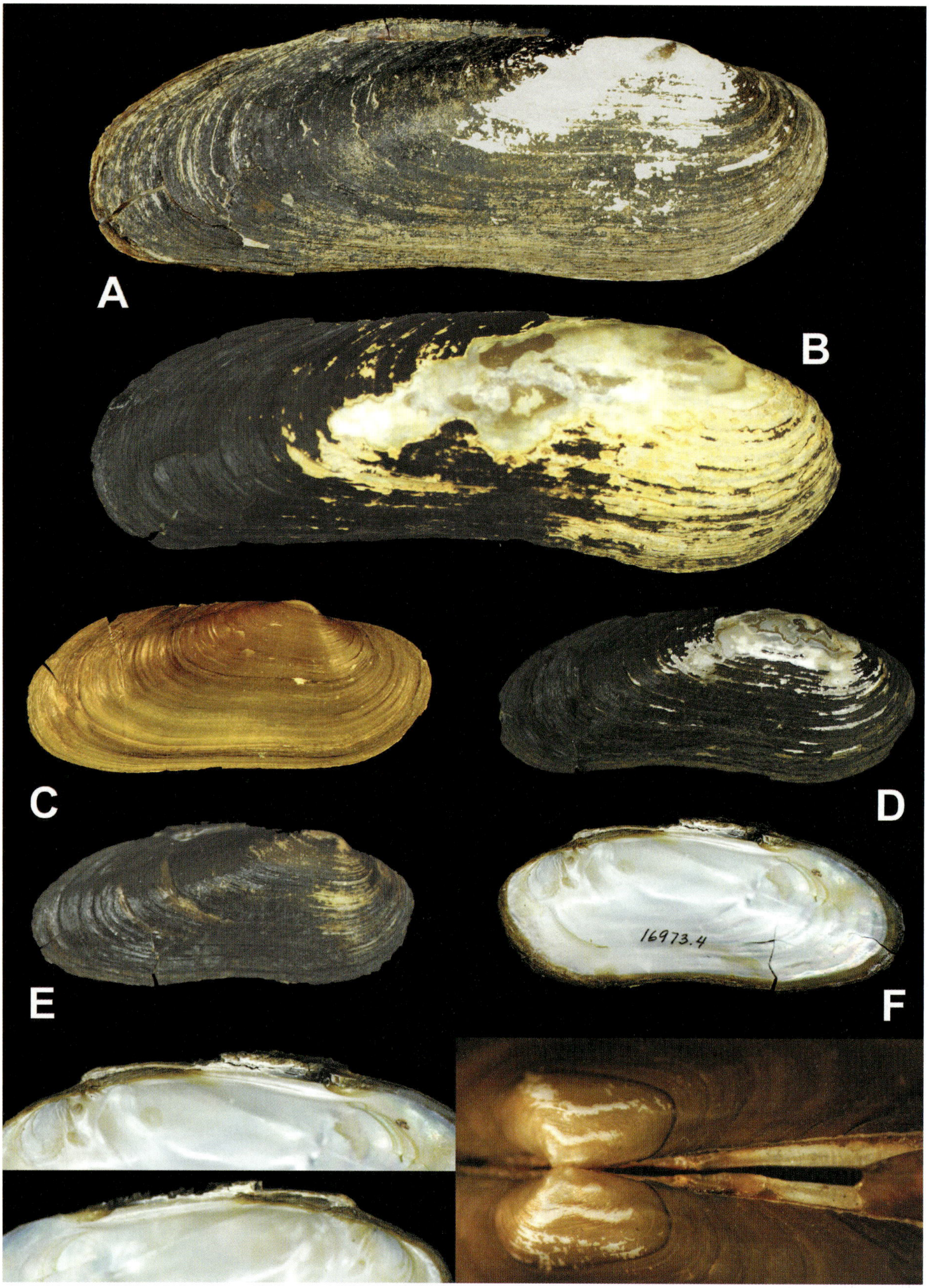

Fig. 18. ***Cumberlandia monodonta*** **(Say, 1829). A.** OSUM 26587.6, Clinch River, Hancock Co., VA. 139 mm. **B.** OSUM 57776, Big Piney River, Pulaski Co., MO. 168 mm. **C, Beak sculpture detail:** OSUM 4915.1, Ohio River, Hamilton Co., OH. 63 mm. **D.** OSUM 10004.2, Ohio River, Hamilton Co., OH. 106 mm. **E, F, Hinge detail:** OSUM 16973.4, Red River, Robertson Co., TN. 104 mm.

Margaritanopsis monodonta (Say, 1829). Smith (2001).
Status. Listed as Extirpated by Ohio Division of Wildlife; listed by U.S. Fish & Wildlife Service as Threatened as of 2009.

ADULT SHELL

Shell. Medium to large size, thin when young becoming subsolid with age, outline unusually elongate, frequently becoming arcuate with age.
Anterior margin. Flattened dorsally becoming uniformly rounded.
Ventral margin. Nearly straight when young becoming emarginate to arcuate with age.
Posterior margin. Gently curved becoming nearly straight dorsally.
Dorsal margin. Very gently curved becoming straight or nearly so in some individuals.
Umbo. Within 25% of the total length from the anterior margin, low, rising only 3–5 mm above the hinge line in oldest specimen, sculpture of 3–5 straight or slightly emarginate subparallel bars passing from anterior slope to posterior slope becoming heavier and parallel to the growth lines.
Ligament. Brownish-black, exposed, heavy, strong, extends the length of the hinge plate becoming gradually heavier posteriorly, extends from umbonal region to the anterior margin of the posterior adductor.
Disc sculpture. Surface smooth from the umbonal sculpture to the margin except for growth-lines and accidental irregularities, anterior and posterior ridges smooth and broadly rounded, the area between becoming very wide, very flat, sulcus scarcely perceptible.
Periostracum. Dusky green when young becoming darker green grading into dark brown or black with age, old specimens typically black. The pronounced tendency of this shell to crack upon drying may be due to shrinkage of a periostracum attached firmly to unusually thin non-shrinking calcareous layers beneath.
Nacre. White, iridescent when and where thin-shelled, Tulberg layers common, indistinct wash of yellow-green frequent in umbonal cavity of young.
Hinge plate. Thin and weak when young becoming moderately stronger with age, elongate, narrow at interdentum, becoming increasingly wider and flattened posteriorly in the region of the lateral teeth, right valve having a single, poorly developed, cardinal tooth beneath and just anterior to the umbo, right lateral tooth single becoming wider posteriorly, left valve with a single poorly developed cardinal tooth beneath and just posterior to the level of the umbo, left laterals poorly developed with one or sometimes absent, older individuals commonly having lateral teeth partially or completely obscured by later nacre deposition.
Muscle scars. Anterior adductor deeply impressed, rough, anterior protractor very close or contiguous, but typically distinct, anterior retractor very close or contiguous, lateral microfossa few or abundant, but rarely absent, dorsal suspensor(s) elongate-simple or abbreviate-multiple, but deeply impressed and extending most of the interdental length posteriorly from the umbonal cavity, posterior adductor lightly impressed, especially in young individuals nearly always confluent with posterior pedal retractor, pallial line complete, subparallel with shell margin, pallial accessory absent.
Umbonal cavity. Shallow.

Glochidia. Depressed subelliptical, with lateral margins equally rounded and a straight hinge line. Literature records (Surber, 1915: 6, pl. 1, fig. 7; Howard, 1915: 6) indicate glochidia dimensions for length (0.05 to 0.07 mm) and height (0.05 to 0.07 mm). Surber (1915: 6) noted, "This minute glochidium suggests *Tritogonia tuberculata* [*T. verrucosa*] in general shape but has less proportionate depth, with longer hinge line." Howard (1915: 6) added, "The glochidia are sub-circular in form, i.e., circular except for the rather short hinge line."

Reproductive biology. Apparently two broods are produced each year, as gravid females were found in May and again in October (Howard, 1915; Gordon & Smith, 1990). Flat, branched conglutinates are produced (Knudsen & Hove, 1997); these are of the amorphous type (Watters, 2008).

Reported potential hosts

Hybopsis amblops (Bigeye Chub)—Baird, 2000 (NI).
Moxostoma macrolepidotum (Shorthead Redhorse)—Baird, 2000 (NI).

Habitat. The Spectaclecase has been found in several different habitats. It often is associated with water willow stands in slow-moving water (Stansbery, 1966a). However, it also occurs in very swift water in sand under large flat rocks. The historic records of this species correlate, in general, with medium to large streams. Within these streams it is rarely found except in close association with the rapid currents associated with riffles and runs. Hildreth (1837) noted of this species that he had "never yet found [it] in a living state, although I have often picked up the shells of the recent dead for the reason that they live in deep water and are buried in mud."

The habitat description above may be based more on inference than experience. Call (1900: 526) reported that "The habits of the animal render it difficult to find and it is not common in even good collections. It buries itself far down in the gravel and mud, on the falls, under

Fig. 19. ***Cumberlandia monodonta*** **(Say, 1829)**
Solid dots—live or freshdead since 1980
Bullseye—all records, any condition

large flat rocks where water circulates freely, or alongside submerged timbers which are well buried in the mud of the river bottom." In the Cumberland River prior to impoundment, Neel & Allen (1964: 432) found *C. monodonta*: "Confined to the main river below the falls; lives under large stones and may be collected only when or where the water is shallow enough to permit wading and moving of such stones; very poorly represented in muskrat piles."

Range. The range of *C. monodonta* is confined to the Mississippi River system. Collection records at OSUM include lots from the following states: Alabama, Iowa, Illinois, Indiana, Kentucky, Minnesota, Missouri, Ohio, Tennessee, Virginia, and Wisconsin. In addition to these vouchered lots, there is a record in the literature from Arkansas (Wheeler, 1918). Parmalee & Bogan (1998) stated that it is known from as far south as the Gulf of Mexico, but this seems unlikely.

It was apparently never common in the Ohio River mainstem above Cincinnati and we have seen no historic records from eastern Kentucky, West Virginia, or Pennsylvania. A single valve from an archaic midden site in the Allegheny River in eastern Pennsylvania suggests a prehistoric range upstream in the Ohio River system well above any historic records.

Range in Ohio. This species has been recorded only from the Ohio River and its tributary streams within the state. The earliest Ohio record appears to be that of the unpublished Hildreth Catalogue (1830: 111, No. 68) (copy at OSUM) with the simple entry of "Ohio." The redescription of *U. monodonta* as *U. soleniformis* Lea, 1831, is based upon a specimen from the "Ohio river. T.G. Lea." Since T.G. Lea was a resident of Cincinnati, Ohio, and sent many unionid specimens to his brother, Isaac Lea, in Philadelphia, it seems likely that the holotype for this newly proposed name came from the Ohio River at or near Cincinnati. Short & Eaton (1831: 79, 80) recorded this species from the "Ohio [River]" "eighty miles due north from Lexington [KY]," which describes the location of Cincinnati. There are additional literature records of this species from Cincinnati over the years. There can be little doubt that the species was represented by a relatively persistent population at least at that site, but records from farther upstream and from streams *within* Ohio are scant. Kirtland's paper (1838: 171) simply noted "in Ohio" while Sullivant's (1838: 10) record gave our only "in-state" record (other than the Ohio River) as "Ohio & Sciota [*sic*] Rivers." Sterki (1907: 393) in his catalogue of Ohio Mollusca reported that *M. monodonta* was in the "Ohio River at Cincinnati (still!), but not common." Five years later Ortmann (1912b: 14) reported collecting it from the "Ohio river at Buffington Island near Portland, Meigs County, Ohio." Finally, a weathered valve was found in 1998 in the Muskingum River at Devola, Washington County. The lack of records either from the Ohio River or from in-state streams in recent years indicates that this species is either very rare or, as we suspect, extirpated from this part of its former range.

Similar species. Adult *C. monodonta* might be confused with several elongate dark-colored species (such as *Elliptio dilatata, Ligumia recta*, etc.) except for its being rounded at both extremities, lacking rays, having an incomplete diaphragm, and having the interlaminar connections of the demibranch developed as continuous septa that pass dorsal to ventral in the anterior direction.

A more frequent problem appearing in collections is the confusion of young *C. monodonta* with young and adult *Simpsonaias ambigua*. There is no problem if the soft parts are present, but the usual specimens are typically only shells that are, at least superficially, quite similar. Both have rounded extremities, both are essentially rayless, both have white nacre, both have reduced cardinal teeth, both have inconspicuous lateral teeth, and both are typically found beneath large flat rocks in riffles and runs.

If the internal structure of the shell is visible, one may identify *C. monodonta* by the presence of lateral teeth, double in the left valve, the elongate dorsal suspensor scar on the underside ("hidden") surface of the interdentum, and lateral microfossae. The beak sculpture, if not eroded away, consists of coarse elongate bars that contrast with the fine double-loops of *S. ambigua*. Other shell differences in both structure and color exist, but they are highly variable and difficult to see—especially in the field or in the case of badly worn shells.

Comments. This species is presumed extirpated from the state since there appear to be no Ohio records of living animals since the Cincinnati record of 1907 and Ortmann's Buffington Island (Portland, Ohio) record of 1912. A weathered valve was recently found in the lower Muskingum River by Watters. The paucity of Ohio material of this species in collections, even old collections, indicates that it was never a common species in the Ohio part of its range. Even so, we should keep in mind that its habitat may render both its limited range and its rarity within that range more apparent than real.

Its continued presence outside Ohio in the upper Tennessee River system, several streams in the Ozark Plateau, and the upper Mississippi River may make possible the reintroduction of the species when (if) the overall requirements for its continued existence here are regained.

Although never an important commercial species, it is the only representative of one of only two families of unionoid molluscs known from Ohio in Recent times. Its absence robs us of any value, academic or economic, it may have, until it can be established here once again.

Family Unionidae Rafinesque, 1820

The family Unionidae is one of several freshwater bivalve groups occurring today that once were believed to be related under the superfamily Unionoida. The number of groups, whether families or subfamilies, is debatable: Unionidae, Margaritiferidae, Hyriidae, Mycetopodidae, Iridinidae, Etheriidae. Several subfamilies have been recognized within the Unionidae as well: Unioninae, Anodontinae, Ambleminae (Pleurobeminae), and Lampsilinae. The Margaritiferidae and Hyriidae are occasionally included as subfamilies there as well.

Actinonaias Crosse & Fischer, 1894

Etymology. Gr. *aktis, aktinos,* ray, beam + Gr. *naias, naiados,* a water nymph of rivers and springs. In the original description, Fischer & Crosse (1894: 557) noted "surface ornamented with numerous narrow rays; . . . ," and this condition is fairly typical within the genus *Actinonaias* as understood today.

Original description

Actinonaias Crosse & Fischer *in* Fischer & Crosse, 1894: 556–557.

Type species. *Unio sapotalensis* Lea, 1841, by subsequent designation of Frierson (1917: 48), supported by the anatomical studies of Ortmann (1912a: 324–327). This designation was followed in the review of Ortmann & Walker (1922: 47).

Synonymy

Actinonaias Fischer & Crosse, 1894: 556–557.

Lampsilis (*Ortmanniana*) Frierson, 1927: 11, 79 [Type species: *Unio carinatus* Barnes, 1823, by original designation].

Geographic range. Central America. If the North American species belong to this genus, which is unlikely, then the Mississippi River drainage must be included.

Comments. It is highly unlikely that the North American species assigned to *Actinonaias* actually belong here, the genus being based on a Central American species. Central and North America share very few genera, and those appear to be derived from North America (*Quadrula couchiana* (Lea, 1860), *Potamilus purpuratus "coloradoensis,"* and several species of *Leptodea*, for example). Those groups have widespread taxa that range from the Mississippi River basin to across the Rio Grande. The North American *Actinonaias* are Mississippian with no such intervening taxa that reach to Central America. Both groups seem to have little more in common than relatively nondescript shells that could place them in any number of equally nondescript genera. Indeed, Simpson (1900a, 1914) seems to have ignored Fischer & Crosse's *Actinonaias*, and placed the southern species in *Nephronaias*, another Central American group. Ortmann (1912a: 324–327) examined the anatomy of *Nephronajas* [*sic*] *ligamentina*, *perdix*, and *sapotalensis*. Of the last, a Mexican species, he remarked that it was "in every respect" like *ligamentina* and *perdix*. However, he also noted that the anatomy of *ligamentina* was "in every particular like that of *Obovaria*." Based on these equivocal comments, there is little justification for identifying *ligamentina* with Central American *Actinonaias*. In all likelihood, based on shell morphology, the North American species should be referred to *Ortmanniana* Frierson, 1927.

Actinonaias ligamentina carinata (Barnes, 1823)

Figures 20, 21

Etymology. L. *carinata,* carinated, ridged.

Vernacular name(s). *Mucket.*

Original description

Unio carinatus Barnes, 1823: 259, pl. 11, fig. 10.

Type locality. Fox River [Wisconsin].

Type material. Apparently lost.

Synonymy. Most workers have not differentiated between the two subspecies presented here, making it difficult to know which taxon was in question. For this reason we present only partial synonymies. For a complete listing as *Actinonaias ligamentina s.l.*, see Bogan & Parmalee, 1998.

Unio carinatus Barnes, 1823: 259, pl. 11, fig. 10.

Mya carinata (Barnes, 1823). Eaton, 1826: 220.

Actinonaias carinata (Barnes, 1823). Ortmann & Walker, 1922: 47.

Lampsilis (*Ortmanniana*) *carinata* (Barnes, 1823). Frierson, 1927: 79.

Actinonaias carinata orbis Morrison, 1942: 361, no figure given [Figured holotype USNM 84998 *fide* Johnson, 1975: pl. 1, fig. 3. Johnson (1975: 52) gave the type locality as "[Tennessee River,] Florence, [Lauderdale Co.,] Alabama."].

Lampsilis (*Ortmanniana*) *carinata carinata* (Barnes, 1823). Murray & Leonard, 1962: 108.

Actinonaias carinata carinata (Barnes, 1823). Murray & Leonard, 1962: 108–111.

Lampsilis (*Ortmanniana*) *carinata orbis* (Morrison, 1942). Haas, 1969: 461.

Actinonaias ligamentina carinata (Barnes, 1823). Oesch, 1984: 146.

Status. Not listed by either Ohio Division of Wildlife or U.S. Fish & Wildlife Service as of 2009 [as *A. l. carinata*].

ADULT SHELL

Shell. Large (to at least 140 mm), moderately heavy,

Fig. 20. ***Actinonaias ligamentina carinata*** **(Barnes, 1823). A.** OSUM 50608, Muskingum River, Washington Co., OH. 109 mm. **B.** OSUM 6615, Muskingum River, Washington Co., OH. 131 mm. **C, F, Hinge detail:** OSUM 17260.5, Walhonding River, Coshocton Co., OH. 109 mm. **D.** OSUM 46559.1, Walhonding River, Coshocton Co., OH. 50 mm. **E.** OSUM 32324.2, Kankakee River, Kankakee Co., IL. 62 mm. **Beak sculpture detail:** OSUM 32324.1, Kankakee River, Kankakee Co., IL. 54 mm total length.

outline oval, juveniles and males more elongate; somewhat laterally compressed.

Anterior margin. Straight for a short distance, then evenly rounded to ventral margin.

Ventral margin. Gently rounded or straight, ending at an acute point posteriorly; females may have a second, obtuse angulation at the beginning of the marsupial region.

Posterior margin. Nearly straight, broader than in *A. l. ligamentina*.

Dorsal margin. Long, straight or somewhat curved, forming an obtuse angle with the posterior margin.

Umbo. Very wide, low, nearly flush with the dorsal margin in many specimens; beak sculpture extremely fine to absent, composed of extremely fine concentric threads on the anterior 25% of the beak; beak sculpture disappears by end of first year's growth.

Ligament. Brown or black, prominent, occupying most of the dorsal slope, partially contained between the valves.

Disc sculpture. Sculptureless except for 1–3 ridges on the posterior slope; the ventral-most of these may form a prominent angle, more pronounced than in *A. l. ligamentina*.

Periostracum. As in *A. l. ligamentina*, but with more of a tendency to be rayed; periostracum often shiny, not matte as in *A. l. ligamentina*.

Nacre. White, iridescent posteriorly; green, orange, or gray Tulberg layers common, sometimes extensive.

Hinge plate. Moderately heavy, more so in old individuals; interdentum short, cord-like; two chunky, triangular cardinal teeth in left valve; one cardinal tooth in right, but often with a prominent anterior accessory tooth; cardinals long, straight, or slightly bowed, two in left, one in right.

Muscle scars. Anterior adductor moderately deep, rough, distinct from large anterior pedal protractor; posterior adductor partially impressed in large specimens, confluent with posterior pedal retractor scar; dorsal suspensors multiple, arranged in a straight or curved line at apex of umbonal cavity from interdentum to side of cardinal fossa; pallial line complete, accessory line present on all but largest specimens, long but not distinctly impressed.

Umbonal cavity. Shallow to moderately deep, wide, open or barely compressed.

Glochidia. Subelliptical, symmetrical, with a straight dorsal margin, gently and evenly arched lateral margins, and a semicircular ventral margin. Beaded exterior valve sculpturing covers the valve, dorsal alae present, and hook absent (Hoggarth, 1999: 68–69).

Literature records (Hoggarth, 1999: 68) indicate glochidial dimensions for length (0.22 mm), height (0.24 mm), and hinge length (0.13 mm). The glochidium of *A. l. carinata* is nearly identical to that of *A. pectorosa*, except it is smaller.

Reproductive biology. Bradytictic. Individuals reach sexual maturity from 4 to 6 years of age (Jirka & Neves, 1992). The posterior half of the outer gills is marsupial (Ortmann, 1912a). Eggs develop in mid-summer (Ortmann, 1919; Jirka & Neves, 1992) and glochidia are present from September to the following May to August (Surber, 1913). Glochidia are released in large (30 x 10 x 2 mm), white, leaf-shaped conglutinates that quickly disassociate (Watters, 2008).

Reported potential hosts

Ambloplites rupestris (Rock Bass)—Lefevre & Curtis, 1910a (LT); Lefevre & Curtis, 1912 (LI); Watters *et al.*, 1998a (LT).

Anguilla rostrata (American Eel)—Coker *et al.*, 1921 (NI).

Campsotoma anomalum (Central Stoneroller)—Watters *et al.*, 1998a (LT).

Cyprinus carpio (Common Carp)—Lefevre & Curtis, 1912 (LI).

Ericymba buccata (Silverjaw Minnow)—Watters *et al.*, 1998a (LT).

Etheostoma tippecanoe (Tippecanoe Darter)—Watters *et al.*, 1998a (LT).

Fundulus diaphanus (Banded Killifish)—Young, 1911 (LT)—exotic.

Lepomis cyanellus (Green Sunfish)—Young, 1911 (LT); Lefevre & Curtis, 1912 (LT); Wilson, 1916 (NI); Coker *et al.*, 1921 (NI).

Lepomis humilis (Orangespotted Sunfish)—Young, 1911 (LT).

Lepomis macrochirus (Bluegill)—Wilson, 1916 (NI); Coker *et al.*, 1921 (NI); Lefevre & Curtis, 1912 (LI).

Micropterus dolomieu (Smallmouth Bass)—Coker *et al.*, 1921 (NT); Howard & Anson, 1923 (NS).

Micropterus salmoides (Largemouth Bass)—Lefevre & Curtis, 1910a (LT); Young, 1911 (LT); Lefevre & Curtis, 1912 (LT); Wilson, 1916 (NI); Reuling, 1919 (LT); Coker *et al.*, 1921 (NI, LT); Howard & Anson, 1923 (LT); Watters *et al.*, 1998a (LT).

Morone chrysops (White Bass)—Surber, 1913 (NI); Wilson, 1916 (NI); Coker *et al.*, 1921 (NI, LT).

Noturus gyrinus (Tadpole Madtom)—Coker *et al.*, 1921 (NI).

Perca flavescens (Yellow Perch)—Lefevre & Curtis, 1910a (LT); Lefevre & Curtis, 1912 (LI); Coker *et al.*, 1921 (NI, LT).

Pomoxis annularis (White Crappie)—Young, 1911 (LT); Lefevre & Curtis, 1912 (LI); Wilson, 1916 (NI); Coker *et al.*, 1921 (NI, LT).

Pomoxis nigromaculatus (Black Crappie)—Howard, 1914a (LI); Coker *et al.*, 1921 (LT).

Rutilus rutilus (Roach)—Lefevre & Curtis, 1912 (LI)—exotic.

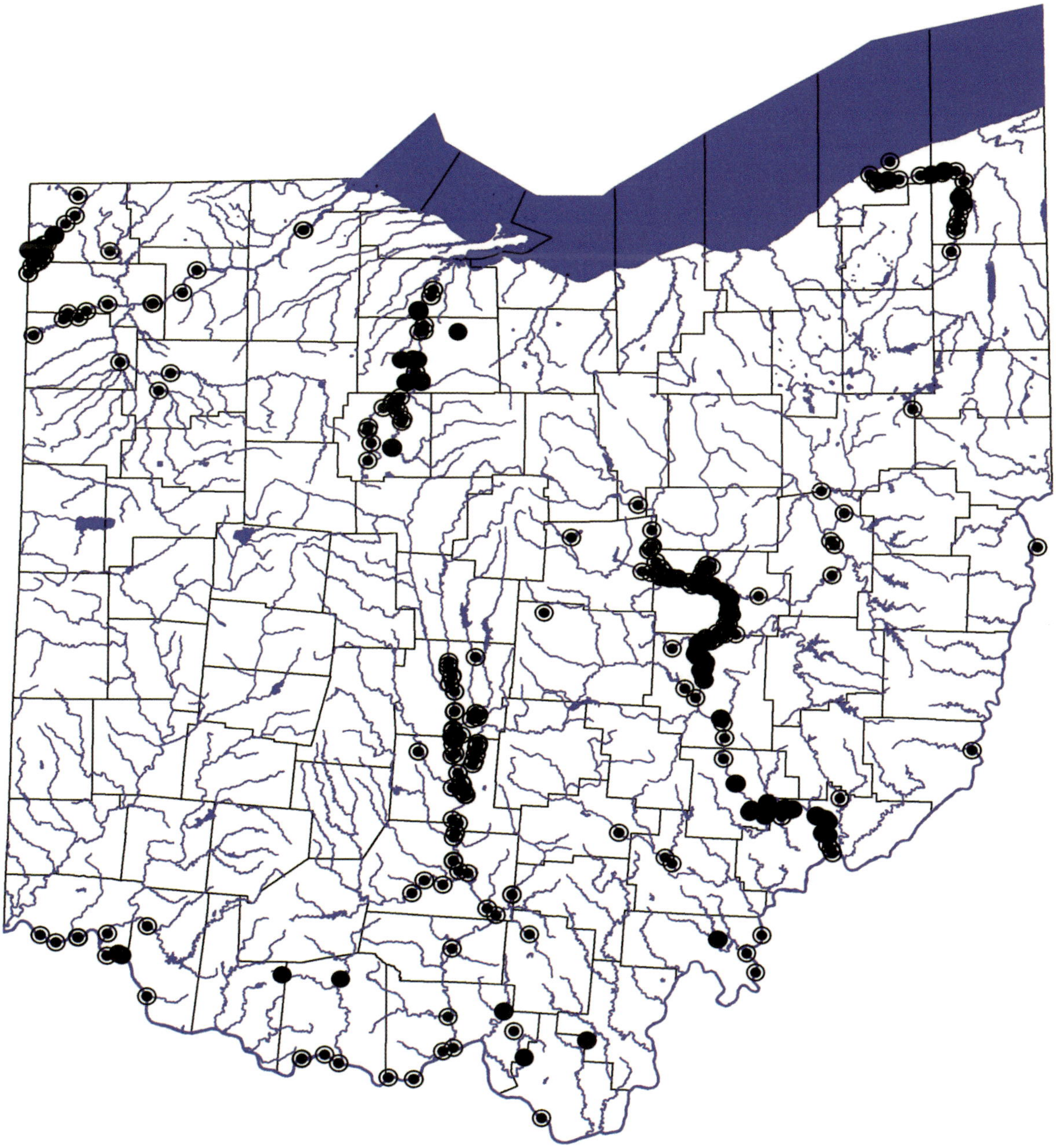

Fig. 21. ***Actinonaias ligamentina carinata*** **(Barnes, 1823)**
Solid dots—live or freshdead since 1980
Bullseye—all records, any condition

Sander canadensis (Sauger)—Coker *et al.*, 1921 (NI); Pearse, 1924 (NI).

Range. Mississippi River drainage and Lawrentian system (Lake Erie, Lake Michigan, and Lake Ontario drainages). The Kansas records of *A. l. carinata* figured by Murray and Leonard (1962: 108, pl. 28, figs. 1–4) are almost certainly *Lampsilis rafinesqueana* Frierson, 1927. The source of these Kansas specimens, the Spring and Neosho rivers, are within the range of *L. rafinesqueana*. Couch (1997) stated that *A. l. carinata* was restricted in Kansas to the Marais des Cygnes River system.

Range in Ohio. Widespread in the larger rivers, but absent from the Great Miami River. Locally abundant in the Sandusky, Grand, and Muskingum drainages. Apparently extirpated from the Scioto River system.

Similar species. The subspecies *ligamentina* differs from *carinata* in having a less carinate dorsal ridge, generally being more oval, and often distinctly humped. Nevertheless, some specimens may be difficult to place.

Comments. This subspecies does not seem to get as large as *A. l. ligamentina*; specimens more than 140 mm in length are rare. Individuals live for up to 25 years.

Actinonaias ligamentina ligamentina (Lamarck, 1819)

Figures 22, 23

Etymology. L. *ligamentum*, a band, tie, bandage. *Actinonaias ligamentina* is a thick-shelled species with a large, strong ligament. Lamarck must have thought this the most striking feature of this species since he discussed it at length in his original description and used this characteristic as his source of the species name.

Vernacular name(s). *Mucket*, Cumberland Mucket.

Original description

Unio ligamentina Lamarck, 1819: 72, species 7, no figure given.

Type locality. "Habite la rivière de l'Ohio. A. Michaud." The type locality was ". . . restricted to the Ohio River, Cincinnati, Ohio" by Johnson, 1969: 53.

Type material. The holotype of *U. ligamentina* is deposited in the Muséum national d'Histoire naturelle, Paris (L. 73 mm, H. 50 mm, W. 32 mm) (Johnson 1969: 53). Johnson (1969: fig. 8) illustrated Lamarck's type of this species.

Synonymy. Most workers have not differentiated between the two subspecies presented here, making it difficult to know which taxon was in question. For this reason we present only partial synonymies. For a complete listing as *Actinonaias ligamentina s.l.*, see Bogan & Parmalee, 1998.

Unio crassus Say, 1817: pl. 1, fig. 8 [*non* Retzius, 1778; Type locality: Ohio River and tributaries].

Unio ligamentina Lamarck, 1819: 72.

Mya crassa (Say, 1817). Eaton, 1826: 217.

Margarita (*Unio*) *crassus* (Say, 1817). Lea, 1836: 24.

Margaron (*Unio*) *crassus* (Say, 1817). Lea, 1852b: 28.

Margaron (*Unio*) *ligamentinus* (Lamarck, 1819). Lea, 1852b: 28.

Lampsilis ligamentinus (Lamarck, 1819). Baker, 1898b: 108.

Lampsilis ligamentina var. *gibbus* Simpson, 1900a: 540 [Type locality: "Ohio River, and southward." Types: Holotype BMNH 1907.10.28.262].

Lampsilis ligamentina gibba Simpson, 1900. Scammon, 1906: 290–291.

Nephronaias ligamentina (Lamarck, 1819). Ortmann, 1912a: 325.

Nephronaias ligamentina gibba (Simpson, 1900). Utterback, 1916a: 341.

Actinonaias ligamentina (Lamarck, 1819). Ortmann, 1919: 232, pl. 14, figs. 5, 6.

Actinonaias ligamentina ligamentina (Lamarck, 1819). Stansbery, 1976: 47.

Unio tecomensis "sp. aff." De Gregorio, 1914: 56. [Type locality?]

Status. Listed as Extirpated by Ohio Division of Wildlife; not listed by U.S. Fish & Wildlife Service as of 2009.

ADULT SHELL

Shell. Large, moderately heavy, outline oval, juveniles and males more elongate; somewhat laterally compressed.

Anterior margin. Straight for a short distance, then evenly rounded to ventral margin.

Ventral margin. Gently rounded or straight, ending at an acute point posteriorly; females may have a second, obtuse angulation at the beginning of the marsupial region.

Posterior margin. Nearly straight.

Dorsal margin. Long, straight or somewhat curved, forming an obtuse angle with the posterior margin.

Umbo. Very wide, low, nearly flush with the dorsal margin in many specimens; beak sculpture extremely fine, composed of 4–10 closely packed concentric threads on the anterior 25% of the beak; beak sculpture disappears by end of first year's growth.

Ligament. Brown or black, prominent, occupying most of the dorsal slope, partially contained between the valves.

Disc sculpture. Sculptureless except for 1–3 ridges on the posterior slope; the ventral-most of these may form a prominent angle.

Fig. 22. ***Actinonaias ligamentina ligamentina*** **(Lamarck, 1819). A.** OSUM 4991, Ohio River, [Co.?], KY. 78 mm. **B.** OSUM 54945.1, Clinch River, Scott Co., VA. 91 mm. **C.** OSUM 9511, Rockcastle River, Laurel Co., KY. 70 mm. **D.** OSUM 35087.6, Clinch River, Anderson Co., TN. 94 mm. **E, F, Hinge detail:** OSUM 45918, Muskingum River, Washington Co., OH. 91 mm. **Beak sculpture detail:** OSUM 35087.2, Clinch River, Anderson Co., TN. 37 mm total length.

Periostracum. Background color of yellowish, greenish, or gray, becoming progressively darker with age in some populations until very old individuals may be nearly black; other populations may remain yellow or green their entire lives; rays vary from absent to bold, wide, dark green rays; rays often discontinuous at annular rings; posterior slope darker in some populations.
Nacre. White, iridescent posteriorly; green, orange, or gray Tulberg layers common, sometimes extensive.
Hinge plate. Moderately heavy, more so in old individuals; interdentum short, cord-like; two chunky, triangular cardinal teeth in left valve; one cardinal tooth in right, but often with a prominent anterior accessory tooth; cardinals long, straight, or slightly bowed, two in left, one in right.
Muscle scars. Anterior adductor moderately deep, rough, distinct from large anterior pedal protractor; posterior adductor partially impressed in large specimens, confluent with posterior pedal retractor scar; dorsal suspensors multiple, arranged in a straight or curved line at apex of umbonal cavity from interdentum to side of cardinal fossa; pallial line complete, accessory line present on all but largest specimens, long but not distinctly impressed.
Umbonal cavity. Shallow to moderately deep, wide, open or barely compressed.

Glochidia. Subelliptical, nearly symmetrical, dorsal margin straight, anterior and posterior margins gently and evenly curved, ventral margin semicircular. Dorsal alae present, beaded exterior valve sculpture covers the surface of the valve, hooks absent (Lea, 1858c: 47, pl. 5, fig. 18; Ortmann, 1911: 364, pl. 89, fig. 16; Hoggarth, 1999).

Literature records (Surber, 1912: 9, pl. 2, fig. 18; Ortmann, 1912a: 326; Utterback, 1916a: 339; Hoggarth, 1999) indicate glochidial dimensions for length (0.22 mm), height (0.24 to 0.26 mm), and hinge (0.13 mm). The glochidium of *A. l. carinata* is nearly identical to that of *A. pectorosa*, except it is smaller.

Reproductive biology. Presumably as in *A. l. carinata*.

Reported potential hosts. Most hosts for *Actinonaias ligamentina* are based on *A. l. carinata*. This subspecies may use the same or similar hosts.

Habitat. The Mucket is widespread and locally abundant. It is most common in cobble and sand in moving water, rarely in ponds or lakes.

Range. Now limited to the Tennessee and Cumberland River systems. Historically also in the lower Ohio River and the lower reaches of its tributaries including: in Illinois the Illinois River, Fox River, and Rock River; in Indiana the Wabash River.

Range in Ohio. The type locality of *A. ligamentina* is the Ohio River and several sizable lots from the Ohio River collected during the middle 1800's (OSUM 19251, 87 specimens; OSUM 47229, 25 specimens) are also of this subspecies. It seems to have become extirpated from Ohio in the 1910's or slightly later, and the last individuals persisted in the Ohio River. Specimens collected in 1909 from the Ohio River at Marietta are *ligamentina* as noted by Ortmann ("with few exceptions = *ligamentina*," label to CMNH 61.10830). At other Ohio River stations *ligamentina* persisted until at least 1911, although *carinata* was already evident as well. Both forms lived in the Tuscarawas River, but *carinata* seems to have been the form found in most other rivers and creeks. Historically it occurred in the Scioto River, Duck Creek (of the Ohio River), and the Little Miami River.

Similar species. The subspecies *ligamentina* differs from *carinata* in having a less carinate dorsal ridge, generally being more oval, and often distinctly humped. Nevertheless, some specimens may be difficult to place.

Comments. The holotype of *A. ligamentina* is surprisingly the same form of this species that Simpson described as "var. *gibba*" in 1900a. This is the high-humped, typically light-colored shell form found commonly today in the middle and upper Cumberland and Tennessee River systems. In the lower parts of the Cumberland and Tennessee River systems, the Ohio River drainage north into the upper Mississippi basin, and the Great Lakes streams, we find today the more elongate, dark-rayed form matching the figured holotype of *Unio carinata* Barnes, 1823 (pl. 11, fig. 10). The Ohio River mainstem apparently retains only a few *A. ligamentina* since its transformation into a series of river-lakes, but those few in collections are the *carinata* subspecies.

St. John (1973), in a comprehensive study of the *A. ligamentina* complex, concluded, after the study of 4,478 specimens from 43 different streams, that there were at least four structurally and geographically definable forms of the *A. ligamentina* complex. The *ligamentina* Lamarck, 1819, form is restricted today to the Cumberland and upper Tennessee River systems although it also occupied all or most of the Ohio River and the lower reaches of at least some of its major tributaries during the pioneer-settlement period (1800–1900). This is clearly documented by early museum records.

The *carinata* Barnes, 1823, form is found broadly distributed from the Lakes Ontario and Erie in eastern New York northwest to Minnesota, southwest to Oklahoma and Missouri, and east to northern Kentucky, western West Virginia, and eastern Pennsylvania. Streams flowing *north* out of the Ozark Plateau also have the *carinata* form.

Fig. 23. ***Actinonaias ligamentina ligamentina*** **(Lamarck, 1819)**
Solid dots—live or freshdead since 1980
Bullseye—all records, any condition

The *gibba* Simpson, 1900, form has as its holotype the specimen figured by Sowerby (1868: fig. 520). This specimen was identified by Sowerby as *Unio crassus* Say, 1817, and its type locality as "Hab. N. America." Simpson (1900a: 540) in using this figure as the holotype gives its range as ". . . the Ohio River southward." Both of these references appear to be more intended as generalized sources or ranges rather than type localities. Morrison (1942: 361) noted that Marshall (1929) selected a neotype for *A. l. gibba* Simpson, 1900, presumably because the specimen cited by Simpson (1900a: 540) and figured by Sowerby and cited by Simpson in the original description either was not considered a legitimate holotype or had become lost. The neotype is USNM 160597 from the Black River, at Black Rock, AR. This specimen is very different from the types of both *A. l. ligamentina* and *A. l. gibba*, and is not a typical *A. l. carinata*. It is unusually dark, heavy, thick, wide, and crudely truncated posteriorly. It has the appearance of being an abnormal specimen of *A. l. carinata*. An examination of the OSUM holdings of Arkansas *A. l.* forms reveals that there are insensible gradations between this form and the common expression of *A. l. carinata* (Barnes, 1823) in the White River system. Marshall's neotype of *A. l. gibba* Simpson form appears to be a variety found only with the parent form within a restricted range. The variation and systematics of the populations of the *A. ligamentina* complex of the Ozark Plateau and its associated streams are in need of study.

Alasmidonta Say, 1818

Etymology. Gr. *a, an,* not, without + Gr. *elasmos*, a thin plate + Gr. *odontos*, tooth. Translation: "without lateral teeth."

Original description

Monodonta Say, 1817: no pagination (under heading "Conchology"), species 9 under Genus *Unio*, pl. 3, fig. 3 [*non Monodonta* Lamarck, 1801 (Gastropoda: Trochidae)].

Alasmidonta Say, 1818: 459 [as replacement name for *Monodonta* Say, 1817].

Type species. *Monodonta undulata* Say, 1817, by monotypy.

Synonymy

Monodonta Say, 1817: no pagination, pl. 3, fig. 3 [*non Monodonta* Lamarck, 1799 (Gastropoda: Trochidae)].

Alasmidonta Say, 1818: 459 [replacement name for *Monodonta* Say, 1817].

Alasmodonta Say, 1819: unpaginated [unjustified emendation of *Alasmidonta* Say, 1817].

Alasmodon [*sic*] Thomson, 1820: 116.

Alasmisodonta [*sic*] de Blainville, 1825: 639.

Decurambis Rafinesque, 1831: 4–5 [Type species: *Alasmodonta scriptum* Rafinesque, 1831 (= *Alasmidonta marginata* Say, 1819), by subsequent designation of Ortmann & Walker, 1922].

Alasmadonta [*sic*] Jay, 1835: 18.

Calceola Swainson, 1840: 382 [*non Calceola* Lamarck, 1799 (Cnidaria). Type species: *Calceola angulata* Swainson, 1840, by monotypy].

Hemiodon Swainson, 1840: 286, 288, 381 [*non* Swainson, 1840: 191 (Gastropoda: Camaenidae) (*nomen nudum*). Type species: *Unio undulata* Say, 1817, by subsequent designation of Herrmannsen, 1847. Four species were listed by Swainson, including "*H. undulatus*," all considered *nomena nuda* by Clarke, 1981a.].

Anelasmodon Agassiz, 1846: ? [unjustified emendation of *Alasmodonta* Say, 1819].

Anelasmodonta Herrmannsen, 1847: 54 [? unjustified emendation of *Anelasmodon* Agassiz, 1846].

Alasmedonta [*sic*] Gray, 1847: 196.

Alasmesodonta [*sic*] Gray, 1847: 196.

Anodontina Mörch, 1853: 48 [*non* Macquart, 1838 (Insecta: Diptera). Unjustified emendation of *Anadontina* Schlüter, 1838].

Alasmidonta (*Bullella*) Simpson, 1900a: 672 [Type species: *Margaritana arcula* Lea, 1838, by original designation].

Alasmidonta (*Pressodonta*) Simpson, 1900a: 667 [Type species: *Unio calceolus* , by original designation].

Alasmidonta (*Rugifera*) Simpson, 1900a: 670 [Type species: *Alasmidonta marginata* Say, 1819, by original designation].

Jugosus Simpson, 1914: 357 [Type species: *Strophitus wrightianus* Walker, 1901, by original designation].

Prolasmidonta Ortmann, 1914: 44 [Type species: *Alasmidonta heterodon* (Lea, 1829) by original designation].

non Uniopsis Swainson, 1840: 289, 392 [Species listed: *U. radiata* Swainson, 1840, and *U. mytiloides* Swainson, 1840; both = *Alasmidonta undulata* (Say, 1817). Agassiz's selection (1852: 49) of *A. edentula* Say, 1829, as type was invalid as that species was not included by Swainson in the genus. This genus was treated as an "Alasmidontine" by nearly all authors, including Swainson himself. But figure 62, which forms the basis for both of Swainson's species *radiata* and *undulata*, clearly shows a rayed unionoid with a massive hinge. The figure appears to represent a non–North American species.].

Geographic range. Eastern North America (Mississippi River system and Atlantic Slope and Gulf drainages from Nova Scotia and New Brunswick to Alabama; absent from peninsular Florida).

Alasmidonta marginata Say, 1818

Figures 24, 25, 26

Etymology. L. *margo, marginis*, edge, border. In the original description Say notes that "The inner margin is of a chalky whiteness, in this respect resembling *Anodonta Mardinata* [*sic*]."

Vernacular name(s). *Elktoe*, Northern Elktoe.

Original description

Alasmodonta marginata Say, 1818: 459–460, no figure given.

Type locality. ". . . by Mr. Isaac Lea, who found it in the Sciota [*sic*] river." The Scioto River is in central and south-central Ohio. Parmalee & Bogan (1998) restricted this to "Scioto River, Chillicothe, Ohio."

Type material. Say's types have apparently been lost (Johnson & Baker, 1973: 146; Clarke, 1981a).

Synonymy

Alasmodonta marginata Say, 1818: 459–460.

Alasmidonta marginata Say, 1818. Rafinesque, 1820: 317.

? *Mya rugulosa* Wood, 1828: 3, pl. 1, fig. 7 [Type locality: "North America, Fresh Water." Type lost *fide* Clarke, 1981].

Alasmodon (*Decurambis*) *scriptum* Rafinesque, 1831: 4 [Type locality: "Green River" [Kentucky]. Type lost *fide* Clarke, 1981].

Margarita (*Margaritana*) *marginata* (Say, 1818). Lea, 1836: 43.

Alasmodon marginata (Say, 1818). Gould, 1841: 116, fig. 77.

Unio marginata (Say, 1818). Hanley, 1842: 211.

Anodonta marginata (Say, 1818). Adams, 1842: 164.

Alasmodonta truncata "Say" Gould, 1841: 171 [*nomen nudum*].

Margaron (*Margaritana*) *marginata* (Say, 1818). Lea, 1852c: 42.

Alasmodonta truncata "Say" Conrad, 1853: 262 [introduced in synonymy of *Alasmidonta marginata* (Say, 1819)].

Baphia marginata (Say, 1818). Adams & Adams, 1857: 500.

Margaritana marginata (Say, 1818). Gould, 1870: 177, fig. 479.

Unio marginatus (Say, 1818). Catlow & Reeve, 1845: 61.

Alasmodon corrugata De Kay, 1843: 198, pl. 24, fig. 259.

Margaritana marginata (Say, 1818). Küster, 1862: 297.

Unio marginata (Say, 1818). Sowerby, 1866: pl. 51, fig. 297.

Margaritana marginata var. *truncata* Wright, 1898: 124 [Type locality: "Clinch River, Virginia." Lectotype USNM 150545 by designation of Clarke, 1981a, 4 paralectotypes MCZ 152253].

Alasmidonta truncata (Wright, 1898). Simpson, 1900a: 671.

Unio calceolus forma sciotincola De Gregorio, 1914: 54, pl. 9, fig. 3 [Type locality: "Sciota" [Scioto River, OH]. Type in Palermo Museum?].

Alasmidonta (*Decurambis*) *marginata* Say, 1818. Ortmann, 1919: 181.

Alasmidonta (*Decurambis*) *marginata susquehannae* Ortmann, 1919: 187, pl. 12, fig. 4 [Type locality: "Susquehanna River, Selinsgrove, Snyder Co., Pennsylvania." Lectotypes by designation of Parodiz (1967) without catalog numbers, later identified as CMNH 61.4679 by Johnson, 1977].

Decurambis marginata (Say, 1818). Frierson, 1927: 21.

Alasmidonta marginata var. *variabilis* Baker, 1928a: 194, pl. 69, figs. 4–9 [Type locality: "Red Cedar River, near Chetek, Barron Co., Wisconsin." Lectotype in Museum of the Wisconsin Geological and Natural History Survey, University of Wisconsin 935a, now UMMZ 209183].

Alasmidonta (*Decurambis*) *marginata* Say, 1818. Clarke, 1981a: 57–68, figs. 19–21.

Status. Listed as a Species of Concern by the Ohio Division of Wildlife; not listed by U.S. Fish & Wildlife Service as of 2009.

ADULT SHELL

Shell. Medium size (to 110 mm), thin but sturdy, approximately twice as long as high, inflated.

Anterior margin. Long and straight before the umbo, broadly rounded to the ventral border.

Ventral margin. Very broadly rounded to straight, occasionally concave in the posterior half.

Posterior margin. Forming an acute point at or near the ventral margin, then obliquely ascending to the dorsal margin. The margin may be rendered undulating by the presence of low projecting ribs on the dorsal slope.

Dorsal margin. Long, straight or gently convex, forming obtuse angle with posterior margin.

Umbo. Very wide and low; umbonal sculpture of 2–3 heavy, broadly "V"-shaped ribs; the distal rib may be nearly straight.

Ligament. Short and prominent, tan to dark brown, not concealed by shells.

Disc sculpture. Sculptureless except for fine oblique corrugations on dorsal slope.

Periostracum. Waxy yellow, green, or tan, becoming darker with age; usually patterned with bold, green rays; rays range in width from very wide to fine lines, occasionally wavy; the rays are overlain by numerous dark green flecks, often smeared distally.

Nacre. White or metallic white, often iridescent posteriorly; rarely flushed with pink or orange posteriorly; commonly with gold Tulberg layers.

Hinge plate. Narrow, short, gently arched; cardinals a single, lamellate, triangular denticle in the right valve, bifid

Fig. 24. ***Alasmidonta marginata*** **Say, 1818. A.** OSUM 57015.3, Sandusky River, Wyandot Co., OH. 90 mm. **B.** OSUM 13192.3, Big Walnut Creek, Franklin Co., OH. 93 mm. **C.** OSUM 58440.1, Little Muskingum River, [Co.?], OH. 46 mm. **D.** OSUM 12205.7, Big Darby Creek, Pickaway Co., OH. 56 mm. **E.** OSUM 55236.2, Greenbrier River, Pocahontas Co., WV. 89 mm. **F.** OSUM 54091, Paint Creek, [Co.?], OH. 79 mm. **G.** OSUM 11652, St. Joseph River, DeKalb Co., IN. 78 mm. **H.** OSUM 12205.9, Big Darby Creek, Pickaway Co., OH. 72 mm. **Hinge detail:** OSUM 12205.8, Big Darby Creek, Pickaway Co., OH. 56 mm total length. **Beak sculpture detail:** OSUM 12205.1, Big Darby Creek, Pickaway Co., OH. 32 mm total length.

in the left valve; laterals absent at all ages; interdentum short and narrow.
Muscle scars. Anterior adductor impressed, at least posteriorly; posterior adductor less so; anterior adductor confluent with large pedal adductor, large posterior pedal retractor confluent with the adductor scar; dorsal suspensors very small, located anteriorly in the umbonal cavity. Pallial line entire, located far from shell margin. Accessory line absent.
Umbonal cavity. Very wide and moderately deep and open.

Glochidia. Pyriform, higher than long, asymmetric, dorsal margin straight, posterior margin broadly curved, anterior margin more broadly curved than the posterior margin, both anterior and posterior margins slightly incurved ventrally producing a narrowly rounded ventral terminus. Exterior surface sculpturing beaded, hooks styliform with as many as 120 microstylets. Microstylets pyramidal, multifaceted and arranged in six to seven proximal rows, reduced to four rows distally (Lea, 1858c: 49, pl. 5, fig. 27; Hoggarth, 1999: figs. 21a–i).

Literature records (Ortmann, 1912a: 297; Surber, 1912: 9, pl. 3, fig. 42; Utterback, 1915: 273; Clarke, 1981a: 60–63, fig. 20; Hoggarth, 1999: figs. 21a–i) indicate glochidial dimensions for length (0.30 to 0.35 mm), height (0.35 to 0.38 mm), and hinge (0.21 to 0.24 mm). Lea (1858) described this glochidium as subtriangular with a long, straight dorsal line and inflated side margins. He described the hook as terminating in an arrowhead point. Clarke (1981a) also described the point of the hook as arrowhead-like. The arrowhead effect is probably due to the collapse of the lateral surfaces of the hook and is therefore an artifact of drying. This glochidium can be distinguished from that of *A. undulata* by differences in exterior valve sculpturing.

Reproductive biology. Bradytictic. Spawning occurs in June; glochidia are found in September, and overwinter in the marsupia until the following June (Baker, 1928a; Ortmann, 1919). However, in Ohio glochidial release may take place as early as May (Watters, unpubl.). No conglutinates are formed (Watters, 2008). All potential hosts are based on a single study that identified natural infestations without evidence of metamorphosis. These "hosts" are questionable at best.

Reported potential hosts

Ambloplites rupestris (Rock Bass)—Howard & Anson, 1923 (NI).
Catostomus commersoni (White Sucker)—Howard & Anson, 1923 (NI).
Hypentelium nigricans (Northern Hogsucker)—Howard & Anson, 1923 (NI).

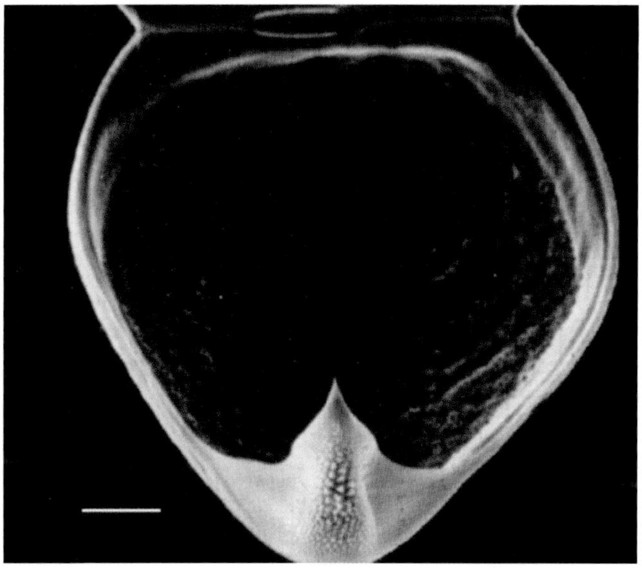

Fig. 25. ***Alasmidonta marginata* Say, 1818.** Glochidium. OSUM MAH 724.1, Fish Creek, Williams Co., OH. Bar = 45 µm. (Hoggarth, 1999).

Lepomis gulosus (Warmouth)—Howard & Anson, 1923 (NI).
Moxostoma macrolepidotum (Shorthead Redhorse)—Howard & Anson, 1923 (NI).

Habitat. This species typically occurs in small rivers and large creeks, in sand and cobble in riffles and runs. It requires high water quality.

Range. Mississippi River system from Minnesota south to Arkansas, including the Tennessee and Cumberland rivers. Lawrentian system except for Lake Superior. On the Atlantic Slope it occurs in the Susquehanna River drainage in Pennsylvania; this form or subspecies has been named *Alasmidonta marginata susquehannae* Ortmann, 1919. Its biological status is uncertain.

Range in Ohio. Widespread but seldom common, in most Ohio River and Lake Erie tributaries. Rare in Lake Erie proper. Rare in unglaciated Ohio.

Similar species. *Strophitus undulatus* and *Anodontoides ferussacianus* may be superficially similar, particularly if rayed. *Alasmidonta marginata* may be distinguished by its corrugated posterior slope and very coarse undulating beak sculpture, and in having the rays flecked with dark spots. The other two species never have a corrugated posterior slope, the beak sculpture is obviously much finer, and although rays may be present, they do not contain dark flecks or spots.

Comments. Individuals grow quickly for 3–4 years, but then abruptly slow. They grow to ca. 110 mm and typically

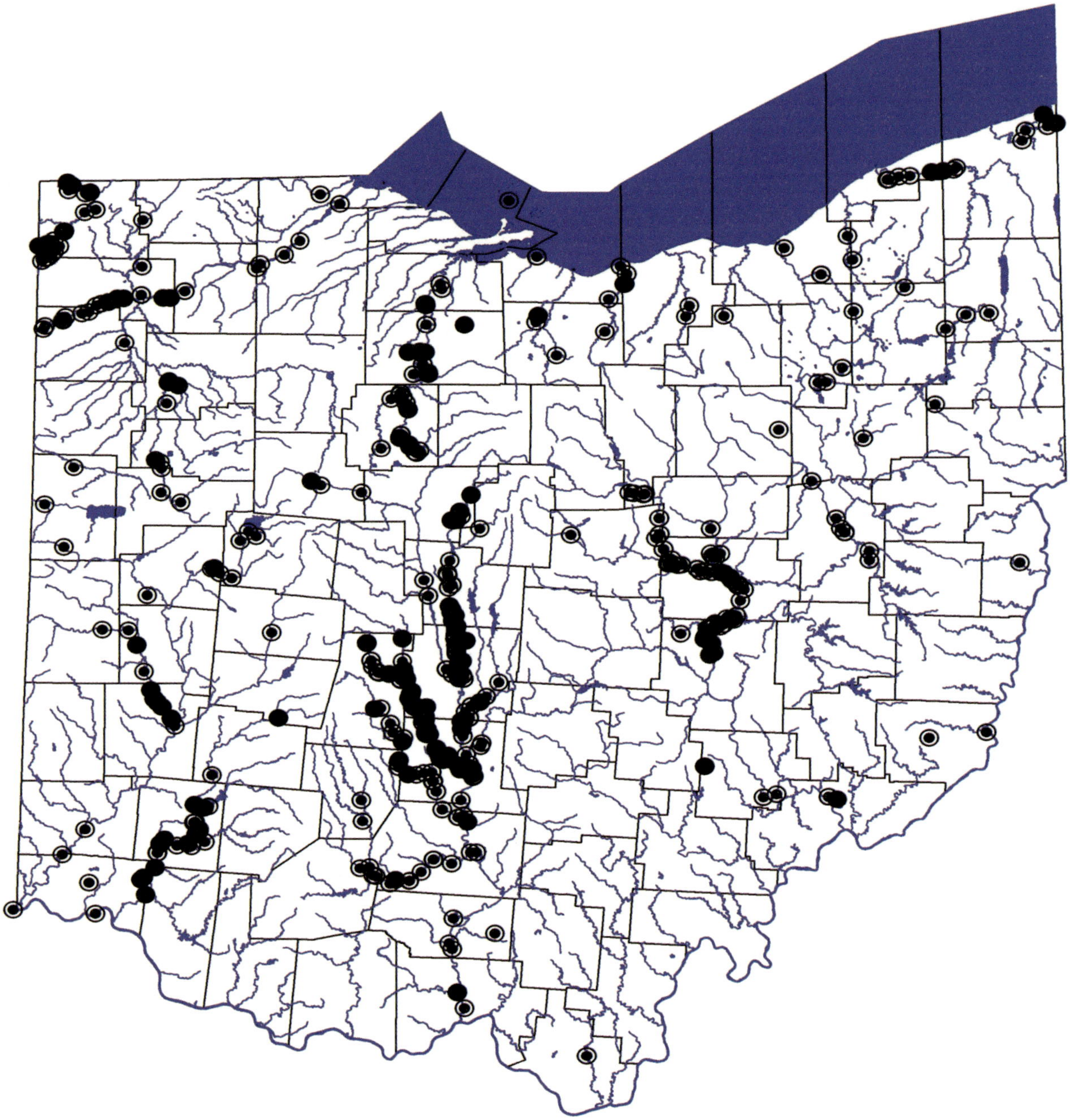

Fig. 26. ***Alasmidonta marginata*** **Say, 1818**
Solid dots—live or freshdead since 1980
Bullseye—all records, any condition

live for ca. 12 years. Some populations tend towards being all green or all tan and are quite uniform in their patterns between specimens.

Alasmidonta viridis (Rafinesque, 1820)

Figures 27, 28, 29

Etymology. L. *viridis*, green. While many unionids have a green or greenish periostracum and/or shell rays, this diminutive headwater species apparently impressed its describer as being especially so.

Vernacular name(s). *Slippershell Mussel*, Green Creek Shell, Little Green.

Original description

Unio (*Elliptio*) *viridis* Rafinesque, 1820: 293, no figure given.

Type locality. "Rare dans l'Ohio, plus commune dans le Kentucky et les petites rivières adjacentes" ["Rare in the Ohio, very common in the Kentucky and the small adjacent rivers"]. Although Johnson & Baker (1973: 174) restricted the type locality to Yellow Creek, near Cincinnati, the neotype (Clarke, 1981a) is from the Ohio River at Cincinnati, Ohio, which becomes the type locality.

Type material. The original type material is apparently lost, but Clarke (1981a: 28) selected a neotype (USNM 8626a). He noted that its locale was "Cincinnati, Ohio." This specimen had previously been in the Isaac Lea collection and was originally collected by Mr. J. Clark on an unspecified date. It is unfortunate that a specimen from a small Kentucky River tributary was not selected since this is distinctly a headwater species. However, the Ohio River was mentioned by Rafinesque as one of his sources, the species is occasionally found in large rivers, and the time of collection of the neotype was surely early in the 19th century.

Synonymy

Unio (*Elliptio*) *viridis* Rafinesque, 1820: 293, no figure given.

Unio calceolus Lea, 1828: 265, pl. 3, fig. 1 [Type locality: "Ohio." Lectotype USNM 86261 by designation of Clarke, 1981].

Lasmigona viridis (Rafinesque, 1820). Rafinesque, 1831: 5.

Lasmigona viridis var. *chloris* Rafinesque, 1831: 5.

Lasmigona viridis var. *radiata* Rafinesque, 1831: 5.

Lasmigona viridis var. *fuscata* Rafinesque, 1831: 5.

Alasmodonta truncata "Say" Conrad, 1834: 73 [*nomen nudum*].

Unio calceola (Lea, 1828). Deshayes, 1835: 546.

Margarita (*Margaritana*) *deltoidea* Lea, 1836: 44 [*nomen nudum*].

Margarita (*Margaritana*) *calceola* (Lea, 1828). Lea, 1836: 45.

Margaritana deltoidea Lea, 1838a: 43, pl. 13, fig. 38 [Type locality: "Scioto River, Ohio." Lectotype USNM 86215 by designation of Clarke, 1981a. Agassiz (1852) believed this was the female of the species.].

Calceola angulata Swainson, 1840: 382 [based on Lea's 1838 figure of *Margaritana deltoidea* (Lea, 1836)].

Unio deltoidea (Lea, 1838). Hanley, 1842: 211.

Margaritana minor Lea, 1845: 163 [Type locality: "N.C." [North Carolina]. Lectotype USNM 86224 by designation of Clarke, 1981a].

Unio deltoideus (Lea, 1836). Catlow & Reeve, 1845: 58.

Uniopsis calceolus (Lea, 1828). Agassiz, 1852: 49.

Margaron (*Margaritana*) *deltoidea* Lea, 1836. Lea, 1852b: 42.

Margaron (*Margaritana*) *minor* (Lea, 1845). Lea, 1852b: 42.

Margaron (*Margaritana*) *calceola* (Lea, 1828). Lea, 1852b: 43.

Strophitus calceolus (Lea, 1828). Conrad, 1853: 262.

Strophitus deltoidea (Lea, 1836). Conrad, 1853: 263.

Strophitus minor (Lea, 1845). Conrad, 1853: 263.

Baphia deltoidea (Lea, 1836). Adams & Adams, 1857: 499.

Baphia minor (Lea, 1845). Adams & Adams, 1857: 499.

Baphia calceola (Lea, 1828). Adams & Adams, 1857: 500.

Margaritana calceola (Lea, 1828). Küster, 1862: 299, pl. 94, fig. 6.

Alasmodonta deltoidea (Lea, 1836). Baker, 1898b: 63–64, pl. 6, fig. 2.

Alasmidonta (*Pressodonta*) *calceola* (Lea, 1828). Simpson, 1900a: 668.

Alasmidonta (*Pressodonta*) *minor* (Lea, 1845). Simpson, 1900a: 668–669.

Alasmidonta calceolus danielsi Baker, 1928a: 187, pl. 69, fig. 2, pl. 72, fig. 9 [Type locality: "Moots Creek, White Co., Ind." Holotype UIMNH Z-14301a].

Alasmidonta calceolus magnalacustris Baker, 1928a: 188, pl. 69, fig. 3, pl. 72, fig. 14 [Type locality: "Sturgeon Bay [of Lake Michigan], Door Co., Wis." Holotype formerly UIMNH 933A, now UMMZ 209215].

Alasmidonta (*Pressodonta*) *viridis* (Rafinesque, 1820). Clarke, 1981a: 17–28, figs. 5–7.

Status. Not listed by either Ohio Division of Wildlife or U.S. Fish & Wildlife Service as of 2009.

ADULT SHELL

Shell. Small (to 60 mm), thin as juveniles becoming thick with age, quadrate, inflated.

Anterior margin. Short and straight before the umbo, broadly rounded to the ventral border.

Ventral margin. Very broadly rounded, straight, or slightly concave posteriorly.

Posterior margin. Forming an acute point near the ventral margin, then ascending straight or slightly concave to the dorsal margin.

Fig. 27. ***Alasmidonta viridis*** **(Rafinesque, 1820). A.** OSUM 42029.2, Little Darby Creek, Madison Co., OH. 27 mm. **B.** OSUM 62666.3, Anderson Fork Caesar Creek, Clinton Co., OH. 51 mm. **C.** OSUM 62728.2, Anderson Fork Caesar Creek, Greene Co., OH. 43 mm. **D.** OSUM 46205, Vermilion River, Huron Co., OH. 35 mm. **E.** OSUM 2756, Little Miami River, Warren Co., OH. 39 mm. **F.** OSUM 9656.17, Clear Fork Mohican River, Ashtabula Co., OH. 44 mm. **G, H, Hinge detail:** OSUM 1457, Big Darby Creek, Madison Co., OH. 37 mm. **Beak sculpture detail:** OSUM 2756.2, Little Miami River, Warren Co., OH. 37 mm total length.

Dorsal margin. Short, straight, forming acute or right angle with posterior margin.
Umbo. Wide and low; umbonal sculpture of 2–3 heavy, irregular ribs; these ribs may be undulating, "W"-shaped, or "V"-shaped with the posterior "V" the most pronounced; the distal portions of the "V" may be nodulose.
Ligament. Short, tan to dark brown, partially concealed by shells.
Disc sculpture. Sculptureless except for 2–3 weak ribs on dorsal slope.
Periostracum. Waxy yellow, green, or tan; rare populations may be blue-green; becoming dark brown with age; usually patterned with bold green rays, often undulating.
Nacre. White, often iridescent posteriorly; commonly with extensive green or gold Tulberg layers.
Hinge plate. Short, narrow; cardinals a single, chunky denticle in right valve, bifid in left valve; laterals absent at all ages; interdentum varies from narrow to thick, occasionally split in half.
Muscle scars. Anterior adductor impressed; posterior adductor less so; anterior pedal protractor small and linear, usually joined anteriorly to the anterior adductor, posterior protractor small and may or may not be confluent with posterior pedal retractor scar; multiple minute dorsal suspensors, located anteriorly in the umbonal cavity. Pallial line entire, located far from shell margin. Accessory line usually present, variable in length.
Umbonal cavity. Very wide and shallow.

Glochidia. Depressed pyriform, slightly asymmetric, dorsal margin straight, posterior margin gently and evenly curved, anterior margin only slightly more produced than the posterior margin, ventral terminus roundly pedicellate. Exterior surface sculpturing loose-looped, hooks styliform, covered with about 30 microstylets. Microstylets lanceolate, arranged in three proximal rows, reduced to two distal rows (Lea, 1858c: 49, pl. 5, fig. 30; Utterback, 1915: 271; Zale & Neves, 1982a: 387, fig. 1; Hoggarth, 1999: figs. 18a–f).

Literature records (Ortmann, 1912a: 295, pl. 19, fig. 4; Surber, 1912: 8, pl. 1, fig. 1; Clarke, 1981a: 22–23, fig. 6; Hoggarth, 1999: figs. 18a–f) indicate glochidial dimensions for length (0.29 to 0.32 mm), height (0.23 to 0.26 mm), and hinge (0.25 to 0.26 mm). This glochidium is distinguished from that of *Strophitus* by its finer exterior valve sculpture and its broadly connected hook.

Reproductive biology. Probably bradytictic. The outer gills are marsupial and turn brown when gravid (Ortmann, 1912a). Gravid females were found in September (Baker, 1928a). No conglutinates are formed (Watters, 2008). The few hosts that have been identified are sculpins and a darter.

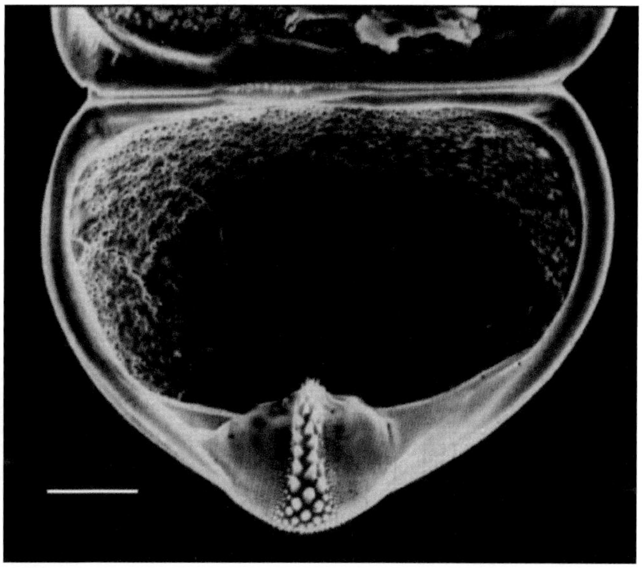

Fig. 28. ***Alasmidonta viridis*** **(Rafinesque, 1820).** Glochidium. OSUM 47518, Horse Lick Creek, Jackson Co., KY. Bar = 45 μm. (Hoggarth, 1999).

Reported potential hosts
Cottus bairdi (Mottled Sculpin)—Morrison *in* Clarke & Berg, 1959 (NS).
Cottus carolinae (Banded Sculpin)—Zale & Neves, 1982a (LT).
Etheostoma nigrum (Johnny Darter)—Morrison *in* Clarke & Berg, 1959 (NS).

Habitat. The slippershell is most commonly found in headwaters. It occurs in a wide variety of substrates, including sandy silt.

Range. Upper Mississippi River system, but primarily the Ohio and upper Tennessee River drainages. Also found in the Lawrentian system exclusive of Lake Superior.

Range in Ohio. Throughout the state but sporadic. Virtually missing from unglaciated and northeastern Ohio. In Lake Erie at the Bass Islands. Its small size probably makes it appear more uncommon than it truly is. It occurs from extreme headwaters to large rivers and lakes. It is not a species of impoundments or ponds.

Similar species. The small size and quadrate shape of most specimens set them apart from other Ohio mussels. Larger, less angular specimens may be confused with *Strophitus undulatus*. That species lacks cardinal teeth, which are at least rudimentary in *A. viridis*.

Comments. Individuals grow quickly for 3 years, but gradually slow. They grow to ca. 60 mm (although most Ohio specimens are much smaller) and typically live for

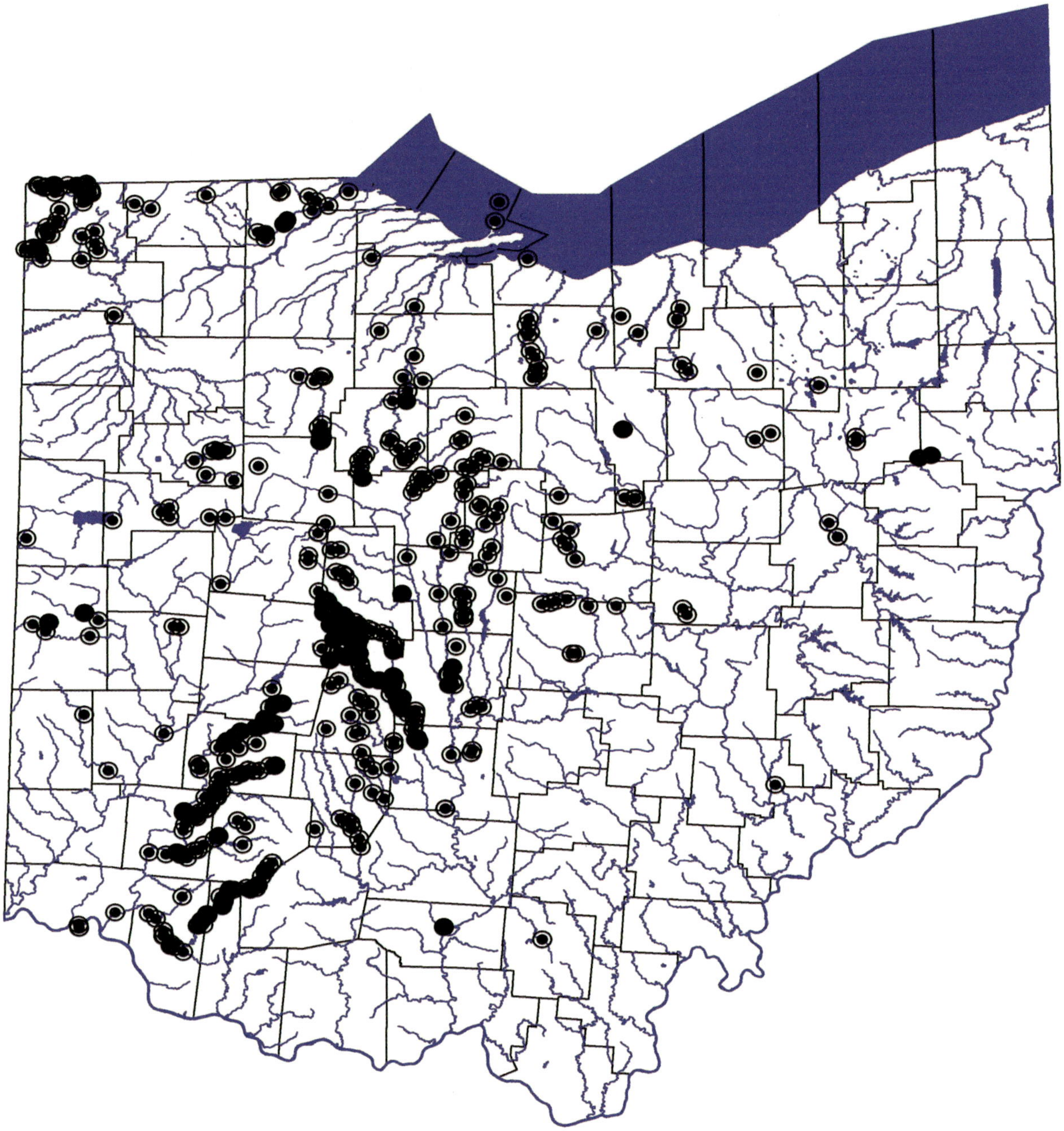

Fig. 29. ***Alasmidonta viridis*** **(Rafinesque, 1820)**
Solid dots—live or freshdead since 1980
Bullseye—all records, any condition

less than 10 years. Within Ohio are two different "forms"—an elongate, *Villosa*-like form and a quadrate form with an enlarged dorsal ridge. These do not appear to be sexual characteristics and perhaps deserve more study. Beyond Ohio, other "forms" are to be found that may constitute separate species.

Amblema Rafinesque, 1820

Etymology. Gr. *amblys*, blunt, dulled, obtuse. Poulson's translation (1832: 54) had Rafinesque describing the genus *Amblema* as having a "shell oval, elliptical or square, very inequilateral; axis lateral, posterior; . . . ." Rafinesque's figures of *Amblema costata* (= *A. p. plicata* (Say, 1817)), (pl. 82, figs. 13, 14) do indeed have blunt margins; all four sides, dorsal, ventral, anterior, and posterior, are relatively straight with the outline approaching a rectangle when viewed laterally. The name, unfortunately, describes the given figures better than it does most specimens of the species.

Original description

Amblema Rafinesque, 1820: 314.

Type species. *Amblema costata* Rafinesque, 1820, by subsequent designation of Frierson (1914a: 7) by elimination. *Amblema costata* Rafinesque is a junior synonym of *U. plicatus* Say, 1817. It should be noted that *Amblema* Rafinesque, 1819 (and its type by monotypy, *A. ovalis* Rafinesque, 1819), was suppressed (Clarke & Clench, 1965: 197).

Synonymy

Amblema Rafinesque, 1820: 314.

Cokeria Marshall, 1916: 133–135 [Type species: *Cokeria southalli* Marshall, 1916, by original designation. Apparently based on a teratological individual of *Amblema plicata plicata* (Say, 1817)].

Geographic range. Mississippi, Mobile, Apalachicola River drainages, and other western Gulf drainages into Texas. Great Lakes.

Amblema plicata plicata (Say, 1817)

Figures 30, 31

Etymology. L. *plicatus*, folded, braided or doubled up; from L. *plico*, to fold or braid. The lateral surface of the shell is nearly always sculptured with a series of oblique folds or ribs which are distinctly more horizontal than vertical.

Vernacular name(s). *Threeridge*, Bluepoint.

Original description

Unio crassus var. *plicata* Say, 1817: no pagination (under heading "Conchology"), no figure.

Type locality. "found by Mr. C. A. Lesueur in Lake Erie." However, the neotype, designated by Haas (1930: 327), is from "Ohio River." This does not meet the qualifying conditions as outlined under ICZN Art. 75.3.6 which calls for the neotype to have come "as nearly as practicable from the original type locality . . . ," and this neotype designation may be invalid. Say's name *plicata* was often used to designate Great Lakes forms, while Rafinesque's *costata* was applied to riverine forms (Ortmann & Walker, 1922). Haas's designation obliterated that historical distinction.

Type material. The original unionoid material of Say has apparently been lost (Johnson & Baker, 1973: 146). Haas (1930: 327) selected SMF No. 4305 as the neotype of this species from the Ohio River.

Synonymy

Unio crassus var. *plicata* Say, 1817: no pagination.

Unio peruvianus Lamarck, 1819: 71 [Type locality: "Habite au Perou, dans la rivières" ("living in Peru, in the rivers"), in error. Holotype in Geneva Museum].

Unio rari-plicata Lamarck, 1819: 71 [Type locality: "Ohio River." Holotype in MNHN *fide* Johnson (1953: 92; 1969: 60)].

Amblema costata Rafinesque, 1820: 315, pl. 82, figs. 13, 14 [Type locality: "de l'Ohio [River]" and "dans les petites rivières du Kentucky, etc." ("in the Ohio" and "in the small rivers of Kentucky"). Lectotype ANSP 20246 by designation of Johnson & Baker (1973: 152) from latter locality].

Unio crassus "Say" Barnes, 1823: 118–120, fig. 1 [*non* Retzius, 1788. Barnes gave eleven varieties (a–l). Type locality: "The Ohio, Mississippi, and the Lakes." Types lost].

Unio plicatus Say, 1817. Barnes, 1823: 120–121.

Mya plicata (Say, 1817). Eaton, 1826: 219.

Unio costatus (Rafinesque, 1820). Say, 1834: unpaginated.

Margarita (*Unio*) *plicatus* (Say, 1817). Lea, 1836: 12.

Unio giganteus Lea, 1838b: 35 [*nomen nudum*].

Unio (*Theliderma*) *plicata* (Say, 1817). Swainson, 1840: 271, fig. 54e.

Unio plicatus Say, 1817. Hanley, 1843: 175, pl. 21, fig. 21.

Unio hippopoeus Lea, 1845: 163 [Type locality: "Lake Erie." Lectotype USNM 84096 *fide* Johnson, 1974: 70].

Complanaria plicatus (Say, 1817). Agassiz, 1852: 49.

Complanaria costatus (Say, 1817). Agassiz, 1852: 49.

Margaron (*Unio*) *plicatus* (Say, 1817). Lea, 1852c: 20.

Margaron (*Unio*) *hippopaeus* [*sic*] (Lea, 1845). Lea, 1852b: 21.

Plectomerus costatus (Rafinesque, 1820). Conrad, 1853: 260.

Plectomerus plicatus (Say, 1817). Conrad, 1853: 261.

Unio pearlensis Conrad, 1855: 256 [Type locality: "Pearl River, Mississippi." Lectotype ANSP 29918a by designation of Johnson & Baker, 1973, corrected to 129918a by Johnson, 1980: 278].

Unio heros Say, 1829. Küster, 1856: 136, pl. 40, figs. 1, 2 [misidentification].

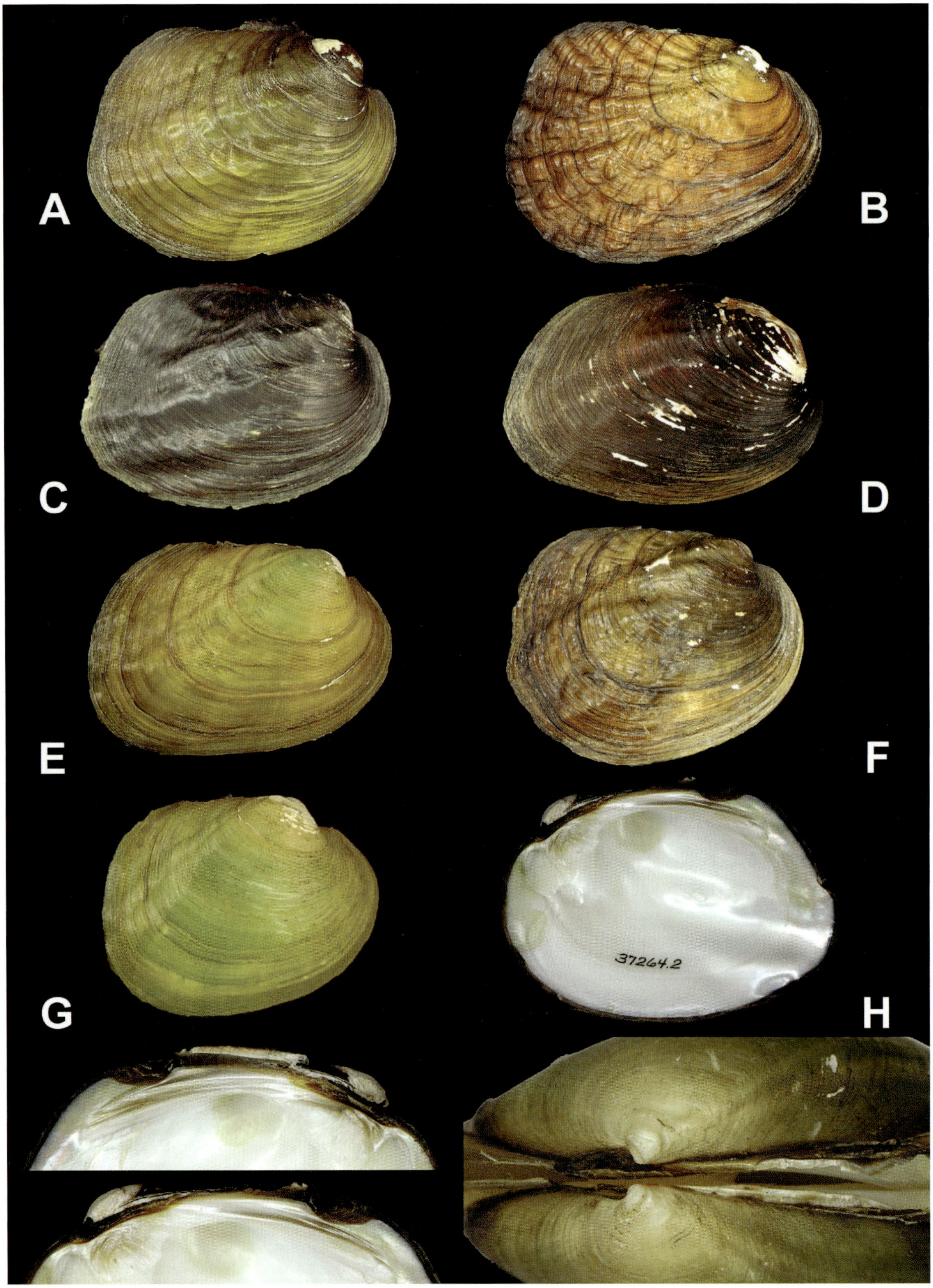

Fig. 30. ***Amblema plicata plicata* (Say, 1817).** **A.** OSUM 25911.2, Lake Erie, Ottawa Co., OH. 66 mm. **B.** OSUM 65075, Black Fork Symmes Creek, Gallia Co., OH. 102 mm. **C, H, Hinge detail:** OSUM 37264.6, Mississippi River, Rock Island Co., IL. 87 mm. **D.** OSUM 25911.7, Lake Erie, Ottawa Co., OH. 81 mm. **E.** OSUM 8818.3, Muskingum River, Morgan Co., OH. 79 mm. **F.** OSUM 9389.1, Big Walnut Creek, Pickaway Co., OH. 83 mm. **G.** OSUM 36859.1, Muskingum River, Washington Co., OH. 26 mm. **Beak sculpture detail:** OSUM 10761.1, Muskingum River, Washington Co., OH. 36 mm total length.

Unio brazosensis Lea, 1868: 144 [Type locality: "Brazos River, Texas." Lectotype USNM 84110].
Unio lincecumii Lea, 1868: 144 [Type locality: [Trinity River] "Dallas Co., Texas." Lectotype USNM 84112].
Margaron (Unio) brazosensis (Lea, 1868). Lea, 1870: 31.
Margaron (Unio) lincecumii (Lea, 1868). Lea, 1870: 31.
Unio pauciplicatus Lea, 1872: 156 [Type locality: "Austin, Texas." Figured lectotype USNM 84111, paralectotypes in ANSP lost *fide* Johnson & Baker, 1973].
Unio quintardii Cragin, 1887: 6 [Type locality: "Salt Creek, a tributary of the Deep Fork of the Canadian River, on the Sac and Fox reservation, Indian Territory." Holotype ANSP 63304 *fide* Johnson & Baker, 1971: 167].
Unio hippopaeus [*sic*] Lea, 1845. Paetel, 1890: 155.
Unio linecurii [*sic*] Lea, 1845. Paetel, 1890: 157.
Unio perlensis [*sic*] Lea, 1845. Paetel, 1890: 163.
Unio pilsbryi Marsh, 1891: 1 [Type locality: "Little Red River, Arkansas." Holotype ANSP 41589a].
Quadrula plicata (Say, 1817). Baker, 1898b: 80–81, pl. 25, fig. 1.
Unio plicata var. *hippopaea* [*sic*] Lea, 1845. Simpson, 1900a: 767.
Quadrula undulata var. *quintardii* (Cragin, 1887). Simpson, 1900a: 768.
Quadrula undulata var. *pilsbryi* (Marsh, 1891). Simpson, 1900a: 769–770.
Quadrula undulata hippopoae (Lea, 1845). Sterki, 1907: 390.
Crenodonta plicata (Say, 1817). Ortmann, 1912a: 246.
Quadrula plicata hippopoea (Lea, 1845). Sterki, 1914: 390.
Quadrula undulata form *hippopoae* (Lea, 1845). Sterki, 1914: 271.
Amblema plicata (Say, 1817). Utterback, 1915: 115.
Amblema peruviana (Lamarck, 1819). Utterback, 1915: 115.
Amblema rariplicata peruviana (Lamarck, 1819). Utterback, 1915: 115.
Amblema perplicata quintardi [*sic*] (Cragin, 1887). Utterback, 1915: 119.
Amblema plicata costata Rafinesque, 1820. Utterback, 1915: 121.
Cokeria southalli Marshall, 1916: 133 [Type locality: "James River, [Beadle Co.] South Dakota." Holotype USNM 215130].
Amblema perplicata quintardii (Cragin, 1887). Utterback, 1916b: 20.
Quadrula peruviana (Lamarck, 1819). Walker, 1918: 168.
Quadrula rariplicata (Lamarck, 1819). Walker, 1918: 168.
Amblema rariplicata (Lamarck, 1819). Baker, 1926: 107.
Amblema plicata hippopaea [*sic*] (Lea, 1845). Frierson, 1927: 61.
Amblema plicata quintardi [*sic*] (Cragin, 1887). Frierson, 1927: 61.
Amblema plicata pilsbryi (Marsh, 1891). Frierson, 1927: 61.
Amblema plicata southalli (Marshall, 1916). Frierson, 1927: 61.
Amblema (Amblema) plicata (Say, 1817). Haas, 1930: 327.
Crenodonta peruviana peruviana (Lamarck, 1819). Murray & Leonard, 1962: 47.
Crenodonta peruviana costata (Rafinesque, 1820). Murray & Leonard, 1962: 49.

Status. Not listed by either Ohio Division of Wildlife or U.S. Fish & Wildlife Service as of 2009.

ADULT SHELL

Shell. Medium to large size (to ca. 170 mm), thick to massive, very variable in shape, from moderately lanceolate to nearly circular, quite compressed to very inflated.

Anterior margin. Very short and straight, then broadly rounded to the ventral border.

Ventral margin. Very rounded to nearly straight.

Posterior margin. Broadly rounded or straight, sometimes forming a blunt acute point at the ventral margin.

Dorsal margin. Long, more or less straight, joining posterior margin with very obtuse angle.

Umbo. Varying from very wide and low to more narrow and elevated; umbonal sculpture of 3–4 concentric ribs, pointed and thickened at the posterior-ventral angle.

Ligament. Long, tan to dark brown, largely concealed by shells.

Disc sculpture. Highly variable depending on habitat. Basic sculpture usually present on all specimens consists of 2–6 oblique, large ribs, beginning posterior of the umbo and running towards the posterior-ventral border. These ribs are not aligned with the growth lines but run obliquely across them. These ribs may be broken into nodules by growth lines. Creek specimens may have numerous secondary ribs and knobs. Lake or large river species typically lack secondary sculpture. The posterior slope may or may not have low radiating ribs as well. The area of the shell anterior to the umbos lacks sculpture.

Periostracum. Yellow, tan, or more rarely greenish as smallest juveniles, becoming dark brown to black with age. Weak rays are rarely seen and never cover more than a small portion of the shell.

Nacre. Porcelain white, often iridescent posteriorly; nearly all specimens seen have extensive gold or brown Tulberg layers.

Hinge plate. Very thick; right cardinal a single, chunky, triangular denticle, often accompanied by an anterior and a posterior accessory denticle; left cardinal bifid; laterals thick, one in right valve, two in left; interdentum short and flattened.

Muscle scars. Anterior adductor impressed; posterior adductor flush; anterior adductor distinct from anterior pedal protractor scar, large posterior adductor confluent

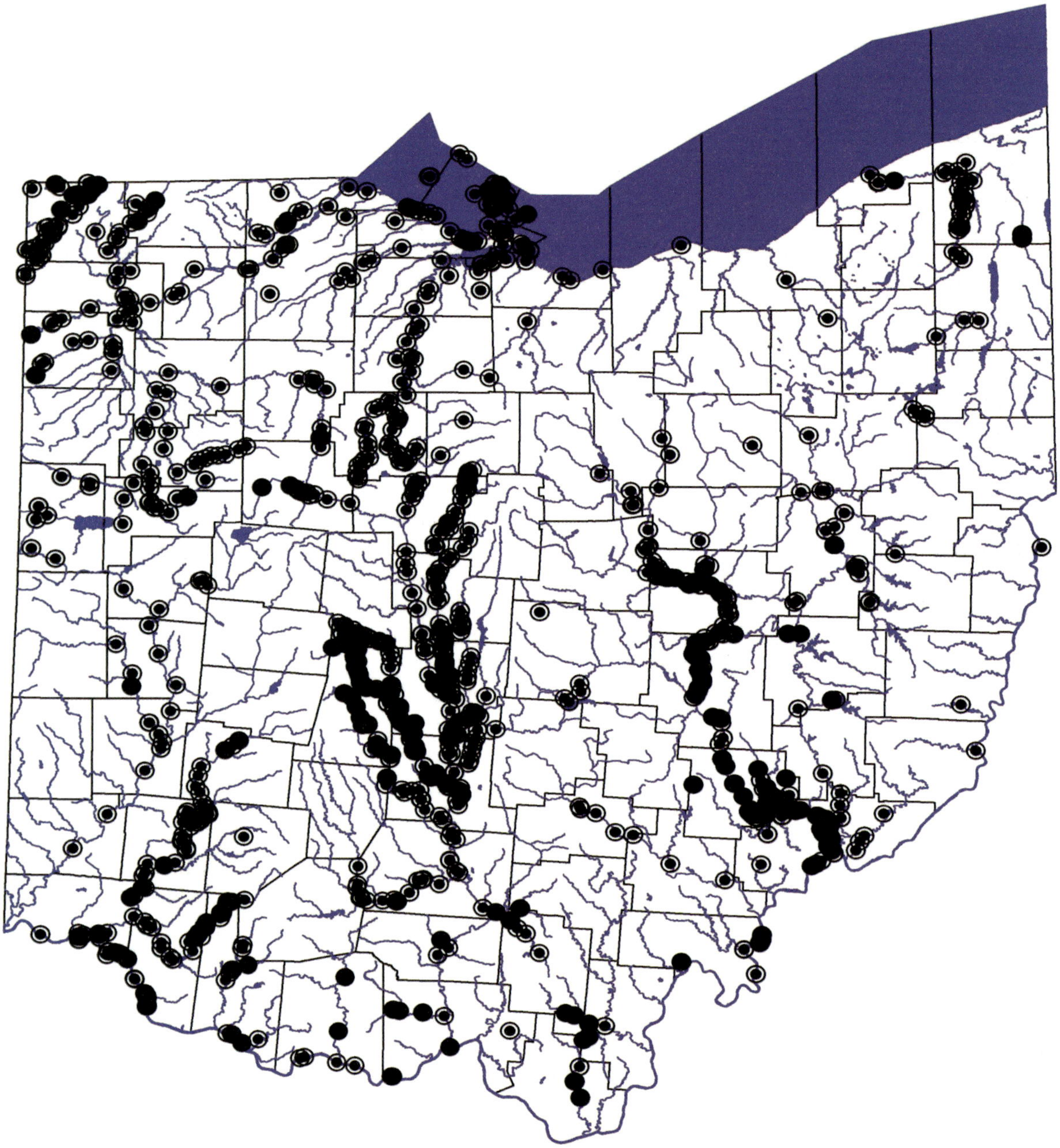

Fig. 31. ***Amblema plicata plicata*** **(Say, 1817)**
Solid dots—live or freshdead since 1980
Bullseye—all records, any condition

with posterior pedal retractor scar; multiple dorsal suspensor located behind anterior adductor and deep within the umbonal cavity. Pallial line entire, located far from shell margin. Accessory line well-developed, reaching nearly half the length of the shell.

Umbonal cavity. Very wide, compressed and deep.

Glochidia. Subelliptical to semicircular, height greater than or equal to length, dorsal margin straight and long, lateral margins equal, ventral margin rounded, hooks absent (Lea, 1858c: 48, pl. 5, fig. 22; Ortmann, 1911: 346, pl. 89, fig. 3; Surber, 1913: 116, pl. 31, fig. 16; Utterback, 1915: 122).

Literature records (Surber, 1912: 10, pl. 2, fig. 34; Ortmann, 1912a: 247; Howard, 1914a: 25, pl. 5, fig. 32; Utterback, 1915: 122) indicate glochidial dimensions for length (0.20 to 0.21 mm) and height (0.20 to 0.22 mm). Ortmann (1912: 247) noted, "Glochidia of suboval shape, without hooks. Lea, 1858, figures them but not quite correct in shape." Howard (1914a: 25) added, "The glochidium is of the hookless type, subsemicircular, elongate and white in color. It possesses a thread gland and larval thread like *Q. heros* [*M. nervosa*]. . . ."

Reproductive biology. Tachytictic. All four gills are marsupial (Utterback, 1915). Females bearing eggs first occur in May, with glochidia developing from June through August (Ortmann, 1919; Baker, 1928a; Heath *et al.*, 1998). Watters (1996d) reported that glochidia were released in Ohio in July when water temperatures reached a sustained 20°C. Utterback (1915) reported fragile, white, leaf-like conglutinates. These are of the amorphous mucus type (Watters, 2008) and quickly disassociate into free glochidia (Watters, unpubl.). Hosts include a wide variety of fishes, including many centrarchids.

Reported potential hosts

Ambloplites rupestris (Rock Bass)—Stein, 1968 (LT).
Aplodinotus grunniens (Freshwater Drum)—Weiss & Layzer, 1995 (NI).
Cyprinella spiloptera (Spotfin Shiner)—Weiss & Layzer, 1995 (NI).
Cyprinella whipplei (Steelcolor Shiner)—Weiss & Layzer, 1995 (NI).
Erimystax dissimilis (Steamline Chub)—Weiss & Layzer, 1995 (NI).
Esox lucius (Northern Pike)—Wilson, 1916 (NI); Coker *et al.*, 1921 (NI).
Hiodon tergisus (Mooneye)—Weiss & Layzer, 1995 (NI).
Hypentelium nigricans (Northern Hogsucker)—Weiss & Layzer, 1995 (NI).
Ictalurus punctatus (Channel Catfish)—Weiss & Layzer, 1995 (NI).
Lepisosteus platostomus (Shortnose Gar)—Coker *et al.*, 1921 (LT); Howard & Anson, 1923 (LI).
Lepomis cyanellus (Green Sunfish)—Stein, 1968 (LT).
Lepomis gibbosus (Pumpkinseed)—Wilson, 1916 (NI); Coker *et al.*, 1921 (NI); Stein, 1968 (LT).
Lepomis gulosus (Warmouth)—Coker *et al.*, 1921 (NI); Pearse, 1924 (NI).
Lepomis macrochirus (Bluegill)—Howard, 1914 (LT); Stein, 1968 (LT); Weiss & Layzer, 1995 (NI).
Micropterus salmoides (Largemouth Bass)—Lefevre & Curtis, 1912 (LT); Howard, 1914 (LT); Coker *et al.*, 1921 (NI, LT).
Morone chrysops (White Bass)—Wilson, 1916 (NI); Coker *et al.*, 1921 (NI).
Moxostoma duquesnei (Black Redhorse)—Weiss & Layzer, 1995 (NI).
Moxostoma erythrurum (Golden Redhorse)—Weiss & Layzer, 1995 (NI).
Notropis atherinoides (Emerald Shiner)—Weiss & Layzer, 1995 (NI).
Perca flavescens (Yellow Perch)—Howard, 1914a (LT); Coker *et al.*, 1921 (LT); Pearse, 1924 (NI); Stein, 1968 (LT).
Percina caprodes (Logperch)—Weiss & Layzer, 1995 (NI).
Pomoxis annularis (White Crappie)—Surber, 1913 (NI); Howard, 1914 (NI, LT); Wilson, 1916 (NI); Coker *et al.*, 1921 (NI).
Pomoxis nigromaculatus (Black Crappie)—Howard, 1914 (NI, LT); Coker *et al.*, 1921 (LT).
Pylodictis olivaris (Flathead Catfish)—Howard, 1914 (LT).
Sander canadensis (Sauger)—Surber, 1913 (NI); Howard, 1914 (NI); Wilson, 1916 (NI); Coker *et al.*, 1921 (NI).

Habitat. It occurs wherever the substrate is firm enough to support it, from creeks to rivers and lakes, in fast currents or slackwater. Once a widespread and abundant species, it is becoming rare and even extirpated in much of its range due to unknown factors.

Range. Mississippi River system from Manitoba to the Gulf of Mexico. Also found in the Great Lakes drainages and in Gulf systems from Texas to Alabama.

Range in Ohio. Once widespread and common, its range has dramatically decreased in the past decades. Rare in northeastern Ohio. Lake Erie in the western basin.

Similar species. The Washboard, *Megalonaias nervosa* (Rafinesque, 1820), is very similar. Individual *M. nervosa* of the same size are not as thick-shelled and have a thinner hinge. The early portions of the shells (to approximately 5 years old) have pronounced divaricate sculpture anterior to the umbo in *M. nervosa*; this is absent in *A. plicata* where all sculpture is posterior to the umbo and consists of undulating ribs (rarely subdivided into pustules).

Comments. Individuals grow at a fairly constant rate for most of their life, only very gradually slowing. However, growth rates vary considerably between populations. A 100-mm specimen may be from 15 to 22 years old. Individuals may grow to ca. 170 mm and live for 50 years.

The great inherent morphological variability of this species, coupled with its wide zoogeographic range, has led to numerous names being erected for what are probably ecophenotypic forms. Mulvey *et al.* (1997) have separated *Unio elliottii* Lea, 1856, as a distinct species from *A. plicata* (Say, 1817) based on genetic differences. *Unio elliotii* occurs in the Alabama-Coosa River system. The subspecies *Amblema plicata perplicata* (Conrad, 1841) has been recognized from the Mobile River system but may not be genetically distinct from *Amblema plicata.*

Anodonta Lamarck, 1799

Etymology. Gr. *a-, an-,* negative prefix expressing want or absence + Gr. *odous (odon), odontos,* tooth. Hinge teeth are usually absent in the *Anodonta* shell. When present the individual teeth are usually very small or the teeth are scarcely perceptible.

Original description

Anodonta Lamarck, 1799: 87, Genus 106, no plates or figures.

Type species. *Mytilus cygneus* Linnaeus, 1758, by monotypy. This is a European species.

Synonymy

Glochidium Rathke, 1797: 166 [Type species: *Glochidium parasiticum* Rathke, 1797, by monotypy but declared invalid by ICZN Opinion 561. This name was bestowed on the larval form, which Rathke believed was a foreign parasite in the gills of *Anodonta.* This generic name is the origin of the vernacular larval name "glochidium."].

Anodonta Lamarck, 1799: 87 [Although *Glochidium* Rathke, 1797, has priority, *Anodonta* was placed on the Official List of Generic Names in Zoology in 1926 (ICZN Opinion 94) and reconfirmed in 1959 (Opinion 561).].

Anodon Oken, 1815: 236 [unjustified emendation of *Anodonta* Lamarck, 1799].

Anodontes Cuvier, 1817: 472 [unjustified emendation of *Anodonta* Lamarck, 1799].

Anodonta (*Anodonta*) Lamarck, 1819. Rafinesque, 1820: 316.

Edentula Nitzsch, 1820: 189 [unnecessary replacement name for *Anodonta*].

Lipodonta Nitzsch, 1820: 189 [unnecessary replacement name for *Anodonta*].

Anodonte Fischer von Waldheim, 1807: 250, 321 [unjustified emendation of *Anodonta* Lamarck, 1799].

Euanodonta Westerlund, 1890: 195 [Type species: many species listed, none designated as type].

Utterbackiana Frierson, 1927: 17 [Type species: *Anodonta suborbiculata* Say, 1831, by original designation].

Anodota [*sic*] Petrbok, 1930: 10.

Geographic range. The morphological boundaries between *Anodonta* and allied genera such as *Utterbackia* and *Pyganodon* have yet to be determined. Furthermore, western North American "anodontas" may be descended from Asian stock that migrated across the Bering Land Bridge, and probably are not related to the eastern taxa, which may be derived from European stock that migrated across Greenland (Watters, 2001). Nominal *Anodonta* occurs in Canada from the Mackenzie River on the Beaufort Sea in the Northwest Territories, south into the United States east of the Rocky Mountains to the Gulf and Atlantic Slope systems, and west into Alaska, Washington, Oregon, and California; also Europe and northern Asia. Whether all of these species belong to *Anodonta,* which is unlikely given their fossil record, or other genera has not been decided.

Anodonta suborbiculata Say, 1831

Figures 32, 33, 34

Etymology. From L. *sub,* under, below, somewhat + L. *orbis,* circle, disc. Shells commonly approach being circular in outline.

Vernacular name(s). *Flat Floater,* Round Floater.

Original description

Anodonta suborbiculata Say, 1831b: not paginated, species no. 2.

Type locality. "Ponds near the Wabash [River near New Harmony, Posey Co., Indiana]" (Say, 1831b: not paginated).

Type material. The holotype of this species is apparently lost (Say, 1840: 4; Johnson & Baker, 1973: 146).

Synonymy

Anodonta suborbiculata Say, 1831b: no pagination [In April, 1831, Say redescribed this species in No. 2 of his *American Conchology.* A plate (pl. 11) was added to this description as well as information concerning its systematic distinctness and habitat.].

Margarita (*Anodonta*) *suborbiculata* (Say, 1831). Lea, 1836: 10.

Anodon suborbiculata (Say, 1831). Catlow & Reeve, 1845: 68.

Margaron (*Anodonta*) *suborbiculata* (Say, 1831). Lea, 1852b: 51.

Fig. 32. ***Anodonta suborbiculata*** **Say, 1831. A.** OSUM 66203, Big Walnut Creek, Franklin Co., OH. 149 mm. **B.** OSUM 54295.3, Mississippi River, Crawford Co., WI. 116 mm. **C.** OSUM 53670, Ohio River, Clermont Co., OH. 113 mm. **D.** OSUM 65042, Sugarcamp Run, Meigs Co., OH. 54 mm. **E, Beak sculpture detail:** OSUM 42481, Black River, Posey Co., IN. 27 mm. **F, Hinge detail:** OSUM 54295.1, Mississippi River, Crawford Co., WI. 98 mm.

Anodon suborbiculatus (Say, 1831). Sowerby, 1867: pl. 5, fig. 11.
Lastena suborbiculata (Say, 1831). Utterback, 1915: 105–109, pl. 4, fig. 19a; pl. 9, fig. 19; pl. 23, figs. 73a–d; page 106, fig. 6; page 107, fig. 7.
Anodonta (*Utterbackiana*) *suborbiculata* Say, 1831. Frierson, 1927: 17.
Anodonta (*Nayadina*) *suborbiculata* Say, 1829. Haas, 1969: 360.
Anodonta (*Utterbackia*) *suborbiculata* Say, 1831, Johnson, 1980: 113–155, pl. 5, pl. 9, fig. 5.

Status. Listed as a Species of Concern by Ohio Division of Wildlife; not listed by U.S. Fish & Wildlife Service as of 2009.

ADULT SHELL

Shell. Medium to very large (to at least 200 mm), thin-shelled as juveniles, becoming thicker with age but always rather fragile, somewhat inflated, especially as adults, oval to nearly round in profile.

Anterior margin. Roundly arched into ventral margin, occasionally angled at mid-point.

Ventral margin. Broadly rounded.

Posterior margin. Ventral and posterior margins form wide, blunt acute angle at or below midline; posterior margin oblique and straight or slightly concave to dorsal margin.

Dorsal margin. Short, straight or gently bowed, particularly in adults, joining posterior margin with very obtuse angle.

Umbo. Flush with dorsal margin, located about 33% to nearly 50% back from anterior margin; umbonal sculpture of 4–6 coarse undulating or nearly linear ridges, vaguely "W"-shaped, occasionally somewhat nodulose on the descending angles.

Ligament. Short and irregular, tan to dark brown, projecting both anterior and posterior of the umbo, partially concealed by shells.

Disc sculpture. Smooth except for dorsal ridge.

Periostracum. Yellow, tan, or greenish as smallest juveniles, becoming dark brown to black with age. Occasionally with weakly defined, numerous fine green or brown rays.

Nacre. White, pinkish, or salmon, very iridescent, particularly posteriorly; nearly all specimens seen have gold or brown Tulberg layers.

Hinge plate. Thin, cord-like. Edendate.

Muscle scars. Anterior adductor very slightly impressed; posterior adductor flush or shallowly impressed; anterior adductor distinct from anterior pedal protractor scar, protractor elongate and parallel to adductor; large posterior adductor confluent and nearly inseparable from posterior pedal retractor scar; minute, multiple dorsal suspensor located within the umbonal cavity except for a much larger anterior one. Pallial line entire, located far from shell margin. Accessory line absent.

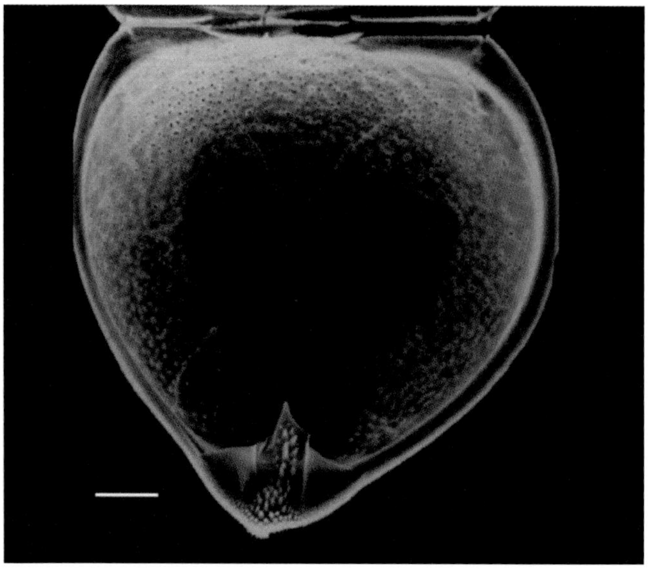

Fig. 33. *Anodonta suborbiculata* Say, 1831. Glochidium. OSUM 13634, Black River, Butler Co., MO. Bar = 40 μm. (Hoggarth, 1999).

Umbonal cavity. Very wide, shallow, and open.

Glochidia. Subtriangular, posterior margin gently curved, anterior margin broadly curved, ventral margin narrowly pointed, dorsal margin straight. Exterior surface sculpturing vermiculate, hooks styliform with about 40 lanceolate microstylets and many micropoints (Hoggarth, 1999: figs. 9a–f).

Literature records (Surber, 1915: 4, pl. 1, fig. 1; Utterback, 1915: 257, fig. 7; Utterback, 1916b: 20; Hoggarth, 1999: figs. 9a–f) indicate glochidial dimensions for length (0.32 to 0.33 mm), height (0.32 to 0.33 mm), and hinge (0.23 to 0.24 mm). Surber (1915: 4) noted, "Glochidium of the *Anodonta* type, large, subtriangular in shape, with spine at tip of each valve; hinge line straight, long; length slightly greater than depth. In general outline *A. suborbiculata* closely resembles *Anodonta grandis* but may be distinguished by its smaller size." Surber (1915: 4) stated that in "general outline [the glochidium of] *suborbiculata* closely resembles *Anodonta grandis* but may be distinguished by its smaller size." He gave measurements of 325 μm x 320 μm for length and height. However, the glochidia of these species are only superficially similar. This glochidium can be distinguished by its exterior valve sculpturing and hook. It resembles *A. cygnea* in shape but it has no other characters to tie it to that species.

Reproductive biology. Bradytictic. Barnhart *et al.* (1995) reported glochidial release from December through February in Missouri. This suggests that the Flat Floater may

Fig. 34. ***Anodonta suborbiculata*** **Say, 1831**
Solid dots—live or freshdead since 1980
Bullseye—all records, any condition

overwinter glochidia on hosts. Glochidia are released in an extensive web of mucus that ensnares hosts (Roberts, 1997; Watters, 2008), mainly catfishes and sunfishes.

Reported potential hosts

Ictalurus punctatus (Channel Catfish)—Howells, 1994a (LT); Howells, 1997 (LT).

Lepomis cyanellus (Green Sunfish)—Howells, 1994a (LT); Howells, 1997 (LT).

Lepomis gulosus (Warmouth)—Barnhart *et al.*, 1995 (LT); Barnhart & Roberts, 1997 (LT).

Lepomis megalotis (Longear Sunfish)—Howells, 1994a (LT); Howells, 1997 (LT).

Micropterus salmoides (Largemouth Bass)—Barnhart *et al.*, 1995 (LT); Barnhart & Roberts, 1997 (LT).

Notemigonus crysoleucas (Golden Shiner)—Barnhart *et al.*, 1995 (LT); Barnhart & Roberts, 1997 (LT).

Pomoxis annularis (White Crappie)—Barnhart & Roberts, 1997 (LT).

Habitat. Originally a species of western bayous and oxbows, it has recently invaded Ohio, where it is found in muddy embayments. In describing the ecology of this species Parmalee (1967: 48) stated, "[Although] least common of the *Anodonta* group in the state [of Illinois], . . . in some local, well-suited habitats (quiet, mud bottom sloughs and river pools), the Heel-splitter may become numerous. In Illinois, this species is restricted to the Mississippi River, its backwater areas, and major tributary systems (Illinois, Sangamon, Kaskaskia)." The typical habitat of this species is in soft stable sediment in pools, backwaters, and low flow reaches of larger rivers. It is perhaps not surprising that the species is extending its range into some of the lower reaches of Ohio's larger rivers. Abundant silt and other fine sediments deposited upon the bottom of the river bed produce favorable habitat for this species.

Range. Parmalee (1967: 48) stated that the species' range was "Southern portion of the Mississippi River system; Nebraska, Iowa and Illinois, south to Louisiana."

Range in Ohio. This species is expanding its range into Ohio. In recent years it has been found on the banks of the Ohio River in Clermont, Scioto, and Meigs counties, the Little Miami River in Warren County, the Whitewater and Great Miami rivers in Hamilton County, the Muskingum River in Washington County, and the Scioto River in Franklin, Pickaway, and Ross counties.

Similar species. The absence of hinge teeth, nearly circular outline, and flattened dorsal margin make this mussel easy to recognize. *Pyganodon grandis grandis* (Say, 1829) is a large, thin-shelled species with edentulous hinge structure and double-looped to nodulous umbo sculpture. However, in *P. grandis* the umbo extends above the dorsal margin, and the outline of the shell is elongate to subrhomboid, inflated. *Utterbackia imbecillis* (Say, 1829) is the only other thin-shelled Ohio species with a flattened umbo that lacks hinge dentition. However, that species is distinctly elongate, usually inflated, and much smaller than adult *A. suborbiculata*.

Comments. Individuals grow very quickly for 2–3 years and then gradually slow. They grow to ca. 200 mm but rarely exceed 10 years.

Anodontoides Simpson *in* Baker, 1898

Etymology. Gr. *an*, not, without + Gr. *odontos*, tooth + Gr. *-oides*, like, resembling. Translation is "resembling *Anodonta*."

Original description

Anodontoides Simpson *in* Baker, 1898b: 72 [replacement name for *Anodontopsis* Simpson *in* Baker, 1898a].

Type species. *Anodonta ferussaciana* Lea, 1834, by subsequent designation of Simpson (1900: 658).

Synonymy

Anodontopsis Simpson *in* Baker, 1898a 76 [*non Anodontopsis* M'Coy, 1851 (Mollusca: Allodesmatidae)].

Anodontoides Simpson *in* Baker, 1898b: 72 [replacement name for *Anodontopsis* Simpson *in* Baker, 1898b].

Geographic range. All described species of this genus are found in eastern and central North America in the Mississippi, St. Lawrence, and Gulf Coast stream systems. The range north includes the interior basin of Canada (Clarke, 1973: 93, map 10).

Comments. One might expect a high number of variant populations in a form that tends to become isolated, or nearly so, in an equal number of headwater habitats. This is the case. Just how many of these populations or population clusters are valid taxa has yet to be determined. Simpson (1914: 467–472) and Haas (1969: 370–372) each list four of these as subspecies or closely related species while Burch (1975: 16) was able to reduce this genus to two species.

Frierson (1927: 17) noted that Simpson's example of *Anodontoides ferussacianus* (Lea, 1834), upon which the genus *Anodontoides* was based, was "abnormal" because of its double-looped beak sculpture and the marsupia occupying all four gills. Typical *A. ferussacianus* has concentric beak sculpture, and the marsupia occur only in the outer gills. This raises the question of upon what species *Anodontoides* is really based.

Anodontoides ferussacianus (Lea, 1834)

Figures 35, 36, 37

Etymology. From Baron de Férussac, André Étienne J. P. J. F. d'A. (1786–1836). This is a patronym in honor of this French malacologist. His paper of 1835 was a sincere effort to bring about the solution of many taxonomic and nomenclatorial problems in this group. Translation is "in honor of Férussac."

Vernacular name(s). *Cylindrical Papershell*, Cylinder Shell, Creek Papershell.

Original description

Anodonta ferussaciana Lea, 1834: 45, pl. 6, fig. 15.

Type locality. "Ohio River, near Cincinnati" [Hamilton Co., Ohio].

Type material. Lectotype USNM 86520 Johnson & Baker (1973: 155), 13 paratypes MCZ 178770. It is indeed strange that this headwater or creek form should have as its type locality a river as large as the Ohio River at Cincinnati. It does occur in large streams, but records indicate it must be rare in such a habitat. Equally peculiar is the fact that this species typically has a basic beak sculpture of single concentric loops whereas the figured holotype has distinct double loops.

Synonymy

Anodonta ferussaciana Lea, 1834: 45, pl. 6, fig. 15.

Margarita (*Anodonta*) *ferussaciana* (Lea, 1834). Lea, 1836: 51.

Margarita (Anodonta) buchanensis Lea, 1836: 54 [*nomen nudum*].

Anodonta buchanensis Lea, 1838a: 47, pl. 14, fig. 43 [Type locality: "Buck Creek, Ohio." Collected by "R. Buchanan, Esq." Lectotype USNM 86719 *fide* Johnson, 1974: 24].

Anodonta subcylindracea Lea, 1838a: 106, pl. 24, fig. 117 [Type locality: "Oak Orchard Creek, Orleans County, New York." Syntypes AMNH 31131, 31131a].

Margarita (*Anodonta*) *subcylindracea* (Lea, 1838). Lea, 1838b: 32.

Anodonta argentea Lea, 1840: 289 [Type locality: "Stone's River, Tenn." Lectotype USNM 86538].

Anodonta ferruginea Lea, 1840: 289 [Type locality: "Simon's Creek, Indiana." Lectotype USNM 86539].

Anodont plicata Haldeman, 1842: 202 [Type locality: "Cumberland River, Tennessee." Lectotype ANSP 41042 by designation of Johnson & Baker (1973: 166)].

Anodon subcylindracea (Lea, 1838). DeKay, 1843: 200, pl. 16, fig. 229.

Anodon ferussaciana (Lea, 1834). DeKay, 1843. 200, pl. 16, fig. 230.

Anodonta oblita Lea, 1852a: 290, pl. 28, fig. 52 [Type locality: "Campbell County, East Tennessee." Lectotype USNM 86536].

Margaron (*Anodonta*) *ferussaciana* (Lea, 1834). Lea, 1852b: 50.

Margaron (*Anodonta*) *ferruginea* (Lea, 1840). Lea, 1852b: 50.

Margaron (*Anodonta*) *argentea* (Lea, 1840). Lea, 1852b: 50.

Margaron (Anodonta) oblita (Lea, 1838). Lea, 1852b: 50.

Margaron (*Anodonta*) *buchanensis* (Lea, 1838). Lea, 1852b: 51.

Margaron (*Anodonta*) *subcylindraceus* (Lea, 1838). Lea, 1852b: 51.

Strophitus argenteus (Lea, 1840). Conrad, 1853: 262.

Anodonta modesta Lea, 1857b: 84 [Type locality: "Pond, near Kalamazoo, Mich." Lectotype USNM 86534].

Anodon modestus (Lea, 1857). Sowerby, 1867: pl. 10, fig. 26.

Anodon subcylindracea (Lea, 1838). Sowerby, 1867: pl. 14, fig. 47.

Margaron (*Anodonta*) *plicata* (Haldeman, 1842). Lea, 1870: 79.

Anodonta subcylindracea (Lea, 1838). Clessin, 1873: 92, pl. 26, figs. 6, 7.

Margaron (*Anodonta*) *modesta* (Lea, 1857). Lea, 1870: 79.

Anodontoides ferussacianus (Lea, 1834). Baker, 1898b: 72, pl. 3, fig. 6, pl. 5, fig. 2.

Anodontoides subcylindraceus (Lea, 1838). Baker, 1898b: 74, pl. 4, fig. 4, pl. 6, fig. 1.

Anodontoides ferussacianus var. *subcylindraceus* (Lea, 1838). Simpson, 1900a: 660.

Anodontoides ferussacianus var. *modestus* (Lea, 1857). Simpson, 1900a: 660.

Anodontoides ferussacianus var. *buchanensis* (Lea, 1857). Simpson, 1914: 469.

Anodontoides modestus (Lea, 1857). Simpson, 1914: 470.

Anodonta subcilindracea [*sic*] Lea, 1838. De Gregorio, 1914: 62.

Anodonta ferrusaceana [*sic*] var. *incertopsis* De Gregorio, 1914: 64, pl. 11, fig. 5 [Type locality: "Sciota [Scioto] R. (Ohio)." Type in Palermo Museum?].

Anodonta ferrusaceana [*sic*] Lea, 1834. De Gregorio, 1914: 71.

Anodontoides birgei Baker, 1923: 123–125 [Type locality: "Sturgeon Bay, Door County, below railroad bridge, Wisconsin." Holotype UMMZ 209188, 4 paratypes UMMZ 209182].

Anodonta (*Anodontoides*) *ferussacianus subcylindracea* (Lea, 1838). Frierson, 1927: 17.

Anodonta (*Anodontoides*) *modesta* (Lea, 1857). Frierson, 1927: 17.

Anodonta (*Anodontoides*) *birgei* (Baker, 1923). Frierson, 1927: 17.

Fig. 35. ***Anodontoides ferussacianus*** **(Lea, 1834). A.** OSUM 61612.2, Sandusky River, Wyandot Co., OH. 101 mm. **B.** OSUM 9083.4, Ohio Canal, Tuscarawas Co., OH. 56 mm. **C.** OSUM 29814, East Branch St. Joseph River, Williams Co., OH. 68 mm. **D, H, Hinge detail:** OSUM 68739.1, Ohio River, Hamilton Co., OH. 73 mm. **E.** OSUM 13216.1, Paint Creek, Fayette Co., OH. 66 mm. **F.** OSUM 65376, Fish Creek, Williams Co., OH. 61 mm. **G.** OSUM S-3507, Hellbranch Run, Franklin Co., OH. 46 mm. **Beak sculpture detail:** OSUM 61202.2, North Fork Deer Creek, Clark Co., OH. 24 mm total length.

Anodontoides ferussacianus buchanensis (Lea, 1838). La Rocque, 1967: 200.
Anodontoides ferussacianus subcylindracea (Lea, 1838). La Rocque, 1967: 200.
Anodontoides ferussacianus modestus (Lea, 1838). La Rocque, 1967: 200.
Anodontoides ferussacianus ferussacianus (Lea, 1834). Haas, 1969: 371.
Anodontoides ferussacianus subcylindraceus (Lea, 1838). Haas, 1969: 371.
Anodontoides ferussacianus birgei Baker, 1923. Haas, 1969: 371.
Anodonta ferussaceana [*sic*] Lea, 1834. Johnson & Baker, 1973: 155.

Status. Not listed by either Ohio Division of Wildlife or U.S. Fish & Wildlife Service as of 2009.

ADULT SHELL

Shell. Small to medium sized (to 100 mm), becoming thicker with age but always rather fragile, usually very inflated, especially as adults, oval to quite elongate in profile.

Anterior margin. Roundly arched into ventral margin.

Ventral margin. Broadly rounded, straight, or slightly concave.

Posterior margin. Ventral and posterior margins form wide, blunt acute angle or squared-off point more or less at the midline; posterior margin oblique and straight or slightly concave to dorsal margin.

Dorsal margin. Long, straight or slightly sinuate, gently bowed, particularly in adults, joining posterior margin with very obtuse angle.

Umbo. Low, broad, located about 25% to nearly 50% back from anterior margin; umbonal sculpture of 2–4 fine angular "V"-shaped ridges, the "V"s tilted anterior and the posterior arm the shorter of the two.

Ligament. Medium length and irregular, tan to dark brown, projecting both anterior and posterior of the umbo, partially concealed by shells.

Disc sculpture. Smooth except for dorsal ridge.

Periostracum. Green, tan, or yellowish, becoming dark brown to black with age. Usually with numerous fine green or brown rays, often merged into broad rays.

Nacre. White, iridescent posteriorly, gold or brown Tulberg layers uncommon.

Hinge plate. Thin, cord-like, slightly sinuate. Edentate.

Muscle scars. Anterior adductor very slightly impressed; posterior adductor flush or shallowly impressed; anterior adductor distinct from anterior pedal protractor scar, protractor elongate and diverging from adductor; large posterior adductor confluent and nearly inseparable from posterior pedal retractor scar; minute, multiple dorsal suspensor located within the umbonal cavity except for 1–3 larger anterior scars. Pallial line entire, located far from shell margin. Accessory line absent.

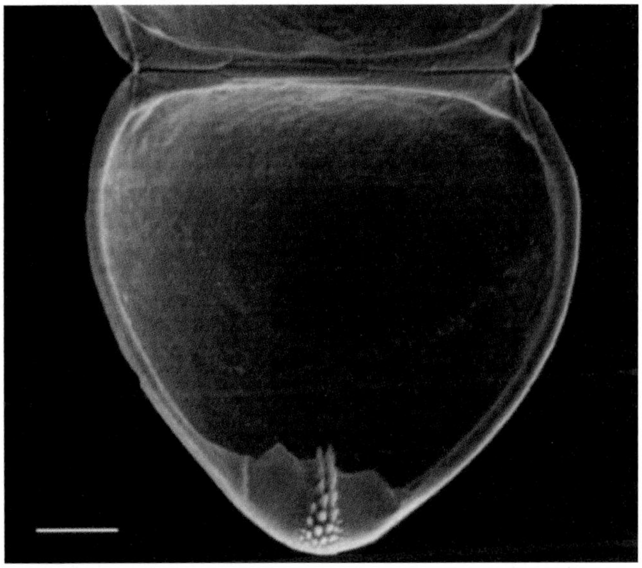

Fig. 36. *Anodontoides ferussacianus* (Lea, 1834). Glochidium. MAH 989.4, Silver Creek, Williams Co., OH. Bar = 50 μm. (Hoggarth, 1999)

Umbonal cavity. Very wide, shallow and open.

Glochidia. Subtriangular, only slightly asymmetric, dorsal margin straight, posterior and anterior margins about equally curved to a bluntly rounded ventral terminus. Exterior surface sculpturing loose-looped, hooks styliform with two rows of microstylets extending distally on the hooks (Lea, 1858c: 50, pl. 5, fig. 35; Hoggarth, 1999: figs. 14a–e).

Literature records (Ortmann, 1912a: 294; Surber, 1912: 8, pl. 3, fig. 43; Utterback, 1915: 269; Hoggarth, 1999: figs. 14a–e) indicate glochidial dimensions for length (0.32 to 0.33 mm), height (0.32 to 0.33 mm), and hinge (0.23 to 0.24 mm). The glochidium of *A. ferussacianus* can be distinguished from above described members of the Anodontinae by its double row of microstylets distally and its symmetry. The hook of this glochidium resembles that of *Strophitus*, but the outline of its valve is essentially like that of *Anodonta*.

Reproductive biology. Bradytictic. The outer gills are marsupial and turn brown when gravid (Ortmann, 1912a). However, Frierson, 1927, remarked that Simpson's specimen of *A. ferussacianus* (if indeed it was this species) was atypical in possessing marsupia in all four gills. Eggs develop in August, and glochidia appear in September. Glochidia are overwintered until the following May (Ortmann, 1919; Clarke, 1981b) and released in mucus strands that entangle hosts (Hove *et al.*, 1995a; Watters, 1995a, 2008). This species appears to be a generalist in its host use.

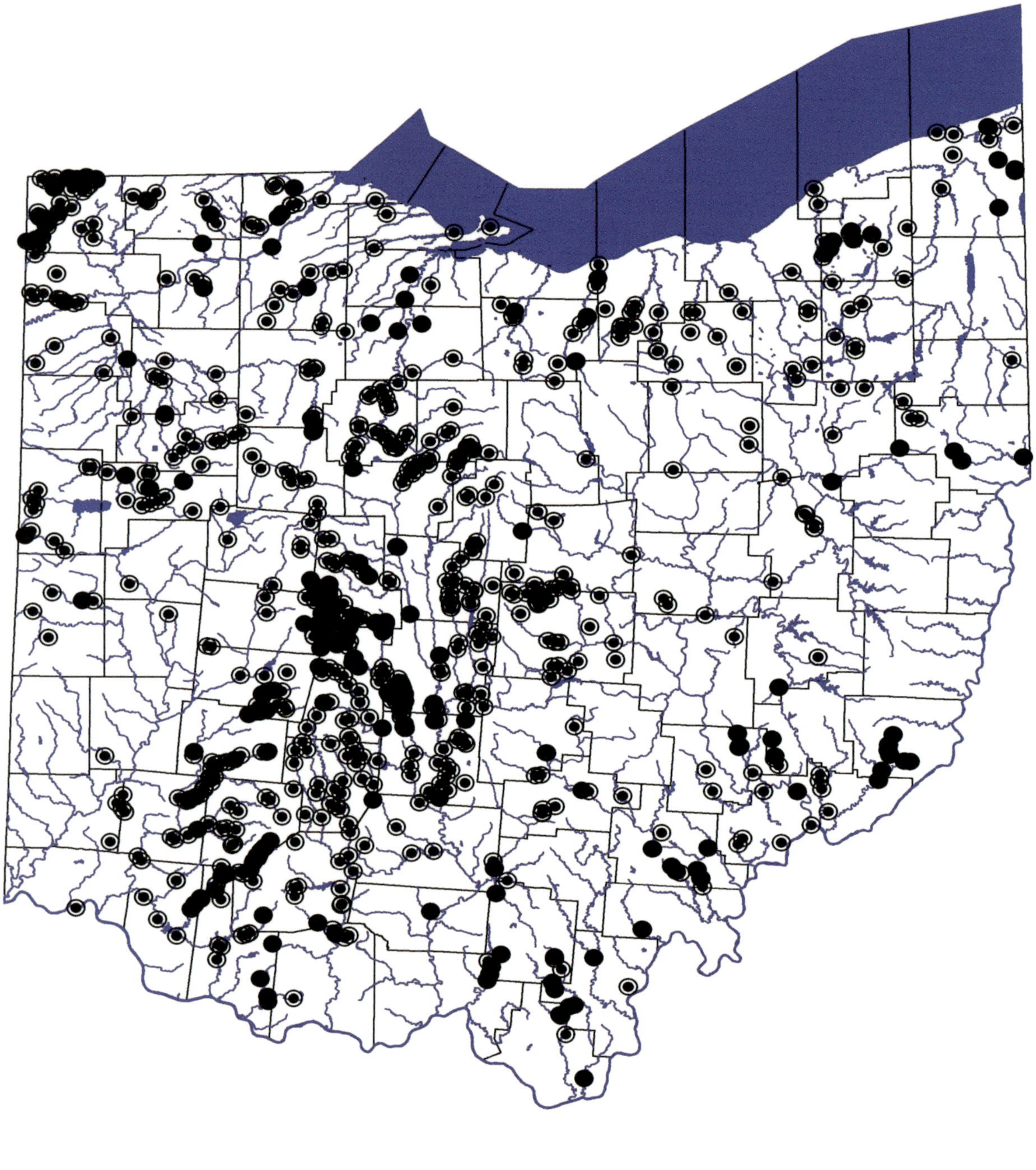

Fig. 37. ***Anodontoides ferussacianus*** **(Lea, 1834)**
Solid dots—live or freshdead since 1980
Bullseye—all records, any condition

Reported potential hosts

Catostomus commersoni (White Sucker)—Kakonge *in* Fuller, 1978 (NS).

Cottus bairdi (Mottled Sculpin)—Morrison *in* Clarke & Berg, 1959 (NS).

Culaea inconstans (Brook Stickleback)—Kakonge *in* Fuller, 1978 (NS).

Cyprinella spiloptera (Spotfin Shiner)—Hove *et al.*, 1995a (LT); Hove *et al.*, 1997 (LT).

Etheostoma exile (Iowa Darter)—Kakonge *in* Fuller, 1978 (NS).

Etheostoma tippecanoe (Tippecanoe Darter)—O'Dee & Watters, 2000 (LT).

Lepomis macrochirus (Bluegill)—Watters, 1995a (LT).

Luxilus cornutus (Common Shiner)—Kakonge *in* Fuller, 1978 (NS).

Micropterus salmoides (Largemouth Bass)—Watters, 1995a (LT); O'Dee & Watters, 2000 (LT).

Notropis heterolepis (Blacknose Shiner)—Kakonge *in* Fuller, 1978 (NS).

Petromyzon marinus (Sea Lamprey)—Wilson & Ronald, 1967 (NI).

Pimephales notatus (Bluntnose Minnow)—Kakonge *in* Fuller, 1978 (NS).

Pimephales promelas (Fathead Minnow)—Kakonge *in* Fuller, 1978 (NS).

Pomoxis nigromaculatus (Black Crappie)—Hove *et al.*, 1995a (LT); Hove *et al.*, 1997 (LT).

Habitat. Primarily a headwater species, occurring farther upstream than any other Ohio unionoid, even in intermittent streams. It lives in a wide variety of substrates, from packed cobble to silty mud and clay. It may be found wedged in the crevices of bedrock-bottomed streams.

Range. Upper Mississippi River, including Missouri, Ohio, and upper Cumberland rivers (but absent from the Tennessee River), east to New York and Pennsylvania, and James and Hudson Bay drainages in Canada. Great Lakes.

Range in Ohio. Very widely distributed in both Ohio River and Lake Erie drainages, particularly in small streams and headwaters. Rare in Lake Erie proper.

Similar species. This species is easily confused with *Pyganodon grandis, Utterbackia imbecillis,* and particularly *Strophitus undulatus.* All of these species occur sympatrically. *Pyganodon grandis* reaches a much greater size and has double-looped, rather than concentrically looped, beak sculpture. *Utterbackia imbecillis* has the beaks flush with the dorsal margin when viewed from the side, in *A. ferussacianus* the beaks obviously project above this margin. *Strophitus undulatus* is so similar to *A. ferussacianus* that some specimens may not be equivocally separated on shell characteristics alone. For specimens of the same size, *S. undulatus* has coarser, larger beak sculpture than does *A. ferussacianus. Anodontoides ferussacianus* usually is more cylindrical in shape, but this varies between populations. *Strophitus undulatus* is more likely to be rayed, but again this varies. Discerning these two species requires comparison with a large number of specimens in collections, and may require a specialist. Finally, both species may actually be complexes of sibling species.

Comments. Individuals grow very quickly for 2–3 years and then gradually slow. They grow to ca. 100 mm but rarely exceed 7 years.

Cyclonaias Pilsbry *in* Ortmann & Walker, 1922

Etymology. Gr. *kyklos*, ring, circle, wheel, sphere + Gr. *naias, naiados*, a water nymph of rivers and springs. The term naiad was used for unionoids by Lamarck as early as 1819 (p. 152). The name literally means "round unionoid," an apt description, although not that distinctive.

Original description

Cyclonaias Pilsbry *in* Ortmann & Walker, 1922: 18–19. [The name *Rotundaria* Rafinesque, 1820, in which the species *tuberculata* Rafinesque, 1820, was described, was given *subrotundus* Rafinesque, 1820, as its type by Herrmannsen (1847: 407). This made *Rotundaria* a synonym of *Obovaria* Rafinesque, 1820, thus unavailable for *tuberculata* Rafinesque, 1820, which is obviously not an *Obovaria. Cyclonaias* was coined by Pilsbry *in* Ortmann & Walker (1822: 18), at the request of Ortmann and Walker, as the generic name for *tuberculata* Rafinesque, 1820.]

Type species. *Obliquaria (Rotundaria) tuberculata* Rafinesque, 1820, by monotypy.

Synonymy

Cyclonaias Pilsbry *in* Ortmann & Walker, 1922: 18–19.

Geographic range. Upper Mississippi River drainage, Lake Erie drainage.

Cyclonaias tuberculata (Rafinesque, 1820)

Figures 38, 39

Etymology. L. *tuberculum*, small swelling or bump; from L. *tuber*, hump, bump, swelling + L. *-ata*, provided with.

Vernacular name(s). *Purple Wartyback*, Purple Pimpleback.

Original description

Obliquaria (*Rotundaria*) *tuberculata* Rafinesque,1820: 308, no figure.

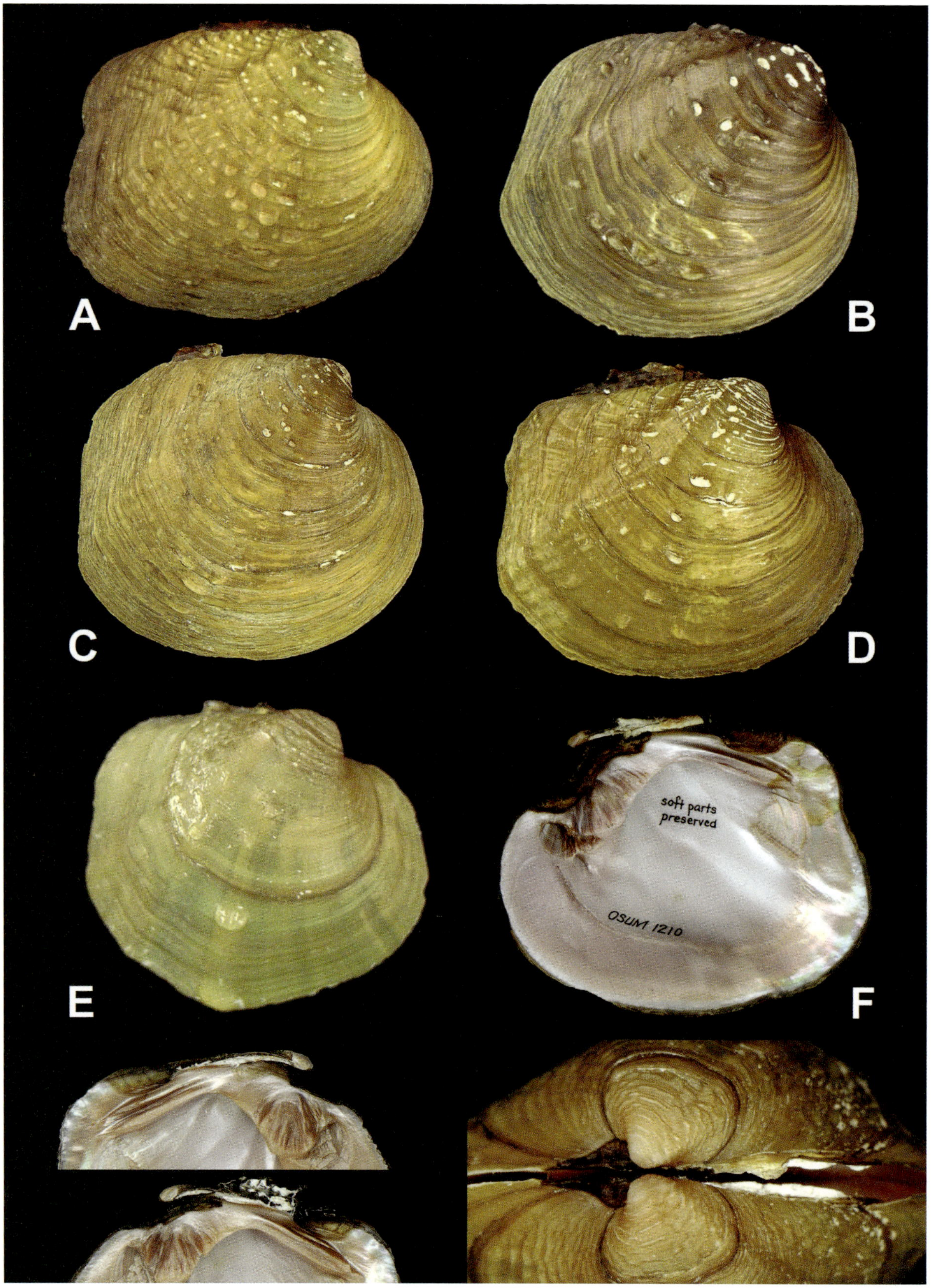

Fig. 38. ***Cyclonaias tuberculata*** **(Rafinesque, 1820). A, F.** OSUM 1210, Fish Creek, Williams Co., OH. 90 mm. **B.** OSUM 16310.8, Ohio River, Hamilton Co., OH. 66 mm. **C.** OSUM 9206.2, Sandusky River, Wyandot Co., OH. 99 mm. **D, Hinge detail:** OSUM 8489.3, Big Darby Creek, Pickaway Co., OH. 69 mm. **E, Beak sculpture detail:** OSUM 8489.1, Big Darby Creek, Pickaway Co., OH. 23 mm.

Type locality. "Très-commune dans l'Ohio et les rivières adjacentes" ["Very common in the Ohio and adjacent rivers"]. The lectotype is from the Ohio River.

Type material. The type mentioned by Vanatta (1915: 558) was selected as lectotype (ANSP 20215 from the Ohio River) by Johnson & Baker (1973: 173).

Synonymy

Obliquaria (Rotundaria) tuberculata Rafinesque, 1820: 308.

Unio verrucosus Barnes, 1823: 123–124, fig. 6 [Type locality: "Ouisconsin [*sic*] River" and "Lake Erie." Types lost].

Mya verrucosa (Barnes, 1823). Eaton, 1826: 216.

Unio tuberculosa [*sic*] (Rafinesque, 1820). Valenciennes, 1827: 232.

Unio verrucosa purpureus Hildreth, 1828: 281 [Type locality: "Muskingum [River, OH]." Type not located].

Unio tuberculatus (Rafinesque, 1820). Conrad, 1836: 43.

Margarita (*Unio*) *verrucosus* (Barnes, 1823). Lea, 1836: 16.

Unio graniferus Lea, 1838a: 69 [Type locality: "Ohio River, Cincinnati, [Hamilton Co.] O[hio]." Lectotype USNM 84296; paralectotype 84293].

Margarita (*Unio*) *graniferus* (Lea, 1838). Lea, 1838b: 15.

Rotundaria tuberculata (Rafinesque, 1820). Agassiz, 1852: 48.

Margaron (*Unio*) *graniferus* (Lea, 1838). Lea, 1852b: 22.

Margaron (*Unio*) *verrucosus* (Barnes, 1823). Lea, 1852b: 22.

Quadrula verrucosus (Barnes, 1823). Baker, 1898b: 85.

Quadrula (*Rotundaria*) *tuberculata* (Rafinesque, 1820). Simpson, 1900a: 795.

Quadrula (*Rotundaria*) *granifera* (Lea, 1838). Simpson, 1900a: 795.

Quadrula granifera var. *pusilla* Simpson, 1900a: 795 [Type locality: "Green River at Mammoth Cave, [Edmonson Co.] Kentucky." Lectotype MCZ 20176].

Plethobasus cooperianus (Lea, 1834). Utterback, 1915: 183, pl. 20, fig. 57 [Misidentification. See *Quadrula* (*Cyclonaias*) *tuberculata utterbackiana* Frierson, 1927, below.].

Cyclonaias tuberculata (Rafinesque, 1820). Ortmann & Walker, 1922: 18.]

Rotundaria granifera (Lea, 1838). Grier, 1922: 11–31.

Cyclonaias granifera (Lea, 1838). Grier & Mueller, 1922: 48.

Quadrula (*Cyclonaias*) *tuberculata* (Rafinesque, 1820). Frierson, 1927: 52.

Quadrula (*Cyclonaias*) *tuberculata granifera* (Lea, 1838). Frierson, 1927: 52.

Quadrula (*Cyclonaias*) *tuberculata pusilla* Simpson, 1900. Frierson, 1927: 52.

Quadrula (*Cyclonaias*) *tuberculata utterbackiana* Frierson, 1927: 52 [Replacement name for *Plethobasus cooperianus* "Lea" Utterback, 1915: 183, pl. 20, fig. 57, *non* Lea, 1834].

Cyclonaias tuberculata granifera (Lea, 1838). Baker, 1928a: 107, pl. 51, figs. 3, 4.

Cyclonaias tuberculata tuberculata (Rafinesque, 1820). Haas, 1969: 606.

Cyclonaias tuberculata pusilla (Simpson, 1900). Haas, 1969: 606.

Cyclonaias tuberculata utterbackiana (Frierson, 1927). Haas, 1969: 607.

Status. Listed as a Species of Concern by Ohio Division of Wildlife; not listed by U.S. Fish & Wildlife Service as of 2009.

ADULT SHELL

Shell. Medium sized (to 120 mm), oval to nearly circular, moderately inflated to compressed; usually with well-developed wing.

Anterior margin. Long, straight to broadly rounded to the ventral border.

Ventral margin. Very broadly rounded or straight, occasionally with a slight concavity at the posterior ridge in old individuals.

Posterior margin. Long, nearly straight or concave to the dorsal margin; sometimes indented at junction of posterior wing.

Dorsal margin. Long, straight to gently arched, joining posterior margin with obtuse or right angle.

Umbo. Wide and low; umbonal sculpture of numerous, irregular concentric threads with prominent sulcus; sculpture may be somewhat pustulose and/or form chevrons.

Ligament. Long and prominent, tan to dark brown, partially concealed by shells. In old individuals ligamental material forms a dense mat over the hinge plate beneath the umbo.

Disc sculpture. Anterior third of shell smooth, remainder sculptured with numerous, low pustules or broken ribs, forming pustulose radiating ribs on the dorsal wing; sulcus often obliterated in adults.

Periostracum. Juveniles are yellow or yellow-green often with vague green rays; adults become a reddish-brown or yellow, often losing the rays.

Nacre. Pale brownish purple, either uniformly colored or restricted to outside of the pallial line, in which case the interior is porcelain white; often iridescent posteriorly; often with gold or brown Tulberg layers.

Hinge plate. Well-developed, massive, angled under umbo; right cardinal a single, massive, triangular denticle, often with small anterior and posterior accessory tooth; left cardinal bifid, the teeth well-separated; laterals well-developed, very short, straight, one in right valve, two in left; interdentum very wide, extending dorsally above cardinals, flattened, long.

Muscle scars. Anterior adductor small and strongly impressed against cardinal teeth; posterior adductor nearly flush; anterior adductor distinct but abutting anterior

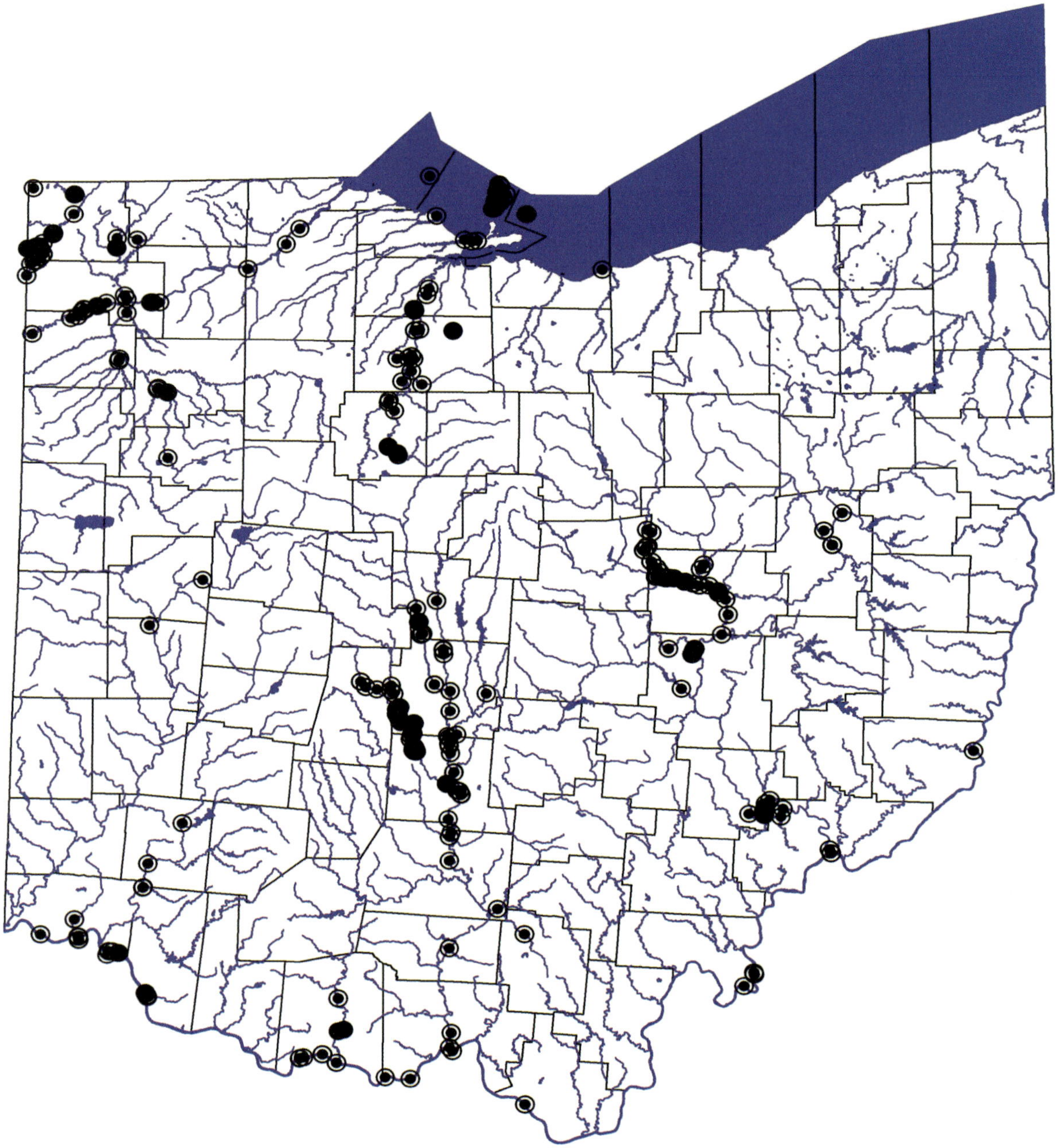

Fig. 39. ***Cyclonaias tuberculata*** **(Rafinesque, 1820)**
Solid dots—live or freshdead since 1980
Bullseye—all records, any condition

pedal protractor scar, large posterior adductor confluent with small posterior pedal retractor scar; puncture-like dorsal suspensors located behind cardinal teeth and interdentum in a line. Pallial line entire, located very far from shell margin. Accessory line present, reaching ¼ the length of the shell.

Umbonal cavity. Very wide (antero-posteriorly), deep, and compressed.

Glochidia. Subelliptical, short, straight dorsal margin, lateral margins equally rounded, ventral margin circular, hooks absent (Utterback, 1915: 151, pl. 1, fig. 4; Ortmann, 1919: 58).

Literature records (Surber, 1912: 9, pl. 2, fig. 19; Howard, 1914a: 16; Utterback, 1915: 151, pl. 1, fig. 4; Utterback 1916b: 20; Ortmann, 1919: 58) indicate glochidial dimensions for length (0.25 to 0.29 mm) and height (0.28 to 0.35 mm). Ortmann (1919: 58) noted, "Glochidia described and figured by Utterback, 1916, pl. 1, fig. 4; They are usually large, 0.267 x 0.325 mm. Surber, 1912, pl. 2, fig. 19 figures the glochidia of the closely allied *R. granifera* [*C. tuberculata*]. They are also large, 0.29 x 0.355 mm., but I have found them in *R. granifera* from the Black River, Arkansas, to be considerably smaller, 0.25 x 0.28 mm."

Reproductive biology. Tachytictic. Ortmann (1919) stated that only the outer gills are used as marsupia. But Frierson (1927: 52) remarked that *C. tuberculata* "is one of those recalcitrant *Naiades* which play havoc with classification based on the number of gills employed as marsupia. Usually, or normally, carrying ova in but two, it sometimes carries them in all four." Individuals reach maturity from 4 to 6 years of age (Jirka & Neves, 1992). Sex ratios are approximately equal (Haggerty *et al.*, 1995). Spermatogenesis occurs in this species mainly between March (Jirka & Neves, 1992) and July (Haggerty *et al.*, 1995). Embryos are brooded in the outer demibranchs between early April and late August, with glochidia formed by May and late August of the same year (Haggerty *et al.*, 1995; Heath *et al.*, 1998). Glochidia are released as white, oblong conglutinates of the simple nonelastic type (Watters, 2008). To date, all identified potential hosts are catfish.

Reported potential hosts

Ameiurus melas (Black Bullhead)—Hove, 1997 (LT); Hove & Kurth, 1997 (LT).
Ameiurus natalis (Yellow Bullhead)—Hove *et al.*, 1994 (LT); Hove & Kurth, 1997 (LT); Hove *et al.*, 1997 (LT).
Ictalurus punctatus (Channel Catfish)—Hove *et al.*, 1997 (LT); Hove & Kurth, 1997 (LT); Hove, 1997 (LT).
Pylodictus olivaris (Flathead Catfish)—Hove, 1997 (LT); Hove & Kurth, 1997 (LT).

Habitat. This species requires high water quality and is usually found buried in coarse sand or gravel in a swift current. It is most common in large creeks. Parmalee (1967: 29) described the preferred habitat of this species as "a gravel/mud bottom, usually in areas of current. . . ." Buchanan (1980: 32) reported that "*Cyclonaias tuberculata* was found in a silt, sand, gravel and cobble or cobble and boulder substrate, but was found most often in a gravel and cobble substrate. It occurred in 1 inch to 5 feet of water in slow (less than 0.2 ft./sec. at the bottom) to moderate (1.4 ft./sec. at the bottom) current."

Range. La Rocque (1967: 151) stated that the species' range was "Mississippi drainage generally; Lake St. Clair drainage; Detroit River and Lake Erie; Ohio River and its drainage." Parmalee (1967: 27) added "Cumberland and Tennessee River systems."

Range in Ohio. The present distribution of this species in Ohio includes both the Ohio River and Lake Erie drainage basins (but it is absent from northeastern Ohio); the western basin of Lake Erie and the Sandusky and Maumee rivers; rare in the Great and Little Miami rivers and Ohio Brush Creek; more common in the upper Muskingum tributaries and the Scioto River.

Similar species. Other pustulate species with circular outlines, such as *Quadrula pustulosa pustulosa* (Lea, 1831), *Plethobasus cooperianus* (Lea, 1834), and *Cyprogenia stegaria* (Rafinesque, 1820), might be superficially similar to this species, but none of these have purple nacre. Other species with purple nacre, such as *Elliptio dilatata* (Rafinesque, 1820), *Elliptio crassidens* (Lamarck, 1819), *Potamilus alatus* (Say, 1817), and *Epioblasma obliquata obliquata* (Rafinesque, 1820), are neither circular in outline nor pustulose and would not be confused with this species. Specimens of *Obovaria retusa* (Lamarck, 1819) are somewhat circular in outline and often have purple nacre inside of the pallial groove, but this species is not pustulose.

Comments. Individuals gradually slow in growth during their entire life. Specimens rarely reach ca. 120 mm and nearly 40 years old.

Cyprogenia Agassiz, 1852

Etymology. Gr. *Kypris*, a name of Aphrodite or *kypros*, henna (*Lawsonia inermis*, an Old-World tropical shrub, the leaves of which yield a reddish dye) + Gr. *genea*, race, generation, offspring. The Greek word *Kypris* (or Latin *Cypris*) has been used to denote beauty, elegance, or grace, qualities possessed by Aphrodite and Venus. Agassiz (1852: 47–48) discussed the spiraling ovisacs of the gravid *Cyprogenia* gill at some length and may have been impressed by their striking appearance. But another possibility is that he was using the word *kypros* to refer to the red color of the eggs (and ovisacs) in a living specimen.

Original description

Cyprogenia Agassiz, 1852: 47–48.

Type species. *Unio irroratus* Lea, 1828, by original designation [= *Obovaria stegaria* Rafinesque, 1820].

Geographic range. Historically two species have been recognized: one east of the Mississippi River (*C. stegaria*) and one west (*C. aberti* Conrad, 1850). *Cyprogenia stegaria* occurs in the Ohio River system. DNA studies suggest that the western populations may comprise more than one species (Serb, 2006).

Cyprogenia stegaria (Rafinesque, 1820)

Figures 40, 41, 42, 43

Etymology. Gr. *stege, stega, stegos*, roof, shelter. Rafinesque (translated by Poulson, 1832: 51) noted that the shell was ". . . slightly imbricated by the large separate wrinkles" [in Rafinesque's French terminology a "wrinkle" is a growth line or annular ring on the shell]. These apparently reminded Rafinesque of overlapping shingles on a roof.

Vernacular name(s). *Fanshell*, Ringed Wartyback.

Original description

Obovaria stegaria (*Unio stegaria*) Rafinesque, 1820: 312, species no. 49, pl. 82, figs. 4, 5.

Type locality. "assez rare dans l'Ohio" ["rather rare in the Ohio (River)"].

Type material. The type mentioned by Vanatta (1915: 554) was selected as lectotype (ANSP 20241 from the Ohio River) by Johnson & Baker (1973: 173). But in Johnson (1980) this specimen is referred to as ANSP 20215.

Synonymy

Obovaria stegaria Rafinesque, 1820: 312, pl. 82, figs. 4, 5.

Obovaria stegaria var. *tuberculata* Rafinesque, 1820: 312 [Type locality: "Ohio River" based on lectotype. Lectotype ANSP 20241 by desigantion of Johnson & Baker, 1973].

Unio irroratus Lea, 1828: 269, pl. 5, fig. 5 [Type locality: "Ohio." Lectotype USNM 84637 and paralectotypes ANSP lost *fide* Johnson, 1980].

Unio irrorata Lea, 1828. Deshayes, 1830: 579.

Unio stegarius (Rafinesque, 1820). Say, 1834: no pagination.

Margarita (*Unio*) *irroratus* (Lea, 1828). Lea, 1836: 16.

Unio (*Theliderma*) *irrorata* (Lea, 1828). Swainson, 1840: 271, fig. 54a.

Cyprogenia irroratus (Lea, 1828). Agassiz, 1852: 47.

Cyprogenia stegurius [*sic*] (Rafinesque, 1820). Agassiz, 1852: 48.

Margaron (*Unio*) *irroratus* (Lea, 1828). Lea, 1852b: 22.

Unio stegarius var. *irroratus* (Lea, 1828). Paetel, 1890: 168.

Cyprogenia irrorata var. *pusilla* Simpson, 1900a: 610 [Type locality: "Green River, Kentucky," restricted to "Mammoth Cave, Edmonson County" by Johnson, 1980. Type not located].

Cyprogenia stegaria (Rafinesque, 1820). Ortmann, 1919: 218, pl. 13, fig. 5.

Cyprogenia irrorata pusilla Simpson, 1900. Frierson, 1927: 66.

Status. Listed as Endangered by both Ohio Division of Wildlife and U.S. Fish & Wildlife Service as of 2009. A Federal recovery plan has been completed (USFWS, 1991a).

ADULT SHELL

Shell. Medium in size (to 80 mm), thick and massive, circular to rather triangular; big river forms have a "humped" appearance; inflated.

Anterior margin. Very short, straight to broadly rounded to the ventral border.

Ventral margin. Very broadly rounded, straight or concave towards the posterior.

Posterior margin. Long, rather straight to the dorsal margin.

Dorsal margin. Long, straight, joining posterior margin with obtuse angle.

Umbo. Wide, prominent; umbos nearly face each other as juveniles but are twisted prominently forward in adults; umbonal sculpture of microscopic wrinkles and a coarse posterior ridge.

Ligament. Very short and prominent, tan to dark brown, not concealed by shells.

Disc sculpture. Disc of shell with coarse, concentric growth ridges overlain by numerous small pustules, particularly on the posterior half; posterior slope with numerous nodulose radiating ridges, at least as juveniles; sculpture strength varies greatly between populations; shell obscurely sulcate, most developed in juveniles.

Periostracum. Surface dull, not shiny; base color yellowish or tan, patterned with radiating rows of green rays composed of microscopic flecks; pustules usually lighter in color.

Nacre. Porcelain white, often iridescent posteriorly; rarely with gold or brown Tulberg spots.

Hinge plate. Well-developed, massive, very short, angled under the umbo; right cardinal a single, massive, triangular denticle, often with small anterior accessory tooth; left cardinal bifid, the teeth well-separated; laterals well-developed, massive, short, straight, one in right valve, two in left; interdentum very wide, thick, short.

Fig. 40. ***Cyprogenia stegaria*** **(Rafinesque, 1820). A.** OSUM 14746, Muskingum River, Washington Co., OH. 82 mm. **B, Hinge detail:** OSUM 5221.7, Muskingum River, Washington Co., OH. 56 mm. **C.** OSUM 102251.1, Salt River, [Co.?], KY. 59 mm. **D.** OSUM 69456.1, Ohio River, Hamilton Co., OH. 41 mm. **E.** OSUM 28322, Tuscarawas River, Tuscarawas Co., OH. 45 mm. **F, H.** OSUM 56379.6, Tippecanoe River, Tippecanoe Co., IN. 63 mm. **G.** OSUM 39552, "Wabash River." 44 mm. **Beak sculpture detail:** OSUM 39861.1, "Ohio River." 25 mm total length.

Muscle scars. Anterior adductor small, "D"-shaped, deeply impressed; posterior adductor nearly flush or impressed; small anterior adductor distinct from anterior pedal protractor scar, posterior adductor distinct from small posterior pedal retractor scar; dorsal suspensors located under interdentum in an irregular line. Pallial line entire, located far from shell margin. Accessory line well-developed, nearly ½ the length of the shell.
Umbonal cavity. Very wide, moderately deep, open.

Glochidia. Elongate-oval, subrotund, dorsal margin straight and long, anterior and posterior margins equal, greatly curved, ventral margin gently curved. Exterior surface sculpturing loose-looped, malleations absent. Dorsal alae absent, hooks absent (Sterki, 1898: 19; Hoggarth, 1999: figs. 39a–d).

Literature records (Sterki, 1898: 19; Surber, 1912: 10, pl. 1, fig. 11; Ortmann, 1912a: 314, pl. 19, fig. 6; Hoggarth, 1999: figs. 39a–d) indicate glochidial dimensions for length (0.20 to 0.21 mm), height (0.16 to 0.19 mm), and hinge (0.11 to 0.14 mm). Ortmann (1912a: 314) stated, "The shape of the glochidia approaches to a degree, that of *Dromus*, but the disproportion between length and height is much less." It should be noted that Ortmann's measurements (0.18 x 0.15 mm) are not included in the ranges given above. This glochidium can be distinguished from other species by its more broadly rounded valve outline.

Reproductive biology. Bradytictic. Eggs are found in September, and glochidia are found in October and November (Ortmann, 1919). The marsupium consists of three to eleven water tubes in the central outer gill. Conglutinates are related to the composite conglutinate B type (Watters, 2008) and project far beyond the distal gill margin, coiling backwards like a watch-spring (Ortmann, 1912a). Females release orange or pink conglutinates 40–50 mm in length coiled like the ovisacs, resembling worms. Each conglutinate contains ca. 8,000 glochidia. These are released through the excurrent siphon (Jones & Neves, 1999). Reported hosts are sculpins and darters.

Reported potential hosts

Cottus bairdi (Mottled Sculpin)—Jones & Neves, 1999 (LT)
Cottus carolinae (Banded Sculpin)—Schulz & Marbain, 1998 (LT); Jones & Neves, 1999 (LT)
Etheostoma aurantiaca (Tangerine Darter)—Jones & Neves, 1999 (LT)
Etheostoma blennioides (Greenside Darter)—Schulz & Marbain, 1998 (LT); Jones & Neves, 1999 (LT)
Etheostoma simoterum (Snubnose Darter)—Jones & Neves, 1999 (LT)
Etheostoma zonale (Banded Darter)—Jones & Neves, 1999 (LT)

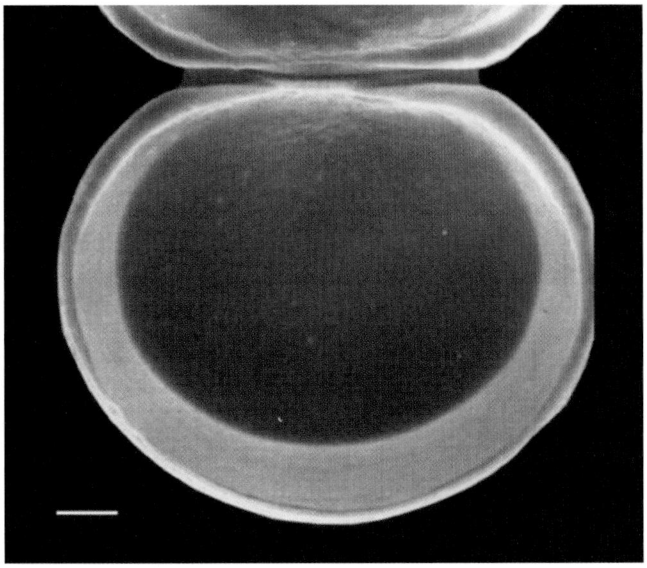

Fig. 41. ***Cyprogenia stegaria*** **(Rafinesque, 1820).** Glochidium. OSUM 6298.21, Green River, Hart Co., KY. Bar = 30 µm. (Hoggarth, 1999).

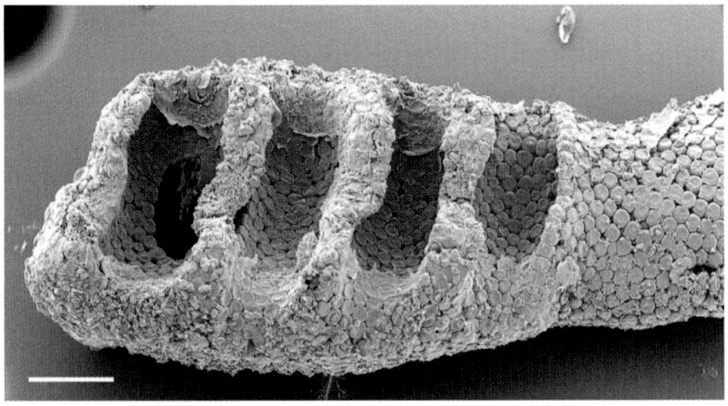

Fig. 42. ***Cyprogenia stegaria*** **(Rafinesque, 1820).** Distal end of conglutinate similar to composite conglutinate B type. OSUM 16757, Clinch River, Hancock Co., TN. Bar = 500 µm. (Watters, 2008).

Percina burtoni (Blotchside Logperch)—Jones & Neves, 1999 (LT)
Percina caprodes (Logperch)—Jones & Neves, 1999 (LT)
Percina roanoka (Roanoke Darter)—Jones & Neves, 1999 (LT)

Habitat. Stable cobble and sand in rivers and large creeks.

Range. Ohio River system, including the Tennessee and Cumberland rivers.

Range in Ohio. Only weathered shells were found in Killbuck Creek, a tributary of the Muskingum River. The only reproducing population in Ohio occurs in the lower Muskingum River (Ecological Specialists, Inc., 1993). In the Ohio River, it lives near Muskingum Island, near the

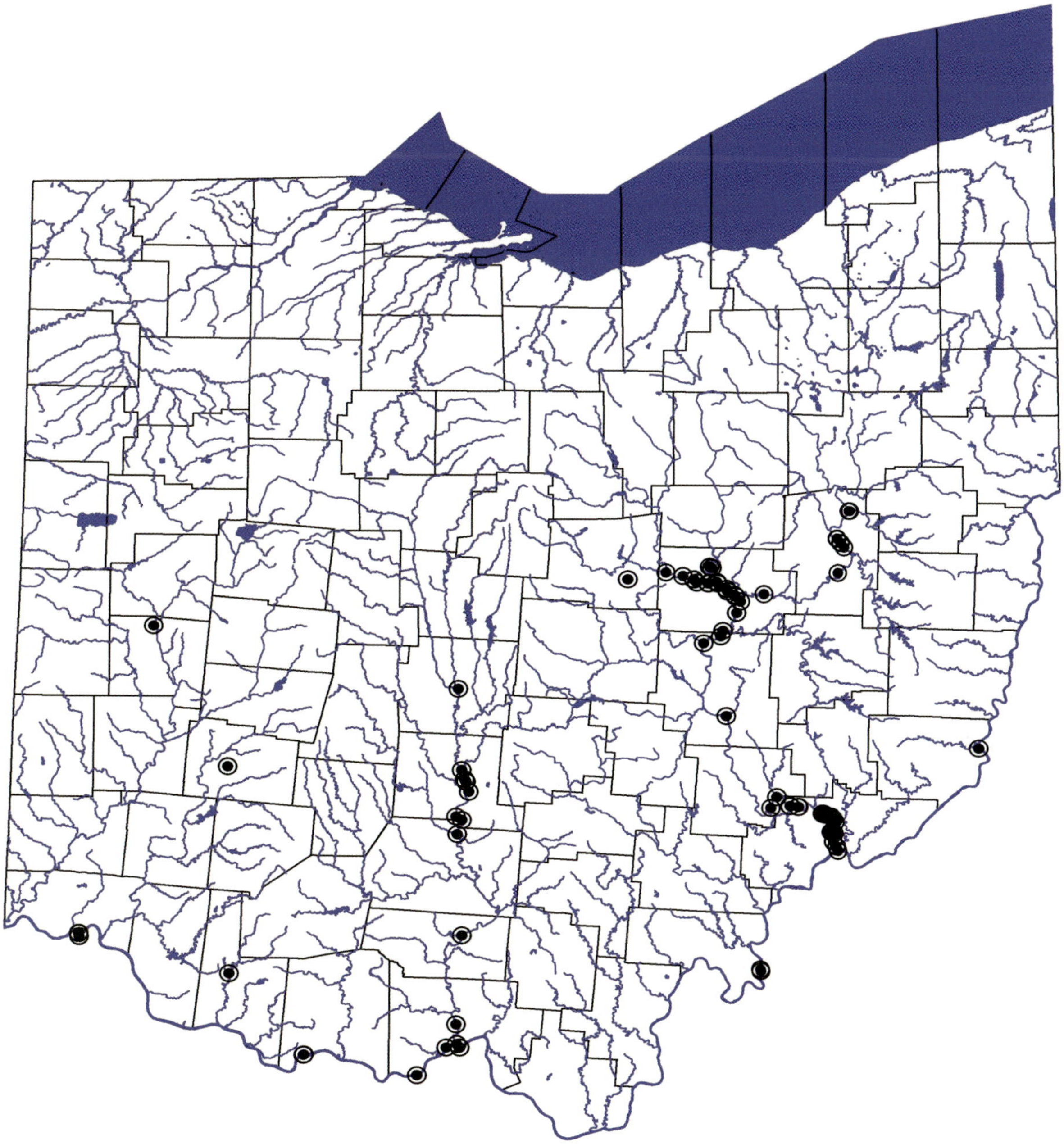

Fig. 43. ***Cyprogenia stegaria*** **(Rafinesque, 1820)**
Solid dots—live or freshdead since 1980
Bullseye—all records, any condition

confluence of the Ohio and Muskingum rivers (Morrison, pers. comm., 1999). Historically it occurred in the Tuscarawas River and in the Ohio River at Portland, Marietta, Clarington, Portsmouth, and Cincinnati (CMNH) and in the Scioto River as far upstream as Columbus. Goodrich found it as far upstream in the Little Miami River as Xenia (CMNH 61.7826, no date). A single record exists from the Great Miami River.

Similar species. Other pustulose species may resemble the Fanshell. *Cyclonaias tuberculata* (Rafinesque, 1820) has purple nacre, never found in *C. stegaria* (although that species may be flushed with pink or salmon). *Quadrula pustulosa* (Lea, 1831) and members of *Plethobasus* all lack the characteristic pattern of fine rays grouped into broad bands across the shell. *Quadrula pustulosa* has only a single green ray or blotch on the early shell. *Plethobasus* species are rayless or rayed only on the early shell and are colored brown, yellow, or tan in contrast to the Fanshell's green base color. The Fanshell lacks the large undulating knobs of *Obliquaria* and some *Epioblasma* species.

Comments. Individuals gradually slow in growth during their entire life. Specimens reach ca. 80 mm and live up to 30 years.

Ellipsaria Rafinesque, 1820

Etymology. Gr. *elleipsis*, a defect, a falling short. The geometric figure was so named because the angle between a conic section (ellipse) and its base "falls short of" (is less than) the inclination of the side of the cone. Rafinesque (1820: 303) described the genus *Ellipsaria* as having a "forme elliptique." Although this was probably an apt description of the species Rafinesque originally included in his new subgenus, it is less appropriate for *E. lineolata* (Rafinesque, 1820), the only species we include in *Ellipsaria* today.

Original description

Obliquaria (*Ellipsaria*) Rafinesque, 1820: 303.

Type species. *Obliquaria* (*Ellipsaria*) *ellipsaria* Rafinesque, 1820 [= *Obliquaria* (*Plagiola*) *lineolata* Rafinesque, 1820], by tautonomy. Rafinesque (1820) described several nominal species that are now believed to be synonymous: *Obliquaria* (*Plagiola*) *depressa*, *O.* (*Ellipsaria*) *ellipsaria*, and *O.* (*Plagiola*) *lineolata*. Conrad (1834: 70), as first revisor, selected *O. lineolata* as the valid name for this species.

Herrmannsen (1847: 279) designated *Obliquaria* (*Plagiola*) *interrupta* Rafinesque, 1820, as the type of *Plagiola*. Simpson (1900a: 603), apparently believing that *interrupta* was not recognizable, subsequently designated *Unio securis* Lea, 1829 (= *Obliquaria* (*Plagiola*) *lineolata* Rafinesque, 1820), as type of the genus *Plagiola*. Simpson's designation was invalid because *U. securis* was not one of the original names included by Rafinesque under *Plagiola*, and because Herrmannsen had already validly designated *O. interrupta* as type. But based on Simpson's action, *Plagiola* Rafinesque, 1820, was long used (though incorrectly) as the generic name for *Obliquaria lineolata* Rafinesque, 1820. Frierson (1914a) selected *Unio fasciolaris* Rafinesque, 1820, as type, but this designation was invalid as that species was not among those listed by Rafinesque. Baker (1964: 140) reviewed this problem and concluded that *Ellipsaria* was "the legal name for '*Plagiola*'," i.e., *Plagiola* was a junior synonym of *Ellipsaria*. He also concluded that the type of *Ellipsaria* was *Obliquaria* (*Ellipsaria*) *ellipsaria* Rafinesque, 1820 (= *O. lineolata*), by absolute tautonomy [ICZN Art. 68.4]. This made the type species of *Plagiola* to be *O. interrupta*, based on Herrmannsen's selection in 1847, and the type species of *Ellipsaria* to be *O. ellipsaria* based on tautonomy. The question was whether *Plagiola* and *Ellipsaria* were synonymous, and that depended on the identity of *Obliquaria* (*Plagiola*) *interrupta* Rafinesque, 1820.

Johnson & Baker (1973) and Johnson (1978) identified *O. interrupta* with *Epioblasma brevidens* (Lea, 1831) and selected ANSP 20257 as a "lectotype." But Bogan (1998), based on an apparent syntype specimen at the Muséum national d'Histoire naturelle, Paris, identified *O. interrupta* as the species now known as *Ptychobranchus fasciolaris* (Rafinesque, 1820), thus making *Plagiola* a senior synonym of *Ptychobranchus*, which was described 80 years after *Plagiola* with *fasciolaris* as its type.

Because *O. ellipsaria* is not a *Ptychobranchus*/*Plagiola*, the correct genus-species combination for this species is *Ellipsaria ellipsaria* (= *Ellipsaria lineolata* based on the action of Conrad (1834) as first revisor). Even though Baker thought *Ellipsaria* and *Plagiola* were synonymous, the selection of *O. ellipsaria* and *O. interrupta*, which are not congeneric, as types, respectively, recreated two separate generic names. Ironically, all of this overlooks the fact that Rafinesque himself in 1831 (p. 3) synonymized *Ellipsaria* with *Elliptio*.

Synonymy

Obliquaria (*Ellipsaria*) Rafinesque, 1820: 303.

Crenodonta Schlüter, 1838: 33 [Type species: *Unio securia* [*sic*] "Deshayes" Lea, 1829, by subsequent designation of Herrmannsen (1852: 38). Simpson's (1900a: 766) later designation of *Unio plicata* Say, 1817, was therefore unnecessary, although because of this *Crenodonta* is often incorrectly used as a synonym of *Amblema*.].

Plagiolopsis Thiele, 1934: 834 [Type species: *Unio securis* Lea, 1829, by monotypy].

Geographic range. Mississippi River system, portions of the Mobile River system.

Ellipsaria lineolata (Rafinesque, 1820)

Figures 44, 45, 46

Etymology. L. *lineola*, a little line + L. *-ata*, provided with. Rafinesque described the typical narrow rays of this species as "brown lines" (1820: 303).

Vernacular name(s). *Butterfly.*

Original description

Obliquaria (*Plagiola*) *lineolata* Rafinesque, 1820: 303, species no. 23, no figure.

Type locality. "Aux chutes de l'Ohio." The Falls of the Ohio are at Louisville, Kentucky. The lectotype is from the Ohio River.

Type material. The type mentioned by Vanatta (1915: 553) was selected as lectotype (ANSP 20242 from the Ohio River) by Johnson & Baker (1973: 161).

Synonymy

Obliquaria (*Plagiola*) *depressa* Rafinesque, 1820: 302, pl. 81, figs. 5–7 [Type locality: "une fois près d'Évamville [*sic*] en Indiana" ("once near Evansville in Indiana"). Neotype ANSP 20251 designated by Johnson & Baker, 1973: 153].

Obliquaria (*Plagiola*) *lineolata* Rafinesque, 1820: 303.

Obliquaria (*Ellipsaria*) *ellipsaria* Rafinesque, 1820: 303 [Type locality: "Dans le Kentuky [River]," but neotype from "Ohio River." Neotype ANSP 20233 by selection of Johnson & Baker (1973), but this action may be invalid as the neotype did not originate from the type locality.].

Unio securis Lea, 1829: 437, pl. 11, fig. 17 [Type locality: "Ohio." Lectotype USNM 84475. Paralectotype in ANSP 59869 not found *fide* Johnson & Baker, 1973].

Unio compressus Rafinesque, 1831: 4 [unnecessary replacement name for *Obliquaria depressa* Rafinesque, 1820, because of presumed priority of *Unio depressa* Lamarck, 1819].

Unio lineolata (Rafinesque, 1820). Say, 1834: unpaginated.

Unio lineolatus (Rafinesque, 1820). Conrad, 1834: 70.

Margarita (*Unio*) *securis* (Lea, 1829). Lea, 1836: 19.

Plagiola lineolata (Rafinesque, 1820). Agassiz, 1852: 48.

Margaron (*Unio*) *securis* (Lea, 1829). Lea, 1852b: 24.

Crenodonta securis (Lea, 1829). Herrmannsen, 1852: 38.

Plagiola lineolata (Rafinesque, 1820). Agassiz, 1852c: 48.

Plagiola securis (Lea, 1829). Smith, 1899: 291, pl. 80.

Plagiolopsis securis (Lea, 1829). Thiele, 1934: 834.

Ellipsaria lineolata (Rafinesque, 1820). Baker, 1964: 141.

Status. Listed as Endangered by Ohio Division of Wildlife; not listed by U.S. Fish & Wildlife Service as of 2009.

ADULT SHELL

Shell. Medium sized (to at least 110 mm); heavy to massive, laterally compressed as a juvenile, outline oval, pointed posteriorly.

Anterior margin. Rounded with short straight segment before umbo.

Ventral margin. Very broadly curved; forming a blunt point at the posterior margin at the base.

Posterior margin. Gently rounded, indistinguishable from the dorsal slope.

Dorsal margin. Very long, continuing the curve of the posterior margin. Dorsal ridge sharp, forming ca. 90° angle.

Umbo. Very wide, low, broadly triangular and flattened; beak sculpture absent but dorsal ridge persists as low radiating rib.

Ligament. Tan to black, short, partially concealed by the valves.

Disc sculpture. Shell sculptureless except for 1–2 very low, broad ribs on the dorsal slope; this sculpture may be absent in some specimens.

Periostracum. Pale yellowish-green to tan, becoming somewhat brown with age. Shells have a characteristic pattern of faint, narrow green rays over the shell. These are overlaid by 7–11 widely spaced, darker green rays, usually composed of dots, dashes, and chevron markings. These markings are often quite bold.

Nacre. White, iridescent posteriorly; often with gold or green Tulberg layers.

Hinge plate. Ca. 60% of shell length, curved, massive; interdentum broad, narrow, thick; cardinal teeth chunky, triangular, two in left valve, one in right with accessory teeth anterior and posterior to cardinal; lateral teeth fairly long, straight or slightly bowed, very thick, two in left, one in right with accessory tooth anterior.

Muscle scars. Adductor muscle scar impressed, posterior adductor much less so, both distinct from protractor and retractor scars; dorsal suspensor scars a long series of pits from behind the cardinals into the umbonal cavity; pallial line entire, accessory line very long, extending to the shell margin.

Umbonal cavity. Wide, open, shallow to moderately deep, somewhat compressed.

Glochidia. Subligulate, much higher than long, dorsal margin slightly curved, anterior and posterior margins more or less straight to slightly incurved dorsally, ventral margin broad and gently curved. Vermiculate sculpture covers the exterior surface of the valve, dorsal alae small and oriented at a 45° angle to the hinge. Micropoints lanceolate and arranged in complete vertical rows. Hooks absent except as small lateral swellings on the ventral margin (Lea, 1858c: 47, pl. 5, fig. 6; Ortmann, 1911: 364, pl. 89, fig. 17; Ortmann, 1912a: 330; Hoggarth, 1999: figs. 45a–f).

Fig. 44. ***Ellipsaria lineolata*** **(Rafinesque, 1820). A.** OSUM 42337.4, Muskingum River, Washington Co., OH. 85 mm. **B.** OSUM 42337.8, Muskingum River, Washington Co., OH. 117 mm. **C.** OSUM 58574.5, "Ohio River." 60 mm. **D.** OSUM 29202, St. Croix River, St. Croix Co., WI. 52 mm. **E.** OSUM 67860.2, Muskingum River, Washington Co., OH. 85 mm. **H.** OSUM 77424.2, Mississippi River, Rock Island Co., IL. 51 mm. **F, G.** OSUM 56173.1, Ohio River, Ballard Co., KY. 76 mm. **Beak sculpture detail:** OSUM 77424.1, Mississippi River, Rock Island Co., IL. 39 mm total length.

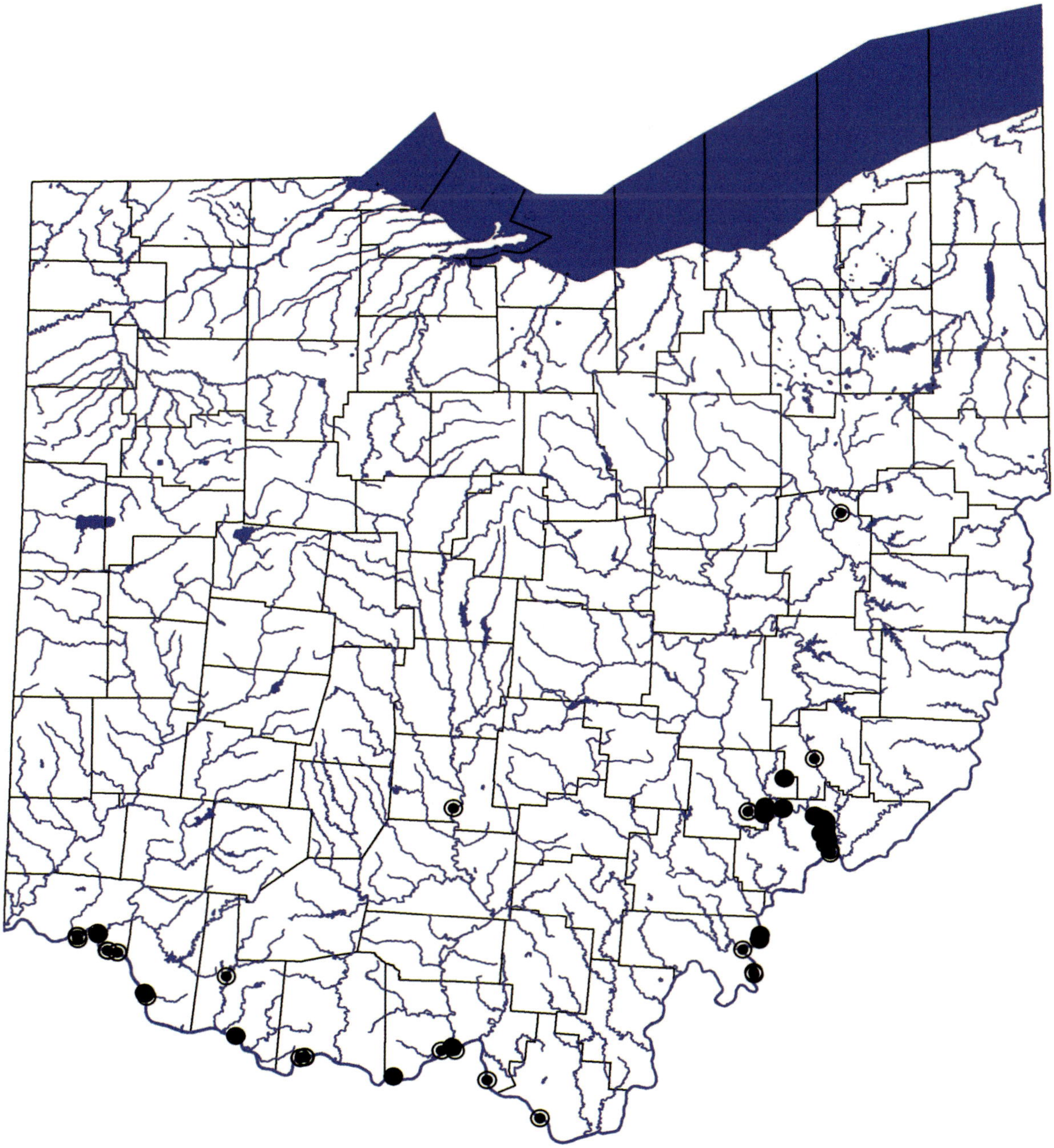

Fig. 45. ***Ellipsaria lineolata*** **(Rafinesque, 1820)**
Solid dots—live or freshdead since 1980
Bullseye—all records, any condition

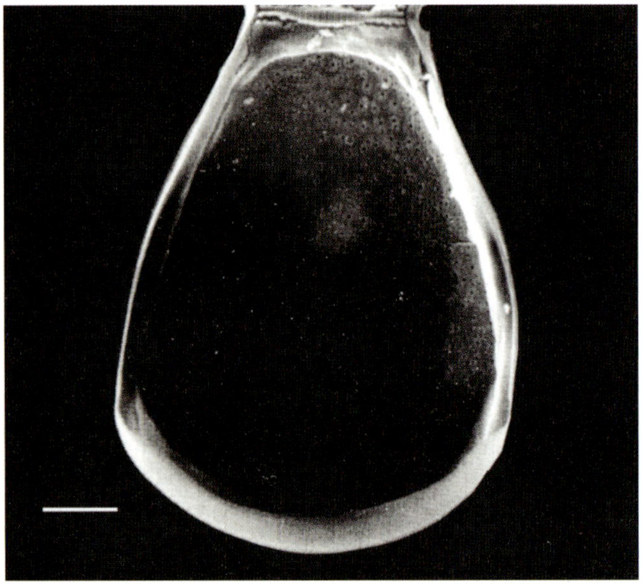

Fig. 46. *Ellipsaria lineolata* (Rafinesque, 1820). Glochidium. OSUM 1984: 14, Ohio River, Campbell Co., KY. Bar = 45 µm. (Hoggarth, 1999).

Literature records (Lefevre & Curtis, 1910a: 97, fig. h; Lefevre & Curtis, 1912: 146, fig. 1h; Surber, 1912: 9, pl. 2, fig. 14; Ortmann, 1912a: 330; Utterback, 1916: 350; Hoggarth, 1999: figs. 45a–f) indicate glochidial dimensions for length (0.23 to 0.26 mm), height (0.31 to 0.35 mm), and hinge (0.09 to 0.10 mm). Ortmann (1912a: 330) noted that the glochidia of this species "are quite large, subspatulate (dilated and rounded off toward the ventral margin), and their anterior and posterior margins are distinctly gaping." This glochidium can be distinguished from all others by its fan-shaped outline, small dorsal alae, exterior valve sculpture and its distinct hook-like folding of the ventral flange.

Reproductive biology. Bradytictic. The posterior half of the outer gills is marsupial (Ortmann, 1912a). Gravid females are present from September to the following July (Ortmann, 1919). Fragile conglutinates are released (Utterback, 1916). Attempts to metamorphose glochidia on a wide range of fishes other than Freshwater Drum have proven unsuccessful (Watters, unpubl.).

Reported potential hosts

Aplodinotus grunniens (Freshwater Drum)—Howard, 1914b (NI); Wilson, 1916 (NI); Coker *et al.*, 1921 (NI, LT); Howard & Anson, 1923 (NS).

Lepomis cyanellus (Green Sunfish)—Surber, 1913 (NI); Wilson, 1916 (NI).

Sander canadensis (Sauger)—Surber, 1913 (NI).

Habitat. Stable cobble and sand in large rivers.

Range. Mississippi River system, including the Ohio, Cumberland, and Tennessee rivers, north to Minnesota and Wisconsin, west in the Missouri, Red, and White rivers. Portions of the Mobile River system.

Range in Ohio. There are records far upstream in the Scioto River at Columbus and in the Tuscarawas River at New Philadelphia. Today it occurs only in the lower Muskingum River and Ohio River, where it is locally common. Absent from Lake Erie drainages.

Similar species. This species may only be confused with some examples of *Truncilla* species. *Ellipsaria lineolata* has a noticeably flattened and very broad umbo—in *Truncilla* these are more inflated. It is elliptical in outline rather than triangular or wedge-shaped. It also grows to a much larger size than any *Truncilla*.

Comments. This species grows fairly rapidly, particularly during the first three years, gradually slowing after that point. Sexual dimorphism in the shell is not pronounced—females tend to be rounder in profile and more inflated. Individuals may exceed 110 mm in length and grow to over 25 years old. Specimens from the Muskingum River are among the largest seen.

Elliptio Rafinesque, 1820

Etymology. Gr. *elleiptikos*, elliptic, having the shape of an ellipse. Nearly all species of *Elliptio* are elongate in lateral view, some scarcely and some extremely so, but this ellipsoidal outline can be considered to justify the name *Elliptio*.

Original description

Unio (*Elliptio*) Rafinesque, 1820: 291.

Type species. *Unio nigra* Rafinesque, 1820 [= *Unio crassidens* Lamarck, 1819, by subsequent designation of Ortmann (1912a). Rafinesque (1820) did not designate a type for his subgenus *Elliptio*, but listed four species: *nigra*, *viridis*, and *fasciata*, all Rafinesque, and *crassa* (Say, 1816) [*non U. crassa* Retzius, 1778]. Simpson (1900a) selected *Unio crassidens* Lamarck, 1819, as type. However, Simpson's designation was invalid, as *U. crassidens* was not one of the four species included by Rafinesque in the original description. Ortmann (1912a) designated *U. nigra* as type, a valid action. *Unio nigra* is a synonym of *U. crassidens* in any event.].

Synonymy

Elliptio Rafinesque, 1819: 426 [*nomen nudum*].

Eurynia Rafinesque, 1819: 426 [*nomen nudum*].

Unio (*Elliptio*) Rafinesque, 1820: 291 [Ortmann & Walker (1922) chose *Elliptio* over *Eurynia* (see next) based on page precedence. Later Rafinesque (1831: 3) commented "*Elliptio* which is the same as *Ellipsaria*."].

Unio (*Eurynia*) Rafinesque, 1820: 297 [Type species: *Unio dilatatus* Rafinesque, 1820, by subsequent designation of Hermannssen (1847)].

Bariosta Rafinesque, 1831: 2 [Rafinesque gave two species in his new genus: *Bariosta ponderosa* Rafinesque, 1831, and *Obliquaria sinuata* Rafinesque, 1820. The type of the former could not be located *fide* Johnson & Baker (1973), and is unidentifiable from the original description *fide* Walker & Ortmann (1922). The latter taxon was identified by Johnson & Baker (1973) with *Elliptio dilatata* (Rafinesque, 1820), although we remain skeptical. Haas (1969) synonymized *Bariosta* with *Amblema*, but without explanation. Inasmuch as *B. ponderosa* cannot be identified, and *O. sinuata* has been referred to *Elliptio*, we place *Bariosta* under *Elliptio*, but with no real conviction. Williams *et al.* (2008) considered *Bariosta* a *nomen oblitum*.].

Cunicula Swainson, 1840: 268, 378 [Type species: *Unio purpurescens* Lamarck, 1819 (= *Unio complanatus* Lightfoot, 1786)].

Curricula Gray, 1847: 196 [unjustified emendation of *Cunicula* Swainson, 1840].

Eurinea [*sic*] Conrad, 1853: 267, 269.

Eurynaia [*sic*] Frierson, 1927: 33.

Geographic range. Most of North America east of the Rocky Mountains and south of James Bay to the Gulf of Mexico. The majority of species occur in the Atlantic Slope and Gulf drainages.

Elliptio complanata (Lightfoot, 1786)

Figures 47, 48

Etymology. L. *com-*, from *cum*, with, together + L. *plana*, to level, make plain, flat, even. So the construction *complanatus* means "flattened together," or compressed. A number of the forms of this complex are very much compressed laterally as are other species in other genera.

Vernacular name(s). *Eastern Elliptio*, Eastern Spike, Flat Ladyfinger, Eastern Ladyfinger, Flat Spike.

Original description

Mya complanata Lightfoot, 1786: 100, lot 2190, no figure [Lister (1685: [60], pl. 150, fig. 5) described one of the many forms of this species as "5. *musculus brevior*," from "Virginiam." This name, being pre-Linnaean, cannot be used, but the text and figure can be made the basis of a name proposed in 1758 or later. The Reverend John Lightfoot, author of *A Catalogue of the Portland Museum* (1786: 100) cited "*Mya . . . Complanata*, List. 150.5" as the name of a listed item therein (Rehder, 1967: 18). This anonymously authored Catalogue was a list of auction material belonging to the late Margaret Cavendish Bentinck (1715–1785), the Dowager Duchess of Portland, that was distributed to potential buyers. Lightfoot was her librarian and chaplain. Because Lightfoot described the entries using binominal nomenclature, the names are available under ICZN rules, even though it was "published" anonymously. It is unlikely that Lightfoot ever believed he was describing new species. In this manner a description and figure published in 1685 became the basis of a name accidentally coined in 1786. In older works these Portland Catalogue species are attributed to Daniel Solander, a student of Linnaeus, who had curated the Duchess's collection. However, research suggests that Lightfoot was the author of the catalogue, not Solander (Kay, 1965).].

Type locality. The source of Lister's (1685: [60], pl. 150, fig. 5) specimens was given as "Virginiam," presumably Virginia, while Johnson (1970: 314) gave "Maryland" as the origin of Lightfoot's type. Regardless, Johnson (1970: 314) restricted the type locality to "the Potomac River, Washington, District of Columbia [approximately opposite Fairfax Co., Virginia]."

Type material. The Oxford Museum has been examined for Lister's specimens but none were found there (Dance, 1966: 292). Johnson (1971) listed the types and figured specimens of the Unionacea in the British Museum of Natural History but apparently did not find the type specimens of either Lister or Lightfoot. Lightfoot's specimens may have sold at the auction. The types are presumed lost.

Synonymy. We have not included a synonymy to this bewildering complex of species that predominately inhabits the Atlantic Coastal Plain. At this time no one has produced a workable breakdown of the many names applied to these taxa, most named by Isaac Lea. Efforts to unravel this morass using genetic markers have suffered from the use of specimens taken from areas other than the type localities that may or may not represent the species in question. We do not see a solution to this problem in the foreseeable future. Bogan (1993) gave the most comprehensive list of names applied to this complex.

Status. Not listed by either Ohio Division of Wildlife or U.S. Fish & Wildlife Service as of 2009.

ADULT SHELL

Shell. Medium to fairly large sized (to 130 mm), moderately thin, elongate, usually rather rectangular; usually quite compressed.

Anterior margin. Long, straight to broadly rounded to the ventral border.

Ventral margin. Very broadly rounded, rarely concave in very old adults.

Posterior margin. Long, rather straight to the dorsal margin.

Dorsal margin. Long, straight, joining posterior margin with obtuse angle.

Fig. 47. *Elliptio complanata* (Lightfoot, 1786). A. OSUM 44015.1, Tuscarawas River, Tuscarawas Co., OH. 85 mm. **B.** OSUM 18351.14, Lake Nipissing, Nipissing Co., Ontario. 66 mm. **C.** OSUM 37586.2, Lake Champlain, Chittenden Co., VT. 83 mm. **D, F, Hinge detail:** OSUM 65760.3, Ottawa River, Pontiac Co., Quebec. 93 mm. **E.** OSUM 18351.1, Lake Nipissing, Nipissing Co., Ontario. 22 mm. **Beak sculpture detail:** OSUM 18351.4, Lake Nipissing, Nipissing Co., Ontario. 23 mm total length.

Fig. 48. ***Elliptio complanata*** **(Lightfoot, 1786)**
Solid dots—live or freshdead since 1980
Bullseye—all records, any condition

Umbo. Wide, very low; umbos nearly face slightly forward; umbonal sculpture of 2–4 angled ridges.
Ligament. Long and prominent, tan to dark brown, not concealed by shells.
Disc sculpture. Disc of shell smooth except for low angulations on posterior slope.
Periostracum. Juveniles may be shiny but adults have a matte finish; juveniles yellow or tan with fine green lines or rays; adults become a reddish-brown or black.
Nacre. Purple, occasionally porcelain white, often iridescent posteriorly; often with extensive gold or brown Tulberg layers.
Hinge plate. Well-developed, very long; right cardinal a single triangular denticle, often with small anterior accessory tooth; left cardinal bifid, the teeth well-separated; laterals well-developed, very long, straight, one in right valve, two in left; interdentum narrow, long.
Muscle scars. Anterior adductor large, oval, impressed; posterior adductor nearly flush; small anterior adductor distinct from anterior pedal protractor scar, posterior adductor confluent with small posterior pedal retractor scar; dorsal suspensors located under interdentum in a short line. Pallial line entire, located far from shell margin. Accessory line well-developed, reaching 50% of the length of the shell.
Umbonal cavity. Very wide, shallow, open.

Glochidia. Subelliptical, dorsal margin straight, lateral margins equally and gently curved, ventral margin broadly rounded, hooks absent (Matteson, 1948: 703–705, fig. 10; Wiles, 1975a: 35, fig. 4; Weir, 1977: 100).

Literature records (Matteson, 1948: 703–705, fig. 10; Weir, 1977: 100) indicate glochidial dimensions for length (0.20 to 0.21 mm) and height (0.19 to 0.20 mm). Matteson (1948) noted, "The glochidium is small in comparison with those of other mussels." However, this observation applies only to the other subelliptical *Elliptio* species.

Reproductive biology. Tachytictic. Only the outer gills are marsupial. Ortmann (1919) reported that in Pennsylvania eggs develop in April, with glochidia appearing from June to August. Fichtel & Smith (1995) reported gravid females from spring to late summer in Vermont. In Michigan, spawning takes place from the end of April to mid-June, when water temperatures reach 20°C (Matteson, 1948). Larvae require about a month to develop into glochidia (Matteson, 1948). Glochidia are discharged as leaf-shaped conglutinates (Ortmann, 1912a). Adults of both sexes may migrate to form aggregates, which may improve fertilization success (Amyot & Downing, 1998). Most identified potential hosts are centrarchids.

Reported potential hosts
Fundulus diaphanus (Banded Killifish)—Young, 1911 (LT); Wiles, 1975b (NI).
Lepomis cyanellus (Green Sunfish)—Young, 1911 (LT).
Lepomis gibbosus (Pumpkinseed)—Watters *et al.*, 2005 (LT).
Lepomis humilis (Orangespotted Sunfish)—Young, 1911 (LT).
Lepomis macrochirus (Bluegill)—Watters *et al.*, 2005 (LT).
Micropterus salmoides (Largemouth Bass)—Young, 1911 (LT).
Perca flavescens (Yellow Perch)—Lefevre & Curtis, 1912 (LT); Matteson, 1948 (LT), 1955 (LT); Tedla & Fernando, 1969 (LT); Weir, 1977 (LT).
Pomoxis annularis (White Crappie)—Young, 1911 (LT).

Habitat. *Elliptio complanata* tolerates many substrates and currents, often occurring in dense beds.

Range. James Bay and Lawrentian drainages, and nearly all of the Atlantic Slope systems to Florida. Absent from peninsular Florida, but may be present in some Gulf drainages (Brim Box & Williams, 2000). The morphological limits of this "species" may never be identified, and many valid sibling and endemic taxa may be involved.

Range in Ohio. There seems little doubt that this species once occurred sporadically in eastern Ohio, although it now appears to be extirpated. Sterki collected it from a millrace on the Tuscarawas River at New Philadelphia in 1898 (CMNH 61.10486), and Ortmann found it in the Grand River at Eagleville in 1919 (CMNH 61.9373).

Similar species. In Ohio this species is most similar to *Elliptio dilatata*. It differs in being less elongate, more compressed, having a coarser periostracum, and lacking the coarse beak sculpture.

Comments. Individuals gradually slow in growth during their entire life. Specimens rarely reach ca. 130 mm and 20 years old.

Elliptio crassidens (Lamarck, 1819)
Figures 49, 50

Etymology. L. *crassus*, solid, thick + L. *dens, dentis*, tooth. This species may well have the thickest shell of any in the genus. Thick shells are typically accompanied by heavy hinge teeth. *Crassidens* is a noun in apposition.
Vernacular name(s). *Elephantear*, Pink Elephantear.

Fig. 49. ***Elliptio crassidens*** **(Lamarck, 1819). A.** OSUM 25410.1, Big Darby Creek, Pickaway Co., OH. 106 mm. **B.** OSUM 54605, Little Darby Creek, Madison Co., OH. 87 mm. **C.** OSUM 8362.2, Big Darby Creek, Pickaway Co., OH. 131 mm. **D.** OSUM 11707.2, Ohio River, Hamilton Co., OH. 67 mm. **E.** OSUM 39845.1, "Ohio River." 27 mm. **F, Hinge detail:** OSUM 11707.5, Ohio River, Hamilton Co., OH. 81 mm. **Beak sculpture detail:** OSUM 39272.1, "Wabash River." 41 mm total length.

Original description

Unio crassidens var. b Lamarck, 1819: 71, no figure given.

Type locality. "Habite l'Amerique septentrionale, dans le Missisipi [*sic*], l'Ohio, et plusieurs lacs" ["Living in northern America, in the Mississippi, the Ohio, and several lakes"]. Two of Lamarck's three unnamed varieties were listed as "du Missisipi [River]" and "du lac Erie." This species does not occur in Lake Erie. The type locality was restricted by Johnson (1969) to "the Ohio River, Cincinnati, Ohio."

Type material. The unnumbered specimen listed as variety (b) by Lamarck is in the Muséum national d'Histoire naturelle, Paris, and was selected as the lectotype by Johnson (1969).

Synonymy

Unio crassidens var. b Lamarck, 1819: 71.

Unio (*Elliptio*) *nigra* Rafinesque, 1820: 291, pl. 80, figs. 1–4 [Type locality: "l'Ohio [River]." Lectotype ANSP 20243 by designation of Johnson & Baker (1973: 163)].

Unio (*Elliptio*) *nigra* var. *fusca* Rafinesque, 1820: 291 [Type locality: none given. Type not located].

Unio (*Elliptio*) *nigra* var. *maculata* Rafinesque, 1820: 291 [Type locality: none given. Type not located].

Unio cuneatus Barnes, 1823: 263–264 [Type locality: "the Ohio [River]." Type lost *fide* Johnson (1970)].

Mya cuneata (Barnes, 1823). Eaton, 1826: 220.

Obliquaria (*Aximedia*) *venus* Rafinesque, 1831: 3 [Type locality: "in the Kentucky and Cumberland [rivers], very rare." Type not located].

Unio niger [*sic*] Rafinesque, 1820. Say, 1834: unpaginated.

Margarita (*Unio*) *crassidens* (Lamarck, 1819). Lea, 1836: 19.

Margaron (*Unio*) *crassidens* (Lamarck, 1819). Lea, 1852b: 24.

Unio discus Lea, 1838. Sowerby, 1868: pl. 62, fig. 310 [misidentification].

Unio crassus Say, 1816. Sowerby, 1868: pl. 95, fig. 520 [misidentification].

Unio lehmanni Wright, 1897: 138 [Type locality: "St. Marys River, [Nassau Co.] Florida." Lectotype USNM 149650 by designation of Johnson (1967)].

Unio danielsii Wright, 1899a: 31 [Type locality: "Spring Creek, Decatur Co., Georgia." Lectotype USNM 168967 by designation of Johnson (1967)].

Unio polymorphus Wright, 1899b: 42 [Type locality: "Spanish Creek, Okefenokee Swamp, Charleton Co., Georgia." Lectotype USNM 152060 by designation of Johnson (1967)].

Unio (*Elliptio*) *crassidens* Lamarck, 1819. Simpson, 1900a: 706–707.

Unio (*Elliptio*) *danielsii* Wright, 1899. Simpson, 1900a: 708.

Elliptio crassidens (Lamarck, 1819). Ortmann, 1912a: 266.

Elliptio niger [*sic*] (Rafinesque, 1920). Ortmann, 1919: 91, pl. 8, fig. 1.

Elliptio (*Elliptio*) *niger* [*sic*] (Rafinesque, 1820). Frierson, 1927: 25.

Elliptio (*Elliptio*) *niger* [*sic*] *danielsi* [*sic*] (Rafinesque, 1820). Frierson, 1927: 25.

Elliptio pachyodon Pilsbry, 1953: 447, pl. 65, fig. 8 [Type locality: "St. Petersburg, Florida. Pliocene." Holotype ANSP 18586].

Elliptio (*Elliptio*) *crassidens* var. *danielsii* (Wright, 1899). Haas, 1969: 215.

Elliptio (*Elliptio*) *crassidens* var. *crassidens* (Lamarck, 1819). Johnson, 1970: 305–307.

Status. Listed as Endangered (as *Elliptio crassidens crassidens*) by Ohio Division of Wildlife; not listed by U.S. Fish & Wildlife Service as of 2009.

ADULT SHELL

Shell. Medium to fairly large in size (to 150 mm), thick and massive, oval to rather elongate, often drawn to point posteriorly and having a "humped" appearance; moderately inflated.

Anterior margin. Short, straight to broadly rounded to the ventral border.

Ventral margin. Very broadly rounded, straight or concave towards the posterior.

Posterior margin. Long, rather straight to the dorsal margin; often merging imperceptibly with the dorsal margin.

Dorsal margin. Long, straight, joining posterior margin with obtuse angle or confluent with it.

Umbo. Very wide, very low; umbos nearly face each other but in some individuals may be twisted prominently forward; umbonal sculpture of 1–2 very weak diagonal ridges.

Ligament. Long and prominent, tan to dark brown, not concealed by shells.

Disc sculpture. Disc of shell smooth except for low angulations on posterior slope; posterior slope weakly corrugated or wrinkled in juveniles; old individuals may become sulcate.

Periostracum. Juveniles may be shiny or have a velvety finish, but adults have a matte or shaggy finish; juveniles yellow or reddish brown with numerous fine green lines or rays; adults become reddish-brown or black and lack rays; posterior slope dark.

Nacre. Deep purple or salmon, very rarely white, often iridescent posteriorly; occasionally with gold or brown Tulberg spots.

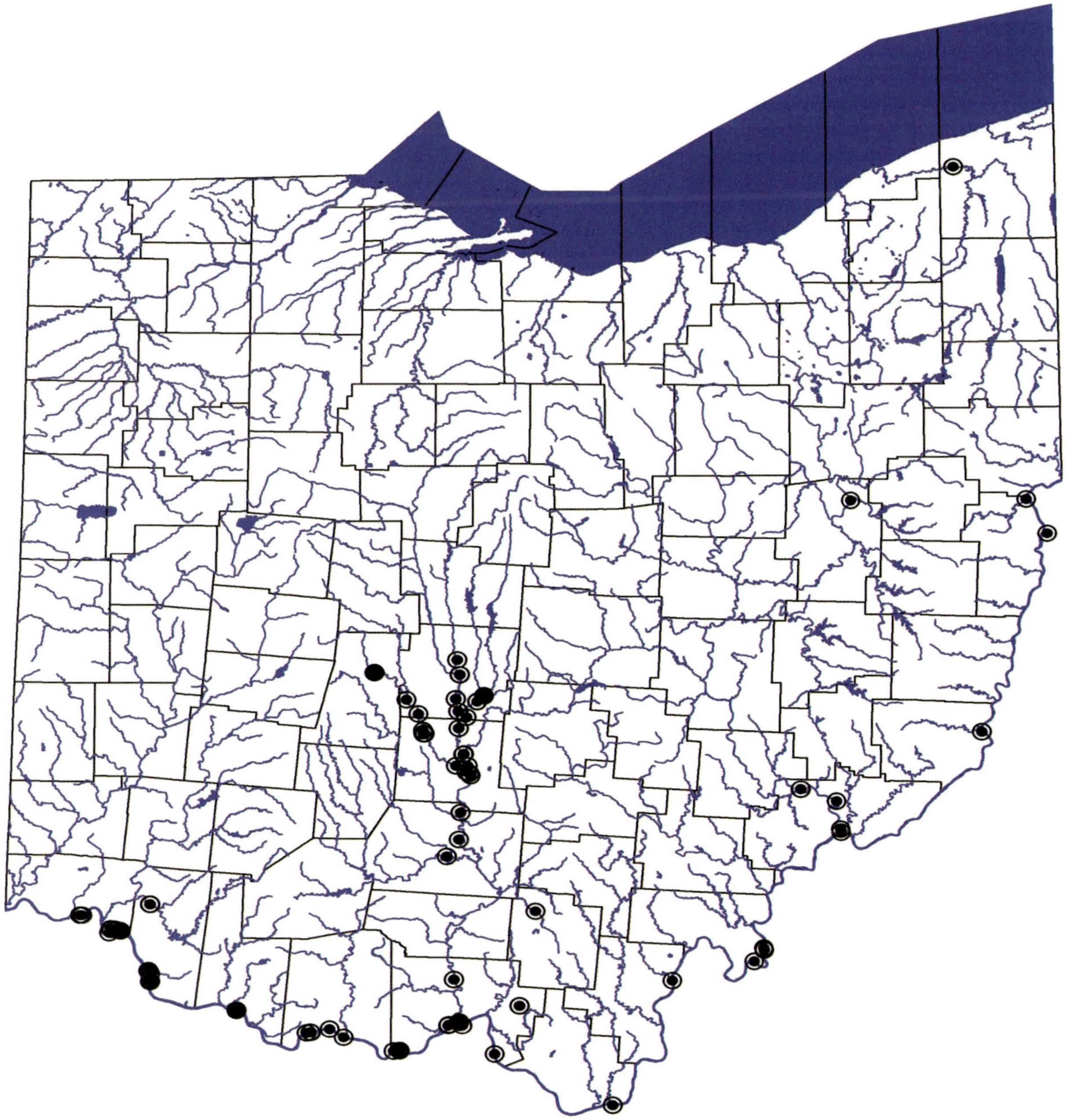

Fig. 50. ***Elliptio crassidens*** **(Lamarck, 1819)**
Solid dots—live or freshdead since 1980
Bullseye—all records, any condition

Hinge plate. Well-developed, massive, long, arched; right cardinal a single, massive, triangular denticle, often with small anterior accessory tooth; left cardinal bifid, the teeth well-separated; laterals well-developed, massive, fairly short, straight, one in right valve, two in left; interdentum wide, thick, long.
Muscle scars. Anterior adductor small, oval, deeply impressed; posterior adductor nearly flush or impressed; small anterior adductor distinct from anterior pedal protractor scar, posterior adductor distinct from small posterior pedal retractor scar; dorsal suspensors located under interdentum in an irregular line. Pallial line entire, located far from shell margin. Accessory line poorly developed, short.
Umbonal cavity. Very wide, shallow, open.

Glochidia. Subtriangular, dorsal margin long and evenly curved, lateral margins rounded, tapering to a pointed ventral terminus, hooks absent (Ortmann, 1911: 364, pl. 89, fig. 6; Surber, 1915: 9, pl. 1, fig. 13).

Literature records (Ortmann, 1912a: 268; Surber, 1915: 9, pl. 1, fig. 13; Utterback, 1915: 200) indicate glochidial dimensions for length (0.13 to 0.15 mm) and height (0.15 to 0.16 mm). Surber (1915: 9) noted, "In this species the glochidium seems to be intermediate in shape between examples of the *Anodonta* group on the one hand and certain species of the *Lampsilis-Quadrula* group on the other, best represented in the former by *Alasmidonta calceola* [*A. viridis*], and in the latter by such species as *Quadrula ebena*, etc. Compared with *Unio gibbosa* [*E. dilatata*] it is more circular in outline, if we except the pointed margin, and has a shorter, more curved hinge line. It is an aberrant type related to *Margaritana margaritifera*."

Reproductive biology. Tachytictic. Outer gills are marsupial. "Breeding season" reported as June and July (Baker, 1928a). Ortmann (1919) found both eggs and developing glochidia in June. Glochidia were released in leaf-shaped conglutinates (Ortmann, 1912a); these are of the simple nonelastic type (Watters, 2008).

Reported potential hosts

Alosa chrysochloris (Skipjack Herring)—Howard, 1914a (NI).

Habitat. In Ohio, in stable cobble and muddy sand in rivers. In southern North America it commonly is found in much smaller creeks.

Range. Mississippi River system, Mobile River system and other Gulf drainages (as *E. crassidens incrassata* Lea, 1840), and the Apalachicola River drainage. Brim Box & Williams (2000) synonymized *E. crassidens incrassata* Lea, 1840 with *E. crassidens*.

Range in Ohio. Ohio River, where it may be locally abundant, and historically from the Scioto River up to Columbus and the Tuscarawas River at New Philadelphia. Occasional specimens have been found as far upstream as Little Darby Creek, but these undoubtedly represent stray occurrences. Does not occur in the Lawrentian drainages—a record from the Grand River is suspect.

Similar species. The Elephant Ear may only be confused with the *Elliptio dilatata* (Rafinesque, 1820), the only other *Elliptio* currently found in Ohio. Other massive shells of similar shape, such as *Actinonaias* species, lack the purple nacre. *Elliptio crassidens* is heavier, larger, and less elongate than *Elliptio dilatata*. In Ohio *E. dilatata* uncommonly occurs in the Ohio River, where *E. crassidens* may be locally abundant. Conversely, *E. dilatata* is common in creeks whereas *E. crassidens* is extremely rare there.

Comments. Individuals gradually slow in growth during their entire life. Specimens reach ca. 150 mm and live over 30 years.

Elliptio dilatata (Rafinesque, 1820)

Figures 51, 52, 53

Etymology. L. *dilato*, spread out, dilate, enlarge, amplify. Rafinesque's concept of width corresponds to the current concept of length. An elongate specimen in today's terminology would be described as wide or dilated by Rafinesque in his time.
Vernacular name(s). *Spike*, Little Spike, Ladyfinger, Bluefinger Spike.
Original description
Unio (*Eurynia*) *dilatata* Rafinesque, 1820: 297, species no. 11, no figure.
Type locality. Rafinesque did not give a type locality, but the title of his paper was "Monographie des Coquilles Bivalves Fluviatiles de la Rivière Ohio, ..." However, as Johnson & Baker (1973: 153) selected a specimen from the Kentucky River as lectotype, that becomes the type locality regardless of Rafinesque's intentions, as clearly stated in ICZN Art. 76.2.
Type material. From the types mentioned by Vanatta (1915: 555), Johnson & Baker (1973: 153) selected ANSP 20248a as lectotype.
Synonymy
Unio nasuta Lamarck, 1819: 75 [*non* Say, 1817. Type locality: "le lac Erié." Type: Muséum national d'Histoire naturelle, Paris, France?].
Unio (*Eurynia*) *dilatata* Rafinesque, 1820: 297.
Obliquaria sinuata Rafinesque, 1820: 320 [Type locality: "Dans le Kentuky [River]." Lectotype ANSP 20252 by

Fig. 51. ***Elliptio dilatata*** **(Rafinesque, 1820). A.** OSUM 6469.2, Olentangy River, Marion Co., OH. 92 mm. **B.** OSUM 38857, Lake Erie, Ottawa Co., OH. 68 mm. **C, F, Hinge detail:** OSUM 61320.1, Meigs Creek, Morgan Co., OH. 86 mm. **D, H.** OSUM 57626.4, Ohio River, [Co.?], OH. 77 mm. **E.** OSUM 55836.1, Little Darby Creek, Madison Co., OH. 61 mm. **G.** OSUM 62335.6, St. Joseph River, Williams Co., OH. 98 mm. **Beak sculpture detail:** OSUM 55836.4, Little Darby Creek, Madison Co., OH. 27 mm total length.

designation of Johnson & Baker (1973). Rafinesque in both 1820 and 1831 described this species with white nacre; the lectotype clearly has purple nacre. However, this does not change the status of the lectotype.].

Unio gibbosus Barnes, 1823: 262–263, fig. 12 [Type locality: "Wisconsan" [*sic*]. Type lost].

Unio mucronatus Barnes, 1823: 266, fig. 13 [Type locality: "the Wisconsan [*sic*] [River]." Type lost].

Mya gibbosa (Barnes, 1823). Eaton, 1826: 220.

Mya mucronatus (Barnes, 1823). Eaton, 1826: 221.

Bariosta sinuata (Rafinesque, 1820). Rafinesque, 1831: 2.

Obliquaria violacea Rafinesque, 1831: 3 [unnecessary replacement name for *Unio dilatata* Rafinesque, 1820, when moved to *Obliquaria*].

Unio (*Eurynia*) *fulvus* Rafinesque, 1831: 3 [also varieties *fuscata* and *rufa*. The former is a senior homonym of *Unio fuscata* Lea, 1843, a common Floridian *Elliptio*. Type locality: "in Green river, Rockcastle river, &c." Type not located.].

Unio bicolor Rafinesque, 1831: 3 [Type locality: "in Kentucky river." Type not located].

Unio euliratus Rafinesque, 1831: 4 [unnecessary replacement name for *Obliquaria sinuata* Rafinesque, 1820, because of presumed priority of *Unio sinuata* Lamarck, 1819].

Unio dilatatus Rafinesque, 1820. Say, 1834: unpaginated.

Margarita (*Unio*) *gibbosus* (Barnes, 1823). Lea, 1836: 38.

Margarita (*Unio*) *arctior* Lea, 1836: 39 [*nomen nudum*].

Unio arctior Lea, 1838a: 10, pl. 4, fig. 10 [Type locality: "Ohio River, near Cincinnati" [Hamilton Co., Ohio]. Holotype USNM 86103 *fide* Johnson, 1974: 15]

Unio stonensis Lea, 1840: 286 [Type locality: "Stones River, Tennessee." Holotype USNM 84969].

Margaron (*Unio*) *stonensis* (Lea, 1840). Lea, 1852b: 27.

Margaron (*Unio*) *gibbosus* (Barnes, 1823). Lea, 1852b: 38.

Margaron (*Unio*) *arctior* (Lea, 1838). Lea, 1852b: 38.

Unio nasutus Lamarck, 1819. Agassiz, 1852: 50.

Unio gibbosus var. *armathwaitensis* Wright, 1898: 123 [Type locality: "a branch of the South Fork of the Cumberland River at Armathwaite, Fentress County, Tennessee." Holotype not located. Paratype MCZ 20167].

Unio (*Elliptio*) *gibbosus* Barnes, 1823. Simpson, 1900a: 703.

Unio (*Elliptio*) *gibbosus* var. *delicatus* Simpson, 1900a: 704 [*non Unio delicatus* Lea, 1863. Type locality: "Entire Mississippi drainage; St. Lawrence and its tributaries; Alabama River system; southeast into Florida; southwest to the Guadelupe River, Texas." Holotype USNM 128797].

Elliptio gibbosus (Barnes, 1823). Ortmann, 1912a: 271.

Unio propeverutus De Gregorio, 1914: 38, pl. 3, figs. 1a–c [Type locality: "Sciota [Scioto] R. (Ohio)." Type in Palermo Museum?].

Elliptio dilatata (Rafinesque, 1820). Utterback, 1915: 201–203.

Elliptio dilatata var. *subgibbosa* (Lea, 1857). Utterback, 1915: 202–203 [misidentification].

Elliptio dilatatus var. *sterkii* Grier, 1918: 9–10 [Type locality: "Big Bend, Presque Isle, Erie Co., Penn." Holotype CMNH 61.4268].

Elliptio dilatatus sterkii Grier, 1918. Ortmann, 1919: 101, pl. 8, fig. 3.

Elliptio (*Elliptio*) *dilatatus* (Rafinesque, 1820). Ortmann & Walker, 1922: 30.

Elliptio (*Eurynia*) *dilatatus* (Rafinesque, 1820). Frierson, 1927: 33.

Elliptio (*Eurynia*) *dilatatus* var. *fulvus* (Rafinesque, 1831). Frierson, 1927: 34.

Elliptio (*Eurynia*) *dilatatus delicatus* (Simpson, 1900). Frierson, 1927: 34.

Elliptio (*Elliptio*) *dilatatus delicatus* (Simpson, 1900). Baker, 1928a: 128, pl. 56, figs. 7, 8.

Elliptio (*Elliptio*) *dilatatus sterkii* Grier, 1918. Baker, 1928a: 130, pl. 56, figs. 3–6.

Elliptio (*Eurynia*) *nasutus nasutus* (Lamarck, 1819). Haas, 1969: 244.

Elliptio (*Eurynia*) *nasutus fulvus* (Lamarck, 1819). Haas, 1969: 244.

Elliptio (*Eurynia*) *nasutus* var. *armathwaitensis* (Wright, 1898). Haas, 1969: 245.

Elliptio (*Eurynia*) *nasutus delicatus* (Simpson, 1900). Haas, 1969: 245.

Elliptio dilatata form *sterkii* Grier, 1918. Strayer & Jirka, 1997: pl. 25, figs. 3, 3a, 4, 4a, 5.

Status. Not listed by either Ohio Division of Wildlife or U.S. Fish & Wildlife Service as of 2009.

ADULT SHELL

Shell. Medium to fairly large sized (to 130 mm), moderately thick, very elongate, usually pointed posteriorly; usually compressed.

Anterior margin. Long, straight to broadly rounded to the ventral border.

Ventral margin. Very broadly rounded to straight, rarely concave in very old adults.

Posterior margin. Long, rather straight to the dorsal margin.

Dorsal margin. Long, straight, joining posterior margin with obtuse angle.

Umbo. Wide, very low; umbos face slightly forward; umbonal sculpture of 2–4 angled or slightly undulating ridges.

Ligament. Long and prominent, tan to dark brown, not concealed by shells.

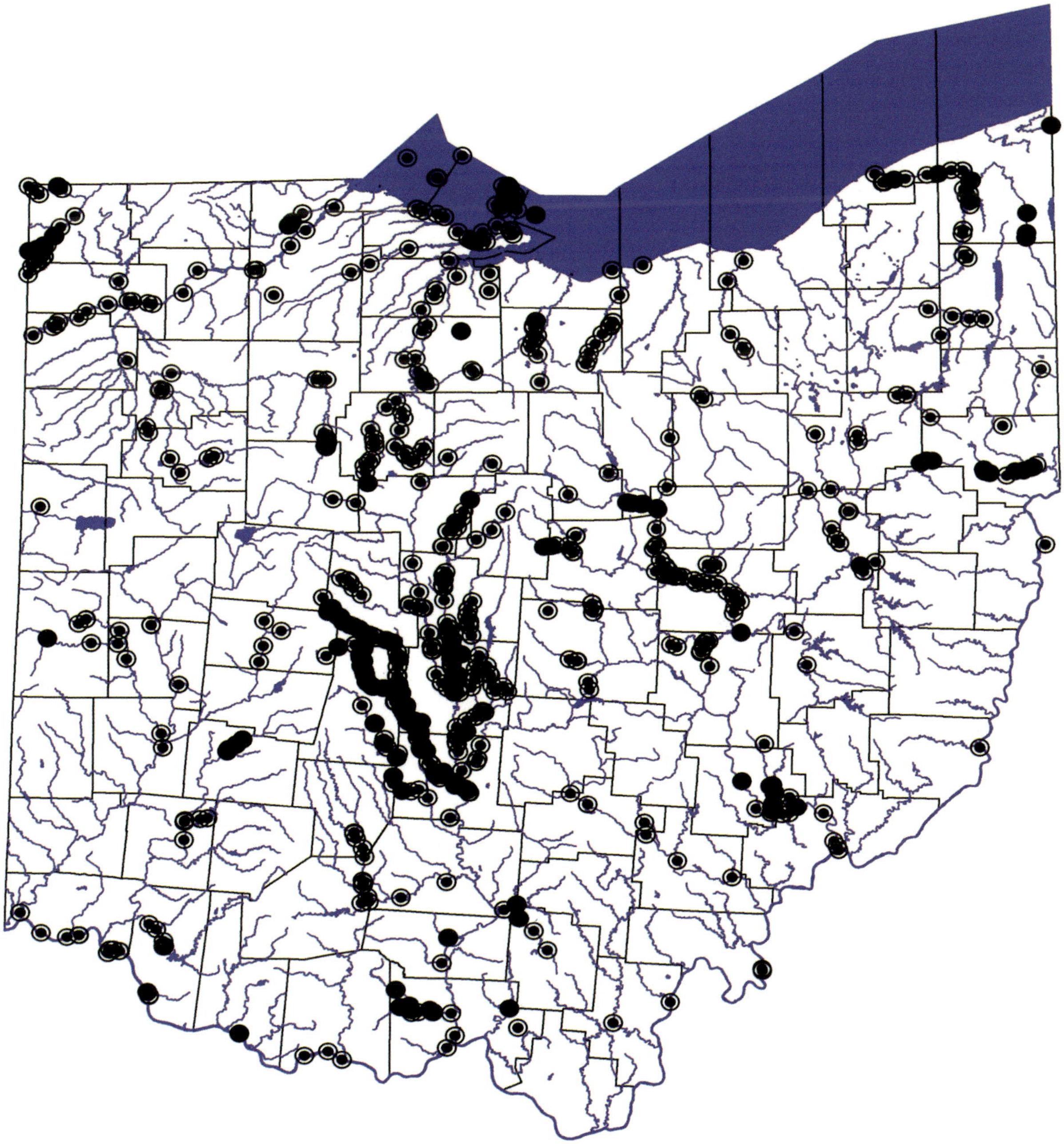

Fig. 52. ***Elliptio dilatata*** **(Rafinesque, 1820)**
Solid dots—live or freshdead since 1980
Bullseye—all records, any condition

Disc sculpture. Disc of shell smooth except for low angulations on posterior slope.
Periostracum. Juveniles may be shiny, but adults have a matte finish; juveniles yellow or tan with fine green lines or rays; adults may become black; dorsal slope usually dark.
Nacre. Usually purple or purple bordered with white; some populations have purple, salmon, and white nacred individuals living sympatrically; often iridescent posteriorly; Tulberg layers rare.
Hinge plate. Well-developed, very long; right cardinal a single, massive, triangular denticle, often with small anterior accessory tooth; left cardinal bifid, the teeth widely separated; laterals well-developed, massive, long, straight, one in right valve, two in left; interdentum impressed, short.
Muscle scars. Anterior adductor medium-sized, oval, impressed; posterior adductor nearly flush to impressed; small anterior adductor distinct from anterior pedal protractor scar, posterior adductor narrowly joined to small posterior pedal retractor scar; dorsal suspensors located under interdentum in a cluster. Pallial line entire, located far from shell margin. Accessory line poorly developed, short.
Umbonal cavity. Very wide, shallow, open.

Glochidia. Subelliptical, dorsal margin straight, ventral margin semicircular, lateral margins about equal, the anterior margin slightly more produced than the posterior margin. Loose-looped sculpture covers the exterior surface of the valve, hooks absent (Lea, 1874b: 69, pl. 1, fig. 13; Ortmann, 1911: 364, pl. 89, fig. 7; Hoggarth, 1999: figs. 34a–f).

Literature records (Lefevre & Curtis, 1910a: 97, fig. n; Lefevre & Curtis, 1912: 146, fig. 1o; Surber, 1912: 10, pl. 2, fig. 38; Ortmann, 1912a: 271; Utterback, 1915: 201; Ortmann, 1919: 111; Hoggarth, 1999: figs. 34a–f) indicate glochidial dimensions for length (0.20 to 0.22 mm), height (0.19 to 0.23 mm), and hinge (0.14 to 0.15 mm). Lea (1863: 54) noted, "Embryonic shell short pouch-shape, white, has no hooks, very nearly the same as *phaseolus* [*P. fasciolaris*]." This glochidium has a much longer hinge line and a more depressed shell than *P. fasciolaris*. The glochidium is more similar to those of the genus *Epioblasma* than any other. However, it can be distinguished from them by the absence of supernumerary hooks in the glochidium of *E. dilatata*.

Reproductive biology. Tachytictic. Individuals reach maturity from 4 to 6 years of age (Jirka & Neves, 1992). The Spike spawns from March to July (Jirka & Neves, 1992), and eggs develop in May and apparently again in July and August (Ortmann, 1919). Although Ortmann (1919: 96) had reported a Spike with eggs in August, and dismissed it as "clearly an abnormal case," there is evidence that the Spike has two breeding periods a year (Watters & O'Dee, 1998c). In their study glochidia were recovered in August and September, and again in November. Wisconsin individuals were found with glochidia from May through July (Heath *et al.,* 1998). In Ohio, individuals overwinter buried in the substrate, but surface in April, presumably to spawn (Watters *et al.,* 2001.). Glochidia are discharged in white, lanceolate conglutinates (Utterback, 1915).

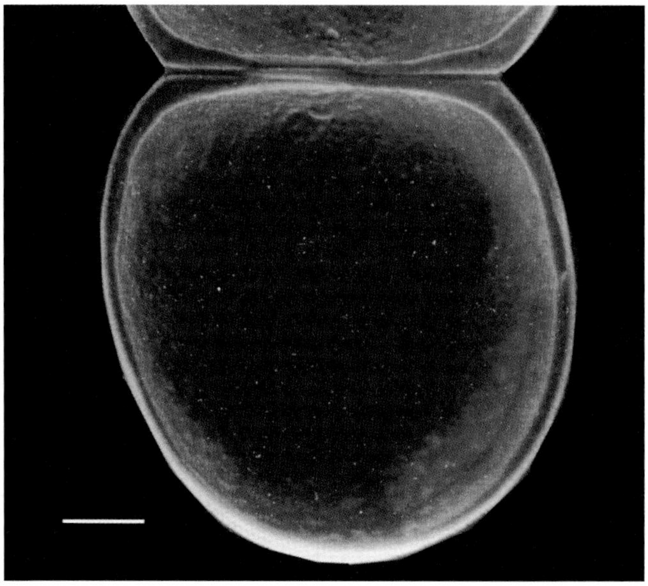

Fig. 53. ***Elliptio dilatata*** **(Rafinesque, 1820).** Glochidium. MAH 946.9, Kalamazoo River, Jackson Co., MI. Bar = 35 µm. (Hoggarth, 1999).

Reported potential hosts
Ambloplites rupestris (Rock Bass)—Luo, 1993 (LT).
Cottus carolinae (Banded Sculpin)—Luo, 1993 (LT).
Dorosoma cepedianum (Gizzard Shad)—Wilson, 1916 (NI).
Etheostoma caeruleum (Rainbow Darter)—Luo, 1993 (LT).
Perca flavescens (Yellow Perch)—Clarke, 1981b (NS).
Pomoxis annularis (White Crappie)—Wilson, 1916 (NI).
Pomoxis nigromaculatus (Black Crappie)—Clarke, 1981b (NS).
Pylodictis olivaris (Flathead Catfish)—Howard, 1914a (LT).
Sander canadensis (Sauger)—Howard, 1914a (NI).

Habitat. A widespread species, the Spike is most common in sand and cobble in creeks. It is rare in large rivers such as the Muskingum and Ohio rivers. Although it occurs in Lake Erie, it is usually not common in other lakes and ponds.

Range. Mississippi River drainage, including the Ohio, Cumberland, and Tennessee rivers. North to Minnesota and Wisconsin. East to Pennsylvania and New York. West in the Missouri, Red, and White rivers. South to Mississippi. Great Lakes drainages.

Range in Ohio. Widespread but declining, most common in the Big Darby Creek system. Western basin of Lake Erie, particularly around the Bass Islands.

Similar species. It is most similar to purple-nacred forms of *Ligumia recta*. That species grows to a much larger size than *E. dilatata* and has fine double-looped beak sculpture. *Elliptio dilatata* has coarse single loops or parallel ribs. It is more elongate and less massive than *Elliptio crassidens*.

Comments. Individuals gradually slow in growth during their entire life. Specimens rarely reach ca. 130 mm and 25 years old.

Epioblasma Rafinesque, 1831

Etymology. Gr. *epi-*, upon, over, above + Gr. *ob-*, toward, to + Gr. *elasma*, metal beaten out, metal plate. Rafinesque (1831: 2) noted that the genus *Epioblasma* "Differs from *Amblema* and *Ellipsaria* by lamellar tooth obliqual, divergent towards the back and straight." In the left valve of *E. flexuosa* (Rafinesque, 1820), the space between the lateral teeth is slightly greater anteriorly than posteriorly; or, in Rafinesque's view, they appear divergent "towards" (*ob-*) the "back" (or umbo) which is positioned upon or above (*epi-*) the shell.

Original description

Epioblasma Rafinesque, 1831: 2.

Type species. *Epioblasma biloba* Rafinesque 1831, by monotypy. Bogan (1998), based on the discovery of a Rafinesque syntype in the Férussac collection, believed that *Epioblasma biloba* Rafinesque, 1831, is the female of what is known as *Epioblasma rangiana* (Lea, 1838).

Synonymy

Epioblasma Rafinesque, 1831: 2.

Dysnomia Agassiz, 1852: 43 [Type species: *Obliquaria (Quadrula) flexuosa* Rafinesque, 1820, by original designation].

Disnomia [*sic*] Bielz, 1869: 43.

Dysonomia [*sic*] Paetel, 1890: 143.

Dysnomia (Truncillopsis) Ortmann & Walker, 1922: 65 [Type species: *Truncilla triqueter* Rafinesque, 1820, by original designation].

Dysnomia (Scalenilla) Ortmann & Walker, 1922: 68 [Type species: *Unio sulcatus* Lea, 1829, by original designation].

Geographic range. Mississippi and Mobile river drainages, Lake Erie and Lake Michigan and their drainages. Absent from the Atlantic Coastal Plain, Florida, and the Apalachicola River.

Epioblasma flexuosa (Rafinesque, 1820)

Figures 54, 55

Etymology. L. *flexuosus*, full of turns or windings, tortuous, flexuous. In his original description (1820: 306) Rafinesque used the word "flexuous" to describe the curvature of both the annulae and the shell margin—appropriately so for this species, and especially so for the female of this species which he had not as yet seen.

Vernacular name(s). *Leafshell*, Northern Leafshell.

Original description

Obliquaria (Quadrula) flexuosa Rafinesque, 1820: 306–307, species no. 33, no figure.

Type locality. Originally "Dans le Kentuky [River], Salt-river et Green-river." The lectotype is from the Kentucky River, which becomes the type locality.

Type material. The type mentioned by Vanatta (1915: 55) was selected as lectotype (ANSP 20249 from the Kentucky River) by Johnson & Baker (1973: 155).

Synonymy

Obliquaria (Quadrula) flexuosa Rafinesque, 1820: 306–307.

Obliquaria (Quadrula) flexuosa var. *bullata* Rafinesque, 1820: 307 [*non Quadrula bullata* Rafinesque, 1820 (see under *Quadrula pustulosa*). Type locality: none given. Type not located].

Unio foliatus Hildreth, 1828: 284, fig. 16 [Type locality: "Ohio." Figured type lost *fide* Johnson, 1978. Agassiz (1852: 43) considered this the female of the species.].

Unio flexuosus (Rafinesque, 1820). Conrad, 1835: 8, pl. 4, fig. 2.

Margarita (Unio) foliatus (Hildreth, 1828). Lea, 1836: 13.

Margaron (Unio) foliatus (Hildreth, 1828). Lea, 1852b: 20.

Dysnomia flexuosa (Rafinesque, 1820). Agassiz, 1852: 43.

Truncilla (Dysnomia) foliata (Hildreth, 1828). Simpson, 1900a: 521.

Dysnomia (Dysnomia) flexuosa (Rafinesque, 1820). Ortmann & Walker, 1922: 70.

Plagiola (Epioblasma) flexuosa (Rafinesque, 1820). Johnson, 1978: 283–287 [in part], pl. 5 [in part], pl. 15, fig. 5.

Status. Listed as Extinct by both Ohio Division of Wildlife and U.S. Fish & Wildlife Service as of 2009.

ADULT SHELL

Shell. Medium size (to 80 mm), rather thick and heavy, bifurcate in profile, particularly in females, moderately inflated. Juveniles much more elongate than adults.

Anterior margin. Flat before the umbo, then broadly rounded to the ventral border.

Fig. 54. ***Epioblasma flexuosa*** **(Rafinesque, 1820). A.** OSUM 39008, "Wabash River." 73 mm. **B.** OSUM 31911.3, Ohio River, Hamilton Co., OH. 57 mm. **C.** OSUM 47313.1, Muskingum River, [Co.?], OH. 59 mm. **D.** OSUM 58869.1, Ohio River, Hamilton Co., OH. 29 mm. **E, H, Hinge detail:** OSUM 47313.2, Muskingum River, [Co.?], OH. 60 mm. **F.** OSUM 31908, Ohio River, Hamilton Co., OH. 47 mm. **G.** OSUM 4050.3, Scioto River, Franklin Co., OH. 55 mm. **Beak sculpture detail:** no specimen available.

Fig. 55. ***Epioblasma flexuosa*** **(Rafinesque, 1820)**
Solid dots—live or freshdead since 1980
Bullseye—all records, any condition

Ventral margin. Gently rounded, descending at ca. 45° to mid-ventral margin. In males the margin forms an obtuse angle or weak digitation, then ascends in a concave line to the posterior angle or digitation. In females the ventral angle forms a distinct wide digitation that projects postero-ventrally, then continues as a short strongly concave margin before forming a second digitation at the posterior angle. In juveniles the ventral and posterior angles are much less pronounced.
Posterior margin. From the second "digit," rounded to dorsal margin.
Dorsal margin. Straight to umbo, forming obtuse angle with posterior margin.
Umbo. Very broad and low; umbonal sculpture eroded from all specimens seen, or perhaps this sculpture is normally absent.
Ligament. Very short, tan, not concealed by shells.
Disc sculpture. Shell sculptureless except for weak radiating striae on the posterior slope; each "digit" is raised above the remaining shell on a broad, weak rib.
Periostracum. Greenish yellow to tan, with numerous radiating, slightly darker green rays; the rays may be wavy; the digits are not darkly colored.
Nacre. White, iridescent posteriorly; rarely with Tulberg layers.
Hinge plate. Heavy and short, interdentum short, wide, and thick; cardinals irregular, with two in the left valve, and one in the right; these teeth may be accompanied by one or more prominent denticles both anterior and posterior to the teeth; lateral very short, straight or concave, two in left valve, one in right.
Muscle scars. Both small, deeply impressed; anterior adductor distinct from anterior pedal adductor, posterior adductor largely confluent with the posterior pedal retractor scar; dorsal suspensors multiple and obscure, located under cardinal fossa, but not on interdentum. Pallial line entire, located far from shell margin. Accessory line extremely long, extending to marsupial region, but often very faint.
Umbonal cavity. Very wide and shallow.

Glochidia. Lea (1863: 79) stated, "Embryonic shell subtriangular, white, has no hooks, nearly the same as *undulatus*, Hild." His figure (Lea, 1858c: 48, pl. 5, fig. 22) showed this shape; however, it is unlikely that this glochidium is subtriangular rather than depressed subelliptical.

Reproductive biology. Reproductive season unknown, presumably bradytictic.

Reported potential hosts. No hosts known.

Habitat. Unknown. There is some evidence it occurred in muddy areas of large rivers.

Range. Large rivers in the Ohio River system, including the Wabash, Tennessee, and Cumberland rivers.

Range in Ohio. Historically from the Ohio River in the vicinity of Cincinnati, and in the Scioto River at Columbus and Chillicothe.

Similar species. Mature specimens of this species can be confused only with *Epioblasma lewisi*. That species differs in having the distal portions of the digitations colored dark green.

Comments. Individuals grew fastest in their second and third years, but not as fast as other species. Sexual dimorphism was apparent by the end of the second year. This species grew to 80 mm and ca. 12 years in age.

Epioblasma lewisi (Walker, 1910)

Figures 56, 57

Etymology. Lewis + L. *-i*, genitive ending; so *E. lewisi* means "the *epioblasma* of Lewis." This is a patronym in honor of James Lewis, M.D., of Little Falls, Herkimer and Mohawk, Herkimer County, N.Y., an early (1800's) student of freshwater molluscs.
Vernacular name(s). *Forkshell*, Southern Leafshell.
Original description
Truncilla lewisii Walker, 1910: 42–44, pl. 3, figs. 3–5.
Type locality. Walker recorded type material "from the Holston River, Tenn. (ex Lewis Coll.), also from the Clinch River, Tenn. (Lewis); Cumberland River, Port Burnside, Ky. (Wetherby) and the Holston River, Knox Co., Tenn. (Andrews)." The specimen selected as lectotype is from the Holston River, Tennessee, which becomes the type locality.
Type material. Johnson (1978: 283) selected the figured female syntype of Walker (1910: pl. 3, fig. 3, UMMZ 91456) as the lectotype of *T. lewisii*.
Synonymy
Truncilla lewisi Walker, 1910: 42–44, pl. 3, figs. 3–5.
Dysnomia (Dysnomia) lewisi (Walker, 1910). Frierson, 1927: 93.
Dysnomia flexuosa lewisi (Walker, 1910). Morrison, 1942: 366.
Dysnomia lewisi (Walker, 1910). Neel & Allen, 1964: 450, figs. 61, 66.
Epioblasma lewisi (Walker, 1910). Stansbery, 1973b: 22.
Plagiola (Epioblasma) flexuosa (Rafinesque, 1820). Johnson, 1978: 283–285 [in part], pl. 15, figs. 6, 8 [misidentification].
Status. Listed as Extinct by Ohio Division of Wildlife; listed as Extinct by U.S. Fish & Wildlife Service as of 2009.

Fig. 56. ***Epioblasma lewisi*** **(Walker, 1910). A.** OSUM 56999, no data. 55 mm. **B.** OSUM 57335, Ohio River, Hamilton Co., OH. 46 mm. **C, D, Hinge detail:** OSUM 50000, "Tennessee River." 49 mm. **Beak sculpture detail:** no specimen available.

ADULT SHELL

Shell. Small size (to 50 mm), thin, bifurcate in profile, particularly in females, not greatly inflated.

Anterior margin. Flat or concave before the umbo, then rounded to the posterior border.

Ventral margin. Straight to slightly concave to the ventral-posterior "digit" of the female; then forming a pronounced projection; then margin briefly but strongly concave, followed by a second "digit" at the posterior angle.

Posterior margin. From the second "digit," rounded to dorsal margin.

Dorsal margin. Straight to umbo.

Umbo. Very broad and low; umbonal sculpture eroded from all specimens seen.

Ligament. Not present on available specimens.

Disc sculpture. Shell sculptureless except for weak radiating striae on the posterior slope; each "digit" is raised above the remaining shell on a broad, weak rib.

Periostracum. Greenish yellow to tan, with numerous obscure, radiating, slightly darker green rays; the "digits" are darker green at the distal margin.

Nacre. White, iridescent posteriorly; gold Tulberg layers may be present.

Hinge plate. Heavy and short, interdentum short, wide,

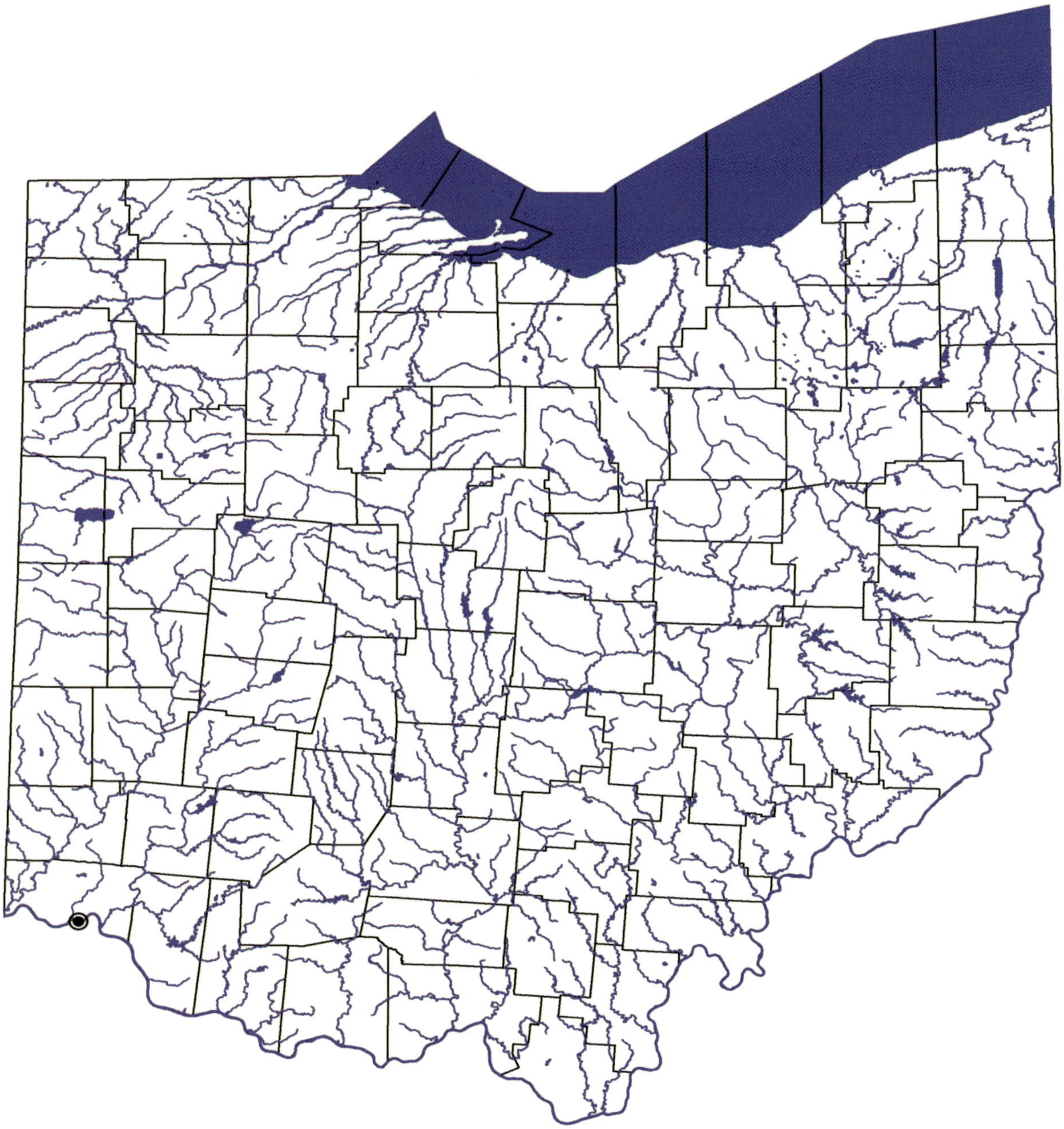

Fig. 57. ***Epioblasma lewisi*** **(Walker, 1910)**
Solid dots—live or freshdead since 1980
Bullseye—all records, any condition

and flattened; cardinals irregular, with two in the left valve, and one in the right; these teeth may be accompanied by one or more prominent denticles both anterior and posterior to the teeth; lateral very short, straight or concave, two in left valve, one in right.
Muscle scars. Anterior adductor small, deep posterior adductor partially confluent with the posterior pedal retractor scar; dorsal suspensors multiple and obscure, located under cardinal fossa, but not on interdentum.
Umbonal cavity. Very wide and shallow.

Glochidia. Unknown.

Reproductive biology. Reproductive season unknown, presumably bradytictic.

Reported potential hosts. No hosts known.

Habitat. Unknown. Apparently it occurred in the shallow riffles of big rivers (Stansbery, 1971: 8).

Range. Simpson (1914: 21) stated that other than the type locality, this species occurred: "Also, Clinch River and Holston River, Knox Co., Tenn., and Cumberland River, Port Burnside, Ky." Museum records at OSUM include a single specimen from the Tennessee River, 3/2 subfossil specimens from Caney Fork River, and one dead specimen from the Ohio River at Cincinnati, Ohio.

Range in Ohio. The species is probably extinct (Stansbery, 1970: 19). It once occurred in the Ohio River at Cincinnati, but it is not found there today. It has not been recorded for any other location in the state.

Similar species. *Epioblasma flexuosa* is very similar but lacks the dark coloration on the distal portion of the digitations.

Comments. Individuals grew fastest in their second and third years, but not as fast as many other species. Sexual dimorphism was apparent by the end of the second year. Apparently not a long-lived species; large (50 mm) specimens are only 7 or 8 years old.

Epioblasma obliquata obliquata (Rafinesque, 1820)

Figures 58, 59

Etymology. L. *obliquatus*, turned, bent, twisted aside or in an oblique direction; from L. *obliquus*, sidelong, slanting, oblique, sideways. Rafinesque apparently had a male specimen before him when he described this species; he mentioned the sulcus on the posterior slope, calling it "une légère dépression longitudinale oblique" (1820: 309). His use of the name *obliquata* again reflects his view of the importance of the orientation of the shell in classification—*E. obliquata* is "oblique" when held with the juvenile shell, or umbo, uppermost.
Vernacular name(s). *Catspaw*, Purple Catspaw.
Original description
Obliquaria (*Scalenaria*) *obliquata* Rafinesque, 1820: 309, species no. 40, no figure.
Type locality. "Dans le Kentucky [River]."
Type material. Johnson & Baker (1973: 163) selected the Rafinesque-Poulson type of Vanatta (1915: 550), from the Kentucky River, as the lectotype, ANSP 20226.
Synonymy
Obliquaria (*Scalenaria*) *obliquata* Rafinesque, 1820: 309.
Unio ridibundus Say, 1829: 308 [Type locality: "Cumberland River." Figured type lost *fide* Johnson, 1978: 278. Agassiz (1852: 43) considered this the female of the species.].
Scalenaria obliquata (Rafinesque, 1820). Agassiz, 1852: 43.
Unio pectitus Conrad, 1853: 255 [*nomen nudum*].
Unio pectitus Conrad, 1854: 297, pl. 27, fig. 4 [Type locality: "Wabash River." Type not found *fide* Johnson & Baker (1973: 165)].
Unio stewardsonii stewardsoni (Lea, 1852). De Gregorio, 1914: 45, pl. 6, fig. 3 [misidentification].
Unio propesulcatus De Gregorio, 1914: 60, pl. 10, fig. 2 [Type locality: "Cumberland River, Tennessee." Type in Palermo Museum?].
Dysnomia sulcata pectita (Conrad, 1853). Frierson, 1927: 96.
Plagiola (*Pilea*) *obliquata* (Rafinesque, 1820). Johnson, 1978: 278–280 [in part], pl. 4 [in part], pl. 14, fig. 10.
Epioblasma obliquata obliquata (Rafinesque, 1820). Stansbery, 1982: 3.
Epioblasma obliquata (Rafinesque, 1820). Parmalee & Bogan, 1998: 97–99 [in part], fig. 31, map 30.
Unio propeculcatus [*sic*] De Gregorio, 1914. Parmalee & Bogan, 1998: 97.
Status. Listed as Endangered by both Ohio Division of Wildlife and U.S. Fish & Wildlife Service as of 2009. A Federal recovery plan has been completed (USFWS, 1992), and it is currently in revision.

ADULT SHELL
Shell. Medium size (to 70 mm), females considerably smaller; thick and heavy, oval in profile, moderately inflated.
Anterior margin. Flat or concave for a short distance before the umbo, then broadly rounded to the ventral border.
Ventral margin. In males, gently rounded to approxi-

Fig. 58. ***Epioblasma obliquata obliquata*** **(Rafinesque, 1820). A.** OSUM 57312.1, Ohio River, Hamilton Co., OH. 27 mm. **B.** OSUM 36899.38, "Ohio River system." 58 mm. **C, F, Hinge detail:** OSUM 60985.2, Killbuck Creek, Coshocton Co., OH. 57 mm. **D.** OSUM 47310, Muskingum River, [Co.?], OH. 49 mm. **E.** OSUM 58867, Ohio River, Hamilton Co., OH. 58 mm. **G, Beak sculpture detail:** OSUM 10372.1, "Ohio River." 40 mm. **H.** OSUM 68408.4, "Ohio River." 38 mm.

Fig. 59. ***Epioblasma obliquata obliquata*** **(Rafinesque, 1820)**
Solid dots—live or freshdead since 1980
Bullseye—all records, any condition

mately 75% back, then forming weak obtuse angle before continuing in a straight or convex line to posterior point; females broadly rounded to posterior margin to form a short broad projection at the postero-ventral margin, then immediately forming the posterior point.
Posterior margin. Rounded to dorsal margin, weakly differentiated from dorsal margin, particularly in males.
Dorsal margin. Rounded or straight to umbo, forming obtuse angle with posterior margin.
Umbo. Very broad and low; umbonal sculpture eroded from all specimens examined, but presumably as in *E. o. perobliqua*.
Ligament. Fairly long, tan to dark brown, not concealed by shells.
Disc sculpture. Shell sculptureless except for weak radiating striae on the posterior slope; the female "digit" is raised above the remaining shell on a broad, weak rib; this protrusion and the posterior point are both serrated along margin.
Periostracum. Yellow to tan or dark brown, with numerous radiating, darker green rays; these are most evident on the posterior half of the shell.
Nacre. Deep purple, often fading to white along the margins, iridescent posteriorly; rarely with Tulberg layers. Some specimens, particularly those from the Ohio River, may have mauve or completely white nacre.
Hinge plate. Heavy and short, interdentum short, wide, and thick; cardinals irregular, with two in the left valve, and one in the right; these teeth may be accompanied by one or more prominent denticles both anterior and posterior to the teeth, with the posterior one positioned on the interdentum; lateral very short, straight or convex, two in left valve, one in right.
Muscle scars. Anterior one deeply impressed; anterior adductor distinct from anterior pedal adductor, posterior adductor varying from entirely confluent with the posterior pedal retractor to nearly separate from the adductor scar; multiple small dorsal suspensors, located in a line behind cardinal fossa and interdentum. Pallial line entire, located far from shell margin. Accessory line extremely long, extending to marsupial region, but often very faint.
Umbonal cavity. Very wide and shallow.

Glochidia. Depressed subelliptical, dorsal margin straight, anterior and posterior margins equally curved, ventral margin broadly curved (Lea, 1858c: 47, pl. 5, fig. 12; Lea, 1863: 78).

Literature records (Surber, 1912: 10, pl. 2, fig. 37) indicate glochidial dimensions for length (0.20 mm) and height (0.21 mm). Lea (1863: 78) noted that the glochidia were "Pouch-shape, white, has no hooks, very nearly the same with *phaseolus*, Hild. [*P. fasciolaris*]." However the glochidium of *E. o. obliquata* is almost equal in length and height, while the glochidium of *P. fasciolaris* is higher than long; the two have very different shapes.

Reproductive biology. Bradytictic. Gravid females were found from September to the following April (Hoggarth *et al.*, 1995; Watters *et al.*, 1998d).

Reported potential hosts
Ambloplites rupestris (Rock Bass)—Watters *et al.*, 1998d (LT).
Cottus bairdi (Mottled Sculpin)—Watters *et al.*, 1998d (LT).
Etheostoma blennioides (Greenside Darter)—Watters *et al.*, 1998d (LT).
Noturus flavus (Stonecat)—Watters *et al.*, 1998d (LT).
Percina maculata (Blackside Darter)—Watters *et al.*, 1998d (LT).
Percina caprodes (Logperch)—Watters *et al.*, 1998d (LT).

Habitat. Although this mussel is usually described as the big river form of the species *obliquata* (Bogan & Parmalee, 1983; Bates & Dennis, 1985), in Ohio it lives in a fairly small stream, Killbuck Creek. In this stream, this mussel can be found in riffle and run habitats with a mixture of sand, gravel, and cobble substrate in fairly fast flowing to almost still water. The mussel lives completely buried in the substrate, although females will move to the surface when they are ready to release their glochidia. These females are found along the margin of the stream in fairly shallow water.

Range. Previously found in the Ohio and Cumberland River systems, but extirpated from all of its range except for (perhaps) the Green River of Kentucky and Killbuck Creek of Ohio.

Range in Ohio. Historically known from the Ohio River at Cincinnati, Portsmouth, and Marietta, the Scioto River at Circleville, and the Walhonding River. Today only known from lower Killbuck Creek.

Similar species. This is the only purple-nacred *Epioblasma* in Ohio. In addition, females resemble female *Epioblasma obliquata perobliqua*, but are larger. Males resemble male *Epioblasma t. rangiana*, but are less sulcate, with the umbos often further forward.

Comments. Male individuals grow fastest in their second and third years, but not as fast as other species. Females grow fast until approximately their third year; subsequent growth is dramatically slower with annuli packed together. Sexual dimorphism is apparent by the end of the second year. Males grow larger than females: an 8-year-old female may be only 26 mm in length while a male of the same age from the same locale is 42 mm. Males reach 70 mm in length and live for 25 years. Females reach 40 mm and 20 years old. This species grows larger than *E. o. perobliqua*.

Epioblasma obliquata perobliqua (Conrad, 1836)

Figures 60, 61

Etymology. L. *per-*, through, thoroughly, very, completely + L. *obliquus*, sidelong, slanting, oblique, sideways. Conrad (1836: 51–52) noted that this species "is much more oblique" than *U. gibbosus* Rafinesque, 1820 (= *Epioblasma torulosa* (Rafinesque, 1820) of which he mistakenly considered *perobliqua* a variety).

Vernacular name(s). *White Catspaw.*

Original description

Unio gibbosus perobliquus Conrad, 1836: 51–52, pl. 27, fig. 2.

Type locality. "Inhabits the Wabash river, Indiana, and Detroit river, Michigan" (Conrad 1836: 52). Repeated efforts to collect *E. o. perobliquus* from the Detroit River have failed. Numerous *Epioblasma t. rangiana* (Lea, 1838) from the Detroit River over the years (Freitag, 1984: 105) suggest that Conrad may have misidentified his specimen from that site and that it may well have been a male individual of the very similar appearing *E. t. rangiana*. Simpson's *Truncilla sulcata delicata* (1900a: 520, no fig.) from "the Detroit River and southern Michigan" is also *E. t. rangiana* (Lea, 1838) (see holotype USNM 128797). Johnson (1978: 261) restricted the type locality of *perobliqua* Conrad, 1836, to the "Wabash River, Indiana."

Type material. Johnson & Baker (1973: 166) were unable to locate the holotype in the collection of the Academy of Natural Sciences of Philadelphia.

Synonymy

Unio gibbosus perobliquus Conrad, 1836: 51–52, pl. 27, fig. 2.

Unio perobliquus Conrad, 1836. Conrad, 1837: back cover.

Dysnomia perobliqua (Conrad, 1836). Agassiz, 1852: 43.

Dysnomia sulcata perobliquus (Conrad, 1836). Stansbery, 1970: 19.

Plagiola (Pilea) obliquata (Rafinesque, 1820). Johnson, 1978: 278–280 [in part], pl. 4 [in part], pl. 14, figs. 11, 12.

Epioblasma obliquata perobliqua (Conrad, 1836). Stansbery, 1982: 3.

Epioblasma obliquata (Rafinesque, 1820). Parmalee & Bogan, 1998: 97–99 [in part].

Status. Listed as Endangered by both Ohio Division of Wildlife and U.S. Fish & Wildlife Service as of 2009. A Federal recovery plan has been completed (USFWS, 1990a), and it is currently in revision.

ADULT SHELL

Shell. Small size (to 55 mm), rather thick and heavy, oval in profile, moderately inflated.

Anterior margin. Flat or concave for a short distance before the umbo, then broadly rounded to the ventral border.

Ventral margin. In males, gently rounded to approximately 66% posterior, then forming weak obtuse angle before continuing in a straight or concave line to posterior point; females broadly rounded to posterior margin to form a short broad projection at the postero-ventral margin, then continuing for a short distance as a concave margin to the posterior point.

Posterior margin. Rounded to dorsal margin, weakly differentiated from dorsal margin.

Dorsal margin. Rounded or straight to umbo, forming obtuse angle with posterior margin.

Umbo. Very broad and low; umbonal sculpture of 3–4 microscopic, very weak "V"-shaped ribs, nearly obsolete in some specimens.

Ligament. Very short, tan to dark brown, not concealed by shells.

Disc sculpture. Shell sculptureless except for weak radiating striae on the posterior slope; the female "digit" is raised above the remaining shell on a broad, weak rib; this protrusion and the posterior point are both serrated along margin.

Periostracum. Greenish yellow to tan, with numerous radiating, darker green rays.

Nacre. White, iridescent posteriorly, rarely with a purple flush below the lateral teeth; rarely with Tulberg layers.

Hinge plate. Heavy and short, interdentum short, wide, and thick; cardinals irregular, with two in the left valve and one in the right; these teeth may be accompanied by one or more prominent denticles both anterior and posterior to the teeth, with the posterior one positioned on the interdentum; lateral very short, straight or convex, two in left valve, one in right.

Muscle scars. Both small, anterior one deeply impressed; anterior adductor distinct from anterior pedal adductor, posterior adductor largely confluent with the posterior pedal retractor scar; dorsal suspensors few and obscure, located in a line behind cardinal fossa and interdentum. Pallial line entire, located far from shell margin. Accessory line extremely long, extending to marsupial region, but often very faint.

Umbonal cavity. Very wide and shallow.

Fig. 60. ***Epioblasma obliquata perobliqua*** **(Conrad, 1836). A.** OSUM 9545.1, St. Joseph River, DeKalb Co., IN. 32 mm. **B.** OSUM 65445, Fish Creek, Williams Co., OH. 49 mm. **C.** OSUM 67902, White River, Jackson Co., IN. 31 mm. **D.** OSUM 68539, Auglaize River, Defiance Co., OH. 34 mm. **E.** OSUM 46602.1, St. Joseph River, Defiance Co., OH. 26 mm. **F, H, Hinge detail, Beak sculpture detail:** OSUM 57277.1, Auglaize River, Defiance Co., OH. 38 mm. **G.** OSUM 57318.3, St. Mary's River, Allen Co., IN. 31 mm.

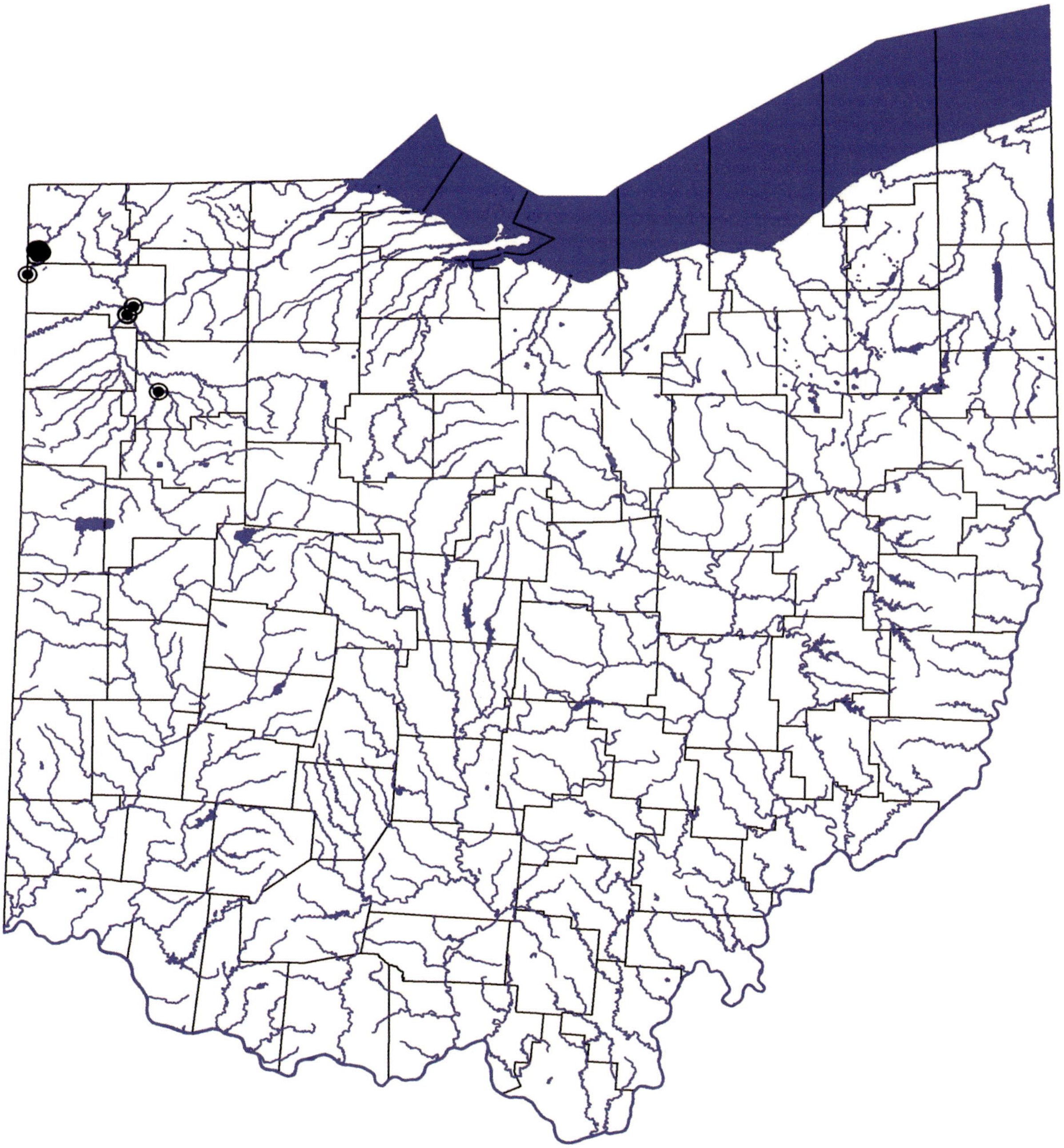

Fig. 61. ***Epioblasma obliquata perobliqua*** **(Conrad, 1836)**
Solid dots—live or freshdead since 1980
Bullseye—all records, any condition

Glochidia. Unknown.

Reproductive biology. Presumably bradytictic.

Reported potential hosts. No hosts known.

Habitat. The few living individuals found in recent years have occurred in stable sand and cobble in moving water of creeks and small rivers.

Range. Once locally common in the Maumee River system, it now is known only from Fish Creek in Williams County, Ohio, and perhaps the St. Joseph River in Fort Wayne, Indiana.

Range in Ohio. Today apparently limited to Fish Creek in Williams County. Historically known from the Maumee and Auglaize rivers.

Similar species. See under *Epioblasma obliquata obliquata.*

Comments. Male individuals grow fastest in their second and third years, but not as fast as other species. Females grow fast until approximately their third year; subsequent growth is dramatically slower with annuli packed together. Sexual dimorphism is apparent by the end of the second year. This species grows to 55 mm, but most are ca. 30 mm in length. They may reach 11 years in age, but most specimens examined were less than 5.

Johnson (1978: 279) stated that "the figured type of *perobliquus* is *P. torulosa.*" However, Johnson had misread the figure numbers on Conrad's plate 27 and reversed the captions for figure 2 and figure 1.

Epioblasma personata (Say, 1829)

Figures 62, 63

Etymology. L. *personatus*, wearing a mask, masked, assumed, pretended, counterfeited. Say described this species from a female specimen(s) and was impressed by the denticulate postventral shell expansion. He compared *personata* to that species, listing similarities, but noted "the form is altogether different" (1829: 309). Say may have been confused, at first, as to the identity of this species; hence the name *personata*.

Vernacular name(s). *Round Combshell.*

Original description

Unio personatus Say, 1829: 309, *Unio* species no. 7, no figure given.

Type locality. Say's original locality was "Inhab. Wabash [River]." However, Johnson & Baker (1973) designated a neotype from the Cumberland River, Tennessee. As stated by ICZN Art. 76.3, the place of origin of the neotype becomes the type locality of the species. But the designation does not fit the qualifying conditions as outlined under ICZN Art. 75.3.6, which calls for the neotype to have come "as nearly as practicable from the original type locality. . . ." This appears to be an invalid neotype designation.

Type material. All or nearly all of the unionoid type material of Say has apparently been lost (Johnson & Baker, 1973: 146). Call (1900: 475) noted, however, "There are larger specimens in the Call Collection, at Harvard, which were labelled by Say, and still have, or had when the collection was transmitted, his original label." Johnson & Baker (1973: 166) reported, "From the four specimens identified by Say . . . is selected a neotype MCZ 25763, from the Cumberland River, Tennessee." Johnson (1978: 276 and caption to pl. 14, fig. 9) gave the neotype as MCZ 5763. As mentioned above, this designation may be invalid.

Synonymy

Unio personatus Say, 1829: 309.

Unio pileus Lea, 1831: 119, pl. 18, fig. 47 [Type locality: "Ohio River, near Cincinnati, Ohio." Figured lectotype USNM 84602a *fide* Johnson, 1978: 276. Agassiz (1852: 43) considered this the male of the species.].

Unio capillaris Lea, 1834: 29, pl. 2, fig. 2 [Type locality: "Ohio." Figured type lost *fide* Johnson, 1978: 276. Agassiz (1852: 43) considered this the female of the species.].

Margarita (*Unio*) *pileus* (Lea, 1834). Lea, 1836: 20.

Margarita (*Unio*) *personatus* (Say, 1829). Lea, 1836: 33.

Margaron (*Unio*) *pileus* (Lea, 1834). Lea, 1852b: 24.

Margaron (*Unio*) *personatus* (Say, 1829). Lea, 1852b: 35.

Scalenaria personata (Say, 1829). Agassiz, 1852: 43.

Truncilla (*Pilea*) *personata* (Say, 1829). Simpson, 1900a: 522.

Dysnomia (*Pilea*) *personata* (Say, 1829). Ortmann, 1925: 361.

Dysnomia personata (Say, 1829). Morrison, 1942: 364.

Plagiola (*Pilea*) *personata* (Say, 1829). Johnson, 1978: 276–278, pl. 6, fig. B, pl. 14, figs. 7–9.

Epioblasma personata (Say, 1829). Stansbery, 1976: 43.

Status. Listed as Extinct by both Ohio Division of Wildlife and U.S. Fish & Wildlife Service as of 2009.

ADULT SHELL

Shell. Small size (to 50 mm), females round in profile, male triangular, both sexes quite inflated.

Anterior margin. Concave for a short distance before the umbo, then broadly rounded to the ventral border.

Ventral margin. In males, broadly rounded to approximately 66% posterior, then forming weak obtuse, rounded angle before continuing in a concave line to posterior point; females broadly rounded to posterior margin to form a very short, very broad projection at the postero-ventral margin, then continuing straight or concave to the posterior point.

Fig. 62. ***Epioblasma personata*** **(Say, 1829). A, F.** OSUM 10370.1, "Ohio River." 45 mm. **B.** OSUM 10379.1, Ohio River, Hamilton Co., OH. 39 mm. **C, Hinge detail:** OSUM 68433, "Ohio." 56 mm. **D.** OSUM 24406, "Ohio River." 46 mm. **E.** OSUM 5118, Scioto River, [Co.?], OH. 49 mm. **Beak sculpture detail:** OSUM 28257. No data. 57 mm total length.

Fig. 63. ***Epioblasma personata*** **(Say, 1829)**
Solid dots—live or freshdead since 1980
Bullseye—all records, any condition

Posterior margin. Rounded to dorsal margin, weakly differentiated from dorsal margin, particularly in males.
Dorsal margin. Rounded or straight in males to umbo, concave in females, forming obtuse angle with posterior margin.
Umbo. Wide and prominently projecting above dorsal margin; umbonal sculpture eroded from all specimens examined.
Ligament. Fairly long, tan to dark brown, not concealed by shells.
Disc sculpture. Shell sculptureless except for weak radiating striae on the posterior slope; the female "digit" is raised above the remaining shell on a broad, weak rib; this protrusion is serrated along its margin; however, this region is very thin and is usually broken.
Periostracum. Greenish, yellow to tan or dark brown, with numerous radiating, darker green rays; these are most evident on the posterior half of the shell.
Nacre. White, iridescent and very thin posteriorly; often with gold Tulberg layers.
Hinge plate. Heavy and short, distinctly angled, interdentum short, wide, and thick; cardinals irregular, with two in the left valve and one in the right; these teeth may be accompanied by one or more prominent denticles both anterior and posterior to the teeth; lateral very short, straight or convex, two in left valve, one in right.
Muscle scars. Anterior adductor very deeply impressed; posterior adductor much less so; anterior adductor distinct from anterior pedal adductor, posterior adductor varying from entirely confluent with the posterior pedal retractor to nearly separate from the adductor scar; multiple small dorsal suspensors, located in a line behind cardinal fossa. Pallial line entire, located far from shell margin. Accessory line long.
Umbonal cavity. Very wide and shallow.

Glochidia. Unknown.

Reproductive biology. Reproductive season unknown, presumably bradytictic.

Reported potential hosts. No hosts known.

Habitat. Unknown. Large rivers.

Range. Large rivers in the Ohio River system: Ohio, Tennessee, Green, and Wabash rivers.

Range in Ohio. Historically from the Ohio River at Cincinnati and the Scioto River at Columbus.

Similar species. The spherical shape of the females of this species most resemble *Quadrula pustulosa*. That species typically has some pustulose sculpture (absent in *E. personata*) and lacks the serrated posterior-ventral margin. The males of *E. personata* may resemble short examples of *Fusconaia ebenus* and the *Pleurobema cordatum*-group. *Epioblasma personata* has a much shallower beak cavity than those species and typically has green rays that persist over the shell, a characteristic usually not found in Ohio *Fusconaia ebenus* or the *Pleurobema cordatum*-group.

Comments. This was a short-lived species (in many ways). They grew quickly for up to 5 years before gradually slowing. Sexual dimorphism was apparent in the third year. Females attained ca. 50 mm in length and lived for 10 years. Males reached 50 mm and lived for 15 years.

Epioblasma phillipsii (Conrad, 1835)

Figures 64, 65

Etymology. Phillips + L. *-i*, genitive ending; so *E. phillipsi* literally means "the *Epioblasma* of Phillips." Conrad (1835: 9, pl. 5, fig. 1) figured this species from a specimen "in the cabinet of Mr. John Phillips, who called only attention to it as a distinct species, and I have dedicated to him . . ."—hence a patronym.
Vernacular name(s). Tubercled Riffleshell.
Original description
Unio phillipsii Conrad, 1835(1): 9, pl. 5, fig. 1.
Type locality. "The specimen figured is from the Wabash [River], and is in the cabinet of Mr. John Phillips, ..."
Type material. Johnson & Baker (1973: 166) identified the figured holotype of *U. phillipsii* Conrad, 1835, as ANSP 19432.
Synonymy
Unio phillipsii Conrad, 1835: 9, pl. 5, fig. 1.
Unio cincinnatiensis Lea, 1840: 285 [Type locality: "Ohio River, Cincinnati, [Hamilton Co.] O[hio]." Possible lectotype USNM 84199].
Margaron (*Unio*) *cincinnatiensis* (Lea, 1840). Lea, 1852b: 22.
Unio phillipsii Reeve, 1864: pl. 4, fig. 15.
Truncilla (*Pilea*) *perplexa* var. *cincinnatiensis* (Lea, 1840). Simpson, 1900a: 523.
Dysnomia (*Torulosa*) *cincinnatiensis* (Lea, 1840). Frierson, 1927: 94.
Dysnomia torulosa cincinnatiensis (Lea, 1840). Morrison, 1942: 365.
Dysnomia (*Pilea*) *torulosa cincinnatiensis* (Lea, 1840). La Rocque, 1967: 283.
Plagiola (*Torulosa*) *torulosa* Rafinesque, 1820. Johnson, 1978: 261–265 [in part], pl. 9, fig. a [in part], pl. 11, fig. 9.
Epioblasma cincinnatiensis (Lea, 1840). Stansbery *et al.*, 1985: 298.

Fig. 64. *Epioblasma phillipsii* (Conrad, 1835). A, B. UMMZ 90598, no data. 43mm. **C, Hinge detail:** UMMZ 2 90598, no data. 39 mm. **Beak sculpture detail:** no specimen available.

Epioblasma phillipsi [*sic*] (Conrad, 1835). Watters, 1993: 41.

Epioblasma torulosa torulosa Rafinesque, 1820. Parmalee & Bogan, 1998: 104–107 [in part].

Status. Listed as Extinct by both Ohio Division of Wildlife and U.S. Fish & Wildlife Service as of 2009.

ADULT SHELL

Shell. Small size (to at least 43 mm), fairly thick and heavy, both sexes similar in outline, males with a concave dorso-ventral margin, with this margin nearly straight in females, both sexes moderately inflated.

Anterior margin. Broadly rounded to the ventral border.

Ventral margin. In males, broadly rounded to approximately 50% posterior, then forming a weak expansion before continuing in a concave line to posterior point; females broadly rounded to 66% posterior, then forming a weak expansion or protrusion that continues, nearly straight, to the dorsal margin.

Posterior margin. Weakly bilobed.

Dorsal margin. Rounded or straight to umbo, forming obtuse angle with posterior margin.

Umbo. Wide and low; umbonal sculpture eroded on all material examined.

Ligament. Fairly short, tan to dark brown, not concealed by shells.

Disc sculpture. Radiating down the center of the shell in a strong descending twist is a single row of elongated pustules that distally divide into two or more rows. A second, more subdued row is present on the dorsal ridge.

Periostracum. Tan, with numerous radiating, darker green rays of various widths.

Nacre. White, iridescent and very thin posteriorly. The nacre becomes abruptly thin at the "marsupial swelling," particularly along the ventral margin.

Hinge plate. Moderately developed and short, distinctly angled, interdentum short, wide, and thick; cardinals irregular, with two in the left valve and one in the right; these teeth may be accompanied by one or more prominent denticles both anterior and posterior to the teeth; the posterior accessory tooth may be positioned on the inter-

Fig. 65. ***Epioblasma phillipsii*** **(Conrad, 1835)**
Solid dots—live or freshdead since 1980
Bullseye—all records, any condition

dentum; lateral very short, straight or convex, two in left valve, one in right.
Muscle scars. Anterior adductor muscle scar small, very deeply impressed; posterior adductor much less so; anterior adductor distinct from anterior pedal adductor, posterior adductor nearly separate from the posterior pedal scar; multiple small dorsal suspensors, located in a line behind cardinal fossa. Pallial line entire, located far from shell margin. Accessory line extremely long, extending to marsupial region.
Umbonal cavity. Very wide and moderately shallow.

Glochidia. Unknown.

Reproductive biology. Reproductive season unknown, presumably bradytictic.

Reported potential hosts. No hosts known.

Habitat. Unknown. Large rivers.

Range. La Rocque (1967: 283) reported that this species was restricted to the Ohio River in the vicinity of Cincinnati. Bogan & Parmalee (1983: 37) reported subfossil specimens removed from archaeological sites in the Pickwick Basin and from Davidson County, Tennessee. Ohio State University Museum records include one dead specimen taken from the Ohio River at Cincinnati, and subfossil specimens taken from Carlston Annis Mound, Kentucky.

Range in Ohio. The species was found in the Ohio River at Cincinnati. It is extinct today.

Similar species. *Epioblasma torulosa torulosa* has fewer knobs in a very different distribution and is generally larger than this species. The female of *E. phillipsii* has a very different marsupial swelling than either *E. t. torulosa* or *E. t. rangiana*. In those taxa the swelling is pronounced, projecting far posterior; in *E. phillipsii* the swelling is scarcely apparent, barely different in outline than the males.

Comments. The identity of this species has escaped many able workers. Simpson (1900a) regarded it as a synonym of *Obliquaria reflexa* Rafinesque, 1820, and Frierson (1927) believed it to be *Plethobasus cyphyus* (Rafinesque, 1820). There are very few good examples of this species in collections.

Based on only a handful of specimens, it appears that this species gradually slowed in growth over time. The largest complete specimen seen is 43 mm in length and only 5 years old.

Epioblasma torulosa rangiana (Lea, 1838)

Figures 66, 67, 68, 69

Etymology. Rang + L. *iana,* adjectival ending. Lea (1838: 96) explained the coining of this patronym: "I name it after an ardent student of the Mollusca, Mons. Sander Rang." P. K. Sander L. Rang (1784–1859) was an early malacologist known to unionoid researchers for his 1835 paper, "Quelques Acéphales d'eau douce du Sénégal."
Vernacular name(s). *Northern Riffleshell,* Tan Riffleshell.
Original description
Unio rangianus Lea, 1838a: 95–96, pl. 18, fig. 56, 3 views. [Bogan (1998) is of the opinion that *Unio rangianus* Lea, 1838, is a junior synonym of *Epioblasma biloba* Rafinesque, 1831; see below.].
Type locality. Lea (1838a: 95) listed his specimens as from "Hab. Ohio River, near Cincinnati [Ohio]. T. G. Lea. [and from] near Poland, Ohio. Dr. Kirtland." The Poland, Ohio, specimen was likely from the Mahoning River near that village and is the type locality.
Type material. Johnson (1974: 122) designated the figured specimen of Lea's original description (1838a: 95) as the lectotype (USNM 84798). This is most appropriate since this specimen was figured and this species is more likely from a stream smaller than the Ohio River at Cincinnati.
Synonymy
Epioblasma biloba Rafinesque, 1831: 2 [Type locality: "Green river and Kentucky river." Johnson (1978: 283) designated a neotype in ANSP 56571 (apparently invalid as the neotype is from the Ohio River), but Bogan (1997) claimed to have located a syntype at the Muséum national d'Histoire naturelle, Paris. If so, this is the senior synonym of *Unio rangianus* Lea, 1838. Parmalee & Bogan (1998) suggested that Rafinesque's name be suppressed. We agree and have maintained the usage of *rangianus* here until the issue is resolved. However, Williams *et al.* (2008) considered *rangianus* only a form of *E. torulosa.*].
Unio rangianus Lea, 1838a: 95–96, pl. 18, fig. 56 [Agassiz (1852: 43) considered this the female of *U. o. perobliquus* Conrad, 1836.].
Margarita (*Unio*) *rangianus* (Lea, 1838). Lea, 1838b: 18.
Margaron (*Unio*) *rangianus* (Lea, 1838). Lea, 1852b: 26.
Truncilla (*Scalenaria*) *sulcata* var. *delicata* Simpson, 1900a: 520 [Type locality: "Detroit River and southern Michigan." Holotype USNM 160853].
Truncilla (*Pilea*) *perplexa* var. *rangiana* (Lea, 1838). Simpson, 1900a: 523.

Fig. 66. ***Epioblasma torulosa rangiana*** **(Lea, 1838). A.** OSUM 9892.5, St. Joseph River, DeKalb Co., IN. 39 mm. **B.** OSUM 46237.5, Sandusky River, Wyandot Co., OH. 41 mm. **C.** OSUM 1434, Big Darby Creek, Pickaway Co., OH. 73 mm. **D.** OSUM 14771, Big Darby Creek, Franklin Co., OH. 51 mm. **E.** OSUM 28280, Walhonding River, [Co.?], OH. 48 mm. **F, Hinge detail:** OSUM 12795.26, Big Darby Creek, Pickaway Co., OH. 63 mm. **G.** OSUM 3925.9, Lake Erie, Ottawa Co., OH. 47 mm. **H.** OSUM 3925.4, Lake Erie, Ottawa Co., OH. 45 mm. **Beak sculpture detail:** OSUM 12795.38, Big Darby Creek, Pickaway Co., OH. 38 mm total length.

Truncilla perplexa rangiana Price, 1900: 79.
Truncilla rangiana (Lea, 1838). Daniels, 1903: 646.
Unio rangianns [*sic*] Lea, 1838. De Gregorio, 1941: 47.
Dysnomia (*Torulosa*) *rangiana* (Lea, 1838). Frierson, 1927: 95.
Dysnomia perplexa rangiana (Lea, 1838). Goodrich, 1932: 113.
Dysnomia rangiana (Lea, 1838). La Rocque & Oughton, 1937: 152.
Dysnomia (*Pilea*) *torulosa rangiana* (Lea, 1838). La Rocque, 1967: 283.
Dysnomia (*Torulosa*) *torulosa rangiana* (Lea, 1838). Haas, 1969: 486.
Plagiola (*Torulosa*) *torulosa* Rafinesque, 1820. Johnson, 1978: 261–264 [in part], pl. 9, fig. a [in part], pl. 11, fig. 10.
Epioblasma torulosa rangiana (Lea, 1838). Stansbery, 1982: 3.
Epioblasma rangiana (Lea, 1838). Cummings & Berlocher, 1990: 92.
Epioblasma torulosa biloba Rafinesque, 1831. Bogan, 1998: 77, 78, 82–84, fig. 3.

Status. Listed as Endangered by both Ohio Division of Wildlife and U.S. Fish & Wildlife Service as of 2009. A Federal recovery plan has been completed (USFWS, 1993).

ADULT SHELL

Shell. Medium size (to 70 mm), females oval in profile, male posteriorly pointed, both sexes moderately inflated.

Anterior margin. Concave for a short distance before the umbo, then broadly rounded to the ventral border.

Ventral margin. In males, broadly rounded to approximately 66% posterior, then forming weak obtuse, rounded angle before continuing in a concave line to posterior point; females broadly rounded to 50% posterior, then forming a very wide, often very prominent expansion or protrusion that continues, broadly rounded, to the dorsal margin.

Posterior margin. In males, rounded to dorsal margin, weakly differentiated from dorsal margin; in females the posterior margin is a continuation of the "marsupial swelling."

Dorsal margin. Rounded or straight in males to umbo, forming obtuse angle with posterior margin.

Umbo. Wide and low; umbonal sculpture of 4–6 microscopic, widely "V"-shaped ridges, the distal-most ones becoming incomplete.

Ligament. Fairly long, tan to dark brown, not concealed by shells.

Disc sculpture. Shell sculptureless except for weak radiating striae on the posterior slope; the annular growth lines become particularly prominent on the marsupial swelling, giving it an undulating aspect.

Periostracum. Greenish, yellow to tan, with numerous radiating, darker green rays; these are most evident on the early portions of the shell; rarely rayless. Rare individuals, particularly from Big Darby Creek, may be colored rose or orange.

Nacre. White, iridescent and very thin posteriorly; very rarely pink or rose; rarely with gold Tulberg layers. The nacre becomes abruptly thin at the marsupial swelling, particularly along the ventral margin.

Hinge plate. Heavy and short, distinctly angled, interdentum short, wide, and thick; cardinals irregular, with two in the left valve and one in the right; these teeth may be accompanied by one or more prominent denticles both anterior and posterior to the teeth; the posterior accessory tooth may be positioned on the interdentum; lateral very short, straight or convex, two in left valve, one in right.

Muscle scars. Anterior adductor very deeply impressed; posterior adductor much less so; anterior adductor distinct from anterior pedal adductor, posterior adductor nearly separate from the posterior pedal scar; multiple small dorsal suspensors, located in a line behind cardinal fossa. Pallial line entire, located far from shell margin. Accessory line extremely long, extending to marsupial region, but often very faint.

Umbonal cavity. Very wide and moderately deep.

Glochidia. Depressed subelliptical, dorsal margin straight; lateral margins and ventral margin are about equally rounded. Exterior surface sculpturing loose-looped, supernumerary hooks present, micropoints blunt (Hoggarth, 1999: figs. 72a–f).

Literature records (Lea, 1858: 48, pl. 5, fig. 21; Ortmann, 1912: 358; Hoggarth, 1999: figs. 72a–f) indicate glochidial dimensions for length (0.24 to 0.26 mm), height (0.21 to 0.24 mm), and hinge (0.16 to 0.19 mm). Hoggarth (1999) noted, "This glochidium can be distinguished by its extreme morphological depression. It is figured by Lea (1858, Pl. 5, Fig. 21, as *Unio perplexus* = *E. rangiana fide* Ortmann & Walker, 1922), who provided length and height measurements of 224 µm x 240 µm. Lea's figure is correctly drawn so it must be assumed that his valve length equals my valve height and that his valve height equals my valve length. His measurements would then agree with mine and Ortmann's (1912); 260 µm x 230 µm."

Reproductive biology. Bradytictic. The posterior half or more of the outer gills is marsupial (Ortmann, 1912a). Gravid females are found from September to the following June (Ortmann, 1919; Watters, unpubl.). Females move to the substrate surface during brooding in the winter and spring. The posterior mantle margin of the displaying female is bright white and can be seen from several meters

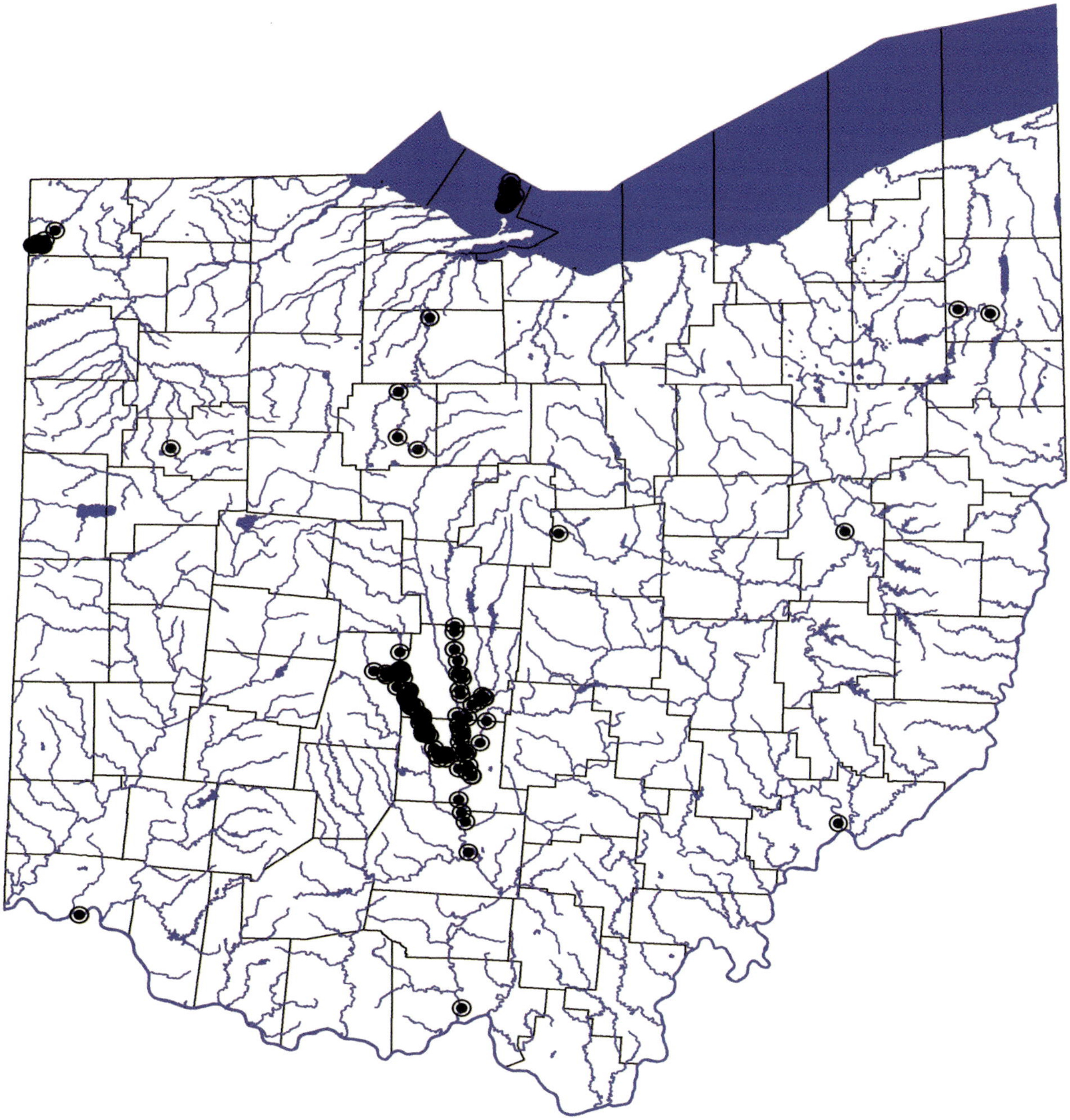

Fig. 67. ***Epioblasma torulosa rangiana*** **(Lea, 1838)**
Solid dots—live or freshdead since 1980
Bullseye—all records, any condition

Fig. 68. *Epioblasma torulosa rangiana* (Lea, 1838). A female has captured a host darter to parasitize.

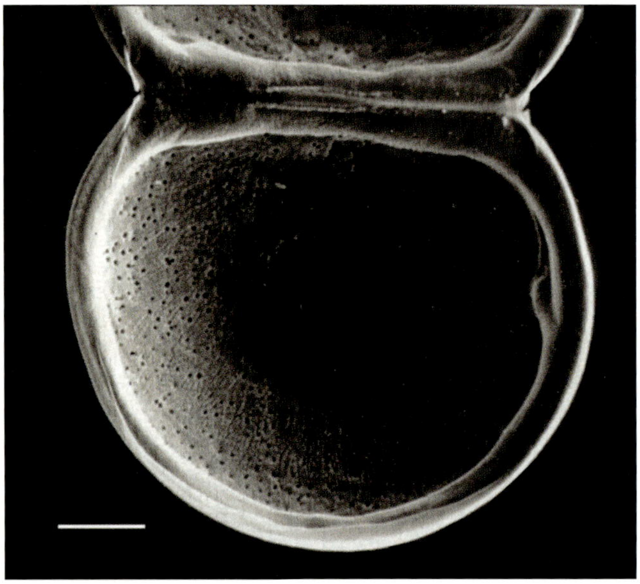

Fig. 69. *Epioblasma torulosa rangiana* (Lea, 1838). Glochidium. MAH 632.1, Big Darby Creek, Pickaway Co., OH. Bar = 35 µm. (Hoggarth, 1999).

away. The female traps the host fish between its shells when the fish inserts its head. The mantle tissue encircles the fish's head, and glochidia are pumped onto the fish for 5–10 minutes before being released. No conglutinate is formed (Watters, 2008). As with most other *Epioblasma* species, the hosts are darters and sculpins. Although Brown Trout was identified as a potential host, it is unlikely that the mussel and trout ever occur together in the wild.

Reported potential hosts

Cottus bairdi (Mottled Sculpin)—Watters, 1996b (LT).
Etheostoma camurum (Bluebreast Darter)—Watters, 1996b (LT); O'Dee & Watters (2000) (LT).
Etheostoma caeruleum (Rainbow Darter)—Watters *et al.*, 2009 (LT).
Etheostoma zonale (Banded Darter)—Watters, 1996b (LT); O'Dee & Watters (2000) (LT).
Salmo trutta (Brown Trout)—Watters, 1996b (LT); O'Dee & Watters (2000) (LT).

Habitat. Stable sand and cobble in the moving water of large creeks. Although known from Lake Erie, it is not a pond or lake species. The Lake Erie specimens apparently occurred in areas with sufficient wave-action to approximate stream conditions.

Range. In Canada it occurs in the Sydenham and Ausable rivers of Ontario (Staton *et al.*, 2000). It was found historically throughout the upper Ohio River system including the Allegheny River. Remnants of an extinct population have been found in rock shelter middens on the Red River of the Kentucky River in Kentucky (Dourson, KY Forest Service, pers. comm., 2000).

Range in Ohio. Historically from the Ohio River and throughout most of the Scioto River and its tributaries, and the Mahoning and Tuscarawas rivers. In the Lawrentian drainage, historically from the Auglaize River and Sandusky River, as well as Lake Erie around the Bass Islands. Currently in Big Darby Creek in central Ohio and Fish Creek in Williams County. Both populations are near extirpation.

Similar species. Females resemble *Epioblamsa torulosa torulosa*. Generally that species is more darkly colored and has prominent, undulating knobs. Male *E. t. rangiana* closely resemble male *E. obliquata s.l.* The latter are more evenly rounded in profile whereas *E. t. rangiana* is concave posteriorly. The sulcus is bounded by more prominent swellings in *E. t. rangiana* than *E. obliquata*. Nevertheless, males of these two species may be difficult to separate. Male *E. t. torulosa* have undulating knobs bordering the anterior of the sulcus—these are reduced or absent in *E. t. rangiana*.

Comments. Individuals grow quickly for the first three years and then gradually slow. Both sexes may grow to 70 mm and live 15 years.

In populations from the Great Lakes drainages the female's marsupial region is distinctly flattened, even concave, while in the Ohio River system females this region is inflated.

Epioblasma torulosa torulosa (Rafinesque, 1820)

Figures 70, 71

Etymology. L. *torus*, swelling, protuberance, knot, bulge + L. *-ula* (diminutive) + L. *-osa*, full of. Rafinesque (1820: 314) apparently named this species for the "quelques nodules," the few irregular nodules or knobs on the median ridge of the shell.

Fig. 70. ***Epioblasma torulosa torulosa*** **(Rafinesque, 1820). A.** OSUM 36904.49, "Tennessee." 64 mm. **B.** OSUM 39011, "Wabash River." 73 mm. **C.** OSUM 36896.3, "Ohio River." 61 mm. **D.** OSUM 67996, Ohio River, Hamilton Co., OH. 59 mm. **E.** OSUM 67995, Ohio River, Hamilton Co., OH. 46 mm. **F, Hinge detail, Beak sculpture detail:** OSUM 57323, Wabash River, Posey Co., IN. 48 mm. **G.** OSUM 47323, Muskingum River, [Co.?], OH. 64 mm. **H.** OSUM 28249, "Ohio River." 49 mm.

Vernacular name(s). *Tubercled Blossom,* Knobbed Riffleshell.

Original description

Amblema torulosa Rafinesque, 1820: 314–315, species no. 55, pl. 82, figs. 11, 12.

Type locality. Originally "Dans l'Ohio [River] et Kentuky [River]." The probable holotype is from the Kentucky River, which becomes the type locality.

Type material. Both Vanatta (1915: 550) and Johnson & Baker (1973: 173) recognized ANSP 20218 from the Kentucky River as the figured holotype, and ANSP 20216 (also Kentucky River) as a paratype.

Synonymy

Amblema torulosa Rafinesque, 1820: 314–315, pl. 82, figs. 11, 12.

Amblema torulosa var. *angulata* Rafinesque, 1820: 315 [Type locality: "l'Ohio et Kentuky [Rivers]." Types lost *fide* Johnson, 1978].

Amblema gibbosa Rafinesque, 1820: 315 [Type locality: "Ohio River" by lectotype designation. Lectotype ANSP 20232 by designation of Johnson & Baker (1973: 156)].

Amblema gibbosa var. *olivacea* Rafinesque, 1820: 315 [Type locality: none given. Type not located].

Amblema gibbosa var. *radiata* Rafinesque, 1820: 315 [Type locality: none given. Type not located].

Amblema gibbosa var. *difformis* Rafinesque, 1820: 315 [Type locality: none given. Type not located].

Truncilla torulosa (Rafinesque, 1820). Ortmann, 1918: 589.

Dysnomia (*Pilea*) *torulosa* (Rafinesque, 1820). Ortmann & Walker, 1922: 69.

Dysnomia torulosa (Rafinesque, 1820). Ortmann, 1926: 182.

Dysnomia (*Torulosa*) *torulosa* (Rafinesque, 1820). Frierson, 1927: 94.

Dysnomia torulosa torulosa (Rafinesque, 1820). Stansbery, 1970: 94.

Epioblasma torulosa torulosa (Rafinesque, 1820). Stansbery, 1976: 51.

Plagiola (*Torulosa*) *torulosa* (Rafinesque, 1820). Johnson, 1978: 261–264 [in part], pl. 9, fig. a [in part], pl. 11, fig. 7, pl. 12, fig. 1.

Epioblasma torulosa (Rafinesque, 1820). Cummings & Berlocher, 1990: 92.

Status. Listed as Extirpated by Ohio Division of Wildlife; listed as Endangered by U.S. Fish & Wildlife Service as of 2009. A Federal recovery plan has been completed (USFWS, 1985a).

ADULT SHELL

Shell. Medium size (to 75 mm), thick and heavy, females oval in profile, male posteriorly pointed, both sexes moderately inflated.

Anterior margin. Broadly rounded to the ventral border.

Ventral margin. In males, broadly rounded to approximately 50% posterior, then forming a weak expansion before continuing in a concave line to posterior point; females broadly rounded to 50% posterior, then forming a very wide expansion or protrusion that continues, broadly rounded, to the dorsal margin.

Posterior margin. In males, rounded to dorsal margin, weakly differentiated from dorsal margin; in females the posterior margin is a continuation of the "marsupial swelling."

Dorsal margin. Rounded or straight to umbo, forming obtuse angle with posterior margin.

Umbo. Wide and low; umbonal sculpture eroded on all material examined, presumably as in *E. t. rangiana.*

Ligament. Fairly long, tan to dark brown, not concealed by shells.

Disc sculpture. Radiating down the center of the shell in a strong descending twist is a single row of large, undulating knobs. These continue to the ventral margin in males, but in females they may become obsolete once the marsupial swelling begins. A second, much more subdued row may be present on the dorsal ridge of some specimens.

Periostracum. Dark green, yellow to tan, with numerous radiating, darker green rays of various widths. As with *E. t. rangiana,* rare pink individuals occur. The distal portion of the marsupial swelling may be dark green. The "valleys" between the medial knobs are often colored dark green.

Nacre. White, iridescent and very thin posteriorly, rarely rose; commonly with gold Tulberg layers. The nacre becomes abruptly thin at the "marsupial swelling," particularly along the ventral margin.

Hinge plate. Heavy and short, distinctly angled, interdentum short, wide, and thick; cardinals irregular, with two in the left valve and one in the right; these teeth may be accompanied by one or more prominent denticles both anterior and posterior to the teeth, and the posterior accessory tooth may be positioned on the interdentum; lateral very short, straight or convex, two in left valve, one in right.

Muscle scars. Anterior adductor small, very deeply impressed; posterior adductor much less so; anterior adductor distinct from anterior pedal adductor, posterior adductor nearly separate from the posterior pedal scar; multiple small dorsal suspensors, located in a line behind cardinal fossa. Pallial line entire, located far from shell margin. Accessory line extremely long, extending to marsupial region.

Umbonal cavity. Very wide and moderately deep.

Glochidia. Unknown.

Reproductive biology. Reproductive season unknown, presumably bradytictic.

Fig. 71. ***Epioblasma torulosa torulosa*** **(Rafinesque, 1820)**
Solid dots—live or freshdead since 1980
Bullseye—all records, any condition

Reported potential hosts. No hosts known.

Habitat. Presumably similar to *Epioblasma torulosa rangiana,* but in larger rivers.

Range. Originally in the Ohio and Tennessee River systems. Primarily a big river species.

Range in Ohio. Historically from the Ohio, Scioto, and Muskingum River mainstems.

Similar species. This species differs from *E. torulosa rangiana* in having more prominent, undulating knobs anterior to the sulcus, and generally being darker in color.

Comments. Individuals do not grow markedly faster as juveniles, but gradually slow. Both sexes may grow to 75 mm and live 15 years.

Epioblasma triquetra (Rafinesque, 1820)

Figures 72, 73, 74

Etymology. L. *triquetrus*, having three corners, triangular. Rafinesque (1820: 300) noted that the shape of this shell is "presque trièdre," or almost trihedral. Together, the posterior slopes of the two valves of an individual of *E. triquetra* (especially a female) more nearly form a plane than in any other species. *Epioblasma triquetra* is also nearly triangular in lateral outline, although the species name apparently refers instead to the three-dimensional shape of the shell.

Vernacular name(s). *Snuffbox*, Northern Arc Shell.

Original description

Truncilla triqueter Rafinesque, 1820: 300–301, species no. 18, pl. 81, figs. 1, 2, 3, 4.

Type locality. "aux chutes de l'Ohio [River]." The Falls of the Ohio are at Louisville, Jefferson Co., Kentucky.

Type material. The type mentioned by Vanatta (1915: 550) was selected as lectotype (ANSP 20231) by Johnson & Baker (1973: 173).

Synonymy

Truncilla triqueter Rafinesque, 1820: 300–301, pl. 81, figs. 1, 2, 3, 4.

Unio triangularis Barnes, 1823: 272–273, fig. 17 [Type locality: "Bois-blanc Isle, Detroit River [MI]." Type in New York Lyceum of Natural History, destroyed by fire *fide* Johnson (1978). Agassiz (1852: 43) considered this the male of the species.].

Unio cuneatus Swainson, 1823: 112 [*non* Barnes, 1823. Type locality: not stated. Type in Manchester Museum?].

Mya triangularis Rafinesque [*sic*]. Eaton, 1826: 221.

Unio triqueter (Rafinesque, 1820). Short & Eaton, 1831: 79.

Unio formosus Lea, 1834: 111, pl. 16, fig. 41 [Type locality: "Ohio River." Type lost *fide* Johnson, 1978. Agassiz (1852: 43) considered this the female of the species.].

Margarita (*Unio*) *triangularis* (Barnes, 1823). Lea, 1836: 18.

Margaron (*Unio*) *triangularis* (Barnes, 1823). Lea, 1852b: 23.

Truncilla (*Truncilla*) *triquetra* Rafinesque, 1820. Simpson, 1900a: 517.

Truncilla triquetra Rafinesque, 1820. Scammon, 1906: 283.

Unio triangularis var. *pergibosus* De Gregorio, 1914: 40, pl. 4, fig. 4 [Type locality: "Sciota [Scioto] River, Ohio." Type in Palermo Museum?].

Unio triangularis var. *longiusculus* De Gregorio, 1914: 40, pl. 4, fig. 5 [Type locality: "Sciota [Scioto] River, Ohio." Type in Palermo Museum?].

Dysnomia (*Truncillopsis*) *triquetra* (Rafinesque, 1820). Ortmann & Walker, 1922: 65–66.

Dysnomia triquetra (Rafinesque, 1820). Danglade, 1922: 5.

Epioblasma triquetra (Rafinesque, 1820). Stansbery, 1973b: 22.

Plagiola (Truncillopsis) triquetra (Rafinesque, 1820). Johnson, 1978: 248–252, pl. 1, pl. 10, figs. 1–4.

Plagiola triquetra (Rafinesque, 1820). Oesch, 1984: 232.

Status. Listed as Endangered by Ohio Division of Wildlife; not listed by U.S. Fish & Wildlife Service as of 2009.

ADULT SHELL

Shell. Medium size (to at least 70 mm), with pronounced sexual dimorphism, heavy for size, males elongate oval, females more angular, both very inflated; marsupial region laterally inflated, angulate along the posterior slope.

Anterior margin. Broadly rounded in both sexes.

Ventral margin. Broadly rounded in male until posterior angulation with posterior margin; straight or rounded in female until marsupial region; marsupial region occupies posterior third of ventral margin to ventral half of posterior margin, laterally inflated, but only slightly expanded beyond normal ventral margin.

Posterior margin. In males, straight for a short distance after posterior angulation, then forming a second angle, then straight to the posterior slope, forming a third angle, then straight again to umbo; females straight to slightly concave from the posterior angulation to the umbo.

Dorsal margin. Gently curved from the posterior slope to under the umbo, then briefly curved up to join the anterior margin.

Umbo. Nearly central to anterior of central, prominent; beak sculpture very fine, varying from closely set "V"-shaped ridges to widely spaced oblique ridges anterior of the umbo; sculpture apparently absent in some individuals; beak sculpture disappears within first year's growth.

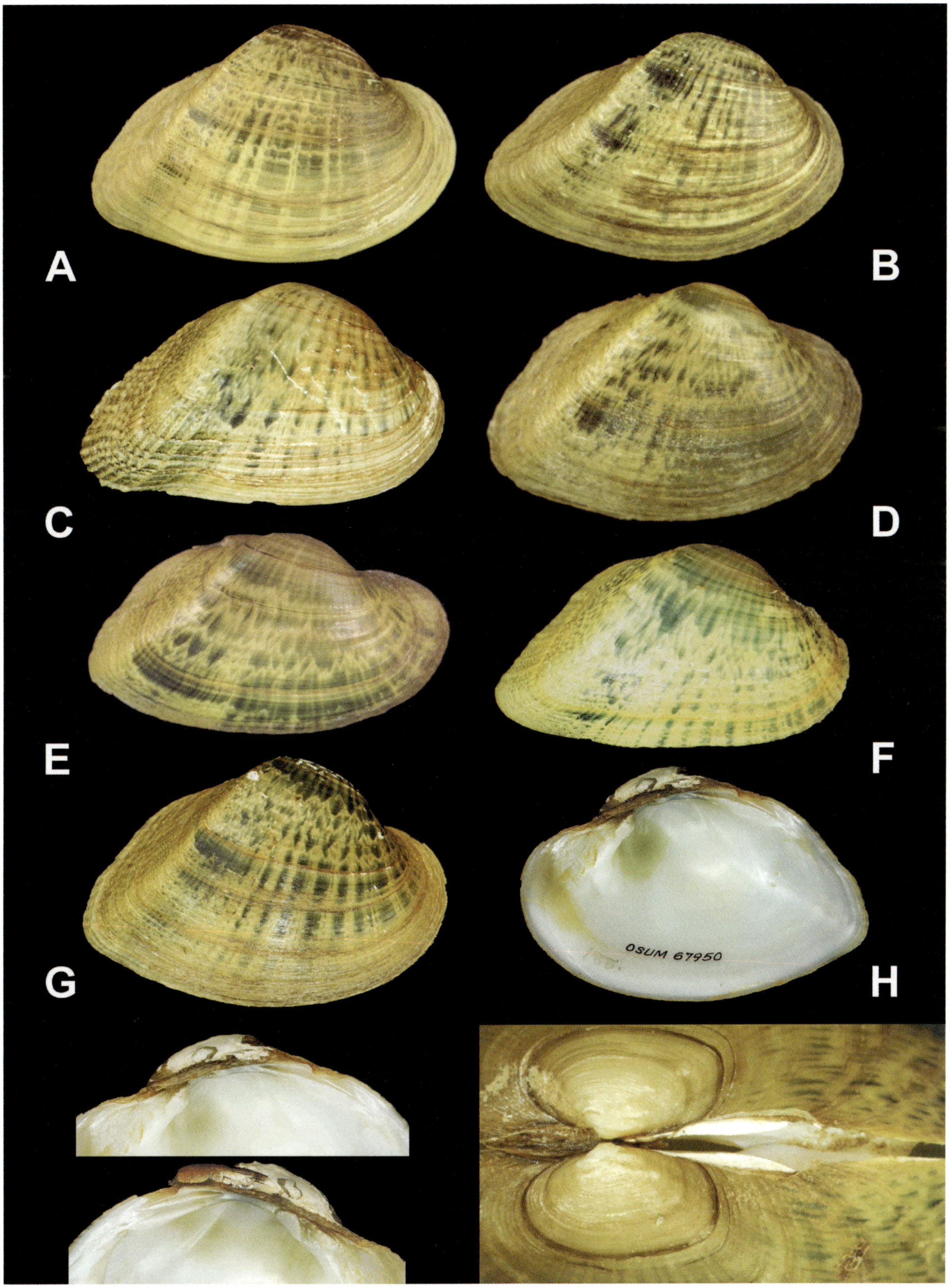

Fig. 72. ***Epioblasma triquetra*** **(Rafinesque, 1820).** **A.** OSUM 4048.6, Big Walnut Creek, Franklin Co., OH. 73 mm. **B.** OSUM 1432, Big Darby Creek, Pickaway Co., OH. 62 mm. **C.** OSUM 45927.2, Lake Erie, Ottawa Co., OH. 35 mm. **D.** OSUM 25971.13, Big Darby Creek, Pickaway Co., OH. 51 mm. **E.** OSUM 18246, Olentangy River, Franklin Co., OH. 46 mm. **F.** OSUM 68052.1, Scioto River, [Co.?], OH. 35 mm. **G, H, Hinge detail:** OSUM 67950, Ohio River, Hamilton Co., OH. 50 mm. **Beak sculpture detail:** OSUM 25971.1, Big Darby Creek, Pickaway Co., OH. 26 mm total length.

Ligament. Brown, fairly small, partially concealed between valves.
Disc sculpture. Male sculptureless except for radiating striae on posterior slope; females sculptureless except for radiating striae on marsupial region, which render the shell margin serrate.
Periostracum. Pale yellow, greenish yellow, or tan; color pattern of dark green, broken radiating rays; these rays are composed of dots, dashes, and chevron-shaped markings; rays may coalesce into irregular dark green blotches; pattern may disappear in old individuals; posterior slope often unpatterned.
Nacre. White; some populations have greenish or grayish Tulberg layers.
Hinge plate. Heavy, interdentum very short, narrow, often with a small denticle or ridge; two lamellate cardinal teeth in left valve, one sub-triangular tooth in right valve, with well-developed anterior denticle; laterals straight to slightly arched, shorter in female, one in right valve, two in left.
Muscle scars. Anterior adductor deep, posterior adductor usually distinct from posterior pedal retractor scar; dorsal suspensors multiple, within umbonal cavity under fossa and along underside of interdentum; pallial line complete, accessory line short and weak.
Umbonal cavity. Wide and deep.

Glochidia. Depressed subelliptical, dorsal margin straight, lateral margins equally and gently curved, ventral margin broadly curved. Loose-looped exterior valve sculpture covers the surface of the valve, supernumerary hooks present, micropoints blunt and irregular in shape (Lea, 1858c: 48, pl. 5, fig. 19; Ortmann, 1911: 347, pl. 89, fig. 24; Hoggarth, 1999: figs. 69a–e).

Literature records (Ortmann, 1912a: 356; Hoggarth, 1999: figs. 69a–e) indicate glochidial dimensions for length (0.21 to 0.22 mm), height (0.21 mm), and hinge (0.15 to 0.16 mm). Ortmann (1912a: 356) noted, "Glochidium of medium size, almost semicircular; length and height 0.21 mm."

Reproductive biology. Bradytictic. The posterior half or more of the outer gills is marsupial (Ortmann, 1912a). Eggs were observed in early September, with glochidia forming in mid-September. Glochidia are overwintered to the following April and May (Ortmann, 1919; Barnhart *et al.*, 1998a). Females move to the surface when ready to release glochidia, but no lures or other host-attracting behaviors and structures have been seen (R. Sherman-Mulcrone, pers. comm., University of Michigan Museum of Zoology, 2000). The female traps the snout of the host fish between its shells when the fish inserts its head (Barnhart *et al.*, 1998a). Hosts are darters and sculpins.

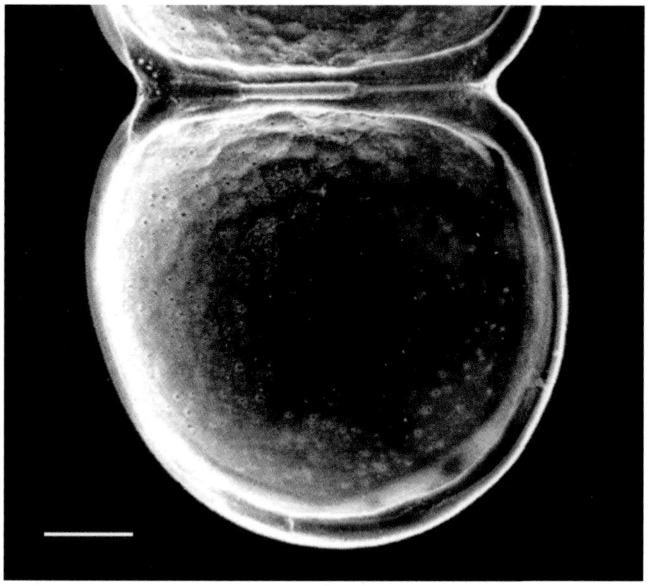

Fig. 73. *Epioblasma triquetra* (Rafinesque, 1820). Glochidium. MAH 588.1, Big Darby Creek, Franklin Co., OH. Bar = 40 µm. (Hoggarth, 1999).

Reported potential hosts

Cottus baileyi (Black Sculpin)—Jones & Neves, 2000 (LT).
Cottus bairdi (Mottled Sculpin)—Watters *et al.*, 2005 (LT).
Cottus carolinae (Banded Sculpin)—Yeager *in* Hill, 1986 (LT); Yeager & Saylor, 1995 (LT).
Cottus hypselurus (Ozark Sculpin)—Barnhart *et al.*, 1998a (LT).
Fundulus olivaceous (Blackspotted Topminnow)—Barnhart *et al.*, 1998a (LT).
Percina caprodes (Log Perch)—Yeager *in* Hill, 1986 (LT); Sherman, 1993 (LT); Yeager & Saylor, 1995 (LT); Hillegass & Hove, 1997 (LT); Barnhart *et al.*, 1998a (LT); Jones & Neves, 1999 (LT); Jones & Neves, 2000 (LT); Hove *et al.*, 2003 (LT).
Percina maculata (Blackside Darter)—Hillegass & Hove, 1997 (LT).
Percina roanoka (Roanoke Darter)—Jones & Neves, 1999 (LT); Jones & Neves, 1999 (LT); Jones & Neves, 2000 (LT).

Habitat. Stable sand and cobble in the moving water of large creeks. It also has been found in impoundments, such as in portions of the Tippecanoe River of Indiana, but this is probably not a preferred habitat. It also is known from Lake Erie. Parmalee (1967: 64) stated, "It usually inhabits bottoms composed of sand and coarse gravel, often in riffles in running water, and most individuals bury themselves deeply in sand. . . ." Buchanan (1980: 63) found this species ". . . in gravel and cobble, cobble, or gravel, cobble and boulder substrates, often entirely buried beneath the substrate. It was collected in 2 inches to 2 feet of water in

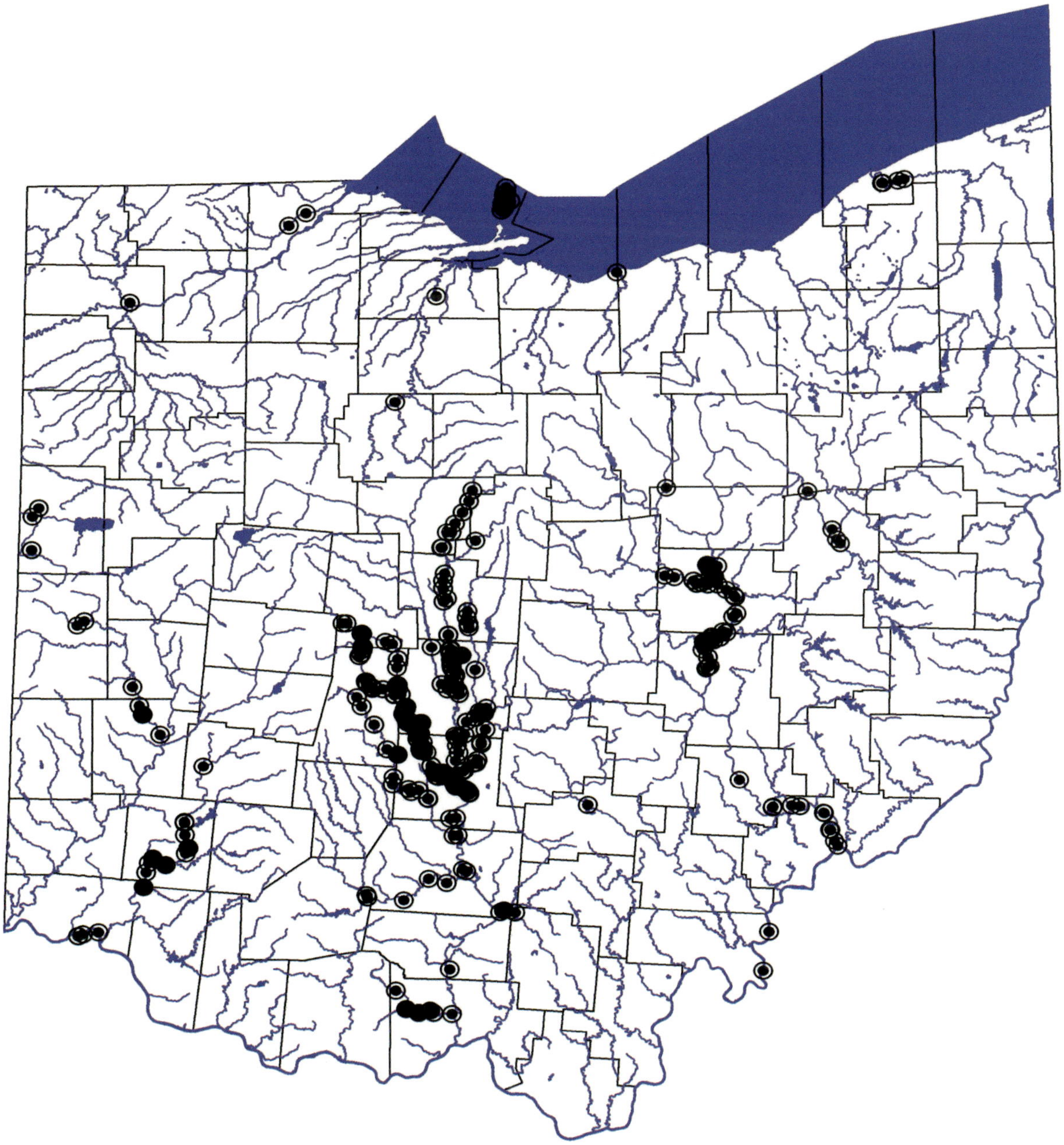

Fig. 74. ***Epioblasma triquetra*** **(Rafinesque, 1820)**
Solid dots—live or freshdead since 1980
Bullseye—all records, any condition

moderate (1.2 ft./sec. at the bottom) to swift (1.7 ft./sec. at the bottom) current."

Range. Simpson (1914: 6) listed this species for the "Ohio River drainage; western New York to southern Michigan; Iowa; eastern Nebraska to Indian Territory." La Rocque (1967: 285) found the general distribution of this species to include "Western New York and southern Ontario west to Nebraska and Kansas, south to West Virginia, Tennessee and northern Alabama, and north to Wisconsin and Michigan. Lake Huron; Lake St. Clair and its drainage; Detroit River; Lake Erie and its drainage."

Range in Ohio. Within the Ohio River drainage basin it has been recorded from the Great and Little Miami rivers, and the Scioto and Muskingum rivers. In the Lake Erie drainage basin it has been found in the Grand, Maumee, and Sandusky rivers, and Swan Creek (OSUM records, La Rocque, 1967: 285; Watters, 1992). In Lake Erie around the Bass Islands.

Similar species. *Truncilla donaciformis* (Lea, 1828) and *Truncilla truncata* Rafinesque, 1820, may be of comparable size and shape, but neither of these species is as inflated as *triquetra*. Neither of those species has a striate posterior slope. The female shell of *triquetra* would not be confused with any other Ohio species.

Comments. Individuals grow very quickly in their second and third year, particularly females, and sexual dimorphism is evident in the second year. Within a given population, females tend to be smaller than males of the same age. Rates of growth between populations are variable: a large individual, 70 mm in length, may be between 10 and 15 years old, depending on habitat. Individuals older than 15 years are rare.

Fusconaia Simpson, 1900

Etymology. L. *fuscus*, dark, dusky, tawny + L. *naias* (from Gr. *naias, naiados*), a water nymph of rivers and springs. In his original description of *Fusconaia*, Simpson (1900: 784) noted ". . . epidermis dark; . . . ," and although some species are rayed, and in others a lighter periostracum of yellowish-brown or tan is common in young individuals, most adults of this genus are a darker shade of brown.

Original description

Quadrula (*Fusconaia*) Simpson, 1900a: 784.

Type species. *Unio trigonus* Lea, 1831 [= *Fusconaia flava* (Rafinesque 1820)] by original designation of Simpson (1900a: 784). He was followed by Haas (1969: 311) who, however, considered *trigonus* to be a subspecies of *Unio undatus* Barnes, 1823. Both *undatus* Barnes and *trigonus* Lea are junior synonyms of *Obliquaria flava* Rafinesque, 1820.

Synonymy

Obovaria Rafinesque, 1819: 426 [Type species: *Obovaria obovalis* Rafinesque, 1820, by subsequent designation of Herrmannsen (1849: 132). *Obovaria obovalis* has been identified with *U. ebenus* Lea, 1831, and a lectotype ANSP 20244 was designated by Johnson & Baker (1973: 163). This name predates Simpson's *Fusconaia* by 81 years. Although *Obovaria obovalis* Rafinesque, 1820, may be petitioned as a *nomen oblitum*, this does not change the identity of *Obovaria*. Hermannsen's designation will need to be suppressed, and a new type species selected, to conserve *Fusconaia* and *Obovaria*. Bogan *et al.* (2006) petitioned the ICZN to conserve the current use of both *Fusconaia* and *Obovaria*, and permission was granted in 2008 (International Commission on Zoological Nomenclature, 2008). See under *Obovaria* for more information.].

Quadrula (*Fusconaia*) Simpson, 1900a: 784.

Quadrula (*Fusconia*) [*sic*] Simpson, 1900. Frierson, 1927: 54.

Geographic range. West of the Rocky Mountains in North America from the Red River of Manitoba throughout the Mississippi River system to Gulf drainages. Great Lakes drainages. Absent from peninsular Florida. Only a single *Fusconaia* inhabits the Atlantic Slope: *F. masoni* (Conrad, 1834).

Fusconaia ebenus (Lea, 1831)

Figures 75, 76

Etymology. L. *ebenus*; from Gr. *ebenos*, the ebony tree or its wood; from Egypt. *hebni*. The periostracum of this species, though bright yellow when very small, is typically dark brown or black. Lea did not name this species *ebeneus*, an adjective meaning ebony-colored or textured, but called it *ebenus*, the name of an object—ebony wood. As such it is a noun in apposition, and its original ending *-us* should not be changed to match the gender of *Fusconaia*. *Fusconaia ebena* is thus an incorrectly formed combination.

Vernacular name(s). *Ebonyshell.*

Original description

Unio ebenus Lea, 1831: 84–85, pl. 9, fig. 14 [Rafinesque named this species twice (*obovalis* and *pachostea*) in 1820, 11 years before Lea's *ebenus*. Rafinesque unnecessarily renamed *pachostea* as *antrosa* when he moved that species to *Amblema*. *Amblema antrosa* was used by Simpson in 1900a and 1914, Frierson in 1927, and Haas in 1969 as a valid name, which does not allow

Fig. 75. ***Fusconaia ebenus*** **(Lea, 1831). A.** OSUM 17233.7, Ohio River, Campbell Co., KY. 90 mm. **B.** OSUM 31462.2, Ohio River, Livingston Co., KY. 85 mm. **C.** OSUM 68571.1, Illinois River, Morgan Co., IL. 65 mm. **D.** OSUM 69010.2, Scioto River, Pike Co., OH. 115 mm. **E.** OSUM 39536.2, "Wabash River." 28 mm. **F, Hinge detail:** OSUM 17233.2, Ohio River, Campbell Co., KY. 79 mm. **Beak sculpture detail:** OSUM 29210.1, Mississippi River, Goodhue Co., MN. 25 mm total length.

it to be declared a *nomen oblitum* by the 50 Year Rule. The names *obovalis* and *pachostea* may be petitioned for *nomena oblita*, but *A. antrosa* will need to be suppressed to preserve the well-known name *Unio ebenus*.]

Type locality. "Ohio river."

Type material. Johnson (1974: 50) recognized the figured specimen, USNM 85792, as the lectotype.

Synonymy

Unio obliqua Lamarck, 1819. Férussac, 1835: 28 [in part, misidentification].

Obovaria obovalis Rafinesque, 1820: 311 [Type locality: "dans l'Ohio [River] et les rivières voisines" ("in the Ohio River and adjacent rivers"). Lectotype ANSP 20244 by selection of Johnson & Baker (1973: 163). Ortmann & Walker (1922: 4) maintained that this name could not be recognized, although stating that "the so-called (Rafinesque-Poulson) 'type' of *O. obovalis* Raf. has been identified as *U. ebenus* Lea" by Vanatta (1915: 558). This is the earliest name for this species, but could be designated a *nomen oblitum*.].

Obovaria pachostea Rafinesque, 1820: 312 [Type locality: "Dans le Kentuky [River]." Johnson & Baker (1973: 164) designated an uncataloged "lectotype" in the Muséum national d'Histoire naturelle, Paris, but Bogan (1998) stated that that is not a type. He maintained that Johnson & Baker's "lectotype" (1973: pl. 10, fig. 2) is a specimen of *Cyclonaias tuberculata* (Rafinesque, 1820) and not *Obovaria pachostea*. If so, then no type of *O. pachostea* has been located. This name could be designated a *nomen oblitum* and *nomen dubium*.].

Amblema antrosa Rafinesque, 1820: 322 [unnecessary new name for *Obovaria pachostea* Rafinesque, 1820, created when Rafinesque moved it to *Amblema*. However, it was used as a valid name for this taxon within the past 50 years, and pre-dates *Unio ebenus* Lea, 1831, by 11 years.].

Unio ebenus Lea, 1831: 84–85, pl. 9, fig. 14.

Unio obovalis (Rafinesque, 1820). Say, 1834: unpaginated.

Margarita (*Unio*) *ebenus* (Lea, 1831). Lea, 1836: 34.

Unio obliquus Lamarck, 1819. Conrad, 1837: 77, pl. 43, fig. 2 [misidentification].

Unio mytiloides Rafinesque, 1820. Swainson, 1840: 270, figs. 52, 53 [misidentification].

Pleurobema obovalis (Rafinesque, 1820). Agassiz, 1852: 49.

Margaron (*Unio*) *ebenus* (Lea, 1831). Lea, 1852b: 35.

Quadrula (*Fusconaia*) *ebenus* (Lea, 1831). Simpson, 1900a: 793.

Quadrula ebena (Lea, 1831). Sterki, 1907: 392.

Fusconaja [*sic*] *ebena* (Lea, 1831). Ortmann, 1912a: 245.

Unio ebenus var. De Gregorio, 1914: 38, 56 [Apparently two different varieties were described. Type locality for both: "Sciota [Scioto] R. (Ohio)." Types in Palermo Museum?].

Quadrula obovalis (Rafinesque, 1820). Vanatta, 1915: 558.

Fusconaia ebena (Lea, 1831). Baker, 1920: 382.

Fusconaia ebenus (Lea, 1831). Ortmann & Walker, 1922: 4.

Quadrula (*Fusconia*) [*sic*] *antrosa* (Rafinesque, 1820). Frierson, 1927: 55.

Fusconaia antrosa (Rafinesque, 1820). Haas, 1969: 314.

Status. Listed as Endangered by Ohio Division of Wildlife; not listed by U.S. Fish & Wildlife Service as of 2009.

ADULT SHELL

Shell. Medium to fairly large sized (to 120 mm), thick, squarish or oval to nearly circular, quite inflated. Individuals become more elongate with age.

Anterior margin. Short and straight, then broadly rounded to the ventral border.

Ventral margin. Broadly rounded in juveniles, with a slight concavity in some very old adults at the dorsal ridge.

Posterior margin. Gently rounded to the dorsal margin.

Dorsal margin. Fairly long, arched, joining posterior margin with slightly obtuse angle.

Umbo. Narrow, prominent, and projecting forward with age; no umbonal sculpture on any specimens examined.

Ligament. Short and prominent, tan to dark brown, not concealed by shells.

Disc sculpture. Disc of shell smooth except for growth lines; posterior slope with two weak angulations.

Periostracum. Tan, yellowish, or reddish brown as juveniles, often becoming black with age; dark annular rings; juveniles may have vague green stripes, but these are lost in adults.

Nacre. Porcelain white, often iridescent posteriorly; occasionally with gold or brown Tulberg layers.

Hinge plate. Well-developed, massive, sharply angled at the umbo; right cardinal a single, massive, triangular denticle with a minute anterior and well-developed posterior accessory tooth; left cardinal bifid, the teeth often far apart; cardinal teeth rotate from a more or less vertical alignment to a horizontal one with age; laterals well-developed, short, arched, one in right valve, two in left; interdentum flattened, broad, very short.

Muscle scars. Anterior adductor strongly impressed; posterior adductor moderately impressed; anterior adductor distinct from anterior pedal protractor scar, large posterior adductor confluent with or distinct from posterior pedal protractor scar; small cluster of dorsal suspensors located behind hinge-plate angulation. Pallial line entire, located far from shell margin. Accessory line weak, reaching 25%–75% the length of the shell.

Umbonal cavity. Very wide, compressed, and deep.

Fig. 76. ***Fusconaia ebenus*** **(Lea, 1831)**
Solid dots—live or freshdead since 1980
Bullseye—all records, any condition

Glochidia. Semicircular, dorsal margin straight but slightly oblique, lateral margins equal, ventral margin rounded, hooks absent (Lea, 1858c: 48, pl. 5, fig. 21; Ortmann, 1911: 346, pl. 89, fig. 1; Surber, 1913: 116, pl. 30, fig. 11; Howard, 1914a: 18, pl. 5, fig. 30; Utterback, 1915: 111).

Literature records (Lefevre & Curtis, 1910a: 97, fig. I; Lefevre & Curtis, 1912: 146, fig. 1i; Surber, 1912: 10, pl. 2, fig. 33; Ortmann, 1912a: 244–245; Howard, 1914a: 18, pl. 5, fig. 30; Utterback, 1915: 111) indicate glochidial dimensions for length (0.13 to 0.16 mm) and height (0.15 mm). Lea (1863: 78) noted, "Embryonic shell white, subrotund, without hooks, very nearly like *perplexus* [*E. t. rangiana*]." Ortmann (1912a: 244) noted, "The glochidia are similar to those of *F. rubiginosa* [*F. flava*], but slightly higher than long." Surber (1912) gave slightly different dimensions: "Dimensions: Height, 0.148 to 0.15 mm; length, 0.153 to 0.16 mm. The dimensions of this shell are near those of *Q. trigona* [*F. flava*] and *Q. solida* [*P. cordatum*] that one is obliged to rely chiefly on the form to distinguish them."

Reproductive biology. Tachytictic. All four gills are marsupial. Individuals become sexually mature at six years (Payne & Miller, 1989). "Breeding season" reported from May to September (Baker, 1928a). Subcylindrical, bright pink conglutinates are released (Utterback, 1915). None of the reported host studies have been carried to metamorphosis, and all "host" identifications should be regarded with skepticism.

Reported potential hosts

Alosa chrysochloris (Skipjack Herring)—Surber, 1913 (NI); Howard, 1914a (NS); Howard, 1916 (LI); Coker *et al.*, 1921 (NI, LI); Howard & Anson, 1923 (NS).

Micropterus salmoides (Largemouth Bass)—Howard, 1914a (NS).

Pomoxis annularis (White Crappie)—Howard, 1914a (NS).

Pomoxis nigromaculatus (Black Crappie)—Howard, 1914a (NS).

Habitat. Sandy mud and gravel of large rivers.

Range. Mississippi River system from Minnesota to northern Louisiana, portions of the Mobile River drainage.

Range in Ohio. Historically in the Ohio River mainstem, the Scioto River up to Columbus, and the lower Muskingum River.

Similar species. *Fusconaia ebenus* may be easily confused with *Fusconaia subrotunda* and *Obovaria olivaria*. Both *F. ebenus* and *F. subrotunda* have deep, compressed beak cavities whereas *O. olivaria* has a shallow cavity. *Fusconaia subrotunda* and *O. olivaria* are often rayed, whereas *F. ebenus* lacks rays except for a yellow dorsal blotch on juvenile shells. Nevertheless, old, eroded specimens of these species may be difficult to discern.

Comments. Individuals gradually slow in growth during their entire life; there is great variation in growth rates among populations. Individuals 60 mm in length may be between 4 and 17 years old. Specimens may reach ca. 120 mm and over 30 years old. The umbos and teeth rotate with age.

Fusconaia flava (Rafinesque, 1820)

Figures 77, 78

Etymology. L. *flavus*, yellow. Poulson (1832: 38–39) in his translation of Rafinesque's original description noted that the shell is "yellowish brown" and the soft parts ". . . of a deep or orange yellow colour," and further that the "*shell* almost yellow when young." While true, the colors "tan to dark brown" for the adult shell would better fit the species.

Vernacular name(s). *Wabash Pigtoe*, Pigtoe.

Original description

Obliquaria (*Quadrula*) *flava* Rafinesque, 1820: 305, pl. 81, figs. 13, 14.

Type locality. "ne se trouvé que dans les petites rivières se jetant dans le Kentuky, Saltrives, et Greenriver" ["found only in the small rivers that flow into the Kentucky, Salt River, and Green River"]. The type locality becomes "small creeks in Kentucky" because of the designation of the lectotype.

Type material. The type mentioned by Vanatta (1915: 557) was selected as lectotype (ANSP 20230, from small creeks in Kentucky) by Johnson & Baker (1973: 155).

Synonymy

Obliquaria (*Quadrula*) *flava* Rafinesque, 1820: 305, pl. 81, figs. 13, 14.

Obliquaria (*Sintoxia*) *lateralis* Rafinesque, 1820: 310 [Type locality: none given, lectotype from "Kentucky River." Lectotype ANSP 20247a, paralectotypes ANSP 20247 and MNHN (unnumbered) by selection of Johnson & Baker (1973: 160). Vanatta (1915: 557) referred this species to *Unio obliqua* Lamarck, 1819.].

Unio undatus Barnes, 1823: 121–122, fig. 4 [Type locality: "Ouisconsin [*sic*] and Fox Rivers." Types lost].

Unio rubiginosus Lea, 1829: 427, pl. 8, fig. 10 [Type locality: "Ohio." Figured possible lectotype USNM 84556. Possible paralectotype ANSP lost *fide* Johnson & Baker (1973: 169)].

Unio trigonus Lea, 1831: 110, pl. 16, fig. 40 [Type locality: "Ohio River, Cincinnati, [Hamilton Co.] Ohio." Figured lectotype USNM 86429. Paralectotypes in ANSP lost *fide* Johnson & Baker (1973: 173)].

Unio flavus (Rafinesque, 1820). Conrad, 1834: 69.

Fig. 77. ***Fusconaia flava*** **(Rafinesque, 1820). A.** OSUM 10033.3, Ohio River, Hamilton Co., OH. 72 mm. **B.** OSUM 41054.8, Lake Erie, Ottawa Co., OH. 57 mm. **C.** OSUM 10901.9, St. Joseph River, Williams Co., OH. 80 mm. **D.** OSUM 7076.1, Deer Creek, Pickaway Co., OH. 71 mm. **E.** OSUM 10642.13, Little Scioto River, Marion Co., OH. 45 mm. **F.** OSUM 49363, Muskingum River, Muskingum Co., OH. 75 mm. **G.** OSUM 15943, Stony Creek, Ionia Co., MI. 89 mm. **H, Hinge detail:** OSUM 19661.11, East Fork Little Miami River, Clermont Co., OH. 54 mm. **Beak sculpture detail:** OSUM 19661.1, East Fork Little Miami River, Clermont Co., OH. 13 mm total length.

Unio undatus (Barnes, 1823). Conrad, 1834: 72.
Margarita (*Unio*) *trigonus* (Lea, 1831). Lea, 1836: 18.
Margarita (*Unio*) *rubiginosus* (Lea, 1829). Lea, 1836: 20.
Unio rubiginosa (Lea, 1829). Deshayes, 1839: 672.
Pleurobema undatus (Barnes, 1823). Agassiz, 1852: 49.
Pleurobema flavus (Rafinesque, 1820). Agassiz, 1852: 49.
Pleurobema rubiginosus (Lea, 1829). Agassiz, 1852: 49.
Margaron (*Unio*) *rubiginosus* (Lea, 1829). Lea, 1852b: 24.
Margaron (*Unio*) *trigonus* (Lea, 1831). Lea, 1852b: 25.
Unio triangularis Say, 1834. Küster, 1852: 56, pl. 15, fig. 5 [misidentification].
Unio pilaris Lea, 1840. Reeve, 1865: pl. 27, fig. 138 [misidentification].
Unio trigonus var. *rubiginosus* (Lea, 1829). Paetel, 1890: 152.
Quadrula trigona (Lea, 1831). Baker, 1898b: 76, pl. 15, fig. 5.
Quadrula rubiginosa (Lea, 1829). Baker, 1898b: 77, pl. 19, fig. 2, pl. 20, fig. 1.
Quadrula (*Fusconaia*) *rubiginosa* (Lea, 1829). Simpson, 1900a: 786.
Quadrula (*Fusconaia*) *trigona* (Lea, 1831). Simpson, 1900a: 787–788.
Quadrula (Fusconaia) obliqua (Lamarck, 1819). Simpson, 1900a: 788 [in part, misidentification].
Fusconaja [*sic*] *undata* (Barnes, 1823). Ortmann, 1912a: 241.
Fusconaja [*sic*] *rubiginosa* (Barnes, 1823). Ortmann, 1912a: 241.
Quadrula (*Fusconaia*) *undata* (Barnes, 1823). Simpson, 1914: 880.
Unio validus var. *continuus* De Gregorio, 1914: 51 [Type locality: "Sciota R." [Scioto River, OH]. Type in Palermo Museum?].
Fusconaia selecta Wheeler, 1914: 76, pl. 4 [Type locality: "Cache River, Arkansas." Type not located].
Fusconaia undata trigonoides Frierson *in* Utterback, 1915: 107, pl. 15, figs. 30a–d, pl. 4, figs. 9a–b [Type locality: "Platte River, Dixon Falls and Garretsburg, Osage River, Schell City and Warsaw, Missouri." Type not located].
Fusconaia flava var. *parvula* Grier, 1918: 11 [Type locality: "Big Bend, Presque Isle Bay, Lake Erie." Holotype CMNH Museum 614513, paratypes ANSP 103017].
Fusconaia flava (Rafinesque, 1820). Ortmann, 1919: 14.
Fusconaia rubiginosa parvula Grier, 1918. Baker, 1922: 19.
Fusconaia undata (Barnes, 1823). Ortmann & Walker, 1922: 6.
Fusconaia flava parvula Grier, 1918. Baker, 1924: 133.
Fusconaia undata trigona (Lea, 1831). Grier & Mueller, 1926: 16, 32.
Quadrula (*Fusconia*) [*sic*] *undata* (Barnes, 1823). Frierson, 1927: 54.
Quadrula (*Fusconia*) [*sic*] *undata trigona* (Lea, 1831). Frierson, 1927: 54.
Quadrula (*Fusconia*) [*sic*] *undata selecta* (Wheeler, 1914). Frierson, 1927: 55.
Quadrula (*Fusconia*) [*sic*] *undata trigonoides* (Frierson *in* Utterback, 1915). Frierson, 1927: 55.
Quadrula (*Fusconia*) [*sic*] *flava* (Rafinesque, 1820). Frierson, 1927: 55.
Fusconaia undata wagneri Baker, 1928a: 64–66, pl. 40, figs. 1–3 [Type locality: "Lake Pepin, near Lake City, Minn." Holotype UMMZ 209194, paratypes UMMZ 209193, IUMNH Z14016-Z14018].
Fusconaia undata undata (Rafinesque, 1820). Haas, 1969: 311.
Fusconaia undata selecta (Wheeler, 1914). Haas, 1969: 312.
Fusconaia undata trigonoides (Frierson, 1915). Haas, 1969: 312.
Fusconaia flava flava (Rafinesque, 1820). Haas, 1969: 313.
Fusconaia flava form *undata trigona* (Lea, 1831). Strayer & Jirka, 1997: pl. 25, figs. 2, 2a, 2b.

Status. Not listed by either Ohio Division of Wildlife or U.S. Fish & Wildlife Service as of 2009.

ADULT SHELL

Shell. Medium size (to 110 mm), rather thick, squarish or triangular, quite inflated in big river and lake populations, compressed in creek forms.

Anterior margin. Short and straight, then broadly rounded to the ventral border.

Ventral margin. Broadly rounded in juveniles, with a slight concavity at the dorsal ridge.

Posterior margin. Gently rounded or straight to the dorsal margin.

Dorsal margin. Rather short, arched or straight, joining posterior margin with slightly obtuse angle.

Umbo. Narrow, low in creek populations, prominent in river and lake forms; umbos face each other; umbonal sculpture of 2–3 minute angulate ridges, usually eroded away.

Ligament. Short and prominent, tan to dark brown, not concealed by shells.

Disc sculpture. Disc of shell smooth except for growth lines; sulcus present; posterior slope with 2–3 weak angulations.

Periostracum. Tan, yellowish, or greenish, becoming darker with age; juveniles may have vague green stripes but these may be lost in adults.

Nacre. Porcelain white, often iridescent posteriorly; occasionally with gold or brown Tulberg layers.

Hinge plate. Well-developed, angled at the umbo; right cardinal a single, massive, triangular denticle, rarely with accessory teeth; left cardinal bifid, the teeth often far apart; laterals well-developed, short, arched or straight, one

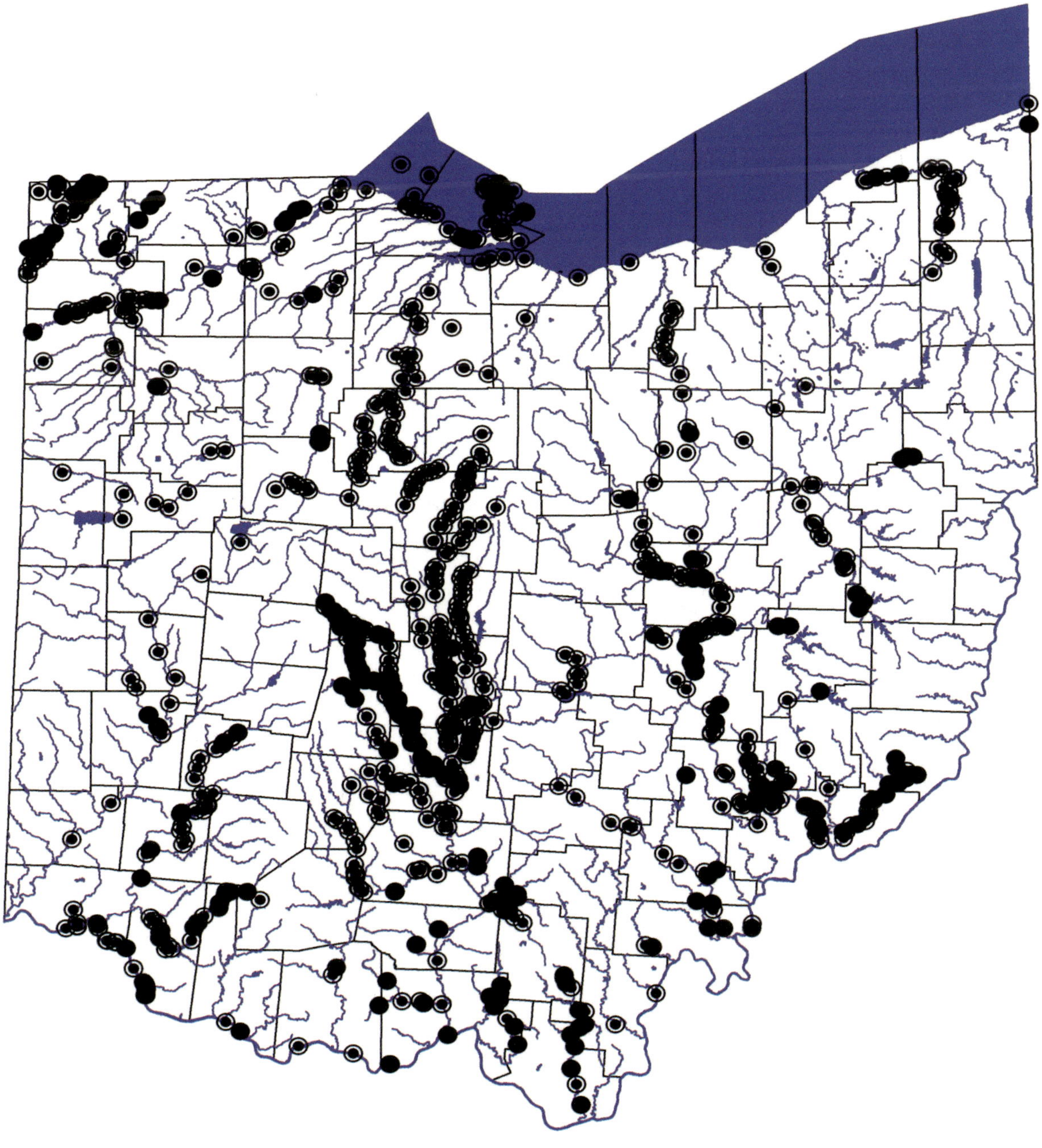

Fig. 78. ***Fusconaia flava*** **(Rafinesque, 1820)**
Solid dots—live or freshdead since 1980
Bullseye—all records, any condition

in right valve, two in left; interdentum flattened, broad, short.
Muscle scars. Anterior adductor strongly impressed; posterior adductor moderately impressed; anterior adductor distinct from anterior pedal protractor scar, large posterior adductor distinct from posterior pedal protractor scar; dorsal suspensors located behind cardinal teeth and interdentum in an irregular line. Pallial line entire, located far from shell margin. Accessory line weak, reaching 25%–75% the length of the shell.
Umbonal cavity. Very wide, rather compressed, and deep.

Glochidia. Semicircular, dorsal margin straight, lateral margins not inflated, more or less straight, ventral margin rounded, hooks absent (Ortmann, 1911: 346, pl. 89, fig. 2; Howard, 1914a: 23–24, pl. 5, fig. 33; Utterback, 1915: 104).

Literature records (Ortmann, 1912a: 243, pl. 2, fig. 36; Surber, 1912: 10, pl. 2, fig. 36; Howard, 1914a: 23–24, pl. 5, fig. 33; Utterback, 1915: 104) indicate glochidial dimensions for length (0.14 to 0.16 mm) and height (0.14 to 0.16 mm). Howard (1914a: 23) noted, "The glochidial shell of this species seems thin and more delicate than in *Q. ebena*. The size and straighter lines of the anterior and posterior edges of the shell should distinguish this glochidium from the similar glochidia of *Q. ebena* and *Q. solida* [*P. cordatum*]."

Reproductive biology. Tachytictic. All four gills are marsupial. Eggs are found in the marsupia beginning in May, and glochidia develop from June to August (Ortmann, 1919; Heath *et al.*, 1998). Individuals from Little Darby Creek released orange-red, vermiform conglutinates in July. Conglutinates from specimens from the Maumee River drainage were bright pink in some individuals, white in others. Specimens collected from Killbuck Creek in May contained primarily bright-pink conglutinates consisting mostly of undeveloped glochidia (Watters, unpubl.). These are of the type composite conglutinate A (Watters, 2008). The color and shape of these conglutinates were due to unfertilized eggs (Barnhart, 1997). Metamorphosis has been reported only for White Crappie, Striped Shiner, and Creek Chub. As with many amblemines, females abort conglutinates (mature or not) with little provocation.

Reported potential hosts

Lepomis macrochirus (Bluegill)—Howard, 1914a (LI); Clarke, 1981b (NS).
Notropis photogenis (Silver Shiner)—Watters & O'Dee, 1997e (LT); O'Dee & Watters, 2000 (LT).
Pomoxis annularis (White Crappie)—Howard, 1914a (LT); Wilson, 1916 (NI); Coker *et al.*, 1921 (NI).
Pomoxis nigromaculatus (Black Crappie)—Surber, 1913 (NI); Wilson, 1916 (NI); Coker *et al.*, 1921 (NI).
Semotilus atromaculatus (Creek Chub)—Watters & O'Dee, 1997e (LT); O'Dee & Watters, 2000 (LT).

Habitat. A widespread species of nearly all substrates that will support it, in fast water or lakes.

Range. Mississippi River system from Minnesota to Louisiana. Red River of Manitoba. Records of this species from the Tombigbee River in Alabama by Parmalee & Bogan (1988) probably represent *F. cerina* (Conrad, 1838).

Range in Ohio. Very widespread in both the Ohio and Lawrentian drainages and the western basin of Lake Erie.

Similar species. *Fusconaia flava* and *Pleurobema sintoxia* may be very similar, particularly headwater specimens. Generally *F. flava* is more sulcate and more trigonal in outline, whereas *P. sintoxia* lacks a sulcus and is rounded in profile. The large river form of *F. flava* resembles members of the *Pleurobema cordatum*-complex but typically is yellowish or tan rather than very dark brown or black. The beaks prominently face each other, rather than forward as in most *P. cordatum*-complex species. *Fusconaia ebenus* and *F. subrotunda* lack a sulcus, which is so prominent in sympatric *F. flava*.

Comments. *Fusconaia flava* var. *parvula* Grier, 1918, is based on stunted Lake Erie specimens.

Individuals gradually slow in growth during their entire life; there is great variation in growth rates among populations. Individuals 60 mm in length may be between 4 and 12 years old. Specimens may reach ca. 110 mm and over 25 years old. Specimens from the Great Lakes and large rivers are usually heavy and trigonal, with prominent umbos. Creek specimens are much more compressed, with low, wide umbos.

Fusconaia subrotunda subrotunda (Lea, 1831)

Figures 79, 80

Etymology. L. *sub-*, somewhat, less than + L. *rotundus*, circular, round.
Vernacular name(s). *Longsolid*, Spotted Pigtoe.
Original description
Unio subrotunda Lea, 1831: 117, pl. 18, fig. 45.
Type locality. "Ohio."
Type material. Possible lectotype USNM 85832 *fide* Johnson, 1974.
Synonymy
? *Obliquaria* (*Rotundaria*) *subrotunda* Var. 1. *Maculata*,

Fig. 79. ***Fusconaia subrotunda subrotunda*** **(Lea, 1831). A.** OSUM 18064.3, Walhonding River, Coshocton Co., OH. 105 mm. **B.** OSUM 49606.8, Muskingum River, Washington Co., OH. 90 mm. **C.** OSUM 17859.1, Mohican River, Coshocton Co., OH. 78 mm. **D.** OSUM 57382.4, Ohio River, Adams Co., OH. 64 mm. **E.** OSUM 61959.2, Mahoning River, Trumbull Co., OH. 23 mm. **F, Hinge detail:** OSUM 49606.1, Muskingum River, Washington Co., OH. 74 mm. **Beak sculpture detail:** OSUM 61959.1, Mahoning River, Trumbull Co., OH. 18 mm total length.

Rafinesque, 1820: 308 [Variety not figured, type apparently lost].
Unio personatus Say, 1829. Conrad, 1834: 71 [misidentification].
Unio subrotunda Lea, 1831: 117, pl. 18, fig. 45.
Unio kirtlandianus Lea, 1834: 98, pl. 14, fig. 41 [Type locality: "Mahoning River, Ohio." Possible lectotype USNM 85815 *fide* Johnson, 1974].
Unio politus Say, 1834: no pagination [*nomen nudum?*]
Margarita (*Unio*) *subrotundus* (Lea, 1831). Lea, 1836: 34.
Margarita (*Unio*) *kirtlandianus* (Lea, 1834). Lea, 1836: 34.
Unio kirklandianus [*sic*] Lea, 1834. Hanley, 1842: 203.
Pleurobema politus Agassiz, 1852: 49 [of Say, 1834?].
Margaron (*Unio*) *subrotundus* (Lea, 1831). Lea, 1852b: 35.
Margaron (*Unio*) *kirtlandianus* (Lea, 1834). Lea, 1852b: 35.
Quadrula (*Fusconaia*) *kirtlandiana* (Lea, 1834). Simpson, 1900a: 791.
Quadrula (*Fusconaia*) *subrotunda* (Lea, 1831). Simpson, 1900a: 791.
Quadrula subrotunda (Lea, 1831). Sterki, 1907: 391.
Quadrula kirtlandianus (Lea, 1834). Sterki, 1907: 391.
Fusconaja [*sic*] *kirtlandiana* (Lea, 1831). Ortmann, 1912a: 245.
Fusconaja [*sic*] *subrotunda* (Lea, 1831). Ortmann, 1912a: 294.
Fusconaja [*sic*] *subrotunda* var. *leucogona* Ortmann, 1913b: 89–90 [Type locality: "Elk River, [Branton Co.] West Virginia." Lectotype Carnegie Museum 61.5239].
Quadrula (*Fusconaia*) *kirtlandiana* (Lea, 1834). Simpson, 1914: 891.
Fusconaia subrotunda (Lea, 1831). Ortmann, 1919: 7, pl. 1, fig. 2.
Fusconaia subrotunda kirtlandiana (Lea, 1834). Ortmann, 1919: 11–14, pl. 1, figs. 3–5.
Obliquaria sintoxia (Rafinesque, 1820). Ortmann & Walker, 1922: 4 [misidentification].
Unio sintoxia Rafinesque, 1820. Frierson, 1927: 56 [misidentification].
Fusconaia kirtlandiana (Lea, 1834). Baker, 1928a: 136.

Status. Listed as Endangered by Ohio Division of Wildlife (as *Fusconaia maculata maculata*); not listed by U.S. Fish & Wildlife Service as of 2009.

ADULT SHELL

Shell. Medium size (to 110 mm), thick, squarish or oval, quite inflated to quite compressed depending on population. Individuals become more elongate with age.
Anterior margin. Straight, then broadly rounded to the ventral border.
Ventral margin. Broadly rounded in juveniles, with a slight concavity in some very old adults at the dorsal ridge.
Posterior margin. Gently rounded to the dorsal margin.
Dorsal margin. Fairly long, arched, joining posterior margin with slightly obtuse angle.
Umbo. Wide, prominent, and projecting somewhat anteriorly; umbonal sculpture of ca. 2 minute angulate ridges, nearly always eroded away.
Ligament. Rather long and prominent, tan to dark brown, not concealed by shells.
Disc sculpture. Disc of shell smooth except for growth lines; posterior slope with 2–3 weak angulations.
Periostracum. Tan, yellowish, or reddish brown as adults; dark annular rings; juveniles usually have a pattern of bold green rays; these are usually lost in the fourth year.
Nacre. Porcelain white, often iridescent posteriorly; occasionally with gold or brown Tulberg layers.
Hinge plate. Well-developed, angled at the umbo; right cardinal a single, massive, triangular denticle with a minute anterior and well-developed posterior accessory tooth; left cardinal bifid, the teeth often far apart; laterals well-developed, short, straight, one in right valve, two in left; interdentum flattened, broad, long to short.
Muscle scars. Anterior adductor strongly impressed; posterior adductor moderately impressed; anterior adductor distinct from anterior pedal protractor scar, large posterior adductor distinct from posterior pedal retractor scar; dorsal suspensors located behind cardinal teeth and interdentum in an irregular line. Pallial line entire, located very far from shell margin. Accessory line weak, reaching 25%–75% the length of the shell.
Umbonal cavity. Very wide, fairly compressed, and deep.

Glochidia. Unknown.

Reproductive biology. Tachytictic. All four gills are marsupial (Ortmann, 1912a, 1919), rarely only two (Frierson, 1927). Gravid females occur in June through August, and eggs may be present at the same time (Ortmann, 1919). Glochidia are released in red or pinkish, cylindrical conglutinates of the composite conglutinate A type (Watters, 2008).

Reported potential hosts. No hosts known.

Habitat. Sandy mud and gravel of large rivers.

Range. Ohio River system, including the Green River and Big Sandy River of Kentucky, White and Wabash rivers of Indiana/Illinois, Kanawha River of West Virginia, Monongahela River and French Creek of Pennsylvania. It does not occur west of the Mississippi River or north into Michigan, Minnesota, or Wisconsin. In the Cumberland and upper Tennessee rivers as *F. s. leseuriana* (Lea, 1840).

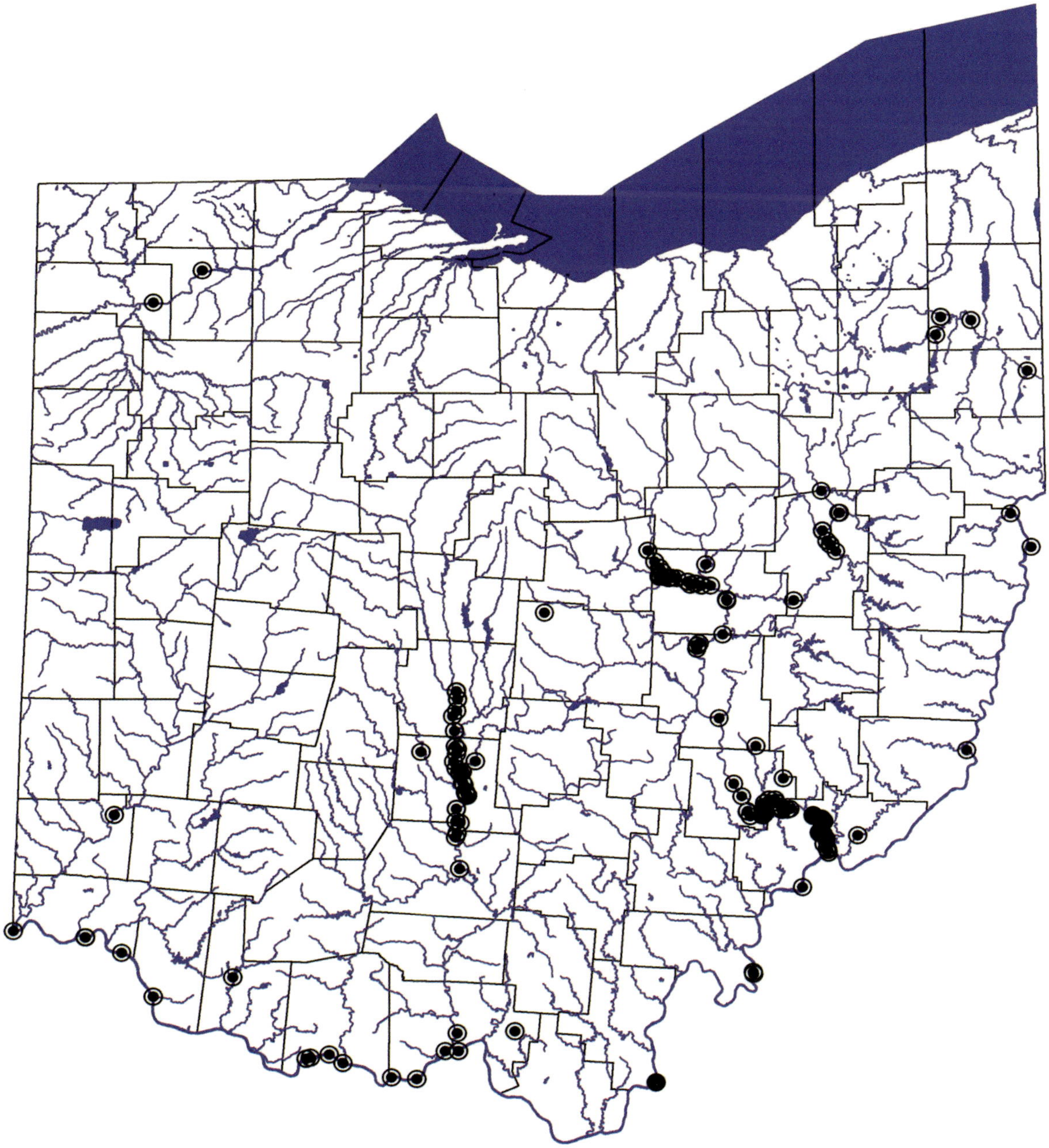

Fig. 80. ***Fusconaia subrotunda subrotunda* (Lea, 1831)**
Solid dots— live or freshdead since 1980
Bullseye—all records, any condition

Range in Ohio. Once common in the Scioto River up to Columbus, it apparently has been extirpated from there. A single record from the Great Miami River. Rare in the Muskingum River, including the Tuscarawas and Walhonding rivers. Unknown from the Lawrentian system until Watters (1988; unpubl.) found it at two sites in the lower St. Joseph River of the Maumee River system in Indiana. It is possible that these shells were brought in by commercial shellers, but the distance between the two collections and the lack of clammers' camps make this a difficult solution. Both specimens were subfossil, and it is more likely that this species once lived in the Maumee system (a carryover from the Wabash River) but is now extirpated from there. Historic records exist for the Mahoning River (CMNH).

Similar species. See under *Fusconaia ebenus* and *F. flava*.

Comments. A peculiar form lives in French Creek in Pennsylvania and Mahoning River in Ohio that is polished and boldly patterned.

Individuals grow quickly for the first 3–4 years and then gradually slow in growth; there is some variation in growth rates among populations. Individuals 55 mm in length may be between 4 and 9 years old. Specimens may reach ca. 110 mm and live to be over 30 years old.

Hemistena Rafinesque, 1820

Etymology. Gr. *hemi-*, half + Gr. *stenos*, narrow. We are uncertain as to what Rafinesque intended to communicate with the name *Hemistena*. We do recognize that this species is elongate and narrow and is about half the height of the other species (*P. ohiensis*) to which the author compared this species.

Original description

Anodonta (*Hemistena*) Rafinesque, 1820: 317 [*Hemistena* Rafinesque, 1820, was proposed as a subgenus of the genus *Anodonta* for species having lamellar teeth, scarcely apparent, along the anterior margin and ligamental material on both sides of the umbo. Rafinesque (1820: 317) stated of *Anodonta lata* that "it must perhaps form another sub-genus *Hemistena*, or be reunited with the first sub-genus (*Lastena*)" [Poulson (1832) translation]. This less than enthusiastic endorsement forms the basis of the genus *Hemistena*.].

Type species. *Anodonta* (*Hemistena*) *lata* Rafinesque, 1820, by monotypy. Herrmannsen's (1849: 577) subsequent designation of *Anodonta ohiensis* Rafinesque, 1820, as type was invalid.

Synonymy

Anodonta (*Hemistena*) Rafinesque, 1820: 317.

Hemilastena [*sic*] Agassiz, 1852: 50.

Sayunio De Gregorio, 1914: 62 [Type species: *Unio* (*Sayunio*) *dehiscens* var. *oriensopsis* De Gregorio, 1914, by monotypy].

Geographic range. The range of *Hemistena* is restricted to the Ohio River system, including Tennessee and Cumberland rivers.

Hemistena lata (Rafinesque, 1820)

Figures 81, 82

Etymology. L. *latus*, broad, wide, wide-spread, extended. Rafinesque's length was measured from the beaks of the umbo across the specimen to the ventral margin opposite. The term *lata* was used to describe the "width" from the anterior to the posterior extremes.

Vernacular name(s). *Cracking Pearlymussel*, Painted Papershell, Fragile Painted Shell.

Original description

Anodonta (*Hemistena*) *lata* Rafinesque, 1820: 317, species no. 59, pl. 82, figs. 17, 18.

Type locality. "dans le Kentuky [River], etc."

Type material. The type mentioned by Vanatta (1915: 554) was selected as lectotype (ANSP 20227 from the Kentucky River) by Johnson & Baker (1973: 160).

Synonymy

Anodonta (*Hemistena*) *lata* Rafinesque, 1820: 317, pl. 82, figs. 17, 18.

Unio dehiscens Say, 1829: 308 [Type locality: "Wabash [River]." Types lost].

Unio oriens Lea, 1831: 68, 73, pl. 6, fig. 5 [Type locality: "Ohio River." Figured lectotype USNM 86271, paralectotype ANSP 41210].

Odatelia radiata Rafinesque, 1832: 154.

Unio latus (Rafinesque, 1820). Conrad, 1834: 70.

Anodonta latus Rafinesque, 1820. Férussac, 1835: 25.

Margarita (*Unio*) *dehiscens* (Say, 1829). Lea, 1836: 35.

Unio hildrethi Delessert, 1841: pl. 19, figs. 4a, b [Type in Geneva Museum?].

Hemilastena dehiscens (Say, 1829). Agassiz, 1852: 50.

Hemilastena oriens (Lea, 1831). Agassiz, 1852: 50.

Hemilastena lata (Rafinesque, 1820). Agassiz, 1852: 50.

Leptodea lata (Rafinesque, 1820). Conrad, 1853: 262.

Margaron (*Unio*) *dehiscens* (Say, 1829). Lea, 1852b: 43.

Baphia dehiscens (Say, 1829). Adams & Adams, 1857: 499.

Anodon lata (Rafinesque, 1820). Sowerby, 1867: pl. 19, fig. 76.

Margaritana dehiscens (Say, 1829). Lewis, 1870: 219.

Anodonta dehiscens (Say, 1829). Paetel, 1890: 178.

Lastena lata (Rafinesque, 1820). Simpson, 1900a: 654.

Unio (*Sayunio*) *dehiscens* var. *oriensopsis* De Gregorio, 1914: 62, pl. 7, figs. 2a, b, pl. 10, figs. 4a–b [Type locality:

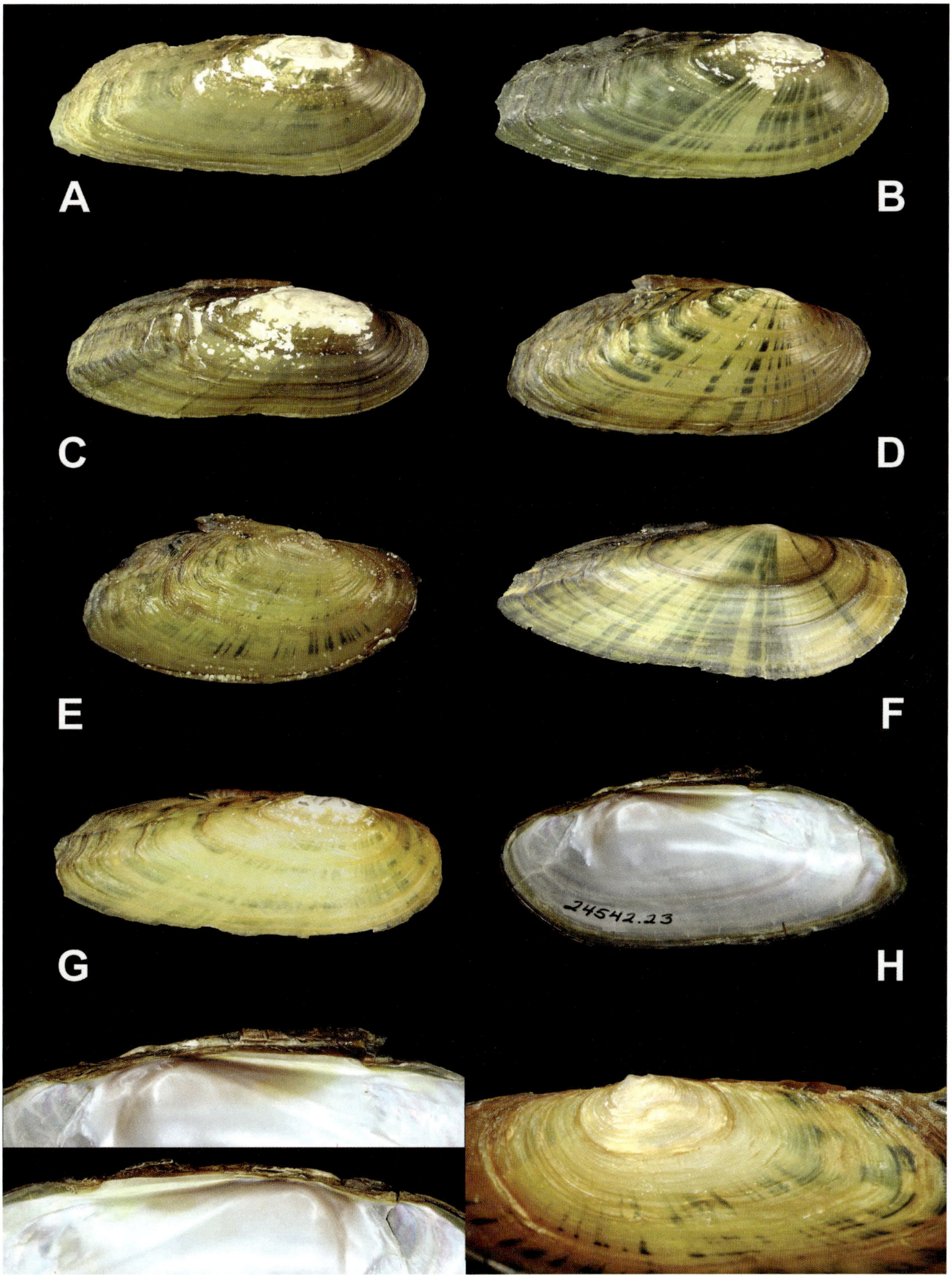

Fig. 81. ***Hemistena lata*** **(Rafinesque, 1820). A.** OSUM 57276.1, Ohio River, Hamilton Co., OH. 53 mm. **B.** OSUM 44226.1, Ohio River, Hamilton Co., OH. 51 mm. **C.** OSUM 44226.2, Ohio River, Hamilton Co., OH. 65 mm. **D, H, Hinge detail:** OSUM 24542.23, Clinch River, Hancock Co., TN. 68 mm. **E, Beak sculpture detail:** OSUM 24542.2, Clinch River, Hancock Co., TN. 32 mm. **F.** OSUM 39072.2, "Wabash River." 80 mm. **G.** OSUM 58534, "Ohio River." 53 mm.

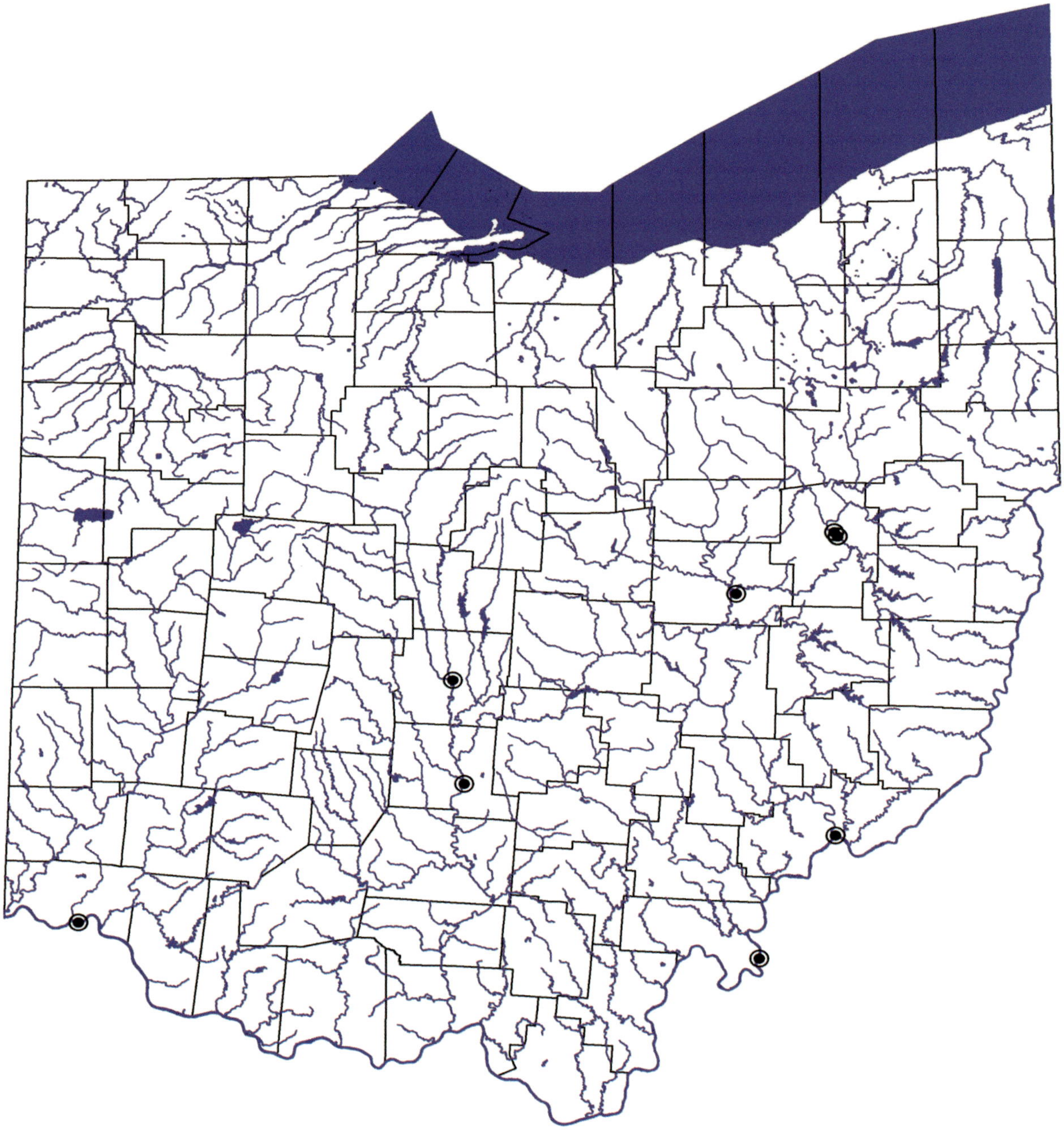

Fig. 82. ***Hemistena lata*** **(Rafinesque, 1820)**
Solid dots—live or freshdead since 1980
Bullseye—all records, any condition

"Sciota river" [Scioto River, OH]. Type in Palermo Museum?].

Hemistena lata (Rafinesque, 1820). Frierson, 1927: 66.

Unio dehiscens orienopsis [*sic*] De Gregorio, 1914. Parmalee & Bogan, 1999: 122.

Status. Listed as Extirpated by Ohio Division of Wildlife; listed as Endangered by U.S. Fish & Wildlife Service as of 2009. A Federal recovery plan has been completed (USFWS, 1990b).

ADULT SHELL

Shell. Medium size (to ca. 95 mm), thin, lanceolate, not inflated.

Anterior margin. Long and straight before the umbo, broadly rounded to the ventral border.

Ventral margin. Very broadly rounded, sinuate, or slightly concave to posterior margin.

Posterior margin. Forming an acute point near the ventral margin, then obliquely ascending to the dorsal margin. The margin may be rendered undulating by the presence of low projecting ribs on the dorsal slope.

Dorsal margin. Long, straight in males, forming obtuse angle with posterior margin.

Umbo. Very wide and low; umbonal sculpture of 2–3 heavy, longitudinal ribs; these ribs may be somewhat undulating.

Ligament. Long, tan to dark brown, not concealed by shells.

Disc sculpture. Sculptureless except for 2–3 weak ribs on dorsal slope.

Periostracum. Waxy yellow or tan, becoming dark brown with age; usually patterned with bold, broken green rays; these may be obliterated on old specimens.

Nacre. White flushed with salmon, purple, or pink in the umbonal cavity, often iridescent posteriorly; commonly with gold Tulberg layers.

Hinge plate. Extremely fine; teeth usually in all but youngest individuals; cardinals at best a single, lamellate, triangular denticle, one in each shell; laterals absent at all ages; interdentum very long and narrow.

Muscle scars. Anterior adductor impressed; posterior adductor less so; anterior adductor distinct from large anterior pedal adductor, large posterior adductor confluent with the posterior pedal retractor scar; two small dorsal suspensors, located in the umbonal cavity. Pallial line entire, located far from shell margin. Accessory line short.

Umbonal cavity. Very wide and moderately deep and open.

Glochidia. Unknown.

Reproductive biology. Presumably tachytictic. The outer gills are marsupial (Ortmann, 1912a). Gravid females were found in May (Ortmann, 1915).

Reported potential hosts

Campostomum anomalum (Central Stoneroller)—Jones & Neves, 1999 (LT); Jones & Neves, 2000 (LT); Jones *et al.*, 2003 (LT).

Cottus carolinae (Banded Sculpin)—Jones & Neves, 1999 (LT); Jones & Neves, 2000 (LT); Jones *et al.*, 2003 (LT).

Cyprinella galactura (Whitetail Shiner)—Jones & Neves, 1999 (LT); Jones & Neves, 2000 (LT); Jones *et al.*, 2003 (LT).

Erimystax dissimilis (Streamline Chub)—Jones & Neves, 1999 (LT); Jones & Neves, 2000 (LT); Jones *et al.*, 2003 (LT).

Etheostoma flabellare (Fantail Darter)—Jones *et al.*, 2003 (LT).

Habitat. Sandy mud, cobble, and gravel of rivers. This species has the ability to pull itself several inches beneath the substrate surface when disturbed.

Range. Ohio River system, including the Green River of Kentucky, Wabash River of Indiana, east to the Allegheny River in Pennsylvania (but extirpated from most of these systems), and the upper Tennessee rivers: Duck, Clinch, Holston, and Elk rivers.

Range in Ohio. Sterki and Ortmann collected this species at the turn of the 20th century in the Tuscarawas River at New Philadelphia and Coshocton, where it was not rare (records from CMNH). It is also known from the Scioto River at Circleville and Columbus, and the Ohio River at Cincinnati and Marietta. Ortmann reported a single specimen from Buffington Island in the Ohio River adjacent to Meigs County in 1911 (CMNH 61.5388)—this apparently was the last record of this species in Ohio.

Similar species. *Hemistena lata* somewhat resembles *Elliptio dilatata* and *Ptychobranchus fasciolaris*. *Elliptio dilatata* never has the broken rayed pattern seen in *H. lata* and has a much thicker hinge and a coarse beak sculpture. *Ptychobranchus fasciolaris* has a color pattern very similar to that of *H. lata*, but internally the shells are very different—*P. fasciolaris* has a massive hinge and never has a purple nacre. Other toothless species lack the color pattern of *H. lata*.

Comments. Individuals grow quickly for 2 years, but gradually slow. Sexual dimorphism is evident by the third year. They grow to ca. 95 mm and typically live for less than 10 years.

Lampsilis Rafinesque, 1820

Etymology. Gr. *elasma*, metal beaten out, metal plate + Gr. *kampsis*, a bending, a curving, from Gr. *kampto*, to bend, bow, curve, turn. Rafinesque (1820: 298) described the lateral tooth of *Lampsilis* as "curved, flexuous" and then goes on to explain, "The name is modified from *Lasmacampsilis*, which means flexuolate blade [= lateral tooth] after the essential character of the genus." He apparently shortened the name to *Lampsilis* for the sake of euphony. The typically smooth shining exterior of *Lampsilis* shells would make the Greek root "*Lampas*, torch, light lamp or *lampein*, to shine," a good name source, but it was apparently not his intent.

Original description

Lampsilis Rafinesque, 1820: 298.

Type species. *Unio ovatus* Say, 1817, by subsequent designation of Herrmannsen (1847: 575).

Synonymy

Lampsilis Rafinesque, 1820: 298.

Aeglia Swainson, 1840: 265, 378 [Type species: *Unio ovatus* Say, 1817, by subsequent designation of Herrmannsen (1846)].

Geographic range. Upper Mackenzie River system in Northwest Territories across Saskatchewan, Alberta, Manitoba, and Ontario in Canada. In the United States generally east of the Rocky Mountains. Eastern Seaboard from Nova Scotia to Georgia, Gulf drainages from Florida Panhandle to Texas. Absent from peninsular Florida. Doubtful into Central America and the Greater Antilles.

Lampsilis abrupta (Say, 1831)

Figures 83, 84, 85

Etymology. L. *abruptus*, broken off from, separated, inaccessible, precipitous, steep; from L. *abrumpo*, to break off. Say (1831a: not paginated) described this species as "Subquadrate, truncate before . . . anterior margin truncated by an almost straight line. . . ." This is an apt description of a female *L. abrupta* if one remembers that Thomas Say had the anterior and posterior ends of the shell reversed.

Vernacular name(s). *Pink Mucket*, Ohio Mucket, Square Mucket.

Original description

Unio abruptus Say, 1831b: no pagination, pl. 17.

Type locality. "It occurs frequently in the Wabash [River]." Parmalee & Bogan's type locality of "Muskingum River, Ohio" is incorrect (1998: 123)—that is the type locality for *Unio orbiculata* Hildreth, 1828, a synonym of *Unio abruptus* Say, 1831.

Type material. The original unionid type material of Say has apparently been lost. Johnson & Baker (1973: 146) stated, "While the Academy is the logical place to turn for the unionid types of Say, there are none there, ..." This species was long referred to in the unionoid literature as *Lampsilis orbiculata* (Hildreth, 1828). An examination of extensive unionoid material from the Muskingum River, Hildreth's original description, especially the dimensions of the holotype, and the Hildreth Collection from Marietta College, Ohio, has revealed that *U. orbiculatus* Hildreth, 1828, is a junior synonym of *Obovaria subrotunda* (Rafinesque, 1820). The earliest name of this species then is *Unio abruptus* Say, 1831.

Frierson (1927: 89–90) commented on this misidentification and claimed it was perpetrated by Lea's "mischievous lead," a direct result of Lea's animosity toward other workers, particularly Rafinesque and Say. Although Lea was aware that Hildreth's name did not apply to *Unio abruptus*, he intentionally misidentified it as that species to save his own name of *Unio circulus* Lea, 1829—a name that would have become a junior synonym if Hildreth's species were correctly identified. As a bonus, Hildreth's misapplied name would have priority over Say's *U. abruptus*. Ironically, neither *U. orbiculatus* nor *U. circulus* has priority for the species now known as *Obovaria subrotunda* (Rafinesque, 1820).

Synonymy

Unio abruptus Say, 1831b: no pagination, pl. 17.

Toxolasma cyclips Rafinesque, 1831: 2 [Type locality: "River Ohio and Mississippi." Type lost *fide* Johnson, 1980. Two varieties also were described: *fuscata* and *lutescens*. Rafinesque remarked, "It is said that this is the *U. abruptus* of Say. I cannot see any thing abrupt in it."].

Unio cyclips (Rafinesque, 1831). Férussac, 1835: 28.

Unio crassus Conrad, 1836: 34, pl. 16 [*non* Retzius, 1778, *nec* Say, 1817, *nec* Nilsson, 1822, *nec* Barnes, 1823. Type locality: "rivers of Ohio, Indiana and Illinois." Type lost *fide* Johnson (1980)].

? *Unio orbicalatus* [*sic*] "Hildreth" De Gregorio, 1914: 70 [from Scioto River, OH].

Lampsilis (*Ortmanniana*) *abrupta* (Say, 1831). Frierson, 1927: 80.

Obovaria retusa (Lamarck, 1819). Frierson, 1927: 89 [in part].

Status. Listed as Endangered (as *Lampsilis orbiculata*) by both Ohio Division of Wildlife and U.S. Fish & Wildlife Service as of 2009. A Federal recovery plan has been completed (USFWS, 1985b).

ADULT SHELL

Shell. Medium size (to 110 mm), thick and very heavy, oval to round in profile, quite inflated.

Anterior margin. Broadly rounded to the ventral border.

Ventral margin. Broadly rounded to posterior margin.

Fig. 83. ***Lampsilis abrupta*** **(Say, 1831).** **A.** OSUM 69472.3, Ohio River, Hamilton Co., OH. 71 mm. **B.** OSUM 44803, Kanawha River, Fayette Co., WV. 103 mm. **C.** OSUM 50728.1, Tennessee River, Hardin Co., TN. 66 mm. **D.** OSUM 68955, Ohio River, Hamilton Co., OH. 68 mm. **E.** OSUM 43980, Muskingum River, Morgan Co., OH. 50 mm. **F, H, Hinge detail:** OSUM 68954, Ohio River, Washington Co., OH. 99 mm. **G.** OSUM 39912, "Ohio." 77 mm. **Beak sculpture detail:** OSUM 75123, Meramec River, St. Louis Co., MO. 56 mm total length.

Posterior margin. In males, forming a broad, rounded point ca. 50% of the way to the dorsal margin; in females abruptly forming obtuse angle at postero-ventral margin, then continuing for a short distance to the broadly rounded posterior point.
Dorsal margin. Rounded or straight in males, forming obtuse angle with posterior margin.
Umbo. Very wide and inflated, but low; umbonal sculpture eroded on all material examined.
Ligament. Long, tan to dark brown, not concealed by shells.
Disc sculpture. Sculptureless except for 2 weak ribs on dorsal slope.
Periostracum. Waxy yellow or tan, faint green rays usually present on juvenile shells. Boldly striped individuals are rare.
Nacre. White, salmon, or pink, often iridescent posteriorly; commonly with gold Tulberg layers.
Hinge plate. Very heavy and short, distinctly angled, interdentum very short, wide, and thick; cardinals irregular, with two in the left valve and one in the right; these teeth may be accompanied by one or more prominent denticles both anterior and posterior to the teeth; the posterior accessory tooth may be positioned on the interdentum; lateral short, straight, two in left valve, one in right.
Muscle scars. Anterior adductor very deeply impressed; posterior adductor much less so; anterior adductor distinct from anterior pedal adductor, posterior adductor nearly separate from the posterior pedal retractor scar; multiple small dorsal suspensors, located in a long line behind cardinal fossa into the umbonal cavity. Pallial line entire, located far from shell margin. Accessory line long and prominent.
Umbonal cavity. Very wide and moderately deep.

Glochidia. Subspatulate, dorsal margin straight, anterior and posterior margins dorsally divergent and ventrally parallel, ventral margin gently curved. Dorsal alae long, exterior surface sculpturing rough. Hooks absent, micropoints lanceolate and arranged in broken vertical rows (Lea, 1858c: 47, pl. 5, fig. 17; Ortmann, 1911: 347, pl. 89, fig. 22; Hoggarth, 1999: figs. 61a–f).

Literature records (Surber, 1912: 9, pl. 2, fig. 21; Ortmann, 1912a: 347, pl. 20, fig. 9; Utterback, 1916: 443; Hoggarth, 1999: figs. 61a–f) indicate glochidial dimensions for length (0.19 to 0.21 mm), height (0.25 to 0.26 mm), and hinge (0.10 to 0.11 mm). Ortmann (1912a: 347) added, "The glochidia are peculiar in so far that I have on my slide two sizes of them. The smaller is more frequent, length 0.19; height, 0.21 mm., and among them are rather rarely larger ones, length 0.20 mm; height 0.25 mm. No intergrades seem to be present." Lea (1863) stated that this glochidium is "almost exactly the same with *multiradiatus*," [= *Lampsilis fasciola fide* Ortmann & Walker, 1922]; however, the glochidium of *L. abrupta* is much smaller than that of *L. fasciola*. Ortmann (1912) gave length and height measurements of 190 µm x 210 µm and 200 µm x 250 µm for the glochidium of *L. orbiculata* [= *L. abrupta*]. Our figures are nearer his second set of measurements, and the glochidia were all about the same size. There is often differential maturation of glochidia, and Ortmann's smaller individuals may have been immature.

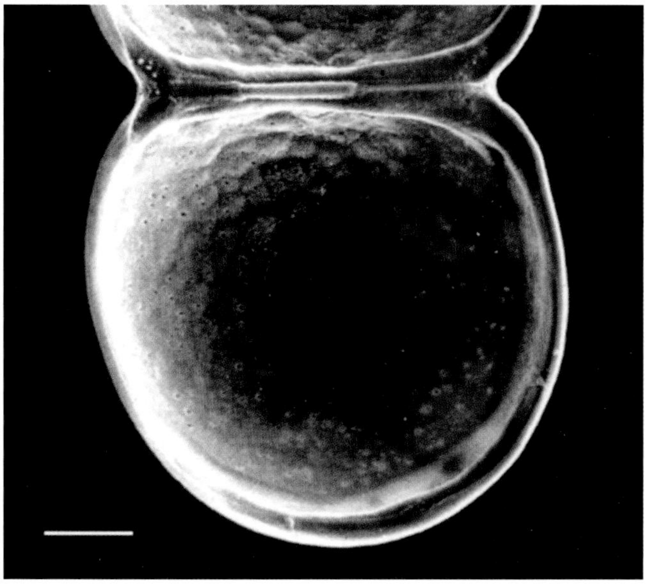

Fig. 84. *Lampsilis abrupta* (Say, 1831). Glochidium. OSUM 38841, Kanawha River, Fayette Co., WV. Bar = 40 µm. (Hoggarth, 1999).

Reproductive biology. Bradytictic. Eggs appear in August, with glochidia formed by September. These are overwintered to the following June (Ortmann, 1919). Reported hosts are bass and Walleye.

Reported potential hosts
Micropterus dolomieui (Smallmouth Bass)—Barnhart *et al.*, 1997 (LT).
Micropterus punctulatus (Spotted Bass)—Barnhart *et al.*, 1997 (LT).
Micropterus salmoides (Largemouth Bass)—Barnhart *et al.*, 1997 (LT).
Sander vitreus (Walleye)—Barnhart *et al.*, 1997 (LT).

Habitat. Sandy mud and gravel of large rivers.

Range. Lower Mississippi River system north to the Ohio River, including the Green River of Kentucky, the Kanawha River of West Virginia, the upper Tennessee and Cumberland rivers, west to the Meramec and Missouri rivers of Missouri, and the White and Red rivers of Arkansas. Replaced by *Lampsilis higginsi* (Lea, 1857) in the upper Mississippi River system.

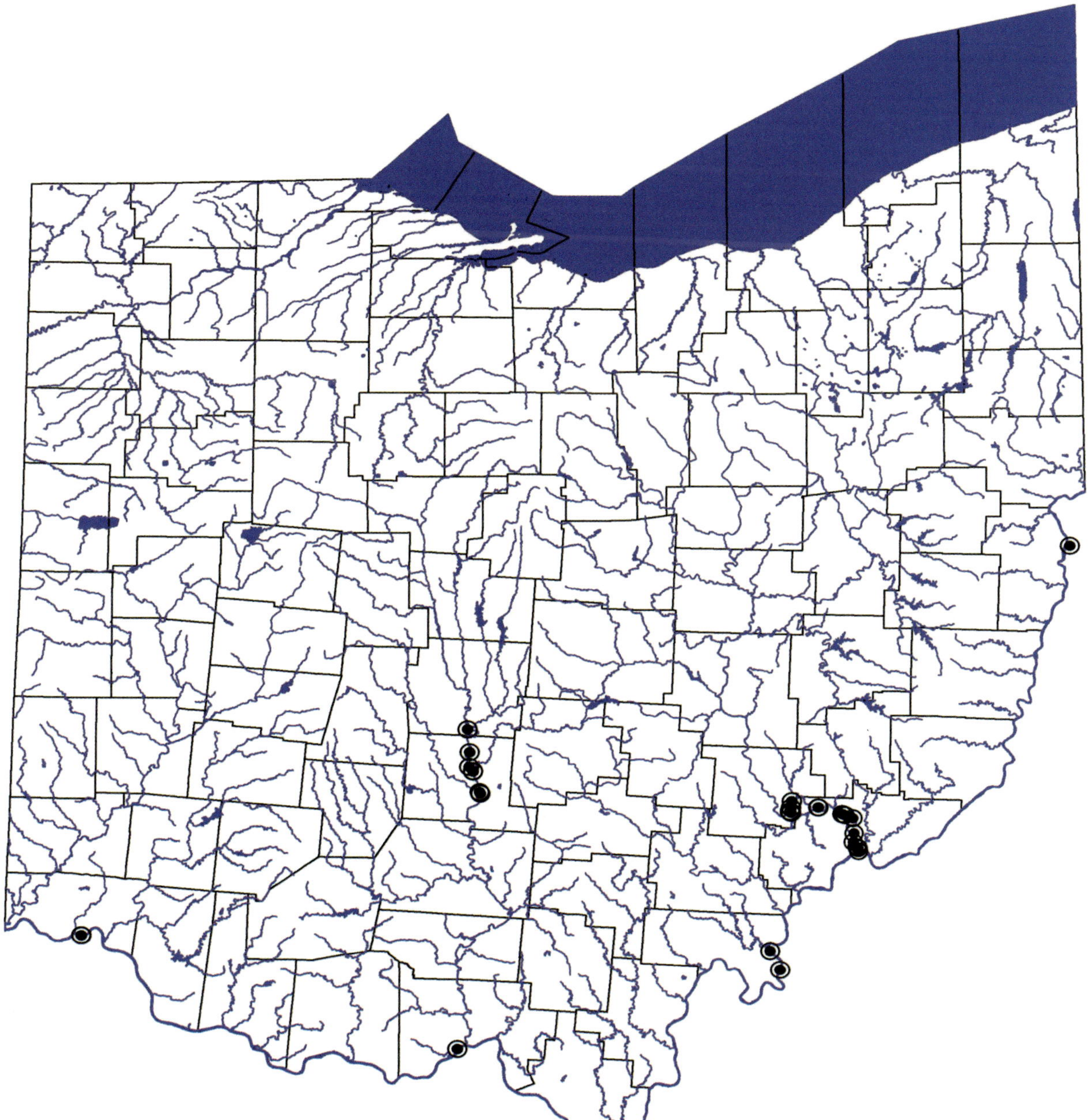

Fig. 85. ***Lampsilis abrupta*** **(Say, 1831)**
Solid dots—live or freshdead since 1980
Bullseye—all records, any condition

Range in Ohio. Currently in the Ohio River mainstem. Extirpated from the Muskingum River and the Scioto River. Living specimens occurred in the lower Muskingum River at Lowell as late as 1971.

Similar species. Female *L. abrupta* resemble female *L. cardium*. The latter grows much larger and has a thinner shell, often with prominent rays, and a thinner hinge. Male *L. abrupta* closely resemble male *L. cardium* and *Actinonaias ligamentina*. Typically male *L. cardium* have a more prominent umbo and thinner hinge. Male *A. ligamentina* often are prominently rayed (*L. abrupta* is rayless or has only vague rays) and have the umbo located more posteriorly than in *L. abrupta*. However, *Actinonaias ligamentina* males may be inseparable from some male *L. abrupta* based solely on the shell.

Comments. Individuals grow quickly for 3–4 years, but gradually slow. Sexual dimorphism is evident by the third year. Both sexes may grow to 110 mm and live for 25 years.

Lampsilis cardium Rafinesque, 1820

Figures 86, 87, 88, 89

Etymology. L. *cardio*, heart, from G. *kardia*, heart. Rafinesque apparently was referring to the heart-shaped cross-section through both valves. *Cardium* is a noun in apposition.

Vernacular name(s). *Plain Pocketbook.*

Original description

Lampsilis cardium Rafinesque, 1820: 298, pl. 80, figs. 16–19.

Type locality. None given. Johnson & Baker (173: 150) gave "Ohio River." Parmalee & Bogan (1998) gave "Wisconsin River; Mississippi, Prairie du Chien, Wisconsin," but that is the type locality for *Unio ventricosus* Barnes, 1823.

Type material. Johnson & Baker (1973: 150) designated ANSP 20210 as lectotype.

Synonymy

Lampsilis cardium Rafinesque, 1820: 298, pl. 80, figs. 16–19.

Unio ventricosus Barnes, 1823: 267–269, fig. 14 [Type locality: "the Wisconsan [River]" and "Mississippi [River], near Prairie du Chien." Type destroyed by fire].

Mya ventricosus (Barnes, 1823). Eaton, 1826: 221.

Unio occidens Lea, 1829: 435 [Type locality: "Ohio." Figured lectotype USNM 84866, paralectotype ANSP lost].

Unio subovatus Lea, 1831: 118, pl. 18, fig. 46 [Type locality: "Ohio River." Figured possible lectotype USNM 84509, possible paralectotype ANSP 42308].

Unio cardium (Rafinesque, 1820). Conrad, 1834: 68.

Unio fasciolus (Rafinesque, 1820). Férussac, 1835: 26 [in part].

Margarita (*Unio*) *subovatus* (Lea, 1831). Lea, 1836: 19.

Margarita (*Unio*) *ventricosus* (Barnes, 1823). Lea, 1836: 23.

Margarita (*Unio*) *occidens* (Lea, 1829). Lea, 1836: 23.

Unio lenis Conrad, 1838: back cover to No. 11 [Type locality: "rivers in North America." Types lost *fide* Johnson & Baker, 1973].

Unio ovata Say, 1816. Deshayes, 1839: 669 [misidentification].

Lampsilis ventricosa (Barnes, 1823). Stimpson, 1851: 14.

Margaron (*Unio*) *subovatus* (Barnes, 1823). Lea, 1852b: 24.

Margaron (*Unio*) *ventricosus* (Barnes, 1823). Lea, 1852b: 26.

Margaron (*Unio*) *occidens* (Lea, 1829). Lea, 1852b: 26.

Unio ovatus Say, 1816. Küster, 1852: 55, pl. 12, fig. 1 [misidentification].

Unio canadensis Lea, 1857c: 85 [Type locality: "St. Lawrence River, near Montreal." Figured possible lectotype USNM 84503, possible paralectotype ANSP lost *fide* Johnson & Baker, 1973].

Unio dolabraeformis Lea, 1838. Sowerby, 1867: 298 [misidentification].

Unio latissimus Rafinesque, 1820. Sowerby, 1868: pl. 66, fig. 337 [misidentification].

Margaron (*Unio*) *canadensis* (Lea, 1857). Lea, 1870: 37.

Unio cardium var. *ventricosus* (Barnes, 1823). Paetel, 1890: 147.

Unio cardium var. *occidens* (Lea, 1829). Paetel, 1890: 147.

Lampsilis ventricosus (Barnes, 1823). Baker, 1898b: 94, pl. 12, figs. 3–5.

Lampsilis (*Lampsilis*) *ventricosus* (Barnes, 1823). Simpson, 1900a: 526.

Lampsilis ventricosa cohongoronta Ortmann, 1912a: 53 [Type locality: "Potomac River, [Hancock, Washington Co.] Maryland." Lectotype CMNH 61.3999, paralectotypes CMNH 61.4000, MCZ 273128 by designation of Parodiz (1967: 28)].

Lampsilis ventricosa var. *lurida* Simpson, 1914: 41 [Type locality: "St. Lawrence drainage." Type lost *fide* Johnson, 1970, 1975].

Lampsilis ventricosa canadensis (Lea, 1857). Ortmann, 1919: 307, pl. 19, figs. 4, 5.

Lampsilis ovata ventricosa (Barnes, 1823). Ortmann, 1913a: 311.

Lampsilis cardia Rafinesque, 1820. Utterback, 1916b: 15.

Lampsilis (*Lampsilis*) *cardium* Rafinesque, 1820. Frierson, 1927: 67.

Lampsilis ventricosa var. *perglobosa* Baker, 1928a: 285, pl. 93, figs. 1–4 [Type locality: "Lake Pepin, near Lake City, Minn." Holotype UMMZ 209195, paratypes UMMZ 209197].

Fig. 86. ***Lampsilis cardium*** **Rafinesque, 1820. A.** OSUM 2784, Lake Erie, Ottawa Co., OH. 82 mm. **B.** OSUM 3712, Big Darby Creek, Pickaway Co., OH. 108 mm. **C.** OSUM 62072, Auglaize River, Allen Co., OH. 83 mm. **D.** OSUM 3394, Big Darby Creek, Pickaway Co., OH. 93 mm. **E.** St. Joseph River, Williams Co., OH. 152 mm. **F, H, Hinge detail:** OSUM 3848.1, Walhonding River, Coshocton Co., OH. 91 mm. **G.** OSUM 52622.3, Caesar Creek, Warren Co., OH. 118 mm. **Beak sculpture detail:** OSUM 5879.2, Big Darby Creek, Pickaway Co., OH. 53 mm total length.

Lampsilis ventricosa occidens (Lea, 1829). Baker, 1928a: 286, pl. 92.
Lampsilis ventricosa lurida (Simpson, 1914). Baker, 1928a: 289, pl. 93, fig. 5, pl. 94, figs. 5, 6.
Lampsilis ventricosa var. *winnebagoensis* Baker, 1928a: 291–293, pl. 94, figs. 1–4 [Type locality: "Winnebago Lake, near Oshkosh, Wis." Holotype UMMZ 209213, paratypes UMMZ 209213, UIMNH Z-12244a, b].
Lampsilis ovata canadensis (Lea, 1857). La Rocque, 1967: 222.
Lampsilis ovata Say, 1816. Strayer & Jirka, 1997: 59 [in part], pls. 1, 2 [misidentification].

Status. Not listed by either Ohio Division of Wildlife or U.S. Fish & Wildlife Service as of 2009.

ADULT SHELL

Shell. Large (to at least 150 mm), thin as a juvenile, becoming very heavy and thick at large sizes, shells sexually dimorphic, females more oval in profile than males; both sexes typically more elongate as juveniles; both sexes may be rather compressed or, more commonly, quite inflated and ventricose.

Anterior margin. Straight or slightly concave before umbo, curving strongly towards the ventral margin.

Ventral margin. Gently rounded in males, more so in females; may be nearly circular in some individuals.

Posterior margin. Males come to a broad point at 33% to 50% the distance to the dorsal margin; a second, weaker angulation may be present at 75% the distance to the dorsal margin; margin straight or gently curved between these points; females have an expanded margin that may have two weak angulations or form an unbroken, nearly circular curve from the ventral to the dorsal margins.

Dorsal margin. Short, particularly in females, straight.

Umbo. Wide and full, height variable but low in most specimens; beak sculpture variable, usually 2–4 "V"-shaped, coarse ridges, the later ones widely separated; in some specimens these are nearly linear or undulating, and some individuals have these ridges broken up into weak pustules.

Ligament. Dark brown, prominent, partially concealed between valves.

Disc sculpture. Shell sculptureless except for a few weak radiating striae on the posterior slope; posterior ridge present and varying from very weak to quite angular.

Periostracum. Pale yellow, yellowish green, olive, tan; old individuals may be dark brown; the surface may be polished, with a waxy appearance, or dull; rays present at least on posterior slope, usually over most of shell to some extent; rays faint to dark green, usually unbroken, often occurring in groups; most specimens have sparse, faint rays, but occasional ones have bold, wide rays; rays may disappear with age; posterior slope rayed and often of a darker base color.

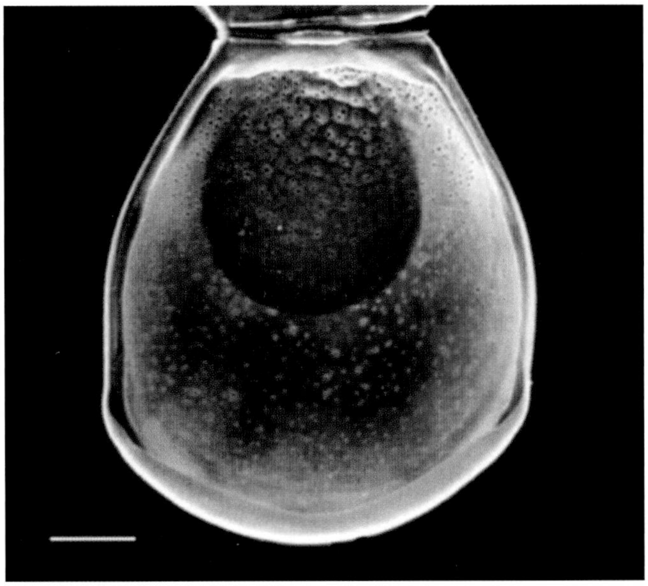

Fig. 87. *Lampsilis cardium* Rafinesque, 1820. Glochidium. OSUM 44619, Green River, Warren Co., KY. Bar = 40 µm. (Hoggarth, 1999).

Nacre. White, occasionally iridescent posteriorly; tan or gold Tulberg layers uncommon; exterior colors show through margin in small specimens.

Hinge plate. Light in juveniles, becoming heavy in old individuals; interdentum long, heavy, and cordlike or flattened; two large cardinals in left valve, one in right, with a weakly developed denticle or lamellum anterior to cardinal; laterals heavy, long, and arched, rarely straight, one lateral in right valve, two in left; very old individuals may have the hinge plate and teeth blunted and contorted, often invaded with periostracum or ligamental material.

Muscle scars. Anterior adductor shallow, becoming somewhat deeper in older individuals; posterior adductor may be confluent with or distinct from the posterior pedal retractor scar; dorsal suspensors multiple, attached in the umbonal cavity to the cardinal fossa and interdentum; pallial line complete, accessory line weak and short, occasionally imperceptible.

Umbonal cavity. Wide, slightly compressed, moderately deep.

Glochidia. Broadly subspatulate, dorsal margin straight, dorsal portion of the lateral margin strongly divergent, the ventral portion of the lateral margin almost parallel, ventral margin gently curved. Dorsal alae long, exterior surface sculpturing rough. Hooks absent, micropoints lanceolate to bluntly pyramidal and arranged in broken vertical

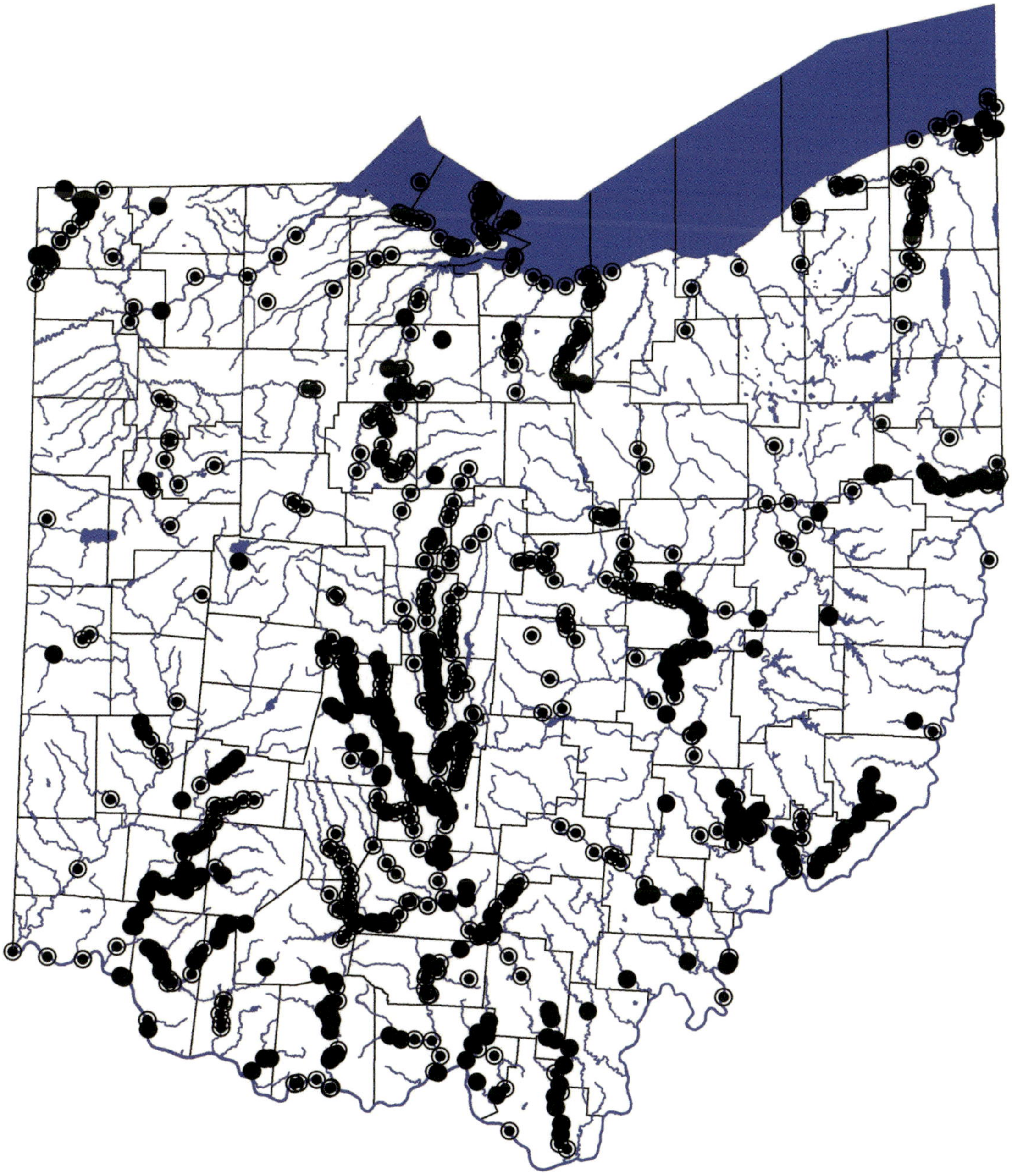

Fig. 88. ***Lampsilis cardium*** **Rafinesque, 1820**
Solid dots—live or freshdead since 1980
Bullseye—all records, any condition

rows (Lea, 1858c: 47, pl. 5, fig. 13; Ortmann, 1911: 347, pl. 89, fig. 23; Hoggarth, 1999: figs. 64a–g).

Literature records (Ortmann, 1912a: 352; Hoggarth, 1999: figs. 64a–g) indicate glochidial dimensions for length (0.24 to 0.26 mm), height (0.27 to 0.29 mm), and hinge (0.10 to 0.12 mm). Other literature records (Surber, 1912: 9, pl. 2, fig. 24; Utterback, 1916a: 450; Waller *et al.*, 1988: 41, figs. 5, 6) indicate different dimensions for the glochidia of this species: length (0.21 to 0.22 mm) and height (0.25 to 0.26 mm). Ortmann (1919: 302) noted, "My measurements are: 0.25 x 0.29 mm.; while Surber gives: 0.205 x 0.255 mm. This is a rather unusual discrepancy." The glochidium of *L. cardium* is longer and higher than that of *L. ovata*, but both species have about the same hinge length. Surber (1912) gave length and height measurements of 200 µm x 250 µm for this glochidium; Ortmann (1912) gave 250 µm x 290 µm; and Waller *et al.* (1988) gave 216 µm x 257 µm. Our measurements are 249 µm x 283 µm. Waller *et al.* and Surber collected their material from the Mississippi River, while Ortmann's and our material came from smaller streams.

Fig. 89. *Lampsilis cardium* Rafinesque, 1820. Mantle display with marsupial portions of gills exposed. Photo courtesy of Tim Revell, *The Columbus Dispatch*.

Reproductive biology. Bradytictic. The posterior half of the outer gills is marsupial (Ortmann, 1912a). Ortmann (1919) believed this species had two broods per year, with egg-bearing or gravid females occurring from July through October, and again from May to July. Holland-Bartels & Kammer (1989) reported the species as spawning in July and August and releasing glochidia in May of the following year. In Ohio, this release may begin in April or in the previous fall if glochidia are overwintered on their hosts. When ready to release glochidia, the marsupia are protruded beyond the posterior shell margins between the mantle flaps. Two types of glochidial release have been seen: white, leaf-shaped conglutinates and free glochidia. The relation of these two methods to the age of the glochidia, or season of the year, has not been identified (Watters, unpubl.). This species can overwinter glochidia on the host for at least six months at 10°C. In Ohio, metamorphosis occurs when spring water temperatures reach a sustained 15°C (Watters & O'Dee, 1999). Most hosts are centrarchids.

Reported potential hosts

Ambystoma tigrinum ssp. (Tiger Salamander)—Watters & O'Dee, 1997a (LT); Watters, 1997a (LT)—exotic.

Betta splendens (Siamese Fighting Fish)—Watters & O'Dee, 1997a (LT); Watters & O'Dee, 1998a (LT)—exotic.

Colisa lalia (Flame Gourami)—Watters & O'Dee, 1997a (LT); Watters & O'Dee, 1998a (LT)—exotic.

Lepomis cyanellus (Green Sunfish)—Watters, 1996a (LT).

Lepomis gibbosus (Pumpkinseed)—Draxler *et al.*, 2006 (LT).

Lepomis macrochirus (Bluegill)—Coker *et al.*, 1921 (LT).

Micropterus dolomieu (Smallmouth Bass)—Coker *et al.*, 1921 (LT).

Micropterus salmoides (Largemouth Bass)—Lefevre & Curtis, 1912 (LT); Reuling, 1919 (LT); Coker *et al.*, 1921 (LT); Waller *et al.*, 1985 (LT); Watters & O'Dee, 1997a (LT); Watters & O'Dee, 1998a (LT); Watters *et al.*, 2005 (LT); Draxler *et al.*, 2006 (LT).

Perca flavescens (Yellow Perch)—Coker *et al.*, 1921 (LT).

Poecilia reticulata (Guppy)—Watters & O'Dee, 1997a (LT); Watters & O'Dee, 1998a (LT)—exotic.

Pomoxis annularis (White Crappie)—Wilson, 1916 (NI); Coker *et al.*, 1921 (NI, LT).

Pomoxis nigromaculatus (Black Crappie)—Draxler *et al.*, 2006 (LT).

Sander canadensis (Sauger)—Wilson, 1916 (NI); Coker *et al.*, 1921 (NI).

Sander vitreus (Walleye)—Waller *et al.*, 1985 (LT).

Trichogaster trichopterus (Lavender Gourami)—Watters & O'Dee, 1997a (LT); Watters & O'Dee, 1998a (LT)—exotic.

Xiphophorus helleri (Painted Sword)—Watters & O'Dee, 1997a (LT); Watters & O'Dee, 1998a (LT)—exotic.

Habitat. The Plain Pocketbook is widespread in creeks, rivers, ponds, and lakes. It tolerates many substrates and water flows.

Range. Red River system in Canada, Great Lakes drainages, Mississippi River system south to northern Louisiana, west to Missouri and Arkansas, east to Pennsylvania.

Range in Ohio. Throughout most of Ohio, including the Ohio River and the western basin of Lake Erie.

Similar species. See comments under *L. abrupta* and *L. ovata*.

Comments. This species may become heavy, large, and very rotund; it is one of Ohio's largest species of mussel. Individuals grow quickly during the first three to four years, but most abruptly slow their growth at this point, probably as sexual maturity is reached. Depending on the environment in which they grew, specimens of very different ages may be the same size. At 90 mm, a small size for this species, individuals may be from 5 to 15 years old. Specimens older than 30 years are rare.

Lampsilis fasciola Rafinesque, 1820

Figures 90, 91, 92, 93

Etymology. L. *fascia*, a band, bandage, zone + L. *-ola* (diminutive). Rafinesque often used the word "bands" when referring to rays on the periostracum of a unionoid shell, as he did in the original description of this species: "épiderme olivâtre, à bandes radiées, flexueuses, inégales, verdâtres" (1820: 299) ["epidermis olivaceous, with radiating, wavy, unequal, green bands"].

Vernacular name(s). *Wavy-rayed Pocketbook*, Fine-rayed Pocketbook, Wavy-rayed Lamp Mussel.

Original description

Lampsilis fasciola Rafinesque, 1820: 299, species no. 16, no figure.

Type locality. "dans le Kentuky [River]."

Type material. As early as 1835, Férussac noted (p. 32) that he had ". . . un bel individu de cette coquille, envoyé par M. Rafinesque, . . ." ["a beautiful individual of this shell, sent by Monsieur Rafinesque"] and identified it as the same species as *U. multiradiatus* Lea, 1829. Since the Rafinesque-Poulson "type" (ANSP 20203) of *Lampsilis fasciola* Rafinesque, 1820, is actually a specimen of *Unio luteola* Lamarck, 1819, the specimen of Férussac, a syntype, appears to be the best candidate as a type for Rafinesque's name. A search of the Muséum national d'Histoire naturelle, Paris, collection for Férussac's specimen failed to find it (Johnson & Baker, 1973: 154–155). We are left only with the original description upon which to identify Rafinesque's species. Fortunately, the species we now know as *L. fasciola* Rafinesque, 1820, is easily identifiable from this description (but see Comments below).

Synonymy

Lampsilis fasciola Rafinesque, 1820: 299.

Unio multi-radiatus Lea, 1829: 434 [Type locality: "Cincinnati, Ohio." Lectotype USNM 84930].

Unio fasciolus (Rafinesque, 1820). Say, 1834: no pagination.

Margarita (*Unio*) *multiradiatus* (Lea, 1829). Lea, 1836: 24.

Margaron (*Unio*) *multiradiatus* (Lea, 1829). Lea, 1852b: 31.

Unio perradiatus Lea, 1858a: 40 [Type locality: [Tennessee River] "Florence, [Lauderdale Co.] Alabama." Lectotype USNM 84501].

Unio altilis Conrad, 1834. Reeve, 1865: pl. 23, fig. 109 [misidentification].

Unio perovalis Conrad, 1834. Sowerby, 1866: pl. 38, fig. 209 [misidentification].

Margaron (*Unio*) *perradiatus* (Lea, 1858). Lea, 1870: 37.

Lampsilis (*Lampsilis*) *multiradiatus* (Lea, 1829). Simpson, 1900a: 532.

Lampsilis multiradiata (Lea, 1829). Sterki, 1907: 388.

Lampsilis (*Lampsilis*) *multiradiata* (Lea, 1829). Simpson, 1914: 55.

Lampsilis (*Lampsilis*) *fasciola* Rafinesque, 1820. Frierson, 1927: 69.

Ligumia fasciola (Rafinesque, 1820). Goodrich, 1932: 109.

Status. Listed as a Species of Concern by Ohio Division of Wildlife; not listed by U.S. Fish & Wildlife Service as of 2009.

ADULT SHELL

Shell. Medium-sized (to ca. 100 mm), thin as a juvenile, heavy in gerontic specimens, outline oval to subcircular; males draw out to a rounded point posteriorly.

Anterior margin. Slightly flattened dorsally in some specimens, broadly rounded in others, strongly curved ventrally.

Ventral margin. Males broadly curved to straight, females broadly curved or straight until the marsupial region, where the margin abruptly bends dorsally.

Posterior margin. Males gently curved towards umbo, females with second angulation approximately 25% distance from dorsal margin, then straight to umbo.

Dorsal margin. Straight over ligament and beneath umbo, gently arched in gerontic specimens; angled down anteriorly.

Umbo. Anterior of center, often greatly so in females; not prominent; beak sculpture of 3–4 weak "V"- or "W"-shaped ridges, often breaking up away from umbo; beak sculpture disappears within first year's growth.

Ligament. Tan to nearly black, prominent, exposed.

Fig. 90. *Lampsilis fasciola* Rafinesque, 1820. A. OSUM 56493.4, Tippecanoe River, Carroll Co., IN. 71 mm. **B.** OSUM 5763.4, Olentangy River, Franklin Co., OH. 85 mm. **C.** OSUM 25970.3, Big Darby Creek, Pickaway Co., OH. 56 mm. **D.** OSUM 25080.3, Blanchard River, Hancock Co., OH. 80 mm. **E.** OSUM 56348.11, Tippecanoe River, Kosciusko Co., IN. 54 mm. **F, G, Hinge detail:** OSUM 25080.2, Blanchard River, Hancock Co., OH. 73 mm. **H.** OSUM 5010, Olentangy River, Marion Co., OH. 60 mm. **Beak sculpture detail:** OSUM 51744.2, Little Darby Creek, Madison Co., OH. 41 mm total length.

Disc sculpture. Disc devoid of sculpture except for 1–2 weak dorsal ridges on posterior slope.
Periostracum. Base color greenish-yellow, yellow, or tan, old individuals darker or gray; in some populations the periostracum is brilliantly shiny with a waxy texture; other populations are duller and have a matte finish; most specimens have numerous dark green rays over the entire shell; these rays are nearly black in some individuals; the pattern typically consists of widely spaced wide rays with intervening narrow ones, although individuals having only fine rays are not uncommon; rays may be grouped in twos and threes; the width of rays and patterns varies from population to population; the rays slightly change their orientation between growth periods, giving the shells a characteristic "wavy rayed" pattern.
Nacre. White, outer margin showing through outer color and pattern; Tulberg layers rare, usually found only in large individuals.
Hinge plate. Moderately heavy, interdentum short, narrow, flat; straight in smaller specimens, more arched in older ones, particularly females; two cardinals in left valve, one in right valve with a well-formed denticle anteriorly; laterals long in males, shorter in females, straight, continuous with interdentum in males, at a distinct angle in females; two laterals in left valve, one in right.
Muscle scars. Anterior adductor deep; posterior adductor adjacent to or confluent with posterior pedal retractor scar; dorsal suspensors multiple, located in umbonal cavity under anterior cardinal fossa and interdentum; pallial line complete, accessory very weak and short.
Umbonal cavity. Fairly deep, compressed, becoming more so with age.

Glochidia. Narrowly subspatulate, dorsal margin straight, posterior margin strongly divergent dorsally, ventrally more or less perpendicular to the hinge, anterior margin less divergent than posterior, ventrally rounded, ventral margin gently and evenly curved. Dorsal alae long, exterior surface sculpturing rough. Hooks absent, micropoints lanceolate and arranged in broken vertical rows (Lea, 1858c: 47, pl. 5, fig. 17; Surber, 1915: 5, fig. 2; Hoggarth, 1999: figs. 68a–f).

Literature records (Ortmann, 1912a: 352; Surber, 1915: 5, fig. 2; Hoggarth, 1999: figs. 68a–f) indicate glochidial dimensions for length (0.23 to 0.25 mm), height (0.29 to 0.30 mm), and hinge (0.11 to 0.12 mm). Surber (1915: 5) noted that the glochidia are of the "*Lampsilis* type; semielliptical; ventral margin rounded, hinge line short and evenly curved; size large. But for the much longer hinge line and larger adductor this could hardly be distinguished from average examples of *Quadrula pustulosa*, as it agrees in size with that form." This glochidium is like the other narrowly subspatulate glochidia except that the anterior margin is slightly more rounded. In this regard

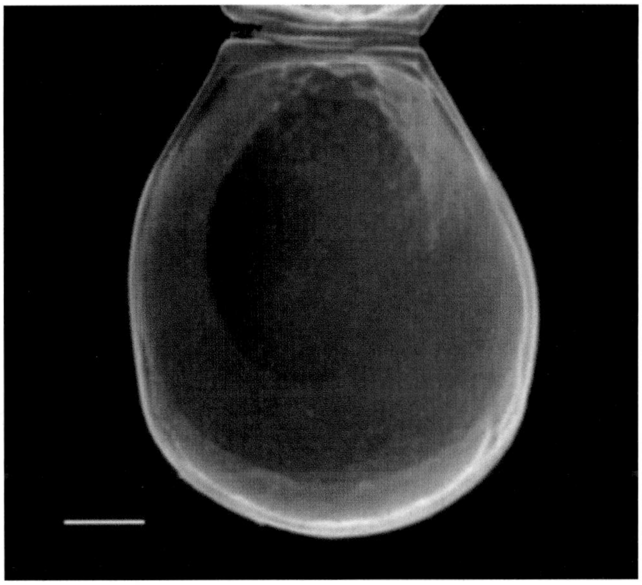

Fig. 91. *Lampsilis fasciola* Rafinesque, 1820. Glochidium. OSUM 55033.2, Clinch River, Russell Co., VA. Bar = 40 µm. (Hoggarth, 1999).

this glochidium resembles that of *Obovaria*, but it will not be confused with *Obovaria*, because of its posterior margin, wide ventral flange, and rough exterior valve sculpture.

Reproductive biology. Bradytictic. The posterior half of the outer gills is marsupial. Zale & Neves (1982c) reported that this species spawned in August and carried glochidia to the following May to August. But Ortmann (1919) suggested that this species has two broods, with gravid females found from August to October, and again from May to August. In support of this, in captivity, females have been found discharging glochidia in both October and April (Watters & O'Dee, 1996). In the laboratory, females "burrowed beneath the surface for two to three days, surfaced and displayed for two days, reburrowed, and emerged in a different part of the aquarium. This cycle was repeated for over two months. By October, both females had stopped displaying and were barren of glochidia. Displays included rhythmic undulations of spotted fish-like mantle flaps. The marsupial regions of the gills protruded between the shells and the mantle flaps. When touched, the flaps were withdrawn, and simultaneously the shells closed on the marsupia, causing them to burst and release a cloud of glochidia" (Watters & O'Dee, 1996: 385). The female will often have a mantle flap that can be observed "flapping" from late July to October. At least two very different mantle flap lures are known from Ohio specimens, even from individuals living side by side. One is a gray lure with brown spots and an eyespot. The other is a bright orange lure with a dark "lateral line" and eyespot. It is possible that this species actually is a series of sibling species.

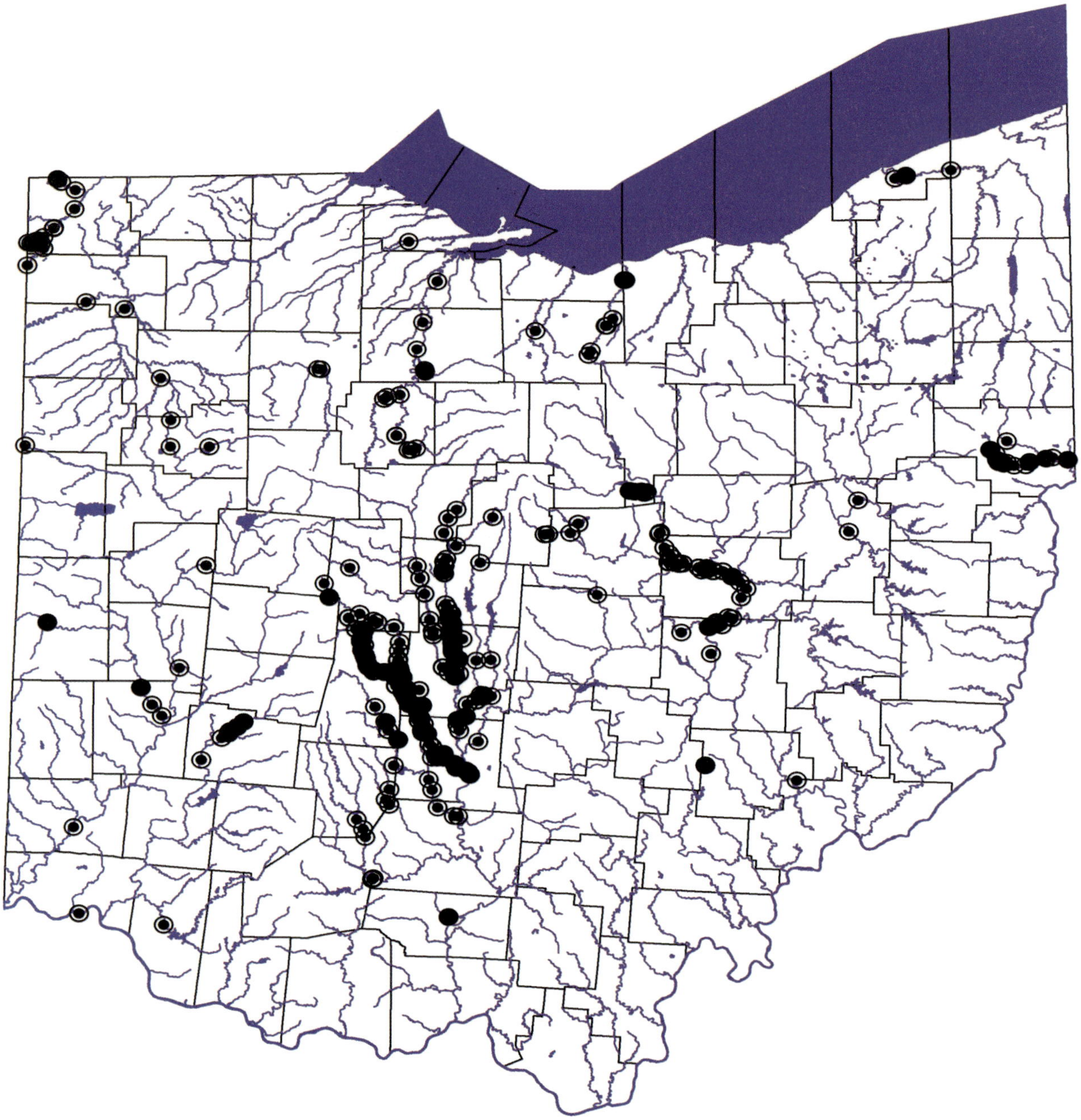

Fig. 92. ***Lampsilis fasciola*** **Rafinesque, 1820**
Solid dots—live or freshdead since 1980
Bullseye—all records, any condition

Fig. 93. *Lampsilis fasciola* Rafinesque, 1820. Mantle lure display.

Reported potential hosts

Lepomis megalotis (Longear Sunfish)—Watters *et al.*, 2009 (LT).

Micropterus dolomieu (Smallmouth Bass)—Zale & Neves, 1982b (LI), 1982c (NS); Neves, 1983 (NS).

Micropterus salmoides (Largemouth Bass)—Watters & O'Dee, 1996 (LT); Watters *et al.*, in press.

Habitat. This is a high-water-quality species found in fast-flowing streams usually in less than 1 m of water on a bottom of cobble to sand. In describing the ecology of this species Parmalee (1967: 67) stated that specimens were found "on a coarse sand and gravel bottom, with little mud, in current, and in less than one foot of water."

Range. Parmalee (1967: 65) stated that the species' range was "Entire Ohio drainage; southern Michigan." La Rocque (1967: 217) added "Great Lakes and their drainage, but not in their entirety." It is absent west of the Mississippi River and does not occur as far north as Minnesota or Wisconsin.

Range in Ohio. Historically this species occurred sporadically in both Ohio River and Lake Erie tributaries but was virtually absent from unglaciated Ohio. Currently much reduced in range, most common in the Scioto River system.

Similar species. This species will not be confused with any other. The combination of size, shape, and especially the density and intensity of the rays is unique to this species in Ohio. Other members of the genus *Lampsilis* are generally larger (*cardium, ovata, teres, abrupta,* and *luteola*), and if rayed, those rays are broader and fewer in number.

Comments. Individuals grow rapidly for the first 4–6 years of life; thereafter, annular growth stages are much narrower, but remain fairly constant in width for the remainder of the animal's life. A large, 100-mm individual may be 10–15 years old. Specimens older than 20 years are rare.

At least three different mantle flap patterns are known (two from Ohio), often found within the same river. Intergrades have not been identified. If *L. fasciola* does indeed have three very different mantle flaps, this would be an extreme case of polymorphism. On the other hand, it is more likely that three sibling species are present, all known under the name *L. fasciola*. Certainly there are differences in the shells of some populations, particularly between the populations of polished shells and matte shells. Whether these shell differences correspond to differences in mantle flaps needs to be studied. If sibling species are present, the question becomes "Which one is the real *L. fasciola* of Rafinesque?" This may never be answered, as the types are apparently lost and Rafinesque's description is not detailed enough to discern the shell differences.

Lampsilis ovata (Say, 1817)

Figures 94, 95, 96

Etymology. L. *ovatus*, egg-shaped, ovate; from L. *ovum*, egg. Say (1817) noted "Shell subovate," and subovate outline is very typical of this species—especially the males and juvenile females.

Vernacular name(s). *Pocketbook*, Ridged Pocketbook.

Original description

Unio ovatus Say, 1817: no pagination, article on "Conchology," Species 3 under Genus *Unio*, pl. 2, fig. 7.

Type locality. "Inhabits the Ohio river and its tributary streams." Unfortunately, the type locality of the neotype designated by Haas (1930: 328) is "U.S.A."

Type material. The original unionid type material of Say has apparently been lost (Johnson, 1970: 386; Johnson & Baker, 1973: 146). Haas (1930: 328) designated SMF No. 4338 as the neotype of *U. ovatus* Say, 1817.

Synonymy

Unio ovatus Say, 1817: no pagination.

Unio ovata Say, 1817. Lamarck, 1819: 75.

Lampsilis ovata (Say, 1817). Rafinesque, 1820: 298.

Mya ovata (Say, 1817). Eaton, 1826: 218.

Margarita (*Unio*) *ovatus* (Say, 1817). Lea, 1836: 19.

Aeglia ovata (Say, 1817). Swainson, 1840: 266, fig. 49.

Margaron (*Unio*) *ovatus* (Say, 1817). Lea, 1852b: 24.

Lampsilis (*Lampsilis*) *ovatus* (Say, 1817). Simpson, 1900a: 530.

Lampsilis (*Lampsilis*) *ovata* (Say, 1817). Simpson, 1914: 48.

Lampsilis ovata (Say, 1817). Frierson, 1927: 67.

Lampsilis (*Lampsilis*) *ovata* (Say, 1817). Johnson, 1970: 386–388 [in part], not pl. 21, figs. 1, 2.

Fig. 94. ***Lampsilis ovata*** **(Say, 1817). A.** OSUM 10317, Muskingum River, [Co.?], OH. 115 mm. **B.** OSUM 33291.1, Green River, Hart Co., KY. 98 mm. **C.** OSUM 39071, "Wabash River." 126 mm. **D.** OSUM 19047.1, Walhonding River, Coshocton Co., OH. 121 mm. **E.** OSUM 46908, Muddy Fork Mohican River, Ashland Co., OH. 122 mm. **F, Hinge detail:** OSUM 44561.1, Elk River, Clay Co., WV. 93 mm. **Beak sculpture detail:** OSUM 18722.1, Paint Rock River, Jackson Co., AL. 31 mm total length.

non Lampsilis ovata (Say, 1817). Strayer & Jirka, 1997: pls. 1, 2 [= *Lampsilis cardium* Rafinesque, 1820].
Status. Listed as Endangered by Ohio Division of Wildlife; not listed by U.S. Fish & Wildlife Service as of 2009.

ADULT SHELL
Shell. Large size (to 150 mm), moderately thick as adults, much thinner as juveniles, oval to somewhat triangular, moderately inflated; females broadly rounded posteriorly, males drawn to a pronounced point.
Anterior margin. Long, straight to broadly rounded to the ventral border.
Ventral margin. Very broadly rounded.
Posterior margin. Very long; in females broadly rounded, forming obtuse angle with ventral margin; in males forming sharp acute angle with ventral margin.
Dorsal margin. Short, straight, joining posterior margin with obtuse angle.
Umbo. Wide and prominent; umbos nearly face each other; umbonal sculpture absent or of very weak undulations.
Ligament. Short and prominent, tan to dark brown, not concealed by shells.
Disc sculpture. Disc of shell smooth except for straight to deeply concave posterior angle.
Periostracum. Shiny to waxy, yellow or tan, often with fine but distinct, widely separated green lines or rays, posterior slope dark.
Nacre. Porcelain white, weakly iridescent posteriorly; often with extensive gold or brown Tulberg layers.
Hinge plate. Well-developed, angled under umbo; right cardinal a single, massive, triangular denticle, usually with a well-developed anterior accessory tooth; left cardinal bifid; laterals well-developed, very short, straight, one in right valve, two in left; interdentum flat, wide, quite long.
Muscle scars. Anterior adductor large, elongate, and impressed; posterior adductor nearly flush; large anterior adductor distinct from anterior pedal protractor scar, large posterior adductor distinct from small posterior pedal retractor scar; cluster of small dorsal suspensors behind cardinal teeth and interdentum in an irregular line. Pallial line entire, located far from shell margin. Accessory line weak to absent.
Umbonal cavity. Very wide, deep, open.

Glochidia. Broadly subspatulate, dorsal margin straight, dorsal portions of the lateral margins strongly divergent, the ventral portions of these margins subparallel, the anterior margin slightly more curved than the posterior margin, ventral margin gently curved. Dorsal alae long, exterior surface sculpturing rough. Hooks absent, micropoints bluntly pyramidal to lanceolate and arranged in broken vertical rows (Lea, 1858c: 47, pl. 5, fig. 15; Ortmann, 1912a: 351; Isom, 1983: 44, fig. 1c; Hoggarth, 1999: figs. 63a–f).

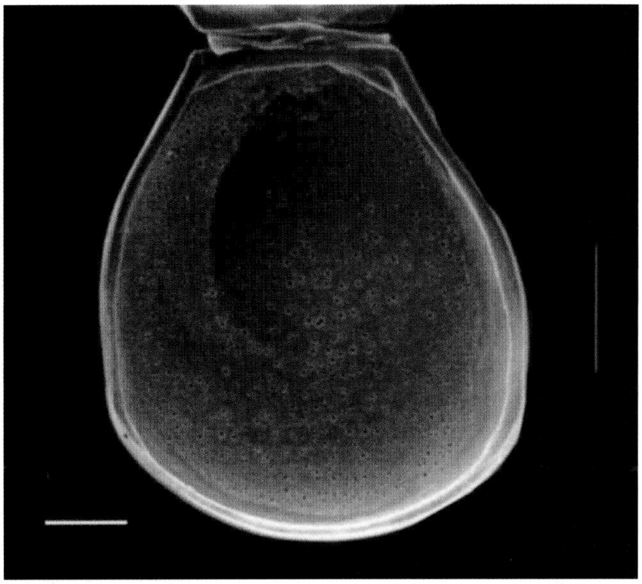

Fig. 95. *Lampsilis ovata* (Say, 1817). Glochidium. OSUM 43164.1, Clinch River, Scott Co., VA. Bar = 40 µm. (Hoggarth, 1999).

Literature records (Ortmann, 1912a: 351; Hoggarth, 1999: figs. 63a–f) indicate glochidial dimensions for length (0.23 to 0.24 mm), height (0.27 to 0.28 mm), and hinge (0.11 to 0.12 mm). Ortmann (1912a: 351) noted, "Glochidia figured by Lea (1858) large, subovate. Here again a mistake occurs in Lea's figure. In fig. 13 of Pl. 5 he figures the glochidium of *occidens* (= *L. cardium*), and the latter is considerably larger than that of *L. ovata*. The fact is, however, that the glochidia of these two forms are practically indistinguishable in size and shape. . . ." The glochidium of *L. ovata* is distinguished from those previously described by its broadly subspatulate shape. However, this shape is also found in the glochidia of the other members of the *L. ovata* complex (i.e., *L. cardium, L. satura,* and *L. ornata*). Relative size may be the only way to distinguish these species.

Reproductive biology. Probably similar to *Lampsilis cardium.* The posterior 1/2–2/3 of the outer gills is marsupial (Ortmann, 1912a). It is "gravid" from August to the following May (Ortmann, 1919).

Reported potential hosts. Records for this species may be confused with those for *Lampsilis cardium.*

Habitat. In Ohio, this is a large creek or riverine species. It occurs in sandy mud or packed sand and cobble, occasionally in riffles.

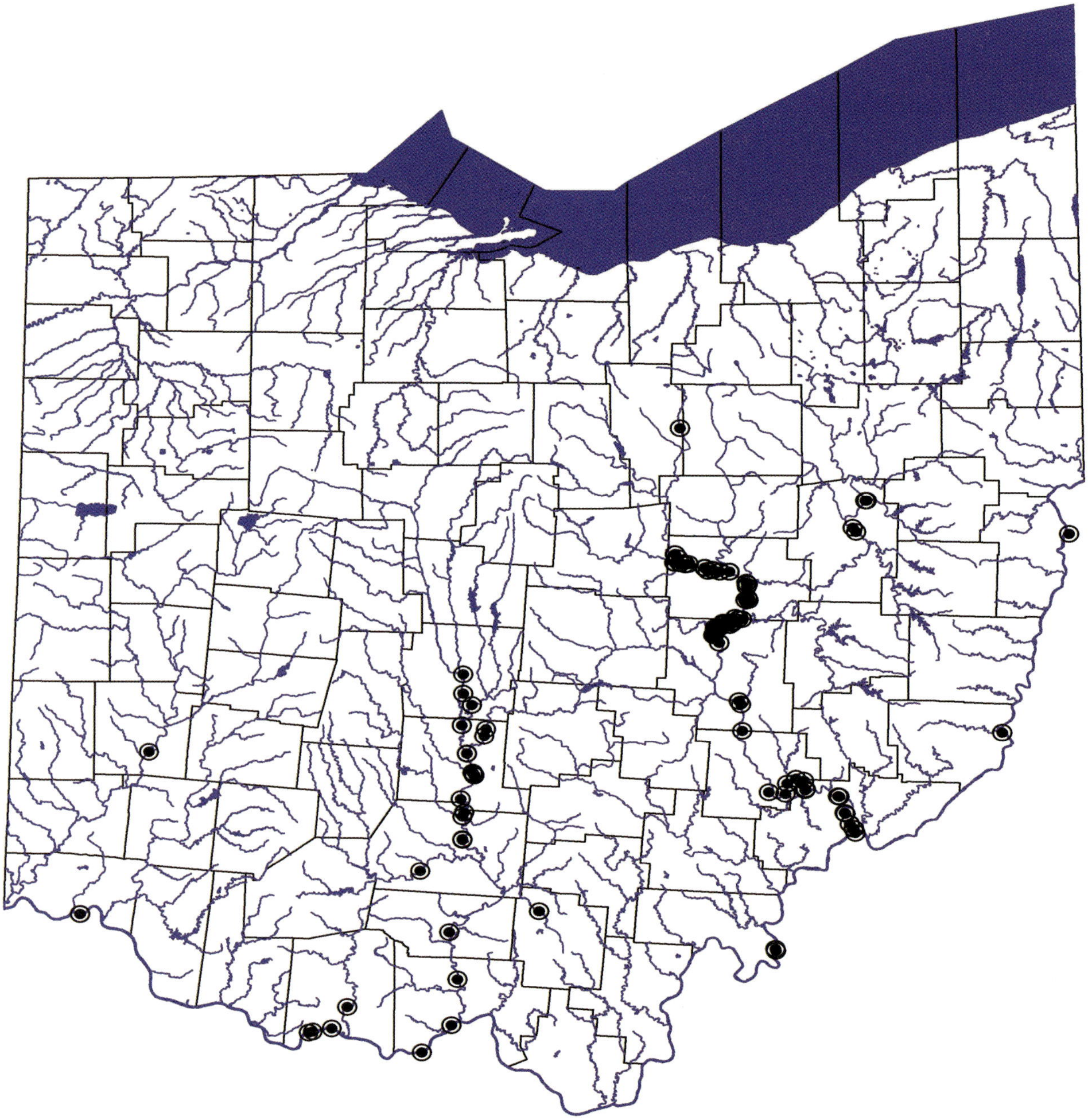

Fig. 96. ***Lampsilis ovata*** **(Say, 1817)**
Solid dots—live or freshdead since 1980
Bullseye—all records, any condition

Range. Because of misidentifications with *L. cardium*, the extent of this species is unclear. Ohio River system, including Tennessee, Cumberland, and Kanawha rivers. West of the Mississippi River in the Red, Missouri, and Arkansas rivers. It does not occur as far north as Minnesota and Wisconsin, and is absent from the Lawrentian system.

Range in Ohio. Historical records exist for the Great Miami River, Ohio Brush Creek, the Scioto and Muskingum rivers, and the Ohio River. Extirpated from most of the state today. Living but uncommon in portions of the Muskingum and Scioto river drainages.

Similar species. This species is very similar to *L. cardium*. Typical examples may be separated from *L. cardium* by the sharp, concave posterior ridge, darkly colored posterior slope, and general lack of rays. However, some specimens closely approach *L. cardium* (or vice versa) and may be impossible to discern based solely on shell characteristics.

Comments. Individuals abruptly slow after 4–5 years. Specimens rarely reach ca. 150 mm and nearly 20 years old.

Lampsilis radiata luteola (Lamarck, 1819)

Figures 97, 98

Etymology. L. *radiatus*, furnished with rays; from L. *radio*, furnish with beams, spokes, make beaming, irradiate. L. *luteus*, yellow + L. *-ola* (diminutive). Lamarck (1819: 79) described *U. luteola* as yellow and rayed, both typical characteristics of this species.

Vernacular name(s). *Fatmucket*, Yellow Mucket

Original description

Unio luteola Lamarck, 1819: 79, species no. 40, no figure given.

Type locality. "Habité la rivière Susquehana et celle Mohancks, dans le États-Unis" ["Living in the Susquehana River and the Mohawk, in the United States"]. Johnson (1969: 54) restricted the type locality to "the Susquehanna River, Columbia, York Co., Pennsylvania."

Type material. Férussac stated that Lamarck's type specimen of *U. luteola* was Say's *U. cariosus*. However, Johnson (1953: 91; 1969: 54, fig. 11) located Lamarck's type of *Unio luteola*, which proved to be a different species. He reported "Holotype Paris Museum, length 69, height 38, width 24.5 mm." There has been some difficulty in determining which of two subspecies, *radiata* Gmelin, 1791, or *siliquoidea* Barnes, 1823, is synonymous with *luteola* Lamarck, 1819. A careful study of Lamarck's original description, the published photograph of the holotype (Johnson, 1969: fig. 11) and numerous collections from across the range of *L. radiata*, and the forms of its complex makes it clear that the holotype of *luteola* is likely either the form otherwise known as *U. siliquoidea* or an intergrade between *U. radiatus* and *U. siliquoidea* that, in its shell characteristics, is closer to the latter. We conclude that *U. siliquoidea* Barnes, 1823, is a junior synonym of *U. luteola* Lamarck, 1819, that intergraded with *L. radiata radiata* in northern New York State in the Finger Lakes, the Erie Canal, the Mohawk, and the upper Susquehanna Rivers. Strayer & Jirka (1997) indicated a particularly large zone of overlap in the Mohawk River basin. It is of interest to note that OSUM has material of additional "western" species, *Alasmidonta marginata* Say, 1818, and *Villosa iris* (Lea, 1829), from the upper Susquehanna River, probably transplanted.

Most recent authors have used the name *siliquoidea* for this taxon, as either a subspecies of *radiata* or a full species. The great similarity of *radiata radiata* and *luteola* and their intergradation where they meet strongly suggest that they are subspecies.

Synonymy

Unio luteola Lamarck, 1819: 79.

Unio inflatus Barnes, 1823: 266–267 [Type locality: "the Wisconsan [*sic*] [River] and Lake Erie." Type lost. A *nomen oblitum*. Contrary to Simpson (1900a: 574) Barnes's name does not preoccupy Lea's 1831 *inflata*, which was described as a *Symphynota*.].

Unio siliquoideus Barnes, 1823: 269–270, fig. 15 [Type locality: "the Wisconsan [*sic*] [River]." Type lost. Agassiz (1852: 43) considered this the male of the species.].

Mya siliquoidea (Barnes, 1823). Eaton, 1826: 221.

Mya inflata (Barnes, 1823). Eaton, 1826: 221.

Margarita (*Unio*) *luteolus* (Lamarck, 1819). Lea, 1836: 25.

Unio childreni Hanley, 1843: 193, pl. 23, fig. 57 [*non* Gray, 1834. Type locality: "North America." Type in BMNH?].

Lampsilis siliquoidea (Barnes, 1823). Stimpson, 1851: 14.

Lampsilis inflatus (Barnes, 1823). Agassiz, 1852: 44 [Agassiz considered this the female of the species.].

Unio siliquoides [*sic*] Barnes, 1823. Küster, 1852: 30, pl. 5, fig. 2.

Margaron (*Unio*) *luteolus* (Lamarck, 1819). Lea, 1852b: 28.

Unio multiradiatus Lea, 1829. Sowerby, 1868: pl. 61, fig. 306 [misidentification].

Unio affinis Lea, 1852. Sowerby, 1868: pl. 69, fig. 307 [misidentification].

Unio hydianus Lea, 1838. Küster, 1861: 201, pl. 67, fig. 1 [misidentification].

Unio distans Anthony, 1865: 156, pl. 13, fig. 2 [Type locality: "Ohio." Holotype MCZ 146978].

Unio superiorensis Marsh, 1897: 103, pls. 1, 2 ,5 [Type locality: "Michipicoton River, upper shore of Lake Superior, Canada." Holotype ANSP 70499a].

Lampsilis luteola (Lamarck, 1819). Baker, 1898b: 103, pl. 11, pl. 37, fig. 12.
Lampsilis superiorensis (Marsh, 1897). Simpson, 1900a: 535.
Lampsilis (*Ligumia*) *fasciata superiorensis* (Marsh, 1897). Frierson, 1927: 72.
Lampsilis siliquoidea pepinensis Baker, 1927b: 223 [Type locality: "Lake Pepin, Lake City, Minn." Lectotype UMMZ 209208, paralectotypes USNM 8509, UIMNH Z-22075].
Lampsilis siliquoidea chadwicki Baker, 1928a: 279–281, pl. 91, figs. 5–8 [Type locality: "Doemel Point, Lake Winnebago, near Oshkosh [WI]." Lectotype UMMZ 209200, paralectotypes UIMNH Z-22076].
Lampsilis radiata siliquoidea (Barnes, 1823). Clarke & Berg, 1959: 59.
Ligumia siliquoidea (Barnes, 1823). Haas, 1969: 433.
Lampsilis radiata luteola (Lamarck, 1819). Starrett, 1971: 336.

Status. Not listed by either Ohio Division of Wildlife or U.S. Fish & Wildlife Service as of 2009.

ADULT SHELL

Shell. Medium to large size (to 130 mm), usually heavy for size, sexually dimorphic, both sexes usually elongate oval, rarely more circular, female more broadly rounded posteriorly; some females are considerably higher posteriorly than anteriorly.
Anterior margin. Straight or concave before the umbo, becoming broadly rounded or somewhat spatulate.
Ventral margin. Males straight or very gently rounded; females straight until the marsupial region, then either broadly rounded or conspicuously inflated.
Posterior margin. Males may have a blunt angulation at the juncture of the posterior and ventral margins, and a second, acute angulation approximately halfway up the margin; margin then straight or broadly rounded to the dorsal margin; females similar in outline, but more inflated in profile.
Dorsal margin. Long and straight to the umbo.
Umbo. Wide and low; more prominent in some ventricose populations; umbonal sculpture of 5–10, closely spaced, fine ridges; the earliest ridges are "V"-shaped; later ridges become wavy and nearly longitudinal; umbonal sculpture disappears by end of first year.
Ligament. Brown to black, prominent, occupying entire length of posterior dorsal margin.
Disc sculpture. Sculptureless except for a few very weak striae on the posterior slope.
Periostracum. Yellow, greenish yellow, rarely tan or brown; patterned with radiating dark green rays; these rays may be broad and widely spaced, narrow and thin, or a combination, often somewhat wavy; rays may be bold or very subdued, and may disappear on old individuals; individuals completely lacking rays are very rare.
Nacre. White, iridescent posteriorly, often with gold or greenish Tulberg layers.
Hinge plate. Fairly light, even in large individuals; interdentum long, narrow, cordlike, straight or slightly arched; two divergent cardinals in left valve; one in right, usually accompanied by a smaller anterior denticle; laterals very long in males, somewhat shorter in females, usually straight, sometimes gently arched.
Muscle scars. Anterior adductor impressed but not deep; posterior retractor may be completely confluent or distinct from posterior pedal retractor scar; dorsal suspensors multiple, attached in umbonal cavity beside interdentum; pallial line continuous, accessory line very weak or imperceptible.
Umbonal cavity. Wide, open, moderately deep.

Glochidia. Subspatulate, dorsal margin straight, anterior and posterior margins dorsally divergent and ventrally parallel, ventral margin gently curved. Dorsal alae long, exterior surface sculpturing rough. Hooks absent, micropoints lanceolate and arranged in broken vertical rows (Lea, 1858c: 47, pl. 5, fig. 10; Arey, 1924b: 334, pl. 1, fig. 1; Hoggarth, 1999).

Literature records (Surber, 1912: 9, pl. 2, fig. 15; Ortmann, 1912a: 349; Utterback, 1916: 447; Hoggarth, 1999) indicate glochidial dimensions for length (0.23 to 0.25 mm), height (0.28 to 0.30 mm), and hinge (0.11 to 0.12 mm). Utterback (1916: 447) noted, "Glochidia rather large (uniformly smaller for lacustrine *luteola*)." Surber (1912) and Ortmann (1912a) gave length and height measurements for the glochidium of *L. r. luteola*; Ortmann's (1912a) figures are 230 µm x 280 µm (almost identical to ours), and Surber (1912) gave 250 µm x 290 µm (almost the same as our measurements for *L. r. radiata*).

Reproductive biology. Bradytictic. The posterior half of the outer gills is marsupial (Ortmann, 1912a). Females release glochidia year-round, with most released from June through August when water temperatures reach a sustained 15°C (Watters, 1996d). Many reported hosts are centrarchids, although this species appears to be a generalist.

Reported potential hosts

Ambloplites rupestris (Rock Bass)—Evermann & Clark, 1918 (LI), 1920 (LI); Weir, 1977 (LT).
Catostomus commersoni (White Sucker)—Kakonge *in* Fuller, 1978 (NS).
Lepisosteus platyrhincus (Florida Gar)— Keller & Ruessler, 1997 (LT).
Lepomis cyanellus (Green Sunfish)—Watters *et al.*, 2005 (LT).

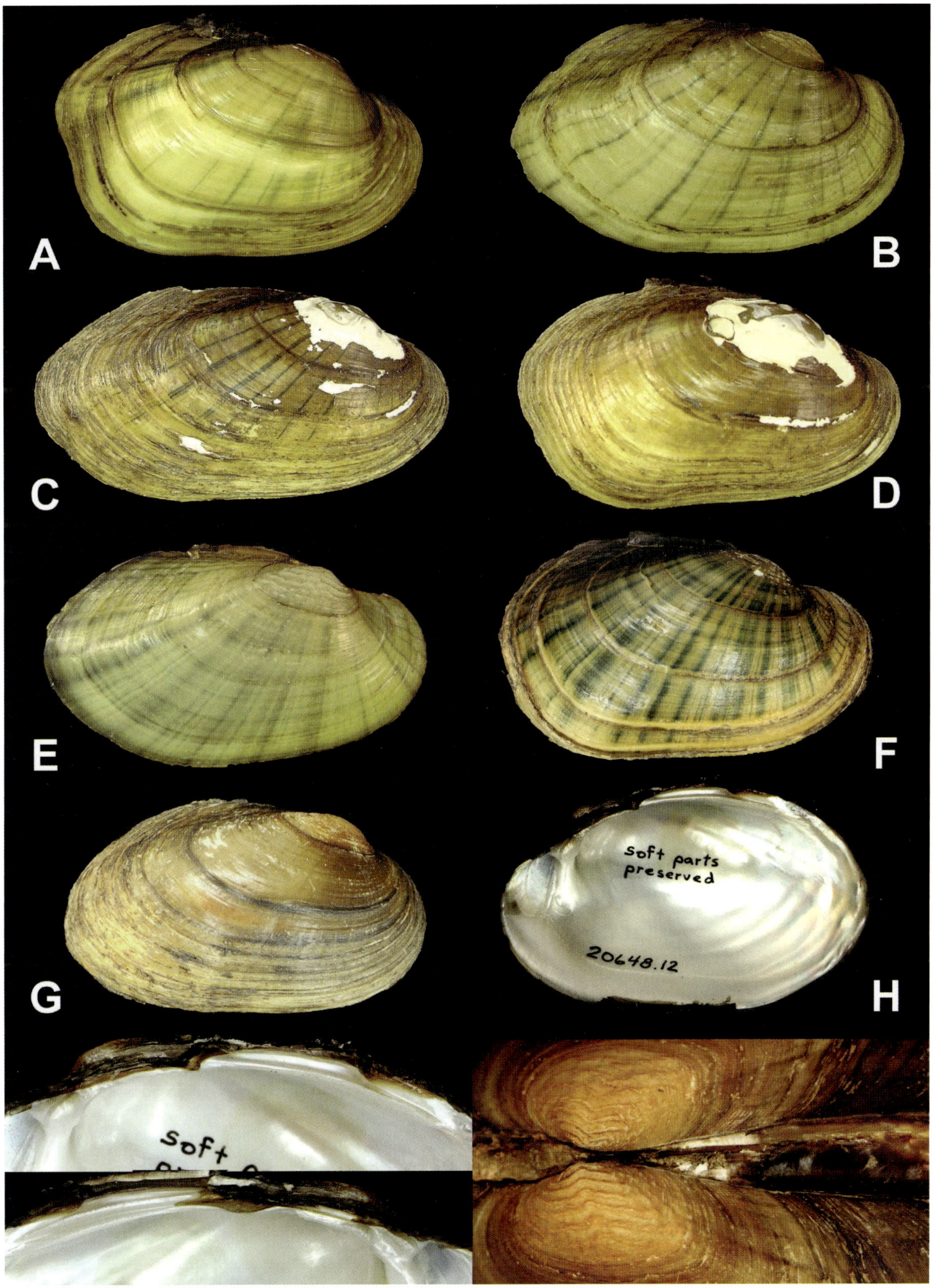

Fig. 97. ***Lampsilis radiata luteola*** **(Lamarck, 1819). A.** OSUM 13110.2, Lake Erie, Ottawa Co., OH. 72 mm. **B.** OSUM 63425.1, Paint Creek, Fayette Co., OH. 69 mm. **C.** OSUM 62944.4, Sunfish Creek, Pike Co., OH. 123 mm. **D.** OSUM 62944.3, Sunfish Creek, Pike Co., OH. 113 mm. **E.** OSUM 26151.1, Alum Creek, Delaware Co., OH. 38 mm. **F.** OSUM 63061, Meigs Creek, Morgan Co., OH. 82 mm. **G.** OSUM 20648.13, Little Darby Creek, Madison Co., OH. 69 mm. **H, Hinge detail:** OSUM 20648.12, Little Darby Creek, Madison Co., OH. 65 mm total length. **Beak sculpture detail:** OSUM 20648.3, Little Darby Creek, Madison Co., OH. 52 mm total length.

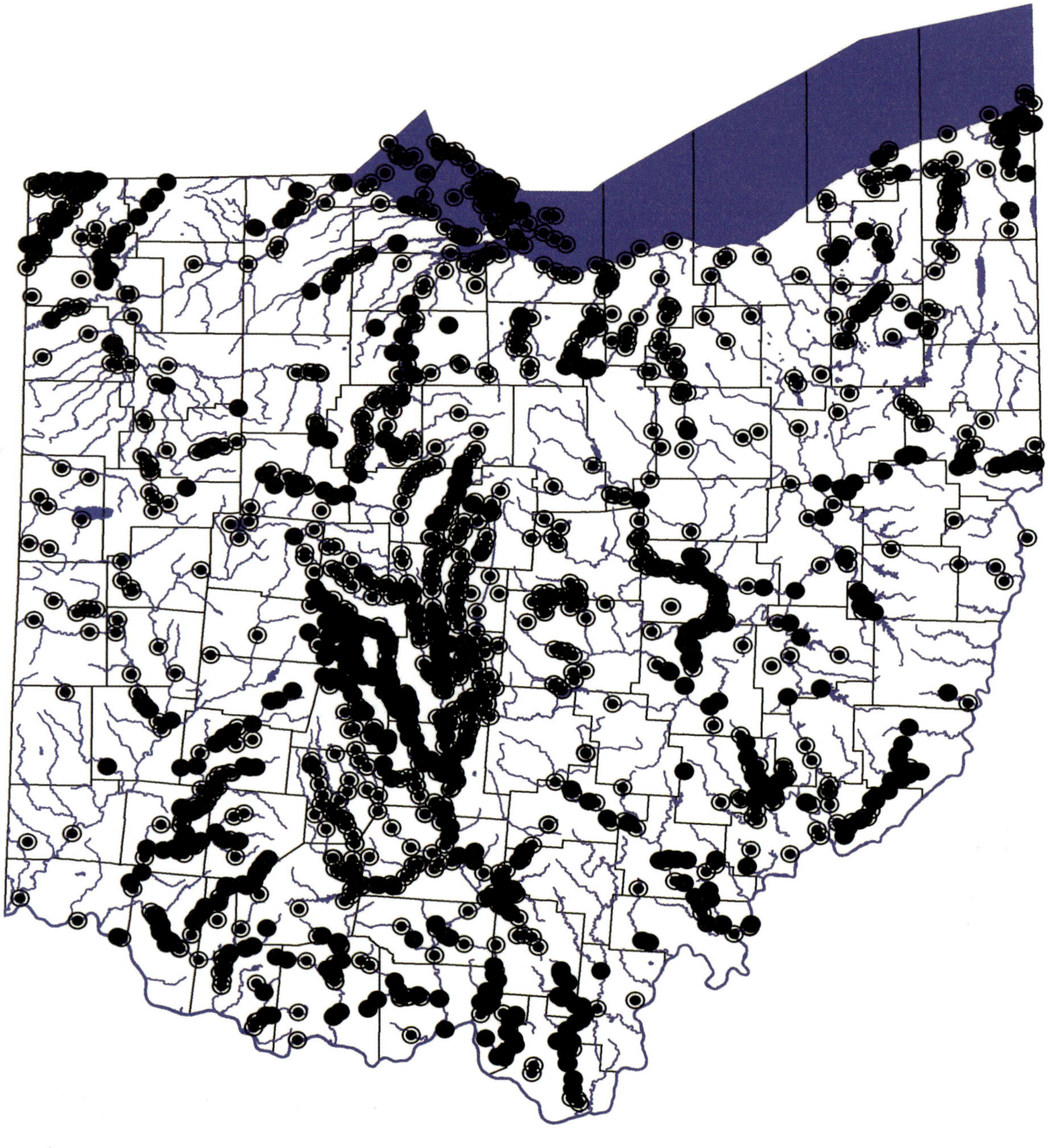

Fig. 98. ***Lampsilis radiata luteola*** **(Lamarck, 1819)**
Solid dots—live or freshdead since 1980
Bullseye—all records, any condition

Lepomis gibbosus (Pumpkinseed)—Trdan, 1981 (NI).
Lepomis gulosus (Warmouth)—Trdan, 1981 (NI).
Lepomis macrochirus (Bluegill)—Evermann & Clark, 1918 (LI); Evermann & Clark, 1920 (LI); Coker *et al.*, 1921 (NI, LT); Howard, 1922 (LT); Weir, 1977 (LT); Trdan, 1981 (NI); Keller & Ruessler, 1997 (LT); Watters & O'Dee, 1997d (LT); O'Dee & Watters, 2000 (LT); Draxler *et al.*, 2006 (LT).
Lepomis megalotis (Longear Sunfish)—Watters & O'Dee, 1997d (LT); O'Dee & Watters, 2000 (LT).
Luxilus cornutus (Common Shiner)—Kakonge *in* Fuller, 1978 (NS).
Micropterus dolomieu (Smallmouth Bass)—Corwin, 1920 (BI); Coker *et al.*, 1921 (LT); Watters & O'Dee, 1997d (LT); O'Dee & Watters, 2000 (LT).
Micropterus salmoides (Largemouth Bass)—Howard, 1914c (LT); Reuling, 1919 (LT); Coker *et al.*, 1921 (LT); Howard, 1922 (LT); Arey, 1932a (LT), 1932b (LT), 1932c (LT); Weir, 1977 (LT); Trdan, 1981 (NI); Keller & Ruessler, 1997 (LT); Watters & O'Dee, 1997d (LT); O'Dee & Watters, 2000 (LT); Watters *et al.*, 2005 (LT); Draxler *et al.*, 2006 (LT).
Morone chrysops (White Bass)—Corwin, 1920 (BI); Coker *et al.*, 1921 (LT).
Notropis ludibundus (Sand Shiner)—Watters & O'Dee, 1997d (LT); O'Dee & Watters, 2000 (LT).
Noturus gyrinus (Tadpole Madtom)—Coker *et al.*, 1921 (NI).
Perca flavescens (Yellow Perch)—Corwin, 1920 (NS); Coker *et al.*, 1921 (NI, LT); Pearse, 1924 (NI); Weir, 1977 (LT); Trdan, 1981 (NI); Draxler *et al.*, 2006 (LT).
Pimephales notatus (Bluntnose Minnow)—Watters & O'Dee, 1997d (LT); O'Dee & Watters, 2000 (LT).
Pomoxis annularis (White Crappie)—Coker *et al.*, 1921 (LT); Howard, 1922 (NI); Trdan, 1981 (NI).
Pomoxis nigromaculatus (Black Crappie)—Coker *et al.*, 1921 (LT); Howard, 1922 (LT); Trdan, 1981 (NI).
Sander canadensis (Sauger)—Corwin, 1920 (NS); Coker *et al.*, 1921 (LT).
Sander vitreus (Walleye)—Corwin, 1920 (NS); Coker *et al.*, 1921 (NI, LT); Corwin, 1921 (LT); Trdan, 1971 (NI); Waller & Mitchell, 1989 (LT).

Habitat. One of the most common species in the state, it tolerates nearly every substrate and flow regime. It occurs from extreme headwaters to ponds, lakes, and rivers, although it is rare in the largest rivers.

Range. One of the most widespread unionoids in the world, ranging from the Upper Mackenzie River system in Northwest Territories across Saskatchewan, Alberta, Manitoba, and Ontario in Canada. Upper Mississippi River drainage, Ohio River drainage (but peculiarly absent from the Cumberland and Tennessee rivers), west of the Mississippi River in the Missouri, White, and Arkansas rivers. Great Lakes drainages.

Range in Ohio. Widespread throughout all of Ohio, from intermittent creeks to large rivers (although rare there), lakes, and ponds.

Similar species. This species is more elongate than either *L. cardium* or *L. fasciola*, and not as elongated as *L. teres*. The low umbo and fine, wavy beak sculpture separate it from all other *Lampsilis*. Males may resemble *Actinonaias ligamentina* but, again, differ in their beak sculpture.

Comments. There is considerable difference in the size of individuals of the same age between populations. For example, medium-sized specimens (ca. 80 mm) may vary in age from 3 to 12 years. Individuals may reach 130 mm in length, but rarely exceed 20 years in age. There also is great variation in the amount of shell sexual dimorphism found in this species. In some populations, differences between males and females may be very subtle, with little evidence of a marsupial region of the shell, while in others the females are greatly inflated. Sexual differences in the shells may not be apparent until the third year of growth. Some females seem to revert to a male, or at least asexual, shell form in later life, perhaps due to senescence.

Lampsilis teres (Rafinesque, 1820)

Figures 99, 100, 101

Etymology. L. *teres*, rounded off, rounded, well-turned, smooth. Individuals of this species are indeed rounded in cross section, often nearly cylindrical in shape.
Vernacular name(s). *Yellow Sandshell*, Bankclimber, Slough Sandshell, Banana Shell.
Original description
Unio (Eurynia) teres Rafinesque, 1820: 321, species no. 66, no figure.
Type locality. "Dans la rivière Wabash." ["In the Wabash River"].
Type material. Conrad (1836: 52–53, pl. 2d) noted under his account of this species, "I have figured a specimen in Mr. Poulson's cabinet, brought from the west and labeled by Rafinesque." Johnson & Baker (1973: 172) interpret the above as designating the type of this species. Conrad's action apparently is valid under ICZN Art. 74.4, 5, and 6, although Conrad did not explicitly state that he was illustrating a type. Should type material be discovered, this figured specimen would become a lectotype. Johnson & Baker (1973: 172) reported that this specimen is lost.
Synonymy
Unio (Eurynia) teres Rafinesque, 1820: 321.

Fig. 99. *Lampsilis teres* (Rafinesque, 1820). A. OSUM 5990, Ohio River, Hamilton Co., OH. 108 mm. **B.** OSUM 45921.1, Muskingum River, Washington Co., OH. 100 mm. **C.** OSUM 58696, Ohio River, Hamilton Co., OH. 85 mm. **D.** OSUM 26849, Green River, Warren Co., KY. 94 mm. **E.** OSUM 12812.2, St. Francis River, Cross Co., AR. 127 mm. **F.** OSUM 44337, Wabash River, Vermillion Co., IN. 131 m. **G, H, Hinge detail:** OSUM 30937, Ohio River, Pulaski Co., IL. 107 mm. **Beak sculpture detail:** OSUM 25474.1, Spoon River, Fulton Co., IL. 57 mm total length.

Unio anodontoides Lea, 1834: 81–82, pl. 8, fig. 11 [Type locality: "Mississippi river. Alabama river. Ohio river." Figured lectotype USNM lost; paralectotype in ANSP lost *fide* Johnson & Baker (1973: 148)].
Margarita (*Unio*) *anodontoides* (Lea, 1834). Lea, 1836: 35.
Lampsilis anodontoides (Lea, 1834). Baker, 1898b: 100, pl. 10, fig. 1.
? *Unio floridensis* Lea, 1852a: 274, pl. 21, fig. 31 [Type locality: "Chácktáhachi river, West Florida," restricted by Clench & Turner (1956: 202) to "Choctawhatchee River, 1 mile west of Caryville, Holmes County, Florida." Lectotype ANSP 42081].
Margaron (Unio) floridensis Lea, 1852b: 39.
Margaron (*Unio*) *anodontoides* (Lea, 1834). Lea, 1852b: 39.
Unio oriens Lea, 1834. Sowerby, 1868: pl. 63, fig. 314 [misidentification].
Lampsilis fallaciosus Smith, 1899: 291, pl. 79 [Type locality: "Ohio River, Sedamsville, Ohio." Possible holotype USNM 30023a. Simpson (1900a: 544) gave himself credit for this species as "Simpson in Smith," but this has not been adopted by later workers.].
Lampsilis (*Eurynia*) *anodontoides* var. *floridensis* (Lea, 1852). Simpson, 1900a: 544.
Lampsilis (*Eurynia*) *fallaciosa* Smith, 1899. Simpson, 1900a: 544.
Lampsilis anadontoides var. *floridensis* (Lea, 1852). Wright & Walker, 1902: 9.
Lampsilis anodontoides fallaciosa Smith, 1899. Grier & Mueller, 1926: 18.
Lampsilis (*Ligumia*) *teres* (Rafinesque, 1820). Frierson, 1927: 70.
Lampsilis (*Ligumia*) *teres floridensis* (Lea, 1852). Frierson, 1927: 70.
Lampsilis (*Ligumia*) *teres fallaciosa* Smith, 1899. Frierson, 1927: 70.
Lampsilis anodontoides floridensis (Lea, 1852). Clench & Turner, 1956: 201–203, pl. 3, fig. 1.
Lampsilis anodontoides anodontoides (Lea, 1834). Murray & Leonard, 1962: 143.
Ligumia teres teres (Rafinesque, 1820). Haas, 1969: 431.
Ligumia teres floridensis (Lea, 1852). Haas, 1969: 432.
Lampsilis anodontoides form *anodontoides* (Lea, 1834). Valentine & Stansbery, 1971: 30.
Lampsilis anodontoides form *fallaciosa* Smith, 1899. Valentine & Stansbery, 1971: 30.
Lampsilis teres teres (Rafinesque, 1820). Oesch, 1984: 207–209.
Lampsilis teres anodontoides (Lea, 1834). Oesch, 1984: 209–212.

Status. Listed as Endangered by Ohio Division of Wildlife; not listed by U.S. Fish & Wildlife Service as of 2009.

ADULT SHELL

Shell. Elongate (to at least 140 mm), very solid, inflated, sexually dimorphic with females having a posterior-ventral swelling.
Anterior margin. Straight before the umbo, then strongly curved towards the ventral margin.
Ventral margin. In males, gently and evenly curved to the posterior point; in females, straight or gently curved, becoming concave before the marsupial region; marsupial region slightly inflated.
Posterior margin. Males come to an abrupt, acute point at or above the middle of the margin, then become straight or curved to the dorsal margin.
Dorsal margin. Long, usually straight.
Umbo. Wide, very flattened, barely protruding above the dorsal margin; beak sculpture extremely fine, composed of 3–5 straight, oblique, parallel threads that begin at mid-shell and run at a posterior-ventral angle to just before the posterior slope; sculpture disappears before end of first year's growth.
Ligament. Tan to dark brown, very long, narrow, partially concealed between valves.
Disc sculpture. Shell sculptureless except for very weak (often absent) radiating threads on posterior slope.
Periostracum. Pale yellow to yellow, tan in larger specimens, usually very polished, with a waxy texture; large specimens may be dull; overlaid with pale green rays, often so faint as to be nearly indistinguishable, rarely completely rayless on disc; posterior slope always with at least one dark ray, occasionally entire slope darkly colored.
Nacre. Porcelaneous white, occasionally flushed with salmon or yellow; golden or tan Tulberg layers rarely present.
Hinge plate. Narrow, interdentum long, cord-shaped; two cardinals in left valve, the posterior one weaker than the anterior; one cardinal in right valve, with a thin, lamellate triangular accessory tooth that is nearly flush with the shell margin; laterals long, straight or very slightly bowed, two in left valve, one in right.
Muscle scars. Adductor scars large, smooth; anterior pedal adductor moderately deep, especially in older individuals; posterior adductor flush with shell posteriorly, slightly indented anteriorly; dorsal suspensors multiple, placed within umbonal cavity and under interdentum; pallial line complete, accessory weak or absent.
Umbonal cavity. Moderately deep, but open.

Glochidia. Subspatulate, dorsal margin straight, anterior and posterior margins equal, dorsally divergent and parallel to slightly convergent ventrally, ventral margin gently curved. Dorsal alae long, exterior surface sculpturing

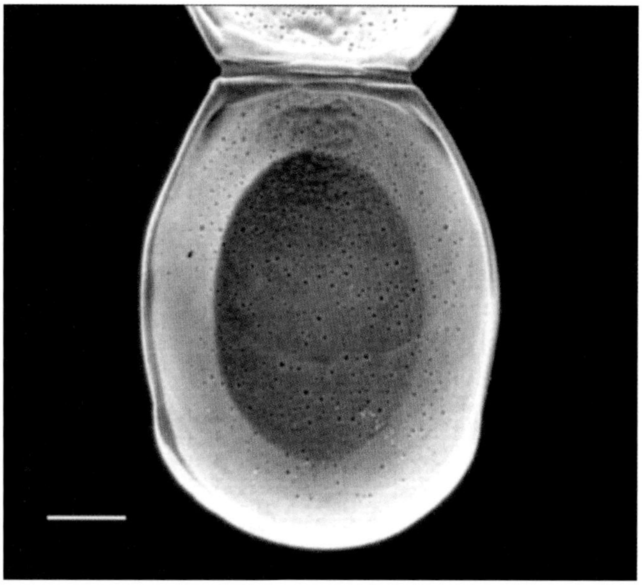

Fig. 100. *Lampsilis teres* (Rafinesque, 1820). Glochidium. OSUM 51669.2, Mississippi River, Crawford Co., WI. Bar = 40 µm. (Hoggarth, 1999).

rough. Hooks absent, micropoints lanceolate and sharply pointed (Hoggarth, 1999: figs. 58a–e).

Literature records (Surber, 1912: 9, pl. 2, fig. 22; Utterback, 1916: 445; Hoggarth, 1999: figs. 58a–e) indicate glochidial dimensions for length (0.19 to 0.20 mm), height (0.24 to 0.27 mm), and hinge (0.10 to 0.12 mm). This glochidium is much smaller than any other recorded for *Lampsilis*, except *L. t.* form *anodontoides*.

Lampsilis t. form *anodontoides*. Subspatulate, dorsal margin straight, anterior and posterior margins equal, dorsally divergent and parallel to slightly convergent ventrally, ventral margin gently curved. Dorsal alae long, exterior surface sculpturing rough. Hooks absent, micropoints lanceolate (Lea, 1858c: 46, pl. 5, fig. 2; Hoggarth, 1999: figs. 59a–e).

Literature records (Surber, 1912: 9, pl. 2, fig. 21; Ortmann, 1912a: 347, pl. 20, fig. 9; Utterback, 1916: 443; Hoggarth, 1999: figs. 59a–e) indicate glochidial dimensions for length (0.19 to 0.20 mm), height (0.24 to 0.26 mm), and hinge (0.11 to 0.12 mm). Hoggarth (1999) noted, "This glochidium is identical to that of *L. t. teres*. Surber (1912) gave length and height measurements of 185 µm x 210 µm for this glochidium and Ortmann (1912) gave 200 µm x 260 µm. Surber's measurements seem too small even for this glochidium and his figure (1912, Pl. 2, Fig. 21) shows the ventral margin as semicircular rather than gently curved. These discrepancies suggest that his material was not mature. This glochidium is also figured by Lea (1858, pl. 5, Fig. 2) and Ortmann (1912, pl. 20, Fig. 9). Lea's figure is much closer to Ortmann's and mine than Surber's figure."

Reproductive biology. Bradytictic. The posterior half of the outer gills is marsupial (Ortmann, 1912a). Gravid females are found in November (Baker, 1928a). White, sole-shaped conglutinates are discharged (Utterback, 1916). Although numerous fishes have been reported as hosts, few of them are based on evidence of metamorphosis.

Reported potential hosts

Ambloplites rupestris (Rock Bass)—Lefevre & Curtis, 1912 (LI).
Atractosteus spatula (Alligator Gar)—Howard, 1914b (NI); Wilson, 1916 (NI); Coker *et al.*, 1921 (LT).
Cyprinus carpio (Common Carp)—Lefevre & Curtis, 1912 (LI).
Etheostoma jordani (Greenthroat Darter)—Prentice *in* Howells, 1994b (LT).
Lepisosteus osseus (Longnose Gar)—Howard, 1914b (NI); Wilson, 1916 (NI); Coker *et al.*, 1921 (NI, LT); Barney, 1922 (LI); Arey, 1932a (LT); Jones, 1950 (LI).
Lepisosteus platostomus (Shortnose Gar)—Howard, 1914b (LI); Wilson, 1916 (NI); Coker *et al.*, 1921 (LT); Barney, 1922 (LI); Howard & Anson, 1923 (LT); Ellis & Ellis, 1926 (LT); Arey, 1932b (LT), 1932c (LT); Jones, 1950 (LI).
Lepomis auritus (Redbreast Sunfish)—Prentice *in* Howells, 1994b (LT).
Lepomis cyanellus (Green Sunfish)—Surber, 1913 (NI).
Lepomis gulosus (Warmouth)—Wilson, 1916 (NI).
Lepomis humilis (Orangespotted Sunfish)—Surber, 1913 (NI).
Lepomis macrochirus (Bluegill)—Lefevre & Curtis, 1912 (LI); Prentice *in* Howells, 1994b (LT).
Micropterus salmoides (Largemouth Bass)—Wilson, 1916 (NI).
Perca flavescens (Yellow Perch)—Lefevre & Curtis, 1912 (LI).
Pomoxis annularis (White Crappie)—Surber, 1913 (NI); Wilson, 1916 (NI).
Pomoxis nigromaculatus (Black Crappie)—Howard, 1913b (NI).
Rutilus rutilus (Roach)—Lefevre & Curtis, 1912 (LI).
Scaphirhynchus platorynchus (Shovelnose Sturgeon)—Surber, 1913 (NI); Wilson, 1916 (NI).

Habitat. Found in sandy mud in large rivers, it rarely strays into smaller streams.

Range. In its various forms, this species is widespread throughout most of the Mississippi River system and Gulf drainages. West to the Missouri, White, Meramec, and St. Francis rivers. North to the upper Mississippi River. East to New York. Absent from the Great Lakes system.

Fig. 101. ***Lampsilis teres*** **(Rafinesque, 1820)**
Solid dots—live or freshdead since 1980
Bullseye—all records, any condition

Range in Ohio. The last living specimen from within the borders of the state was found in lower Ohio Brush Creek in 1988 (Watters, 1988). Common in the Ohio River immediately below the Gallipolis Lock and Dam in Gallia County (Watters, unpubl.), but rare elsewhere along the Ohio shore. Absent from the Great Lakes drainages. Historically from the lower Scioto, Great Miami, and Muskingum rivers.

Similar species. The extremely elongated, waxy yellow shell can be confused only with some extreme examples of *L. r. luteola*. That species has a fine, wavy beak sculpture rather than the parallel threads found in *L. teres*.

Comments. Despite its size, this is a relatively short-lived mussel. Individuals grow rapidly most of their life, only appreciably slowing their growth at 7 years or more. Female shell characteristics are apparent by age 3. Large specimens (130 mm or more) may be less than 10 years old, and few are found older than this.

Three subspecies or forms are often recognized based primarily on size, coloration, and distribution. *Lampsilis teres teres*, more commonly known as *Lampsilis teres fallaciosa* (Smith, 1899), tends to be smaller than the other taxa and is rayed. *Lampsilis teres anodontoides* (Rafinesque, 1820) is large and nearly uniformly yellow. However, there appears to be a continuum between the two forms. *Lampsilis t. anodontoides* may be the big river form of the species (it is this form that occurs in Ohio). We do not recognize the distinction here. Williams *et al.* (2008) recognized the third subspecies or form, *L. t.* form *floridensis* (Lea, 1852), as a distinct species.

Lasmigona Rafinesque, 1831

Etymology. Gr. *elasma*, metal beaten out, metal plate + Gr. *gonia*, corner, angle. Rafinesque used the word *elasma* to refer to a unionoid lateral tooth, which is often thin and plate-like. Of *Lasmigona*, he notes (1831: 4) that the usual lateral tooth is replaced by "an horizontal angular projection, flat above." Although he doesn't call it by name, Rafinesque is actually describing a claustrum.

Original description

Lasmigona Rafinesque, 1831: 4–5, no figure.

Type species. *Alasmidonta costata* Rafinesque, 1820, by subsequent designation of Simpson (1900a: 664). Rafinesque gave *Alasmidonta marginata* Say, 1817, and *Alasmidonta costata* Rafinesque, 1820, as examples of his *Lasmigona*.

Synonymy

Lasmigona Rafinesque, 1831: 4–5, no figure.

Sulcularia Rafinesque, 1831: 4–5 [Type species: *Sulcularia badium* Rafinesque, 1831, by subsequent designation of Frierson, 1914a].

Amblasmodon Rafinesque, 1831: 4–5 [Type species: *Amblasmodon hians* Rafinesque, 1831 (= *Alasmidonta costata* Rafinesque, 1820), by original designation].

Pterosyna Rafinesque, 1831: 5 [Type species: *Alasmidonta complanata* "Say" (error for Barnes, 1823) by original designation].

Alasmodon [*sic*] Rafinesque, 1831: 5.

? *Gonamblus* Rafinesque, 1831: 5 [conditionally proposed name for *Alasmodon ponderosum* Rafinesque, 1831 (= ?)].

Amblamadon [*sic*] Rafinesque, 1831: 5.

? *Megadomus* Swainson, 1840: 265, 275, 378 [Type species: *Megadomus gigas* Swainson, 1840, by original designation. Swainson mentions only a single lateral tooth in each valve.].

Complanaria Swainson, 1840: 228, 290, 382 [Type species: *Complanaria gigas* Swainson, 1840 (= *Alasmidonta complanata* Barnes, 1823), by original designation].

Elasmogona [*sic*] Agassiz, 1846: 583.

Elasmigena [*sic*] Herrmannsen, 1852: 49.

Symphynota (*Symphynota*) Simpson, 1900a: 662 [*non Symphynota* Lea, 1829. Type species: *Symphynota compressa* Lea, 1829].

Symphynota (*Pterosygna*) [*sic*] Simpson, 1900a: 665.

Symphynota (*Alasminota*) Ortmann, 1914: 42 [Type species: *Unio holstonia* Lea, 1838, by original designation].

Platynaias Walker, 1918: 2 [Type species: *Symphynota compressa* Lea, 1829, by original designation].

Geographic range. North America east of the Rocky Mountains from southern Ontario to the Gulf of Mexico. On the Atlantic Slope from New Hampshire to South Carolina.

Comments. This genus is unique in possessing a composite conglutinate C composed of glochidia tethered to a core of unfertilized eggs (Watters, 2008).

Lasmigona complanata (Barnes, 1823)

Figures 102, 103, 104

Etymology. L. *complanatus*, flattened. Most individuals of this species are laterally flattened or compressed.

Vernacular name(s). *White Heelsplitter*, Elephant Ear.

Original description

Alasmodonta complanata Barnes, 1823: 278, fig. 22.

Type locality. "Fox River" and "Wisconsan [*sic*]."

Type material. The types of Barnes (1823) are apparently lost.

Synonymy

Alasmodonta complanata Barnes, 1823: 278–279, fig. 22.

? *Unio gigas* Swainson, 1824: 15–17 [Type locality: "Oronokoo [Orinoco] River [Venezuela]" in error. Type lost].

Fig. 102. ***Lasmigona complanata*** **(Barnes, 1823). A.** OSUM 56322, Walnut Creek, Pickaway Co., OH. 155 mm. **B.** OSUM 63462, Lake Erie, Ottawa Co., OH. 121 mm. **C.** OSUM 52619.3, Caesar Creek, Warren Co., OH. 162 mm. **D, F.** OSUM 46538, Little Muskingum River, Washington Co., OH. 101 mm. **E.** OSUM 41353.1, Paint Creek Reservoir, Ross Co., OH. 44 mm. **Hinge detail:** OSUM 49577.2, Cowan Creek, Clinton Co., OH. 83 mm total length. **Beak sculpture detail:** OSUM 49577.1, Cowan Creek, Clinton Co., OH. 46 mm total length.

Mya complanata (Barnes, 1823). Eaton, 1826: 222.
Symphynota complanata (Barnes, 1823). Lea, 1829: 448.
Pterosyna complanata "Say." Rafinesque, 1831: 5.
Margarita (*Margaritana*) *complanata* (Barnes, 1823). Lea, 1836: 43.
Unio katherinae Lea, 1838a: 143 [Type locality: "Lake Superior." Type in private collection of Lady Katherine Douglas, St. Mary's Isles, Scotland, *fide* Lea, 1838b: 143; presumed lost].
Complanaria gigas Swainson, 1840: 258, 290, 382 [based on Sowerby, 1839, fig. 141 = *Lasmigona complanata* (Barnes, 1823)].
? *Megadomus gigas* Swainson, 1840: 265, 378.
Alasmodon complanatus (Barnes, 1823). Sowerby, 1842: 61, fig. 140.
Complanaria complanata (Barnes, 1823). Agassiz, 1852: 49.
Baphia complanata (Barnes, 1823). Adams & Adams, 1857: 500.
Margaritana complanata (Barnes, 1823). Calkins, 1874: 46.
Symphynota (*Pterosygna* [*sic*]) *complanata* (Barnes, 1823). Simpson, 1900a: 665–666.
Symphynota (*Pterosygna* [*sic*]) *complanata* var. *katherinae* (Lea, 1838). Simpson, 1900a: 666.
Lasmigona (*Lasmigona*) *complanata complanata* (Barnes, 1823). Clarke, 1985: 25–36, figs. 7–9.

Status. Not listed by either Ohio Division of Wildlife or U.S. Fish & Wildlife Service as of 2009.

ADULT SHELL

Shell. Large (to 190 mm), thin-shelled, particularly as juveniles, quite compressed, oval or triangular in profile but usually with a pronounced wing.
Anterior margin. Broadly rounded into ventral margin.
Ventral margin. Nearly straight to broadly rounded, forming oblique angle in females posteriorly.
Posterior margin. Forms 1–2 broad angulations ventrally and then in a straight line obliquely to form the posterior margin of the wing.
Dorsal margin. Forming oblique or right angle with posterior margin of wing, long, more or less straight to umbo.
Umbo. Low, pointed, minute, located far anterior of the midline; umbonal sculpture of 3–4 coarse, broad "W"-shaped ridges, the anterior loop rounded, the posterior one pointed.
Ligament. Long, tan to dark brown, almost completely concealed by shells.
Disc sculpture. Smooth or with a few irregular radiating grooves, or malleated posteriorly; the wing is invariably corrugated.
Periostracum. Green or tan as juveniles; adults usually have the earlier portions somewhat abraded to yellow or white and the newer shell dark brown; often with numerous fine rays.

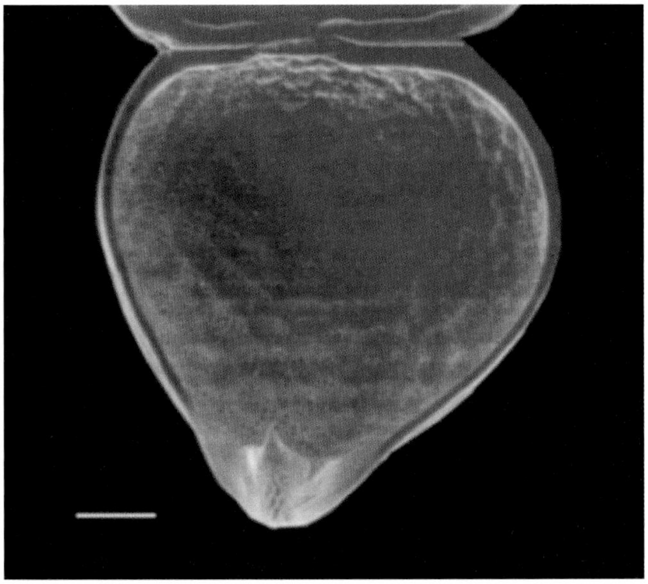

Fig. 103. *Lasmigona complanata* (Barnes, 1823). Glochidium. MAH 278.2, Big Darby Creek, Pickaway Co., OH. Bar = 45 µm. (Hoggarth, 1999).

Nacre. Porcelain white, often iridescent posteriorly; many specimens have irregular gold or brown Tulberg layers.
Hinge plate. Very thick; right cardinal a single, chunky, triangular denticle, often accompanied by an anterior and a posterior accessory denticle, and often radially subdivided; left cardinal bifid or subdivided, extending across the interdentum; laterals thick but not well-developed, becoming obsolete in adults, one in each valve; interdentum short, wide, and flattened.
Muscle scars. Anterior adductor large and shallowly impressed; posterior adductor large and flush; anterior adductor distinct from anterior pedal protractor scar, large posterior adductor confluent with posterior pedal retractor scar; multiple dorsal suspensor located deep within the umbonal cavity. Pallial line entire, located far from shell margin. Accessory line weak, short.
Umbonal cavity. Very wide, shallow, and open.

Glochidia. Pyriform, almost symmetrical, dorsal margin straight, anterior and posterior margins equal, ventral terminus narrowing and furnished with a styliform hook. Exterior surface sculpturing rosette. Microstylets (about 100) lanceolate, multifaceted, and arranged in six proximal rows reduced to four rows distally (Lea, 1858c: 49, pl. 5, fig. 29; Lefevre & Curtis, 1910a: 97, fig. a; Ortmann, 1911: 346, pl. 89, fig. 11; Hoggarth, 1999: figs. 28a–f).

Literature records (Lefevre & Curtis, 1912: 146, fig. 1a; Surber, 1912: 9, pl. 1, fig. 6; Ortmann, 1912a: 283; Utterback, 1915: 250; Clarke, 1985: 28–31, fig. 8; Hoggarth,

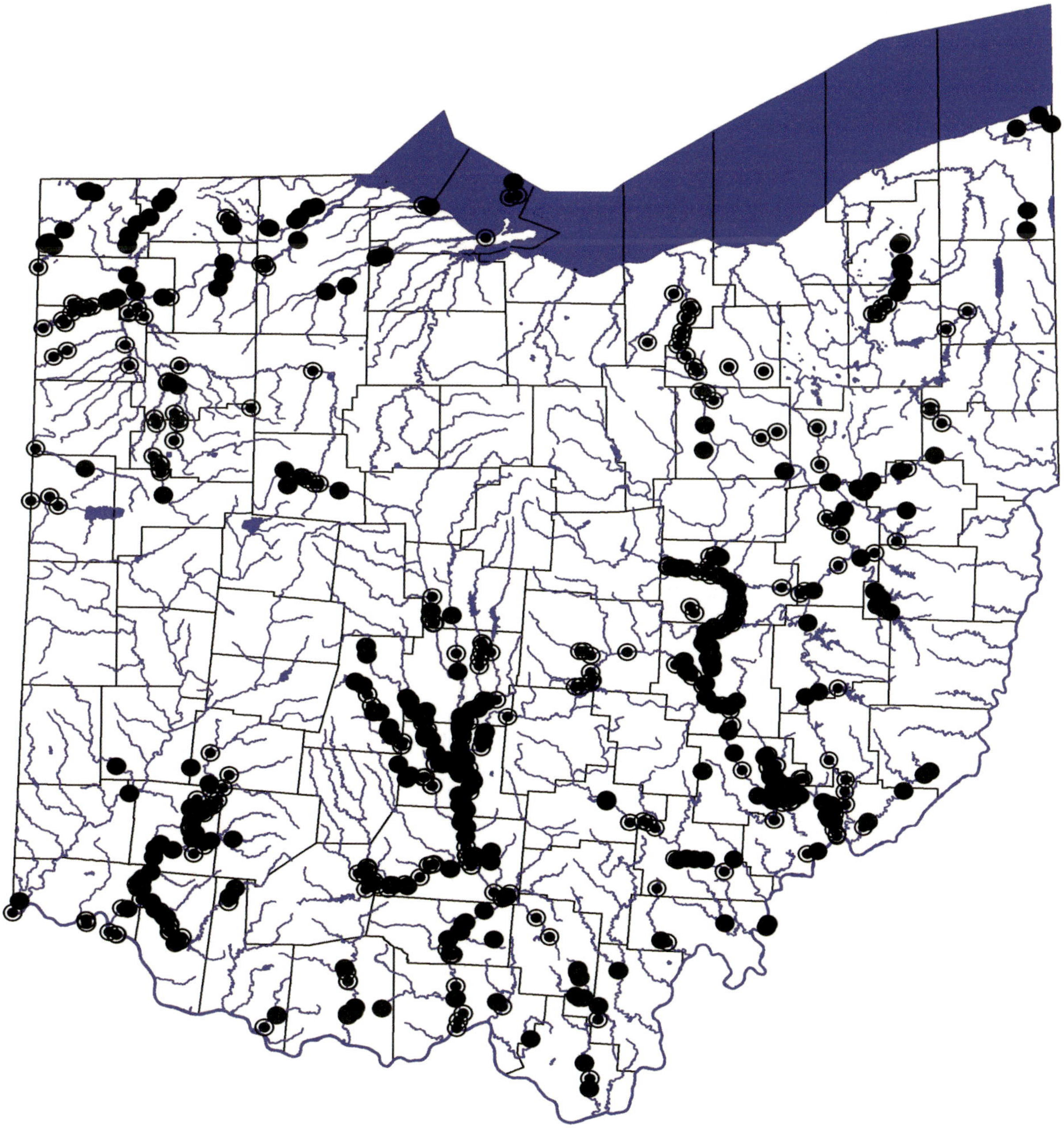

Fig. 104. ***Lasmigona complanata*** **(Barnes, 1823)**
Solid dots—live or freshdead since 1980
Bullseye—all records, any condition

1999: figs. 28a–f) indicate glochidial dimensions for length (0.29 to 0.34 mm), height (0.29 to 0.34 mm), and hinge (0.19 to 0.23 mm). This glochidium can be distinguished by its nearly equal length and height, few micropoints, and exterior valve sculpture.

Reproductive biology. Bradytictic. Marsupia are formed of the entire outer gills (Utterback, 1915). Gravid females occur from September to the following May (Ortmann, 1919). Conglutinates of the composite C type are formed (Watters, 2008). Most reported hosts are centrarchids.

Reported potential hosts

Cyprinus carpio (Common Carp)—Lefevre & Curtis, 1910a (LT); Lefevre & Curtis, 1912 (LI)—exotic.
Dorosoma cepedianum (Gizzard Shad)—Weiss & Layzer, 1995 (NI).
Fundulus diaphanus (Banded Sunfish)—Young, 1911 (LT).
Lepisosteus osseus (Longnose Gar)—Weiss & Layzer, 1995 (NI).
Lepomis cyanellus (Green Sunfish)—Young, 1911 (LT); Lefevre & Curtis, 1912 (LT).
Lepomis humilis (Orangespotted Sunfish)—Young, 1911 (LT).
Lepomis megalotis (Longear Sunfish)—Watters (unpubl.) (LT).
Micropterus salmoides (Largemouth Bass)—Lefevre & Curtis, 1910a (LT); Young, 1911 (LT).
Moxostoma carinatum (River Redhorse)—Weiss & Layzer, 1995 (NI).
Perca flavescens (Yellow Perch)—Watters (unpubl.) (LT).
Pomoxis annularis (White Crappie)—Young, 1911 (LT); Coker *et al.*, 1912 (LT); Lefevre & Curtis, 1912 (LT).
Pomoxis nigromaculatus (Black Crappie)—Watters (unpubl.) (LT).
Sander canadensis (Sauger)—Weiss & Layzer, 1995 (NI).

Habitat. This species is widespread, often found in sluggish water in sandy mud and silt. It occurs in creeks, rivers, and lakes. Because of its ability to parasitize Common Carp, it often is common below sewage outfalls and impoundments.

Range. Mississippi River system from Saskatchewan to Ontario and south, Great Lakes drainages. Absent from the Atlantic Slope. Clarke (1985) designated the southern populations as *L. complanata alabamensis,* but this appears to be an endemic Mobile River species apart from *L. complanata* (Williams *et al.*, 2008).

Range in Ohio. Widespread across the state from small creeks to rivers and lakes; western basin of Lake Erie.

Similar species. The greatly compressed, plate-like shells with a corrugated posterior wing and coarse beak sculpture readily identify this species. Other alate species such as *Potamilus alatus* and. *P. ohiensis* lack a corrugated wing and have pink or purple nacre. *Leptodea fragilis* is consistently yellow and also lacks a corrugated wing. No other alate Ohio mussel has coarse beak sculpture.

Comments. The degree of alateness varies considerably. In high energy environments the wings may be nearly nonexistent, while other populations have well-developed wings. Individuals grow quickly for 3 years and then slow. They grow to 190 mm in length and rarely live more than 12 years.

Lasmigona compressa (Lea, 1829)

Figures 105, 106, 107, 108

Etymology. L. *com-* from *cum*, with, together + L. *pressus*, crowd, bear down upon. Individuals of this species are typically compressed (laterally flattened) individuals.
Vernacular name(s). *Creek Heelsplitter*, Flat Creek Shell.
Original description
Symphynota compressa Lea, 1829: 450–451, pl. 12, fig. 22.
Type locality. Lea gave "Hab. Ohio. T. G. Lea" and "Norman's Kiel, near Albany [NY]." The figured lectotype USNM 83961 restricted the type locality to the Ohio specimen noted by Johnson & Baker (173: 151) as from "[Yellow Creek, near Cincinnati, Hamilton Co.] Ohio." The bracketed information is here presumed to be from the original label.
Type material. Johnson & Baker (1973: 151) recognized the figured specimen, USNM 83961, as the "holotype," but Clarke (1985) correctly designated this same specimen as the lectotype.
Synonymy
Symphynota compressa Lea, 1829: 450–451, pl. 12, fig. 22.
Unio alasmodontina "Barnes" Lea, 1829: 451 [unavailable, introduced in synonymy of *Symphynota compressa* Lea, 1829].
Unio compressus (Lea, 1829). Conrad, 1834: 68.
Margarita (*Unio*) *compressus* (Lea, 1829). Lea, 1836: 11.
Unio pressus Lea, 1842a: 237 [replacement for *Symphynota compressa* Lea, 1829, when transferred to *Unio* by Lea, 1836, to avoid conflict with *Unio compressus* Sowerby, 1828. This action became unnecessary when transferred back to *Symphynota* by Simpson (1900).].
Unio compressus var. *plebius* Adams, 1842: 16 [Type locality: "a small brook in Middlebury [VT]." Lectotype MCZ 154361, 3 paralectotypes MCZ 154366 by selection of Johnson, 1956].

Fig. 105. ***Lasmigona compressa*** **(Lea, 1829). A.** OSUM 9895.3, Cuyahoga River, Portage Co., OH. 86 mm. **B.** OSUM 44509.2, Raccoon Creek, Licking Co., OH. 107 mm. **C.** OSUM 21917.2, Big Darby Creek, Madison Co., OH. 85 mm. **D.** OSUM 24897.2, Norman's Kill, Albany Co., NY. 71 mm. **E, H.** OSUM 40824.1, Lake Fork Licking River, Licking Co., OH. 77 mm. **F.** OSUM 5596, Muskingum River, Muskingum Co., OH. 86 mm. **G, Hinge detail, Beak sculpture detail:** OSUM 1206.1, Fish Creek, Williams Co., OH. 88 mm.

Complanaria alasmodontina Stimpson, 1851: 14 [*nomen nudum*].
Complanaria compressus (Lea, 1829). Agassiz, 1852: 49.
Margaron (*Unio*) *pressus* (Lea, 1842). Lea, 1852b: 19.
Complanaria compressa (Lea, 1829). Conrad, 1853: 261.
Unio alasmodontinas Paetel, 1890: 144 [*nomen nudum*].
Microcondylaea compressa (Lea, 1829). Paetel, 1890: 175.
Alasmodonta pressa (Lea, 1842). Baker, 1898b: 58–60, pl. 6, fig. 3.
Symphynota (*Symphynota*) *compressa* Lea, 1829. Simpson, 1900a: 662–663.
Symphynota (*Symphynota*) *compressa* var. *plebius* (Adams, 1842). Simpson, 1900a: 663.
Unio compressa var. *lindus* De Gregorio, 1914: 44, pl. 6, figs. 1a–d [Type locality: "Sciota" [Scioto River, OH]. Type in Palermo Museum?].
Lasmigona (*Platynaias*) *compressa* (Lea, 1829). Clarke, 1985: 40–50, figs. 12–14.

Status. Listed as a Species of Concern by Ohio Division of Wildlife; not listed by U.S. Fish & Wildlife Service as of 2009.

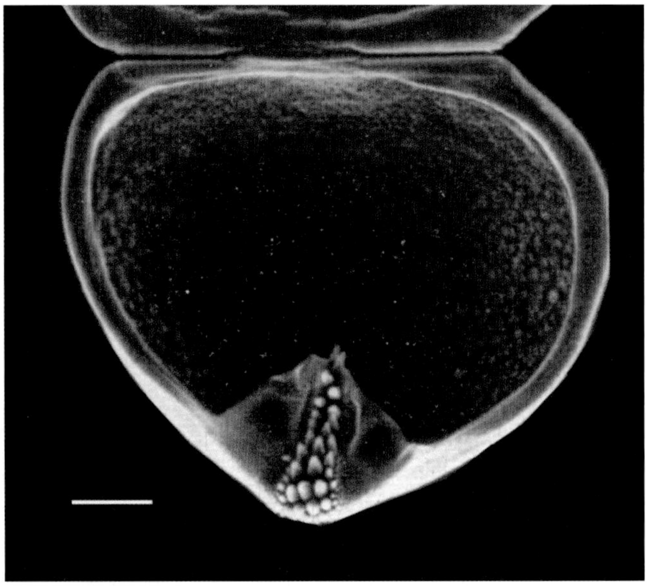

Fig. 106. *Lasmigona compressa* (Lea, 1829). Glochidium. OSUM 23179.1, Little Darby Creek, Madison Co., OH. Bar = 45 µm. (Hoggarth, 1999).

ADULT SHELL

Shell. Medium size (to 110 mm), thin-shelled as juveniles, much thicker with age, somewhat compressed, rather rectangular in profile.
Anterior margin. Roundly arched into ventral margin.
Ventral margin. Nearly straight to broadly rounded.
Posterior margin. Ventral and posterior margins form wide, blunt acute angle, often somewhat expanded ventrally; posterior margin oblique and straight to dorsal margin.
Dorsal margin. Long, more or less straight, joining posterior margin with very obtuse angle.
Umbo. Low, pointed, located about 33% back from anterior margin; umbonal sculpture of 3–4 coarse, undulating or nearly linear ridges, vaguely "W"-shaped.
Ligament. Very long, tan to dark brown, partially concealed by shells.
Disc sculpture. Smooth except for coarse growth lines.
Periostracum. Yellow, tan, or greenish as smallest juveniles, becoming dark brown to black with age. Usually with numerous fine green or brown rays, particularly posteriorly.
Nacre. Porcelain white, often iridescent posteriorly, occasionally with a faint salmon, yellow, or greenish flush beneath the hinge; nearly all specimens seen have gold or brown Tulberg layers.
Hinge plate. Thick; right cardinal a single, chunky, triangular denticle; left cardinal very broadly bifid, the anterior portion very weak, the posterior forming a wide tongue-shaped tooth or interdental projection beneath the interdentum; laterals straight or gently arched, two in left valve, one in right; interdentum wide and flattened.
Muscle scars. Anterior adductor impressed; posterior adductor flush or shallowly impressed; anterior adductor distinct from anterior pedal protractor scar, large posterior adductor confluent with posterior pedal retractor scar; multiple dorsal suspensor located behind cardinal tooth and within the umbonal cavity. Pallial line entire, located far from shell margin. Accessory line weak, short.
Umbonal cavity. Very wide, shallow, and open.

Glochidia. Depressed pyriform, longer than high, and strongly asymmetric, dorsal margin straight, posterior margin strongly curved, anterior margin greatly inflated dorsally, becoming more gently curved ventrally. Posterior and anterior margins join at a gently rounded, nipple-like ventral terminus. Exterior surface sculpturing loose-looped, hooks styliform with about 25 stout microstylets arranged in three proximal rows reduced to two widely offset distal rows (Lea, 1858c: 48, pl. 5, fig. 23; Ortmann, 1911: 346, pl. 89, fig. 10; Hoggarth, 1999: figs. 24a–f).

Literature records (Ortmann, 1912a: 282; Surber, 1912: 8, pl. 3, fig. 44; Tompa, 1979: 188; Clarke, 1985: 44–46, fig. 13; Hoggarth, 1999: figs. 24a–f) indicate glochidial dimensions for length (0.32 to 0.35 mm), height (0.26 to 0.31 mm), and hinge (0.23 to 0.24 mm). This glochidium can be distinguished by its widely offset double row of microstylets and its wide unsculptured distal hook margin.

Reproductive biology. Bradytictic. The outer gill is marsupial. Ortmann (1912, 1919) reported this species as hermaphroditic. Eggs develop in August, and glochidia appear in September. Glochidia are released the following May (Ortmann, 1919; Baker, 1928a). Ortmann (1912a)

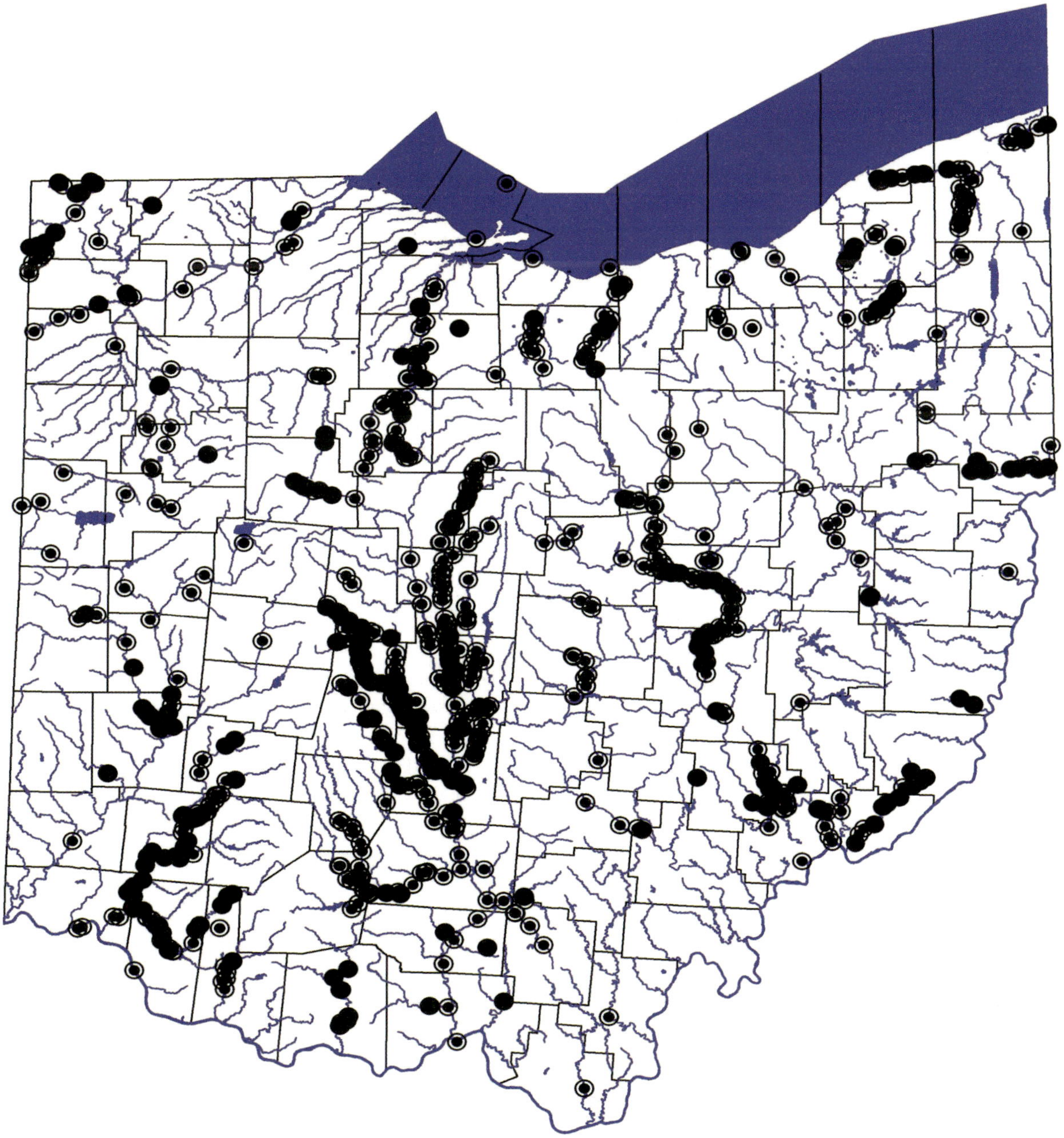

Fig. 107. ***Lasmigona compressa*** **(Lea, 1829)**
Solid dots—live or freshdead since 1980
Bullseye—all records, any condition

Fig. 108. *Lasmigona compressa* (Lea, 1829). Portion of the composite conglutinate type C. Glochidia are tethered to a core of structural programmed eggs. Bar = 500 µm. (Watters, 2008).

stated that this species did not produce a conglutinate, but such is not the case. Conglutinates of the composite C type are formed (Watters, 2008).

Reported potential hosts

Ameiurus melas (Black Bullhead)—McGill *et al.*, 2002 (LT).
Ameiurus natalis (Yellow Bullhead)—McGill *et al.*, 2002 (LT).
Cottus cognatus (Slimy Sculpin)—Hove *et al.*, 1995b (LT).
Culaea inconstans (Brook Stickleback)—McGill *et al.*, 2002 (LT).
Cyprinella spiloptera (Spotfin Shiner)—Hove *et al.*, 1995b (LT); McGill *et al.*, 2002 (LT).
Dorosoma cepedianum (Gizzard Shad)—McGill *et al.*, 2002 (LT).
Hybognathus hankinsoni (Brassy Minnow)—McGill *et al.*, 2002 (LT).
Lebistes reticulatus (Guppy)—Tompa, 1979 (LT)—exotic.
Lepisosteus platostomus (Shortnose Gar)—McGill *et al.*, 2002 (LT).
Lepomis cyanellus (Green Sunfish)—McGill *et al.*, 2002 (LT).
Lepomis humilis (Orangespotted Sunfish)—McGill *et al.*, 2002 (LT).
Lepomis macrochirus (Bluegill)—McGill *et al.*, 2002 (LT).
Micropteris dolomieu (Smallmouth Bass)—McGill *et al.*, 2002 (LT).
Notropis atherinoides (Emerald Shiner)—McGill *et al.*, 2002 (LT).
Notropis volucellus (Mimic Shiner)—McGill *et al.*, 2002 (LT).
Perca flavescens (Yellow Perch)—Hove *et al.*, 1995b (LT).
Pomoxis nigromaculatus (Black Crappie)—Hove *et al.*, 1995b (LT); McGill *et al.*, 2002 (LT).
Pylodictis olivaris (Flathead Catfish)—McGill *et al.*, 2002 (LT).
Rhinichthys cataractae (Longnose Dace)—McGill *et al.*, 2002 (LT).
Rhinichthys sp.—McGill *et al.*, 2002 (LT).
Semotilus atromaculatus (Creek Chub)—McGill *et al.*, 2002 (LT).

Habitat. A high-water-quality species, the Creek Heelsplitter is found in clean creeks in sand and cobble, in a current or in slackwater.

Range. Upper Mississippi River system, Ohio River system (but absent from the Tennessee and Cumberland rivers), Great Lakes drainages. On the Atlantic Slope in the Hudson River and several tributaries of the St. Lawrence River.

Range in Ohio. Widespread but sporadic, rarely in large rivers. Very rare in Lake Erie.

Similar species. The peculiar tongue-like interdental projection of the hinge is a diagnostic feature of this species. *Lasmigona costata* is very similar, but the posterior slope is coarsely corrugated; *L. compressa* possesses lateral teeth, which are only rudimentary in *L. costata*. Small specimens of *L. complanata* may be similar but are more circular and also have a corrugated posterior wing. *Alasmidonta marginata* also has a corrugated posterior slope, and the rays are often spotted.

Comments. Individuals grow very fast for the first 2–3 years before dramatically slowing. This species may grow to 110 mm and live 13 years.

Lasmigona costata (Rafinesque, 1820)

Figures 109, 110, 111, 112

Etymology. L. *costatus*, having rib-like lines or ridges; from L. *costa*, rib. The posterior slope of individuals of this species often has a series of prominent costae.

Vernacular name(s). *Fluted Shell*, Sandshell, Fluted Riffleshell, Hackle-back, Ribbed Creek Shell, Ribbed Riffleshell.

Original description

Alasmidonta costata Rafinesque, 1820: 318, pl. 82, figs. 15, 16.

Type locality. "elle a été recueillié dans la rivière Kentuky, où elle parait être rare" ["it has been collected in the Kentucky (River), where it is seems to be rare"].

Type material. Johnson & Baker (1973) did not include this species in the Rafinesque-Poulson unionoids examined at the Academy of Natural Sciences, Philadelphia. Type lost according to Clarke (1985: 12).

Synonymy

Alasmidonta costata Rafinesque, 1820: 318, pl. 82, figs. 15, 16.

Fig. 109. ***Lasmigona costata*** **(Rafinesque, 1820). A.** OSUM 47467, Muskingum River, Muskingum Co., OH. 140mm. **B.** OSUM 22153, Mohican River, Coshocton Co., OH. 113 mm. **C.** OSUM 58645.2, Lake Erie, [Co.?], OH. 63 mm. **D.** 62934.4, Dyes Fork, Morgan Co., OH. 116 mm. **E.** OSUM 20705.2, Little Darby Creek, Madison Co., OH. 93 mm. **F.** OSUM 21820.2, Deer Creek, Union Co., OH. 53 mm. **G, H. Hinge detail:** OSUM 46228.3, Sandusky River, Wyandot Co., OH. 106 mm. **Beak sculpture detail:** OSUM 43758.1, Little Darby Creek, Madison Co., OH. 64 mm total length.

Alasmidonta rugosa Barnes, 1823: 278, fig. 21 [Type locality: "Wisconsan" [*sic*] and "Fox River." Type lost *fide* Clarke, 1985].

Mya rugosa (Rafinesque, 1820). Eaton, 1826: 222.

Alasmodon hians Rafinesque, 1831: 5 [Type locality: "River Tennessee." Type lost *fide* Johnson & Baker, 1973, and Clarke, 1985].

Lasmigona rugosum (Barnes, 1823). Rafinesque, 1831: 5.

Alasmodonta costata Rafinesque, 1820. Say, 1834: unpaginated.

Alasmodonta hians (Rafinesque, 1831). Férussac, 1835: 25.

Margarita (*Margaritana*) *rugosa* (Barnes, 1823). Lea, 1836: 44.

Unio rugosa (Barnes, 1823). Hanley, 1842: 211.

Alasmodon rugosa (Barnes, 1823). DeKay, 1843: 196, pl. 14, fig. 226.

Unio rugosus (Barnes, 1823). Catlow & Reeve, 1845: 63.

Complanaria rugosa (Barnes, 1823). Stimpson, 1851: 14.

Margaron (*Margaritana*) *rugosa* (Barnes, 1823). Lea, 1852b: 42.

Complanaria costata (Rafinesque, 1820). Agassiz, 1852: 49.

Baphia rugosa (Barnes, 1823). Adams & Adams, 1857: 500.

Margaritana rugosa (Barnes, 1823). Calkins, 1874: 46.

Symphynota (*Lasmigona*) *costata* (Rafinesque, 1820). Simpson, 1900a: 665.

Lasmigona costata var. *eriganensis* Grier, 1918: 10 [Type locality: "Big Bend, Presque Isle Bay, Lake Erie." Lectotype CMNH 209131].

Lasmigona costata pepinensis Baker, 1928a: 144 [Type locality: "Lake Pepin" [WI]. Holotype UMMZ 209133].

Lasmigona costata nuda Baker, 1928a: 145 [Type locality: "Red Cedar River, west of Chetek, Barron Co." [WI]. Holotype UMMZ 209131].

Lasmigona (*Lasmigona*) *costata* (Rafinesque, 1820). Clarke, 1985: 12–24; figs. 4–6.

Status. Not listed by either Ohio Division of Wildlife or U.S. Fish & Wildlife Service as of 2009.

ADULT SHELL

Shell. Large size (to 150 mm), thin-shelled as juveniles, much thicker with age, somewhat compressed, rather rectangular in profile.

Anterior margin. Roundly arched into ventral margin.

Ventral margin. Nearly straight to broadly rounded, often indented near the midline.

Posterior margin. Ventral and posterior margins form wide, blunt acute angle, often somewhat expanded ventrally; posterior margin oblique and straight to dorsal margin.

Dorsal margin. Long, more or less straight, joining posterior margin with very obtuse angle.

Umbo. Low, pointed, located about 33% back from anterior margin; umbonal sculpture of 3–4 coarse undulating or nearly linear ridges, vaguely "W"-shaped.

Ligament. Very long, tan to dark brown, partially concealed by shells.

Disc sculpture. Smooth except for the corrugated posterior slope; these corrugations are most prominent on juveniles and may become quite obsolete in old individuals. Rare individuals also corrugated on posterior half of shell.

Periostracum. Yellow, tan, or greenish as smallest juveniles, becoming dark brown to black with age. Usually with numerous fine green or brown rays, particularly posteriorly.

Nacre. Porcelain white, often iridescent posteriorly, occasionally with a faint salmon, yellow, or greenish flush beneath the hinge; nearly all specimens seen have gold or brown Tulberg layers.

Hinge plate. Thick; right cardinal a single, chunky, triangular denticle; left cardinal very broadly bifid, the anterior portion very weak, the posterior forming a wide "V"-shaped tooth or interdental projection beneath the interdentum; laterals obsolete at all stages of life; interdentum wide and flattened.

Muscle scars. Anterior adductor impressed; posterior adductor flush or shallowly impressed; anterior adductor distinct from anterior pedal protractor scar, large posterior adductor confluent with posterior pedal retractor scar; multiple dorsal suspensor located behind cardinal tooth and within the umbonal cavity. Pallial line entire, located far from shell margin. Accessory line weak, short.

Umbonal cavity. Very wide, shallow, and open.

Glochidia. Pyriform, asymmetric, dorsal margin straight, posterior and anterior margins broadly rounded dorsally and slightly incurved ventrally, ventral margin broadly pedicellate. Exterior surface sculpturing densely beaded, hooks styliform, arising from the ventral terminus as a broadly incurved triangular plate. Microstylets lanceolate, multifaceted, and arranged in about seven proximal rows, reduced to five distal rows and numbering about 100 (Lefevre & Curtis, 1910a: 97, fig. B; Arey, 1924b: 334, pl. 1, fig. 2; Hoggarth, 1999: figs. 27a–e).

Literature records (Lea, 1858c: 48, pl. 5, fig. 26; Lefevre & Curtis, 1912: 146, fig. 1b; Ortmann, 1912a: 283; Surber, 1912: 9, pl. 1, fig. 7; Utterback, 1915: 252; Clarke, 1985: 16–19, fig. 5; Hoggarth, 1999: figs. 27a–e) indicate glochidial dimensions for length (0.34 to 0.39 mm), height (0.36 to 0.40 mm), and hinge (0.23 to 0.25 mm). The glochidium of *L. costata* can be distinguished by its distinctly pear-shaped outline, its exterior valve sculpture, and hook structure.

Reproductive biology. Bradytictic. Eggs appear in August (Ortmann, 1919). Females bear glochidia from August to the following May (Clarke, 1985). Two types of

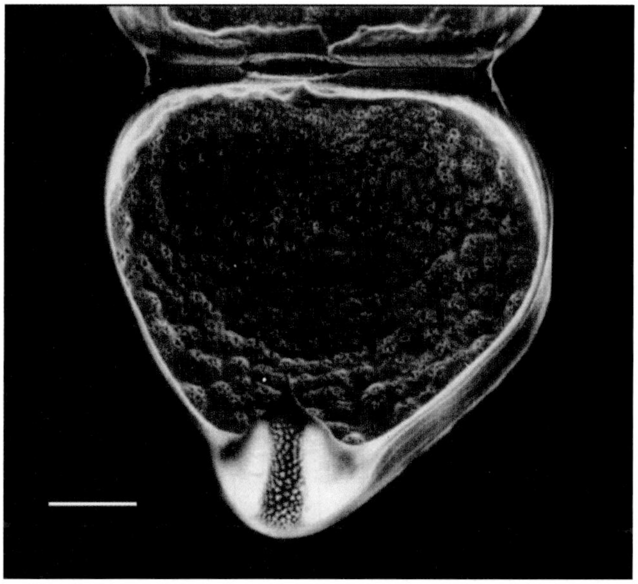

Fig. 110. *Lasmigona costata* (Rafinesque, 1820). Glochidium. MAH 585. Big Darby Creek, Franklin Co., OH. Bar = 50 µm. (Hoggarth, 1999).

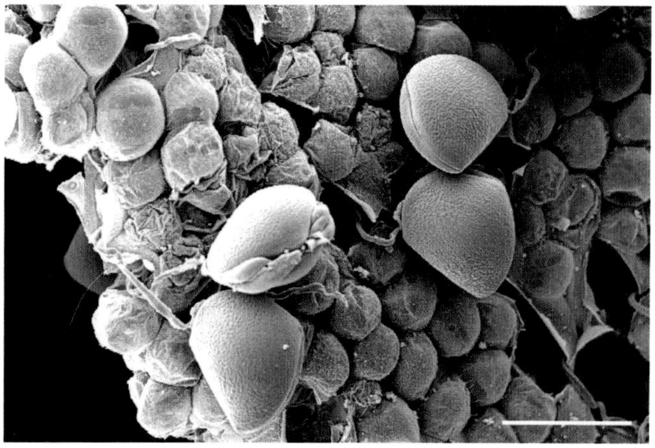

Fig. 111. *Lasmigona costata* (Rafinesque, 1820). Portion of the composite conglutinate type C. Glochidia are tethered to a core of structural programmed eggs. Bar = 500 µm. (Watters, 2008).

glochidia have been found within the marsupia. The first is a triangular, strongly hooked infective glochidium. The second is a round, weakly hooked or unhooked glochidium. Individuals of this second type had obviously grown shell material while in the marsupium, and appear to be non-parasitic. This species, like its congener *Lasmigona subviridis*, may be able to forego the parasitic stage. It is possible that the two types of glochidia are formed at different times of the year (Watters *et al.*, 1998b).

Reported potential hosts

Ambloplites rupestris (Rock Bass)—Luo, 1993 (LT).
Ameiurus nebulosus (Brown Bullhead)—Luo, 1993 (LT).
Amia calva (Bowfin)—Hove *et al.*, 1994 (LT).
Campostoma anomalum (Central Stoneroller)—Watters *et al.*, 2005 (LT).
Carrasius auratus (Goldfish)—Watters *et al.*, 2005 (LT)—exotic.
Cottus carolinae (Banded Sculpin)—Luo, 1993 (LT).
Cyprinus carpio (Common Carp)—Lefevre & Curtis, 1910a (LI); 1912 (LI)—exotic.
Dorosoma cepedianum (Gizzard Shad)—Weiss & Layzer, 1995 (NI).
Esox lucius (Northern Pike)—Hove *et al.*, 1994 (LT).
Etheostoma caeruleum (Rainbow Darter)—Luo, 1993 (LT).
Etheostoma flabellare (Fantail Darter)—Watters *et al.*, 1998d (LT).
Etheostoma virgatum (Striped Darter)—Luo, 1993 (LT).
Etheostoma zonale (Banded Darter)—Watters *et al.*, 1998c (LT); Watters *et al.*, 1998d (LT).
Fundulus catenatus (Northern Studfish)—Luo, 1993 (LT).
Hypentelium nigricans (Northern Hogsucker)—Watters *et al.*, 1998b (LT), 1998d (LT).
Lepomis cyanellus (Green Sunfish)—Luo, 1993 (LT).
Lepomis gibbosus (Pumpkinseed)—Watters *et al.*, 1998d (LT).
Lepomis macrochirus (Bluegill)—Watters *et al.*, 2005 (LT).
Lepomis megalotis (Longear Sunfish)—Luo, 1993 (LT).
Micropteris dolomieu (Smallmouth Bass)—Luo, 1993 (LT).
Micropteris salmoides (Largemouth Bass)—Watters *et al.*, 1998d (LT).
Moxostoma carinatum (River Redhorse)—Weiss & Layzer, 1995 (NI).
Perca flavescens (Yellow Perch)—Hove *et al.*, 1994 (LT).
Rhinichthys cataractae (Longnose Dace)—Watters *et al.*, 1998b (LT); 1998d (LT).
Sander vitreus (Walleye)—Hove *et al.*, 1994 (LT).
Semotilus atromaculatus (Creek Chub)—Watters *et al.*, 2005 (LT).

Habitat. This is a species of large creeks, rivers, and lakes in sandy mud and cobble, often in moving water. Locally abundant (i.e., Stillwater River).

Range. Mississippi River system from southern Ontario to Arkansas, Great Lakes drainage and St. Lawrence River. Hudson River.

Range in Ohio. Widespread across the state; western basin of Lake Erie.

Similar species. The compressed shell, green rays, and coarsely corrugated slope are unique among Ohio mussels. *Alasmidonta marginata* also has corrugations and rays,

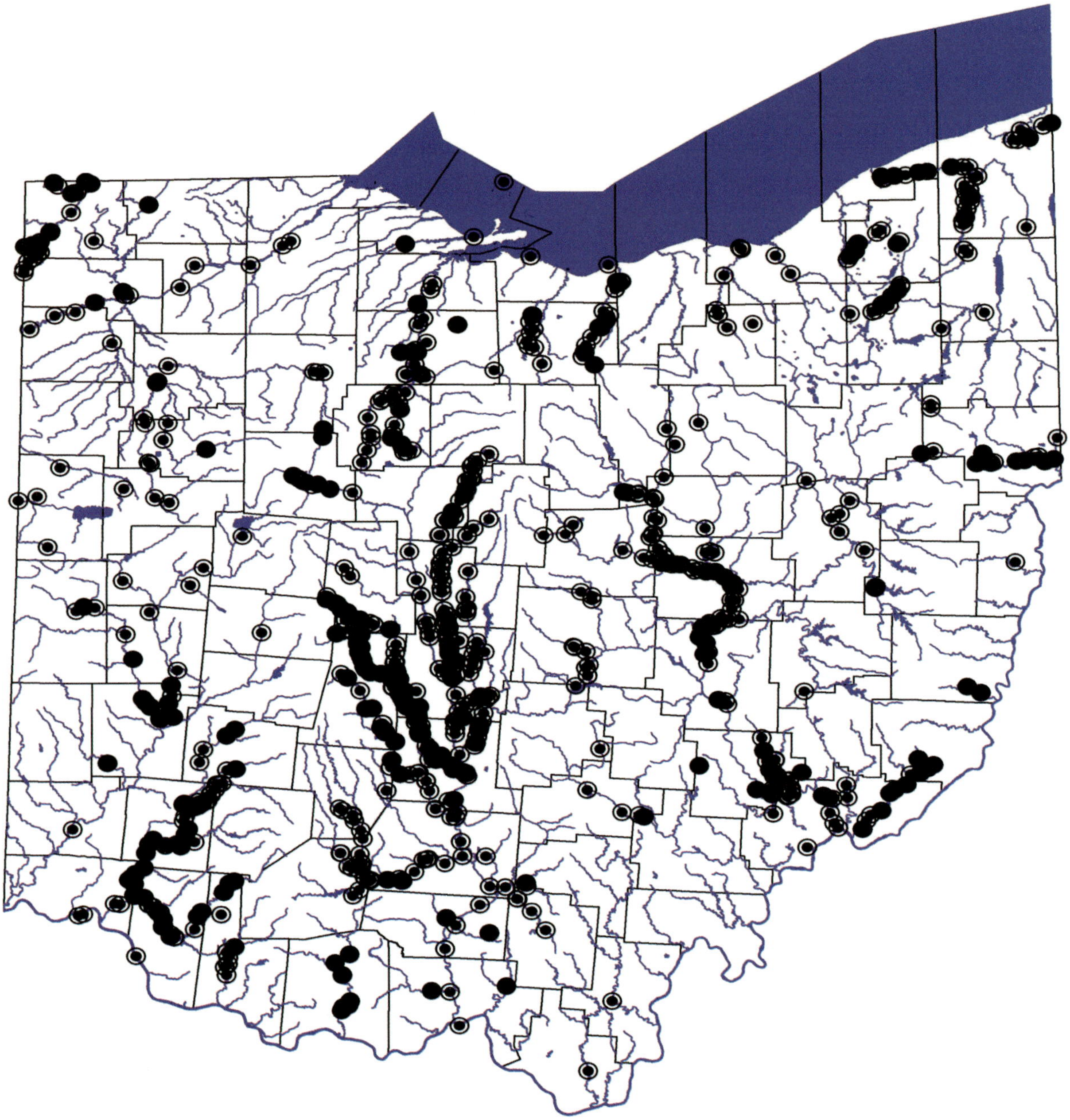

Fig. 112. ***Lasmigona costata*** **(Rafinesque, 1820)**
Solid dots—live or freshdead since 1980
Bullseye—all records, any condition

but the corrugations are much finer and the rays are often spotted. *Lasmigona complanata* is also corrugated, but the shell is much more circular, usually with a distinct posterior wing. *Lasmigona compressa* is very similar to *L. costata* but lacks the corrugations and has a tongue-like structure on the left valve hinge on the interdentum; *L. compressa* possesses lateral teeth, which are only rudimentary in *L. costata*.

Comments. Individuals grow quickly for 2–3 years and then gradually slow. They grow to ca. 150 mm and live for 20 years.

Leptodea Rafinesque, 1820

Etymology. Gr. *leptos*, fine, thin, delicate + Gr. *odous* (*odon*), *odontos*, tooth. In his original description, Rafinesque (1820: 295) writes of the cardinal teeth of *Leptodea*: "Bilobed tooth entire and smooth; that of the right valve simple. Contour not thickened."

Original description

Unio (*Leptodea*) Rafinesque, 1820: 295.

Type species. *Unio fragilis* Rafinesque, 1820, by subsequent designation of Herrmannsen (1847: 584). See also Utterback (1916b: 151) and Haas (1969: 418) for a contradictory opinion.

Synonymy

Unio (*Leptodea*) Rafinesque, 1820: 295.

Lasmonos Rafinesque, 1831: 1, 5–6 [Type species: *Unio fragilis* Rafinesque, 1820, by original designation].

Paraptera Ortmann, 1911: 368 [Type species: *Unio gracilis* Barnes, 1823, by original designation].

Geographic range. Mississippi River system and Gulf drainages from Alabama to Texas. Great Lakes drainages. Species from Central America assigned to *Leptodea* require futher study.

Leptodea fragilis (Rafinesque, 1820)

Figures 113, 114, 115

Etymology. L. *fragilis*, easily broken, brittle, fragile, weak. Rafinesque (1820: 295) described the shell of *U. fragilis* as "very thin and fragile," which is typical of this species.

Vernacular name(s). *Fragile Papershell.*

Original description

Unio (*Leptodea*) *fragilis* Rafinesque, 1820: 295, species no. 6, no figure.

Type locality. Rafinesque did not give a type locality. The lectotype is from "creeks in Kentucky," and this becomes the type locality.

Type material. The type mentioned by Vanatta (1915: 552) was selected as lectotype (ANSP 20209 from "creeks in Kentucky") by Johnson & Baker (1973: 155).

Synonymy

Unio (*Leptodea*) *fragilis* Rafinesque, 1820: 295.

Unio fragilis Rafinesque, 1820. Swainson, 1823: pl. 171.

Unio planus Barnes, 1823: 272, fig. 16 [Type locality: "the Wisconsan [*sic*] [River]." Type lost].

Unio gracilis Barnes, 1823: "174" [274] [Type locality: "the Wisconsan [*sic*] [River]" and "the [Great] Lakes." Type lost].

Mya plana (Barnes, 1823). Eaton, 1826: 221.

Mya gracilis (Barnes, 1823). Eaton, 1826: 222.

Symphynota gracilis (Barnes, 1823). Lea, 1829: 452.

Unio (?*Metaptera*) *diaphana* Rafinesque, 1831: 3 [also variety *lineolata*. Type locality: "in small streams in Kentucky." Type not located].

Lasmonos fragilis (Rafinesque, 1820). Rafinesque, 1831: 5–6.

Symphynota fragilis (Rafinesque, 1820). Férussac, 1835: 25.

Margarita (*Unio*) *gracilis* (Barnes, 1823). Lea, 1836: 11.

Unio (*Niaa*) *atrata* Swainson, 1841: pl. 171 [*non* Sowerby, 1839].

Unio atratus Swainson, 1841. Hanley, 1842: 199.

Unio fragilis var. *fragilis* (Rafinesque, 1820). Jay, 1850: 59.

Unio leptodon var. *planus* Barnes, 1823. Jay, 1850: 61.

Metaptera gracilis (Barnes, 1823). Stimpson, 1851: 14.

Metaptera fragilis (Rafinesque, 1820). Lapham, 1852: 369.

Margaron (*Unio*) *gracilis* (Barnes, 1823). Lea, 1852b: 19.

Unio fragilis var. *gracilis* Barnes, 1823. Paetel, 1890: 153.

Lampsilis gracilis (Barnes, 1823). Baker, 1898b: 99, pl. 19, fig. 1.

Lampsilis simpsoni Ferriss, 1900: 38 [Type locality: "Spring River, Hardy, [Sharp Co.] Arkansas." Possible holotype FMNH, number not given].

Lampsilis (*Proptera*) *gracilis* (Barnes, 1823). Simpson, 1900a: 573.

Paraptera gracilis (Barnes, 1823). Ortmann, 1911: 334.

Paraptera fragilis (Rafinesque, 1820). Ortmann, 1918: 572.

Leptodea fragilis (Rafinesque, 1820). Ortmann & Walker, 1922: 82.

Paraptera gracilis lacustris Baker, 1922: 131 [Type locality: "off Plumbers Point, Lake Butte des Morts, Wisconsin." Holotype UMMZ 209186].

Leptodea fragilis lacustris Baker, 1924: 132.

Lampsilis (*Leptodea*) *fragilis* (Rafinesque, 1820). Frierson, 1927: 82.

Lampsilis (*Leptodea*) *fragilis simpsoni* Ferriss, 1900. Frierson, 1927: 82.

Proptera gracilis (Barnes, 1823). Haas, 1969: 416.

non Leptodea fragilis (Rafinesque, 1820). Strayer & Jirka, 1997: pl. 11 [= *Potamilus ohiensis* Rafinesque, 1820].

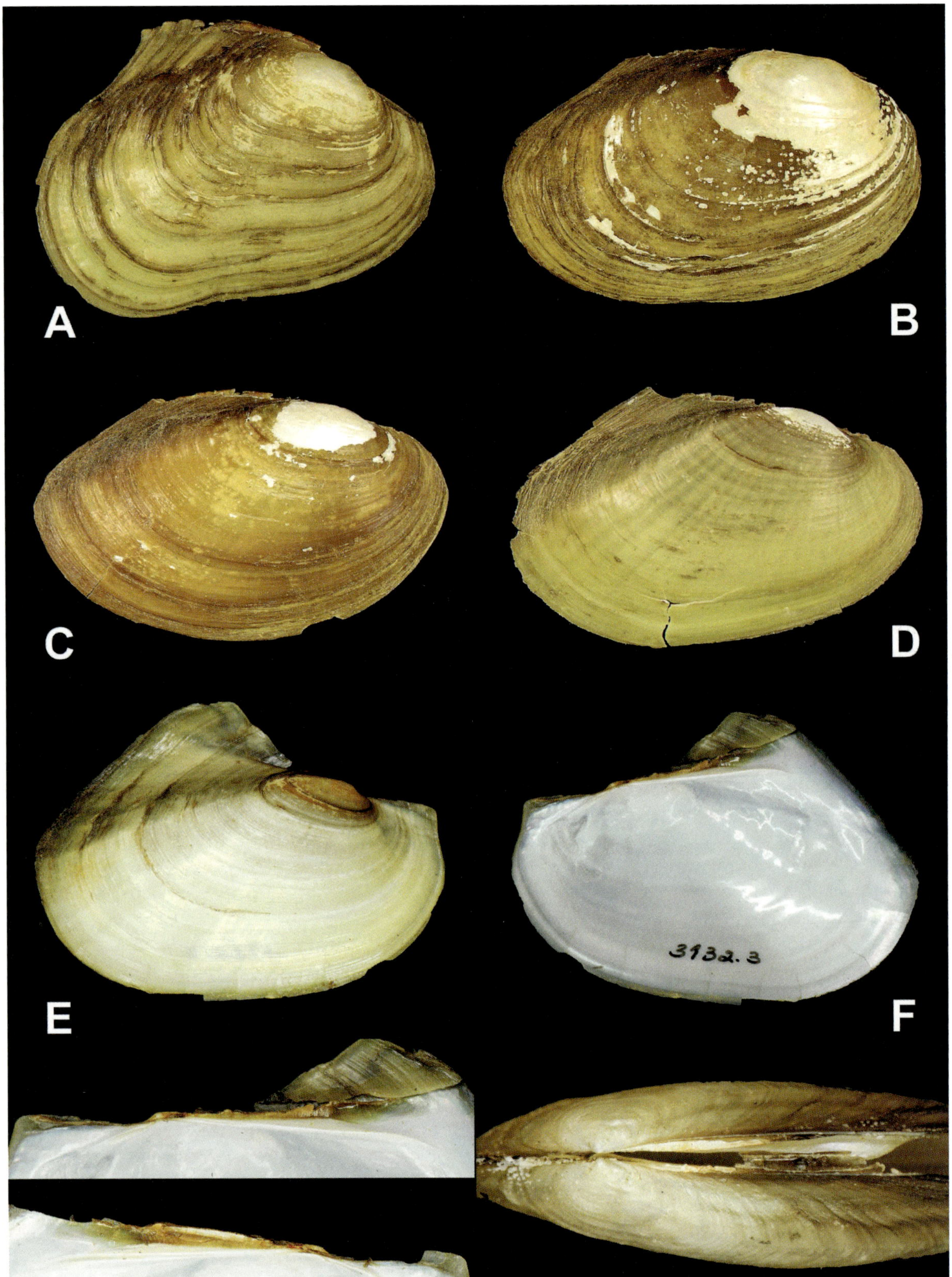

Fig. 113. ***Leptodea fragilis*** **(Rafinesque, 1820). A.** OSUM 3932.15, Lake Erie, Ottawa Co., OH. 91 mm. **B.** OSUM 65751.1, Killbuck Creek, Coshocton Co., OH. 113 mm. **C.** OSUM 61353.2, Wills Creek, Coshocton Co., OH. 139 mm. **D.** OSUM 61833, Clear Fork, Monroe Co., OH. 86 mm. **E, F, Hinge detail:** OSUM 3932.3, Lake Erie, Ottawa Co., OH. 61 mm. **Beak sculpture detail:** OSUM 51691.1, Lake Erie, Ottawa Co., OH. 30 mm total length.

Status. Not listed by either Ohio Division of Wildlife or U.S. Fish & Wildlife Service as of 2009.

ADULT SHELL

Shell. Large (to 160 mm), compressed, thin (pliable when wet), with pedal and "siphonal" gape, oval to somewhat elongate.

Anterior margin. Straight, concave, or convex, sometimes forming a minute wing on the dorsal margin before abruptly curving ventrally.

Ventral margin. Males gently rounded; females straight or slightly convex until the marsupial region; marsupial region expanded to various degrees, usually not as prominent as in *Potamilus, Villosa,* or *Lampsilis.*

Posterior margin. Broadly rounded, rarely coming to a blunt point; then in a straight line to the dorsal wing.

Dorsal margin. The degree of development of the wing is quite variable; generally smaller individuals and lake specimens have a more extensive wing (specimens from the Bass Islands area of Lake Erie often have well-developed wings); most adult riverine specimens have a very modest wing or none at all; the dorsal margin is straight, concave, or serrate.

Umbo. Very broad and low; beak sculpture extremely fine, consisting of 2–4 "V"-shaped ridges pointing toward the umbo, occasionally somewhat nodulose; beak sculpture disappears early in the first year of growth.

Ligament. Large, brown, occupying most of dorsal margin; if a wing is present, the ligament is contained within it; if not, the ligament is usually prominent.

Disc sculpture. Sculptureless except for 1–3 radiating ridges on the dorsal slope.

Periostracum. Pale yellow or green to tan, not appreciably darkening with age; most specimens have faint, diffuse greenish rays; the dorsal slope always has one or two dark rays or may be uniformly darkly colored; rays continuous throughout the life of the mussel; annular rings dark brown.

Nacre. White, iridescent posteriorly, usually flushed with pink on the dorsal 1/3 or more; gold Tulberg layers are common.

Hinge plate. Very light, interdentum long but barely perceptible; cardinal teeth lamellar, nearly obsolete, two in left valve, one in right with minute accessory tooth; lateral teeth long, bowed, one in left valve, two in right; in old individuals the teeth may become lost or rudimentary.

Muscle scars. Adductors large, flush or nearly so with shell; anterior adductor removed from large anterior pedal protractor; posterior retractor present as a long scar basally attached to posterior pedal retractor scar; numerous suspensors arranged in a straight or arched oblique row, extending ventrally from the umbonal cavity; pallial line complete, indented at juncture with posterior adductor; accessory line absent.

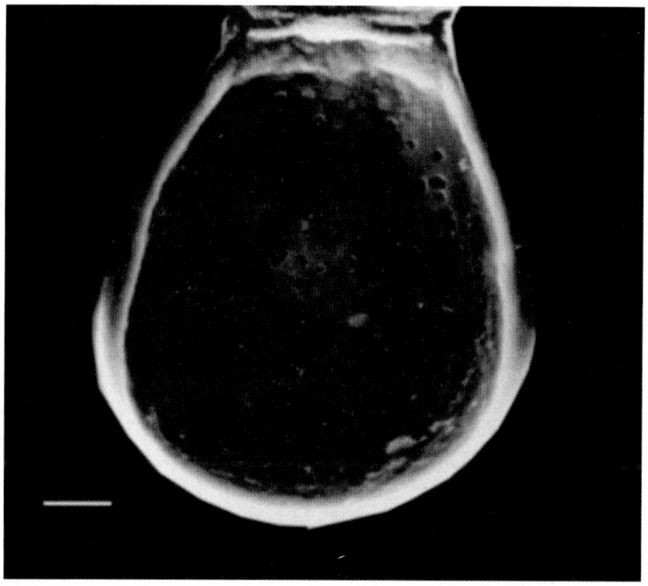

Fig. 114. ***Leptodea fragilis*** **(Rafinesque, 1820).** Glochidium. MAH 626.1, Big Darby Creek, Pickaway Co., OH. Bar = 10 µm. (Hoggarth, 1999).

Umbonal cavity. Wide but shallow.

Glochidia. Subelliptical, dorsal margin straight, anterior and posterior margins about equally divergent dorsally and ventrally curved, ventral margin broadly rounded. Exterior surface lightly covered with loose-looped valve sculpture, dorsal alae very small, lateral valve gape present. Hooks absent, micropoints wide, lamellate plates arranged in complete vertical rows (Ortmann, 1911: 347, pl. 89, fig. 19; Coker & Surber, 1911: 180, pl. 1, figs. 2, 2a; Hoggarth, 1999: figs. 46a–e).

Literature records (Lefevre & Curtis, 1910a: 97, fig. k; Lefevre & Curtis, 1912: 146, fig. 1k; Surber, 1912: 9, pl. 2, fig. 28; Ortmann, 1912a: 331; Utterback, 1916a: 353; Hoggarth, 1999: figs. 46a–e) indicate glochidial dimensions for length (0.07 to 0.08 mm), height (0.08 to 0.10 mm), and hinge (0.03 to 0.04 mm). Coker & Surber (1911: 180) noted, "The glochidium of *gracilis* [*L. fragilis*] is oval pear shaped and very similar to that of *ligamentinus* [*A. ligamentina*] or *ventricosus* [*L. cardium*], or to what we generally regard as a typical form." Utterback (1916a: 353) added, "Glochidia among the very smallest, belonging to the *Lampsilis* type." This glochidium will not be confused with any other. Its size, rather round subelliptical shape, small dorsal alae, and lamellate micropoints will distinguish it from all others.

Reproductive biology. Bradytictic. The posterior half of the outer gills is marsupial (Ortmann, 1912a). Gravid females are found from September to the following July (Ortmann, 1919). Glochidia are released from September through November, probably resulting in glochidia

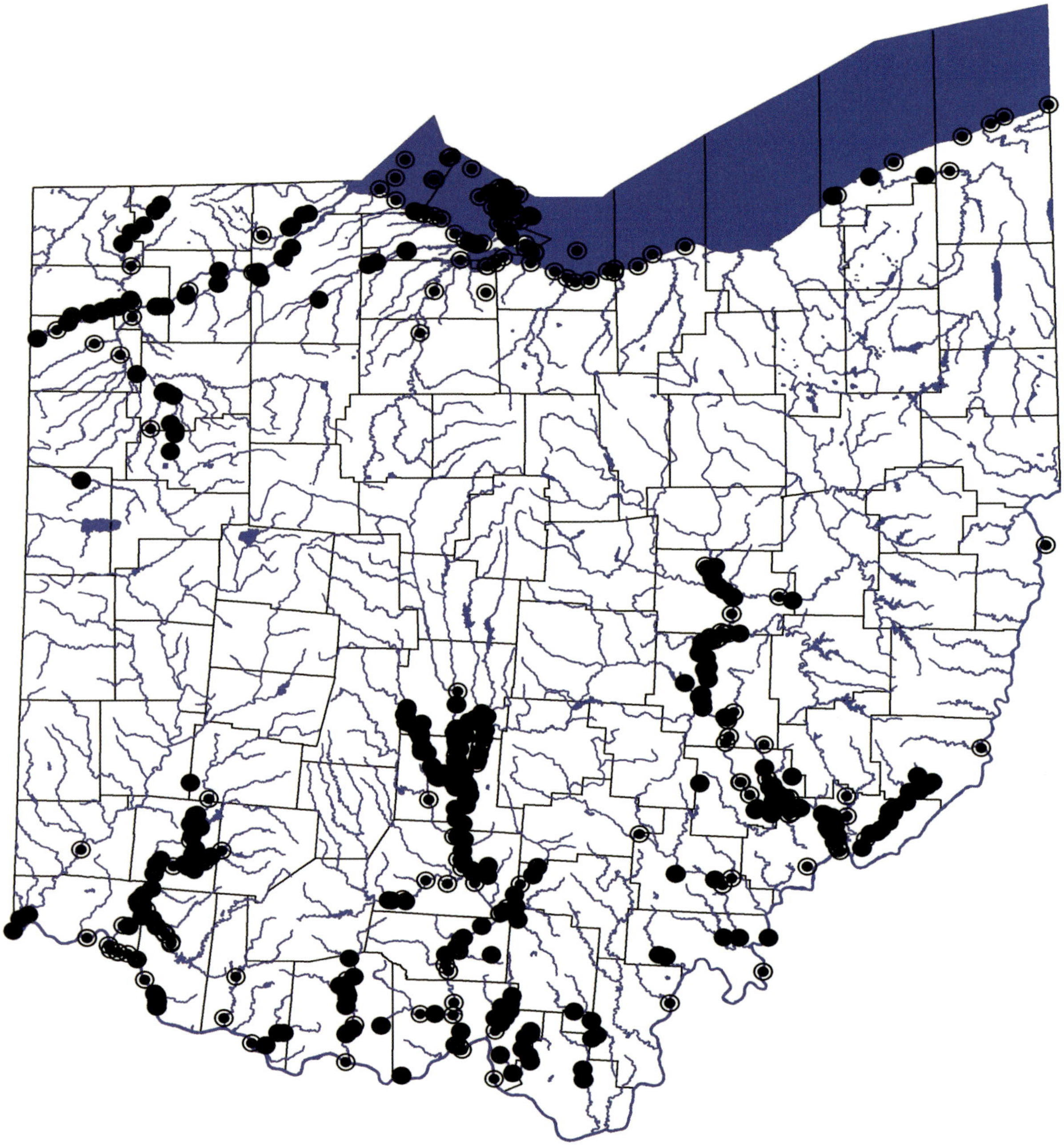

Fig. 115. ***Leptodea fragilis*** **(Rafinesque, 1820)**
Solid dots—live or freshdead since 1980
Bullseye—all records, any condition

overwintering on their hosts (Watters, unpubl.). Glochidia are released in fragile, white, leaf-like conglutinates, which quickly disassociate into free glochidia (Watters, 2008). Despite reports of Freshwater Drum as host, no metamorphosis has been reported, but attempts to obtain metamorphosis on a wide range of fishes other than Freshwater Drum also have proven unsuccessful (Watters, unpubl.). The range of this species is limited by man-made obstacles to its fish hosts, such as low-head dams (Watters, 1996c).

Reported potential hosts

Aplodinotus grunniens (Freshwater Drum)—Howard, 1913 (NI); Wilson, 1916 (NI); Howard & Anson, 1923 (NI); Cummings & Mayer, 1993 (NI).

Habitat. *Leptodea fragilis* occurs in large streams, rivers, and lakes. It may be found in many substrates, from sandy mud to packed cobble. It is one of the most active mussels in Ohio and often is found crawling over the surface of the substrate during periods of low water. Individuals of this species transplanted to Leading Creek in southern Ohio had moved several hundred feet within a few months.

Range. Mississippi River system from Ontario to southern Louisiana, Gulf drainages from Alabama to Texas, Great Lakes drainages.

Range in Ohio. Widespread and common in the Little Miami, Scioto, and Muskingum rivers, less so in other Ohio River tributaries. In the Lawrentian drainage it is common within Lake Erie but occurs in relatively few tributaries—the Maumee, Sandusky, and lower Grand rivers are exceptions. A river and lake species, seldom found in creeks.

Similar species. *Leptodea fragilis* usually will not be confused with any other species. Outwardly it is of a different color than other large, alate species such as *Lasmigona complanata* or members of *Potamilus*. Its nacre is never uniformly deep purple or pink as in *Potamilus*, and it has much weaker hinge than in *Lasmigona complanata*. *Leptodea leptodon* is similar but more elongate and lacks beak sculpture.

Comments. *Leptodea fragilis* is a fast-growing mussel for the first two to four years of life, though the growth rate is highly variable. For example, specimens 90 mm long (medium-sized) may be between 3 and 8 years old. Very large specimens may attain 160 mm or more in length. Nevertheless, it is a fairly short-lived species; individuals over 12 years old are rare.

Leptodea leptodon (Rafinesque, 1820)

Figures 116, 117

Etymology. Gr. *leptos*, fine, thin, delicate + Gr. *odous* (*odon*), *odontos*, tooth. Rafinesque (1820: 295) described the lateral tooth of this species as "lamellar tooth thin and long," hence the name *leptodon*. *Leptodon* is a noun in apposition.

Vernacular name(s). *Scaleshell.*

Original description

Unio (*Leptodea*) *leptodon* Rafinesque, 1820: 295, species no. 5, pl. 80, figs. 5–7.

Type locality. "Assez commune dans les parties inférieures de l'Ohio" ["Rather common in the lower parts of the Ohio"]. The neotype is from the Kentucky River. However, as this specimen is not from the Ohio River, this designation is probably invalid.

Type material. A specimen identified by Rafinesque and mentioned by Vanatta (1915: 551) was selected as neotype (ANSP 20214 from the Kentucky River) by Johnson & Baker (1973: 160).

Synonymy

Unio (*Leptodea*) *leptodon* Rafinesque, 1820: 295, pl. 80, figs. 5–7.

Anodon purpurescens Swainson, 1823: pl. 160 [Type probably in Museum of University of Manchester, but merged with general collection (Kohn, 1992)].

Symphynota tenuissima Lea, 1829: 453, pl. 11, fig. 21 [Type locality: "Ohio." Type not located].

Unio velum Say, 1829: 293 [Type locality: "Kentucky River." Type not located].

Unio leptodon Rafinesque, 1820. Say, 1834: no pagination.

Symphynota leptodon (Rafinesque, 1820). Férussac, 1835: 25.

Margarita (*Unio*) *tenuissimus* (Lea, 1829). Lea, 1836: 38.

Unio tenuissimus (Lea, 1829). Hanley, 1843: 206, pl. 20, fig. 42.

Alasmodonta leptodon (Rafinesque, 1820). Lapham, 1852: 370.

Metaptera leptodon (Rafinesque, 1820). Agassiz, 1852: 46.

Margaron (*Unio*) *tenuissimus* (Lea, 1829). Lea, 1852b: 38.

Leptodea leptodon (Rafinesque, 1820). Conrad, 1853: 262.

Lampsilis (*Proptera*) *leptodon* (Rafinesque, 1820). Simpson, 1900a: 575.

Lampsilis blatchleyi Daniels, 1902: 13, pl. 2 [Type locality: "Wabash River, Section 32, Linn [Lynn] Township, Posey Co., Indiana." Holotype UMMZ 82960].

Proptera leptodon (Rafinesque, 1820). Sterki, 1907: 393.

Lampsilis (*Proptera*) *blatchleyi* Daniels, 1902. Simpson, 1914: 190.

Unio (*Symphynota*) *tenuissimus* (Lea, 1829). De Gregorio, 1914: 36.

Fig. 116. ***Leptodea leptodon*** **(Rafinesque, 1820). A, H, Hinge detail, Beak sculpture detail:** OSUM 18498, East Fork Little Miami River, [Co.?], OH. 90 mm. **B.** OSUM 57257.2, Ohio River, Hamilton Co., OH. 72 mm. **C.** OSUM 57257.1. Ohio River, Hamilton Co., OH. 53 mm. **D.** OSUM 13398.3, Gasconade River, Maries Co., MO. 47 mm. **E.** OSUM 24542.2, Clinch River, Hancock Co., TN. 26 mm. **F.** OSUM 54583.2, Meramec River, Jefferson Co., MO. 44 mm. **G.** OSUM 10256.4, Ohio River, Hamilton Co., OH. 75 mm.

Fig. 117. ***Leptodea leptodon*** **(Rafinesque, 1820)**
Solid dots—live or freshdead since 1980
Bullseye—all records, any condition

Unio shepardianus forma duttonianus Lea, 1841. De Gregorio, 1914: 55 [misidentification].
Lampsilis leptodon (Rafinesque, 1820). Vanatta, 1915: 551.
Lasmonos leptodon (Rafinesque, 1820). Utterback, 1916b: 388.
Paraptera leptodon (Rafinesque, 1820). Ortmann, 1918: 571.
Lampsilis (Leptodea) leptodon (Rafinesque, 1820). Frierson, 1927: 82.
Leptodea blatchleyi (Daniels, 1902). Goodrich & van der Schalie, 1944: 316.

Status. Listed as Extirpated by Ohio Division of Wildlife; listed as Endangered by U.S. Fish & Wildlife Service as of 2009.

ADULT SHELL

Shell. Medium-sized (to 100 mm); thin and fragile, outline elongated, umbo 25% from anterior margin.
Anterior margin. Narrowly curved.
Ventral margin. Very broadly curved, or nearly straight in females; females may have a thin, scale-like extension that projects from the posterior at an angle. This projection may be fluted or irregularly folded.
Posterior margin. Gently rounded or straight.
Dorsal margin. Long, straight or slightly convex.
Umbo. Narrow, low, barely rising above the hinge line; beak sculpture absent.
Ligament. Tan to brown, very long, largely concealed by the valves.
Disc sculpture. Shell sculptureless except for 2–3 low, narrow ridges.
Periostracum. Pale yellowish-green to tan, becoming olive green with age; shells usually have numerous faint green rays over the shell. The scale-like extension on female shells may be brown, contrasting with the remainder of the shell. Periostracum shiny.
Nacre. White, often flushed with pink or salmon, particularly under the umbos, entire shell iridescent; often with gold or green Tulberg layers.
Hinge plate. Very long, straight, very narrow, interdentum nearly absent; cardinal teeth rudimentary, stubby, often absent in large specimens, one in each valve; lateral teeth long, rudimentary, straight, very narrow, one in each valve (left valve may have additional very weak claustrum).
Muscle scars. Anterior adductor muscle scars impressed, posterior adductor scar flush with shell surface, both indistinct from protractor and retractor scars; dorsal suspensor scars a series of elongated scars oriented obliquely under the interdentum; pallial line entire, accessory line absent.
Umbonal cavity. Wide, open, shallow, nearly nonexistent.

Glochidia. Not described.

Reproductive biology. Bradytictic. Females with early embryos were found in August, with mature glochidia present by the following April (Barnhart *et al.*, 1998b).

Reported potential hosts
Aplodinotus grunniens (Freshwater Drum)—Barnhart *et al.*, 1998b (LT).

Habitat. Sandy mud and cobble in rivers.

Range. Upper Mississippi River system, including the Missouri, Ohio, Tennessee, and Cumberland rivers. West in the Gasconade, Meramec, Red, and White rivers.

Range in Ohio. Extirpated. Historically in the Ohio River at Cincinnati and Marietta, the lower Muskingum River, and the Scioto River at Columbus and Circleville.

Similar species. This species can be confused with *Leptodea fragilis*. That species typically is less elongate, being oval in profile. *Leptodea leptodon* apparently lacks beak sculpture, which is present (but very fine) in *L. fragilis*. Like *L. fragilis*, *H. lata* also is a nearly toothless, thin-shelled species, but has a characteristic pattern of broken rays, absent in *L. leptodon*.

Comments. This species is fast growing and short-lived. Growth is accelerated for the first two years, then slightly slowing for the remainder of the animal's life. Sexual dimorphism in the shell is pronounced. Females are more broadly rounded posteriorly and may have a unique scale-like extension on the posterior margin. Females appear to be much rarer than males. Individuals may exceed 100 mm in length, but rarely reach 8 years old.

The scale-like extensions of the female shell are unique among North American unionids. The orientation, contrasting color, and shape resemble the mantle flap lures of the genus *Lampsilis*. It is possible that these extensions are static lures formed by the shell rather than the living mantle. If so, then these represent an amazing example of convergence between *Lampsilis* and *Leptodea*.

Ligumia Swainson, 1840

Etymology. L. *legumen, leguminis*, bean, leguminous plant, seed or pod of a legume. Swainson described *ligumia* (1840: 378) as "very long and pod-shaped; . . . ," so it seems more probable that he was using the word for "legume."
Original description
Ligumia Swainson, 1840: 268, 274, 378.
Type species. *Unio recta* Lamarck, 1819, by original designation.

Synonymy

Ligumia Swainson, 1840: 268, 274, 378.
Ligumea [*sic*] Swainson, 1840: 263.
Legumia [*sic*] Agassiz, 1852: 45.
Ligumina [*sic*] Philippi, 1853: 465.

Geographic range. North America east of the Rocky Mountains, from the Red River of Canada to the Gulf drainages, the Atlantic Slope from New England to Virginia.

Comments. Smith (2000a) recently added the eastern *Unio ochraceus* Say, 1817, to this genus based on mantle morphology and glochidia. If this is correct, it is quite unlike other *Ligumia* in shell morphology.

Ligumia nasuta (Say, 1817)

Figures 118, 119

Etymology. L. *nasutus*, having a large nose, large-nosed. Say (1817: not paginated), in his original description, noted that the shell of *L. nasuta* is "thin, oblong, compressed, rostrated, ..." The pointed posterior of this species apparently reminded Say of a rostrum or nose.

Vernacular name(s). *Eastern Pondmussel.*

Original description

Unio nasutus Say, 1817: no pagination, article on "Conchology," species no. 6 under Genus *Unio*, pl. 4, fig. 1.

Type locality. "Very common in the Delaware and Schuylkill [rivers]." Johnson (1970: 380) added "near Philadelphia, Philadelphia County, Pennsylvania."

Type material. The original unionoid material of Say has apparently been lost (Johnson & Baker 1973: 146), and Haas (1930: 322), in selecting types for many of Say's species, somehow overlooked *U. nasutus* Say, 1817. Johnson (1970: 380) stated that the type of this species was lost from ANSP, but Johnson & Baker (1973: 162) recognized ANSP 41461a as the "figured specimen" of *U. nasutus* Say, 1817. However, there are two shells illustrated in Conrad's figure 1. The upper figure is *Unio nasutus* Say, 1817, but the lower is clearly an individual of the *Unio fisherianus* Lea, 1833, complex established by Simpson (1900a). Johnson & Baker referred to the second figured specimen from the Chester River, Maryland, as the type.

Synonymy

Unio nasutus Say, 1817: unpaginated.
Unio nasuta Say, 1817. Lamarck, 1819: 75.
Obliquaria attenuata Rafinesque, 1820: 304 [Type locality: "la fleuve Hudson" ("the Hudson River"). Type lost *fide* Johnson, 1970].
Mya nasuta (Say, 1817). Eaton, 1826: 218.
Unio rostrata Valenciennes, 1827: 233, pl. 53, fig. 3 [*non* Lamarck, 1819. Type locality: "Philadelphia vicinis" ("vicinity of Philadelphia"). Types in MNHN?].
Margarita (*Unio*) *nasutus* (Say, 1817). Lea, 1836: 37.
Eurynia nasuta (Say, 1817). Agassiz, 1852: 45.
Margaron (*Unio*) *nasutus* (Say, 1817). Lea, 1852b: 37.
Unio vaughanianus Sowerby, 1868: pl. 61, fig. 308 [*non* Lea, 1838].
Lampsilis nasuta (Say, 1817). Simpson, 1900a: 545–546.
Eurynia nasuta (Say, 1817). Ortmann, 1919: 271, pl. 16, figs. 10, 11.
Ligumia nasuta (Say, 1817). Clarke & Berg, 1959: 51, fig. 54.

Status. Listed as Endangered by Ohio Division of Wildlife; not listed by U.S. Fish & Wildlife Service as of 2009.

ADULT SHELL

Shell. Medium to fairly large sized (to 120 mm), sturdy but not heavy, elongated.

Anterior margin. Straight, then evenly rounded to the ventral margin.

Ventral margin. Broadly rounded in males, meeting the posterior margin in an acute angulation about halfway between the dorsal and ventral margins; females with conspicuous marsupial region, ventral margin straight or concave anterior to marsupia, forming an obtuse angle with the posterior margin at the base.

Posterior margin. Long, straight or concave in males; females straight from the ventral angulation, then forming an acute angle approximately 33% down from the dorsal margin, then straight to the dorsal margin. Females may be produced posteriorly into a pronounced "snout."

Dorsal margin. Long, straight or somewhat convex.

Umbo. Low, wide, barely rising above the hinge line; beak sculpture of 5–7 fine "W"-shaped ridges, pointed towards the umbo, the peaks aligned obliquely toward the ventral-posterior border; this sculpture abruptly disappears by the end of the first year's growth.

Ligament. Brown, long, occupying most of the dorsal margin, partially enclosed by the valves.

Disc sculpture. Shell smooth except for broad dorsal ridge.

Periostracum. Base color yellow or greenish; with age, the older portions of the shell become dark brown, leaving the newer annuli lighter; shell usually with well-defined, dark green rays; darkest on the posterior half; rays continuous or broken across annuli, rarely wavy.

Nacre. Pearly white, very iridescent posteriorly; green or gold Tulburg layers common and extensive.

Hinge plate. Very light, interdentum long, almost imperceptible in many examples; cardinals lamellate; two in left valve, but posterior one may be very weak; one in right valve, but with well-developed accessory tooth anterior to cardinal nearly confluent with shell margin; laterals very long, straight or slightly bowed, two in left, one in right.

Muscle scars. Anterior adductor barely indented, smooth, well-separated from anterior pedal protractor; posterior adductor flush with shell, partially confluent with posterior pedal retractor scar; dorsal suspensors few, inside

Fig. 118. ***Ligumia nasuta*** **(Say, 1817). A.** OSUM 54723.2, Cuyahoga River, Portage Co., OH. 74 mm. **B.** OSUM S-2852, Lake Erie, Ottawa Co., OH. 93 mm. **C.** OSUM 9872, Lake Erie, Erie Co., OH. 83 mm. **D.** OSUM 43990, ? Muskingum River, Washington Co., OH. 68 mm. **E, Beak sculpture detail:** OSUM 2299, "Lake Erie." 18 mm. **F.** OSUM 5162.7, Muzzy Lake, Portage Co., OH. 82 mm. **G, H, Hinge detail:** Lake St. Clair, Kent Co., Ontario. 76 mm.

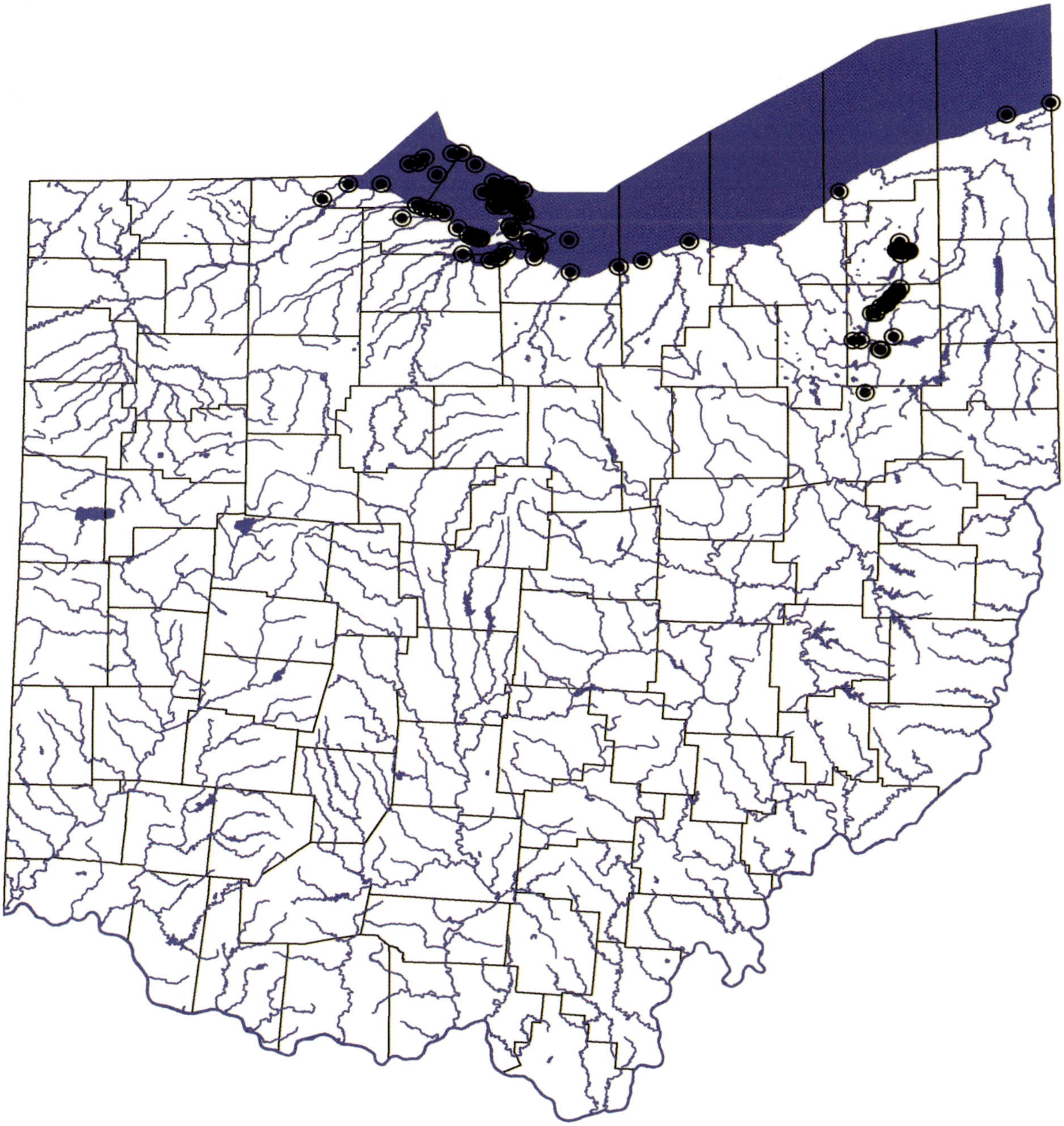

Fig. 119. ***Ligumia nasuta*** **(Say, 1817)**
Solid dots—live or freshdead since 1980
Bullseye—all records, any condition

umbonal cavity; pallial line entire, accessory very long but nearly imperceptible.
Umbonal cavity. Wide, open, moderately deep.

Glochidia. Subelliptical, dorsal margin short and straight, anterior and posterior margins equally curved, ventral margin rounded, hooks absent (Lea, 1863: 40; Lea, 1874b: 68, pl. 21, fig. 2; Reardon, 1929: 12, pl. 2, fig. 6).

Literature records (Ortmann, 1912a: 343, pl. 20, fig. 8) indicate glochidial dimensions for length (0.25 mm) and height (0.29 mm). Lea (1863: 40) noted, "Embryonic shell short, pouch-shape, very nearly like that of *phaseolus* [*P. fasciolaris*]."

Reproductive biology. Bradytictic. Marsupium formed by the posterior half to three-fourths of the outer gill (Ortmann, 1912a). Ortmann (1919) reported gravid females from August to the following May, but Connor (1907) believed that this species bred year round.

Reported potential hosts. No hosts known.

Habitat. Sandy mud, often in ponds or slackwater areas.

Range. St. Lawrence River to James River in Virginia on the Atlantic Slope. Mohawk River to Lake Erie. Conneaut Outlet, French Creek in Pennslyvania. Upper Cuyahoga River in Ohio.

Range in Ohio. Upper Cuyahoga River. Bass Islands of Lake Erie and most of Lake Erie proper. A record from the Muskingum River is probably erroneous.

Similar species. This species resembles *Ligumia recta*, *Ligumia subrostrata*, and even *Elliptio dilatata*. Both *L. nasuta* and *L. subrostrata* are thinner and smaller than *L. recta* and tend towards having a rudimentary wing. *Elliptio dilatata* differs from all *Ligumia* in having coarse, nearly linear beak sculpture, whereas *L. recta* has 3–5 oblique ridges, *L. subrostrata* has 5–7 inverted "V"-shaped ridges, and *L. nasuta* has double-looped beak sculpture.

Comments. Specimens may live for 10 years or more and attain a length of 120 mm, although most are ca. 90 mm or less. Shell growth gradually slows over the short life span of the animal.

Ligumia recta (Lamarck, 1819)

Figures 120, 121, 122

Etymology. L. *rectus*, straight, upright; from L. *rego*, to keep straight, to lead straight, guide, direct, rule. Lamarck chose an appropriate name for this species, since it is one of only a few elongate species that have the dorsal and ventral margins parallel and nearly straight.
Vernacular name(s). *Black Sandshell.*
Original description
Unio recta Lamarck, 1819: 74, species no. 19, no figure given.
Type locality. "Habite le lac Erie. Michaud" ["living in Lake Erie"] was included with the original description. Johnson (1953: 92–93; 1969: 60, fig. 13) noted the label with the possible holotype as saying "de la [vici] ite de Niaga [*sic*]." We interpret this information as a restriction of the type locality being Lake Erie in the vicinity of Niagara Falls, New York.
Type material. Holotype: MNHN *fide* Johnson, 1953: 92–93. The holotype was illustrated by Johnson (1969: 60, fig. 13) and its "length 100, height 41, width 31 mm" also recorded. The possible holotype thus is a juvenile.
Synonymy
Unio recta Lamarck, 1819: 74.
Unio (Eurynia) latissima Rafinesque, 1819: 426 [*nomen nudum*].
Unio (*Eurynia*) *latissima* Rafinesque, 1820: 297, pl. 80, figs. 14, 15 [Type locality: "dans le fleuve Susquehannah," but lectotype from "Ohio River." Lectotype ANSP 20212 by designation of Johnson & Baker (1973: 160), but this may be invalid.].
Unio praelongus Barnes, 1823: 261–262, fig. 11 [Type locality: "Fox River." Types lost].
Mya praelonga (Barnes, 1823). Eaton, 1826: 220.
Margarita (*Unio*) *rectus* (Lamarck, 1819). Lea, 1836: 34.
Unio rectus Lamarck, 1819. Conrad, 1836: 33, pl. 15.
Unio sageri Conrad, 1836: 53 [Type locality: "Detroit River, Michigan. Also, Red River of the North." Types not found *fide* Johnson & Baker (1973: 169)].
Unio (*Ligumia*) *recta* Lamarck, 1819. Swainson, 1840: 267, 271, fig. 55.
Eurynea [*sic*] *praelonga* (Barnes, 1823). Stimpson, 1851: 13.
Eurynia recta (Lamarck, 1819). Agassiz, 1852: 45.
Margaron (*Unio*) *rectus* (Lamarck, 1819). Lea, 1852b: 35.
Unio arquatus Conrad, 1854: 297, pl. 26, fig. 8 [Type locality: "Wabash River." Figured type in ANSP lost *fide* Johnson & Baker, 1973].
Unio leprosus Miles, 1861: 240 [Type locality: "Huron River, Livingston Co., Michigan." Syntypes ANSP 42392, MCZ 151653].
Lampsilis rectus (Lamarck, 1819). Smith, 1899: 290, fig. 78.
Lampsilis recta (Lamarck, 1819). Sterki, 1907: 389.
Eurynia (*Eurynia*) *recta* (Lamarck, 1819). Ortmann, 1912a: 344, fig. 24.
Lampsilis recta var. *sageri* (Conrad, 1836). Simpson, 1914: 96.

Fig. 120. ***Ligumia recta*** **(Lamarck, 1819).** **A.** OSUM 6339, Muskingum River, Washington Co., OH. 165 mm. **B.** OSUM 23598, Rock River, Winnebago Co., IL. 129 mm. **C.** OSUM S-1076, Lake Erie, Ottawa Co., OH. 104 mm. **D.** OSUM 63612, Ohio River, Campbell Co., KY. 143 mm. **E, Beak sculpture detail:** OSUM 51605.1, Mississippi River, Crawford Co., WI. 48 mm. **F.** OSUM 60724, Ohio River, Meigs Co., OH. 136 mm. **G.** OSUM 62282, St. Joseph River, Williams Co., ON. 87 mm. **H, Hinge detail:** OSUM 45920.1, Muskingum River, Washington Co., OH. 87 mm.

Unio angustatus var. *cuniculus* "Desf" [?] De Gregorio, 1914: 52, pl. 10, fig. 1 [Type locality: "Sciota R." [Scioto River, OH]. Type in Palermo Museum?].
Eurynia (*Eurynia*) *recta latissima* (Rafinesque, 1820). Ortmann, 1919: 276, pl. 16, figs. 12, 13.
Eurynia recta (Lamarck, 1819). Baker, 1920: 383.
Ligumia recta (Lamarck, 1819). Ortmann & Walker, 1922: 59.
Ligumia recta latissima (Rafinesque, 1820). Ortmann & Walker, 1922: 59.
Ligumia (*Ligumia*) *recta* (Lamarck, 1819). Baker, 1928a: 255, pl. 87, figs. 4, 5.
Ligumia (*Ligumia*) *recta latissima* (Rafinesque, 1820). Baker, 1928a: 257, pl. 87, figs. 1–3, pl. 86, figs. 13, 14.
Lampsilis (*Ligumia*) *recta latissima* (Rafinesque, 1820). Frierson, 1927: 70.
Lampsilis (*Ligumia*) *recta* (Lamarck, 1819). Haas, 1930: 328.
Ligumia recta recta (Lamarck, 1819). Haas, 1969: 430.

Status. Listed as Threatened by Ohio Division of Wildlife; not listed by U.S. Fish & Wildlife Service as of 2009.

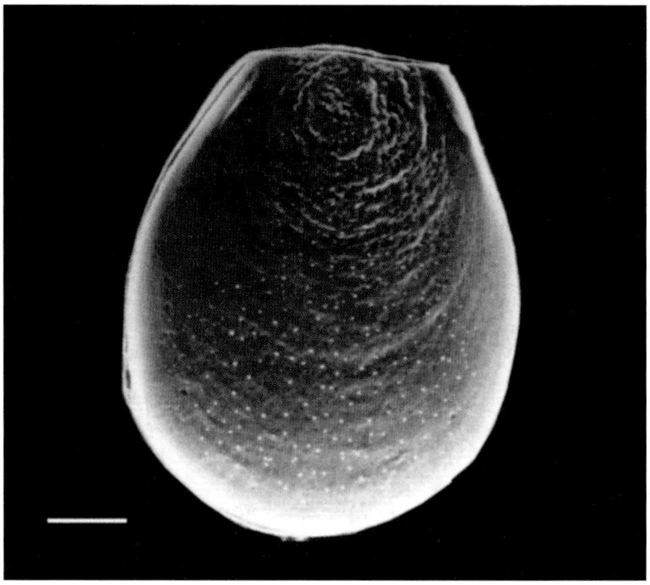

Fig. 121. *Ligumia recta* (Lamarck, 1819). Glochidium. OSUM: 1984, 2, Mississippi River, Crawford Co., WI. Bar = 35 µm. (Hoggarth, 1999).

ADULT SHELL

Shell. Very large (to 180 mm), very elongate, massive as an adult.

Anterior margin. Straight or convex, becoming abruptly but evenly rounded.

Ventral margin. Very broadly and evenly rounded in males; females are straight or slightly concave until the marsupial region.

Posterior margin. Males come to a single or double acute angulation approximately 50% or less from the dorsal margin; females variably shaped, some broadly rounded posteriorly, reaching a point at the dorsal margin; others have the marsupial region directed ventrally with the angulation located at the dorso-ventral margin.

Dorsal margin. Nearly straight in both sexes.

Umbo. Low, broad, pointed; beak sculpture of 3–5 straight ridges on the posterior portion of the beaks, oriented obliquely towards the posterior margin; 3–7 extremely fine concentric threads may be present on the anterior third of the beak; sculpture disappears during first year of growth.

Ligament. Occupies 66% to 75% of the anterior portion of the dorsal margin, low, partially concealed by the valves.

Disc sculpture. Sculptureless except for one or two weak longitudinal ridges on the posterior slope.

Periostracum. When young, green (rarely tan), with numerous broad, diffuse dark green rays; adults are progressively darker, sometimes becoming nearly uniformly black; rays are continuous, rarely wavy.

Nacre. White, occasionally flushed with pink or purple, rarely uniformly colored pink or purple; golden or gray Tulberg layers are common, particularly in the umbonal cavity.

Hinge plate. Very long, narrow; interdentum short, very narrow, flattened; two chunky cardinals in left valve, one in right, accompanied anteriorly by a weak accessory tooth; laterals very long, straight except for largest individuals (bowed).

Muscle scars. Anterior adductor moderately deep, smooth, well-separated from anterior pedal protractor; posterior adductor shallow, smooth, confluent with posterior pedal retractor scar; numerous dorsal suspensors, running in an arc from beneath the interdentum onto the outside of the cardinal fossa; pallial line complete; accessory line present but very short.

Umbonal cavity. Shallow to deep, but wide and open.

Glochidia. Subelliptical, dorsal margin slightly curved, anterior and posterior margins equally curved, ventral margin rounded. Dorsal alae long, rough exterior surface sculpturing covers the valve. Hooks absent, micropoints numerous, lanceolate (Lea, 1858c: 47, pl. 5, fig. 11; Ortmann, 1911: 347, pl. 89, fig. 21; Isom & Hudson, 1982: 147, fig. 1; Isom, 1983: 44, figs. 1a–b; Hoggarth, 1999: figs. 52a–e).

Literature records (Lefevre & Curtis, 1910a: 97, fig. l; Lefevre & Curtis, 1912: 146, fig. 11; Surber, 1912: 9, pl. 2, fig. 17; Ortmann, 1912a: 345; Utterback, 1916a: 440; Hoggarth, 1999: figs. 52a–e) indicate glochidial dimensions

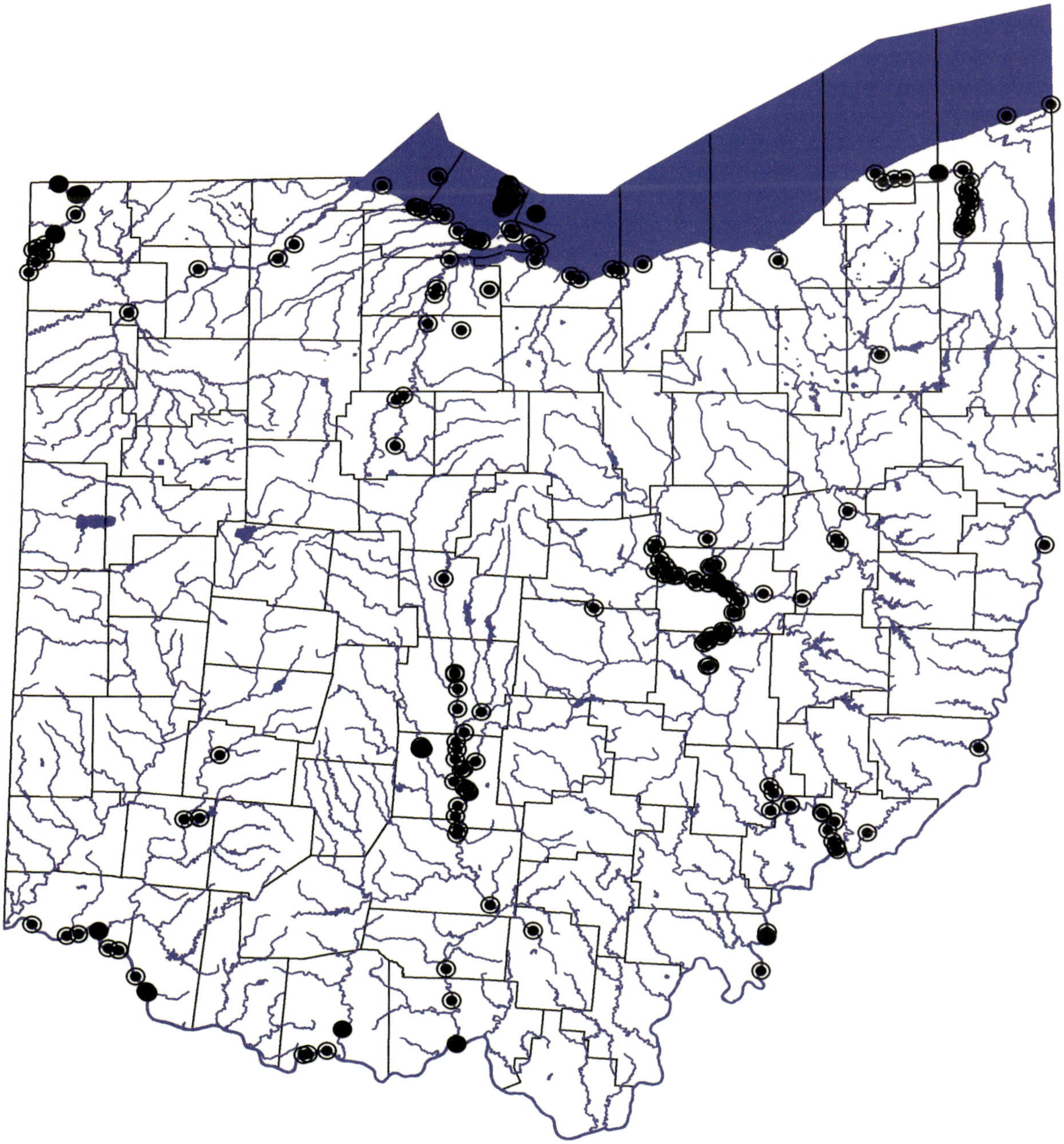

Fig. 122. ***Ligumia recta*** **(Lamarck, 1819)**
Solid dots—live or freshdead since 1980
Bullseye—all records, any condition

for length (0.20 to 0.22 mm), height (0.24 to 0.28 mm), and hinge (0.11 to 0.12 mm). This glochidium has the same valve outline as that of *A. pectorosa*. However, the glochidium of *L. recta* can be distinguished from that glochidium by its larger size, broader ventral flange, and longer dorsal alae.

Reproductive biology. Bradytictic. Marsupia occupy the posterior half or less of the outer gills (Ortmann, 1912a). Ortmann (1919) found eggs in August, with glochidia developing by September. These overwinter until the following July. Females display a line of long, dark or light tentacles along most of the ventral and posterior border that may serve to attract hosts; an eyespot is also present. White conglutinates are discharged (Utterback, 1916a), but these are not durable (Watters, 2008).

Reported potential hosts

Ambloplites rupestris (Rock Bass)—Lefevre & Curtis, 1912 (LI); Steg & Neves, 1997 (LT).
Anguilla rostrata (American Eel)—Coker *et al.*, 1921 (NI).
Campostoma anomalum (Central Stoneroller)—Watters *et al.*, 1999 (LT).
Cichlasoma nigrofasciatum (Convict Cichlid)—Steg & Neves, 1997 (LT)—exotic.
Cyprinus carpio (Common Carp)—Lefevre & Curtis, 1912 (LI).
Fundulus diaphanus (Banded Killifish)—Young, 1911 (LT).
Lepomis auritus (Redbreast Sunfish)—Steg & Neves, 1997 (LT).
Lepomis cyanellus (Green Sunfish)—Young, 1911 (LT); Steg & Neves, 1997 (LT).
Lepomis gibbosus (Pumpkinseed)—Watters *et al.*, 1999 (LT).
Lepomis humilis (Orangespotted Sunfish)—Young, 1911 (LT).
Lepomis macrochirus (Bluegill)—Surber, 1913 (NI); Wilson, 1916 (NI); Coker *et al.*, 1921 (NI); Lefevre & Curtis, 1921 (LI); Hove *et al.*, 1994 (LT); Hove *et al.*, 1998a (LT); Watters *et al.*, 1999 (LT); Khym & Layzer, 2000 (LT).
Lepomis megalotis (Longear Sunfish)—Watters *et al.*, 1999 (LT).
Micropterus salmoides (Largemouth Bass)—Young, 1911 (LT); Lefevre & Curtis, 1912 (LI); Steg & Neves, 1997 (LT); Hove *et al.*, 1998a (LT); Watters *et al.*, 1999 (LT); Khym & Layzer, 2000 (LT); Watters *et al.*, 2009.
Morone americana (White Perch)—Steg & Neves, 1997 (LT).
Notropis rubellus (Rosyface Shiner)—Watters *et al.*, 1999 (LT).
Perca flavescens (Yellow Perch)—Lefevre & Curtis, 1912 (LI); Steg & Neves, 1997 (LT).
Pomoxis annularis (White Crappie)—Young, 1911 (LT); Wilson, 1916 (NI); Coker *et al.*, 1921 (NI); Khym & Layzer, 2000 (LT).
Pomoxis nigromaculatus (Black Crappie)—Khym & Layzer, 2000 (LT).
Rutilus rutilus (Roach)—Lefevre & Curtis, 1912 (LI).
Sander canadensis (Sauger)—Pearse, 1924 (NI); Khym & Layzer, 2000 (LT).
Sander vitreus (Walleye)—Hove *et al.*, 1994 (LT); Hove *et al.*, 1998a (LT).
Xiphophorus maculatus (Southern Platyfish)—Steg & Neves, 1997 (LT)—exotic.

Habitat. Widespread but sporadic in rivers and lakes, less commonly in streams. It may occur in soft or hard substrates. In describing the ecology of this species Parmalee (1967: 75) stated, "This mussel is usually found in areas of strong current, on coarse sand or gravel bottoms, and at depths varying from a few inches to 3–4 feet." However, in the Walhonding River this species was found in shallow water near the margin of the river in areas of swift current to slackwater habitats. The specimens were usually only partly buried in the substrate. In the St. Joseph River and the West Branch of the St. Joseph River this species was found in 3–5 feet of water in moderate current almost completely buried in the sand and gravel substrate.

Range. Parmalee (1967: 74) gives the range of this species as "Mississippi River drainage; western New York west to South Dakota and Kansas, north to Minnesota; Manitoba, Ontario, and Quebec, Canada; south to Louisiana, Alabama, and Georgia."

Range in Ohio. Records indicate that this species once occurred in the Maumee, Sandusky, and Grand rivers of Lake Erie and the lake proper. In the Ohio River drainages in the Little Miami, Scioto, and Muskingum rivers as well as the Ohio River. Today the species appears to be limited to the St. Joseph River, the West Branch of the St. Joseph River, Big Darby Creek, the Muskingum River, and the Walhonding River within Ohio.

Similar species. *Lampsilis teres* (Rafinesque, 1820) approaches the shape and size of this species, but that species is generally yellow in color and has a shining, waxy periostracum. *Elliptio dilatata* (Rafinesque, 1820) may be confused with smaller specimens of this species, but that species has coarse beak sculpture and a more cloth-like periostracum, and is more elliptical rather than rectangular. See also under *Ligumia nasuta*.

Comments. This species grows rapidly for up to six or

seven years before abruptly slowing. Female shell characteristics become apparent in the third year. Growth rates are fairly uniform across populations, and individuals may reach 180 mm or more and live for at least 30 years.

Megalonaias Utterback, 1915

Etymology. Gr. *megas, megalo-*, large, great + Gr. *naias, naiados*, a water nymph of rivers and springs. This genus has the largest heavy-shelled unionid in North America. The term *naias* has been used in reference to a number of freshwater invertebrate animals and plants inhabiting springs, brooks, and pools.

Original description

Megalonaias Utterback, 1915: 41 [The first name used for this genus was *Magnonaias*, also proposed by Utterback on page 7 in the same publication. A request to suppress *Magnonaias*, despite page priority, for the more commonly used name, *Megalonaias*, was granted by the International Commission on Zoological Nomenclature (Bogan & Williams 1986: 273; Bull. ICZN 1988: 159).].

Type species. *Unio heros* Say, 1829 (= *Unio nervosa* Rafinesque, 1820), by monotypy.

Synonymy

Magnonaias Utterback, 1915: 7.
Megalonaias Utterback, 1915: 41.

Geographic range. Mississippi River system from Minnesota to the Gulf of Mexico. Gulf drainages from Mobile River to Texas, Mexico, and Guatemala.

Megalonaias nervosa (Rafinesque, 1820)

Figures 123, 124, 125

Etymology. L. *nervosus*, full of sinews, nerves, or fibers. French *nervure*, vein, rib, raised band. Rafinesque may have drawn upon his native French as well as classical Latin in order to describe the sculpture of the figured shell (Rafinesque, 1820: pl. 80, figs. 8–10).

Vernacular name(s). *Washboard*, Black Rock Shell, Fine-ridged Rock Shell, Big Black.

Original description

Unio (*Leptodea*) *nervosa* Rafinesque, 1820: 296, pl. 80, figs. 8–10.

Type locality. "Je l'ai trouvée aux rapides de l'Ohio" ["I have found it at The Falls of the Ohio"]. The Falls of the Ohio River are between Louisville, Jefferson County, Kentucky and the Falls City area of Floyd County, Indiana.

Type material. The type was described by Rafinesque as "Larguer un pouce et demi"—only 1.5 inches in length, obviously a juvenile. The type is apparently lost, as Johnson & Baker (1973) in their search for potential type material in the Rafinesque-Poulson Collection at the ANSP failed to locate such a specimen.

Synonymy

Unio (*Leptodea*) *nervosa* Rafinesque, 1820: 296, pl. 80, figs. 8–10.
Unio giganteus Barnes, 1823: 119 [Conditionally proposed under *Unio crassus* var. *i*—"might perhaps constitute a separate species under the designation *Unio Giganteus*." Type locality: "obtained by the N. W. Expedition." Type "in Dr. Mitchell's cabinet," presumably lost. Second specimen mentioned, from "Detroit," must be *Amblema costata*.].
Unio undulatus Barnes, 1823: 120, fig. 2 [Type locality: "in the Ohio [River]." Types lost].
Unio heros Say, 1829: 291 [Type locality: "Fox River of the Wabash [River]." Neotype SMF 4307 by selection of Haas (1930: 328)].
Mya undulata (Barnes, 1823). Eaton, 1826: 219.
Unio undulata Barnes, 1823. Valenciennes, 1827: 229, pl. 54, figs. 3, 3a, 3b.
Unio multiplicatus Lea, 1831: 70, pl. 4, fig. 2 [Type locality: "Tennessee and Ohio Rivers." Type not located].
Margarita (*Unio*) *undulatus* (Barnes, 1823). Lea, 1836: 12.
Margarita (*Unio*) *multiplicatus* (Lea, 1831). Lea, 1836: 12.
Unio atrocostatus Lea, 1848: 70, pl. 1, fig. 5 [Type locality: "Claiborne, Ala., Tuscaloosa, Ala., Alexandria, La." Lectotype USNM 84015].
Complanaria heros (Say, 1829). Agassiz, 1852: 49.
Complanaria atrocostatus (Lea, 1848). Agassiz, 1852: 49.
Margaron (*Unio*) *atrocostatus* (Lea, 1848). Lea, 1852b: 20.
Margaron (*Unio*) *multiplicatus* (Lea, 1831). Lea, 1852b: 20.
Margaron (*Unio*) *undulatus* (Barnes, 1823). Lea, 1852b: 29.
Unio eightsii Lea, 1860b: 306 [Type locality: "Sabinas River, [Nuevo Leon] Mexico." Lectotype USNM 83991].
Unio atrocostatus Lea, 1848. Reeve, 1864: pl. 4, fig. 13 [Changed in errata to *Unio heros* Say, 1829].
Margaron (*Unio*) *eightsii* (Lea, 1860). Lea, 1870: 29.
Unio eighti [*sic*] (Lea, 1831). Paetel, 1890: 151.
Unio heros var. *multiplicatus* (Lea, 1831). Paetel, 1890: 155.
Quadrula undulata (Barnes, 1823). Baker, 1898b: 82–83, pl. 22, figs. 1, 2, pl. 12, fig. 1.
Quadrula triumphans Wright, 1898: 101 [Type locality: "Coosa River, St. Clair Co., Alabama." Possible holotype USNM 150554].
Quadrula heros (Say, 1829). Simpson, 1900a: 770.
Crenodonta undulata (Barnes, 1823). Ortmann, 1912a: 246.
Crenodonta heros (Say, 1829). Ortmann, 1912a: 248.
Magnonaias heros (Say, 1829). Utterback, 1915: 47.

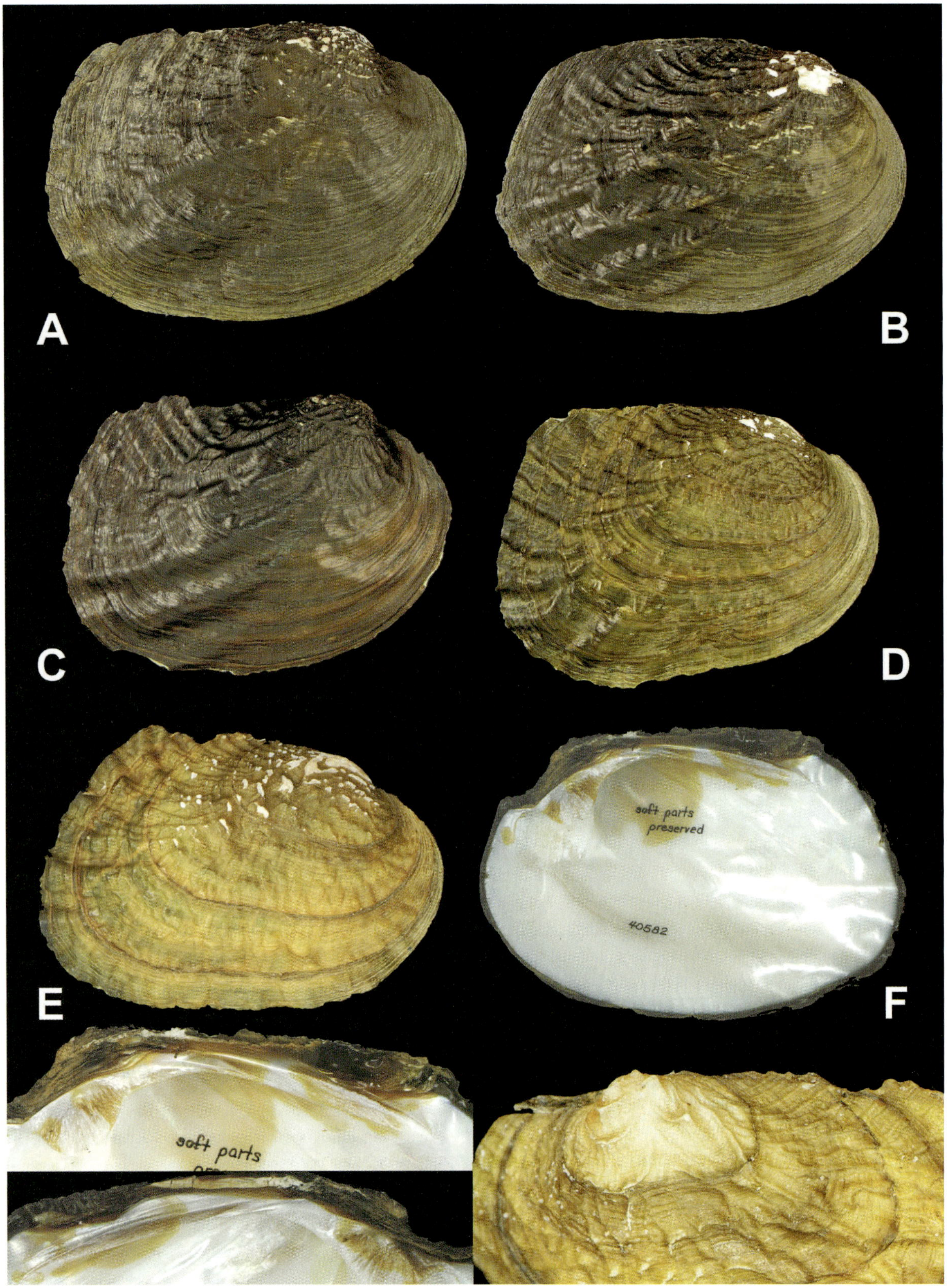

Fig. 123. ***Megalonaias nervosa*** **(Rafinesque, 1820). A.** OSUM 7869, Big Walnut Creek, Franklin Co., OH. 161 mm. **B.** OSUM 44846, Illinois River, Morgan Co., IL. 180 mm. **C.** OSUM 39434, Little Miami River, [Co.?], OH. 116 mm. **D.** OSUM 10055, "Ohio River." 98 mm. **E.** OSUM 58418, "Ohio River." 57 mm. **F, Hinge detail:** OSUM 40582, Mississippi River, Crawford Co., WI. 111 mm. **Beak sculpture detail:** OSUM 45445.1, Mississippi River, Crawford Co., WI. 35 mm total length.

Megalonaias heros (Say, 1829). Utterback, 1915: 125–129, pl. 7, fig. 16, pl. 17, figs. 48a–f.
Megalonaias gigantea (Barnes, 1823). Ortmann & Walker, 1922: 7–8.
Amblema (*Megalonaias*) *gigantea* (Barnes, 1823). Frierson, 1927: 62.
Amblema (*Megalonaias*) *triumphans* (Wright, 1898). Frierson, 1927: 62.
Amblema gigantea (Barnes, 1823). Strecker, 1931: 34–35.
Megalonaias nervosa (Rafinesque, 1820). Morrison, 1969: 24.
Megalonaias gigantea gigantea (Barnes, 1823). Haas, 1969: 284–285.
Megalonaias triumphans (Wright, 1898). Haas, 1969: 285.
Status. Listed as Endangered by Ohio Division of Wildlife; not listed by U.S. Fish & Wildlife Service as of 2009.

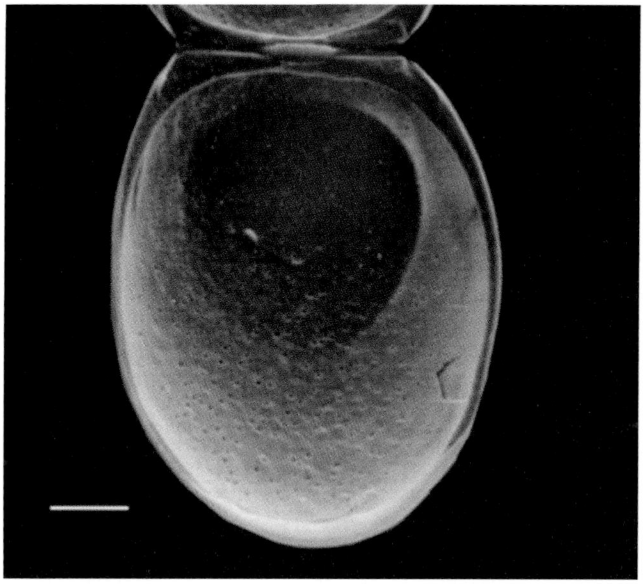

Fig. 124. *Megalonaias nervosa* (Rafinesque, 1820). Glochidium. OSUM 54178, Mississippi River, Pike Co., IL. Bar = 50 µm. (Hoggarth, 1999).

ADULT SHELL

Shell. Very large size (to 210 mm), thick to massive, oval to rectangular in profile, rather inflated.
Anterior margin. Very short and straight, then broadly rounded to the ventral border.
Ventral margin. Very rounded to nearly straight, occasionally somewhat concave in large individuals.
Posterior margin. Broadly rounded or straight, usually forming a blunt acute point at the ventral margin.
Dorsal margin. Long, gently arched, joining posterior margin with very obtuse angle.
Umbo. Very wide and low; umbonal sculpture of 3–5 concentric double-looped ridges, usually quite prominent, the posterior loops tending to become nodulose or weakly pointed.
Ligament. Long, tan to dark brown, partially concealed by shells.
Disc sculpture. During the first three years of growth the shell is covered with a complex series of anastomosing chevrons, overlain by larger, oblique ribs running from the midline to the ventral margin; posterior slope strongly corrugated; the oblique ribs and posterior corrugations persist on the adult shell, but the chevron sculpture is usually lost, particularly anterior to the mid-line.
Periostracum. Juveniles are olive or yellowish with dark annular rings; past three years old most specimens become dark brown to black; juveniles may have very vague green rays.
Nacre. Porcelain white, often iridescent posteriorly; often with gold or brown Tulberg layers.
Hinge plate. Well-developed; right cardinal a single, massive, triangular denticle with a much smaller anterior ridge; left cardinal bifid, the teeth often far apart; laterals well-developed, one in right valve, two in left; interdentum flattened, rather long.
Muscle scars. Anterior adductor impressed; posterior adductor flush; anterior adductor distinct from anterior pedal protractor scar, large posterior adductor confluent with posterior pedal retractor scar; dorsal suspensors located behind anterior adductor and in a line inside of interdentum. Pallial line entire, located far from shell margin. Accessory strong, reaching 33% the length of the shell.
Umbonal cavity. Very wide, open, and deep.

Glochidia. Subelliptical, dorsal margin straight, lateral margins gently curved but unequal, ventral margin narrowly rounded. Tight-looped sculpture covers the exterior surface of the valve, coiled larval thread present. Hooks absent; lanceolate micropoints occur in broken vertical rows on a narrow ventral flange (Lea, 1858c: 46, pl. 5, fig. 3; Hoggarth, 1999: figs. 29a–g).

Literature records (Surber, 1912: 9, pl. 2, fig. 32; Howard, 1914a: 28–30, pl. 5, fig. 35; Surber, 1915: 7, pl. 1, fig. 10; Utterback, 1915: 126, figs. 3a–b; Hoggarth, 1999: figs. 29a–g) indicate glochidial dimensions for length (0.25 to 0.28 mm), height (0.34 to 0.38 mm), and hinge (0.15 to 0.16 mm). Surber (1915: 7) noted, "Notwithstanding its great variation in size, and even outline, this species can not be readily confused with any other, even though the larval gland may have been absorbed, which occurs during metamorphosis. This coiled gland is a prominent feature in the mature glochidium."

Reproductive biology. Bradytictic. All four gills are marsupial (Utterback, 1915). Eggs develop from June to September (Bingham, 1968). Gravid females are found in winter, and glochidia are released by the following April (Utterback, 1915). Utterback (1915) reported that whole, sole-shaped conglutinates were released, but specimens

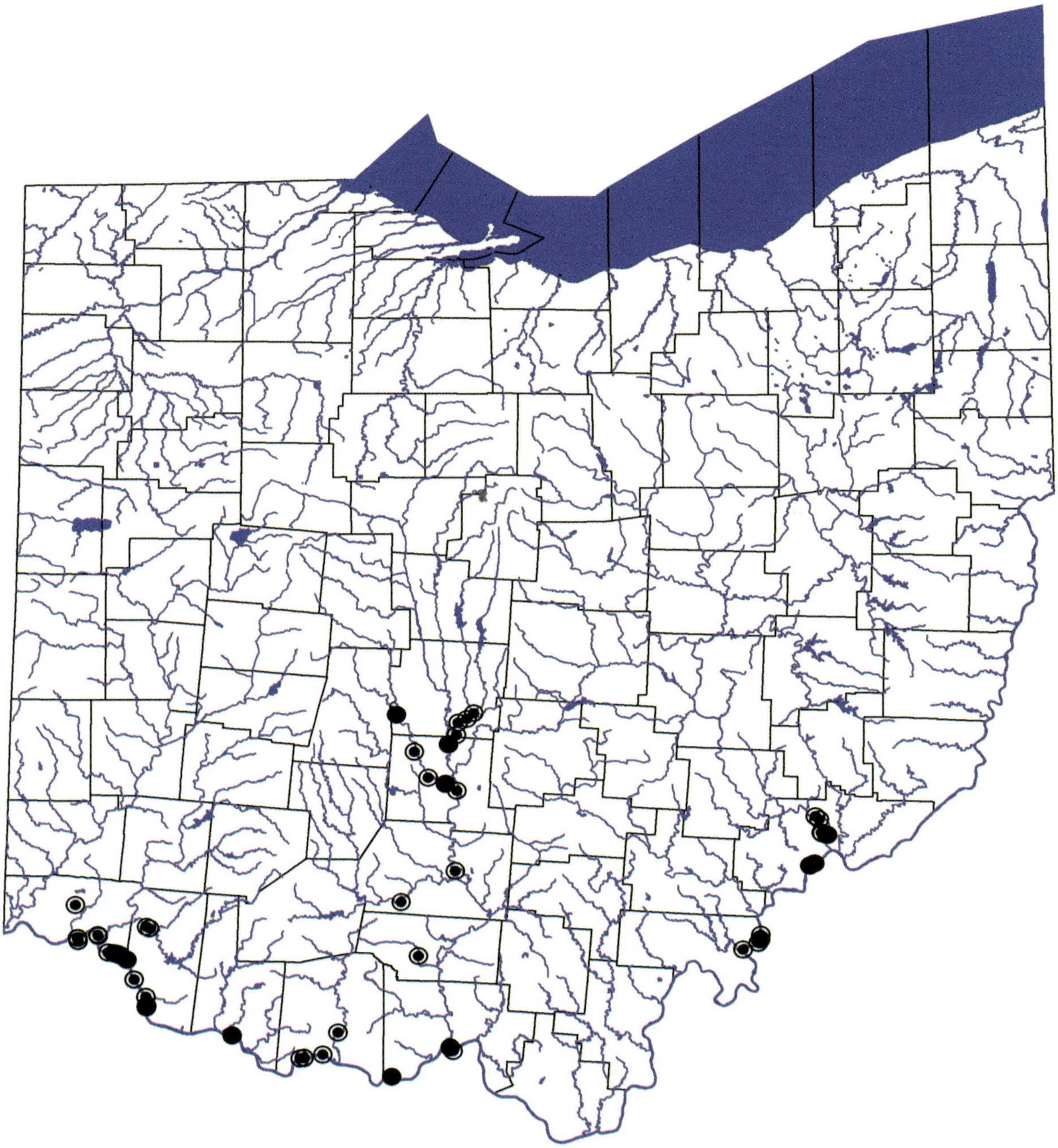

Fig. 125. ***Megalonaias nervosa*** **(Rafinesque, 1820)**
Solid dots—live or freshdead since 1980
Bullseye—all records, any condition

observed in captivity released numerous glochidia in large masses of mucus (Watters, unpubl.). This species may be a generalist in its host use.

Reported potential hosts

Alosa chrysochloris (Skipjack Herring)—Wilson, 1916 (NI); Coker *et al.*, 1921 (NI).

Ameiurus melas (Black Bullhead)—Howard, 1914a (LI); Coker *et al.*, 1921 (LT); Woody & Holland-Bartels, 1993 (LT); Watters *et al.*, 2009 (LT).

Ameiurus natalis (Yellow Bullhead)—Watters *et al.*, 2009 (LT).

Ameiurus nebulosus (Brown Bullhead)—Coker *et al.*, 1921 (LT).

Amia calva (Bowfin)—Howard, 1914a (NI).

Anguilla rostrata (American Eel)—Surber, 1915 (NI); Wilson, 1916 (NI); Coker *et al.*, 1921 (NI).

Aplodinotus grunniens (Freshwater Drum)—Surber, 1913 (NT); Howard, 1914a (LI); Surber, 1915 (NI); Wilson, 1916 (NI); Coker *et al.*, 1921 (NI, LT); Weiss & Layzer, 1995 (NI).

Campsotoma anomalum (Central Stoneroller)—Watters & O'Dee, 1998b (LT).

Carpiodes velifer (Highfin Carpsucker)—Howard, 1914a (LI).

Dorosoma cepedianum (Gizzard Shad)—Howard, 1914a (NI); Coker *et al.*, 1921 (NI); Weiss & Layzer, 1995 (NI).

Fundulus catenatus (Northern Studfish)—Watters *et al.*, 2009 (LT).

Ictalurus punctatus (Channel Catfish)—Howard, 1914a (LI); Coker *et al.*, 1921 (LT); Woody & Holland-Bartels, 1993 (LT).

Lepisosteus osseus (Longnose Gar)—Weiss & Layzer, 1995 (NI); Watters & O'Dee, 1998b (LT); O'Dee & Watters, 2000 (LT).

Lepomis cyanellus (Green Sunfish)—Coker *et al.*, 1921 (LT); Howard, 1914a (LT); Woody & Holland-Bartels, 1993 (LT); Watters *et al.*, 2009 (LT).

Lepomis gulosus (Warmouth)—Weiss & Layzer, 1995 (NI).

Lepomis macrochirus (Bluegill)—Howard, 1914a (LI); Coker *et al.*, 1921 (LI, NT); Watters & O'Dee, 1998b (LT); O'Dee & Watters, 2000 (LT).

Lepomis megalotis (Longear Sunfish)—Watters & O'Dee, 1998b (LT); O'Dee & Watters, 2000 (LT).

Micropterus punctulatus (Spotted Bass)—Weiss & Layzer, 1995 (NI).

Micropterus salmoides (Largemouth Bass)—Howard, 1914a (LI); Keller & Ruessler, 1997 (LT); Watters & O'Dee, 1998b (LT); O'Dee & Watters, 2000 (LT); Watters *et al.*,2009 (LT).

Morone chrysops (White Bass)—Howard, 1914a (NI, LT); Wilson, 1916 (NI); Coker *et al.*, 1921 (NI); Weiss & Layzer, 1995 (NI).

Necturus maculosus (Mudpuppy)—Howard, 1914d (NI); Surber, 1915 (NI).

Notemigonus crysoleucas (Golden Shiner)—Watters *et al.*, 2009 (LT).

Noturus gyrinus (Tadpole Madtom)—Coker *et al.*, 1921 (NI).

Perca flavescens (Yellow Perch)—Watters & O'Dee, 1998b (LT); O'Dee & Watters, 2000 (LT).

Percina caprodes (Logperch)—Watters & O'Dee, 1998b (LT); O'Dee & Watters, 2000 (LT).

Percina phoxocephala (Slenderhead Darter)—Watters & O'Dee, 1998b (LT); O'Dee & Watters, 2000 (LT).

Pomoxis annularis (White Crappie)—Howard, 1914a (LI); Coker *et al.*, 1921 (NI, LT); Weiss & Layzer, 1995 (NI).

Pomoxis nigromaculatus (Black Crappie)—Howard, 1914a (LI); Coker *et al.*, 1921 (LT).

Pylodictis olivaris (Flathead Catfish)—Howard, 1914a (NI); Coker *et al.*, 1921 (NI, NT); Weiss & Layzer, 1995 (NI).

Sander canadensis (Sauger)—Howard, 1914a (LI).

Scaphirhynchus platorynchus (Shovelnose Sturgeon)—Watters *et al.*, 2009 (LT).

Habitat. The Washboard is a species of large rivers and impoundments of these rivers. As such, it can tolerate a wide variety of substrates and water flows.

Range. Mississippi River system. Gulf drainages from the Mobile River to Texas. If *Unio eightsii* Lea, 1860, from the Rio Sabinas in Mexico is this species, then it is one of the few North American mussels to occur in Central America as well (Rio Sabinas is a Mexican tributary of the Rio Grande). Specimens from the Ochlockonee and Apalachicola rivers were named *M. boykiniana* (Lea, 1840), considered by Mulvey *et al.* (1997) and Brim Box & Williams (2000), as a synonym of *M. nervosa*.

Range in Ohio. The only reproducing population is in the lower Muskingum River and the Ohio River mainstem. Historically in the Scioto River into Franklin County, and the lower Little Miami River. Rare strays may occur in smaller streams, such as Big Darby Creek. It is absent from the Lake Erie drainages (a record from the Maumee River is undoubtedly in error).

Similar species. In Ohio *M. nervosa* can be confused only with *Amblema plicata*. For specimens of the same size, *Megalonaias* is thinner than *Amblema*. On *Megalonaias* there is divaricate sculpture on the early shell, with sculpture anterior to the umbo. On *Amblema* there are only oblique ribs, which do not reach forward of the umbo.

Comments. Individuals grow very quickly for the first

2–3 years and then slow. Individuals grow to at least 210 mm in length and over 30 years in age.

Obliquaria Rafinesque, 1820

Etymology. L. *obliquus,* sidelong, slanting, oblique, sideways. Rafinesque (1820: 301) explained the use of the name *Obliquaria* for his new genus by noting that "it differs from *Unio* or *Elliptio* principally by its shape, by the oblique ligament and lamellar tooth, etc." When Rafinesque described a specimen, he held it with the beaks of the umbo at the very uppermost central position (anterior end up). When viewed this way, the various species that Rafinesque included in *Obliquaria,* including *reflexa* Rafinesque, 1820, do indeed have "oblique" ligaments and lateral teeth.

Original description

Obliquaria Rafinesque, 1820: 301.

Type species. *Obliquaria reflexa* Rafinesque, 1820, by subsequent designation of Simpson (1900a: 610).

Synonymy

Obliquaria Rafinesque, 1820: 301.

Oblicaria [*sic*] d'Orbigny, 1846: 709.

Geographic range. Mississippi River system from Minnesota to southern Louisiana. Mobile River system.

Obliquaria reflexa Rafinesque, 1820

Figures 126, 127, 128, 129

Etymology. L. *reflexus,* bent back, turned back, reflected, from L. *reflecto,* to bend or turn back. Rafinesque, in his original description (1820: 306) of *O. reflexa,* noted "lower margin reflected with a posterior sinus; . . . ." By "posterior sinus," Rafinesque was probably referring to the slight emargination of the post-ventral margin, a typical characteristic of this species. At the point where a new knob is being formed, the mid-ventral margins of the shell typically appear "bent," or in Rafinesque's words, "reflected," one laterally and the other mesially, toward the valve on which the new knob is developing.

Vernacular name(s). *Threehorn Wartyback,* Horny-Back, Three Knot, Three Dot.

Original description

Obliquaria (*Quadrula*) *reflexa* Rafinesque, 1820: 306, species no. 31, no figure.

Type locality. Rafinesque gave two localities: "Dans le Kentuky et aux rapides de Letart" ["In the Kentucky River and in the Rapids of Letart" (of the Ohio River, Meigs Co., Ohio)]. The type locality became the "rapides de Letart" when Johnson & Baker (1973) designated a lectotype from that locale.

Type material. From the types mentioned by Vanatta (1915: 554), Johnson & Baker (1973: 168) selected ANSP 20206a, from the Ohio River at Letart Falls, as lectotype, and ANSP 20206 as paralectotype.

Synonymy

Obliquaria (*Quadrula*) *reflexa* Rafinesque, 1820: 306.

Unio cornutus Barnes, 1823: 122–123, fig. 5 [Type locality: "Fox River." Types lost].

Mya cornuta (Barnes, 1823). Eaton, 1826: 216.

Unio torulosus Rafinesque, 1820. Short & Eaton, 1831: 75 [misidentification].

Unio reflexa (Rafinesque, 1820). Say, 1834: no pagination.

Margarita (*Unio*) *cornutus* (Rafinesque, 1820). Lea, 1836: 15.

Theliderma cornuta (Barnes, 1823). Swainson, 1840: 269.

Quadrula reflexus (Rafinesque, 1820). Agassiz, 1852: 48.

Margaron (*Unio*) *cornutus* (Rafinesque, 1820). Lea, 1852b: 22.

Obliquaria reflexa Rafinesque, 1820. Baker, 1898b: 89, pl. 14, fig. 5, pl. 20, fig. 2.

Obliquaria reflexa var. *conradi* Frierson, 1927: 65 [Type locality: "Alabama River." Lectotype ANSP 56695a *fide* Johnson, 1972: 142].

Status. Listed as Threatened by Ohio Division of Wildlife; not listed by U.S. Fish & Wildlife Service as of 2009.

ADULT SHELL

Shell. Medium-sized (to 80 mm); very heavy for its size, outline rounded, pointed posteriorly, umbo just anterior of center.

Anterior margin. Broadly curved.

Ventral margin. Very broadly curved anteriorly, then straight or concave to acute or 90° point at the posterior margin.

Posterior margin. Short, straight, forming distinct obtuse angle with dorsal slope.

Dorsal margin. Short, straight; dorsal ridge sharp, forming ca. 90° angle or less, concave or straight.

Umbo. Wide, low; beak sculpture microscopic, of fine concentric striae, two irregular knobs on the posterior slope, and a single medial knob. This medial knob is the first of several that will persist on the shell as adult sculpture; except for this knob, the beak sculpture disappears early in the first year's growth.

Ligament. Brown to black, short, partially concealed by the valves.

Disc sculpture. Shell sculpture varies greatly even within a single population. All specimens have a single medial row of knobs radiating from the umbo to the ventral margin. The knobs alternate in position between left and right valves. No more than four knobs are made per side (not including the beak sculpture), but either side may make the final one. Occasional specimens have additional wavy,

Fig. 126. ***Obliquaria reflexa*** **Rafinesque, 1820. A.** OSUM 56771.5, Ohio River, Hamilton Co., OH. 56 mm. **B.** OSUM 5893.2, Muskingum River, Washington Co., OH. 49 mm. **C.** OSUM 39267.1, "Wabash River." 33 mm. **D.** OSUM 49364, Muskingum River, Muskingum Co., OH. 67 mm. **E, F, Hinge detail:** OSUM 68464.32, Ohio River, Clermont Co., OH. 49 mm. **G.** OSUM 25868, Lake Erie, Ottawa Co., OH. 43 mm. **H.** OSUM S-152, Lake Erie, Ottawa Co., OH. 50 mm. **Anterior view:** note knobs alternating from side to side. **Beak sculpture detail:** OSUM 68464.6, Ohio River, Clermont Co., OH. 30 mm.

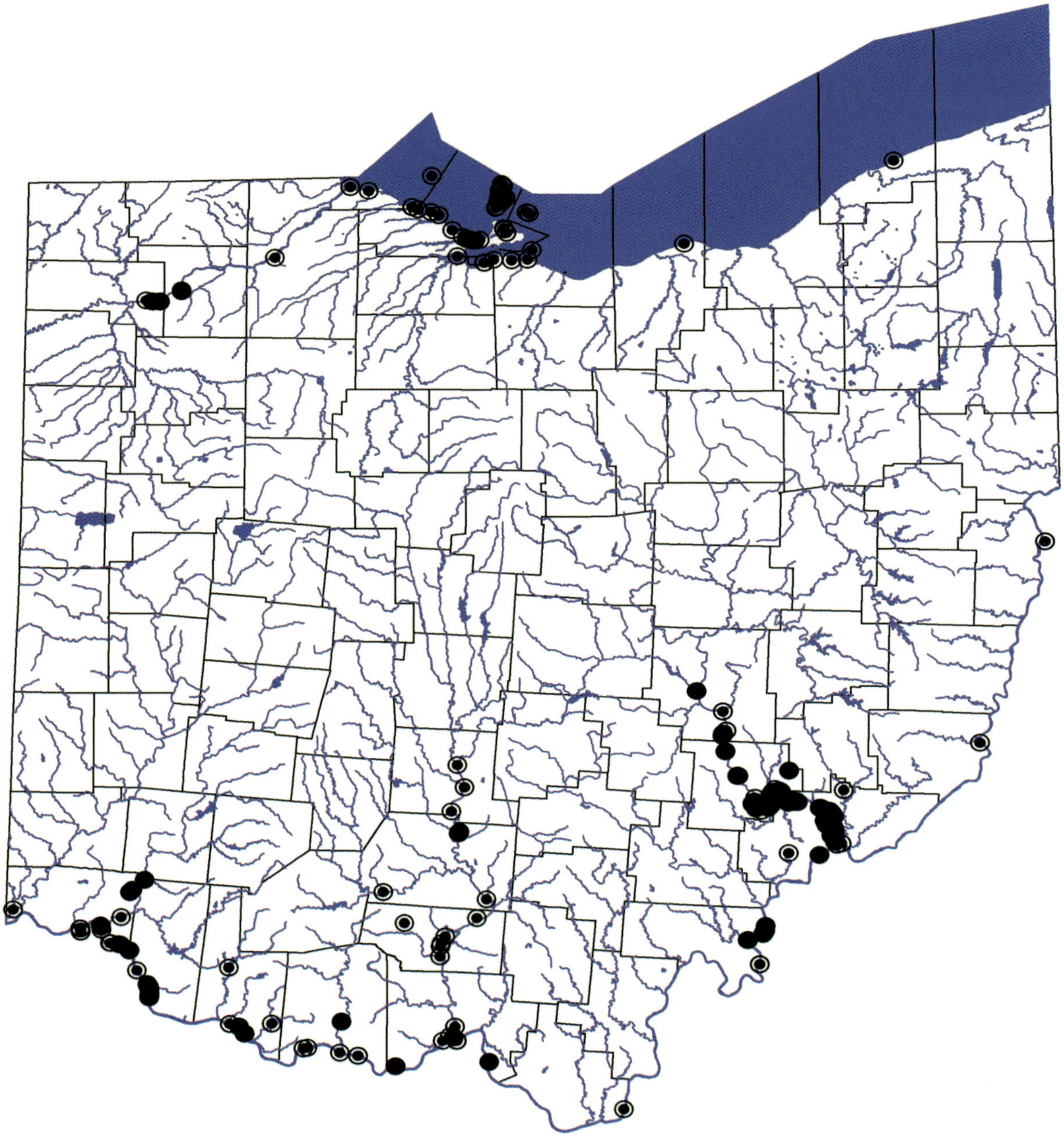

Fig. 127. ***Obliquaria reflexa*** **Rafinesque, 1820**
Solid dots—live or freshdead since 1980
Bullseye—all records, any condition

teardrop-shaped, or pustulose sculpture scattered anywhere on the shell disc. Radiating narrow ribs generally, but not always, are present on the posterior slope.
Periostracum. Extremely variable in color. Pale yellow, orangish, green, or tan, becoming darker with age. Shells vary from rayless to boldly patterned. Rayless individuals tend to be yellow; boldly patterned ones tend to be dark green. Most specimens have numerous fine green rays. A single wide green ray may radiate down through the row of knobs.
Nacre. White, rarely pink or rose, iridescent posteriorly; rarely with gold or green Tulberg layers.
Hinge plate. 60% of shell, strongly curved or angled, fairly thick; interdentum wide, thick; cardinal teeth chunky, triangular, oriented vertically, two in left valve, one in right, often with weak anterior and/or posterior accessory teeth; lateral teeth short, straight or gently curved, narrow, two in left, one in right.
Muscle scars. Anterior adductor muscle scar impressed, posterior adductor scar flush with shell surface, anterior pedal protractor scar distinct, posterior pedal retractor scar narrowly confluent or separate; dorsal suspensor scars a series of irregular pits within the umbonal cavity; pallial line entire, accessory line extremely long, extending to the shell margin almost to the posterior of the shell.
Umbonal cavity. Wide, open, shallow.

Glochidia. Subrotund, symmetrical, dorsal margin slightly curved outward, lateral and ventral margins more or less round in outline. Dorsal alae absent, hooks absent, micropoints bluntly pyramidal (Ortmann, 1912a: 311, pl. 20, fig. 1; Hoggarth, 1999: figs. 38a–e).

Literature records (Lefevre & Curtis, 1910a: 97, fig. m; Lefevre & Curtis, 1912: 146, fig. 1n; Surber, 1912: 10, pl. 2, fig. 39; Ortmann, 1912a: 311, pl. 20, fig. 1; Utterback, 1916a: 319; Hoggarth, 1999: figs. 38a–e) indicate glochidial dimensions for length (0.21 to 0.23 mm), height (0.21 to 0.24 mm), and hinge (0.12 to 0.13 mm). Ortmann (1912a: 311) stated, "The glochidia are of medium size, almost subcircular (their shape may be best compared with a circle a small section of which is cut off)." This glochidium can be distinguished by its shape and micropoints.

Reproductive biology. Tachytictic. Four to eight water tubes in the central portion of the outer gill form the marsupium (Ortmann, 1912a). Eggs are formed in May, and glochidia appear from June to August (Ortmann, 1919). In Ohio, infective conglutinates were released in mid-June (Watters, unpubl.). Glochidia are released in solid, white, club-shaped conglutinates (12 mm long x 1.5 mm wide), composed completely of glochidia, each encased in a thick, somewhat elastic matrix (Watters, 2008). There are no accessory unfertilized eggs (Watters, unpubl.). Suggestions that it may not require a host to complete development (Utterback, 1916a) have not been substantiated.

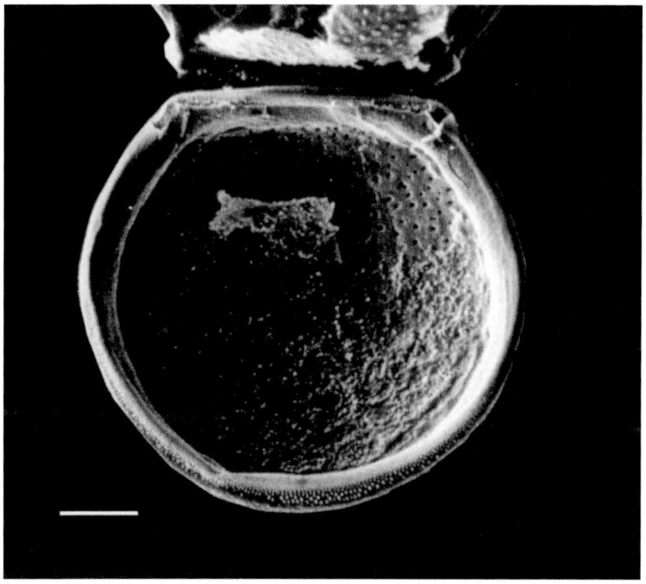

Fig. 128. ***Obliquaria reflexa*** **Rafinesque, 1820.** Glochidium. OSUM 54361, Mississippi River, Jackson Co., IA. Bar = 35 µm. (Hoggarth, 1999).

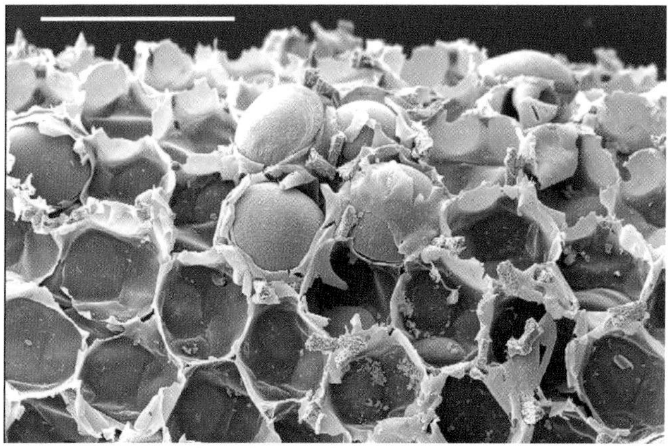

Fig. 129. ***Obliquaria reflexa*** **Rafinesque, 1820.** Portion of elastic simple conglutinate with glochidia embedded in attached egg membranes. Bar = 500 µm. (Watters, 2008).

Reported potential hosts

Ericymba buccata (Silverjaw Minnow)—Watters *et al.*, 1998c (LT).
Luxilus chrysocephalus (Common Shiner)—Watters *et al.*, 1998c (LT).
Rhinichthys cataractae (Longnose Dace)—Watters *et al.*, 1998c (LT).

Habitat. A river and lake species, tolerating slackwater or swift currents, muddy sand, or cobble. Rarely found in smaller creeks as strays.

Range. Mississippi River system from Minnesota to southern Louisiana, Texas, Kansas, and Oklahoma. East to western Pennsylvania. Mobile River system.

Range in Ohio. Abundant in the lower Muskingum River, but as rare strays elsewhere. Historically from the lower Little Miami River, the Scioto River up to Circleville, and the Muskingum River to Zanesville. In Lake Erie, particularly in the western basin around the Bass Islands. Sporadic in the Maumee River.

Similar species. Among pustulose species only *Obliquaria* has a single row of large knobs arranged across the shells that alternate in position from the left to right valves.

Comments. This is a fairly slow growing and long lived species. Growth is accelerated for the first 2–3 years and then abruptly slows. Sexual dimorphism in the shell is not pronounced and often not discernible. Females are more rounded than males posteriorly—males tend to have a concave posterior margin. This species reaches 80 mm and grows to 18 years in age.

Obovaria Rafinesque, 1819

Etymology. L. *ob*, toward, to + L. *ovatus*, egg-shaped. An apt description of many of the shells of this genus, particularly *Obovaria subrotunda* (Rafinesque, 1820), the type of the genus.

Original description

Obovaria Rafinesque, 1819: 426 [*Obovaria* was described without any species assigned to it. The type species became *Obliquaria obovalis* Rafinesque, 1820, by subsequent designation of Herrmannsen (1849: 132). But *Obovaria obovalis* has been identified with *U. ebenus* Lea, 1831, and a lectotype ANSP 20244 was designated by Johnson & Baker (1973: 163), making *Obovaria* the senior synonym of the widely used *Fusconaia*. The next available name is *Rotundaria* Rafinesque, 1820, which has as its type *Obliquaria subrotundus* Rafinesque, 1820. Simpson's (1900a: 599) designation of *Unio retusa* Lamarck, 1819, as the type of *Obovaria* was invalid because of Herrmannsen's previous action. The only way to preserve *Obovaria* as the valid name for this genus is to suppress Hermannsen's type designation in favor of Simpson's and suppress *O. obovalis* as the type species. This action was taken by the ICZN in Opinion 2188 (2008), thus conserving the current concepts of both *Fusconaia* and *Obovaria*.].

Type species. As mentioned above, the type of *Obovaria*, *Obliquaria obovalis* Rafinesque, 1820, was suppressed in favor of *Unio retusa* Lamarck, 1819.

Synonymy

Obliquaria (*Rotundaria*) Rafinesque, 1820: 308 [Type species: *Obliquaria* (*Rotundaria*) *subrotunda* Rafinesque, 1820, by subsequent designation of Herrmannsen (1847: 407)].

Obovaria (*Pseudoon*) Simpson, 1900a: 601 [Type species: *Unio ellipsis*, by original designation].

Geographic range. Mississippi River system, Great Lakes and St. Lawrence drainages, Mobile River system.

Obovaria olivaria (Rafinesque, 1820)

Figures 130, 131

Etymology. L. *olivaria*, of or belonging to olives, from L. *oliva*, an olive. Rafinesque (1820: 314) described the periostracum of this species as "olivaceous," referring to its typical olive-green-tan color.

Vernacular name(s). *Hickorynut*, Eggshell, Glassyback.

Original description

Amblema olivaria Rafinesque, 1820: 314, species no. 53, no figure.

Type locality. "Dans le Kentuky [River]."

Type material. From the types mentioned by Vanatta (1915: 553), Johnson & Baker (1973: 164) selected ANSP 20251a from the Kentucky River as lectotype. These authors also recognized specimens from the Férussac Collection (1835: 34) in the Muséum national d'Histoire naturelle, Paris, as paralectotypes of this species.

Synonymy

Unio brevialis Lamarck, 1819. Sowerby & Sowerby, 1823: fig. ? [misidentification].

Amblema olivaria Rafinesque, 1820: 314.

Unio ellipsis Lea, 1828: 268, pl. 4, fig. 4 [Type locality: "Ohio." Syntypes ANSP 56614 *fide* Johnson & Baker (1973: 154)].

Unio olivarius (Rafinesque, 1820). Conrad, 1834: 70.

Margarita (*Unio*) *ellipsis* (Lea, 1828). Lea, 1836: 22.

Diplasma vitrea Rafinesque, 1820. Potiez & Michaud, 1844: 150, pl. 58, fig. 1 [misidentification].

Margaron (*Unio*) *ellipsis* (Lea, 1828). Lea, 1852b: 26.

Unio taitianus Lea, 1834. Sowerby, 1868: pl. 67, fig. 338 [misidentification].

Unio pealei Lea, 1871: 191 [Type locality: "Topeka, Kansas." Figured lectotype USNM 85745].

Lampsilis ellipsis (Lea, 1828). Kelly, 1899: 401.

Obovaria (*Pseudoon*) *ellipsis* (Lea, 1828). Simpson, 1900a: 602.

Obovaria ellipsis (Lea, 1828). Scammon, 1906: 305.

Unio pearlii [*sic*] Lea, 1871. Scammon, 1906: 305.

Obovaria olivaria (Rafinesque, 1820). Vanatta, 1915: 553.

Obovaria (*Pseudoon*) *olivaria* (Rafinesque, 1820). Ortmann, 1919: 229.

Fig. 130. ***Obovaria olivaria*** **(Rafinesque, 1820). A.** OSUM 69404.2, Ohio River, Meade Co., KY. 52 mm. **B.** OSUM 47237.12, Muskingum River, [Co.?], OH. 57 mm. **C, F, Hinge detail:** OSUM 29394.3, Wabash River, Tippecanoe Co., IN. 65 mm. **D.** OSUM 58576.1, Ohio River, Hamilton Co., OH. 19 mm. **E.** OSUM 59110.2, Bad River, Ashland Co., WI. 60 mm. **Beak sculpture detail:** OSUM 29394.1, Wabash River, Tippecanoe Co., IN. 49 mm total length.

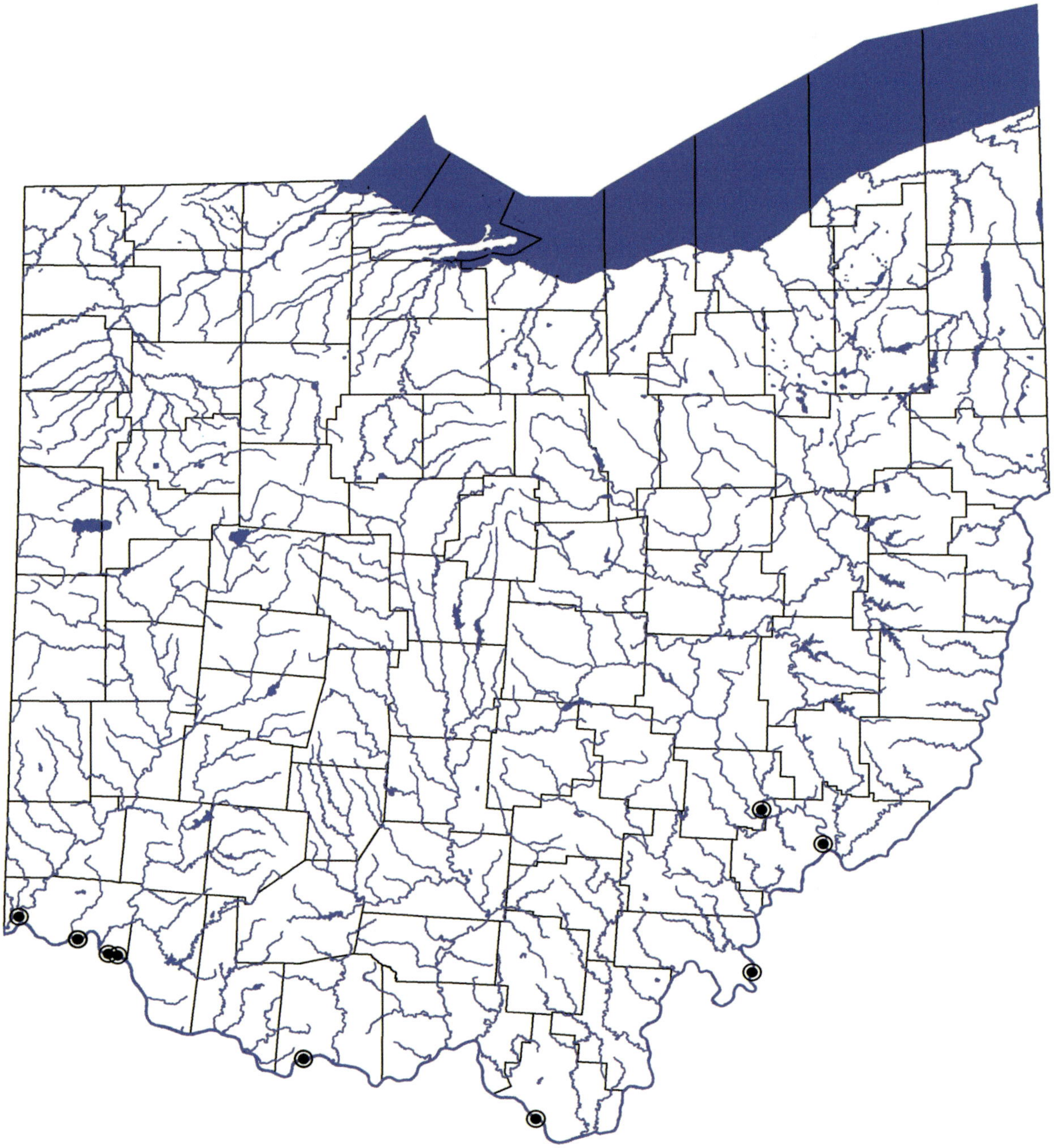

Fig. 131. ***Obovaria olivaria*** **(Rafinesque, 1820)**
Solid dots—live or freshdead since 1980
Bullseye—all records, any condition

Status. Listed as Extirpated by Ohio Division of Wildlife; not listed by U.S. Fish & Wildlife Service as of 2009.

ADULT SHELL

Shell. Medium-sized (to at least 100 mm); heavy to massive, outline oval.

Anterior margin. Straight or rounded.

Ventral margin. Very broadly curved or straight in large specimens, angled at the posterior-ventral junction in females.

Posterior margin. Gently rounded, nearly indistinguishable from the dorsal slope in males, straight or rounded, forming a broad angle with the posterior slope in females.

Dorsal margin. Short, straight, becoming convex in large individuals.

Umbo. Wide, prominent but low; beak sculpture very fine, of 2–4 double-looped ridges; beak sculpture disappears early in the first year's growth.

Ligament. Tan to black, short, partially concealed by the valves.

Disc sculpture. Shell sculptureless except for 2–4 low, broad ribs on the dorsal slope, divided by shallow grooves; this sculpture may be nearly lost in some specimens.

Periostracum. Greenish-yellow with pale green rays as juveniles, yellow to tan in older specimens with persistent rays, brown to reddish-brown in very old individuals. Periostracum velvety.

Nacre. White, iridescent posteriorly; rarely with gold Tulberg layers.

Hinge plate. Short, distinctly curved, massive; interdentum very broad, short, thick; cardinal teeth chunky, triangular, two in left valve, one in right with an accessory tooth anterior and posterior to cardinal; cardinal teeth rotated to orient longitudinally in old individuals; lateral teeth short, slightly bowed, very thick, two in left, one in right flanked with 1–2 accessory teeth.

Muscle scars. Both adductor muscle scars impressed, anterior protractor distinct from anterior pedal adductor, posterior retractor confluent ventrally with posterior pedal retractor scar; dorsal suspensor scars a series of numerous pits lying within deepest portion of umbonal cavity; pallial line entire, accessory line very long, extending to the shell margin.

Umbonal cavity. Wide, open, shallow, nearly nonexistent in large specimens.

Glochidia. Subelliptical, slightly asymmetric, dorsal margin slightly curved, posterior margin mostly straight-sided with a bend about half the distance from dorsal to ventral, anterior margin gently curved, ventral margin semicircular. Dorsal alae present; loose-looped sculpture covers the exterior surface of the valve. Hooks absent, micropoints lanceolate (Hoggarth, 1999: figs. 42e–h).

Literature records (Surber, 1912: 9, pl. 2, fig. 25; Ortmann, 1912a: 324, pl. 19, fig. 11; Utterback, 1916a: 325; Hoggarth, 1999: figs. 42e–h) indicate glochidial dimensions for length (0.20 to 0.21 mm), height (0.25 to 0.27 mm), and hinge (0.10 to 0.12 mm). Ortmann (1912a: 324) noted, "Glochidia similar to those of the other species [of *Obovaria*]." This glochidium is very similar to that of *O. retusa*. However, it is smaller and it has a different exterior valve sculpturing. Surber gives length and height measurements of 210 µm x 265 µm, and Ortmann (1912) records length and height measurements of 190 µm x 220 µm. Ortmann's (1912) numbers are considerably smaller than all others recorded in the literature and are not included in the ranges above.

Reproductive biology. Bradytictic. The posterior half of the outer gill is marsupial (Ortmann, 1912a). "Breeding" has been reported from August to the following June (Baker, 1928a). Ortmann (1919) and Bingham (1968) reported finding eggs from June through August, and glochidia in September. These overwintered until the following June. Utterback (1916a) reported that glochidia are released in loose masses, but we have found small, white, sole-shaped conglutinates as well (Watters, unpubl.). Sturgeon are the only reported hosts.

Reported potential hosts

Acipenser fulvescens (Lake Sturgeon)—Brady *et al.*, 2005 (LT).

Scaphirhynchus platorynchus (Shovelnose Sturgeon)—Howard, 1914b (NI); Coker *et al.*, 1921 (NI, LT).

Habitat. Muddy sand or gravel in rivers and lakes.

Range. St. Lawrence River system and Great Lakes drainages of Lake Ontario, Lake Huron, and Lake Erie. Mississippi River system from Minnesota to Louisiana.

Range in Ohio. Ohio River near Cincinnati. Historically in the lower Muskingum River and Ohio River to Marietta.

Similar species. The heavy, egg-shaped shell is characteristic. Other species of *Obovaria* are less elongate, with more central umbos. See also under *Fusconaia ebenus*.

Comments. This species is slow growing and lives many years. Growth is accelerated for the first 2–5 years and then abruptly slows. Specimens of the same size may vary greatly in age—an 80-mm individual may be from 3 to 10 years old. Sexual dimorphism in the shell is not pronounced, females being more pointed posteriorly than males. Individuals may exceed 100 mm in length and grow to over 30 years old.

Obovaria retusa (Lamarck, 1819)

Figures 132, 133, 134

Etymology. L. *retusa*, dulled, made blunt, from L. *re-*, back, again + L. *tundo*, beat, strike. In his original description, Lamarck (1819: 72) wrote "with blunt, eroded umbos." The umbos of *O. retusa* are in fact rounded and, in that manner, blunted, but Lamarck may have been using *retusa* in the more strictly literal sense of "beat back," referring to the strong anterior rotation of the umbos, a striking feature of this species.

Vernacular name(s): *Ring Pink*, Golf Stick.

Original description

Unio retusa Lamarck, 1819: 72, species no. 9, no figure given.

Type locality. "Habite les rivières de la Nouvelle Écosse" ["Lives in the rivers of Nova Scotia"]. Johnson (1969: 60) corrected the type locality to "The Ohio River, Cincinnati, Ohio," since it is generally agreed that Nova Scotia, far beyond the known range of this species, must be an error.

Type material. The presumed holotype of *U. retusa* is in the Muséum national d'Histoire naturelle, Paris (Johnson 1953: 93), and has been measured (length 44 mm, height 45 mm, width 30 mm) and illustrated by Johnson (1969: 60, 58, fig. 12).

Synonymy

Unio retusa Lamarck, 1819: 72.

Obovaria torsa Rafinesque, 1820: 311, pl. 82, figs. 1–3 [Type locality: "rare dans l'Ohio, plus commune dans les petites rivières" ("rare in the Ohio River, more common in the small rivers"). Holotype ANSP 20256 *fide* Johnson & Baker, 1993].

Unio retusus Lamarck, 1819. Conrad, 1836: 19, pl. 8.

Margarita (*Unio*) *retusus* (Lamarck, 1819). Lea, 1836: 34.

Unio torsus (Rafinesque, 1820). Potiez & Michaud, 1844: 149, pl. 57, figs. 1, 2.

Margaron (*Unio*) *retusus* (Lamarck, 1819). Lea, 1852b: 35.

Obovaria (*Obovaria*) *retusa* (Lamarck, 1819). Simpson, 1900a: 599.

Status. Listed as Extirpated by the Ohio Division of Wildlife; listed as Endangered by U.S. Fish & Wildlife Service as of 2009. A Federal recovery plan has been completed (USFWS, 1991b).

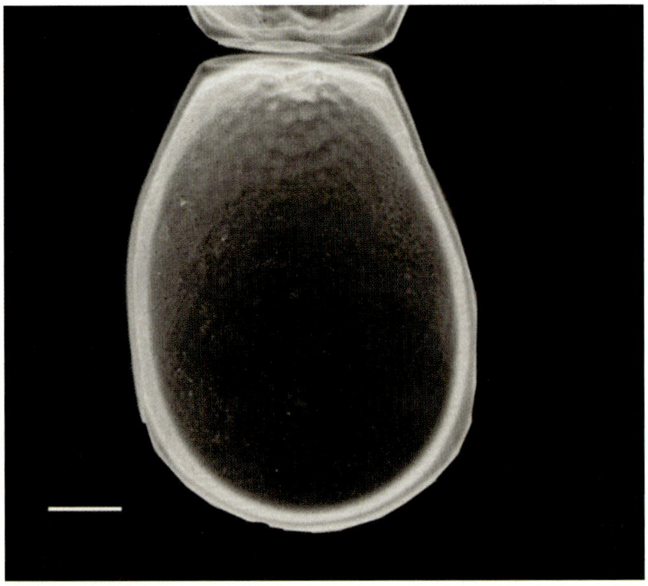

Fig. 132. *Obovaria retusa* (Lamarck, 1819). Glochidium. UMMZ uncataloged, Tennessee River, Humphreys Co., TN. Bar = 40 µm. (Hoggarth, 1999).

ADULT SHELL

Shell. Medium-sized (to 80 mm); heavy to massive, outline rounded or square with prominent umbos.

Anterior margin. Roundly curved.

Ventral margin. Broadly rounded in both sexes, but ending in broad, blunt point at posterior-ventral margin in females.

Posterior margin. Rounded in males; in females the posterior margin is vertically straight or concave between two swellings—one the posterior-ventral margin, the other the posterior ridge.

Dorsal margin. Short, straight or gently convex.

Umbo. Very wide and prominent, distinctly twisted anteriorly; the umbo becomes increasingly more anteriorly located with age, often at or beyond the anterior margin. No apparent beak sculpture. Umbos usually eroded, even in small specimens.

Ligament. Tan to brown, very short, not concealed by valves.

Disc sculpture. Shell sculptureless except for 0–4 low, broad ribs on the dorsal slope, divided by shallow grooves; this sculpture may be nearly lost in some specimens.

Periostracum. Pale yellowish-green to tan, becoming darker with age; posterior 33% of shell often lighter colored than anterior 66%. Rayless or with vague green rays. Periostracum satiny or matte.

Nacre. Pale to dark purple, abruptly changing to white at or just beyond the pallial line, iridescent posteriorly, hinge and teeth purple; rarely with gold Tulberg layers.

Hinge plate. Short, distinctly angled at ca. 90°, massive; interdentum very short, broad, thick; cardinal teeth chunky, triangular, two in left valve, one in right with accessory teeth anterior and posterior to cardinal; lateral teeth short, slightly bowed, very thick, two in left, one in right often flanked with two weak accessory teeth.

Muscle scars. Both adductor muscle scars impressed, both distinct from protractor and retractor scars; dorsal suspensor scars a series of pits behind the cardinal teeth just within the umbonal cavity; pallial line entire, accessory line long but faint, extending to the shell margin.

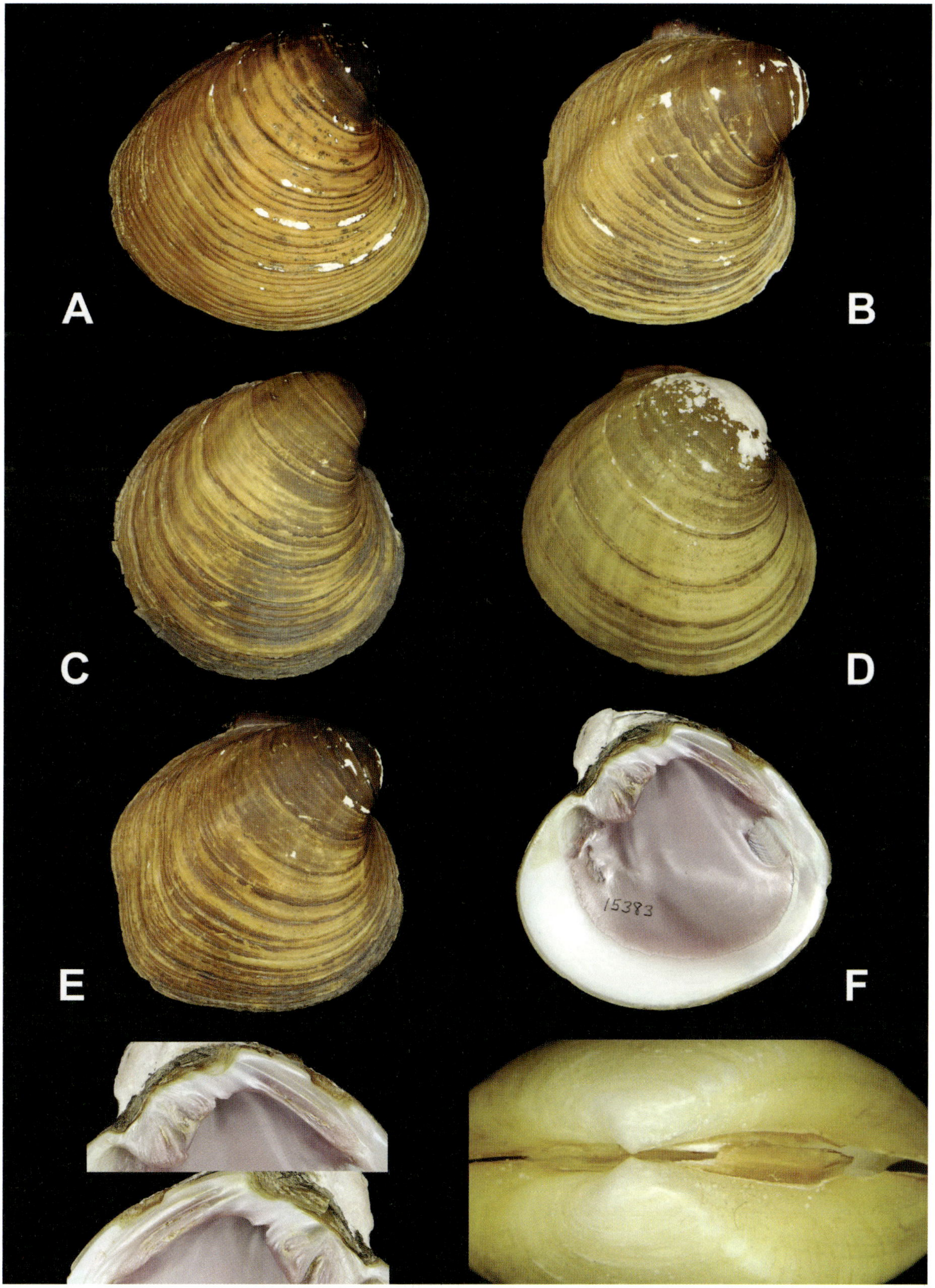

Fig. 133. ***Obovaria retusa*** **(Lamarck, 1819). A, F, Hinge detail:** OSUM 15383, Tennessee River, Hardin Co., TN. 83 mm. **B.** OSUM 45452.8, Ohio River, Hamilton Co., OH. 51 mm. **C.** OSUM 68360, Ohio River, Union Co., KY. 63 mm. **D.** OSUM 46876.1, Muskingum River, Washington Co., OH. 35 mm. **E.** OSUM 65030, Ohio River, Scioto Co., OH. 49 mm. **Beak sculpture detail:** OSUM 68291.2, Wabash River, [Co.?], IN. 13 mm total length.

Fig. 134. ***Obovaria retusa*** **(Lamarck, 1819)**
Solid dots—live or freshdead since 1980
Bullseye—all records, any condition

Umbonal cavity. Deep, compressed.

Glochidia. Subelliptical, slightly asymmetric, dorsal margin slightly curved, dorsal half of posterior margin straight, oblique, ventral half of posterior margin gently curved, anterior margin gently curved, ventral margin semicircular. Dorsal alae present; beaded to loose-looped sculpture covers the exterior surface of the valve. Hooks absent, micropoints lanceolate and arranged in broken vertical rows (Lea, 1858c: 47, pl. 5, fig. 7; Lea, 1863: 69; Hoggarth, 1999: figs. 42a–d).

Literature records (Surber, 1912: 9, pl. 3, fig. 47; Ortmann, 1912a: 322, pl. 19, fig. 9; Hoggarth, 1999: figs. 42a–d) indicate glochidial dimensions for length (0.22 to 0.24 mm), height (0.27 to 0.30 mm), and hinge (0.11 to 0.12 mm). In the case of *O. retusa* the moderately sized dorsal alae and valve shape will distinguish the glochidium from other species of *Obovaria*. Its beaded to loose-looped sculpture will distinguish this species from the other members of the genus.

Reproductive biology. Probably bradytictic. Marsupium formed in the posterior half of the outer gill (Ortmann, 1912a). Ortmann (1919) found eggs in August and September, and glochidia in late September.

Reported potential hosts. No hosts known.

Habitat. Sand and cobble of small and large rivers.

Range. Ohio River system, including the Tennessee, Cumberland, and Green rivers.

Range in Ohio. Extirpated. At the beginning of the 20th century Sterki and Ortmann found it in the Ohio River at Toronto, Clarington, Marietta, Portland, and Portsmouth (CMNH). At several of these locations it was fairly common. Also known historically from the Scioto River at Circleville and the Muskingum River.

Similar species. The vertically elongated shell, with its prominent umbo and purple nacre, is unique. Extreme specimens of *O. subrotunda* may be very similar but lack the pronounced dorsal ridge and typically have a smaller umbo. Purple nacre rarely occurs in *O. subrotunda*. The interdentum of *O. retusa* is very wide, whereas the interdentum of *O. subrotunda* is narrow.

Comments. This species is slow growing and lives many years. Growth is accelerated for the first 2–4 years, but not as dramatically as in other species. Sexual dimorphism in the shell is pronounced, with females having a truncated posterior margin. Individuals may reach 80 mm in length and commonly grow to over 25 years old—but some slower growing individuals may reach this age at 45 mm.

Obovaria subrotunda (Rafinesque, 1820)

Figures 135, 136, 137

Etymology. L. *sub-*, under, rather, somewhat + L. *rotundus*, wheel-shaped, round, circular, spherical, from L. *rota*, a wheel. Since in his original description (1820: 308) Rafinesque described this species as "perfectly rounded," it is uncertain why he used the modifier *sub-* except that in reality few individuals are perfectly round.

Vernacular name(s). *Round Hickorynut.*

Original description

Obliquaria (*Rotundaria*) *subrotunda* Rafinesque, 1820: 308, species no. 38, pl. 81, figs. 21–23.

Type locality. Rafinesque gave his source as "très commune dans l'Ohio et toutes les rivières qui s'y jettent" ["Very common in the Ohio [River] and all of the rivers that flow to it"]. But the type locality became the "Kentucky River" when Johnson & Baker (1973) designated a neotype from that locale.

Type material. The specimen mentioned by Vanatta (1915: 552) was selected as neotype (ANSP 20254 from the Kentucky River) by Johnson & Baker (1973: 171).

Synonymy

Unio (Aximedia) levigata Rafinesque, 1820: 296–297, pl. 80, figs. 11–13 [Type locality: "Dans le Kentucky [River]." Lectotype ANSP 20255a by selection of Johnson & Baker (1973: 160)].

Obliquaria (*Rotundaria*) *subrotunda* Rafinesque, 1820: 308, pl. 81, figs. 21–23.

Obovaria striata Rafinesque, 1820: 311.

Unio orbiculatus Hildreth, 1828: 284 [Type locality: "Muskingum [River]." Type lost. See under *Lampsilis abrupta* for the use of this name.].

Unio circulus Lea, 1829: 433, pl. 9, fig. 14 [Type locality: "Ohio [River], Cincinnati; Monongahela [River], Pittsburgh; Tennessee [River], Nashville." Figured USNM syntype and ANSP syntype 56602 lost *fide* Johnson & Baker (1973: 150)].

Unio lens Lea, 1831: 80, pl. 8, fig. 10 [Type locality: "Ohio and Tennessee." Type not located].

Unio laevigatus [*sic*] Rafinesque, 1820. Say, 1834: no pagination.

Margarita (*Unio*) *circulus* (Lea, 1829). Lea, 1836: 33.

Margarita (*Unio*) *lens* (Lea, 1831). Lea, 1836: 33.

Margaron (*Unio*) *circulus* (Lea, 1829). Lea, 1852b: 34.

Margaron (*Unio*) *lens* (Lea, 1831). Lea, 1852b: 34.

Mya rotunda [*sic*] (Rafinesque, 1820). Wood, 1856: 199, pl. 1, suppl. fig. 1.

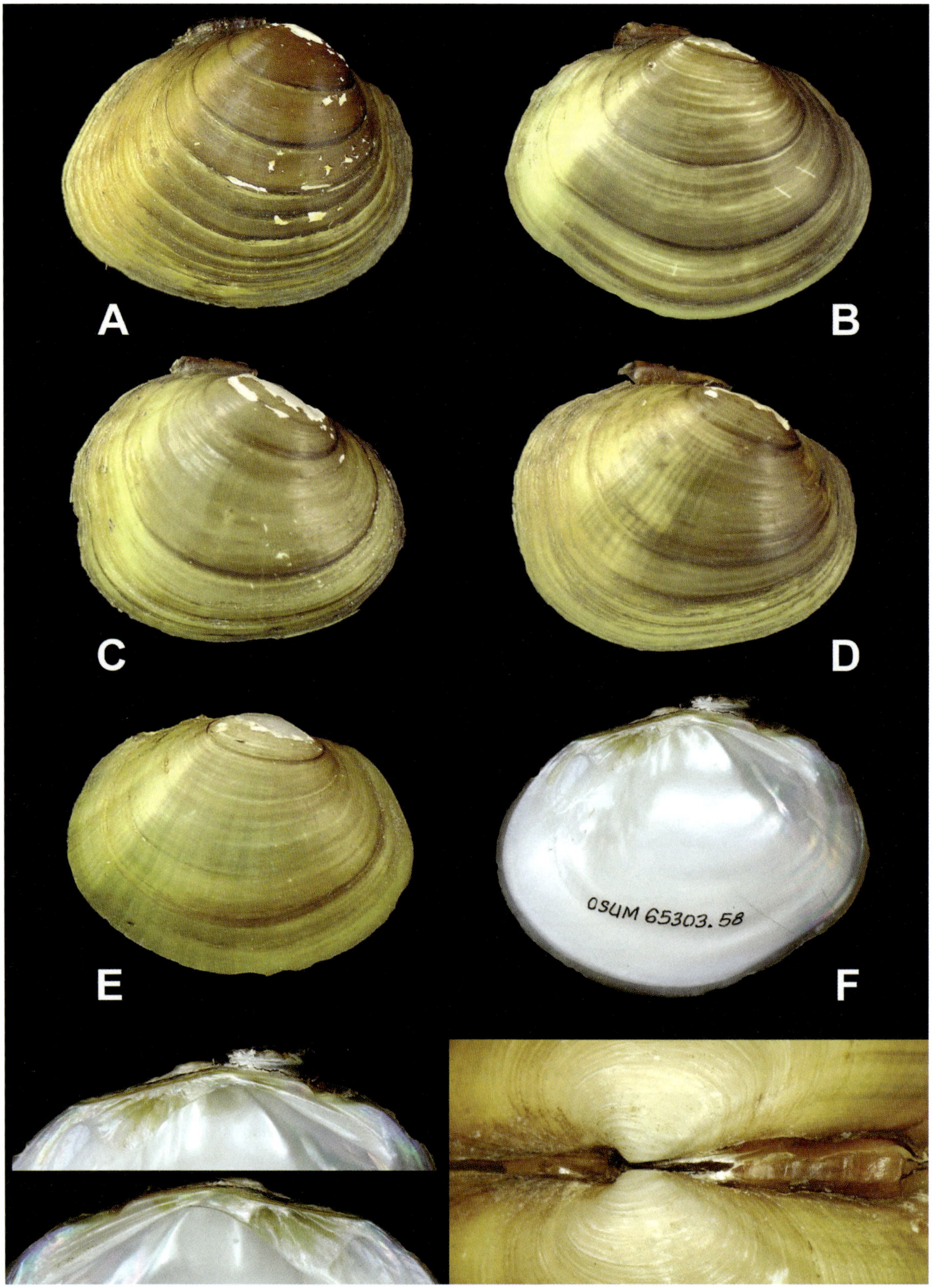

Fig. 135. ***Obovaria subrotunda*** **(Rafinesque, 1820). A.** OSUM 28185, Lake Erie, Ottawa Co., OH. 39 mm. **B.** OSUM 65303.58, Middle Fork Salt Creek, Ross Co., OH. 35 mm. **C.** OSUM 20815.12, Grand River, Ashtabula Co., OH. 32 mm. **D.** 65303.50, Middle Fork Salt Creek, Ross Co., OH. 33 mm. **E.** 65303.1, Middle Fork Salt Creek, Ross Co., OH. 21 mm. **F, Hinge detail:** OSUM 65303.58, Middle Fork Salt Creek, Ross Co., OH. **Beak sculpture detail:** OSUM 58573.1, Ohio River, Hamilton Co., OH. 35 mm total length.

Unio leibii Lea, 1862: 168 [Type locality: "Erie Co., Michigan." Lectotype USNM 84425].
Unio depygis Conrad, 1866: 107, pl. 10, fig. 1 [Type locality: "Harpeth River, Tennessee." Figured holotype lost *fide* Johnson & Baker, 1973: 153].
Margaron (*Unio*) *leibii* (Lea, 1862). Lea, 1870: 36.
Unio subrotundus var. *circulus* Lea, 1829. Paetel, 1890: 168.
Obovaria circulus (Lea, 1829). Simpson, 1900a: 600.
Obovaria lens (Lea, 1831). Simpson, 1900a: 600.
Obovaria lens var. *depygis* (Conrad, 1866). Simpson, 1900a: 601.
Obovaria leibii (Lea, 1862). Simpson, 1900a: 601.
Obovaria circulus leibii (Lea, 1862). Sterki, 1907: 390.
Obovaria lens var. *parva* Simpson, 1914: 294 [Type locality: "Maumee and Wabash Rivers; Sandusky, Ohio; Tennessee." Lectotype from the Duck River, Tennessee, USNM 150457].
Obovaria lens var. *elongata* Simpson, 1914: 294–295 [Type locality: "Ohio River." Lectotype USNM 35734 *fide* Johnson, 1975].
Obovaria subrotunda (Rafinesque, 1820). Vanatta, 1915: 552.
Obovaria levigata (Rafinesque, 1820). Vanatta, 1915: 552.
Obovaria (*Obovaria*) *subrotunda* (Rafinesque, 1820). Ortmann, 1918: 567.
Obovaria (*Obovaria*) *subrotunda levigata* (Rafinesque, 1820). Ortmann, 1918: 568.
Obovaria circula (Lea, 1829). Baker, 1920: 383.
Obovaria subrotunda depygis (Conrad, 1866). Frierson, 1927: 90.
Obovaria levigata (Rafinesque, 1820). Frierson, 1927: 90.
Obovaria subrotunda leibii (Lea, 1862). Goodrich, 1932: 103.
Obovaria subrotunda globula Morrison, 1942: 360 [Type locality: "[Tennessee River] Tuscumbia, Alabama." Holotype USNM 85789].
Obovaria subrotunda lens (Lea, 1831). Goodrich & van der Schalie, 1944: 318.
Obovaria subrotunda levigata (Rafinesque, 1820). La Rocque, 1967: 242.
Obovaria (*Obovaria*) *levigata* (Rafinesque, 1820). Haas, 1969: 422.
Quadrula subrotunda globula (Morrison, 1942). Johnson, 1980: 96.
Obovaria subroturda [*sic*] (Rafinesque, 1820). Hoggarth, 1999: 73.

Status. Not listed by either Ohio Division of Wildlife or U.S. Fish & Wildlife Service as of 2009.

ADULT SHELL

Shell. Small to medium-sized (to 75 mm), nearly circular, or obliquely circular in profile; rarely transverse; moderately to very inflated.

Anterior margin. Straight, then abruptly but evenly rounded to ventral margin.

Ventral margin. Evenly rounded, females slightly straighter.

Posterior margin. Rounded in females; having a blunt angulation in the ventral half of the margin in males.

Dorsal margin. Short, straight.

Umbo. Broad, prominent; beak sculpture of 4–6 "V"-shaped or wavy ridges; the posterior arm of the "V" is shorter than the anterior arm and may become disconnected and nodulose; in some populations the ridges are nearly linear; beak sculpture disappears at or before first annular ring.

Ligament. Large, brown, occupying most of the dorsal margin, scarcely concealed by the valves.

Disc sculpture. Absent except for 1–2 inconspicuous ridges on the dorsal slope.

Periostracum. Yellowish, tan, chestnut brown on disc of shell, posterior slope often lighter; rays may be present in some populations; these are usually broad and diffuse, rarely bold; rays continuous or broken at annular rings; posterior slope rarely rayed.

Nacre. White, rarely salmon or purple, iridescent posteriorly; gold Tulberg layers rare. The purple nacre fades with time.

Hinge plate. Moderately heavy, short, strongly bowed; interdentum very short, laterally flattened, rarely with a small tooth; one cardinal in left valve, but often accompanied by either or both anterior and posterior accessory teeth; two cardinal teeth in right valve, lamellar when young, becoming more massive with age; lateral teeth short, straight, two in right valve, one in left, often accompanied by a second smaller tooth ventrally.

Muscle scars. Anterior adductor deep, rough, distinct from anterior pedal protractor; posterior adductor slightly impressed, partially or not at all confluent with posterior pedal retractor scar; dorsal suspensors multiple, situated well within the umbonal cavity in a line from the interdentum to under the anterior cardinal fossa; pallial line complete, accessory line long, curving posteriorly.

Umbonal cavity. Deep, wide, somewhat laterally compressed.

Glochidia. Subelliptical, dorsal margin slightly curved, posterior margin straight and oblique dorsally, ventrally curved, anterior margin gently curved, ventral margin broadly rounded. Dorsal alae present; loose-looped valve sculpture covers the exterior surface of the valve. Hooks absent, micropoints lanceolate and arranged in broken vertical rows (Ortmann, 1911: 364, pl. 89, fig. 15; Hoggarth, 1999: figs. 43a–e).

Literature records (Ortmann, 1912a: 323; Surber, 1915: 7, pl. 1, fig. 8; Hoggarth, 1999: figs. 43a–e) indicate

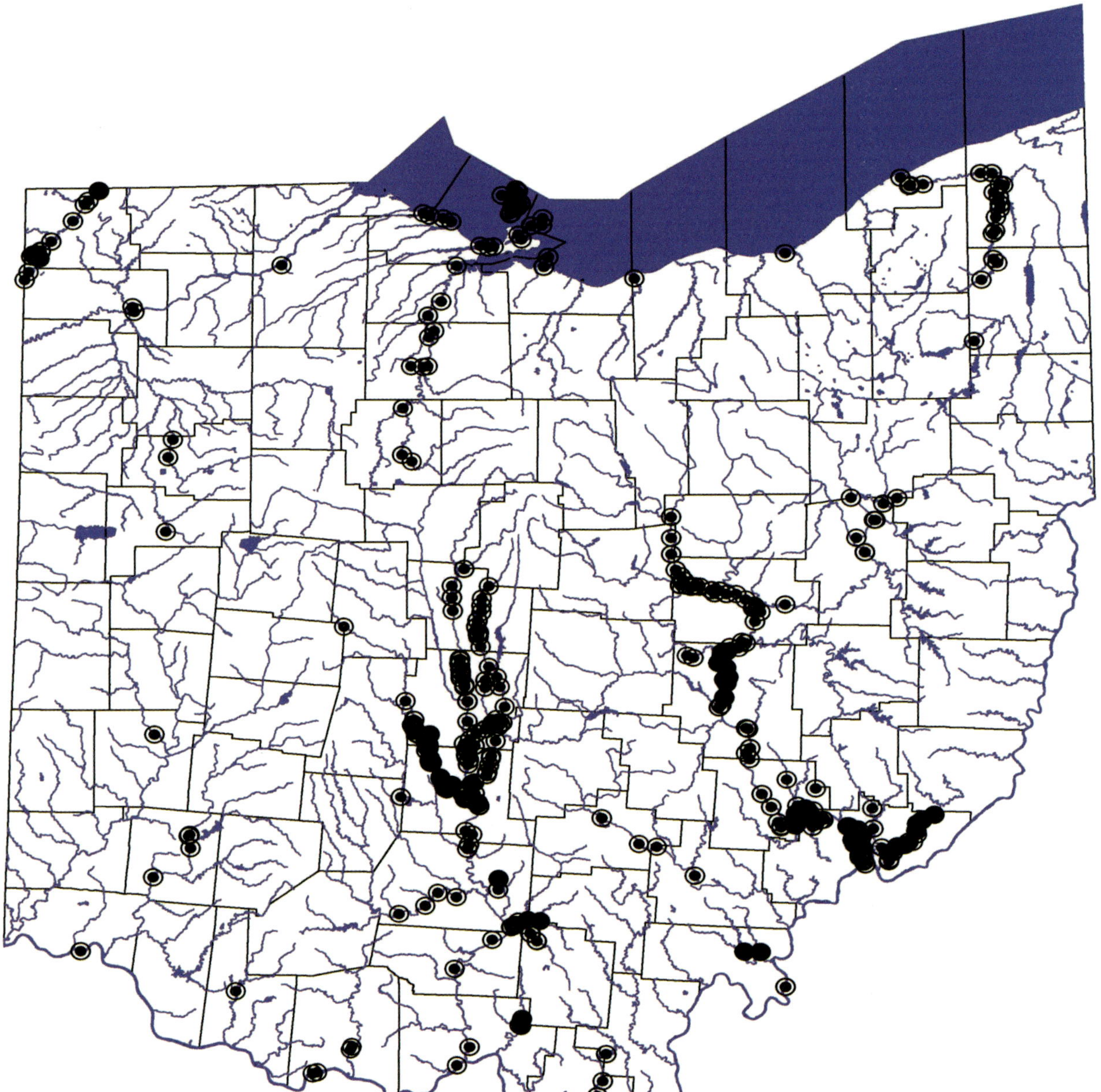

Fig. 136. ***Obovaria subrotunda*** **(Rafinesque, 1820)**
Solid dots—live or freshdead since 1980
Bullsey— all records, any condition

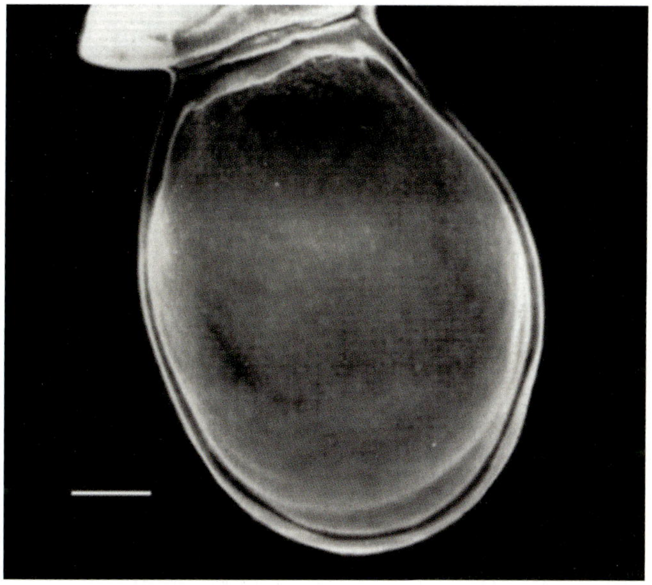

Fig. 137. *Obovaria subrotunda* (Rafinesque, 1820). Glochidium. MAH 805.1, Fish Creek, Williams Co., OH. Bar = 35 µm. (Hoggarth, 1999).

glochidial dimensions for length (0.17 to 0.20 mm), height (0.20 to 0.23 mm), and hinge (0.09 to 0.10 mm). Surber (1915: 7) noted, "This glochidium is apparently intermediate in form between *retusa* and *ellipsis* [*O. olivaria*], but rather nearer the latter, yet it differs from both in being of smaller size. In *ellipsis* the hinge line is straight, while in *retusa* the same line is undulate." This glochidium shares moderately long dorsal alae with *O. retusa*, and moderately dense loose-looped sculpture with *O. olivaria*.

Reproductive biology. Bradytictic. Egg-bearing females are found in September, and glochidia in late September, with glochidia overwintering in the marsupia until the following May and June (Ortmann, 1919). No durable conglutinates formed (Watters, 2008).

Reported potential hosts. No hosts known.

Habitat. This uncommon species occurs in a variety of habitats. In the Big Darby Creek system it is associated with riffles and runs. But in Shade River and Salt Creek, it occurs in slow-moving water in sandy mud.

Range. Ohio River system, including Tennessee and Cumberland rivers; drainages of Lake Erie and Lake St. Clair.

Range in Ohio. Widely but sporadically distributed. Locally common in the Shade River and Middle Fork Salt Creek, uncommon in Big Darby and Fish creeks, rare in the Muskingum River and several small southern Ohio streams. Historically throughout the Scioto and Hocking rivers; rare in the Great and Little Miami rivers. Historically from Lake Erie, particularly the western basin. Also in the Grand, Sandusky, and Maumee River systems.

Similar species. See under *O. retusa*.

Comments. Rates of growth vary between populations, with some growing at a nearly constant rate and others growing quickly for the first 4–5 years before slowing. At 55 mm in length, individuals may be from 4 to 10 years old. Female shell characteristics are apparent by the third year. Exceptionally large specimens may be 75 mm long. Few individuals live longer than 12 years.

Plethobasus Simpson, 1900

Etymology. Gr. *plethos*, multitude, magnitude, mass, main body, largest part, OR Gr. *plethora*, fullness + Gr. *basis*, step, base, pedestal, foundation. Simpson (1900a: 764) described the genus *Plethobasus* as ". . . somewhat suddenly swollen at the posterior base; . . . ." This ventral fullness is especially apparent in the two species he included in this genus: *P. cyphyus* (Rafinesque, 1820) (as *Pleurobema aesopus* (Green, 1827)) and *P. cicatricosus* (Say, 1829).

Original description

Pleurobema (*Plethobasus*) Simpson, 1900a: 764 [Simpson established the Section *Plethobasus* to receive *Pleurobema* species that were sculptured with tubercles. Ortmann (1912a: 259) elevated the group to generic level.].

Type species. *Unio aesopus* Green, 1827 [= *Obliquaria cyphya* Rafinesque, 1820] by original designation.

Synonymy

Pleurobema (*Plethobasus*) Simpson, 1900a: 764 [Type species: *Unio aesopus* Green, 1827, by original designation].

Plethobasus Simpson, 1900. Ortmann, 1912a: 259.

Quadrula (*Striata*) Frierson, 1927: 50–51 [*non Striata* Boettger, 1878 (Mollusca: Clausiliidae). Type species: *Quadrula striata* Rafinesque, 1820, by original designation].

Quadrula (*Luteacarnea*) Frierson, 1927: *errata et corrigenda* [replacement name for *Striata* Frierson, 1927].

Geographic range. Ohio River system, including the Tennessee and Cumberland rivers.

Plethobasus cicatricosus (Say, 1829)

Figures 138, 139

Etymology. L. *cicatrix*, a scar + L. *-osus*, full of. The lower median disc of this species usually has a sculpture consisting of a number of relatively large welts or flattened elongate raised areas.

Fig. 138. *Plethobasus cicatricosus* (Say, 1829). A. OSUM 44083, Tennessee River, Hardin Co., TN. 100 mm. **B.** OSUM 57472, Ohio River, Adams Co., OH. 93 mm. **C.** OSUM 69462, Ohio River, Hamilton Co., OH. 78 mm. **D, F.** OSUM 68131.3, Ohio River, Hamilton Co., OH. 56 mm. **E.** OSUM 10131, "Wabash River." 110 mm. **G, Hinge detail:** OSUM 68230.1, Ohio River, Hamilton Co., OH. 42 mm. **Beak sculpture detail:** no specimen available.

Vernacular name(s). *White Wartyback,* Welted Pigtoe.

Original description

Unio cicatricosus Say, 1829: 292–293, no figure given.

Type locality. "Inhabits Wabash [River]."

Type material. The original unionoid type material of Say has apparently been lost (Johnson & Baker, 1973: 146). Haas (1930: 327) designated SMF No. 4539 as neotype.

Synonymy

Unio cicatricosus Say, 1829: 292–293.

Unio varicosus Lea, 1829: 424 [*non* Lamarck, 1819. Type locality: "Ohio River." Holotype USNM 84320].

Unio cyphia (Rafinesque, 1820). Conrad, 1834: 68 [in part].

Margarita (*Unio*) *varicosus* (Lea, 1829). Lea, 1836: 17.

Margaron (*Unio*) *varicosus* (Lea, 1829). Lea, 1852b: 23.

Unio cicatricosus var. *varicosus* (Lea, 1829). Paetel, 1890: 148.

Pleurobema (*Plethobasus*) *cicatricosum* (Say, 1829). Simpson, 1900a: 765.

Pleurobema cicatricosa (Say, 1829). Baker, 1906: 78.

Quadrula cicatricosa (Say, 1829). Sterki, 1907: 392.

Unio detectus Frierson, 1911: 52, pl. 2, lower fig. [Type locality: "Ohio River." Holotype UMMZ 87517].

Unio cicatricoides Frierson, 1911: 53, pl. 2, upper fig. [Type locality: "Wabash River." Holotype UMMZ 87570].

Plethobasus cicatricosa (Say, 1829). Ortmann, 1919: 61.

Plethobasus cyphyus (Rafinesque, 1820). Ortmann & Walker, 1922: 19 [in part].

Pleurobema cicatricoideum (Frierson, 1911). Frierson, 1927: 45–46.

Plethobasus pachosteus (Rafinesque, 1820). Morrison, 1969: 24 [misidentification].

Plethobasus cicatricoideus (Frierson, 1911). Morrison, 1969: 24.

Plethobasus cicatricosus form *detectus* (Frierson, 1911). Warren, 1975: 66.

Status. Listed as Extirpated by the Ohio Division of Wildlife; listed as Endangered by U.S. Fish & Wildlife Service as of 2009. A Federal recovery plan has been completed (USFWS, 1984b).

ADULT SHELL

Shell. Medium to fairly large sized (to 120 mm), moderately thick, oval to somewhat triangular, moderately inflated; old individuals may become abruptly elongate.

Anterior margin. Long, straight to broadly rounded to the ventral border.

Ventral margin. Very broadly rounded, rarely concave in very old adults.

Posterior margin. Short, rounded to the dorsal margin.

Dorsal margin. Long, straight, joining posterior margin with obtuse angle.

Umbo. Wide and prominent; umbos nearly face each other (ca. 10°) but often project beyond anterior margin; umbonal sculpture eroded away on all specimens examined.

Ligament. Long and prominent, tan to dark brown, not concealed by shells.

Disc sculpture. Disc of shell smooth except for row of low, almost imperceptible, concentrically oriented knobs or series of pustules; sulcus very weak or absent; slope with 2–3 weak angulations.

Periostracum. Juveniles shiny, yellow or yellow-green with fine green lines or rays, posterior slope lighter; adults become a reddish-brown but are still rather shiny; rays in adults.

Nacre. Porcelain white, often iridescent posteriorly; often with gold or brown Tulberg layers.

Hinge plate. Well-developed, angled under umbo; right cardinal a single, massive, triangular denticle, often with small anterior accessory tooth, posterior accessory low but nearly as large as cardinal; left cardinal bifid, the teeth well-separated; laterals well-developed, short to medium-length, straight, one in right valve, two in left; interdentum flat, wide, short.

Muscle scars. Anterior adductor small, elongate, and strongly impressed; posterior adductor nearly flush; large anterior adductor distinct from anterior pedal protractor scar, large posterior adductor distinct from small posterior pedal retractor scar; cluster of small dorsal suspensors located behind cardinal teeth and interdentum in an irregular line. Pallial line entire, located far from shell margin. Accessory line weak, reaching 25% the length of the shell.

Umbonal cavity. Very wide, deep, compressed.

Glochidia. Unknown.

Reproductive biology. Probably tachytictic.

Reported potential hosts. No hosts known.

Habitat. Large rivers.

Range. Ohio River system, including the Wabash, Kanawha, Tennessee, and Cumberland rivers.

Range in Ohio. Extirpated. Sterki and Ortmann found it in the Ohio River at Marietta and Clarington at the turn of the 20th century (CMNH). Other historical records include the Ohio River at Cincinnati and Manchester, and the lower Muskingum River.

Similar species. This species (and *Plethobasus* in general) resembles a member of the *Pleurobema cordatum* complex with very low pustules. The pustules are arranged down the center of the shell and run confluent with the lines of

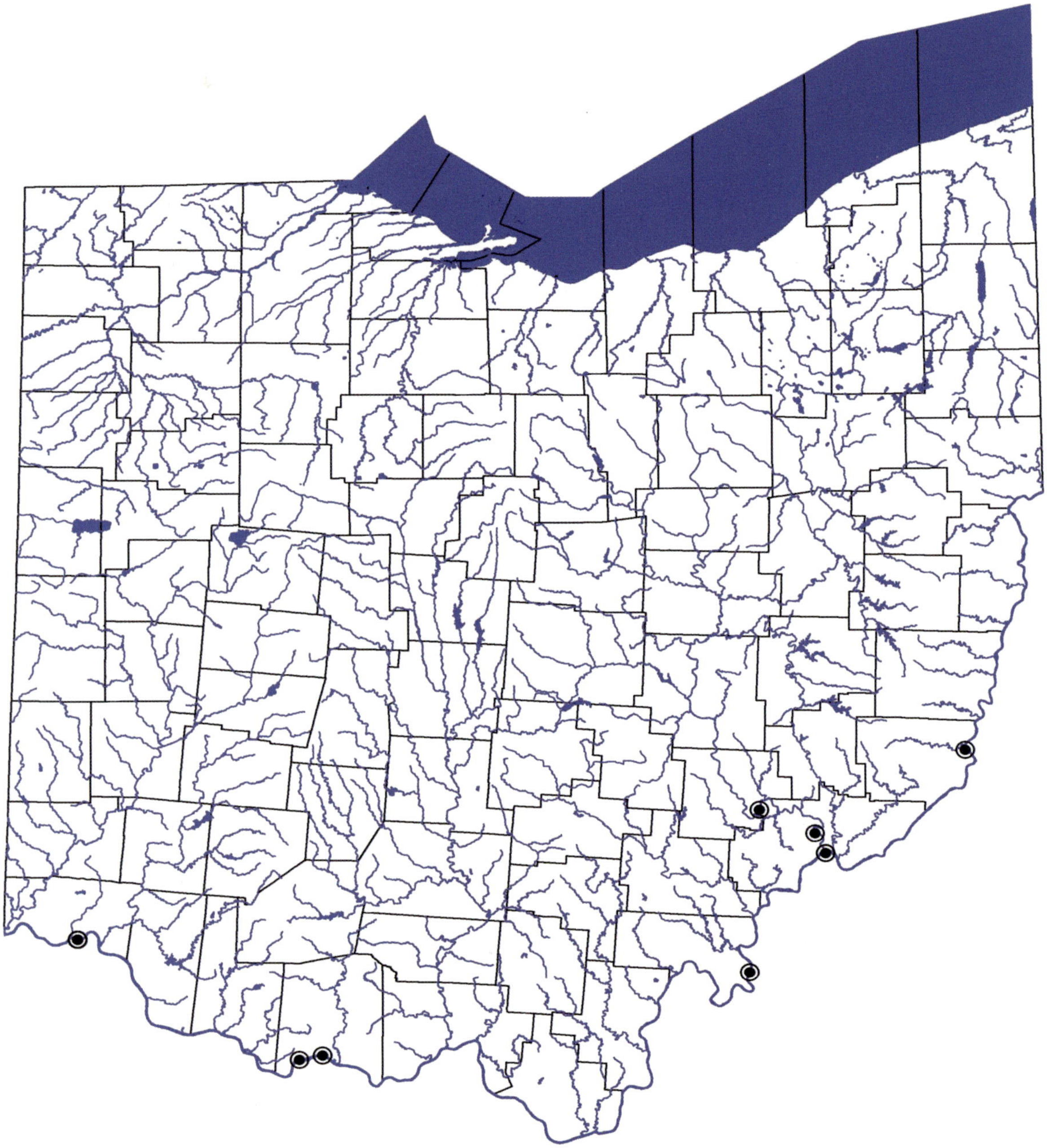

Fig. 139. ***Plethobasus cicatricosus*** **(Say, 1829)**
Solid dots—live or freshdead since 1980
Bullseye—all records, any condition

growth. *Quadrula pustulosa* has a green ray or blotch on the umbo which is lacking in *P. cicatricosus*. *Plethobasus cyphyus* is very similar but typically is more yellow and waxy (at least in younger specimens), and the ventral margin is more angled. *Plethobasus cooperianus* has more pustules distributed over more than just the center of the shell; the pustules are not aligned in a single row. The single row of pustules in *Obliquaria* is much more pronounced and alternates in position from left to right valves.

Comments. Individuals gradually slow in growth during their entire life. Specimens rarely reach ca. 120 mm and nearly 25 years old.

Plethobasus cooperianus (Lea, 1834)

Figures 140, 141

Etymology. The name honors "William Cooper, Esq.," who had lent Lea the first specimens.
Vernacular name(s). *Orangefoot Pimpleback.*
Original description
Unio cooperianus Lea, 1834: 61, pl. 8, fig. 21.
Type locality. "Ohio River."
Type material. The type has not been found and is presumed lost.
Synonymy
Unio cooperianus Lea, 1834: 61, pl. 8, fig. 21 [Type locality: "Ohio River." Type not located].
Margarita (Unio) cooperianus (Lea, 1834). Lea, 1836: 16.
Margaron (Unio) cooperianus (Lea, 1834). Lea, 1852b: 22.
Unio striatus (Rafinesque, 1820). Reeve, 1864: pl. 8, fig. 30.
Unio cooperianus (Lea, 1834). Reeve, 1864: *errata.*
Quadrula cooperianus (Lea, 1834). Simpson, 1900a: 781.
Pleurobema cooperianus (Lea, 1834). Ortmann, 1910: 117.
Plethobasus cooperianus (Lea, 1834). Ortmann, 1912a: 261.
Quadrula cooperiana (Lea, 1834). Walker, 1918: 166.
Quadrula striata striata (Rafinesque, 1820). Frierson, 1927: 50.
Quadrula (Luteacarnea) striata (Rafinesque, 1820). Haas, 1969: 296.
Plethobasus striatus (Rafinesque, 1820). Morrison, 1969: 24.
Status. Listed as Extirpated by the Ohio Division of Wildlife; listed as Endangered by U.S. Fish & Wildlife Service as of 2009. A Federal recovery plan has been completed (USFWS, 1984a).

ADULT SHELL
Shell. Medium size (to 90 mm), moderately thick, oval to somewhat triangular, moderately inflated.
Anterior margin. Long, straight to broadly rounded to the ventral border.
Ventral margin. Very broadly rounded or straight, often with a slight concavity at the posterior ridge.
Posterior margin. Long, nearly straight to the dorsal margin.
Dorsal margin. Short, straight, joining posterior margin with obtuse angle.
Umbo. Wide and prominent; umbos nearly face each other (ca. 10°); umbonal sculpture eroded on all examples seen.
Ligament. Long and prominent, tan to dark brown, not concealed by shells.
Disc sculpture. Disc of shell with irregular pustules, often concentrically elongated, on the posterior 50%–66% of shell. Posterior slope with a few curving pustulose ribs.
Periostracum. Juveniles are yellow or tan; adults become a reddish-brown; juveniles have very numerous fine green rays, but these may be lost in adults.
Nacre. Porcelain white, often iridescent posteriorly; often with gold or brown Tulberg layers. Rarely with a pinkish hue on the hinge.
Hinge plate. Well-developed, massive, angled under umbo; right cardinal a single, massive, triangular denticle, often with small anterior accessory tooth, posterior accessory rare; left cardinal bifid, the teeth well-separated; laterals well-developed, short, arched or straight, one in right valve, two in left; interdentum cord-like, narrow, short.
Muscle scars. Anterior adductor small and strongly impressed; posterior adductor moderately impressed; anterior adductor distinct from or confluent with anterior pedal protractor scar, large posterior adductor distinct from small posterior pedal retractor scar; cluster of small dorsal suspensors located behind cardinal teeth and interdentum in an irregular line. Pallial line entire, located far from shell margin. Accessory line weak, reaching 25% the length of the shell.
Umbonal cavity. Very wide, deep, and compressed.

Glochidia. Unknown.

Reproductive biology. Probably tachytictic. The outer gills are used as marsupia (Ortmann, 1919).

Reported potential hosts. No hosts known.

Habitat. Large rivers.

Range. Ohio River system, including the Wabash, Cumberland, and Tennessee rivers.

Range in Ohio. Extirpated. Records of Sterki and Ortmann from the turn of the 20th century indicate that it was

Fig. 140. ***Plethobasus cooperianus*** **(Lea, 1834). A.** OSUM 39827.2, Ohio River, Hamilton Co., OH. 78 mm. **B.** OSUM 69461, Ohio River, Hamilton Co., OH. 84 mm. **C.** OSUM 10130, "Ohio River." 80 mm. **D, F, Hinge detail:** OSUM 2325.2, Tennessee River, Hardin Co., TN. 71 mm. **E.** OSUM 57252.1, "Tennessee." 34 mm. **Beak sculpture detail:** OSUM 57262. No data. 27 mm total length.

Fig. 141. ***Plethobasus cooperianus*** **(Lea, 1834)**
Solid dots—live or freshdead since 1980
Bullseye—all records, any condition

fairly common in the Ohio River at Marietta, Clarington, and Portland (CMNH). Additional records exist for Cincinnati (CMNH). Although a valve from the Hildreth collection was labeled "Muskingum River," other evidence suggested that it was not collected there. A subfossil valve found by Watters in the lower Muskingum River at Devola is the only authenticated record of this species from that river. There is a single historical record for the lower Scioto River.

Similar species. *Quadrula pustulosa* and *Cyclonaias tuberculata* both resemble this species. *Quadrula pustulosa* has a more distinct umbo that bears a green ray or blotch. *Cyclonaias tuberculata* has purple nacre and is sulcate as a juvenile. See also under *P. cicatricosus*.

Comments. This species is occasionally cited as *Plethobasus striatus* (Rafinesque, 1820). That species has been shown to be a synonym of *Obovaria subrotunda;* the designation of Johnson & Baker (1973: 171) is invalid (A. Bogan, in lit., 2009).

This species grows to 90 mm and lives to at least 30 years old. As the name suggests, the foot of the living animal is orange.

Plethobasus cyphyus (Rafinesque, 1820)

Figures 142, 143, 144

Etymology. Gr. *kyphos*, humped, sloped, humpbacked, curved. The central ridge of this species is typically sculptured with a radical row of humps or swellings.

Vernacular name(s). *Sheepnose*, Bullhead, Yellow Knob Shell.

Original description

Obliquaria (*Quadrula*) *cyphya* Rafinesque, 1820: 305, species no. 29, no figure.

Type locality. "Aux chutes de l'Ohio." The Falls of the Ohio is located at Louisville, Kentucky, on the Ohio River. However, the lectotype is from an unspecified site in the Ohio River.

Type material. The type mentioned by Vanatta (1915: 556) was selected as lectotype (ANSP 20239 from the Ohio River) by Johnson & Baker (1973: 152).

Synonymy

Obliquaria (*Quadrula*) *cyphya* Rafinesque, 1820: 305.
Unio aesopus Green, 1827: 46, fig. 3 [Type not located].
Unio cyphius [*sic*] (Rafinesque, 1820). Say, 1834: no pagination.
Unio cyphia [*sic*] (Rafinesque, 1820). Conrad, 1834: 68.
Margarita (*Unio*) *aesopus* (Green, 1827). Lea, 1836: 17.
Quadrula cyphius [*sic*] (Rafinesque, 1820). Agassiz, 1852: 48.
Margaron (*Unio*) *aesopus* (Green, 1827). Lea, 1852b: 23.
Unio scyphius [*sic*] (Rafinesque, 1820). Küster, 1861: 181, pl. 57, fig. 2.
Unio cyphias [*sic*] var. *aesopus* Green, 1827. Paetel, 1890: 150.
Pleurobema (*Plethobasus*) *aesopus* (Green, 1827). Simpson, 1900a: 764.
Pleurobema aesopus (Green, 1827). Baker, 1905: 255.
Unio compertus Frierson, 1911: 53, pl. 3, middle and lower figs. [Type locality: "Clinch and Holston Rivers." Lectotype UMMZ 87769].
Plethobasus aesopus (Green, 1827). Ortmann, 1912a: 260, fig. 8.
Pleurobema (*Plethobasus*) *compertus* (Frierson, 1911). Simpson, 1914: 809.
Plethobasus cyphyus compertus (Frierson, 1911). Ortmann, 1918: 544.
Plethobasus cyphyus (Rafinesque, 1820). Ortmann, 1919: 65, pl. 5, fig. 6, pl. 6, figs. 1–3.
Pleurobema (*Plethobasus*) *cyphyum* (Rafinesque, 1820). Frierson, 1927: 45.
Plethobasus cyphyum (Rafinesque, 1820). Haas, 1969: 250.
Plethobasus compertus (Frierson, 1911). Haas, 1969: 251.

Status. Listed as Endangered by Ohio Division of Wildlife; not listed by U.S. Fish & Wildlife Service as of 2009.

ADULT SHELL

Shell. Medium to fairly large sized (to 130 mm), moderately thick, oval to somewhat triangular, moderately inflated.

Anterior margin. Long, straight to broadly rounded to the ventral border.

Ventral margin. Very broadly rounded or straight, often with a slight concavity at the posterior ridge.

Posterior margin. Long, nearly straight to the dorsal margin.

Dorsal margin. Short, straight, joining posterior margin with obtuse angle.

Umbo. Wide and prominent; umbos nearly face each other (ca. 10°); umbonal sculpture of 2–3 widely spaced, undulating ridges.

Ligament. Long and prominent, tan to dark brown, not concealed by shells.

Disc sculpture. Disc of shell smooth except for row of low, almost imperceptible knobs, one per year, bordering the anterior of the sulcus; weak sulcus present; posterior slope with 2–3 weak angulations.

Periostracum. Juveniles are shiny, waxy, and yellow or yellow-green; adults become a reddish-brown but are still rather shiny; juveniles rarely have very vague, washed-out green stripes, but these may be lost in adults.

Fig. 142. ***Plethobasus cyphyus*** **(Rafinesque, 1820). A.** OSUM 6267, Muskingum River, Washington Co., OH. 110 mm. **B.** OSUM 11082, Muskingum River, Washington Co., OH. 98 mm. **C.** OSUM 69446, Ohio River, Hamilton Co., OH. 68 mm. **D, F, Hinge detail:** OSUM 5411.5, Wabash River, Lawrence Co., IL. 73 mm. **E.** OSUM 6182.1, Muskingum River, Washington Co., OH 47 mm. **Beak sculpture detail:** OSUM 24538.1, Clinch River, Hancock Co., TN. 17 mm total length.

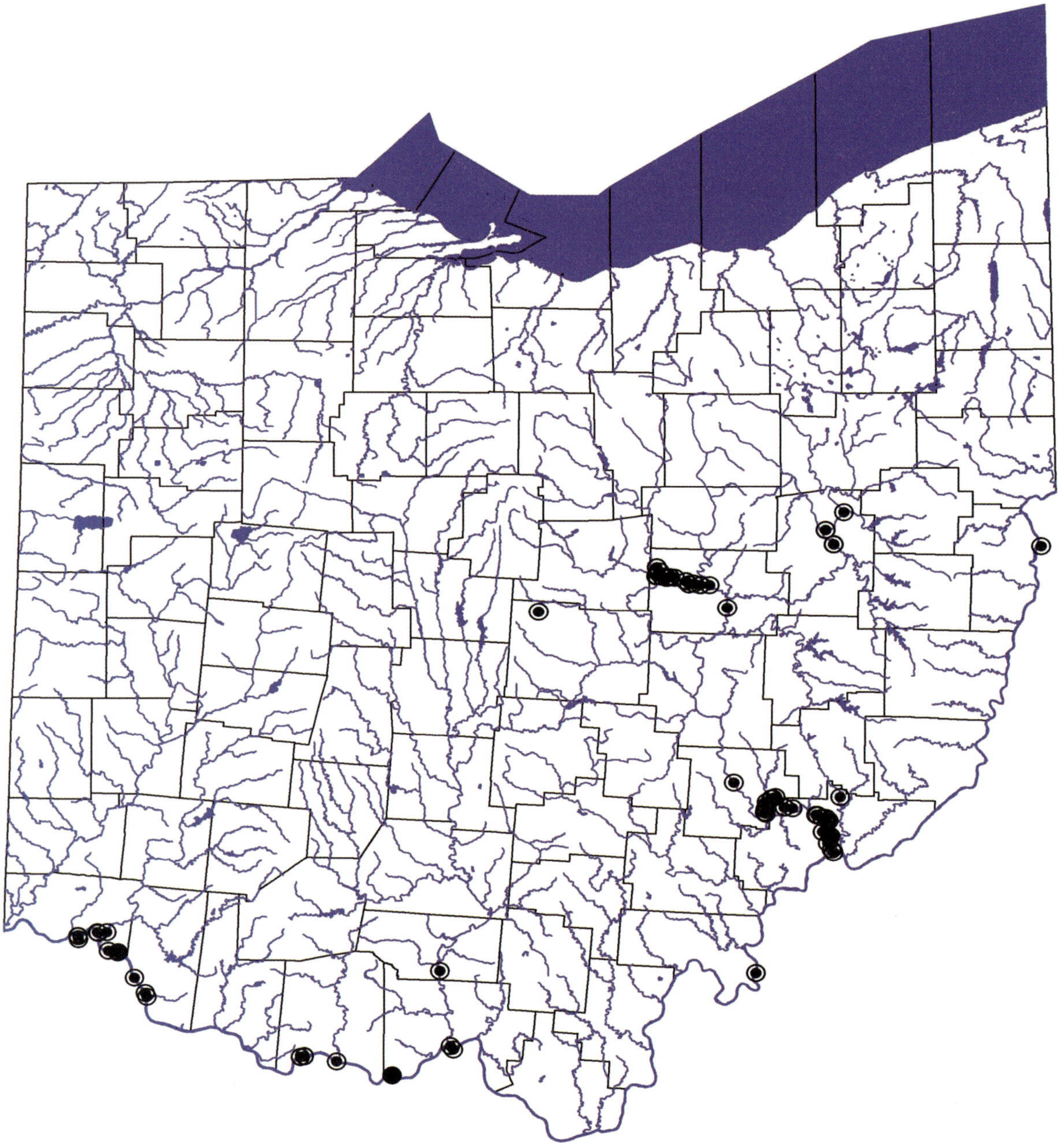

Fig. 143. ***Plethobasus cyphyus*** **(Rafinesque, 1820)**
Solid dots—live or freshdead since 1980
Bullseye—all records, any condition

Fig. 144. ***Plethobasus cyphyus*** **(Rafinesque, 1820).** Portion of non-elastic simple conglutinate. Bar = 500 µm. (Watters, 2008).

Nacre. Porcelain white, often iridescent posteriorly; occasionally with gold or brown Tulberg layers.

Hinge plate. Well-developed, massive, angled under umbo; right cardinal a single, massive, triangular denticle, often with small anterior accessory tooth, posterior accessory rare; left cardinal bifid, the teeth well-separated; laterals well-developed, short, arched or straight, one in right valve, two in left; interdentum cord-like, narrow, short.

Muscle scars. Anterior adductor small and strongly impressed; posterior adductor moderately impressed; anterior adductor distinct from or confluent with anterior pedal protractor scar, large posterior adductor distinct from small posterior pedal retractor scar; cluster of small dorsal suspensors located behind cardinal teeth and interdentum in an irregular line. Pallial line entire, located far from shell margin. Accessory line weak, reaching 25% the length of the shell.

Umbonal cavity. Very wide, open, and shallow.

Glochidia. Depressed semicircular, length greater than height, dorsal margin straight, lateral margins equally rounded, ventral margin obliquely rounded, hooks absent (Ortmann, 1912a: 261; Surber, 1913: 116, pl. 31, fig. 17; Utterback, 1915: 185; Ortmann, 1919: 66).

Literature records (Surber, 1912: 10, pl. 1, fig. 12; Utterback, 1915: 185; Ortmann, 1919: 66) indicate glochidial dimensions for length (0.21 to 0.22 mm) and height (0.20 mm). Ortmann (1912a: 261) gave measurements for immature glochidia (0.18 mm x 0.18 mm) and noted, "Glochidia of the usual shape, rather small, semioval, without hooks." Later, Ortmann (1919: 66) gave dimensions for the mature glochidium of this species more in line with those of other investigators.

Reproductive biology. Tachytictic. Only the outer gills are marsupial. Eggs develop in June and glochidia appear in July (Ortmann, 1919). Glochidia are discharged in pink or red, lanceolate conglutinates (Ortmann, 1912a; Utterback, 1915). These are non-elastic simple conglutinates lacking unfertilized eggs (Watters, 2008). So far only the Central Stoneroller has been confirmed as a host.

Reported potential hosts

Campostoma anomalum (Central Stoneroller)—Watters *et al.*, 2005 (LT).

Sander canadensis (Sauger)—Surber, 1913 (NI); Wilson, 1916 (NI).

Habitat. Sandy mud and gravel in rivers.

Range. Ohio River system, including Tennessee and Cumberland rivers. North as far as Minnesota and Wisconsin, east to the Monongahela River.

Range in Ohio. Muskingum River, historically as far upstream as the Licking and Tuscarawas rivers; lower Scioto River. Ohio River mainstem. Absent from Lake Erie drainages.

Similar species. See under *P. cicatricosus*.

Comments. Individuals gradually slow in growth during their entire life. Specimens rarely reach ca. 130 mm and nearly 30 years old.

Pleurobema Rafinesque, 1819

Etymology. Gr. *pleura*, the side, or Gr. *pleuron*, a rib + Gr. *bema*, step, pace, raised place, base, pedestal. In his original description Rafinesque (1820: 313) gave us no clue as to his choices of word roots to form the name *Pleurobema*. He may have been referring to the step-like areas (= "ribs"?) commonly found between the annulae on the valves (sides) of many unionoids. These ridges are often pronounced in *Pleurobema clava* (Lamarck, 1819) (= *P. mytiltoides* Rafinesque, 1820), and the visual effect of stair-steps is enhanced by the interrupted ray pattern of this species.

Original description

Pleurobema Rafinesque, 1819: 427.

Type species. *Pleurobema mytiloides* Rafinesque, 1820 [= *Unio clava* Lamarck, 1819], by subsequent designation of Herrmannsen (1847: 291–292). Two species were included under *Pleurobema* in 1819: *mytiloides* and *conica*, both *nomen nuda* and therefore unavailable. In 1820 Rafinesque described and gave *mytiloides* and *cuneata* as examples. Because the 1819 taxa were unavailable, the first available names were Rafinesque's 1820 taxa. Nevertheless, the genus name *Pleurobema* must date from 1819, not 1820, even though none of the included nominal species were available at that time.

Pleurobema clava is quite unlike most other species assigned to this genus. Most have white viscera and foot, but in *P. clava* these structures are orange. *Pleurobema clava* and its clavate relatives *P. decisum* (Lea, 1831), *P. chattanoogaense* (Lea, 1858), and a few others probably constitute a separate group apart from other *Pleurobema*.

Synonymy

Pleurobema Rafinesque, 1819: 427.

Obliquaria (*Scalenaria*) Rafinesque, 1820: 309 [Type species: *Obliquaria* (*Scalenaria*) *scalenia* Rafinesque, 1820, by subsequent designation of Herrmannsen (1847: 527)].

Obliquaria (*Sintoxia*) Rafinesque, 1820: 310 [Type species: *Obliquaria* (*Sintoxia*) *sintoxia* Rafinesque, 1820, by subsequent designation of Herrmannsen (1847: 527)].

Geographic range. Mississippi River from Minnesota to the Gulf of Mexico, Gulf drainages from peninsular Florida to Texas. Great Lakes drainages. The Atlantic Slope species "*Canthyria*" *collina* (Lea, 1836) is now considered to be a *Pleurobema* (Boss & Clench, 1967; Turgeon *et al.*, 1996).

Pleurobema clava (Lamarck, 1819)

Figures 145, 146

Etymology. L. *clava*, staff, club, cudgel. The shell of *P. clava* typically has its greatest height and width very near the anterior end and gradually becomes lower and narrower moving posteriorly. The resulting form is that of a tapered wedge or club. *Clava* is a noun in apposition.

Vernacular name(s). *Clubshell*, Northern Club Shell, Ohio Club Shell, Ohio River Club Shell, Common Club Shell.

Original description

Unio clava Lamarck, 1819: 74, no figure given [Frierson (1927: 40) argued that Lamarck's name could not be positively identified with any known species. He advocated dropping *clava* in favor of *Pleurobema mytiloides* Rafinesque, 1820.].

Type locality. Lamarck gave two sources for his specimen(s): "Habite dans le lac Erie" ["living in Lake Erie"] and "La varieté (b) vit dans la rivière de la Nouvelle Écosse" ["variety b living in the river in Nova Scotia"]. Ortmann & Walker (1922: 25–26) believed that both of these localities were erroneous. This species has not, to our knowledge, been recorded in any form from Nova Scotia since Lamarck described the shells delivered to him by the younger Michaux sometime before 1819. However, the University of Michigan Museum of Zoology has a specimen labeled "Lake Erie," so that part of the original type locality cannot be ruled out. Although apparently absent from Lake Erie in recent times, *P. clava* has been recorded from the Maumee River of the lake's drainage basin. Johnson (1969: 53) therefore unnecessarily "corrected" the type locality to "the Ohio River, Cincinnati, Ohio." Most recent collections are from much smaller rivers and creeks than the Ohio River.

Type material. The holotype of *U. clava* is deposited in the Muséum national d'Histoire naturelle, Paris, as is a paratype described as variety (b) in the original description, and is figured by Johnson (1969, fig. 9).

Synonymy

Unio clava Lamarck, 1819: 74.

Unio (*Aximedia*) *elliptica* Rafinesque, 1820: 296 [Type locality: "vue de Louisville et de Maysville" ("seen from Louisville and Maysville [KY]"). Lectotype ANSP 20213 by selection of Johnson & Baker (1973: 154)].

Obliquaria (*Scalenaria*) *scalenia* Rafinesque, 1820: 309–310, pl. 81, figs. 24, 25 [Type locality: "dans le Kentuky [River]," but the neotype is from "Ohio River." Neotype ANSP 20229 by selection of Johnson & Baker (1973: 169). This designation may be invalid.].

Pleurobema mytiloides Rafinesque, 1820: 313, pl. 82, figs. 8–10 [Type locality: "dans le Wabash [River]." Labeled "type" ANSP 20223 is actually *Pleurobema cordatum* (Rafinesque, 1820) and does not match the figures, *fide* Johnson & Baker (1973: 162). Location of actual type is unknown.].

Pleurobema cuneata Rafinesque, 1820: 313 [Type locality: "assez commune dans l'Ohio près de Steubenville, Marietta, etc." ("rather common in the Ohio River near Steubenville, Marietta, etc. [OH]"). The lectotype is from "Ohio River." Lectotype ANSP 20288 by designation of Johnson & Baker (1973: 152)].

Unio mytiloides (Rafinesque, 1820): 269.

Unio patulus Lea, 1829: 441, pl. 12, fig. 20 [Type locality: "Ohio." Figured lectotype USNM 84760, paralectotypes ANSP 129871].

Unio cuneatus (Rafinesque, 1820). Say, 1834: unpaginated.

Margarita (*Unio*) *mytiloides* (Rafinesque, 1820). Lea, 1836: 21.

Margarita (*Unio*) *patulus* (Lea, 1829). Lea, 1836: 22.

Margarita (*Unio*) *clavus* (Lamarck, 1819). Lea, 1836: 22.

Unio bournianus Lea, 1840: 288, no figure [Type locality: "Sciota [Scioto] River near Chillicothe, Ohio." Holotype USNM 84683 *fide* Johnson, 1974: 22. "One probable paratype MCZ 178833" *fide* Johnson, 1956: 108].

Cunicula patula (Lea, 1829). Swainson, 1840: 378.

Unio clavus Lamarck, 1819. Reeve, 1841: 117, pl. 88, fig. 3.

Pleurobema clava (Lamarck, 1819). Agassiz, 1852: 49.

Margaron (*Unio*) *mytiloides* (Rafinesque, 1820). Lea, 1852b: 25.

Fig. 145. ***Pleurobema clava*** **(Lamarck, 1819).** **A.** OSUM 34785.2, "Ohio River." 72 mm. **B.** OSUM 10904.6, St. Joseph River, Williams Co., OH. 72 mm. **C.** OSUM 47189.5, Muskingum River, [Co.?], OH. 38 mm. **D.** OSUM 27207, Big Darby Creek, Franklin Co., OH. 48 mm. **E.** OSUM 57251.1, Mahoning River, [Co.?], OH. 33 mm. **F, H, Hinge detail, Beak sculpture detail:** OSUM 9758.1, Scioto River, [Co.?], OH. 42 mm. **G.** OSUM 31763.4, Pymatuning Creek, Ashtabula Co., OH. 81 mm.

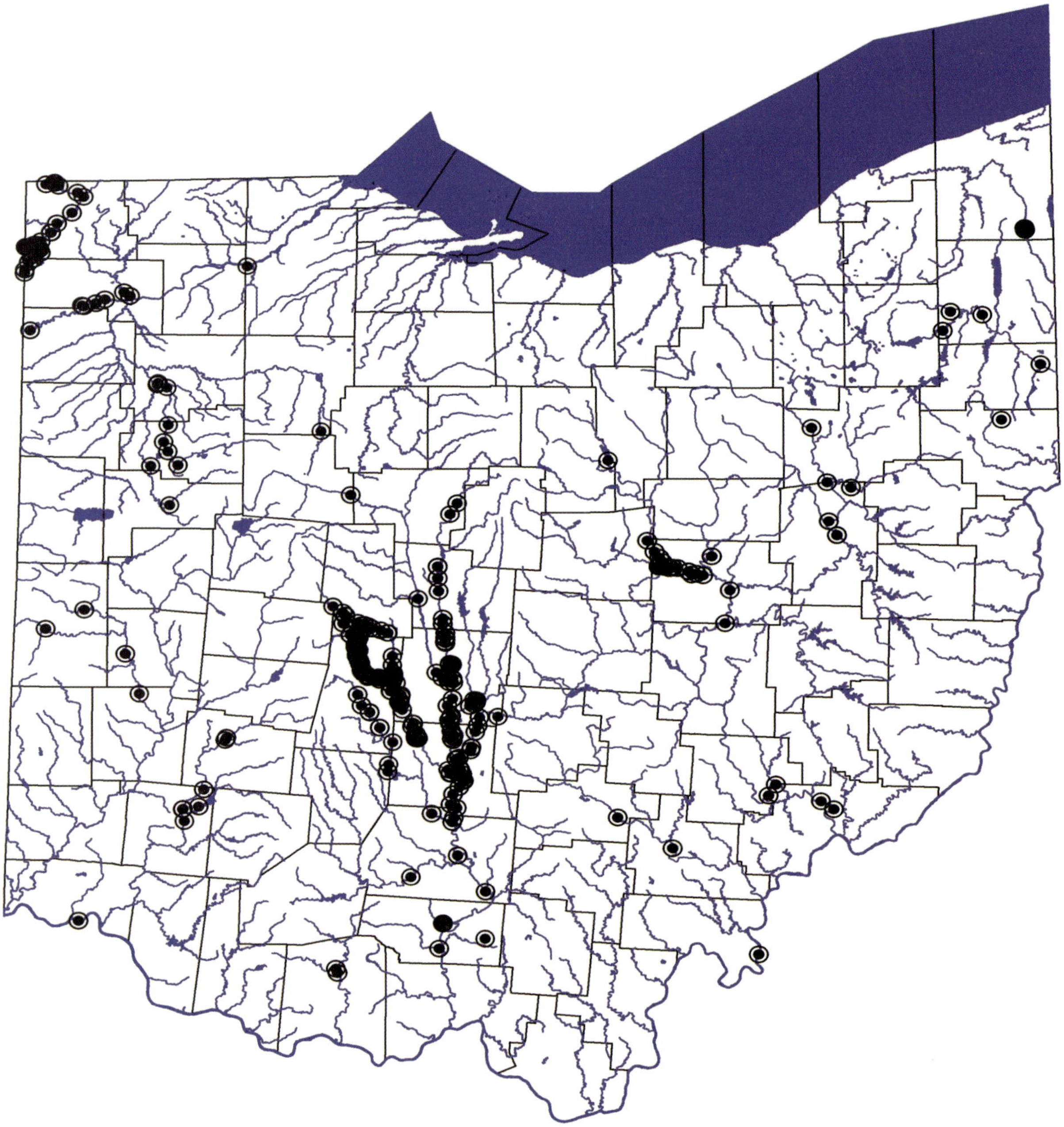

Fig. 146. ***Pleurobema clava*** **(Lamarck, 1819)**
Solid dots—live or freshdead since 1980
Bullseye—all records, any condition

Margaron (*Unio*) *patulus* (Lea, 1829). Lea, 1852b: 26.
Margaron (*Unio*) *clavus* (Lamarck, 1819). Lea, 1852b: 26.
Unio consanguineus De Gregorio, 1814: 46 [*non* Lea, 1863. Type locality: "Sciota [Scioto] R. (Ohio)." Type in Palermo Museum?].
Pleurobema bournianum (Lea, 1840). Simpson, 1900a: 747.
Unio anaticulus var. *ohiensis* De Gregorio, 1914: 51, pl. 10, fig. 3. [Type locality: "Sciota. [Scioto River] Ohio." Type in Palermo Museum?].
Pleurobema mytiloides patulum (Lea, 1829). Frierson, 1927: 40.
Pleurobema clava clava (Lamarck, 1819). Haas, 1969: 252.
Pleurobema clava patulum (Lea, 1829). Haas, 1969: 252.
non Unio mytiloides (Rafinesque, 1820). Swainson, 1840: 270, figs. 52, 53 [= *Fusconaia ebenus* (Lea, 1831)].

Status. Listed as Endangered by both Ohio Division of Wildlife and U.S. Fish & Wildlife Service as of 2009. A Federal recovery plan exists for this species (USFWS, 1993).

ADULT SHELL

Shell. Small to medium size (to 80 mm), rather thick, elongate triangular, moderately inflated.

Anterior margin. Short, nearly vertical, straight or broadly rounded to the ventral border.

Ventral margin. Very broadly rounded or straight, rarely with a slight concavity at the posterior ridge.

Posterior margin. Short, gently rounded or straight to the dorsal margin.

Dorsal margin. Long, arched or straight, joining posterior margin with very obtuse angle.

Umbo. Wide to rather narrow, low in creek populations, prominent in river forms; umbos face forward ca. 30°; umbonal sculpture of 2–3 angulate ridges, sometimes prominently developed posteriorly.

Ligament. Short and prominent, tan to dark brown, not concealed by shells.

Disc sculpture. Disc of shell smooth except for growth lines; weak sulcus present; posterior slope with 2–3 weak angulations.

Periostracum. Tan, yellowish, or greenish, becoming darker with age; juveniles have bold green stripes, often interrupted into a checkerboard pattern, but these may be lost in adults.

Nacre. Porcelain white, often iridescent posteriorly; occasionally with gold or brown Tulberg layers.

Hinge plate. Well-developed, gently arched; right cardinal a single, small, triangular denticle, rarely with accessory teeth; left cardinal bifid, the teeth not far apart and often joined dorsally; cardinals tend to rotate to a horizontal position with age; laterals well developed, short, arched or straight, one in right valve, two in left; interdentum flattened, broad, long.

Muscle scars. Anterior adductor strongly impressed; posterior adductor moderately impressed; anterior adductor distinct from anterior pedal protractor scar, large posterior adductor distinct from small circular posterior pedal retractor scar; dorsal suspensors located behind cardinal teeth and interdentum in an irregular line. Pallial line entire, located far from shell margin. Accessory line rudimentary, often absent, reaching 24% the length of the shell.

Umbonal cavity. Very wide, open, and shallow.

Glochidia. Semicircular, dorsal margin straight, lateral margins equal, ventral margin rounded, hooks absent (Ortmann, 1911: 364, pl. 89, fig. 5; Ortmann, 1912a: 265).

Literature records (Ortmann, 1912a: 265) indicate glochidial dimensions for length (0.16 mm) and height (0.16 mm). Ortmann (1912a: 265) noted, "Glochidia (see Ortmann, 1911, Pl. 89, fig. 5) of small size, subovate, without hooks. Their length and height about the same."

Reproductive biology. Tachytictic. Eggs appear in May, and glochidia develop in June and July (Ortmann, 1919). In Ohio, females release fragile, white, non-elastic conglutinates (Watters, 2008), and are barren by the end of June (Watters, unpubl.).

Reported potential hosts

Campostoma anomalum (Central Stoneroller)—Watters & O'Dee, 1997e (LT); O'Dee & Watters, 2000 (LT).
Luxilus chrysocephalus (Striped Shiner)—Watters & O'Dee, 1997e (LT); O'Dee & Watters, 2000 (LT).
Percina caprodes (Logperch)—Watters & O'Dee, 1997e (LT); O'Dee & Watters, 2000 (LT).
Percina maculata (Blackside Darter)—Watters & O'Dee, 1997e (LT); O'Dee & Watters, 2000 (LT).

Habitat. The clubshell occurs in clean, coarse sand and cobble in a current, where it may live several inches beneath the surface. It is most common in the downstream ends of riffles and islands. Adults caught and released will burrow out of sight within 24 hours.

Range. Ohio River system, including the Tennessee, Cumberland, Kanawha, and Wabash river drainages. Maumee River system of Lake Erie. Although it was originally described from Lake Erie, we have found only one record of its occurrence there (UMMZ).

Range in Ohio. Once widespread and common, this species is now nearly extirpated. Living populations remain only in Pymatuning Creek, Little Darby Creek, Fish Creek, and the St. Joseph River. The Pymatuning Creek population consists of large (> 80 mm long), old (> 20 years) individuals that apparently are not reproducing. Recorded only from the Maumee River system for the Lawrentian systems.

Similar species. The extreme forward placement of the umbo, often beyond the anterior border, the bold green rays, and the elongated wedge shape all identify this shell. Rare examples of *Pleurobema sintoxia* may approximate this species, but usually lack the rays and extreme umbonal position.

Comments. Individuals gradually slow in growth during their entire life; there is great variation in growth rates among populations. Individuals 60 mm in length may be between 4 and 12 years old. Specimens rarely reach ca. 80 mm and over 20 years old; most individuals are half this size and age. Specimens from large rivers usually have prominent umbos, often projecting past the anterior margin.

Unio bournianus Lea, 1840 (the Scioto Pigtoe), is tentatively considered here a teratological specimen of *P. clava*. It is known only from the type locality of the Scioto River. It has not been collected in well over 150 years. Only the two specimens Lea had at hand are known of this species. However, R. Anderson (USGS, Pittsburgh, pers. comm., 2000) found listings for this species in the Earlham College collection at Richmond, Indiana, but the specimens have not been located and no locality data were evident. Additional listings were found by Anderson for the Call collection, but again the specimens have not been found. If a valid species, the Scioto Pigtoe is probably extinct.

Pleurobema cordatum (Rafinesque, 1820)

Figures 147, 148, 149

Etymology. L. *cor, cordis*, heart + L. *-atus*, provided with; i.e., heart-shaped. Most *Pleurobema* species, when viewed from the anterior, are cordate or heart-shaped, and the well-developed adult of *P. cordatum* is especially so.

Vernacular name(s). *Ohio Pigtoe*, Ohio River Pigtoe, River Pigtoe, Common Pigtoe, Big River Pigtoe.

Original description

Obovaria cordata Rafinesque, 1820: 312–313, species no. 50, pl. 82, figs. 6, 7.

Type locality. "dans l'Ohio [River], etc."

Type material. The type mentioned by Vanatta (1915: 558) was selected as lectotype (ANSP 20221 from the Ohio River) by Johnson & Baker (1973: 151).

Synonymy

? *Unio obliqua* Lamarck, 1819: 72 [*nomen dubium*. Type locality: "Ohio River." Holotype lost *fide* Johnson, 1953: 91].

Obovaria cordata Rafinesque, 1820: 312–313, pl. 82, figs. 6, 7.

? *Obovaria cordata* var. *rosea* Rafinesque, 1820: 312 [Type locality: "dans l'Ohio [River]." Type not located].

Unio mytiloides Rafinesque, 1820. Short & Eaton, 1831: 74 [misidentification].

Unio cordatus (Rafinesque, 1820). Conrad, 1836: 48.

Margarita (*Unio*) *obliqua* (Lamarck, 1819). Lea, 1836: 20.

Unio obliquus (Lamarck, 1819). Hanley, 1842: 186.

Margarita (*Unio*) *obliquus* (Lamarck, 1819). Lea, 1838b: 17.

Margaron (*Unio*) *obliquus* (Lamarck, 1819). Lea, 1852b: 25.

Quadrula obliqua (Lamarck, 1819). Simpson, 1900a: 788.

Pleurobema obliquum (Lamarck, 1819). Ortmann, 1912a: 264.

Pleurobema obliquum cordatum (Rafinesque, 1820). Ortmann, 1918: 548.

Pleurobema cordatum (Rafinesque, 1820). Ortmann, 1924: 34.

Quadrula (*Obliquata*) *cordata* (Rafinesque, 1820). Frierson, 1927: 53.

Pleurobema cordata (Rafinesque, 1820). Morrison, 1942: 354.

Quadrula (*Obliquata*) *cordata cordata* (Rafinesque, 1820). Haas, 1969: 297.

Pleurobema cordatum cordatum (Rafinesque, 1820). Burch, 1973: 15, fig. 38.

Status. Listed as Endangered by Ohio Division of Wildlife; not listed by U.S. Fish & Wildlife Service as of 2009.

ADULT SHELL

Shell. Medium to fairly large sized (to 130 mm), rather thick, triangular, moderately inflated; individuals become more elongate with age.

Anterior margin. Long, nearly vertical, straight to the ventral border.

Ventral margin. Very broadly rounded or straight, with a concavity at the posterior ridge.

Posterior margin. Short, broadly arched at the dorsal margin.

Dorsal margin. Long, arched, joining posterior margin with very obtuse angle or joining imperceptibly.

Umbo. Rather narrow, very prominent; umbos twisted forward at least 45°; projecting beyond anterior margin, particularly in older individuals; umbonal sculpture of 2–3 minute oblique ridges, sometimes prominently developed posteriorly.

Ligament. Long and prominent, tan to dark brown, not concealed by shells.

Disc sculpture. Disc of shell smooth except for growth lines; prominent sulcus present; posterior slope with 2–3 weak angulations.

Fig. 147. ***Pleurobema cordatum*** **(Rafinesque, 1820). A.** OSUM 58322, Ohio River, Meade Co., KY. 68 mm. **B.** OSUM 40285, Muskingum River, Washington Co., OH. 84 mm. **C.** OSUM 42336.7, Muskingum River, Washington Co., OH. 106 mm. **D.** OSUM 51297.6, Muskingum River, Washington Co., OH. 47 mm. **E, Beak sculpture detail:** OSUM 19291.1, Clinch River, Hancock Co., TN. 22 mm. **F, Hinge detail:** OSUM 42336.2, Muskingum River, Washington Co., OH. 69 mm.

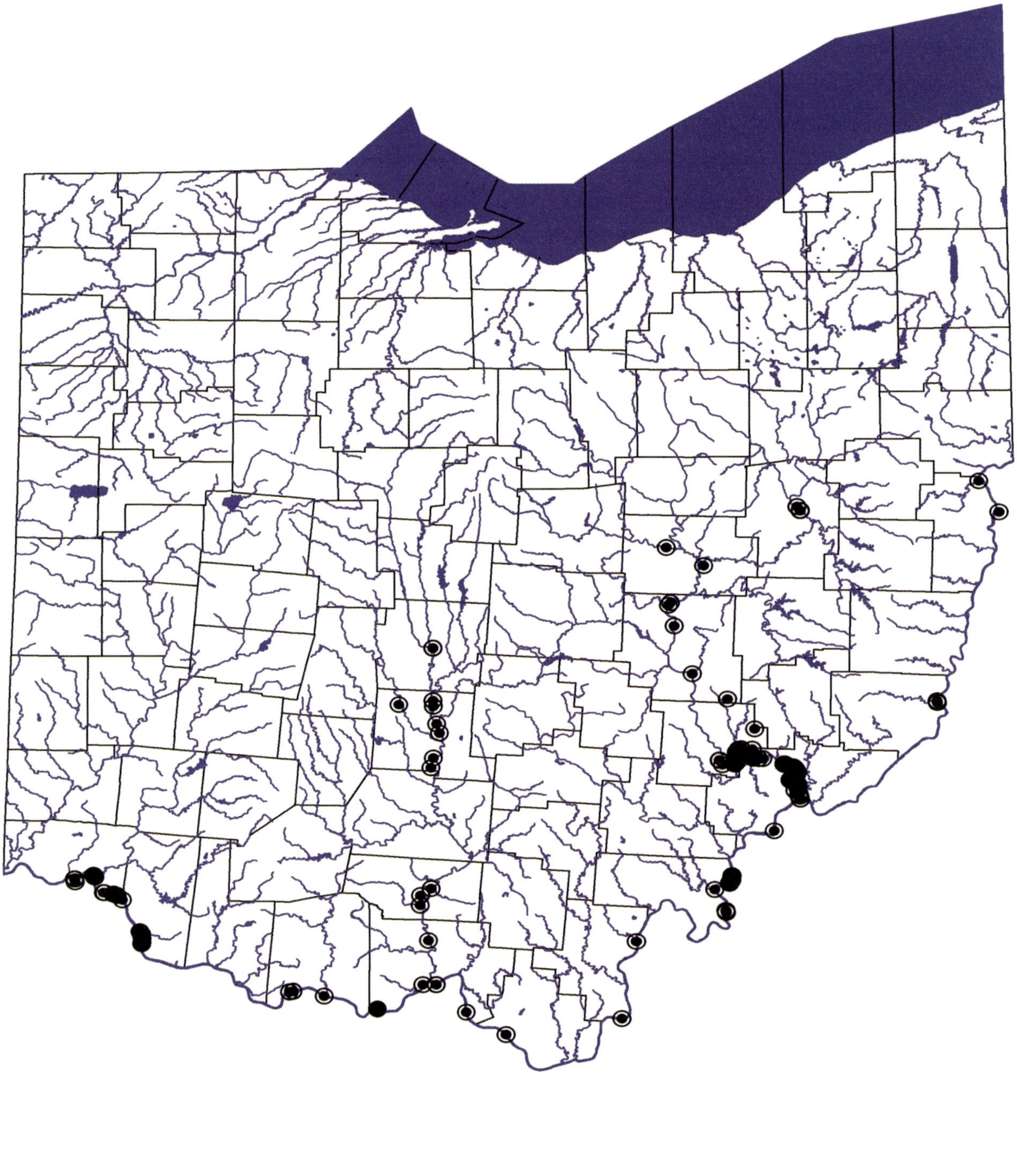

Fig. 148. ***Pleurobema cordatum*** **(Rafinesque, 1820)**
Solid dots—live or freshdead since 1980
Bullseye—all records, any condition

Periostracum. Shiny, tan, yellowish, or reddish brown, becoming darker with age; juveniles have green stripes, but these may be lost in adults.

Nacre. Porcelain white, very rarely with rose flush, often iridescent posteriorly; occasionally with gold or brown Tulberg layers.

Hinge plate. Well-developed, angled under umbo; right cardinal a single, massive, triangular denticle, often with small anterior accessory tooth, posterior accessory rare; left cardinal bifid, the teeth well-separated; laterals well-developed, short, arched, one in right valve, two in left; interdentum flattened, broad, short.

Muscle scars. Anterior adductor small and strongly impressed; posterior adductor moderately impressed; anterior adductor distinct or just touching anterior pedal protractor scar, large posterior adductor distinct from small posterior pedal retractor scar; small dorsal suspensors located behind cardinal teeth and interdentum in an irregular line. Pallial line entire, located far from shell margin. Accessory line distinct but short, reaching 25% the length of the shell.

Umbonal cavity. Very wide, compressed, and deep.

Glochidia. Subelliptical, dorsal margin straight, lateral margins equal, evenly rounded, ventral margin rounded, hooks absent (Surber, 1915: 8, pl. 1, fig. 12; Utterback, 1916b: 20).

Literature records (Surber, 1912: 10, pl. 2, fig. 35; Howard, 1914a: 23, pl. 5, fig. 34; Surber, 1915: 8, pl. 1, fig. 12; Utterback, 1915: 190, 193; Utterback, 1916b: 20; Yokley, 1972: 359, figs. 12, 13) indicate glochidial dimensions for length (0.14 to 0.17 mm) and height (0.15 to 0.18 mm). Surber (1915: 8) noted, "This glochidium seems to be rather intermediate in shape between *Q. trigona* [*F. flava*] and *Unio gibbosa* [*E. dilatata*], agreeing with the former in length, but with less depth, *trigona* being longer than deep, and *obliqua* [*P. cordatum*] deeper than long. While agreeing with *gibbosa* in shape, except somewhat longer hinge line, it is much smaller and is not likely to be confused with it."

Reproductive biology. Tachytictic. Only the outer gills are marsupial (Ortmann, 1912a). Eggs are present in April, and females are gravid in May through September (Ortmann, 1919; Scruggs, 1960; Yokley, 1972). Conglutinates are of the non-elastic type without unfertilized eggs (Watters, 2008).

Reported potential hosts

Culaea inconstans (Brook Stickleback)—Watters & Kuehnl, 2004 (LT); Watters *et al.*, 2005 (LT).

Lebistes reticulata (Guppy)—Watters & Kuehnl, 2004 (LT); Watters *et al.*, 2005 (LT)—exotic.

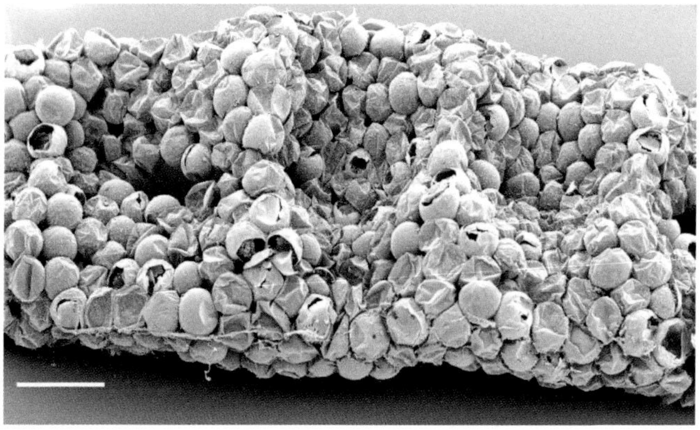

Fig. 149. *Pleurobema cordatum* (Rafinesque, 1820). Portion of simple non-elastic conglutinate. Bar = 500 µm. (Watters, 2008).

Lythrurus ardens (Rosefin Shiner)—Yokley, 1972 (LT); Fuller, 1974 (NS).

Lepomis macrochirus (Bluegill)—Fuller, 1974 (NS).

Semotilus atromaculatus (Creek Chub)—Watters & Kuehnl, 2004 (LT); Watters *et al.*, 2005 (LT).

Habitat. The Ohio Pigtoe occurs in stable, muddy sand and cobble in large rivers and some of their impoundments.

Range. Mississippi River system, including Ohio, Cumberland, and Tennessee rivers, upper St. Lawrence River, but absent from the Great Lakes drainages. Specimens from west of the Mississippi River may represent a different species.

Range in Ohio. Previously in the large Ohio rivers: Scioto River to Columbus, and Muskingum River. Today it lives in the lower Muskingum River, where it is common, and Big Darby Creek, where it is very rare, although additional specimens were transplanted there within very recent times. Historical records from the turn of the 20th century include the Tuscarawas River (CMNH).

Similar species. This species belongs to a complex of very similar-appearing taxa. The differences between them are often subtle and difficult to discern. They are briefly compared below. Ohio has four members of the complex; all historically occurred in the lower Muskingum River and Ohio River at Marietta, as well as the Tuscarawas River at New Philadelphia. A fifth species, *Pleurobema taitianum* (Lea, 1834), occurs in the southeastern United States.

Pleurobema cordatum—prominent sulcus, umbos prominent and facing forward, umbos near anterior margin, elongated only as old individuals, nacre white; riverine.

Pleurobema plenum—no sulcus, umbos prominent and facing each other, umbos not near anterior margin, not elongated; nacre white, periostracum rather shaggy or matte; riverine.

Pleurobema rubrum—prominent sulcus, umbos very prominent and facing forward, umbos at or beyond anterior margin, often quite elongated, nacre pink or white; riverine.

Pleurobema sintoxia—sulcus weak or absent, umbos prominent in river forms, not in creek forms, facing each other; umbos not before anterior margin, elongated only as old individuals, nacre pink, salmon, or white; widely distributed in creeks and rivers.

Comments. Individuals gradually slow in growth during their entire life. Specimens reach ca. 130 mm and nearly 30 years old.

Pleurobema plenum (Lea, 1840)

Figures 150, 151

Etymology. L. *plenus*, full, stout, plump. Lea noted (1840: 286), in his original description of this species, that it is "*inflata*" or wide, and later (Lea 1842: 211) compared it with *P. cordatum* and *F. flava* in other regards. Specimens of this species typically appear "full, stout or plump."

Vernacular name(s). *Rough Pigtoe*, Ridged Pigtoe, Full Pigtoe.

Original description

Unio plenus Lea, 1840: 286, no figure. Redescribed in Lea, 1842: 211–212, pl. 14, fig. 26.

Type locality. "Ohio River, near Cincinnati."

Type material. Johnson (1974: 114) recognized the figured specimen (USNM 84677) as the lectotype.

Synonymy

Unio plenus Lea, 1842b: 211–212, pl. 14, fig. 26.

Margaron (*Unio*) *plenus* (Lea, 1842). Lea, 1852b: 25.

Quadrula plena (Lea, 1840). Simpson, 1900a: 790.

Quadrula obliqua (Lamarck, 1819). Daniels, 1903: 652.

Pleurobema obliquum (Lamarck, 1819). Ortmann, 1910: 117.

Quadrula cordata (Rafinesque, 1820). Vanatta, 1915: 558 [misidentification].

Pleurobema obliquum var. *plenum* (Lea, 1842). Utterback, 1915: 188.

Pleurobema obliquum cordatum (Rafinesque, 1820). Ortmann, 1918: 548 [in part, misidentification].

Pleurobema plenum (Lea, 1840). Ortmann & Walker, 1922: 23.

Pleurobema cordatum var. *plenum* (Lea, 1840). Ortmann, 1925: 339.

Quadrula (*Obliquata*) *cordata plena* (Lea, 1840). Frierson, 1927: 53.

Fusconaia plena (Lea, 1840). Morrison, 1942: 354.

Pleurobema cordatum (Rafinesque, 1820). Bickel, 1968: 20 [in part, misidentification].

Pleurobema premorsa (Rafinesque, 1820). Morrison, 1969: 23 [misidentification].

Quadrula (*Obliquaria*) *cordata* var. *plena* (Lea, 1840). Haas, 1969: 298.

Status. Listed as Extirpated by the Ohio Division of Wildlife; listed as Endangered by U.S. Fish & Wildlife Service as of 2009.

ADULT SHELL

Shell. Medium size (to 90 mm), rather thick, triangular, moderately inflated.

Anterior margin. Long, nearly vertical, straight to the ventral border.

Ventral margin. Very broadly rounded or straight, often with a slight concavity at the posterior ridge.

Posterior margin. Long, nearly straight to the dorsal margin.

Dorsal margin. Short, arched, joining posterior margin with obtuse angle.

Umbo. Rather narrow, prominent; umbos nearly face each other ca. 10°; umbonal sculpture of 2–3 minute angulate ridges, sometimes prominently developed posteriorly.

Ligament. Short and prominent, tan to dark brown, not concealed by shells.

Disc sculpture. Disc of shell smooth except for growth lines; weak sulcus present; posterior slope with 2–3 weak angulations.

Periostracum. Coarse, with a satin finish, tan, yellowish, or reddish brown, becoming darker with age; juveniles have green stripes, but these may be lost in adults.

Nacre. Porcelain white, rarely with rose flush, often iridescent posteriorly; occasionally with gold or brown Tulberg layers.

Hinge plate. Well-developed, massive, angled 90° under umbo; right cardinal a single, massive, triangular denticle, often with small anterior accessory tooth, posterior accessory rare; left cardinal bifid, the teeth well-separated; laterals well-developed, short, arched or straight, one in right valve, two in left; interdentum flattened, broad, very short.

Muscle scars. Anterior adductor small and strongly impressed; posterior adductor moderately impressed; anterior adductor distinct from anterior pedal protractor scar, large posterior adductor distinct from small posterior pedal retractor scar; cluster of small dorsal suspensors located behind cardinal teeth and interdentum in an irregular line. Pallial line entire, located far from shell margin. Accessory line rudimentary, often absent, reaching 25% the length of the shell.

Umbonal cavity. Very wide, somewhat compressed, and deep.

Fig. 150. *Pleurobema plenum* (Lea, 1840). A. OSUM 17243, Ohio River, Campbell Co., KY. 72 mm. **B, F.** OSUM 44609.5, Green River, Warren Co., KY. 73 mm. **C.** OSUM 68351.3, Clinch River, Union Co., TN. 60 mm. **D.** OSUM 68351.2, Clinch River, Union Co., TN. 44 mm. **E.** OSUM 68308, Ohio River, Hamilton Co., OH. 43 mm. **Hinge detail:** OSUM 44609.1, Green River, Warren Co., KY. 63 mm total length. **Beak sculpture detail:** OSUM 20356, Clinch River, Hancock Co., TN. 24 mm total length.

Fig. 151. ***Pleurobema plenum*** **(Lea, 1840)**
Solid dots—live or freshdead since 1980
Bullseye—all records, any condition

Glochidia. Unknown.

Reproductive biology. Probably tachytictic. Frierson (1927: 53) stated that the marsupia may occasionally occupy all four gills.

Reported potential hosts. No hosts known.

Habitat. This Federally endangered species occurs in stable, muddy sand and cobble in large rivers and their impoundments.

Range. Ohio River system, including the Green River of Kentucky, White River of Indiana, Wabash, Cumberland and Tennessee rivers; east to the Monongahela River (but extirpated from most of these rivers). Does not occur in the upper Mississippi River or the Great Lakes. Specimens from west of the Mississippi River may represent a different species.

Range in Ohio. Extirpated. Subfossil shells occur in the lower Muskingum River and the Scioto River below Columbus. Historical records from the 1900's exist for the Ohio River from Buffington Island to Cincinnati.

Similar species. See under *Pleurobema cordatum*.

Comments. Individuals gradually slow in growth during their entire life. Specimens rarely reach ca. 90 mm and nearly 30 years old.

Pleurobema rubrum (Rafinesque, 1820)

Figures 152, 153

Etymology. L. *ruber*, red, ruddy. Rafinesque may have used the term "ruber" to refer to the pink nacre frequently found in the species or to the reddish brown periostracum that typically accompanies it. The nacre, however, is white-iridescent in many individuals.

Vernacular name(s). *Pyramid Pigtoe*, Pink Pigtoe.

Original description

Obliquaria rubra Rafinesque, 1820: 314, species no. 54, no figure. Although described under the section on *Amblema*, Rafinesque referred this to *Obliquaria*.

Type locality. "Dans le Kentuky [River]." Parmalee & Bogan's (1988: 192) type locality of "Ohio" was incorrect.

Type material. The type mentioned by Vanatta (1915: 557) was selected as lectotype (ANSP 20237) by Johnson & Baker (1973: 169).

Synonymy

Obliquaria rubra Rafinesque, 1820: 314.

Unio cardiacea Guérin-Méneville, 1828: pl. 28, fig. 7 [Type not located].

Unio mytiloides Rafinesque, 1820. Deshayes, 1830: 586, pl. 249, fig. 4 [misidentification].

Unio pyramidatus Lea, 1831: 109, pl. 16, fig. 39 [Type locality: "Ohio." Syntypes in ANSP 41726 not found *fide* Johnson & Baker (1973: 167)].

Margarita (*Unio*) *pyramidatus* (Lea, 1831). Lea, 1836: 21.

Pleurobema ruber [*sic*] (Rafinesque, 1820). Agassiz, 1852: 49.

Margaron (*Unio*) *pyramidatus* (Lea, 1831). Lea, 1852b: 25.

Unio ruber [*sic*] (Rafinesque, 1820). Conrad, 1853: 257.

? *Unio obliqua* "Lamarck" Wood, 1856: 200 [See under *Pleurobema cordatum* (Rafinesque, 1820).].

Unio mytiloides var. *pyramidatus* Lea, 1831. Paetel, 1890: 160.

Quadrula pyramidatum (Lea, 1831). Simpson, 1900a: 790.

Pleurobema pyramidatum (Lea, 1834). Ortmann, 1912a: 264.

Unio plenus var. *interduos* De Gregorio, 1914: 48 [Type locality: "Sciota R." [Scioto River, OH]. Type in Palermo Museum?].

Quadrula rubra (Rafinesque, 1820). Vanatta, 1915: 557.

Pleurobema obliquum rubrum (Rafinesque, 1820). Ortmann, 1918: 550.

Quadrula (*Obliquata*) *obliquata* Rafinesque, 1820. Frierson, 1927: 52 [misidentification].

Unio mytilloides [*sic*] Lea [*sic*]. Frierson, 1927: 52 [misidentification].

Pleurobema cordatum var. *pyramidatum* (Lea, 1831). Goodrich & van der Schalie, 1932: 309.

Pleurobema cordatum pyramidatum (Lea, 1831). Murray & Leonard, 1962: 76.

Pleurobema rubrum (Rafinesque, 1820). Stansbery, 1976: 50.

Status. Listed as Endangered by Ohio Division of Wildlife; not listed by U.S. Fish & Wildlife Service as of 2009.

ADULT SHELL

Shell. Medium size (to 90 mm), rather thick, triangular, moderately inflated.

Anterior margin. Long, nearly vertical or slanted posteriorly, straight to the ventral border.

Ventral margin. Very broadly rounded or straight, often with a slight concavity at the posterior ridge.

Posterior margin. Long, nearly straight to the dorsal margin.

Dorsal margin. Short, arched, joining posterior margin with very obtuse angle.

Umbo. Rather narrow, prominent; umbos twisted forward and angled 45°–60°; umbonal sculpture of 2–3 minute angulate ridges, tending towards double-looped, sometimes prominently developed posteriorly.

Fig. 152. ***Pleurobema rubrum*** **(Rafinesque 1820). A.** OSUM 68310.2, Ohio River, Hamilton Co., OH. 78 mm. **B, F, Hinge detail:** OSUM 34790.2, Tuscarawas River, Tuscarawas Co., OH. 76 mm. **C.** OSUM 46942, Muskingum River, Washington Co., OH. 71 mm. **D.** OSUM 68310.1, Ohio River, Hamilton Co., OH. 29 mm. **E.** OSUM 68270, Ohio River, Hamilton Co., OH. 90 mm. **Beak sculpture detail:** OSUM 20355.3, Clinch River, Hancock Co., TN. 23 mm total length.

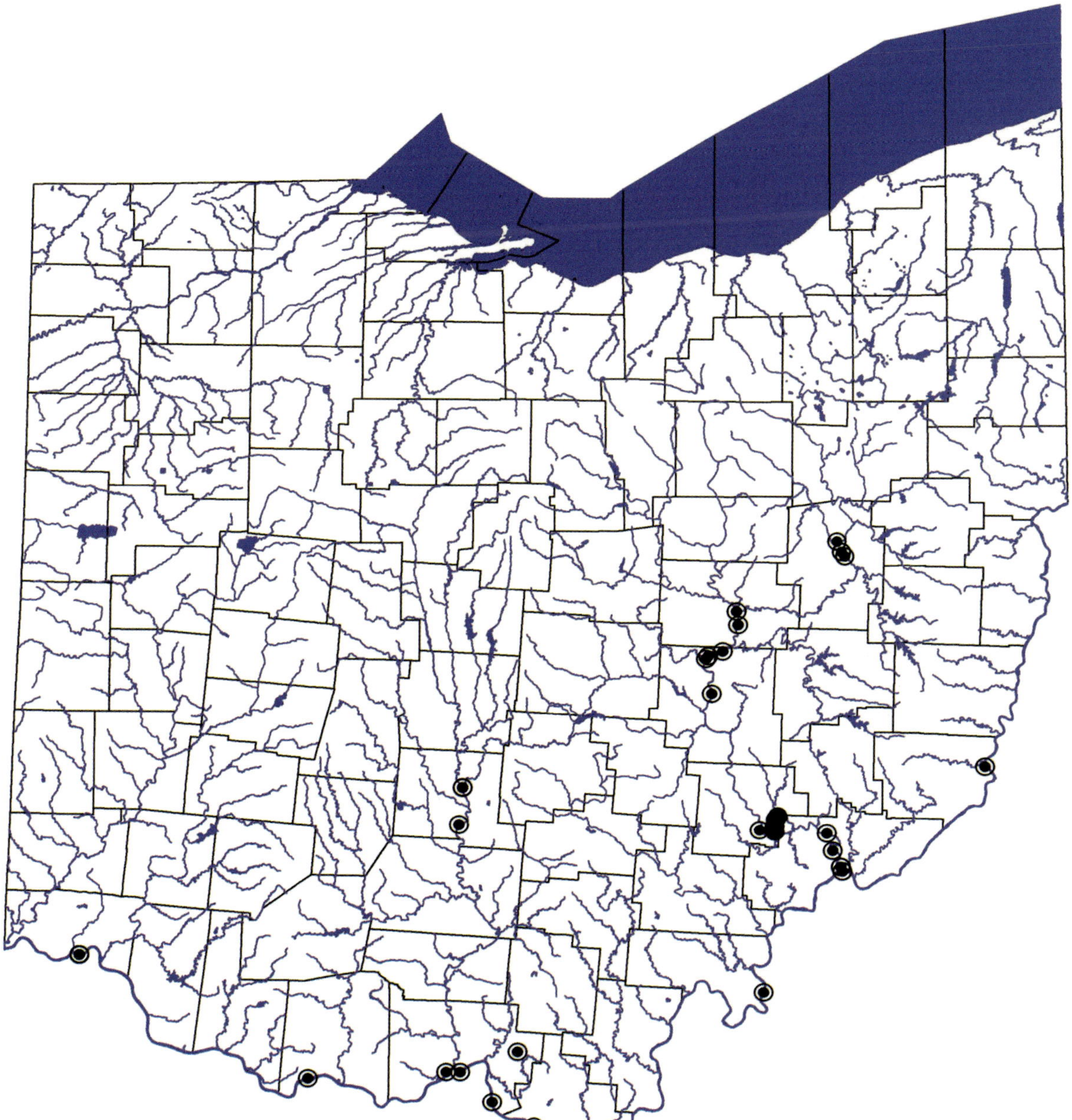

Fig. 153. ***Pleurobema rubrum*** **(Rafinesque, 1820)**
Solid dots—live or freshdead since 1980
Bullseye—all records, any condition

Ligament. Moderately long and prominent, tan to dark brown, not concealed by shells.
Disc sculpture. Disc of shell smooth except for growth lines; usually with well-developed sulcus; posterior slope with 2–3 weak angulations.
Periostracum. Coarse, dull, tan, yellowish, or usually reddish brown, becoming darker with age; juveniles have green stripes, but these may be lost in adults.
Nacre. Porcelain white or with rose flush (ca. 50/50 in specimens examined), often iridescent posteriorly; occasionally with gold or brown Tulberg layers.
Hinge plate. Well-developed, angled 90° under umbo; right cardinal a single, massive, triangular denticle, often with small anterior and posterior accessory tooth; left cardinal bifid, the teeth well-separated; laterals well-developed, short, arched or straight, one in right valve, two in left; interdentum flattened, broad, short.
Muscle scars. Anterior adductor small and strongly impressed; posterior adductor moderately impressed; anterior adductor distinct from or just touching anterior pedal protractor scar, large posterior adductor distinct from small posterior pedal retractor scar; small dorsal suspensors located behind cardinal teeth and interdentum in an irregular line. Pallial line entire, located far from shell margin. Accessory line weak, reaching 25% the length of the shell.
Umbonal cavity. Very wide, open to very compressed, and shallow to deep.

Glochidia. Unknown.

Reproductive biology. Tachytictic. Females bearing eggs were found in May and July (Ortmann, 1919).

Reported potential hosts
Cyprinella spilopetra (Spotfin Shiner)—Culp *et al.*, 2006 (LT).
Erimystax dissimilis (Streamline Chuub)—Culp *et al.*, 2006 (LT).
Lythurus fasciolaris (Scarlet Shiner)—Culp *et al.*, 2006 (LT).
Notropis photogenis (Silver Shiner)—Culp *et al.*, 2006 (LT).

Habitat. The Pyramid Pigtoe is found in stable, muddy sand and cobble in large rivers and their impoundments.

Range. Mississippi River system from Arkansas to Wisconsin, Cumberland and Tennessee rivers. Specimens from west of the Mississippi River probably represent a different species.

Range in Ohio. Living individuals are found only in the lowest reach of the Muskingum River (Ecological Specialists, Inc., 1993), where it is very rare. Historically, it lived in the Ohio River from Marietta to Cincinnati, the Muskingum River to the Tuscarawas River, and the Scioto River in Pickaway County.

Similar species. See under *Pleurobema cordatum.*

Comments. Individuals gradually slow in growth during their entire life. Specimens rarely reach ca. 90 mm and nearly 30 years old.

Pleurobema sintoxia (Rafinesque, 1820)

Figures 154, 155

Etymology. L. *semi-*, half; small, light + Gr. *toxon*, bow, arch. Rafinesque (1820: 310), in characterizing his subgenus *Sintoxia*, noted that both the lateral tooth and ligament are curved. Although true, the widespread occurrence of this character in many unionid species makes it of little value as a species character. *Sintoxia* is a noun in apposition.
Vernacular name(s). *Round Pigtoe*, Flats Pigtoe, Solid Pigtoe.
Original description
Obliquaria (*Sintoxia*) *sintoxia* Rafinesque, 1820: 310, species no. 44, no figure.
Type locality. "Dans l'Ohio [River]." The lectotype is from the Ohio River. Parmalee & Bogan's type locality of "Mahoning River near Pittsburgh, Pennsylvania" was incorrect (1988: 194)—that is the type locality of the synonymous *Unio coccineus* Conrad, 1836.
Type material. The type mentioned by Vanatta (1915: 558) was selected as lectotype (ANSP 20208 from the Ohio River) by Johnson & Baker (1973: 170).
Synonymy
Obliquaria (*Sintoxia*) *sintoxia* Rafinesque, 1820: 310.
Unio rubens Menke, 1828: 90 [Type: Menke's collection was dispersed and lost *fide* Dance (1966: 294), but Zilch (1967: 54) gave SMF 10950 as the lectotype.].
Unio coccineus Conrad, 1836: 29 [Type locality: "Mahoning River near Pittsburgh [PA]." Figured holotype ANSP 2104 lost *fide* Johnson & Baker (1973: 151)].
Unio catillus Conrad, 1836: 30 [Type locality: "Scioto River, Ohio." Holotype ANSP 2103 lost *fide* Johnson & Baker (1973: 150)].
Margarita (*Unio*) *solidus* "Conrad" Lea, 1836: 20 [*nomen nudum*].
Margarita (*Unio*) *coccineus* (Conrad, 1836). Lea, 1836: 34.
Unio solidus Lea, 1838a: 13, pl. 5, fig. 13 [Type locality: "Ohio River, Cincinnati; Mahoning River, Ohio." Types lost *fide* Johnson, 1974].
Unio cuneus Conrad, 1838: No. 11 back cover [Type

Fig. 154. ***Pleurobema sintoxia*** **(Rafinesque, 1820). A.** OSUM 56037, Mississippi River, Houston Co., MN. 67 mm. **B.** OSUM 14165.19, Lake Erie, Ottawa Co., OH. 58 mm. **C.** OSUM 18071, Walhonding River, Coshocton Co., OH. 101 mm. **D.** OSUM 29891.2, Fish Creek, Williams Co., OH. 77 mm. **E, Hinge detail:** OSUM 249.2, Fish Creek, DeKalb Co., IN. 75 mm. **F.** OSUM 249.1, Fish Creek, DeKalb Co., IN. 62 mm. **G.** OSUM 53687, Ohio River, Campbell Co., KY. 70 mm. **H.** OSUM 6896.3, Big Darby Creek, Pickaway Co., OH. 79 mm. **Beak sculpture detail:** OSUM 43245.1, Big Darby Creek, Pickaway Co., OH. 35 mm total length.

locality: "Little Red River, Arkansas." Figured holotype ANSP 41645].
Unio gouldianus "Ward" Jay, 1839: 24 [*nomen nudum*].
Unio cardiacea Guérin-Méneville, 1828. Deshayes, 1839: 19, pl. 31, figs. 1, 2 [misidentification].
Unio fulgidus Lea, 1845: 164 [Type locality: "Alexandria, Louisiana." Lectotype USNM 84649].
Margaron (*Unio*) *solidus* (Conrad, 1836). Lea, 1852b: 25.
Margaron (*Unio*) *coccineus* (Conrad, 1836). Lea, 1852b: 35.
Unio obovalis (Rafinesque, 1820). Conrad, 1853: 253 [misidentification].
Unio catilus [*sic*] Conrad, 1836. Wright, 1888: no pagination.
Quadrula coccinea (Conrad, 1836). Baker, 1898b: 79, pl. 14, fig. 1, pl. 19, fig. 3.
Quadrula coccinea var. *paupercula* Simpson, 1900a: 789 [Type locality: "Niagra R. 3 mi. above Niagra Falls." Lectotype USNM 134834].
Quadrula (*Fusconaia*) *solida* Simpson, 1900a: 789.
Pleurobema missouriensis Marsh, 1901: 74 [Type locality: "Black River near Poplar Bluff, Missouri." Holotype UMMZ 38855].
Quadrula coccinea paupercula (Simpson, 1900). Sterki, 1907: 392.
Pleurobema coccineum (Conrad, 1836). Ortmann, 1912a: 263.
Quadrula (*Fusconaia*) *coccinea* var. *magnalacustris* Simpson, 1914: 884 [unnecessary replacement name for *Quadrula coccinea* var. *paupercula* Simpson, 1900a].
Pleurobema obliquum catillus (Conrad, 1836). Utterback, 1915: 190, pl. 20, figs. 62a, b.
Pleurobema obliquum coccineum (Conrad, 1834). Utterback, 1915: 191, pl. 20, figs. 61a–d.
Pleurobema catillus (Conrad, 1836). Utterback, 1915: 193, pl. 20, figs. 59a, b.
Quadrula cocinea [*sic*] (Conrad, 1836). Walker, 1918: 167.
Pleurobema obliquum coccineum (Conrad, 1836). Ortmann, 1919: 78.
Pleurobema obliquum pauperculum (Simpson, 1900). Ortmann, 1919: 83, pl. 7, fig. 6.
Fusconaja [*sic*] *solida* (Lea, 1838). Baker, 1920: 382.
Pleurobema coccineum pauperculum (Simpson, 1900). Ortmann & Walker, 1922: 24.
Pleurobema cordatum pauperculum (Simpson, 1900). Goodrich, 1932: 90.
Quadrula (*Obliquata*) *coccinea* (Conrad, 1834). Frierson, 1927: 53.
Quadrula (*Obliquata*) *coccinea* var. *cuneus* (Conrad, 1838). Frierson, 1927: 53.
Quadrula (*Obliquata*) *coccinea magnalacustris* (Simpson, 1914). Frierson, 1927: 53.
Quadrula (*Obliquata*) *catillus* (Conrad, 1836). Frierson, 1927: 54.
Quadrula (*Obliquata*) *missouriensis* (Marsh, 1901). Frierson, 1927: 54.
Pleurobema coccineum var. *catillus* (Conrad, 1836). Baker, 1928a: 117, pl. 55, figs. 1–3.
Pleurobema coccineum solida (Lea, 1838). Baker, 1928a: 118, pl. 52, pl. 53, fig. 6.
Pleurobema coccineum var. *mississippiensis* Baker, 1928a: 121, pl. 53, figs. 1–5. [Type locality: "Lake Pepin. Mississippi River." Holotype University of Illinois Museum of Natural History 14019].
Pleurobema catillus var. *coccinea* (Conrad, 1836). Richardson, 1928: 458.
Pleurobema catillus var. *solida* (Lea, 1838). Richardson, 1928: 458.
Pleurobema cordatum var. *coccineum* (Conrad, 1836). Goodrich, 1932: 90.
Pleurobema cordatum paupercula (Simpson, 1900). Goodrich, 1932: 90.
Pleurobema cordatum var. *catillus* (Conrad, 1836). Goodrich, 1932: 90.
Quadrula solida (Lea, 1838). La Rocque, 1953: 98.
Pleurobema cordatum coccineum (Conrad, 1836). Murray & Leonard, 1962: 71.
Pleurobema cordatum catillus (Conrad, 1836). Murray & Leonard, 1962: 74.
Quadrula (*Obliquata*) *coccinea coccinea* (Conrad, 1836). Haas, 1969: 298.
Pleurobema coccineum forma solida (Lea, 1838). Starrett, 1971: 309.
Pleurobema sintoxia (Rafinesque, 1820). Oesch, 1984: 122.
Pleurobema cordatum form *coccineum* (Conrad, 1836). Strayer & Jirka, 1997: pl. 21, figs. 4, 4a.
Pleurobema cordatum form *pauperculum* (Simpson, 1900). Strayer & Jirka, 1997: pl. 21, fig. 5.

Status. Listed as a Species of Concern by the Ohio Division of Wildlife; not listed by U.S. Fish & Wildlife Service as of 2009.

ADULT SHELL

Shell. Medium-sized (to 110 mm), thin to rather thick, triangular to oval, moderately inflated in big river forms, compressed in creek populations.

Anterior margin. Short, straight or broadly rounded to the ventral border.

Ventral margin. Very broadly rounded or straight, rarely with a slight concavity at the posterior ridge.

Posterior margin. Short, gently rounded or straight to the dorsal margin.

Dorsal margin. Long, arched or straight, joining posterior margin with very obtuse angle.

Umbo. Wide to rather narrow, low in creek populations, prominent in river and lake forms; umbos face forward ca. 20°–30°; umbonal sculpture of 2–3 angulate ridges,

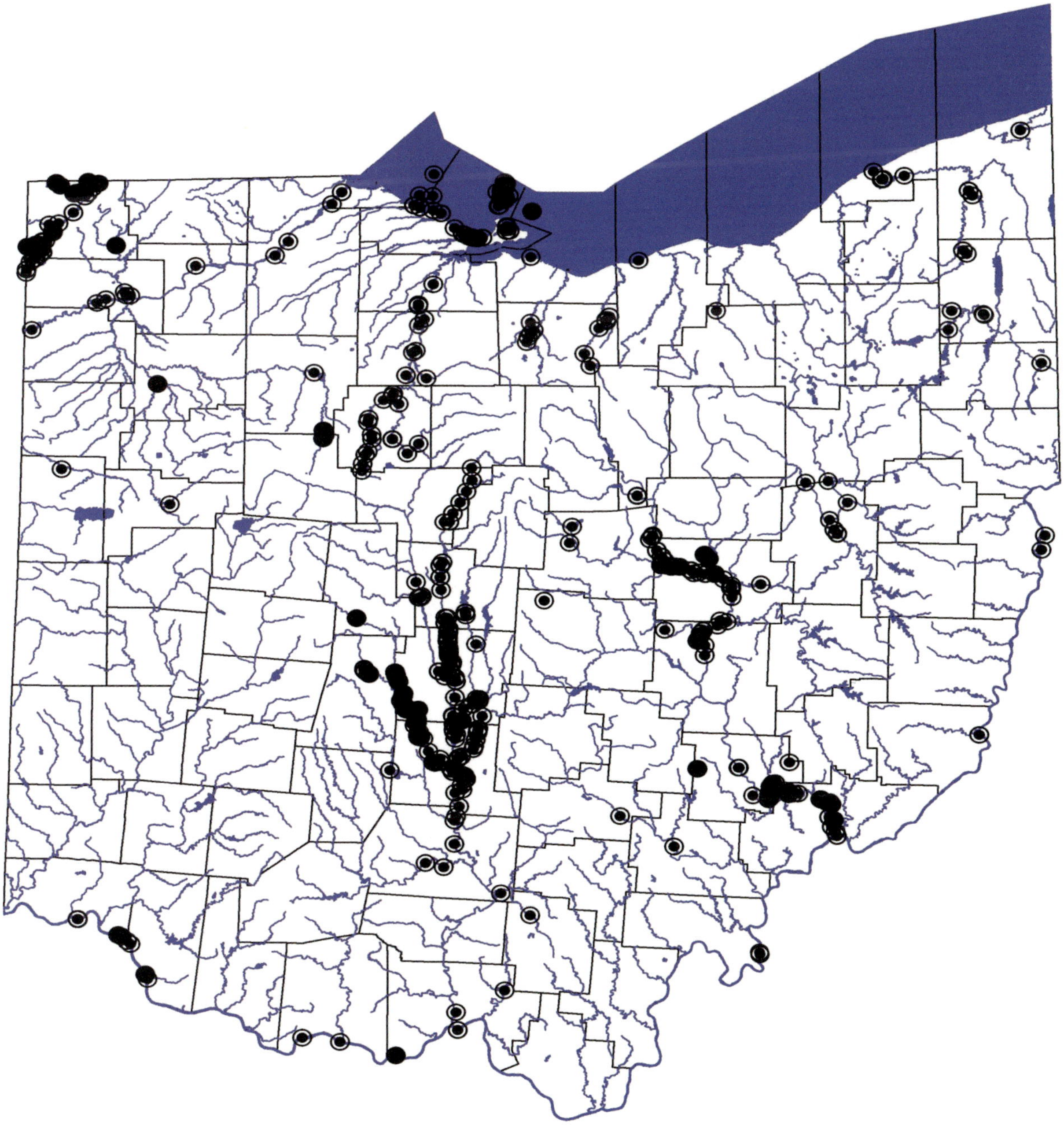

Fig. 155. ***Pleurobema sintoxia*** **(Rafinesque, 1820)**
Solid dots—live or freshdead since 1980
Bullseye—all records, any condition

sometimes approaching double-looped, sometimes prominently developed posteriorly.

Ligament. Short and prominent, tan to dark brown, not concealed by shells.

Disc sculpture. Disc of shell smooth except for growth lines; sulcus absent; posterior slope with 2–3 weak angulations.

Periostracum. Tan or yellowish, becoming darker with age; individuals with rose nacre often have a reddish-brown periostracum; juveniles have vague, washed-out green stripes, but these may be lost in adults.

Nacre. Porcelain white, rose, or salmon, often iridescent posteriorly; occasionally with gold or brown Tulberg layers.

Hinge plate. Well-developed, gently arched; right cardinal a single, small, triangular denticle, rarely with well-developed anterior and posterior accessory teeth; left cardinal bifid, the teeth widely separated; laterals well-developed, short to moderately long, arched or straight, one in right valve, two in left; interdentum flattened, broad or narrow, short.

Muscle scars. Anterior adductor strongly impressed; posterior adductor moderately impressed; anterior adductor distinct from anterior pedal protractor scar, large posterior adductor distinct from small circular posterior pedal retractor scar; a cluster of small dorsal suspensors located behind cardinal teeth and interdentum. Pallial line entire, located far from shell margin. Accessory line rudimentary, often absent, reaching 25% the length of the shell.

Umbonal cavity. Very wide, open to slightly compressed, and shallow to moderately deep.

Glochidia. Semicircular, dorsal margin straight, lateral margins equal, ventral margin rounded, slightly oblique, hooks absent (Ortmann, 1911: 362, pl. 89, fig. 4; Ortmann, 1912a: 263; Surber, 1915: 7, pl. 1, fig. 9).

Literature records (Ortmann, 1912a: 263; Surber, 1915: 7, pl. 1, fig. 9; Utterback, 1915: 192, 194) indicate glochidial dimensions for length (0.15 to 0.16 mm) and height (0.15 to 0.16 mm). Surber (1915: 7) noted, "The glochidium is of the type represented by *ebena*, *plicata*, etc., and very closely resembles *solida* [*P. sintoxia*], with which it agrees in size but differs in the obliquely rounded ventral margin."

Reproductive biology. Tachytictic. The outer gills are marsupial. Eggs appear in May, and glochidia are developed from May through July (Ortmann, 1919; Heath *et al.*, 1998). Ohio specimens had developing glochidia in early June. White, non-elastic, leaf-shaped conglutinates (12 x 8 x 1 mm) are released (Watters, 2008). These have a barred appearance across their width (Watters, unpubl.). Cyprinids are the most commonly reported hosts.

Reported potential hosts

Campostoma anomalum (Central Stoneroller)—Watters *et al.*, 2005 (LT).

Cyprinella spiloptera (Spotfin Shiner)—Hove, 1995b (LT); Hove *et al.*, 1997 (LT).

Lepomis macrochirus (Bluegill)—Surber, 1913 (NI); Coker *et al.*, 1921 (NI); Watters *et al.*, 2005 (LT).

Phoxinus erythrogaster (Southern Redbelly Dace)—Hove, 1995b (LT).

Phoxinus eos (Northern Redbelly Dace)—Hove, 1995b (LT); Hove *et al.*, 1997 (LT).

Phoxinus erythrogaster (Southern Redbelly Dace)—Watters *et al.*, 2005 (LT).

Pimephales notatus (Bluntnose Minnow)—Hove, 1995b (LT); Hove *et al.*, 1997 (LT).

Habitat. A sporadic but widespread species occurring in creeks, rivers, and lakes. It is most commonly found in moving water in sandy cobble. Buchanan (1980: 35) stated that this species "was found in a variety of substrates from silt to gravel, cobble and boulder, but occurred most commonly in a gravel and cobble substrate. It was found in 1 inch to 5 feet of water in standing (0.0 ft./sec. at the bottom) to moderately-flowing (1.5 ft./sec. at the bottom) water."

Range. Baker (1928a: 115) stated that this species was distributed in "the upper Mississippi drainage and St. Lawrence drainage. Its range is apparently bounded by western New York on the east, Iowa and Kansas on the west, Michigan and Wisconsin on the north, and Alabama and Arkansas on the south."

Range in Ohio. This species was once widely distributed within the Ohio River and Lake Erie drainage basins in the state, but curiously absent from the Great and Little Miami rivers. The range of this species has severely declined, with recent records from only a few streams including the St. Joseph River and its tributaries in Williams County, the Olentangy River, Big Darby Creek, Little Darby Creek, Caesar Creek, Walhonding River, the lower Muskingum River (rare), and the western basin of Lake Erie.

Similar species. See under *Pleurobema cordatum*. *Fusconaia flava* (Rafinesque, 1820), particularly headwater specimens, may look very much like this species but generally is more yellow, has a deeper umbo cavity, is usually more triangular, and possesses a sulcus. *Fusconaia subrotunda* has a much deeper and more compressed beak cavity.

Comments. Individuals gradually slow in growth during their entire life; there is great variation in growth rates among populations. Individuals 70 mm in length may be between 8 and 16 years old. Specimens reach ca. 110 mm and nearly 30 years old. Specimens from large rivers usually have more prominent umbos and are more inflated than creek individuals.

Potamilus Rafinesque, 1818

Etymology. Gr. *potamos*, a river or stream. Rafinesque could not have chosen a more appropriate name for his first described unionoid genus.

Original description

Potamilus Rafinesque, 1818: 355.

Type species. *Unio alatus* Say, 1817, by subsequent monotypy. Rafinesque's original description of *Potamilus* (1818: 355) included a number of *nomina nuda* and "*P. alatus*." If this listed species is *U. alatus* Say, 1817, then it is the only recognizable species given by Rafinesque. Despite the fact that ICZN Art. 69.4 clearly states, "Elimination of all but one of the originally included nominal species from a nominal genus or subgenus does not in itself constitute type fixation," because the other names are unavailable *nomina nuda*, *P. alatus* becomes the type by subsequent monotypy (ICZN Art. 69.3).

Synonymy

Potamilus Rafinesque, 1818: 355.

Proptera Rafinesque, 1819: 426 [Type species: *Unio alatus* Say, 1817, by subsequent designation of Herrmannsen, 1846].

Metaptera Rafinesque, 1820: 299 [Invalid emendation of *Proptera*, brought about when Rafinesque realized he had mistakenly confused the anterior and posterior ends of the shells].

Anodonta (*Lastena*) Rafinesque, 1820: 316–317 [Type species: *Anodonta* (*Lastena*) *ohiensis* Rafinesque, 1820, by monotypy. See under *Hemistena*.].

Symphynota Lea, 1829: 424, 442 [Type species: *Unio alatus* Say, 1817, withdrawn as type apparently to avoid conflict with Rafinesque's *Proptera*, which had the same type (and priority). However, no replacement type was designated, nor would such an action have been valid.].

Naidea Swainson, 1840: 378.

Lymnadia Sowerby, 1839: 60.

Lymnadea [*sic*] Swainson, 1840: 379.

Symphinota [*sic*] Villa & Villa, 1841: 43.

Lostena Gray, 1847: 196 [unjustified emendation of *Lastena* Rafinesque, 1820].

Noidea [*sic*] Schaufuss, 1869: 21.

Symphionata [*sic*] Philippi, 1853: 450.

Symphyonota [*sic*] Philippi, 1853: 492.

Lastenes [*sic*] Desmarest, 1859: 212.

Geographic range. Mississippi River system from the Red River of Canada to Louisiana, Ohio, Tennessee, and Cumberland rivers. Gulf drainages from Florida Panhandle to Texas. Rio Grande drainage.

Potamilus alatus (Say, 1817)

Figures 156, 157, 158

Etymology. L. *alatus*, winged; from L. *ala*, wing. Say was obviously impressed with the large postdorsal wing so typical of this species and so well-developed in most Lake Erie specimens. He wrote "hinge margin remarkably oblique, rising . . . into an alated projection" (1817: not paginated). However, some specimens of *P. alatus* typically have two wings; the smaller dorsal wing anterior to the umbo may not have been preserved on Say's specimen, and it is entirely missing in Say's figure (1817: pl. 4, fig. 2).

Vernacular name(s). *Pink Heelsplitter.*

Original description

Unio alatus Say, 1817: no pagination, article on "Conchology," pl. 4, fig. 2.

Type locality. Say, in his original description, noted, "Mr. Lesuer found this species in Lake Erie" but goes on to say, "The specimen from which this description is taken, . . . [was] purchased by Mr. Joseph Watson . . . and presented to the Academy of Natural Sciences of Philadelphia." The type locality was not "Ohio River" as stated by Williams *et al.* (2008).

Type material. The original unionoid type material of Say has apparently been lost (Johnson & Baker, 1973: 148). Haas (1930: 328) designated SMF No. 4349 as the neotype of this species.

Synonymy

Unio alatus Say, 1817: no pagination, article on "Conchology," pl. 4, fig. 2.

Unio alata Say, 1817. Lamarck, 1819: 76.

Megaptera megaptera Rafinesque, 1820: 300, pl. 80, figs. 20–22 [Type locality: "dans l'Ohio [River]." Lectotype ANSP 20211 by designation of Johnson & Baker (1973: 161)].

Mya alata (Say, 1817). Eaton, 1826: 219.

Symphynota alata (Say, 1817). Lea, 1829: 448.

Margarita (*Unio*) *alatus* (Say, 1817). Lea, 1836: 11.

Lymnadia alata (Say, 1817). Swainson, 1840: 265, fig. 48.

Mysca alata (Say, 1817). Swainson, 1841: 28.

Metaptera alata (Say, 1817). Stimpson, 1851: 14.

Margaron (*Unio*) *alatus* (Say, 1817). Lea, 1852b: 19.

Lampsilis alatus (Say, 1817). Baker, 1898b: 97.

Lampsilis (*Proptera*) *alata* (Say, 1817). Simpson, 1900a: 567.

Lampsilis alata (Say, 1817). Scammon, 1906: 299–300.

Proptera alata (Say, 1817). Ortmann, 1912a: 333.

Fig. 156. ***Potamilus alatus*** **(Say, 1817).** **A.** OSUM 27784, Lake Erie, Ottawa Co., OH. 74 mm. **B.** OSUM 53429.1, Ohio River, Wood Co., WV. 112 mm. **C.** OSUM 59154, Hocking River, Athens Co., OH. 150 mm. **D.** OSUM 46703.1, Scioto Brush Creek, Scioto Co., OH. 115 mm. **E, Beak sculpture detail:** OSUM 44293, Mississippi River, Crawford Co., WI. 22 mm. **F, Hinge detail:** OSUM 11698.1, Ohio River, Carroll Co., KY. 90 mm.

Unio (*Symphynota*) *alatus* Say, 1817. De Gregorio, 1914: 37.
Potamilus alatus (Say, 1817). Morrison, 1969: 24.
Status. Not listed by either Ohio Division of Wildlife or U.S. Fish & Wildlife Service as of 2009.

ADULT SHELL

Shell. Large (to at least 190 mm), thin, complanate (females more inflated), subcircular, often with a dorsal wing; narrow pedal and "siphonal" gapes are present.
Anterior margin. Straight in most specimens, forming a small, often inconspicuous wing, then strongly rounded toward the ventral margin.
Ventral margin. In males, very slightly curved; in females, straight or concave to the marsupial region.
Posterior margin. Males broadly rounded, terminating in two angulations at the dorso-ventral margin; females more inflated in marsupial region, which may protrude slightly beyond the ventral margin, posterior margin rising vertically and then forming two angulations.
Dorsal margin. Convex, usually expanded into a wing; this wing is variable in size and shape; it may be half the height of the shell disc; it may be bent or curved to either side; its dorsal margin may be straight, convex, or prominently serrate.
Umbo. Low, wide, pointed; beak sculpture extremely fine, consisting of 2–3 very small "V"-shaped ridges, pointing toward the umbo; beak sculpture disappears very early in life, before the end of the first year.
Ligament. Short, stocky, contained within the wing; portions of the ligament may extend for some distance in the wing as a thin layer.
Disc sculpture. Sculptureless except for 1–3 raised threads radiating from the umbo, delineating the posterior slope.
Periostracum. Small specimens are tan, yellow, or green; larger ones become progressively darker until nearly a uniform dark brown or black; most specimens diffusely rayed, at least as juveniles; usually with a combination of very broad rays and much finer ones; rays are unbroken and persist throughout life, but may be concealed by the dark coloration in older specimens.
Nacre. Purple, less commonly pinkish or copper, either uniformly colored or fading to white around the shell margin; green or gray Tulberg layers uncommon.
Hinge plate. Delicate in small specimens, becoming chunky in old individuals; interdentum short, flattened, bowed, becoming longer in large specimens as the lateral teeth fuse anteriorly; teeth lamellar in small individuals, more robust in older ones; two small cardinal teeth in left valve, one in right with a small accessory tooth anterior; laterals bowed, becoming progressively more so with age; one lateral in the right valve, two in left; the left laterals have a tendency to fuse anteriorly with age.

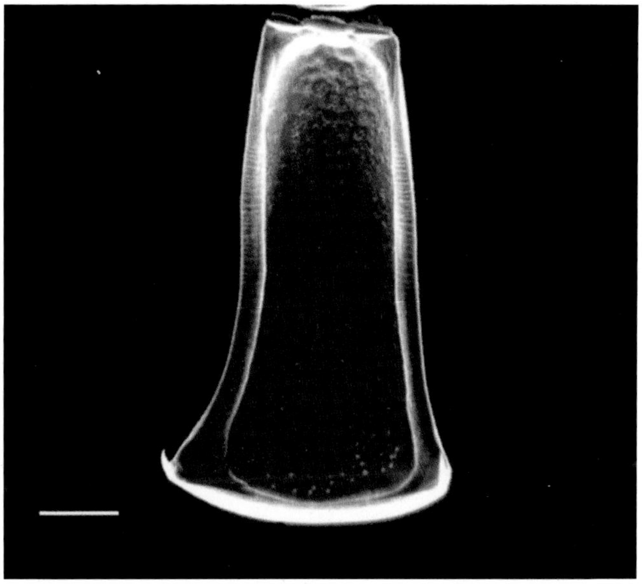

Fig. 157. *Potamilus alatus* (Say, 1817). Glochidium. OSUM 55465, Clinch River, Scott Co., VA. Bar = 50 μm. (Hoggarth, 1999).

Muscle scars. Adductor muscle scars smooth and flush with shell when young; the anterior adductor becomes moderately deep with age; anterior pedal protractor well-separated from anterior adductor; posterior retractor confluent with posterior pedal retractor; numerous suspensors are aligned in an arc from the interdentum ventrally for some distance; pallial line complete, accessory line long, present as a single groove curving ventrally and then posteriorly from the anterior adductor.
Umbonal cavity. Wide, open, generally shallow.

Glochidia. Ligulate, ax-head-shaped, dorsal margin straight, although appearing slightly curved in exterior view due to the deep umbo cavity; anterior and posterior margins equal, parallel-sided dorsally and gently outwardly curved ventrally, ventral margin slightly curved. Dorsal alae small, lateral valve gape extreme; exterior surface is covered with vermiculate sculpture. Lanceolate hooks present at the lateral margins of the ventral flange, micropoints lanceolate and arranged in complete vertical rows (Lea, 1858c: 48, pl. 5, fig. 25; Ortmann, 1911: 347, pl. 89, fig. 18; Coker & Surber, 1911: 179, pl. 1, fig. 3; Hoggarth, 1999: figs. 50a–f).

Literature records (Lefevre & Curtis, 1910a: 97, fig. d; Lefevre & Curtis, 1912: 146, fig. 1d; Surber, 1912: 9, pl. 1, fig. 8; Ortmann, 1912a: 333; Utterback, 1916a: 390; Hoggarth, 1999: figs. 50a–f) indicate glochidial dimensions for length (0.21 to 0.23 mm), height (0.37 to 0.41 mm), and hinge (0.10 to 0.11 mm). This glochidium can be distinguished from that of other species by its size, shape, and micropoint structure. As noted by Ortmann (1912a), the glochidium of *P. capax* is similar to that of *P. alatus*, but it

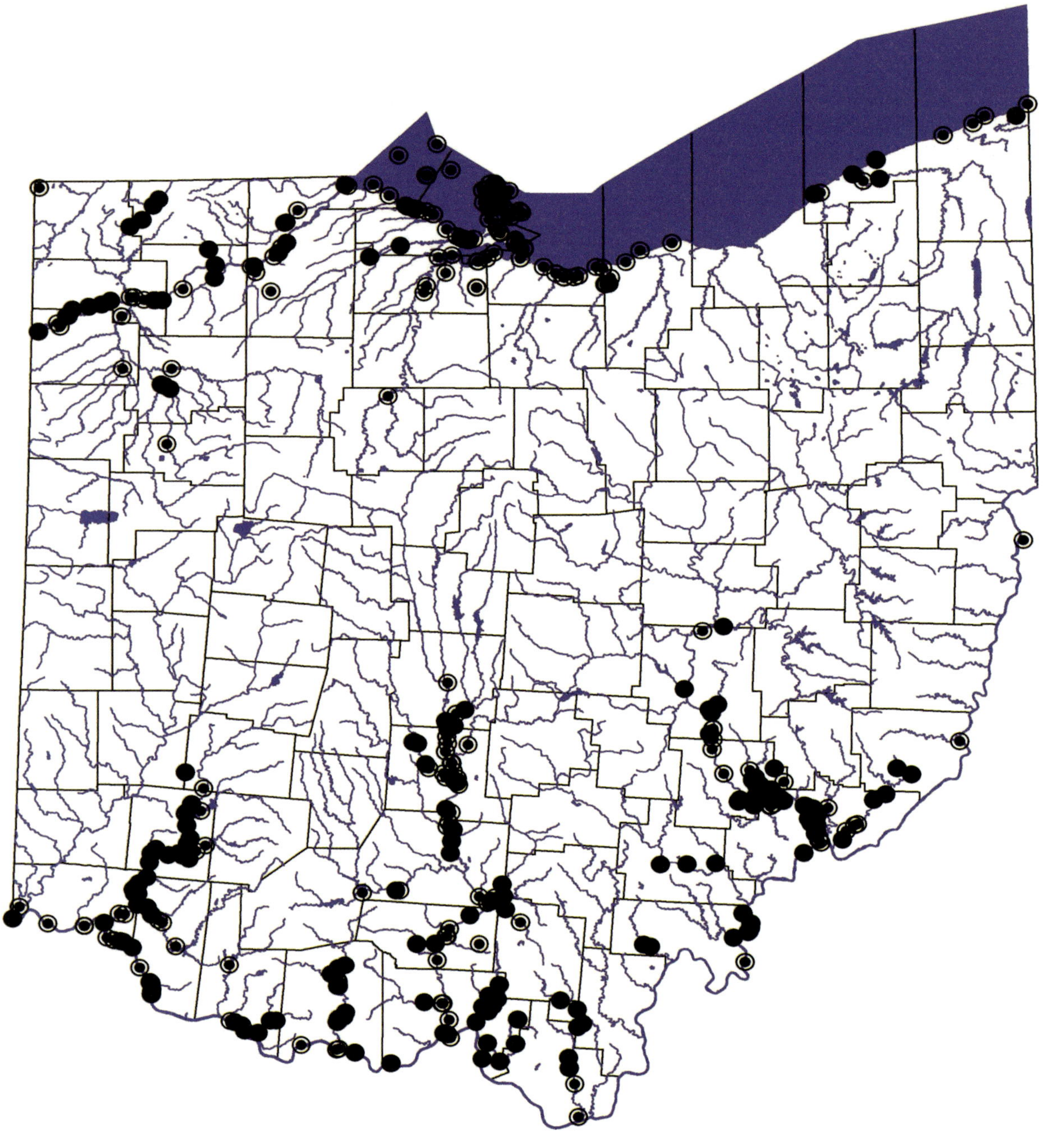

Fig. 158. ***Potamilus alatus*** **(Say, 1817)**
Solid dots—live or freshdead since 1980
Bullseye—all records, any condition

is much smaller, approaching the dimensions of *P. ohiensis* instead of *P. alatus*.

Reproductive biology. Bradytictic. The posterior part of the outer gills is marsupial (Ortmann, 1912a). Ortmann (1919) believed that this species bred year round, with overlapping broods. He found gravid females from June to October, and again from May to July. Holland-Bartels & Kammer (1989) also found gravid females from May to July. Glochidia are released in large conglutinates (Utterback, 1916a) that disassociate (Watters, 2008). Although Freshwater Drum is often reported as the only host, metamorphosis has not been observed on any host.

Reported potential hosts

Aplodinotus grunniens (Freshwater Drum)—Howard, 1913 (NI); Wilson, 1916 (NI); Howard & Anson, 1923 (NS); Cummings & Mayer, 1993 (NI); Weiss & Layzer (NI); Brady *et al.*, 2004 (LT).

Habitat. The Pink Heelsplitter is widespread in lakes, impoundments, and rivers, rarely in smaller systems, often found in slackwater in silty sand and mud.

Range. Mississippi River system from the Red River of Canada to Arkansas and northern-most Alabama. Great Lakes drainages and upper St. Lawrence River.

Range in Ohio. Widespread in most large creeks and rivers. Most Ohio River tributaries of any appreciable size. Lake Erie, and Maumee and Sandusky rivers, occasionally in smaller Lake Erie tributaries.

Similar species. The large, compressed, alate shell and purple nacre readily identify this species, but it may resemble several others, particularly *Potamilus ohiensis*. The umbos of *P. ohiensis* are usually lighter colored than the remainder of the shell—this is not the case for *P. alatus*. The shell of *P. alatus* has a matte periostracum, whereas the periostracum of *P. ohiensis* is often quite polished. *Potamilus ohiensis* tends towards having alae both before and after the umbo, but *P. alatus* rarely has anterior alae. The nacre of *P. alatus* is a solid deep purple to pink. The nacre of *P. ohiensis* is a metallic pink or purple, not as deeply colored as in *P. alatus*. *Leptodea fragilis* differs in having a yellow or tan shell, with a silvery nacre flushed with pink or purple in the umbonal cavity. *Lasmigona complanata* has a corrugated posterior wing, heavy hinge, and white nacre.

Comments. There is great variation in the strength of the posterior dorsal wing. Individuals living in high-energy environments often have the wing reduced (by abrasion?) to the point where it is barely perceptible. Those living in low-energy backwater or lacustrine situations may produce very extensive wings.

Individuals grow rapidly for 2–4 years, when they abruptly slow their growth. Female shell characteristics are apparent in the second year. There is great variability in growth rates: individuals 80 mm long may be from 2 to 6 years old. Very large individuals may reach over 190 mm in length and be 15 years old.

Potamilus ohiensis (Rafinesque, 1820)

Figures 159, 160, 161

Etymology. Ohio [River] + L. *-ensis*, of that place. In his original description of this species, Rafinesque (1820: 317) noted that it was "Très-commune dans l'Ohio et toutes les rivières adjacentes."

Vernacular name(s). *Pink Papershell*, Fragile Heelsplitter.

Original description

Anodonta (*Lastena*) *ohiensis* Rafinesque, 1820: 316–317, species no. 58, no figure. Ortmann & Walker (1922: 37) argued that this species and the genus *Lastena*, which is based on it, cannot be recognized from the description.

Type locality. "Très-commune dans l'Ohio et toutes les rivieres adjacentes" ["Very common in the Ohio [River] and all of the adjacent rivers"].

Type material. Rafinesque's material of this species has not, with certainty, been located (Johnson & Baker, 1973: 164). A replacement type has not been selected.

Synonymy

Anodonta (*Lastena*) *ohiensis* Rafinesque, 1820: 316–317.
Symphynota laevissima Lea, 1829: 444, pl. 13, fig. 23 [Type locality: "Ohio"; lectotype label reads "Cincinnati." Lectotype USNM 83938.].
Unio ohioensis [*sic*] (Rafinesque, 1820). Say, 1834: no pagination.
Unio laevissimus (Lea, 1829). Say, 1834: no pagination.
Unio laevissima (Lea, 1829). Deshayes, 1835: 558.
Margarita (*Unio*) *laevissimus* (Lea, 1829). Lea, 1836: 11.
Symphynota ohioensis [*sic*] (Rafinesque, 1820). Férussac, 1835: 25.
Metaptera ohioensis [*sic*] (Rafinesque, 1820). Agassiz, 1852: 46.
Margaron (*Unio*) *laevissimus* (Lea, 1829). Lea, 1852b: 19.
Metaptera ohiensis (Rafinesque, 1820). Conrad, 1853: 260.
Unio discoideus (Lea, 1834). Sowerby, 1866: pl. 53, fig. 275 [misidentification].
Lampsilis (*Proptera*) *laevissima* (Lea, 1829). Simpson, 1900a: 574.
Proptera laevissima (Lea, 1829). Ortmann, 1912a: 334.
Leptodea laevissima (Lea, 1829). Goodrich & van der Schalie, 1944: 316.
Potamilus ohiensis (Rafinesque, 1820). Morrison, 1969: 24.
Leptodea fragilis (Rafinesque, 1820). Strayer & Jirka, 1997: pl. 11 [misidentification].

Fig. 159. ***Potamilus ohiensis*** **(Rafinesque, 1820).** **A.** OSUM 55993, Muskingum River, Muskingum Co., OH. 136 mm. **B, Hinge detail:** OSUM 53135.14, Mississippi River, Allamakee Co., IA. 77 mm. **C.** OSUM 31889, Scioto River, Pickaway Co., OH. 88 mm. **D, F.** OSUM 65.4, Scioto River, Pike Co., OH. 125 mm. **E.** OSUM 28994.8, Meramec River, Jefferson Co., MO. 19 mm. **Beak sculpture detail:** 28994.3, Meramec River, Jefferson Co., MO. 19 mm total length.

Lastena ohiensis (Rafinesque, 1820). Hoggarth, 1999: 81.
Status. Not listed by either Ohio Division of Wildlife or U.S. Fish & Wildlife Service as of 2009.

ADULT SHELL

Shell. Medium size, very thin, oval to elongate, often with a prominent posterior wing and a smaller anterior one; usually compressed.
Anterior margin. Very short (to 170 mm), straight to broadly rounded to the ventral border; in some specimens the margin forms an anterior wing.
Ventral margin. Very broadly rounded.
Posterior margin. Long, rather straight to the dorsal margin.
Dorsal margin. Long, straight, joining posterior margin in a wing.
Umbo. Narrow, very low, placed forward of midline; umbos face slightly forward; umbonal sculpture a weak microscopic ridge.
Ligament. Long, tan to dark brown, completely concealed by posterior wing.
Disc sculpture. Disc of shell smooth except for low angulations on posterior slope.
Periostracum. Shiny even as adults; juvenile shell grayish-green, very rarely with vague rays, adult shell darker brown or greenish, annular rings dark.
Nacre. Uniformly iridescent pale pink or copper; rarely with Tulberg layers.
Hinge plate. Well-developed but narrow and delicate; right cardinal a single, lamellar, triangular denticle, often with small anterior accessory tooth; left cardinal with single poorly developed triangular tooth; laterals well-developed, thin, long, curved, one in right valve, two in left; interdentum very narrow, long.
Muscle scars. Anterior adductor large, oval, flush with shell; posterior adductor large, flush; large anterior adductor distinct from anterior pedal protractor scar, small posterior adductor confluent with small posterior pedal retractor scar; large dorsal suspensors located under umbo in a diagonal line. Pallial line entire, located far from shell margin. Accessory line absent or weakly impressed.
Umbonal cavity. Very wide, shallow, open.

Glochidia. Ligulate, ax-head-shaped, parallel-sided dorsally, becoming very broad ventrally, dorsal margin straight and very short, anterior and posterior margins equal; straight-sided dorsally and outwardly crescent-shaped ventrally, ventral margin rounded. Dorsal alae small, lateral valve gape extreme; loose-looped exterior sculpture covers exterior surface of valve. Hooks absent, micropoints lamellate and arranged in broken vertical rows (Ortmann, 1912a: 334; Arey, 1924b: 334, pl. 1, fig. 3; Hoggarth, 1999: figs. 46a–e).

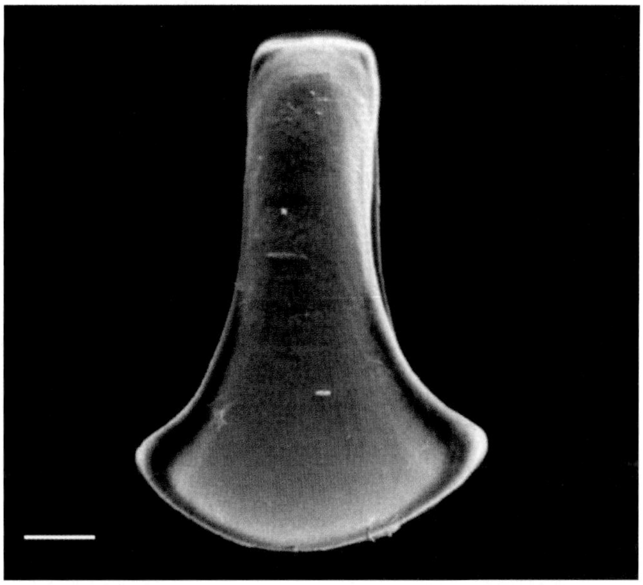

Fig. 160. *Potamilus ohiensis* (Rafinesque, 1820). Glochidium. OSUM 54520.1, Big Blue River, Gage Co., NE. Bar = 30 µm. (Hoggarth, 1999).

Literature records (Lea, 1858c: 48, pl. 5, fig. 24; Coker & Surber, 1911: 179, pl. 1, figs. 1, 1a; Surber, 1912: 9, pl. 1, fig. 10; Ortmann, 1912a: 334; Utterback, 1916a: 393; Hoggarth, 1999: figs. 46a–e) indicate glochidial dimensions for length (0.10 to 0.13 mm), height (0.16 to 0.19 mm), and hinge (0.04 to 0.05 mm). Ortmann (1912a: 334) noted, "The glochidia are of the *Proptera*-type [*Potamilus*], but they differ distinctly from those of *P. alata* in being considerably smaller, and in having the ventral margin broader and more curved." Lea (1858, 1863) described this glochidium as wedge-shaped with a hook-like process at each corner of the ventral margin. However, Coker & Surber (1911: 179) reported that the spines or hooks were absent, and Surber (1912: 9) reported, "Glochidium without spines (?)." This glochidium can be distinguished from that of most other members of the genus *Potamilus* by the absence of spines. They are further distinguished by micropoint structure and exterior valve sculpture and shape, especially the roundness of the ventral margin. This glochidium shares these characters with *Leptodea fragilis*, but can be distinguished from that species by its ax-head shape.

Reproductive biology. Bradytictic, although Surber (1913) found gravid females year round. The posterior part of the outer gills is marsupial (Utterback, 1916). Glochidia are discharged in fragile, white conglutinates that quickly break apart.

Reported potential hosts

Aplodinotus grunniens (Freshwater Drum)—Coker &

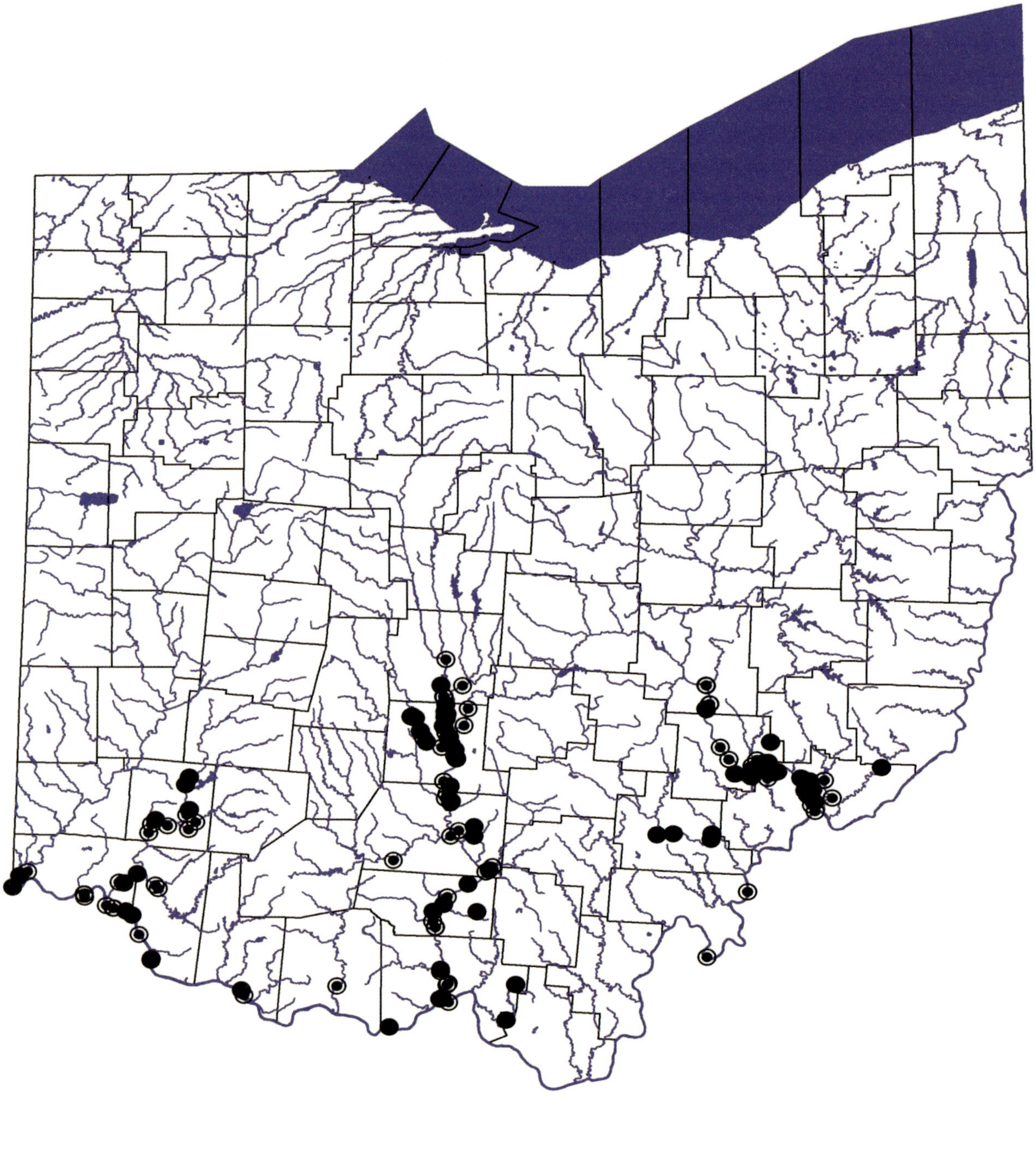

Fig. 161. ***Potamilus ohiensis* (Rafinesque, 1820)**
Solid dots—live or freshdead since 1980
Bullseye—all records, any condition

Surber, 1911 (NI); Howard, 1913 (NI); Surber, 1913 (NI); Wilson, 1916 (NI); Howard & Anson, 1923 (NI); Cummings & Mayer, 1993 (NI); Brady *et al.*, 2004 (LT).
Pomoxis annularis (White Crappie)—Wilson, 1916 (NI).

Habitat. This species is widespread in lakes, impoundments, and rivers, rarely in smaller systems. It is often found in slackwater in silty sand. Like other species of *Potamilus*, it may invade and prosper in impoundments.

Range. Mississippi River system from Minnesota to Louisiana, into Missouri, Kansas, and Colorado. Absent from the Great Lakes drainages.

Range in Ohio. Most of the larger Ohio River tributaries: Great Miami, Scioto, and Muskingum rivers. The dredging of the Hocking River rendered it a suitable habitat for this once-rare species, and it is now common there. Rarely straying into smaller creeks. This species does not occur in Lake Erie or its tributaries.

Similar species. See under *Potamilus alatus*.

Comments. A large fast-growing species, attaining 170 mm in length but rarely living more than 7 years.

Ptychobranchus Simpson, 1900

Etymology. Gr. *ptychodes*, in folds or layers + Gr. *branchion*, gill. Simpson (1900a: 612) noted that "embryos are contained in distinct ovisacs with rounded bases, occupying the entire outer gills, which, when gravid, consist of a series of folds."

Original description

Ptychobranchus Simpson, 1900a: 79.

Type species. *Unio phaseolus* Hildreth, 1828 [= *Obliquaria fasciolaris* Rafinesque, 1820] by original designation.

Synonymy

Truncilla (*Plagiola*) Rafinesque, 1820: 302 [Bogan (1998) has shown that *Obliquaria interrupta* Rafinesque, 1820, the type species of *Plagiola*, is actually the species now known as *Ptychobranchus fasciolaris* (Rafin–esque, 1820). This makes *Plagiola* the senior synonym of *Ptychobranchus*. Williams *et al.* (2008) considered *Plagiola* a *nomen oblitum*. We agree with this proposal and have maintained *Ptychobranchus* in our account.].

Ptychobranchus Simpson, 1900a: 79 [Type species: *Unio phaseolus* Hildreth, 1828, by original designation].

Ptychobranchus (*Subtentus*) Frierson, 1927: 10, 65 [Type species: *Unio subtentus* Say, 1825, by original designation].

Geographic range. Mississippi River system, including the Ohio, Cumberland, and Tennessee rivers. West of the Mississippi River in Missouri, Arkansas, and Oklahoma. Mobile River system. Great Lakes.

Ptychobranchus fasciolaris (Rafinesque, 1820)

Figures 162, 163, 164, 165, 166, 167

Etymology. L. *fascia*, a band, bandage, zone + L. *-ola* (diminutive) + L. *-aris*, pertaining to. Rafinesque described the striking rays of this species as "oblique brown bands" (1820: 303), although they are more often green.

Vernacular name(s). *Kidneyshell.*

Original description

Obliquaria (*Ellipsaria*) *fasciolaris* Rafinesque, 1820: 303–304, species no. 25, no figure.

Type locality. "Assez commune dans l'Ohio, le Wabash, Kentuky, etc." ["Rather common in the Ohio, Wabash, Kentucky (Rivers), etc."]. The lectotype is from the Kentucky River, which becomes the type locality. Parmalee & Bogan's (1998: 201) type locality of the "Muskingum River" was in error.

Type material. The type mentioned by Vanatta (1915: 554) was selected as lectotype ANSP 20253, from the Kentucky River, by Johnson & Baker (1973: 155).

Synonymy

Potamilus fasciolaris Rafinesque, 1818: 355 [*nomen nudum*].

Plagiola fasciolaris Rafinesque, 1819: 426 [*nomen nudum*].

Obliquaria (*Plagiola*) *interrupta* Rafinesque, 1820: 302 [Type locality: "Dans le Kentuky et Ohio [Rivers]." "Lectotype" ANSP 20257 is a substituted specimen of *Epioblasma brevidens* (Lea, 1831) and not from the type lot of this species; a true paralectotype is in MNHN, *fide* Bogan, 1998.].

Obliquaria (*Ellipsaria*) *fasciolaris* Rafinesque, 1820: 303–304.

Obliquaria (Ellipsaria) fasciolaris var. *interrupta* Rafinesque, 1820: 303 [*non* Rafinesque, 1820: 302. Type locality: none given. Type not located].

Obliquaria (*Ellipsaria*) *fasciolaris* var. *fuscata* Rafinesque, 1820: 304 [Type locality: none given. Type not located].

Obliquaria (*Ellipsaria*) *fasciolaris* var. *obliterata* Rafinesque, 1820: 304 [Type locality: none given. Type not located].

Obliquaria (*Ellipsaria*) *fasciolaris* var. *longa* Rafinesque, 1820: 304 [Type locality: none given. Type not located].

Unio phaseolus Hildreth, 1828: 283 [Type locality: "Muskingum [River]." Type not located].

Unio planulatus Lea, 1829: 431, pl. 9, fig. 13 [Type locality: "Cincinnati, Ohio." Holotype USNM 86059].

Fig. 162. ***Ptychobranchus fasciolaris*** **(Rafinesque, 1820). A.** OSUM 6114, Tennessee River, Lauderdale Co., AL. 130 mm. **B.** OSUM 60871, Walhonding River, Coshocton Co., OH. 130 mm. **C, H, Hinge detail:** OSUM 10911, St. Joseph River, Williams Co., OH. 105 mm. **D.** OSUM 3792.7, Lake Erie, Ottawa Co., OH. 57 mm. **E.** OSUM 4204, Olentangy River, Franklin Co., OH. 83 mm. **F.** OSUM 6272.1, Muskingum River, Muskingum Co., OH. 75 mm. **G.** OSUM 14306.2, Olentangy River, Franklin Co., OH. 104 mm. **Beak sculpture detail:** OSUM 5692.1, Deer Creek, Fayette Co., OH. 36 mm total length.

Unio fasciolaris (Rafinesque, 1820). Conrad, 1834: 69.
Unio camelus Lea, 1834: 102, pl. 15, fig. 45 [Type locality: "Ohio River, Cincinnati." Holotype USNM 84486].
Margarita (*Unio*) *camelus* (Lea, 1834). Lea, 1836: 19.
Margarita (*Unio*) *phaseolus* (Hildreth, 1828). Lea, 1836: 38.
Unio planulata Lea, 1829. Deshayes, 1839: 672.
Unio compressissimus Lea, 1845: 163 [Type locality: "Holston River, Tennessee." Lectotype USNM 84474].
Margaron (*Unio*) *camelus* (Lea, 1834). Lea, 1852b: 24.
Margaron (*Unio*) *compressissimus* (Lea, 1845). Lea, 1852b: 24.
Margaron (*Unio*) *phaseolus* (Hildreth, 1828). Lea, 1852b: 38.
? *Unio arquatus* Conrad, 1854: 297, pl. 26, fig. 8 [Type locality: "Wabash River." Figured type lost *fide* Johnson & Baker, 1973. Possibly = *Ligumia recta*].
Ptychobranchus phaseolus (Hildreth, 1828). Simpson, 1900a: 612.
Ellipsaria fasciolaris (Rafinesque, 1820). Frierson, 1914a: 7.
Ptychobranchus phaseolus var. *camelus* (Lea, 1834). Simpson, 1914: 335.
Unio imperitus De Gregorio, 1914: 45–46, pl. 9, fig. 1 [Type locality: "Sciota [Scioto] R. (Ohio)." Type in Palermo Museum?]
Unio lanceolatus var. *blandus* De Gregorio, 1914: 52, pl. 8, fig. 2 [Type locality: "Sciota" [Scioto River, OH]. Type in Palermo Museum?].
Unio compressissimus var. *performosus* De Gregorio, 1914: 53–54 [Type locality: "Sciota" [Scioto River, OH]. Type in Palermo Museum?].
Ptychobranchus fasciolare (Rafinesque, 1820). Ortmann & Walker, 1922: 42.
Ptychobranchus (*Ptychobranchus*) *fasciolare* (Rafinesque, 1820). Frierson, 1927: 64.
Ptychobranchus (*Ptychobranchus*) *fasciolare longum* (Rafinesque, 1820). Frierson, 1927: 64.
Ptychobranchus (*Ptychobranchus*) *fasciolare arquatum* (Conrad, 1854). Frierson, 1927: 64.
Ptychobranchus fasciculare [*sic*] (Rafinesque, 1820). Baker, 1928a: 486.
Ptychobranchus fasciolaris lacustris Baker, 1928b: 52 [Type locality: "Assembly grounds, Chautauqua Lake, New York." Holotype University of Illinois Museum of Natural History, Urbana, Z-23799].
Ptychobranchus fasciolaris (Rafinesque, 1820). Goodrich & van der Schalie, 1939: 319.

Status. Listed as a Species of Concern by the Ohio Division of Wildlife; not listed by U.S. Fish & Wildlife Service as of 2009.

ADULT SHELL

Shell. Medium- to large-sized (to 130 mm), heavy to massive, outline elongated.
Anterior margin. Straight, then broadly curved.

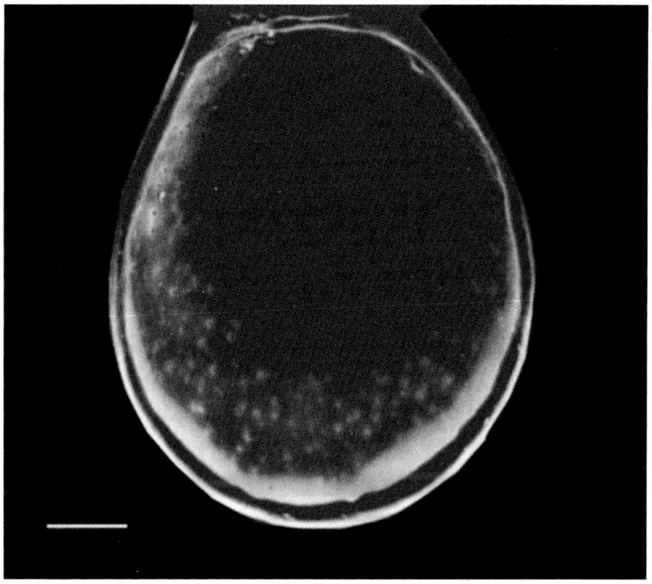

Fig. 163. *Ptychobranchus fasciolaris* (Rafinesque, 1820). Glochidium. MAH 640.1, Little Darby Creek, Madison Co., OH. Bar = 25 µm. (Hoggarth, 1999).

Ventral margin. Very broadly curved or straight, slightly concave in females; forming a blunt point at the posterior margin at the base.
Posterior margin. Gently rounded, nearly indistinguishable from the dorsal slope.
Dorsal margin. Very long, straight or slightly convex.
Umbo. Very wide, low, barely rising above the hinge line; beak sculpture nearly obsolete, of 2–4 "V"-shaped or double-looped ridges, pointing towards the umbo; occasionally only the distal arms of the "V" are apparent; beak sculpture disappears early in the first year's growth.
Ligament. Tan to brown, very long, largely concealed by the valves.
Disc sculpture. Shell sculptureless except for 2–4 low, broad ribs on the dorsal slope, divided by shallow grooves; this sculpture may be nearly lost in some specimens.
Periostracum. Pale yellowish-green to tan, becoming somewhat darker with age; shells have a characteristic pattern of broad, dark green, widely separated, broken rays (rarely continuous), at least on the juvenile shell; these rays may have a "smeared" appearance, or form a distinct checkerboard pattern aligned concentrically; on close examination, the broad rays will be seen to be composed of numerous finer rays smeared together. The dorsal slope is usually devoid of these broad rays, being either patternless, or with several very fine rays; we have not seen any completely unrayed specimens.
Nacre. White, iridescent posteriorly; often with gold or green Tulberg layers.
Hinge plate. Short, distinctly angled, massive; interdentum very broad, arched, lamellar over dorsal suspensor scar; cardinal teeth chunky, triangular, two in left valve,

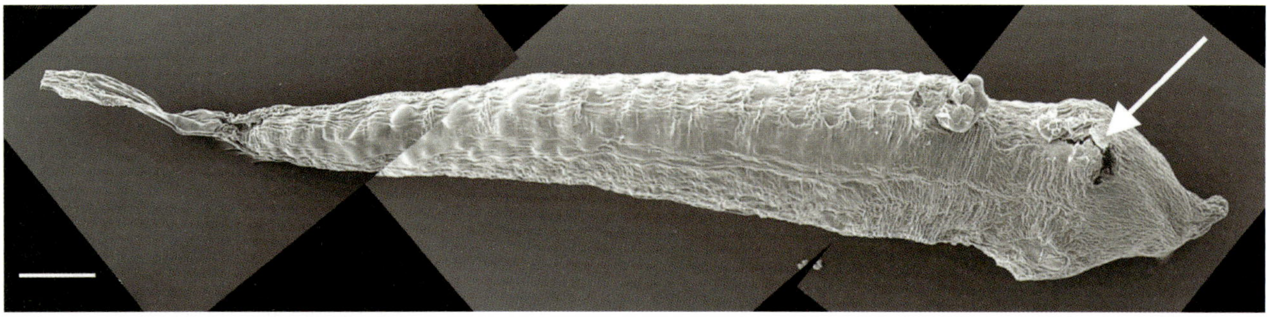

Fig. 164. ***Ptychobranchus fasciolaris*** **(Rafinesque, 1820).** Scanning electron micrograph of metaconglutinate. "Tail" at left is adhesive. "Head" at right contains "eye-spots" (arrow) through which glochidia are released under pressure. Bar = 500 µm. (Watters, 2008).

Fig. 165. ***Ptychobranchus fasciolaris*** **(Rafinesque, 1820).** Metaconglutinates mimicking fish fry.

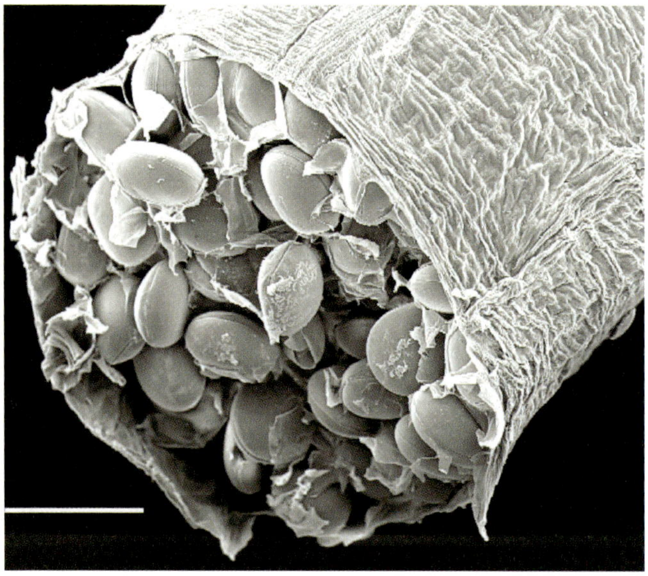

Fig. 166. ***Ptychobranchus fasciolaris*** **(Rafinesque, 1820).** Cross-section through metaconglutinate showing central core of glochidia. (Watters, 2008).

one in right with an accessory tooth anterior to cardinal; lateral teeth short, straight or slightly bowed, very thick, two in left, one in right.

Muscle scars. Both adductor muscle scars impressed, both distinct from protractor and retractor scars; dorsal suspensor scars usually fused into a single elongated deep scar under the interdentum; pallial line entire, accessory line short, extending to the shell margin.

Umbonal cavity. Wide, open, shallow, nearly nonexistent. Radiating from the umbonal cavity to a point just below the posterior adductor is a distinct groove. Below this, obliquely radiating across the center of the shell, is a hollowed-out area that corresponds to the marsupia of the females.

Glochidia. Subelliptical, dorsal margin slightly curved, anterior and posterior margins equally curved, ventral margin broadly rounded. Dorsal alae absent, and loose-looped sculpturing covers the exterior surface of the valve; hooks absent (Ortmann, 1911: 346, pl. 89, fig. 14; Hoggarth, 1999: figs. 35a–e).

Literature records (Ortmann, 1912a: 307; Hoggarth, 1999: figs. 35a–e) indicate glochidial dimensions for length (0.17 to 0.18 mm), height (0.18 to 0.20 mm), and hinge (0.08 to 0.09 mm). Ortmann (1912a: 307) stated, "The poor quality of Lea's figures [1858: 47, pl. 5, fig. 12] of the glochidia is clearly shown in this instance. The figure of the glochidium of *phaseolus* [*P. fasciolaris*] (fig. 12) stands next to that of *Eurynia recta* [*Ligumia recta*] (fig. 11), and is distinctly larger than the latter, while actually the glochidium of *E. recta* is by far the larger of the two." This glochidium can be distinguished by its small size, central ligament position, and very simple ventral margin.

Reproductive biology. Bradytictic. The outer gills act as marsupia and consist of peculiar curtain-like folds. Conglutinates appear first in the anterior-most end of the gill and eventually mature posteriorly (Watters, unpubl.). Eggs appear in August, with glochidia developing by September and overwintering till the following year (Ortmann, 1919). Ortmann (1919) reported that glochidia were released

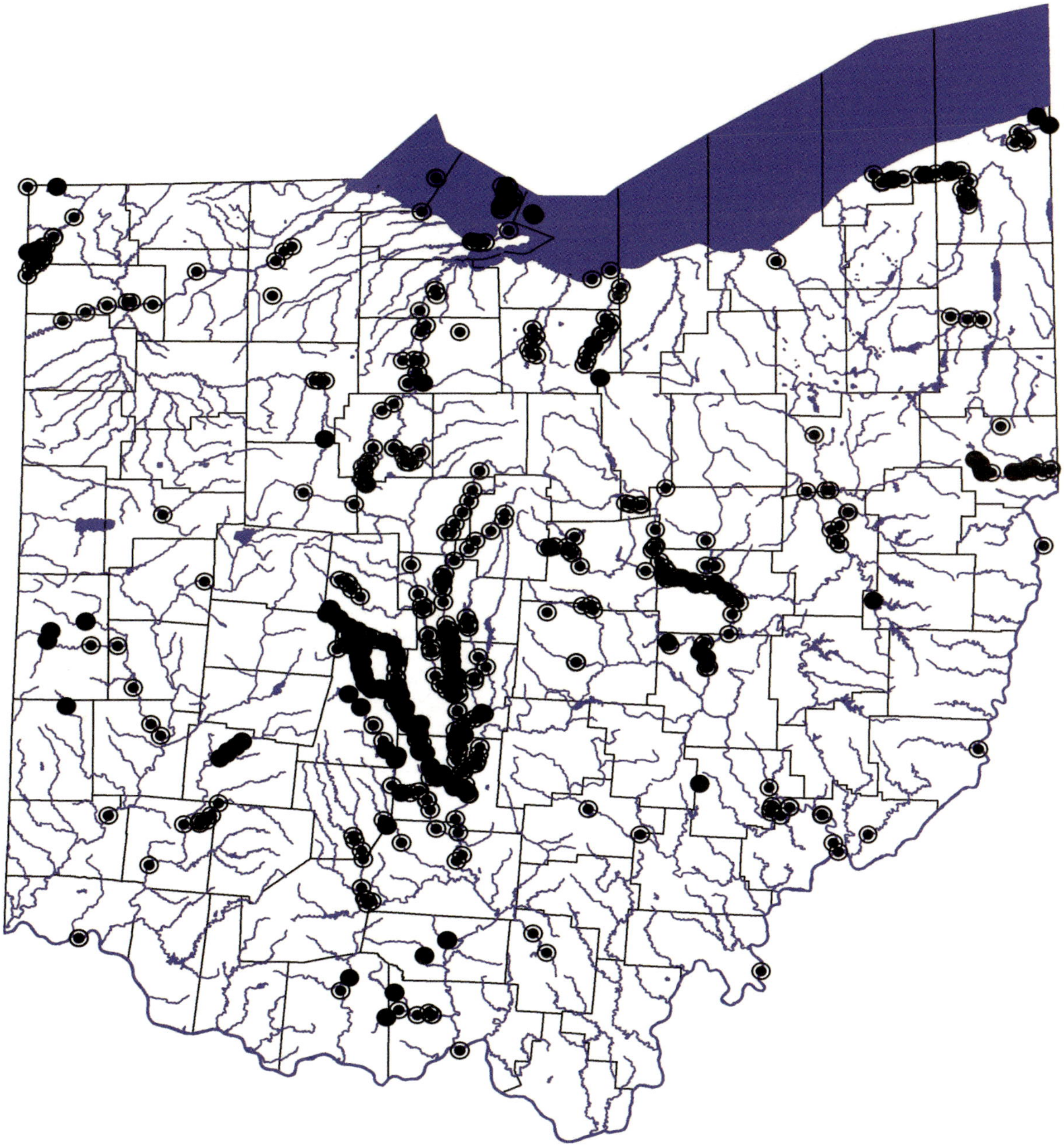

Fig. 167. ***Ptychobranchus fasciolaris*** **(Rafinesque, 1820)**
Solid dots—live or freshdead since 1980
Bullseye—all records, any condition

from June to August, but in Ohio this species releases in April and May (Watters, unpubl.). Two types of conglutinates have been seen in the same population, perhaps associated with age. Both are torpedo-shaped packages containing several thousand glochidia. One type has a bright red apical spot with poorly defined "eyes" and apparently mimics insect larvae. The second type has no red spot but distinct "eyes" and mimics fish larvae. In both types the "tail" is adhesive. This conglutinate was described in detail by Watters (1999a) and termed a metaconglutinate (Watters, 2008). Conglutinates are discharged individually through the distal edge of the gill, being "shot" out of the excurrent siphon, and may travel 6 inches or more in calm water. When released, the conglutinates stick to rocks, twigs, etc., and wiggle in the current. These conglutinates are avidly consumed by fishes, resulting in glochidial attachment to gill filaments (Watters, unpubl.).

Reported potential hosts

Culaea inconstans (Brook Stickleback)—Watters *et al.*, 2005 (LT).
Etheostoma caeruleum (Rainbow Darter)—Watters *et al.*, 2009 (LT).
Etheostoma flabellare (Fantail Darter)—Watters *et al.*, 2009 (LT).

White *et al.* (1996) suggested the following hosts based on the genetics of glochidia on naturally infected fish:

Etheostoma blennioides (Greenside Darter)
Etheostoma flabellare (Fantail Darter)
Etheostoma nigrum (Johnny Darter)
Etheostoma zonale (Banded Darter)

Habitat. A high-water-quality species occurring from creeks to rivers, and in Lake Erie. It is usually associated with sand and cobble in moving water.

Range. Ohio River system, including Cumberland and Tennessee rivers, and Great Lakes drainages.

Range in Ohio. Widespread and sporadic, but locally common. Western basin of Lake Erie. Abundant in Fish Creek and Big and Little Darby creeks.

Similar species. The combination of elongate shell, yellow color with broken rays and lines, white nacre, and short massive hinge easily identifies this species.

Comments. This species is slow growing, but reaches a large size and lives many years. Growth is accelerated for the first 3 years, but not as dramatically as in other species. Sexual dimorphism in the shell is not at all pronounced and is most noticeable from the marsupial groove on the inside of the female shell. Individuals may exceed 130 mm in length and commonly grow to over 30 years old.

Pyganodon Fischer & Crosse, 1893

Etymology. From G. *pyge, pyganos*, rump, buttocks, nates + G. *an*, not, without + G. *odontos*, tooth. Shells of this genus commonly have swollen umbos, resembling nates or buttocks, while lacking hinge teeth in the same (dorsal) region of the shell. Literally "buttocks without teeth."

Original description

Anodonta (*Pyganodon*) Fischer & Crosse, 1893: 518.

Type species. Five species were listed in the original description of the genus *Pyganodon*. The first listed, *Anodonta globosa* Lea, 1841, was selected as the type species by Frierson (1927: 9, *errata & corrigenda*). All five species are Mexican.

Synonymy

Anodonta (*Pyganodon*) Fischer & Crosse, 1893: 518.
Anodonta (*Nayadina*) De Gregorio, 1914: 64 [*non* Munier-Chlamas, 1864 (Mollusca: Ostreidae). Type species: *Anodonta* (*Nayadina*) *venusta* De Gregorio, 1914, by monotypy (= *Anodonta grandis* Say, 1829, *fide* Walker (1918))].

Geographic range. *Pyganodon* forms are widespread in North America and occur south into Central America through Mexico to Nicaragua. They are found from the Rocky Mountains (Henderson, 1924: 85–87) to the Atlantic Seaboard, in Gulf of Mexico drainages, and from Mexico (Fischer & Crosse, 1893: 518–537; Martens, 1901: 653–654) north to tributaries of the Arctic Ocean (Clarke, 1981b).

Pyganodon cataracta marginata (Say, 1817)

Figures 168, 169, 170

Etymology. G. *kataraktes*, waterfall; L. *cataracta*, waterfall. L. *marginus*, border, edge, margin.

Vernacular name(s). *Eastern Floater*, Gaspé Floater.

Original description

Anodonta marginata Say, 1817: no pagination, article on "Conchology," species no. 2 under *Anodonta*, pl. 3, fig. 5.

Type locality. ". . . , it is very common in our rivers." A specific geographic locality was not given. Johnson (1970: 356) subsequently restricted the type locality to "near Philadelphia, Philadelphia Co., Pennsylvania."

Synonymy

Anodonta marginata Say, 1817: no pagination, pl. 3, fig. 5.

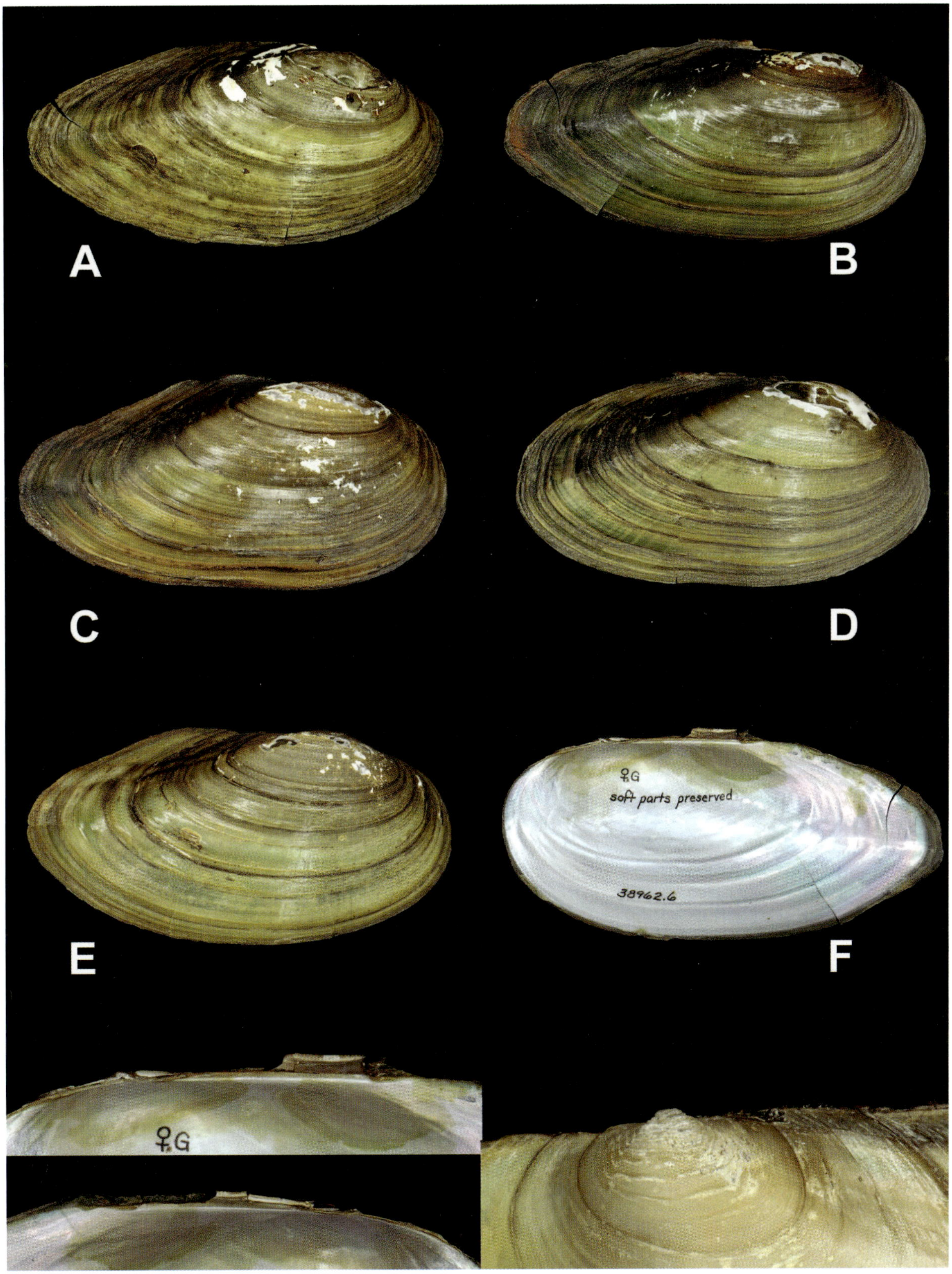

Fig. 168. ***Pyganodon cataracta marginata*** **(Say, 1817).** **A.** OSUM 45600.4, Lac Rond, Maskinongee Co., Quebec. 85 mm. **B, F, Hinge detail:** OSUM 38962.6, Black Moshannon Lake, Centre Co., PA. 107 mm. **C.** OSUM 48106, stream, Dutchess Co., NY. 93 mm. **D.** OSUM 56555, Pine Lake, Rusk Co., WI. 87 mm. **E.** OSUM 26123.9, Bostwick Lake, Kent Co., MI. 75 mm. **Beak sculpture detail:** OSUM 6758.8, Lac à Foin, Matane Co., Quebec. 61 mm total length.

Anodonta marginatus Say, 1817. Hanley, 1843: 217.
Status. Not listed by either Ohio Division of Wildlife or U.S. Fish & Wildlife Service as of 2009.

ADULT SHELL

Shell. Medium to large size (to 140 mm), very thin, elongate, bluntly pointed posteriorly; somewhat compressed to inflated.
Anterior margin. Very short, straight to broadly rounded to the ventral border.
Ventral margin. Very broadly rounded.
Posterior margin. Long, rather straight to the dorsal margin.
Dorsal margin. Long, straight, joining posterior margin with very obtuse angle.
Umbo. Narrow, very low, placed far forward; umbos nearly face each other; umbonal sculpture of 5–7 ridges, the earliest double-looped but becoming concentric further from the umbo.
Ligament. Long, poorly developed, tan to dark brown, not concealed by shells.
Disc sculpture. Disc of shell smooth except for low angulations on posterior slope.
Periostracum. Shiny even as adults; base color yellow, green, or tan with or without vague green rays, dorsal slope darker, annular rings dark.
Nacre. Iridescent white, particularly posteriorly, occasionally flushed in the umbonal region with salmon or light purple; often with extensive gold or brown Tulberg layers; commonly the nacre shows signs of disease, parasites, or damage.
Hinge plate. Edentulous.
Muscle scars. Anterior adductor large, "D"-shaped, flush with shell; posterior adductor large, flush; large anterior adductor distinct from anterior pedal protractor scar, small posterior adductor confluent with small posterior pedal retractor scar; dorsal suspensors located under umbo in cluster, line, or semicircle. Pallial line entire, located far from shell margin. Accessory line absent.
Umbonal cavity. Very wide, shallow, open.

Glochidia (*P. cataracta s.l.*). Subtriangular and asymmetric with a straight dorsal margin, posterior margin gently and evenly curved and anterior margin broadly curved to its maximum inflation at about 70% from dorsal to ventral. Coarse loose-looped sculpturing covers the exterior surface of the valve. Hook styliform, with about 20 lanceolate microstylets. Microstylets arranged in four rows near the ventral terminus, reduced to a single row distally and clustered near the tip of the hook (Hoggarth, 1999: 25–26, figs. 11a–e).

Literature records (Ortmann, 1912a; Lefevre & Curtis, 1910a: fig. C; Lefevre & Curtis, 1912, fig. 1c; Calloway & Turner, 1979: pl. 3, figs. 1, 3, 5, 7, 8; Wiles, 1975a: figs. 3, 7; Rand & Wiles, 1982: figs. 1–4; Hoggarth, 1999: 25–26, figs. 11a–e) provided dimensions for length (0.36 to 0.38 mm), height (0.35 to 0.38 mm), and hinge length (0.28 to 0.29 mm). The glochidium of *P. cataracta* can be distinguished from that of *P. grandis* by its longer central ligament. The glochidia of both *P. c. cataracta* and *P. c. marginata* have longer anterior ligaments than any species in the Anodontinae.

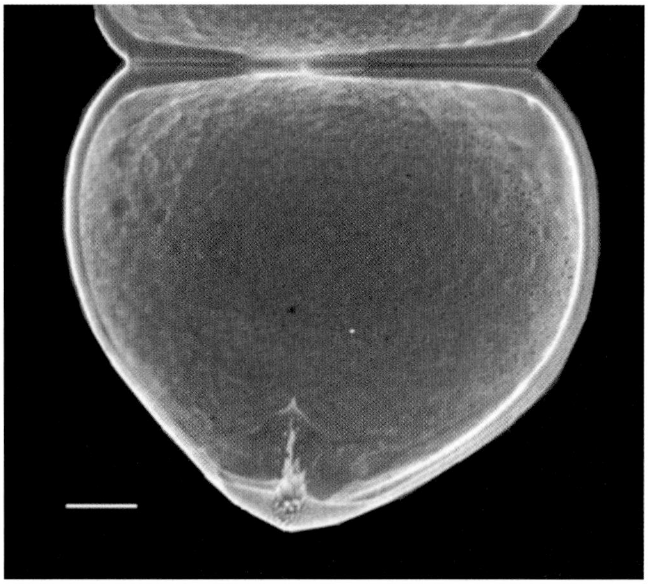

Fig. 169. *Pyganodon cataracta cataracta* (Say, 1817). Glochidium. OSUM 52462.35, Great Herring Pond, Barnstable Co., MA. Bar = 35 µm. (Hoggarth, 1999).

Reproductive biology (for *Pyganodon cataracta s.l.*). Bradytictic. Eggs were found in August (Ortmann, 1912a), and glochidia were discharged the following April–May (Ortmann, 1919; Fichtel & Smith, 1995). During gravid periods the marsupial portion of the gills develop secondary water tubes for respiration, which disappear when not gravid (Tankersley & Dimock, 1992). Metamorphosis has been observed on White Sucker, Rock Bass, and Pumpkinseed.

Reported potential hosts (for *Pyganodon cataracta s.l.*)
Ambloplites rupestris (Rock Bass)—Gray *et al.,* 1999 (LT).
Catostomus commersoni (White Sucker)—Wiles, 1975a (NI); Gray *et al.*, 1999 (LT).
Cyprinus carpio (Common Carp)—Lefevre & Curtis, 1910a (LI); Lefevre & Curtis, 1912 (LI).
Gasterosteus aculeatus (Threespine Stickleback)—Wiles, 1975a (NI); Threlfall, 1986 (NI).
Lepomis gibbosus (Pumpkinseed)—Conner, 1905 (NI); Gray *et al.*, 1999 (LT).

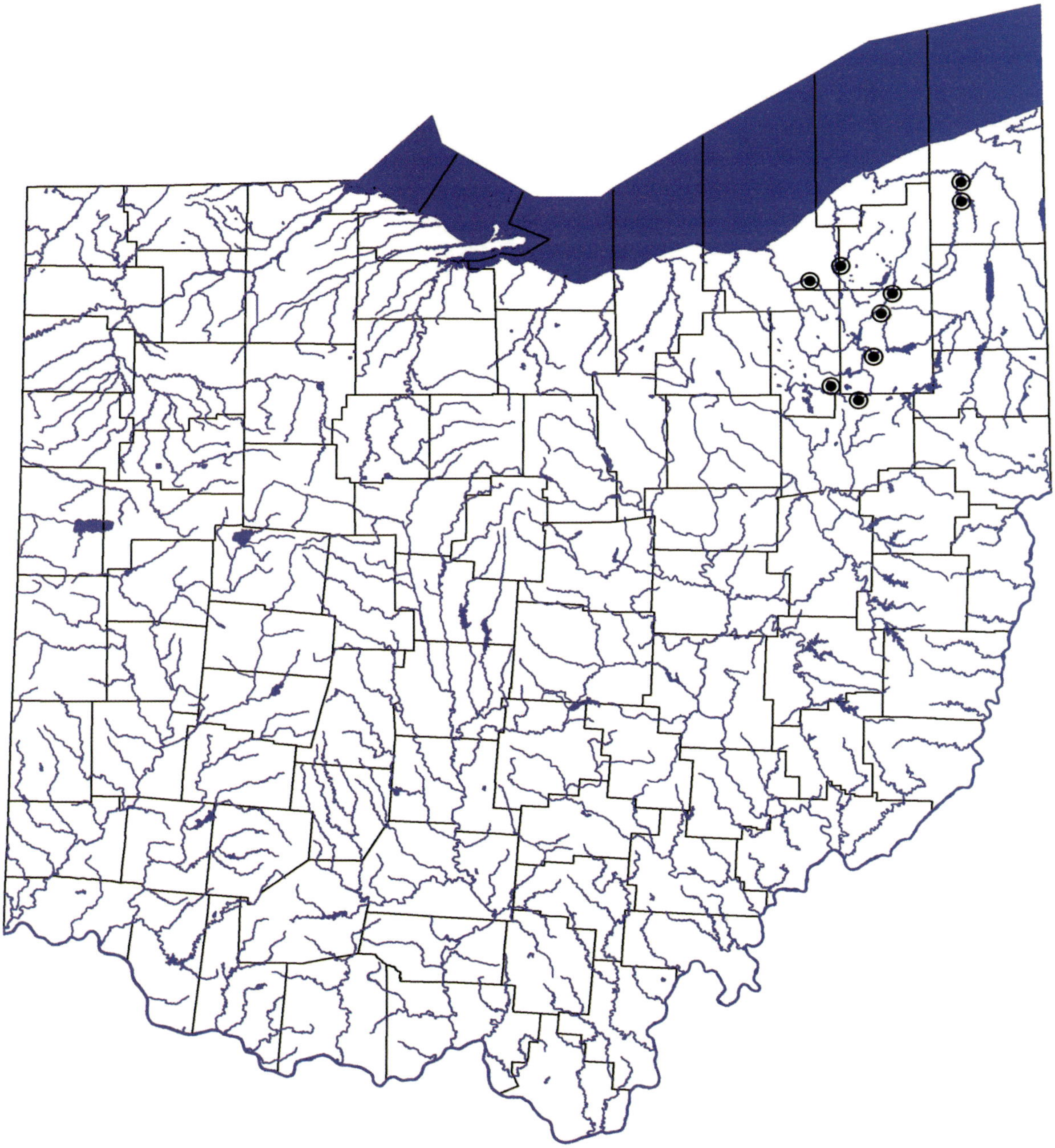

Fig. 170. ***Pyganodon cataracta marginata*** **(Say, 1817)**
Solid dots—live or freshdead since 1980
Bullseye—all records, any condition

Habitat (for *Pyganodon cataracta s.l.*). Most common in lakes and ponds and in sand, silty sand, and sandy gravel (Fichtel & Smith, 1995).

Range. Simpson (1900a: 633) gave the distribution of this form as "St. Lawrence drainage," and records exist from Wisconsin (Morrison, 1929: 421; 1932: 363, 379), Michigan (Goodrich & van der Schalie, 1939: 37; van der Schalie 1938: 52), Pennsylvania (Rhoads, 1899: 136; Ortmann 1909b: 184), New York (Jay, 1852: 72; Letson, 1905: 86; Baker, 1916: 253; Clarke & Berg, 1959: 36, 38), and Canada (Robertson & Blakeslee, 1948: 100; La Rocque, 1953: 88). Evidence of this form from south of the Ohio River has not been found, and no lots bearing this name from Ohio are in the OSUM collection (but see next).

Range in Ohio. Several accounts of Ohio unionids (Sterki, 1907: 394; Zimmerman, 1960: 20; La Rocque, 1960: 134–136) have included Recent or Pleistocene records of *Pyganodon cataracta marginata* (as *Anodonta marginata* Say, 1817). During the preparation of this book, specimens from the Carnegie Museum of Natural History were identified as this species. These specimens are all from northeastern Ohio Lake Erie drainages (Cuyahoga River, Little Cuyahoga River, Tinkers Creek, Chagrin River, and Grand River) or the Mahoning River—all were collected prior to 1920. They are elongate individuals with the umbos far forward and the beak sculpture varying from double-looped through longitudinal lines, with no apparent separation.

Similar species. Both *Pyganodon lacustris* (Lea, 1857) (see "Questionable and Potential Ohio Occurrences") and *Pyganodon cataracta marginata* are elongate forms having low umbos distinctly closer to the anterior margin of the shell than *P. c. cataracta*. Both have modified double-loop umbonal sculpture that is distinctly more delicate than that of either *P. c. cataracta* or *P. g. grandis*. The fine double-loops of *P. lacustris* become flattened moving ventrally down the shell away from the umbonal beaks, and the paired loops sometimes merge, becoming sets of scarcely perceptible, nearly linear ridges parallel to the hinge. The fine-sculptured linear or near-linear ridges common to what we recognize as *P. c. marginata* may yet be found to intergrade with those of *P. lacustris*. At present these forms appear to be either sibling species or subspecies of *P. c. cataracta*. Shell forms from Ohio streams currently thought to be modified elongate expressions of *P. grandis* may, in fact, belong to the *P. c. marginata–P. lacustris* complex.

Comments. The validity of this named species and its range have long been uncertain. Hoeh & Burch (1989: 263–265) reviewed aspects of this problem as appropriate in the recognition of *Pyganodon lacustris* (Lea, 1857) (as *Anodonta lacustris*) as a valid species. This latter name was long considered a junior synonym of *A. marginata*. They concluded, "The identity of Say's *A. marginata* should be determined if possible because of the name's priority, but whatever the species turns out to be, we are confident that it will not be the same as *A. lacustris*." We concur in spite of the similarity of shell characters.

Individuals grow quickly for 3–4 years before slowing. Exceptional specimens reach ca. 140 mm and 17 years; most are ca. 80 mm and 9 years old.

Pyganodon grandis grandis (Say, 1829)

Figures 171, 172, 173, 174

Etymology. From L. *grandis*, grand, large, imposing. This species has individuals with dimensions that rank it among the largest unionoids known.

Vernacular name(s). *Giant Floater*, Big Floater, Large Floater, Stout Floater, Fat Floater.

Original description

Anodonta grandis Say 1829: 341, species no. 3, no plates or figures.

Type locality. "Inhab. Fox River of the Wabash."

Type material. The holotype of this species is apparently lost (L. Say, 1840: 4; Johnson & Baker, 1973: 157). Simpson (1900a: 642, footnote) recorded "I have seen Say's type of *A. grandis*" but fails to say where. Haas (1930: 326) selected SMF No. 4300 as the neotype of *Anodonta* (*Anodonta*) *grandis* Say, 1829.

Synonymy. The list below includes names that are believed to be synonyms based upon the morphological intergradation and distributional patterns of the described forms. We anticipate that additional described names will be added to this list as additional intergrading forms are collected and added to museum research collections. Knowing that some valid species overlap in their major morphological features, it will also not be surprising to learn that some presumed synonyms may eventually be found to be distinct sibling species in their own right.

Anodonta lugubris Say, 1829: 340 [Type locality: "Cumberland River." Types lost].

Anodonta grandis Say, 1829: 341.

Anodonta stewartiana Lea, 1834: 47, pl. 6, fig. 17 [Type locality: "River Teche, Louisiana." Lectotype USNM 86664].

Anodonta palna [*sic*] Lea, 1834: 48, pl. 7, fig. 18 [Incorrect original spelling of *plana*. Type locality: "Bear Grass Creek near Louisville [KY]." Lectotype USNM 86626].

Symphynota benedictensis Lea, 1834: 104, pl. 16, fig. 48

Fig. 171. ***Pyganodon grandis grandis* (Say, 1829). A.** OSUM 11941.4, St. Joseph River, DeKalb Co., IN. 129 mm. **B.** OSUM 305, Lake Erie, Ottawa Co., OH. 68 mm. **C, H, Hinge detail:** OSUM 5420, Olentangy River, Franklin Co., OH. 124 mm. **D.** OSUM 17552.1, Cowan Lake, Clinton Co., OH. 180 mm. **E.** 66877.2, Olentangy River, Marion Co., OH. 80 mm. **F.** OSUM 45527, Ohio River, Hamilton Co., OH. 72 mm. **G.** OSUM 21585.2, Mill Creek, Ashtabula Co., OH. 76 mm. **Beak sculpture detail:** OSUM 13214, Scioto River, Delaware Co., OH. 54 mm total length.

[Although named for "Professor Benedict," the spelling lapse must stand. Type locality: "Lake Champlain [VT/NY]." Lectotype USNM 86381].

Anodonta corpulenta Cooper, 1834: 154 [Type locality: "Upper Missouri River." Type not located].

Anodonta declivus Conrad, 1834: 341, pl. 1, fig. 11 [Type locality: "Flint River, Morgan Co., Alabama." Figured holotype lost *fide* Johnson & Baker, 1973].

Anodonta benedictensis (Lea, 1834). Férussac, 1835: 25.

Margarita (*Anodonta*) *benedictensis* (Lea, 1834). Lea, 1836: 28.

Margarita (*Anodonta*) *salmonia* Lea, 1836: 51 [*nomen nudum*].

Margarita (*Anodonta*) *grandis* (Say, 1829). Lea, 1836: 52.

Margarita (*Anodonta*) *gigantea*. Lea, 1836: 52 [*nomen nudum*].

Margarita (*Anodonta*) *plana* (Lea, 1834). Lea, 1836: 52.

Margarita (*Anodonta*) *stewartiana* (Lea, 1834). Lea, 1836: 52.

Margarita (*Anodonta*) *decora* Lea, 1836: 52 [*nomen nudum*].

Margarita (*Anodonta*) *ovata* Lea, 1836: 52 [*nomen nudum*].

Anodonta gigantea Lea, 1838a: 1, pl. 1, fig. 1 [Type locality: "near Port Gibson [Claiborne Co., MS]"]. Lectotype USNM 86659].

Anodonta ovata Lea, 1838a: 2, pl. 2, fig. 2 [Type locality: "near Marietta, Ohio." Lectotype USNM 86615].

Anodonta salmonia Lea, 1838a: 45, pl. 14, fig. 41 [Type locality: "near Poland, Ohio." Lectotype USNM 86450].

Anodonta decora Lea, 1838a: 64, pl. 20, fig. 63 [Type locality: "canal near Cincinnati, Ohio." Lectotype USNM 86633].

Anodonta pepinianus Lea, 1838a: 96, pl. 16, fig. 51 [Type locality: "Lake Pepin, Portage County, Ohio." Lectotype USNM 86440].

Margarita (*Anodonta*) *pepiniana* Lea, 1838b: 30 [*nomen nudum*].

Anodonta pepiniana Lea, 1838b: 154.

Anodonta footiana Lea, 1840: 289 [Type locality: "vicinity of Fort Winnebago [WI]." Lectotype USNM 86468].

Anodonta harpethensis Lea, 1840: 289 [Type locality: "Harpeth River, Tennessee." Lectotype USNM 86657].

Anodonta maryattiana Lea, 1840: 289 [Type locality: "vicinity of Fort Winnebago [WI]." Lectotype USNM 86706].

Anodonta marryattana Lea, 1842: 226, pl. 20, fig. 45 [emendation of *Anodonta maryattiana* Lea, 1840].

Anodon plana (Lea, 1834). DeKay, 1843: 201, pl. 17, fig. 232.

Anodon benedictensis (Lea, 1834). Catlow & Reeve, 1845: 66.

Anodon decora (Lea, 1838). Catlow & Reeve, 1845: 66.

Anodon grandis (Say, 1829). Catlow & Reeve, 1845: 67.

Anodon ovata (Lea, 1838). Catlow & Reeve, 1845: 67.

Anodon stewardtiana (Lea, 1834). Catlow & Reeve, 1845: 68.

Anodon salmonia (Lea, 1838). Catlow & Reeve, 1845: 68.

Anodonta opaca Lea, 1852a: 285, pl. 25, fig. 46 [Type locality: "near New Orleans, Louisiana." Lectotype USNM 128835].

Anodonta linneana Lea, 1852a: 289, pl. 27, fig. 51 [Type locality: "Lake Concordia, Louisiana." Lectotype USNM 86702].

Anodonta virens Lea, 1852a: 290, pl. 18, fig. 53 [Type locality: "Red River, Alexandria, Louisiana." Lectotype USNM 86666].

Margaron (*Anodonta*) *benedictensis* (Lea, 1834). Lea, 1852b: 47.

Margaron (*Anodonta*) *pepiniana* (Lea, 1838). Lea, 1852b: 49.

Margaron (*Anodonta*) *footiana* (Lea, 1840). Lea, 1852b: 49.

Margaron (*Anodonta*) *gigantea* (Lea, 1834). Lea, 1852b: 50.

Margaron (*Anodonta*) *plana* (Lea, 1834). Lea, 1852b: 50.

Margaron (*Anodonta*) *decora* (Lea, 1838). Lea, 1852b: 50.

Margaron (*Anodonta*) *ovata* (Lea, 1838). Lea, 1852b: 50.

Margaron (*Anodonta*) *salmonia* (Lea, 1838). Lea, 1852b: 50.

Margaron (*Anodonta*) *harpethensis* (Lea, 1840). Lea, 1852b: 50.

Margaron (*Anodonta*) *opaca* (Lea, 1852). Lea, 1852b: 50.

Margaron (*Anodonta*) *stewartiana* (Lea, 1834). Lea, 1852b: 51.

Margaron (*Anodonta*) *maryattana* (Lea, 1840). Lea, 1852b: 51.

Margaron (*Anodonta*) *linneana* (Lea, 1852). Lea, 1852b: 51.

Margaron (*Anodonta*) *virens* (Lea, 1852). Lea, 1852b: 51.

Anodonta nilssonii Küster, 1853: 61, pl. 17, figs. 3, 4 [Type not located].

Anodonta lewisii Lea, 1857b: 84 [Type locality: "Erie Canal and Mohawk River, Herkimer County, New York." Lectotype USNM 86595].

Anodonta danielsii Lea, 1858b: 139 [Type locality: "Topeka, Kansas." Lectotype USNM 86426].

Anodonta texasensis Lea, 1859a: 113 [Type locality: "Texas." Lectotype USNM 86680].

Anodonta simpsoniana Lea, 1861b: 56 [Type locality: "Great Slave Lake, Fort Rae, Arctic America." Lectotype USNM 86434].

Anodonta leonensis Lea, 1862: 169 [Type locality: "Leon, Texas." Lectotype USNM 87806].

Anodonta dallasiana Lea, 1863: 190 [Type locality: "Saskatchewan River at mouth, Winnepeg, Canada." Lectotype USNM 86443].

Anodonta bealei Lea, 1863: 194 [Type locality: "Leon, Texas; Verdigris River." Lectotype USNM 86690].

Anodon subangulata Anthony, 1865: 158, pl. 13, fig. 1

[Type locality: "Shears Lake, Michigan." Lectotype MCZ 150645].

Anodon imbricata Anthony, 1865: 159, pl. 14, fig. 1 [Type locality: "Camp Lake, Michigan." Lectotype MCZ 161860].

Anodon opalina Anthony, 1865: 159, pl. 14, fig. 2 [Type locality: "Shear's Lake, Michigan." Lectotype MCZ 150633].

Anodon subinflata Anthony, 1865: 160, pl. 15, fig. 1 [Type locality: "Michigan." Lectotype MCZ 161863].

Anodon micans Anthony, 1865: 162, pl. 16, fig. 1 [Type locality: "Texas." Lectotype MCZ 187290].

Anodon mcnielii Anthony, 1866: 144, pl. 6, fig. 1 [Type locality: "Michigan." Lectotype MCZ 150644].

Anodon subgibbosa Anthony, 1866: 144, pl. 6, fig. 2 [Type locality: "Black Lake, Michigan." Lectotype MCZ 161854].

Anodon inornata Anthony, 1866: 145 [Type locality: "Slawson's Lake, Michigan." Lectotype MCZ 161876].

Anodon gigantea (Lea, 1834). Sowerby, 1867: pl. 8, fig. 18.

Anodon footiana (Lea, 1840). Sowerby, 1867: pl. 14, fig. 48.

Anodon edentulus (Say, 1829). Sowerby, 1867: pl. 17, fig. 60 [misidentification].

Anodonta houghtonensis Currier, 1868: 11 [*nomen nudum*].

Anodon harpethensis (Lea, 1840). Sowerby, 1869: pl. 21, fig. 82.

Margaron (*Anodonta*) *lewisii* (Lea, 1857). Lea, 1870: 70.

Margaron (*Anodonta*) *benedictii* Lea, 1870: 75 [emendation of *Symphynota benedictensis* Lea, 1834; named for "Professor Benedict"].

Margaron (*Anodonta*) *danielsii* (Lea, 1858). Lea, 1870: 78.

Margaron (*Anodonta*) *leonensis* (Lea, 1862). Lea, 1870: 78.

Margaron (*Anodonta*) *dallasiana* (Lea, 1862). Lea, 1870: 78.

Margaron (*Anodonta*) *micans* (Anthony, 1865). Lea, 1870: 78.

Margaron (*Anodonta*) *texasensis* (Lea, 1859). Lea, 1870: 81.

Margaron (*Anodonta*) *bealei* (Lea, 1863). Lea, 1870: 81.

Margaron (*Anodonta*) *corpulenta* (Lea, 1834). Lea, 1870: 81.

Margaron (*Anodonta*) *subglobosa* [*sic*] (Anthony, 1866). Lea, 1870: 81.

Anodon subgibbosus Anthony, 1866. Sowerby, 1870: pl. 28, fig. 107.

Anodon marryattana [*sic*] (Lea, 1840). Sowerby, 1870: pl. 28, fig. 111.

Anodon corpulenta (Cooper, 1834). Sowerby, 1870: pl. 32, fig. 129.

Anodon stewartianus (Cooper, 1834). Sowerby, 1870: pl. 32, fig. 133.

Anodon virens (Lea, 1852). Sowerby, 1870: pl. 34, fig. 138.

Anodon lewisii (Lea, 1857). Sowerby, 1870: pl. 35, fig. 142.

Anodon linneana (Lea, 1852). Sowerby, 1870: pl. 35, fig. 144.

Anodon texasensis (Lea, 1859). Sowerby, 1870: pl. 36, fig. 146.

Anodon pepinianus (Lea, 1838). Sowerby, 1870: pl. 36, fig. 150.

Anodon giganteus (Lea, 1834). Sowerby, 1870: pl. 37, fig. 152.

Anodonta sulcata Küster, 1853: 62, pl. 18, fig. 1.

Anodonta hockingensis "Moores" Call, 1880: 530 [*nomen nudum*].

Anodonta somersii "Moores" Call, 1880: 530 [*nomen nudum*].

Anodonta houghtonensis Currier *in* DeCamp, 1881: 14, pl. 1, fig. 2 [Type locality: "Houghton's Lake, Roscomon Co., Michigan." Figured holotype ANSP 126581a].

Anodonta benedictii (Lea, 1870). Latchford, 1882: 55.

Anodonta imbricata (Anthony, 1865). Wright, 1888: no pagination.

Anodonta micans (Anthony, 1865). Wright, 1888: no pagination.

Anodonta subangulata (Anthony, 1865). Wright, 1888: no pagination.

Anodonta subinflata (Anthony, 1865). Wright, 1888: no pagination.

Anodonta subgibbosa (Anthony, 1866). Wright, 1888: no pagination.

Margaritana danielsii (Lea, 1858). Paetel, 1890: 173.

Anodonta grandis var. *footiana* (Lea, 1840). Simpson, 1900a: 642.

Anodonta marryattiana [*sic*] (Lea, 1840). Simpson, 1900a: 642.

Anodonta grandis var. *gigantea* (Lea, 1834). Simpson, 1900a: 643.

Anodonta grandis var. *leonensis* (Lea, 1862). Simpson, 1900a: 643.

Anodonta grandis var. *benedictensis* (Lea, 1834). Simpson, 1900a: 644.

Anodonta kennicottii Lea, 1861. Simpson, 1900a: 647 [in part, misidentification].

Anodonta (*Nayadina*) *venusta* De Gregorio, 1914: 65, pl. 12, fig. 2 [Type locality "Sciota [Scioto] R. (Ohio)." Type in Palermo Museum?].

Anodonta grandis gigantea Lea, 1834. Baker, 1905: 254.

Anodonta grandis footiana Lea, 1840. Chadwick, 1906: 94.

Anodonta grandis decora (Lea, 1838). Sterki, 1907: 394.

Anodonta dakota Frierson, 1910: 113, pl. 10 [Type locality: "Ulvers point, Clear Lake, Deuel County, South Dakota." Holotype UMMZ 87528].

Anodonta dakotana Frierson, 1910. Utterback, 1915: 265, pl. 24, figs. 77a, b [unjustified emendation for *Anodonta dakota* Frierson, 1910, but *non* Frierson, 1910, *fide* Frierson, 1927: 15, misidentification].

Anodonta (*Pyganodon*) *grandis* Say, 1829. Frierson, 1927: 14.
Anodonta (*Pyganodon*) *grandis bealei* Lea, 1863. Frierson, 1927: 15.
Anodonta (*Pyganodon*) *grandis lugubris* Say, 1829. Frierson, 1927: 14.
Anodonta (*Pyganodon*) *grandis benedictensis* (Lea, 1834). Frierson, 1927: 15.
Anodonta (*Pyganodon*) *grandis pepiniana* Lea, 1838. Frierson, 1927: 15.
Anodonta (*Pyganodon*) *grandis dakota* Frierson, 1910. Frierson, 1927: 15.
Anodonta (*Pyganodon*) *grandis dakotana* Utterback, 1915. Frierson, 1927: 15.
Anodonta salmonea [*sic*] Lea, 1838. Baker, 1928a: 155.
Anodonta grandis plana (Lea, 1834). Baker, 1928a: 155, pl. 60, fig. 5, pl. 61, figs. 3–6.
Anodonta grandis lugubris Say, 1829. Strecker, 1931: 10.
Anodonta grandis bealei Lea, 1863. Strecker, 1931: 10.
Anodonta (*Anodonta*) *grandis* Say, 1829. Clarke & Berg, 1959: 36, fig. 38.
Nayadina venusta (De Gregorio, 1914). Haas, 1969: 359.
Anodonta grandis corpulenta Cooper, 1834. Starrett, 1971: 319.
Anodonta grandis simpsoniana Lea, 1861. Clarke, 1973: 85, pl. 5, figs. 5–10, map 9.
Anodonta bealii [*sic*] Lea, 1863. Johnson, 1980: 92.
Anodonta grandis grandis Say, 1829. Oesch, 1984: 40.
Pyganodon grandis (Say, 1829). Hoeh, 1990: 63.

Status. Not listed by either Ohio Division of Wildlife or U.S. Fish & Wildlife Service as of 2009.

ADULT SHELL

Shell. Medium to very large (to 210 mm), thin-shelled as juveniles, becoming thicker with age but always rather fragile, somewhat inflated to quite inflated. Usually elongate, but some specimens more oval in profile.

Anterior margin. Roundly arched into ventral margin, occasionally angled with ventral margin.

Ventral margin. Nearly straight to broadly rounded.

Posterior margin. Ventral and posterior margins form wide, blunt acute angle at or below midline; posterior margin oblique and straight to dorsal margin.

Dorsal margin. Moderately long, straight or gently bowed, joining posterior margin with very obtuse angle.

Umbo. Low, pointed, located about 33% back from anterior margin in the typical form, high, narrow, and more centrally located in the form *corpulenta;* umbonal sculpture of 2–5 double-looped ridges, occasionally nodulose.

Ligament. Long, tan to dark brown, extending anterior of the umbo, partially concealed by shells.

Disc sculpture. Smooth except for posterior ridge.

Periostracum. Yellow, tan, or greenish as smallest juveniles, becoming dark brown to black with age. Occasionally with numerous fine green or brown vague rays.

Nacre. Porcelain white, often iridescent posteriorly, occasionally with a faint salmon, copper, or yellow flush; nearly all specimens seen have gold or brown Tulberg layers.

Hinge plate. Thin, cord-like, edentulous, gently bowed.

Muscle scars. Anterior adductor barely impressed; posterior adductor flush or shallowly impressed; anterior adductor large and distinct from anterior pedal protractor scar, large posterior adductor confluent and nearly indistinguishable from posterior pedal retractor scar; minute, multiple dorsal suspensor located within the umbonal cavity, the anterior-most scar much larger. Pallial line entire, located far from shell margin. Accessory line absent.

Umbonal cavity. Very wide, shallow, and open.

Glochidia. Subtriangular and asymmetric, posterior margin gently curved, anterior margin broadly curved, ventral margin broadly rounded, dorsal margin straight. Exterior surface sculpturing coarse, loose-looped; hooks styliform with about 20 microstylets (Lea, 1858c: 49, pl. 5, figs. 32–34; Hoggarth, 1999: figs. 10a–f).

Literature records (Ortmann, 1912a: 292; Surber, 1912: 8, pl. 3, fig. 45; Utterback, 1915: 263; Tucker, 1928: 123; Hoggarth, 1999: figs. 10a–f) indicate glochidial dimensions for length (0.35 to 0.41 mm), height (0.34 to 0.42 mm), and hinge (0.25 to 0.26 mm). Ortmann (1912a: 292) noted, "Glochidia very large, the largest known to me, even larger than those of *A. cygnea*, sub-triangular, slightly higher than long, with hooks. The glochidia figured by Lea differ somewhat from each other, while they actually should be all alike." Ortmann (1912) gave 360 µm x 370 µm for length and height of the glochidium of *P. g. grandis,* and Tucker (1928) reported the following ranges: length, 350–398 µm; height, 343–390 µm. However, Surber (1912) gave measurements of 410 µm x 420 µm. This extremely large range in size led Tucker (1928) to suggest that it was the result of examining the glochidia of a far-ranging, variable species.

Reproductive biology. Bradytictic. The outer gills are marsupial (Ortmann, 1912a). Gravid females were recorded from August to the following May (Ortmann, 1919), but Watters & O'Dee (1998c) found that host-overwintering glochidia were released from October to February. The glochidia are released in a tangle of mucus and larval threads (Utterback, 1915). This species is a generalist in host use.

Reported potential hosts

Alosa chrysochloris (Skipjack Herring)—Surber, 1913 (NI); Wilson, 1916 (NI).

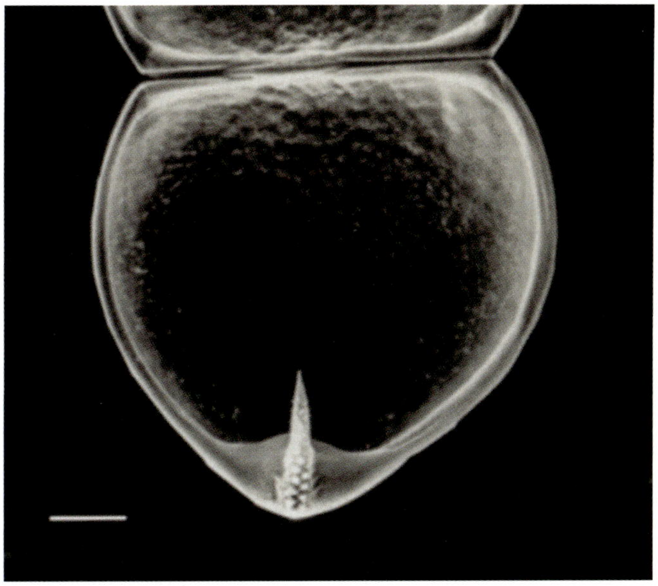

Fig. 172. ***Pyganodon grandis grandis*** **(Say, 1829).** Glochidium. OSUM 38467.10, Great Miami River, Montgomery Co., OH. Bar = 35 μm. (Hoggarth, 1999).

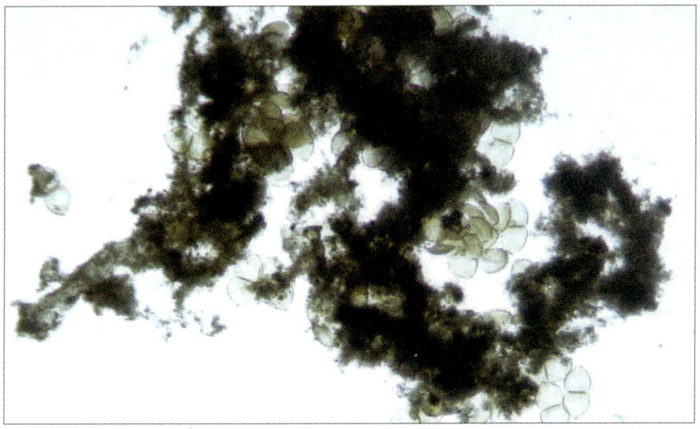

Fig. 173. ***Pyganodon grandis grandis*** **(Say, 1829).** Glochidia entrained in mucus strand released by female to entangle host fish.

Ambloplites rupestris (Rock Bass)—Lefevre & Curtis, 1910a (NI); Tucker, 1928 (LT); Trdan & Hoeh, 1982 (NI, LT).

Ameiurus natalis (Yellow Bullhead)—Wilson, 1916 (NI).

Aplodinotus grunniens (Freshwater Drum)—Wilson, 1916 (NI).

Campostoma anomalum (Central Stoneroller)—Trdan & Hoeh, 1982 (NI, LT).

Carpiodes carpio (River Carpsucker)—Morrison *in* Clarke & Berg, 1959 (NS).

Carrasius auratus (Gold Fish)—Watters *et al.*, 2005 (LT)—exotic.

Catostomus commersoni (White Sucker)—Kakonge *in* Fuller, 1978 (NS).

Cichlosoma cyanoguttatum (Rio Grande Cichlid)—Howells, 1997 (LT).

Culaea inconstans (Brook Stickleback)—Morrison *in* Clarke & Berg, 1959 (NS); Trdan & Hoeh, 1939 (LT).

Cyprinus carpio (Common Carp)—Lefevre & Curtis, 1910a (NI); Morrison *in* Clarke & Berg, 1959 (NS).

Dorosoma cepedianum (Gizzard Shad)—Wilson, 1916 (NI).

Etheostoma caeruleum (Rainbow Darter)—Trdan & Hoeh, 1982 (NI, LT).

Etheostoma exile (Iowa Darter)—Morrison *in* Clarke & Berg, 1959 (NS); Trdan & Hoeh, 1982 (NI, LT).

Etheostoma nigrum (Johnny Darter)—Hankinson, 1908 (NI); Morrison *in* Clarke & Berg, 1959 (NS); Trdan & Hoeh, 1982 (NI, LT).

Fundulus chrysotus (Golden Topminnow)—Penn, 1939 (LT).

Fundulus diaphanus (Banded Killifish)—Trdan & Hoeh, 1982 (LT).

Labidesthes sicculus (Brook Silverside)—Trdan & Hoeh, 1982 (NI, LT).

Lepisosteus osseus (Longnose Gar)—Trdan & Hoeh, 1982 (LT).

Lepomis cyanellus (Green Sunfish)—Wilson, 1916 (NI); Tucker, 1928 (LT); Trdan & Hoeh, 1982 (NI).

Lepomis gibbosus (Pumpkinseed)—Trdan & Hoeh, 1982 (LT); Watters *et al.*, 2005 (LT).

Lepomis humilis (Orangespotted Sunfish)—Arey, 1932a (LT).

Lepomis macrochirus (Bluegill)—Lefevre & Curtis, 1910a (NI); Wilson, 1916 (NI); Penn, 1939 (LT); Morrison *in* Clarke & Berg, 1959 (NS); Trdan & Hoeh, 1982 (NI, LT); Watters *et al.*, 2005 (LT).

Lepomis megalotis (Longear Sunfish)—Penn, 1939 (LT).

Luxilus chrysocephalus (Striped Shiner)—Lefevre & Curtis, 1910 (NI); Read & Oliver, 1953 (NI); Trdan & Hoeh, 1982 (LT).

Luxilus cornutus (Common Shiner)—Kakonge *in* Fuller, 1978 (NS); Trdan & Hoeh, 1982 (NI, LT).

Lythrurus umbratilis (Redfin Shiner)—Trdan & Hoeh, 1982 (LT).

Margariscus margarita (Pearl Dace)—Trdan & Hoeh, 1982 (NI).

Micropterus salmoides (Largemouth Bass)—Wilson, 1916 (NI); Penn, 1939 (LT); Morrison *in* Clarke & Berg, 1959 (NS); Trdan & Hoeh, 1982 (NI); Watters *et al.*, 2005 (LT).

Morone chrysops (White Bass)—Wilson, 1916 (NI).

Neogobius melanostomus (Round Goby)—Watters *et al.*, 2005 (LT)—exotic.

Notemigonus crysoleucas (Golden Shiner)—Trdan & Hoeh, 1982 (LT).

Notropis heterodon (Blackchin Shiner)—Trdan & Hoeh, 1982 (NI, LT).

Notropis heterolepis (Blacknose Shiner)—Trdan & Hoeh, 1982 (NI, LT).

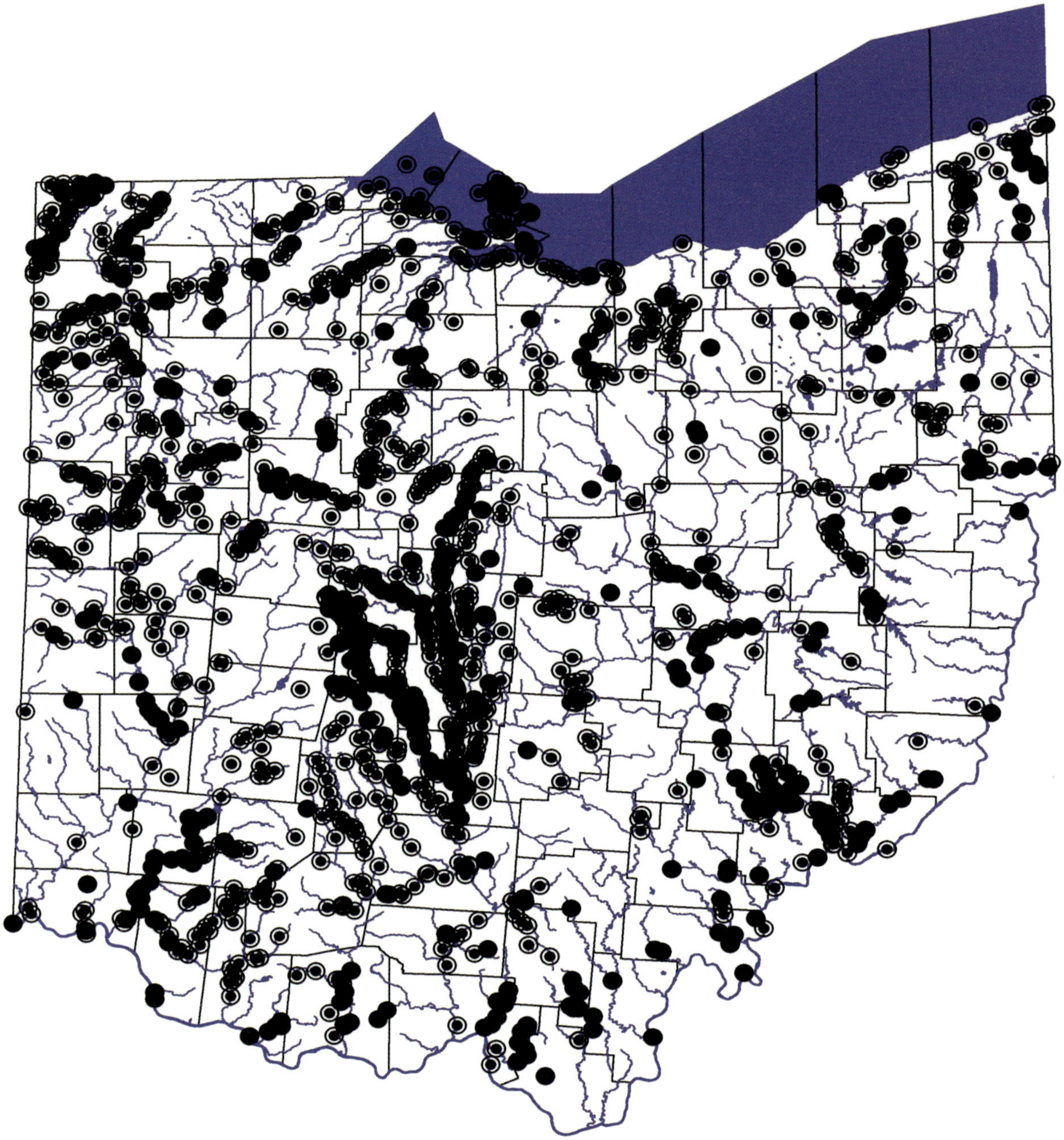

Fig. 174. ***Pyganodon grandis grandis*** **(Say, 1829)**
Solid dots—live or freshdead since 1980
Bullseye—all records, any condition

Perca flavescens (Yellow Perch)—Lefevre & Curtis, 1910a (NI); Trdan & Hoeh, 1982 (NI, LT); Jansen, 1991 (NI); Jansen & Hanson, 1991 (NI).

Pimephales notatus (Bluntnose Minnow)—Trdan & Hoeh, 1982 (NI, LT).

Poecilia reticulata (Guppy)—Watters *et al.*, 2005 (LT)—exotic.

Pomoxis annularis (White Crappie)—Lefevre & Curtis, 1910a (NI); Wilson, 1916 (NI); Morrison *in* Clarke & Berg, 1959 (NS); Watters *et al.*, 2005 (LT).

Pomoxis nigromaculatus (Black Crappie)—Wilson, 1916 (NI); Trdan & Hoeh, 1982 (NI, LT).

Rhinichthys atratulus (Blacknose Dace)—Trdan & Hoeh, 1982 (LT).

Rutilus rutilus (Roach)—Lefevre & Curtis, 1910a (NI)—exotic.

Semotilus atromaculatus (Creek Chub)—Trdan & Hoeh, 1982 (LT); Watters *et al.*, 2005 (LT).

Habitat. A widespread and common species found in nearly every type of substrate and water flow.

Range. Widespread from the mouth of the Mackenzie River in the Northwest Territories across Canada to the Great Lakes drainages, St. Lawrence River system, Mississippi River system, Gulf drainages from Ochlockonee and Apalachicola rivers to Texas and northern Mexico. Not found on the Atlantic Slope or peninsular Florida. Brim Box & Williams (2000) synonymized *A. gesnerii* Lea, 1858, and *A. hallenbeckii* Lea, 1858, both from Uphapee Creek, Alabama, with *P. grandis*.

Range in Ohio. The most common species in the state, it is widespread and common in ponds, impoundments, and slack-water reaches of rivers and creeks. It often invades farm ponds where it may become abundant. Nearly ubiquitous in the state and probably more common now than ever before due to the presence of impoundments.

Similar species. See under *Pyganodon cataracta marginata*. This species also resembles *Anodontoides ferussacianus* and *Strophitus undulatus*. Both of those species have single-looped beak sculpture.

Comments. A form or subspecies is sometimes recognized: *Pyganodon grandis corpulenta* (Cooper, 1834). It can be much larger than typical *P. grandis*, is greatly inflated, and often has a peculiar copper-purple nacre. This form turns up across the range of *P. grandis* in isolated colonies and appears to be an ecophenotype. In Ohio a population occurs in Cowan Lake.

Individuals grow quickly for 2–3 years and then gradually slow. They grow to ca. 210 mm (particularly the form *corpulenta*). Even the largest individuals rarely live for more than 10 years.

Quadrula Rafinesque, 1820

Etymology. L. *quadrula*, a little square; from L. *quadra*, a square, + L. *-ula* (diminutive). Not all *Quadrula* species are quadrate. Many are rectangular, and many are round or nearly so. The roundish species do approach being equilateral and, in so doing, also approach, in the broadest sense, being quadrate.

Original description

Obliquaria (*Quadrula*) Rafinesque, 1820: 305 [*Quadrula* was proposed as a subgenus of *Obliquaria* to receive those species that were essentially square but somewhat longer than high and rounded anteriorly. It originally included pustulose and apustulose species.].

Type species. *Obliquaria* (*Quadrula*) *quadrula* Rafinesque, 1820, by absolute tautonomy and subsequent designation of Herrmannsen (1848: 384). Simpson's (1900a: 765) later selection of *Quadrula metanevra* Lea, 1820, is invalid.

Synonymy

Obliquaria (*Quadrula*) Rafinesque, 1820: 305.

Theliderma Swainson, 1840: 378 [Type species: *Unio lacrymosa* , by subsequent designation of Simpson (1900: 775)].

Telederma Paetel, 1875: 203 [Unjustified emendation of *Theliderma* Swainson, 1840].

Orthonymus Agassiz, 1852: 48 [Type species: *Unio cylindricus* Say, 1816, by original designation].

Amphinaias Crosse & Fischer, 1894: 556 [Type species: *Unio couchiana* Lea, 1860, by original designation].

Quadrula (*Bullata*) Frierson, 1927: 48 [*non* Jousseaume, 1875 (Mollusca: Marginellidae). Type species: *Unio pustulosa* Lea, 1829, by original designation].

Geographic range. Mississippi River system, Mobile River system and Gulf drainages from the Mobile to Texas. No *Quadrula* occur on the Atlantic Slope.

Comments. Frierson (1927: 52) remarked that some of the species in this genus "constitute a chain of such marked family likeness that they provide an almost ideal proof of evolution."

Serb *et al.* (2003) examined the phylogeny of *Quadrula* using the mitochondrial ND1 gene. They resolved the species into four groups (*pustulosa*, *quadrula*, *Tritogonia verrucosa*, and *metanevra*) and included all within the genus *Quadrula*, thus subordinating *Tritogonia* to a synonym of *Quadrula* despite the substantial conchological differences between the two. Although not named in Serb *et al.* (2003), all four groups had been previously recognized: *Bullata* Frierson, 1927, for their *pustulosa* group; *Quadrula*

Rafinesque, 1820, for their *quadrula* group; *Tritogonia* Agassiz, 1852; and *Orthonymus* Agassiz, 1852, for their *metanevra* group. Given the genetic and conchological differences, we would advocate that all four be raised to genus-level status (although a new name would need to be coined for Frierson's preoccupied *Bullata*). However, we await further studies before either raising or sinking these generic names.

Quadrula cylindrica cylindrica (Say, 1817)

Figures 175, 176

Etymology. L. *cylindrus*, cylinder, roller, from Gr. *kylindros*. It would appear that the type specimen, given to Say by Professor Barton, and now lost, was the cylindrical downstream form rather than the relatively compressed form found upstream.

Vernacular name(s). *Rabbitsfoot*, Cob Shell.

Original description

Unio cylindricus Say, 1817: no pagination, article on "Conchology," pl. 4, fig. 3.

Type locality. Say gave the type locality as "From . . . Professor [Benjamin Smith] Barton; it is said to be found in the river Wabash." Unfortunately, the type locality becomes "Nordamerika" because of the selection of Haas's neotype.

Type material. The original unionoid material of Say has apparently been lost (Johnson & Baker, 1973: 146). Haas (1930: 327) selected SMF No. 4310 as a neotype and No. 4540 as a neoparatype. Both specimens are from "Nordamerika."

Synonymy

Unio cylindricus Say, 1817: no pagination, pl. 4, fig. 3.

Unio (*Eurynia*) *solenoides* Rafinesque, 1820: 298 [Type locality: "dans la partie supérièure de l'Ohio" ("in the upper part of the Ohio River"). Lectotype ANSP 20204 by selection of Johnson & Baker (1973: 170)].

Unio naviformis Lamarck, 1819: 75 [Type locality: "Ohio River." Holotype MNHN unnumbered *fide* Johnson, 1963].

Mya cylindricus (Say, 1817). Eaton, 1826: 219.

Margarita (*Unio*) *cylindricus* (Say, 1817). Lea, 1836: 17.

Unio (*Theliderma*) *cylindrica* (Say, 1817). Swainson, 1840: 271, fig. 54c.

Margaron (*Unio*) *cylindricus* (Say, 1817). Lea, 1852b: 23.

Orthonymus cylindricus (Say, 1817). Agassiz, 1852: 48.

Unio rugosus Chenu, 1859: 138, fig. 668 [*non* Barnes, 1823, *nec* Poiret, 1801].

Quadrula cylindrica (Say, 1817). Lewis, 1870: 218.

Quadrula (*Quadrula*) *cylindrica* (Say, 1817). Simpson, 1900a: 773.

Unio cilindricus [*sic*] var. *propetipicus* De Gregorio, 1914: 41–42, pl. 4, fig. 1 [Type locality: "Scieta [Scioto] R." [Ohio]. Type in Palermo Museum?].

Unio cilindricus [*sic*] var. *acrispatus* De Gregorio, 1914: 41, pl. 4, fig. 2 [Type locality: "Scieta [Scioto] R." [Ohio]. Type in Palermo Museum?].

Quadrula cylindrica cylindrica (Say, 1817). Oesch, 1984: 91.

Status. Listed as Endangered by Ohio Division of Wildlife; not listed by U.S. Fish & Wildlife Service as of 2009.

ADULT SHELL

Shell. Medium to fairly large sized (to 130 mm), thick, very elongate, squared-off posteriorly and rounded anteriorly, compressed to nearly cylindrical.

Anterior margin. Short and straight, then broadly rounded to the ventral border, often produced slightly dorsally of the dorsal margin.

Ventral margin. Rounded to the sulcus, then slightly concave (rarely straight) to dorsal ridge.

Posterior margin. The posterior slope is extended into a pointed, broad flange; concave dorsally to the dorsal margin.

Dorsal margin. Very long, forming a well-developed wing, straight or arched, joining posterior margin with slightly obtuse or right angle.

Umbo. Broad and low; umbonal sculpture ca. 5 irregular chevron ridges, usually nodulose, particularly posteriorly; chevrons broken into individual pustules anteriorly or forming irregular parallel ridges; sculpture continues across weak sulcus.

Ligament. Long and prominent, tan to dark brown, not concealed by shells.

Disc sculpture. Sculpture varies from densely pustulose and irregularly ribbed to nearly smooth; all specimens have a heavy nodulose rib bounding the posterior edge of the sulcus; dorsal slope forming a wing; wing with oblique, irregular cords which may or may not overlay a series of low, undulating ribs.

Periostracum. The base color of most specimens is greenish or yellowish overlain with abundant green, inverted tent-shape markings, some large, some minute; in adults these markings may change to rays or streaks of green; pustules, particularly on juveniles, may be lighter colored; in some populations the shell is unicolored yellow, green, or tan.

Nacre. Porcelain white, often iridescent posteriorly; occasionally flushed with pink on the hinge and in umbonal cavity; often with gold or brown Tulberg layers.

Hinge plate. Well-developed; right cardinal a single, massive, triangular denticle with a much smaller anterior ridge; left cardinal bifid, the teeth often far apart; laterals well-developed, very long, straight or barely arched, short,

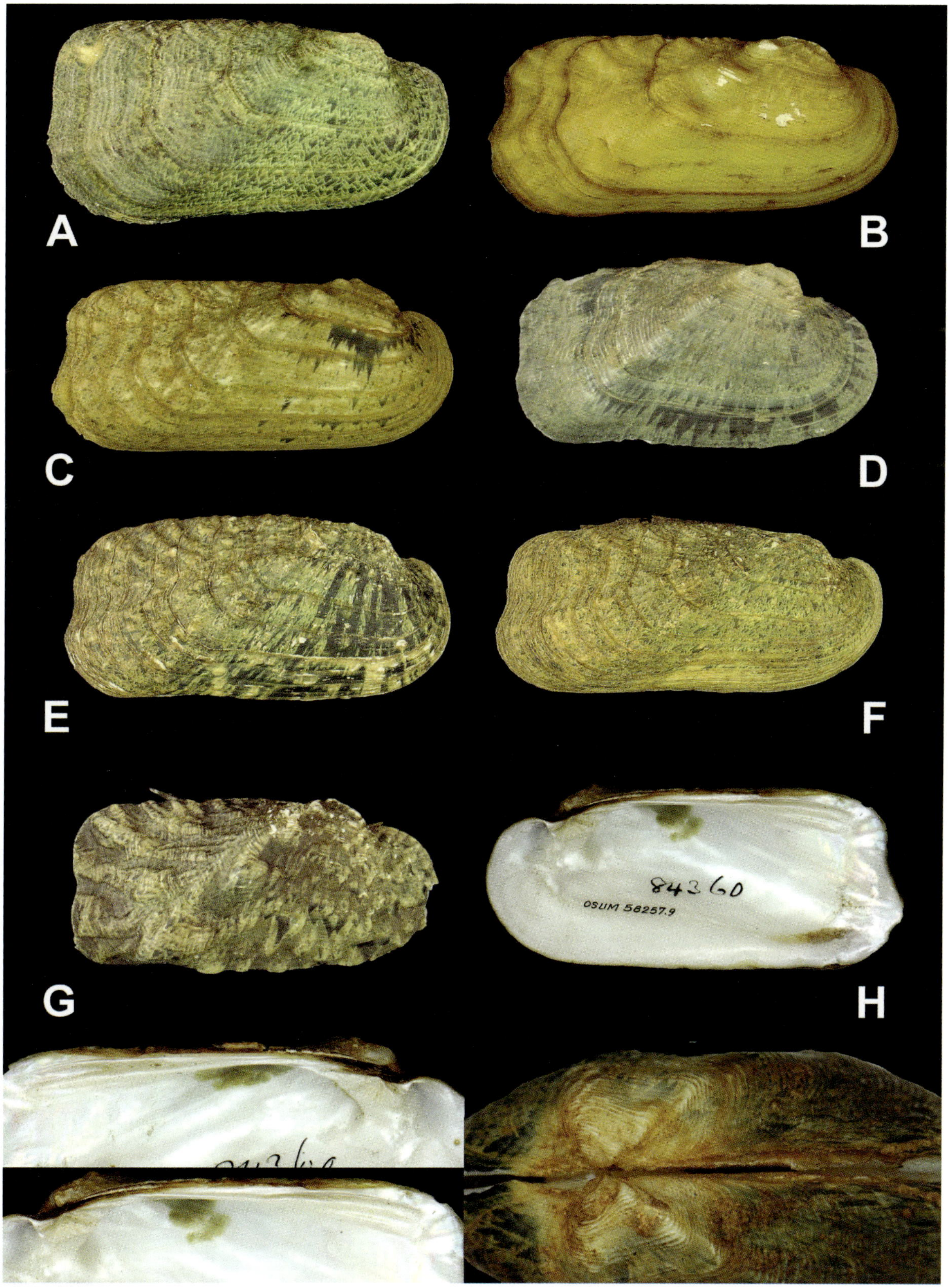

Fig. 175. ***Quadrula cylindrica cylindrica*** **(Say, 1817). A.** OSUM 18066, Walhonding River, Coshocton Co., OH. 75 mm. **B.** OSUM 33613.1, St. Francis River, Wayne Co., AR. 80 mm. **C, H, Hinge detail:** OSUM 58257.9, Ohio River, Hamilton Co., OH. 81 mm. **D, Beak sculpture detail:** OSUM 58257.3, Ohio River, Hamilton Co., OH. 34 mm. **E.** OSUM 31308.4, Fish Creek, Williams Co., OH. 103 mm. **F.** OSUM 7524.2, Big Darby Creek, Pickaway Co., OH. 110 mm. **G.** OSUM 52494.4, Duck River, Marshall Co., TN. 41 mm.

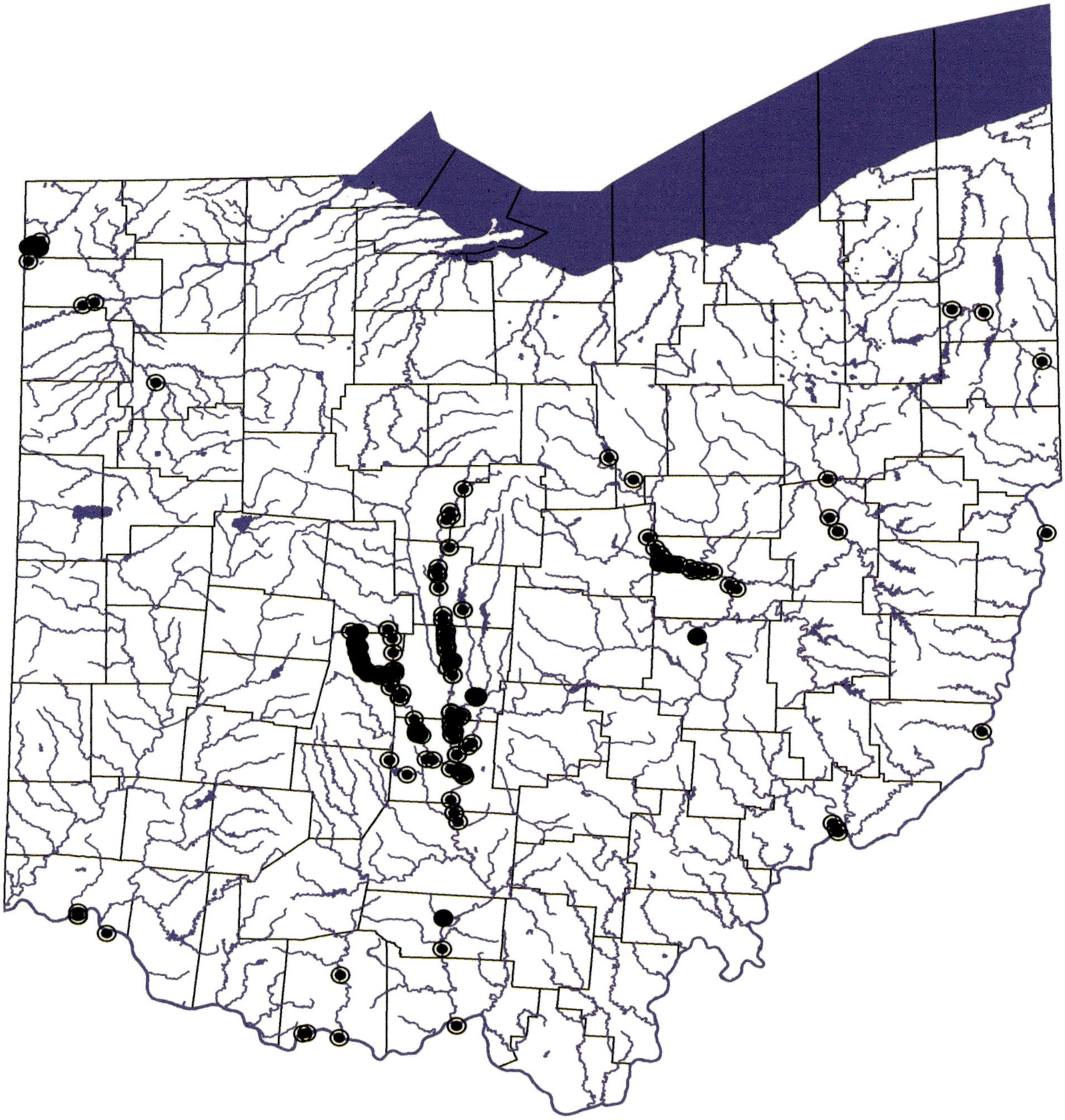

Fig. 176. ***Quadrula cylindrica cylindrica*** **(Say, 1817)**
Solid dots—live or freshdead since 1980
Bullseye—all records, any condition

one in right valve, two in left; interdentum cord-like, narrow, long.
Muscle scars. Anterior adductor impressed; posterior adductor flush; anterior adductor distinct from anterior pedal protractor scar, large posterior adductor attached but rarely confluent with posterior pedal retractor scar; dorsal suspensors located behind anterior adductor and in a line inside of interdentum, anterior- and posterior-most suspensor scars the largest. Pallial line entire, located far from shell margin. Accessory line weak, reaching 25% to 75% the length of the shell.
Umbonal cavity. Very wide, compressed, and deep.

Glochidia. Subcircular with equal length and height, lateral margins equal, dorsal margin straight, ventral margin subcircular, without hooks (Ortmann, 1919: 54).

Literature records (Ortmann, 1919: 54) indicate glochidial dimensions for length (0.19 mm) and height (0.19 mm). Ortmann (1919: 54) stated, "Glochidia have been observed in specimens from the Holston River in Tennessee [*Q. c. strigillata*]. Their shape is subcircular; length and height about the same. Thus they resemble those of *Q. metanevra* (0.18 x 0.19), but are more nearly circular."

Reproductive biology. Tachytictic [Ortmann's (1919: 54) statement that it is bradytictic clearly is a typographic error]. Females develop eggs from June to August, and glochidia are released in August (Utterback, 1915). All four gills are used as marsupia, and glochidia are discharged in tan to orange, lanceolate conglutinates (Ortmann, 1919).

Reported potential hosts
Etheostoma caeruleum (Rainbow Darter)—Watters *et al.*, 2009 (LT).
Luxilus chrysocephalus (Striped Shiner)—Watters *et al.*, 2009 (LT).

The following are host identifications for the Cumberlandian subspecies *Quadrula cylindrica strigillata* (Wright, 1898).

Cyprinella galactura (Whitetail Shiner)—Jenkinson, 1983 (LT); Yeager & Neves, 1986a (LT).
Cyprinella spiloptera (Spotfin Shiner)—Yeager & Neves, 1986a (LT); Yeager & Neves, 1986b (LT).
Hybopsis amblops (Bigeye Chub)—Yeager & Neves, 1986a (LT); Yeager & Neves, 1986b (LT).

Habitat. This now rare species occurs in high-water-quality creeks and small rivers, often found unburied along the water's edge. It will usually bury itself only under drought conditions.

Range. Mississippi River drainage from Minnesota to northern Louisiana, including Tennessee and Cumberland rivers. West to Arkansas and Missouri. Western Lake Erie drainages.

Range in Ohio. Sporadic and rare, extirpated from most of its range. Little Darby Creek (extremely rare in Big Darby Creek), Fish Creek, Walhonding River. Historically throughout the Scioto drainage, the upper Muskingum drainages, Maumee and Auglaize rivers, Mahoning River, Ohio Brush Creek.

Similar species. The elongated, pustulose shell is unique among Ohio mussels. *Tritogonia verrucosa* is similar but lacks the tent-like markings.

Comments. The shape of the shell and coloration of the animal do not suggest a close affinity with any other *Quadrula* with the possible exception of *Quadrula metanevra* (Rafinesque, 1820). Agassiz (1852) erected the genus *Orthonymus* for this species, setting it apart from other *Quadrula*. This action was not without merit as this species is certainly unique.

Quadrula cylindrica strigillata (Wright, 1898) occurs in the Tennessee River system. It differs in being more compressed and heavily sculptured, although some creek forms of *Q. c. cylindrica* are extremely similar.

Individuals gradually slow in growth during their entire life. Specimens may reach ca. 130 mm and 20 years old. The coarseness of the sculpture varies between populations; some may be quite nodulose whereas others are nearly smooth. However, all individuals are sculptured as juveniles.

Quadrula fragosa (Conrad, 1835)
Figures 177, 178

Etymology. L. *fragosus-*, broken, rough, uneven. This is an obvious reference to the coarse sculpture of the shell.
Vernacular name(s). *Winged Mapleleaf*, False Mapleleaf, Weltwing Mapleleaf.
Original description
Unio fragosus Conrad, 1835: 12, pl. 6, fig. 2.
Type locality. "Scioto river, Ohio."
Type material. Figured type apparently lost (Johnson & Baker, 1973).
Synonymy
Unio rugosus Barnes, 1823: 126–127, fig. 9 [*non Unio rugosus* Poiret, 1801 = *Margaritifera crassa* (Retzius, 1778) *fide* Simpson, 1900a: 678; 1914: 518. Type locality: "Ohio." Type lost].

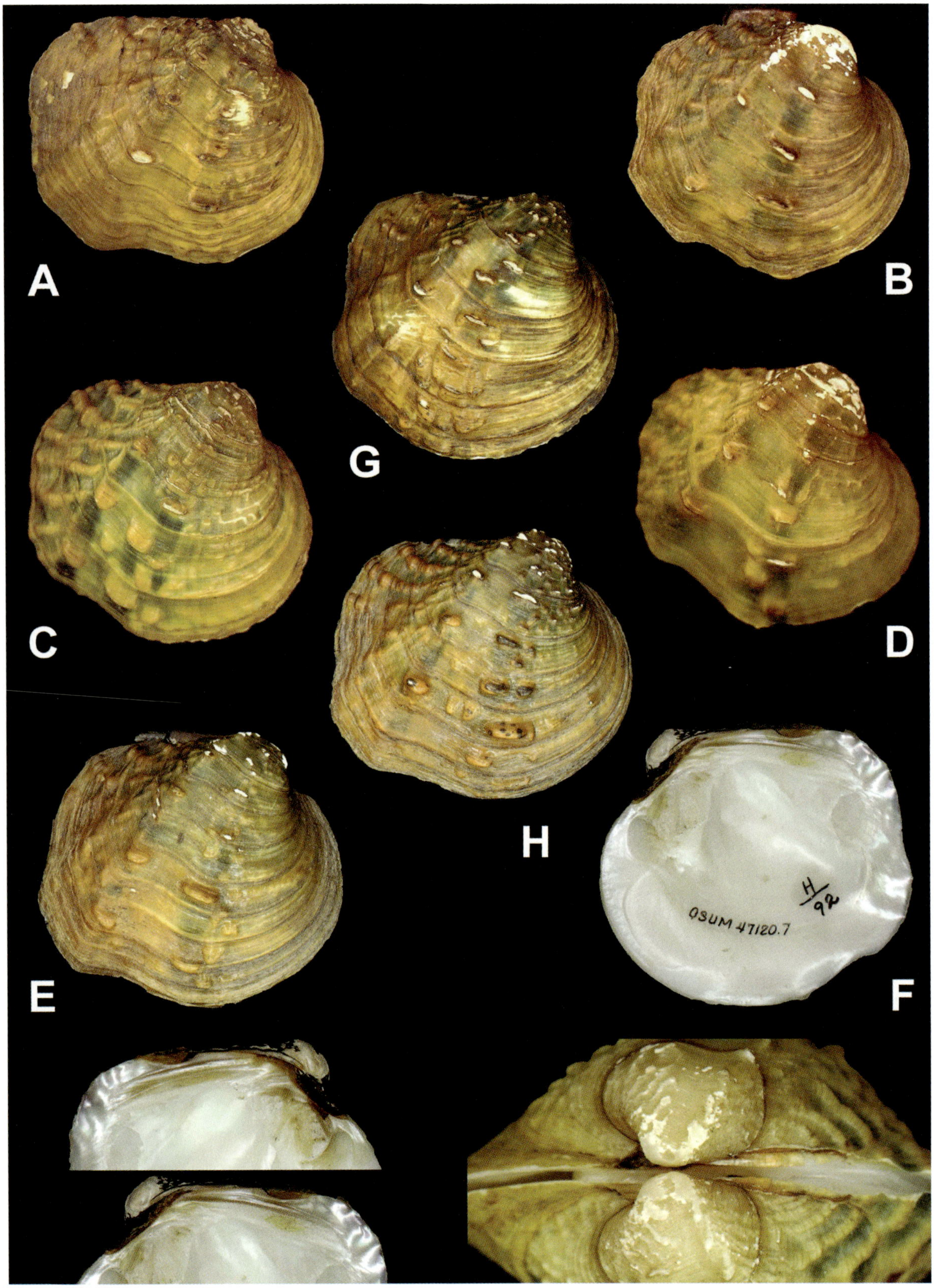

Fig. 177. ***Quadrula fragosa*** **(Conrad, 1835). A.** OSUM 47120.8, Scioto River, [Co.?], OH. 76 mm. **B.** OSUM 56855.2, Ohio River, Hamilton Co., OH. 52 mm. **C.** OSUM 39038, Wabash River, [Co.?], IN. 45 mm. **D.** OSUM 47120.1, Scioto River, [Co.?], OH. 24 mm. **E.** OSUM 30123.4, St. Croix River, Chisago Co., MN. 60 mm. **F, Hinge detail:** OSUM 47120.7, Scioto River, [Co.?], OH. 61 mm. **G.** OSUM 10081, "Ohio." 66 mm. **H.** OSUM 57453, Mississippi River, Scott Co., IA. 62 mm. **Beak sculpture detail:** 47120.2, Scioto River, [Co.?], OH. 28 mm total length.

Unio fragosus Conrad, 1835: 12, pl. 6, fig. 2.
Margarita (*Unio*) *fragosus* (Conrad, 1835). Lea, 1836: 14.
Unio tragosus [*sic*] Conrad, 1835. Hanley, 1843: 178, pl. 20, fig. 40.
Unio fragosa Conrad, 1835. Catlow & Reeve, 1845: 59.
Quadrula rugosus (Barnes, 1823). Agassiz, 1852: 48.
Quadrula fragosus (Conrad, 1835). Agassiz, 1852: 48.
Margaron (*Unio*) *fragosus* (Conrad, 1835). Lea, 1852b: 22.
Quadrula (*Quadrula*) *fragosa* (Conrad, 1835). Simpson, 1900a: 777–778.
Quadrula fragosa (Conrad, 1835). Chadwick, 1906: 96.
Quadrula quadrula fragosa (Conrad, 1835). Ortmann, 1925: 335.
Quadrula (*Quadrula*) *quadrula fragosa* (Conrad, 1835). Frierson, 1927: 47.

Status. Listed as Extirpated by the Ohio Division of Wildlife; listed as Endangered by U.S. Fish & Wildlife Service as of 2009.

ADULT SHELL

Shell. Medium size (to 90 mm), thick, squarish in profile, rather inflated.
Anterior margin. Short and straight, then broadly rounded to the ventral border.
Ventral margin. Rounded to the sulcus, then quite concave to dorsal ridge.
Posterior margin. The posterior slope is extended into a blunt, broad flange; concave dorsally to the dorsal margin.
Dorsal margin. Very long, forming a moderately well developed wing, straight or arched, joining posterior margin with slightly obtuse or right angle.
Umbo. Rather narrow and projecting well above the dorsal margin; umbonal sculpture has approximately 5 chevron ridges, usually nodulose, particularly posteriorly, chevrons broken into individual pustules anteriorly, with well-defined sulcus.
Ligament. Short and prominent, tan to dark brown, not concealed by shells.
Disc sculpture. Sulcus bounded anteriorly and posteriorly by heavy, nodulose ribs; sulcus may have pustules; anterior disc of shell covered with dense, broken, pustulose chevrons, particularly as juveniles; these rarely persist past ca. 3 years in age. Dorsal slope with nodulose, widely separated ribs, rarely broken into a series of radiating pustules.
Periostracum. Green or yellow with lighter pustules as juveniles, reddish brown or yellow-green as adults, often with vague green stripes, particularly down the anterior-most row of knobs.
Nacre. Porcelain white, often iridescent posteriorly; often with gold or brown Tulberg layers.
Hinge plate. Well-developed; right cardinal a single, massive, triangular denticle with a much smaller anterior (and occasionally posterior) ridge; left cardinal bifid, the teeth often far apart; laterals well-developed, straight or barely arched, short, one in right valve, two in left; interdentum flattened, broad, rather long.
Muscle scars. Anterior adductor impressed; posterior adductor flush; anterior adductor distinct from anterior pedal protractor scar, large posterior adductor attached to but rarely confluent with posterior pedal retractor scar; dorsal suspensors located behind anterior adductor and in a line inside of interdentum, anterior- and posterior-most suspensor scars the largest. Pallial line entire, located far from shell margin. Accessory line strong, reaching 25% to 75% the length of the shell.
Umbonal cavity. Very wide, somewhat compressed, and deep.

Glochidia. Unknown.

Reproductive biology. In Wisconsin, gravid females were found only in September and early October. Conglutinates were 10 x 5 mm tapered rods (Heath *et al.*, 1998).

Reported potential hosts
Ictalurus furcatus (Blue Catfish)—Steingraeber *et al.*, 2004 (LT).
Ictalurus punctatus (Channel Catfish)—Hove *et al.*, 2000 (LT), 2001 (LT); Steingraeber *et al.*, 2004 (LT).

Habitat. Muddy sand in slow-moving rivers.

Range. Upper Mississippi River system, but extirpated from most of this range. Possibly still present in the Ohio River based on a specimen found in the 1980's. Specimens from west of the Mississippi River may represent a different species.

Range in Ohio. Historically from the Scioto River (the type locality) at Columbus and Circleville and the Ohio River at Cincinnati. Presumably extirpated. The last collected specimen was from Raccoon Creek, now a creek severely polluted by acid mine runoff. Absent from Lake Erie and its tributaries.

Similar species. *Quadrula fragosa* is obviously closely related to *Quadrula quadrula*. It differs from that species in having a more pronounced dorsal wing and a much coarser overall sculpture. On the posterior wing the sculpture of *Q. fragosa* consists of a few irregular nodulose ribs, whereas the same region in *Q. quadrula* has numerous beaded cords. *Quadrula fragosa* tends towards dark brown and green coloration; *Q. quadrula* is often more yellowish or tan.

Fig. 178. ***Quadrula fragosa*** **(Conrad, 1835)**
Solid dots—live or freshdead since 1980
Bullseye—all records, any condition

Comments. Individuals gradually slow in growth during their entire life. Specimens may reach ca. 90 mm and 20 years old. It does not get as large as *Quadrula quadrula* and is always heavily sculptured.

Quadrula metanevra (Rafinesque, 1820)

Figures 179, 180

Etymology. Gr. *meta-*, in the midst of, between, after, behind + Gr. *neura*, string or cord of sinew. The posterior slope of individuals of this species are almost always sculptured with a series of curved, sinew-like ribs from the posterior ridge to the post-dorsal margin. Poulson's translation (1832: 40) of Rafinesque's (1820: 305–306) original description of *U. metanevra* noted that the shell is "nerved posteriorly" and that there are nerves "obliquely curved on the dilated posterior edge." The name *metanevra* literally means "ribbed behind."

Vernacular name(s). *Monkeyface*, Knobbed Rock Shell.

Original description

Obliquaria (*Quadrula*) *metanevra* Rafinesque, 1820: 305–306, pl. 81, figs. 15, 16.

Type locality. Rafinesque's material came from "dans le Kentuky [River]." However, the neotype of Johnson & Baker's (1973) designation was from the Ohio River. This designation may be invalid.

Type material. Johnson & Baker (1973: 162, pl. 3, fig. 4) selected type material from the ANSP collection. "From the specimens identified by Rafinesque, mentioned as types by Vanatta, 1915, PANSP 67:556, is here selected a neotype ANSP 20238a from the Ohio River." They also cited additional Rafinesque-identified specimens: ANSP 20239 and MNHN (uncatalogued).

Synonymy

Obliquaria (Quadrula) metanevra Rafinesque, 1820: 305–306, pl. 81, figs. 15–16.

Unio nodosus Barnes, 1823: 124–125, fig. 7 [Type locality: "Oiusconsin [*sic*]." Types lost].

Mya nodosus (Barnes, 1823). Eaton, 1826: 216.

Unio metanever [*sic*] (Rafinesque, 1820). Short & Eaton, 1831: 76.

Unio metaneverus [*sic*] (Rafinesque, 1820). Say, 1834: no pagination.

Margarita (*Unio*) *metanever* [*sic*] (Rafinesque, 1820). Lea, 1836: 15.

Unio (*Theliderma*) *metanevra* (Rafinesque, 1820). Swainson, 1840: 268, figs. 50, 54b.

Unio meteniver [*sic*] (Rafinesque, 1820). Catlow & Reeve, 1845: 61.

Unio metaneurus [*sic*] (Rafinesque, 1820). Küster, 1852: 50, pl. 10, fig. 4.

Quadrula metanever [*sic*] (Rafinesque, 1820). Agassiz, 1852: 48.

Quadrula nodosus (Barnes, 1823). Agassiz, 1852: 48.

Margaron (*Unio*) *metanevrus* (Rafinesque, 1820). Lea, 1852b: 22.

Unio wardii Lea, 1861d: 392 [Type locality: "Walhonding River, Ohio. Wassepinicon River, Iowa. Coal River, Logan co., Va"; actual type locality is probably Coal River *fide* Lea *in litt. in* Johnson, 1956. Lectotype USNM 84237].

Margaron (*Unio*) *wardii* (Lea, 1861). Lea, 1870: 33.

Quadrula (*Quadrula*) *metanevra* (Rafinesque, 1820). Simpson, 1900a: 774.

Quadrula (*Quadrula*) *metanevra* var. *wardii* (Lea, 1861). Simpson, 1900a: 774.

Quadrula metanevra (Rafinesque, 1820). Ortmann, 1912a: 255, figs. 6, 6a.

Quadrula (*Orthonymus*) *metanevra* (Rafinesque, 1820). Frierson, 1927: 51.

Quadrula (*Orthonymus*) *metanevra wardii* (Lea, 1861). Frierson, 1927: 52.

Quadrula metanevra wardii (Lea, 1861). La Rocque, 1967: 136.

Orthonymus metanevrus metanevrus (Rafinesque, 1820). Haas, 1969: 309.

Orthonymus metanevrus wardii (Lea, 1861). Haas, 1969: 310.

Status. Listed as Endangered by Ohio Division of Wildlife; not listed by U.S. Fish & Wildlife Service as of 2009.

ADULT SHELL

Shell. Medium size (to 110 mm), thick, rectangular to squarish in profile, rather inflated.

Anterior margin. Broadly rounded to the ventral border.

Ventral margin. Rounded to the sulcus, then quite concave to dorsal ridge.

Posterior margin. The posterior slope is extended into a blunt, broad flange; concave dorsally to the dorsal margin.

Dorsal margin. Very long, forming a moderately well developed wing, straight or arched, joining posterior margin with slightly obtuse or right angle.

Umbo. Rather narrow and projecting well above the dorsal margin; umbonal sculpture with several very weak chevron or angular swellings, most developed posteriorly where the posterior ridge forms.

Ligament. Very short and prominent, tan to dark brown, not concealed by shells.

Disc sculpture. Sulcal area with a wide, prominent rib bearing longitudinally compressed, protruding knobs; central disc with coarse, chevron-shaped pustules; anterior third of shell usually smooth; posterior slope and wing

Fig. 179. ***Quadrula metanevra*** **(Rafinesque, 1820). A.** OSUM 57574.6, Ohio River, Hamilton Co., OH. 65 mm. **B, F, Hinge detail:** OSUM 41079.2, Ohio River, Spencer Co., IN. 52 mm. **C.** OSUM 23248, Wabash River, Tippecanoe Co., IN. 65 mm. **D.** OSUM 51943.1, Muskingum River, Washington Co., OH. 57 mm. **E.** OSUM 39400.1, "Wabash River." 25 mm. **G.** OSUM 60641, Mississippi River, Houston Co., MN. 83 mm. **H.** OSUM 41454.7, Meramec River, Crawford Co., MO. 80 mm. **Beak sculpture detail:** OSUM 39400.2, "Wabash River." 32 mm total length.

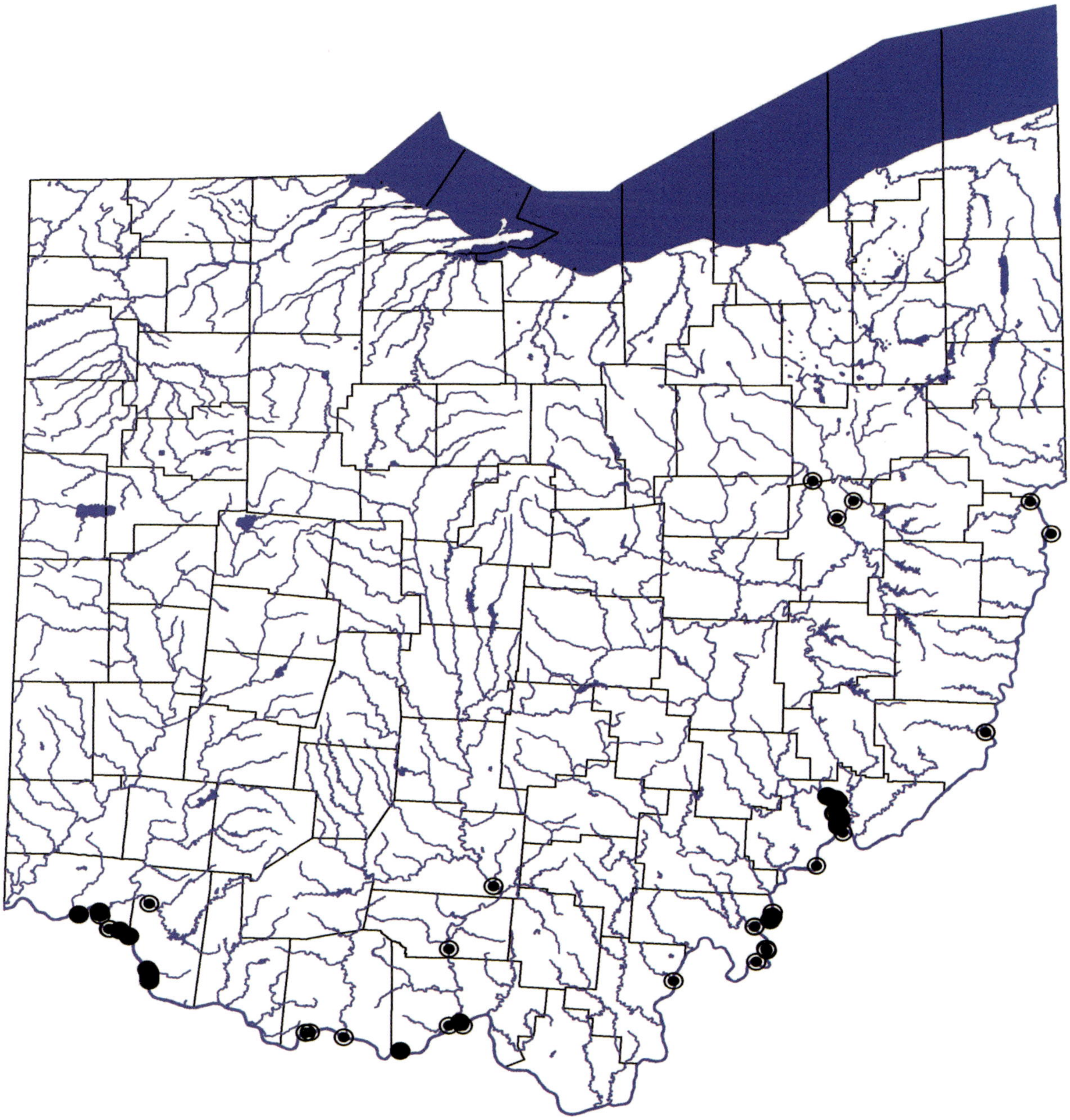

Fig. 180. ***Quadrula metanevra*** **(Rafinesque, 1820)**
Solid dots live or freshdead since 1980
Bullseye—all records, any condition

with oblique, coarse cords or ribs, sometimes nodulose.

Periostracum. Green, tan, or yellow, occasionally with lighter pustules, often with bold green, inverted tent-like markings of various sizes; in some individuals the markings become rays or streaks; some specimens nearly patternless; old individuals are usually reddish-brown.

Nacre. Porcelain white, often iridescent posteriorly; often with gold or brown Tulberg layers.

Hinge plate. Massive; right cardinal a single, massive, triangular denticle with a much smaller anterior ridge; left cardinal bifid, the teeth often far apart; laterals well-developed, very short, straight, one in right valve, two in left; interdentum flattened, broad, short.

Muscle scars. Anterior adductor impressed; posterior adductor flush or slightly impressed; anterior adductor distinct or just touching anterior pedal protractor scar, large posterior adductor confluent with posterior pedal retractor scar; dorsal suspensors located behind anterior adductor and in a line inside of interdentum, anterior- and posterior-most suspensor scars the largest. Pallial line entire, located far from shell margin. Accessory line strong, reaching 75% the length of the shell.

Umbonal cavity. Wide, very compressed, and deep.

Glochidia. Subelliptical, slightly higher than long, lateral margins equal, dorsal margin slightly curved, ventral margin subcircular, without hooks (Howard, 1914a: 17, fig. 31; Ortmann, 1919: 48–49; Utterback, 1915: 146).

Literature records (Lefevre & Curtis, 1910a: 97, fig. e; Lefevre & Curtis, 1912: 146, fig. 1e; Surber, 1912: 9, pl. 2, fig. 26; Howard, 1914a: 17, fig. 31; Ortmann, 1919: 48–49; Utterback, 1915: 146) indicate glochidial dimensions for length (0.18 to 0.20 mm) and height (0.19 to 0.22 mm). Utterback (1915: 146) noted, "Glochidia average, semi-elliptical, ventral margin rounded, spineless, medium size, hinge line undulate." This glochidium appears to be intermediate in size between the smaller *Q. quadrula* and *Q. cylindrica* and the larger *Q. nodulata* and *Q. p. pustulosa*.

Reproductive biology. Tachytictic. All four gills are marsupial (Utterback, 1915). Gravid females were reported from May to August (Heath *et al.*, 1998), but eggs were also present from May through July (Ortmann, 1919; Bingham, 1968). Fragile, leaf-shaped, whitish conglutinates are produced (Ortmann, 1912a; Utterback, 1915) of the amorphous mucus type (Watters, 2008).

Reported potential hosts

Cyprinella spiloptera (Spotfin Shiner)—Crownheart *et al.*, 2006 (LT).

Lepomis cyanellus (Green Sunfish)—Surber, 1913 (NI); Wilson, 1916 (NI); Coker *et al.*, 1921 (NI).

Lepomis macrochirus (Bluegill)—Surber, 1913 (NI); Howard, 1914 (NI); Wilson, 1916 (NI); Coker *et al.*, 1921 (NI).

Pimephales notatus (Bluntnose Minnow)—Crownheart *et al.*, 2006 (LT).

Rhinichthys atratulus (Eastern Blacknose Dace)—Crownheart *et al.*, 2006 (LT).

Sander canadensis (Sauger)—Howard, 1914 (NI); Coker *et al.*, 1921 (NI); Pearse, 1924 (NI).

Semotilus atromaculatus (Creek Chub)—Crownheart *et al.*, 2006 (LT).

Habitat. The Monkeyface occurs in moving water in muddy sand and cobble in large rivers.

Range. Mississippi River system from Minnesota to Louisiana, into Arkansas and Missouri. Absent from the Great Lakes drainages. Mobile River system.

Range in Ohio. Lower Muskingum River and Ohio River. Historically as far upstream as the Tuscarawas River and the lower Scioto River.

Similar species. *Quadrula metanevra* shares the peculiar tent-like markings with *Quadrula cylindrica*, but it is immediately separated by its squarish, rather than very elongate, shape. *Quadrula quadrula* and *Q. fragosa* are superficially similar, but where they have a sulcus, *Q. metanevra* has a row of pronounced knobs. *Obliquaria reflexa* also has a row of knobs, but these alternate in position from left to right valves. *Obliquaria reflexa* also lacks the tented color pattern.

Comments. Individuals gradually slow in growth during their entire life. Specimens may reach ca. 110 mm and 30 years old.

Quadrula nodulata (Rafinesque, 1820)

Figures 181, 182

Etymology. L. *nodulus*, a little knot + *-ata*, provided with. Individuals of this species are typically provided with a prominent row of nodules as external sculpture down both the central disc and posterior ridges of each valve.

Vernacular name(s). *Wartyback*, Winged Orb Shell, Winged Pimpleback.

Original description

Obliquaria (*Quadrula*) *nodulata* Rafinesque, 1820: 307, pl. 81, figs. 17, 18.

Type locality. "dans le Kentuky [River]."

Type material. The Rafinesque-Poulson Collection deposited at ANSP contained specimens of this species from the Kentucky River (Vanatta, 1915: 557). From this

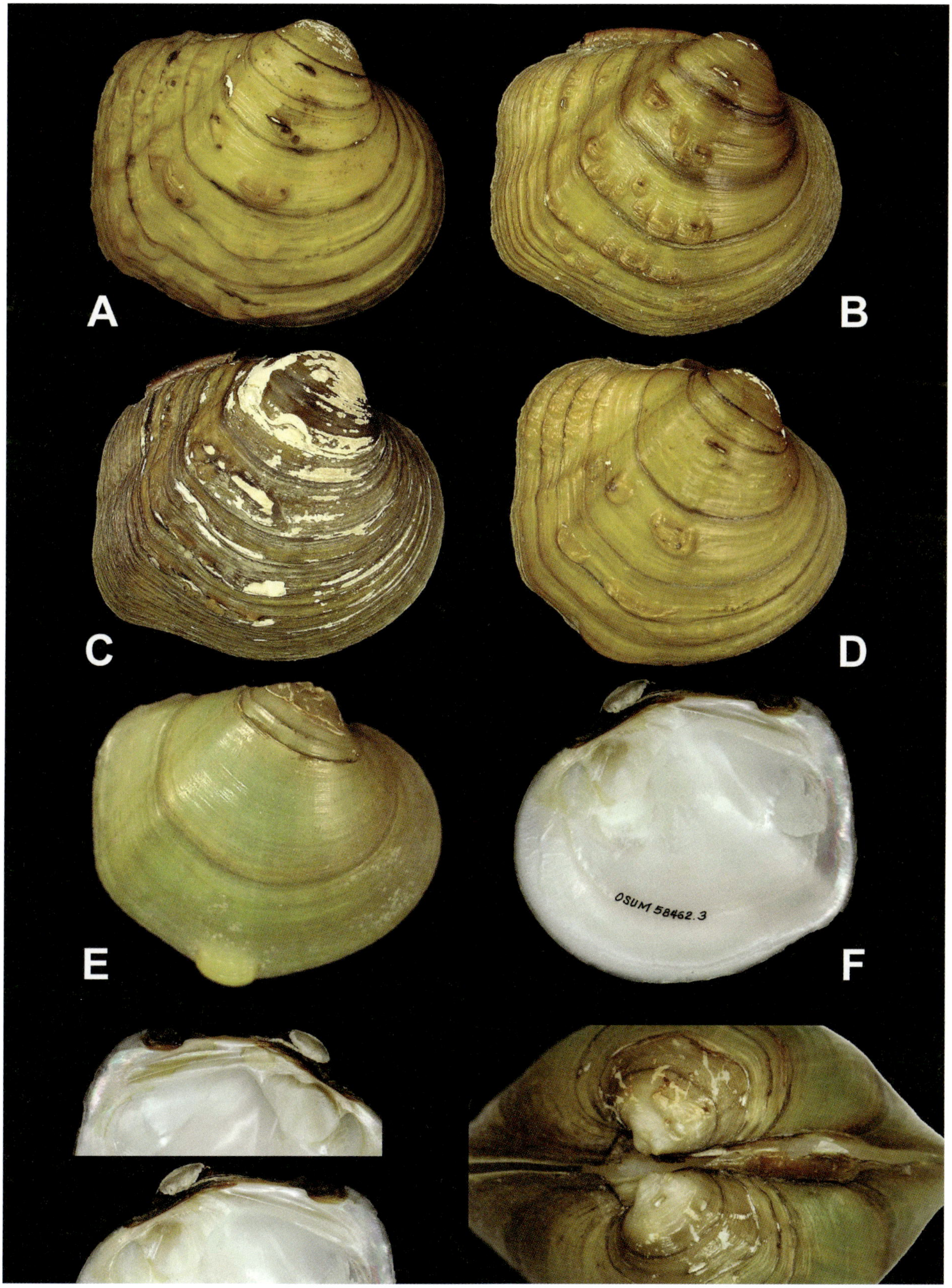

Fig. 181. ***Quadrula nodulata*** **(Rafinesque, 1820). A, F, Hinge detail:** OSUM 58462.3, Ohio River, Hamilton Co., OH. 61 mm. **B.** OSUM 55581, Wabash River, Wabash/Gibson Co., IL/IN. 63 mm. **C.** OSUM 44745, Salt River, Spencer Co., KY. 85 mm. **D.** OSUM 58462, Ohio River, Hamilton Co., OH. 57 mm. **E, Beak sculpture detail:** OSUM 12525, Mississippi River, Henderson Co., IL. 27 mm.

material, Johnson & Baker (1973: 163) selected a lectotype ANSP 20225a from the Kentucky River.

Synonymy

Obliquaria (*Quadrula*) *nodulata* Rafinesque, 1820: 307, pl. 81, figs. 17, 18.

Unio pustulatus Lea, 1831: 79, pl. 7, fig. 9 [Type locality: "Ohio; Tennessee." Lectotype USNM 84200].

Unio nodulatus (Rafinesque, 1820). Say, 1834: no pagination.

Margarita (*Unio*) *pustulatus* (Lea, 1831). Lea, 1836: 15.

Quadrula nodulatus (Rafinesque, 1820). Agassiz, 1852: 48.

Margaron (*Unio*) *pustulatus* (Lea, 1831). Lea, 1852b: 22.

Unio nodulatus var. *pustulatus* Lea, 1831. Paetel, 1890: 161.

Quadrula (*Quadrula*) *pustulata* (Lea, 1831). Simpson, 1900a: 781.

Quadrula pustulata (Lea, 1831). Baker, 1905: 256.

Quadrula nodulata (Rafinesque, 1820). Utterback, 1915: 139.

Quadrula (*Bullata*) *nodulata* (Rafinesque, 1820). Frierson, 1927: 49.

Quadrula (*Pustulosa*) *nodulata* (Rafinesque, 1820). Haas, 1969: 295.

Status. Listed as Endangered by Ohio Division of Wildlife; not listed by U.S. Fish & Wildlife Service as of 2009.

ADULT SHELL

Shell. Medium size (to 80 mm), thick, squarish to nearly circular, quite inflated.

Anterior margin. Short and straight, then broadly rounded to the ventral border.

Ventral margin. Broadly rounded in juveniles, with a slight concavity in most adults at the dorsal ridge.

Posterior margin. The posterior slope is weakly extended into a blunt, broad flange; concave dorsally to the dorsal margin.

Dorsal margin. Short, with a moderately well developed wing, sinuous under the umbo, arched, joining posterior margin with slightly obtuse or right angle.

Umbo. Somewhat narrow and projecting well above the dorsal margin; umbonal sculpture of a few very weak double-looped ridges, nodulose on the ventral portion of the loop; these nodules persist onto the adult shell; broken, irregular threads anteriorly on some specimens, no sulcus.

Ligament. Very short and prominent, tan to dark brown, not concealed by shells.

Disc sculpture. Disc of shell smooth except for two radiating rows of longitudinally compressed, prominent nodules, well-separated from each other, the posterior row on the posterior ridge; wing with irregular radiating, nodulose ribs or often without sculpture.

Periostracum. Green or yellow as juveniles, becoming brown or reddish-brown as adults; pustules often darker; dark annular rings; juveniles may have vague green stripes, but these are lost in adults.

Nacre. Porcelain white, often iridescent posteriorly; often with gold or brown Tulberg layers.

Hinge plate. Well-developed, angled at the umbo; right cardinal a single, massive, triangular denticle with a much smaller anterior (and occasionally posterior) ridge; left cardinal bifid, the teeth often far apart; laterals well-developed, short, arched, one in right valve, two in left; interdentum flattened, broad, very short.

Muscle scars. Anterior adductor impressed; posterior adductor flush or nearly so; anterior adductor distinct from anterior pedal protractor scar, large posterior adductor near but unconnected to posterior pedal retractor scar; dorsal suspensors located behind anterior adductor and in a line inside of interdentum, anterior- and posterior-most suspensor scars the largest. Pallial line entire, located far from shell margin. Accessory line strong, reaching 75% the length of the shell.

Umbonal cavity. Very wide, somewhat compressed, and deep.

Glochidia. Subelliptical, hinge straight, ventral margin rounded, lateral margin equal, hooks absent (Surber, 1912: 9, pl. 2, fig. 27).

Literature records (Surber, 1912: 9, pl. 2, fig. 27) indicate glochidial dimensions for length (0.20 to 0.23 mm) and height (0.25 to 0.29 mm). Surber (1912: 9) noted, "The glochidial characters are the same [as those of *Q. pustulosa*], except a difference in size, the glochidium of *nodulata* being larger."

Reproductive biology. Tachytictic. Baker (1928a) gave the "breeding" season as June to July. Metamorphosis has not been observed on any host.

Reported potential hosts

Ictalurus punctatus (Channel Catfish)—Wilson, 1916 (NI); Coker *et al.*, 1921 (NI).

Lepomis macrochirus (Bluegill)—Howard, 1914 (LI).

Micropterus salmoides (Largemouth Bass)—Howard, 1914 (LI).

Pomoxis annularis (White Crappie)—Surber, 1913 (NI); Wilson, 1916 (NI); Coker *et al.*, 1921 (NI).

Pomoxis nigromaculatus (Black Crappie)—Howard, 1914 (LI).

Pylodictis olivaris (Flathead Catfish)—Coker *et al.*, 1921 (NI).

Habitat. This species occurs in moving water in muddy sand and cobble in large rivers.

Range. Mississippi River system from Minnesota and Wisconsin to Louisiana, including the Cumberland and

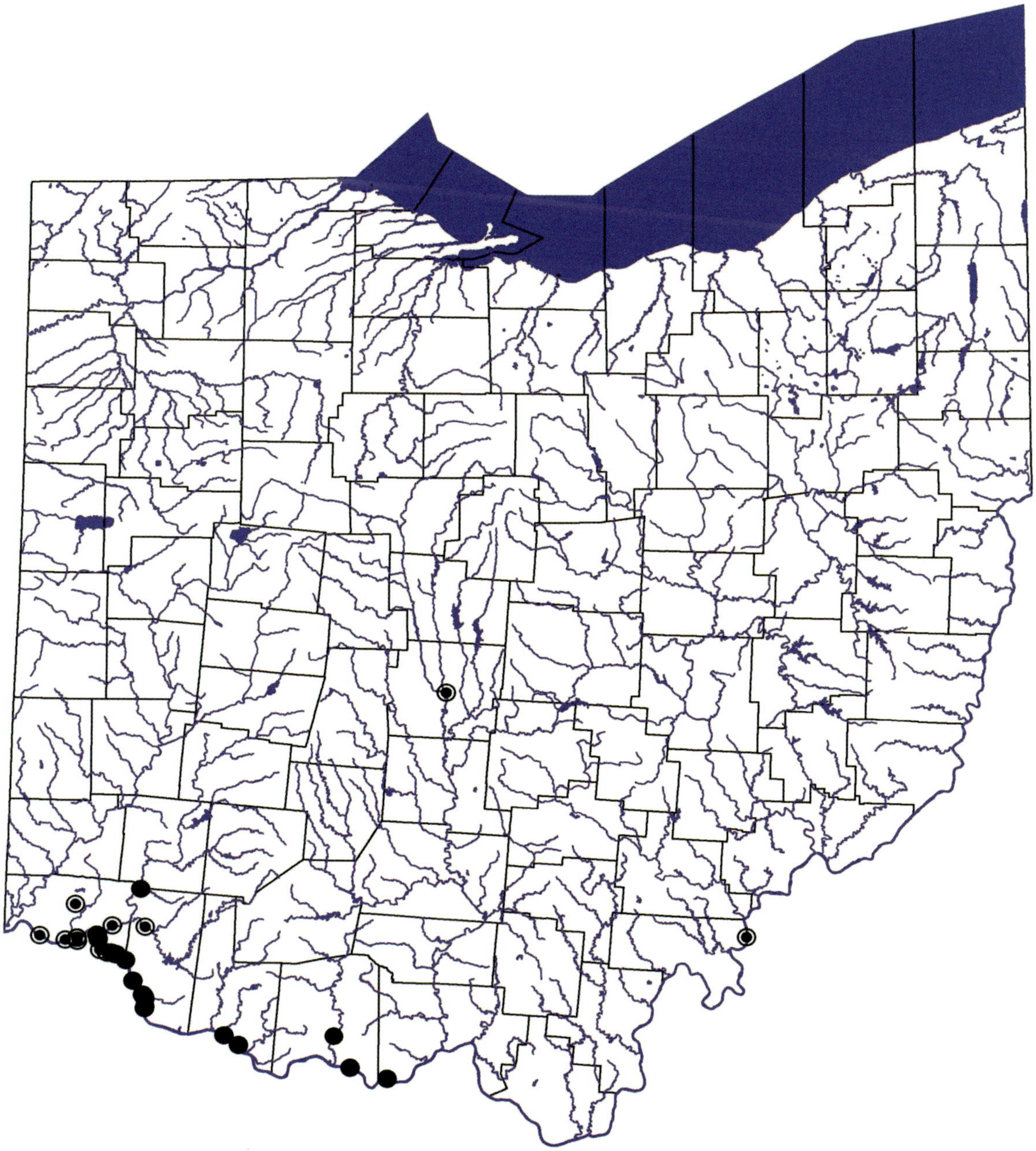

Fig. 182. ***Quadrula nodulata*** **(Rafinesque, 1820)**
Solid dots—live or freshdead since 1980
Bullseye—all records, any condition

Tennessee rivers. West of the Mississippi River to Kansas and Texas.

Range in Ohio. Currently in the Ohio River at Cincinnati and in Ohio Brush Creek. Historically as far upstream in the Ohio as Portland. Extirpated from the lower Great Miami River. A single stray record exists for the Scioto River at Columbus.

Similar species. This species is characterized by its tan, inflated shell and two radiating rows of pustules or knobs. It lacks the green ray or blotch of *Quadrula pustulosa*. *Obliquaria reflexa* has a single row of knobs. Unlike *Quadrula quadrula* and *Q. fragosa*, it lacks a sulcus.

Comments. Individuals gradually slow in growth during their entire life. Specimens may reach ca. 80 mm and 13 years old. The sculpture is very consistent among populations.

Quadrula pustulosa pustulosa (Lea, 1831)

Figures 183, 184

Etymology. L. *pustula*, a pimple or blister + *-osa*, full of. Individuals of this species very frequently develop pustules, a few or many, on the surface of the disc.

Vernacular name(s). *Pimpleback*, White Pimpleback, Round Pimpleback, Warty Pigtoe, Wartyback. Specimens of this species without pustules have been referred to as Smooth Pimpleback.

Original description

Unio pustulosus Lea, 1831: 76–77, pl. 7, fig. 7.

Type locality. "Hab. Ohio, T. G. Lea. Alabama river, Judge Tait" (Lea 1831: 76). The Alabama River specimen was not *Q. pustulosa* but its sibling species *Quadrula asperata* (Lea, 1861), common in the Alabama River.

Type material. The figured type of this species should have been in the Isaac Lea Collection that was deposited in the National Museum of Natural History (Smithsonian Institution). A search in 1973–1974 did not reveal its presence there. The figured type is apparently lost. Johnson (1974: 119) noted "syntypes ANSP 43058 from Ohio mentioned as in ANSP." It may well be that the type material of this species was part of the ANSP collection rather than the Isaac Lea Collection.

Synonymy

Obliquaria (*Quadrula*) *bullata* Rafinesque, 1820: 307–308 [*non Obliquaria flexuosa* var. *bullata* Rafinesque, 1820: 307 = *Epioblasma flexuosa* (Rafinesque, 1820). Type locality: "aux chutes de l'Ohio," but neotype from "Kentucky River" by Johnson & Baker (1973: 149), which may be an invalid designation. Rafinesque described *Obliquaria bullata* and *Obliquaria flexuosa bullata* in the same publication (1820). Although *Obliquaria flexuosa bullata* has line precedence, according to ICZN Art. 57.7, the species-group use of *bullata* takes precedence over the subspecies group use. This rule renders *Obliquaria bullata* available and *Obliquaria flexuosa bullata* unavailable by homonymy because subspecific and specific names compete for availability. Therefore Vanatta (1915: 557) was incorrect in believing *Obliquaria bullata* was not available. Indeed, as Morrison (1969: 24) pointed out, this is the earliest name for this taxon. Even Agassiz (1852: 48) mentioned *Quadrula bullata* in his list of *Quadrula*, but not *pustulosa*. The above conclusion hinges on the correct identification of Rafinesque's species with the taxon known as *Unio pustulosa* Lea, 1831. Johnson & Baker (1973: 149) designated a species of what is now known as *pustulosa* (ANSP 20250) from the Kentucky River as neotype of *O. bullata* Rafinesque, 1820. But there are nagging doubts that Rafinesque's name is another species. As pointed out by Ortmann & Walker (1922: 15), Rafinesque's description mentions an oblique furrow and rosy nacre—characteristics not found in any *U. pustulosa*. Zoological nomenclature would be best served by not applying this name to the species commonly known as *Quadrula pustulosa* (Lea, 1831), a widespread and important commercial species. We believe *Obliquaria bullata* Rafinesque, 1820, is a *nomen dubium* and suggest that the Commission be petitioned to suppress the name.].

Obliquaria (*Quadrula*) *retusa* Rafinesque, 1820: 306, pl. 81, figs. 19–20 [Type locality: "dans l'Ohio et le Kentuky [Rivers]," but the neotype is from the "Green River, Kentucky." Neotype ANSP 20220 by Johnson & Baker (1973: 168). This name had not been applied to this taxon with any certainty until Johnson & Baker's selection of a neotype. It is a *nomen oblitum*.].

Unio verrucosus Barnes, 1823. Valenciennes, 1827: 231, pl. 53, fig. 2 [misidentification].

Unio pustulosus Lea, 1831: 76, pl. 7, fig. 7.

Unio premorsus Rafinesque, 1831: 4 [Unnecessary replacement name for *Obliquaria retusa* Rafinesque, 1820, because of presumed priority of *Unio retusa* Lamarck, 1819].

Unio nodulosus (Wood, 1815). Say, 1834: no pagination [misidentification].

Unio bullatus (Rafinesque, 1820). Conrad, 1834: 68.

Unio prasinus Conrad, 1834: 44, pl. 3, fig. 1 [Type locality: given as "Scioto river, Indiana" (error for Ohio) but based on same specimen as *Unio schoolcraftensis* Lea, 1834 *fide* Frierson (1927); see below. Obviously

Fig. 183. ***Quadrula pustulosa pustulosa*** **(Lea, 1831). A.** OSUM 57573.2, Ohio River, Hamilton Co., OH. 66 mm. **B.** OSUM 7059.1, Scioto River, Ross Co., OH. 95 mm. **C.** OSUM 44252.3, Muskingum River, Washington Co., OH. 50 mm. **D.** OSUM 25910.8, Lake Erie, Ottawa Co., OH. 61 mm. **E.** OSUM 6834.3, Sandusky River, Seneca Co., OH. 59 mm. **F, Hinge detail:** OSUM 25910.2, Lake Erie, Ottawa Co., OH. 48 mm. **G.** OSUM 23401.1, Muskingum River, Washington Co., OH. 27 mm. **H.** OSUM 15355.2, Wabash River, Tippecanoe Co., IN. 67 mm. **Beak sculpture detail:** OSUM 6964, Olive Green Creek, Washington Co., OH. 15 mm total length.

a single specimen cannot originate from two places, indicating that the type locality for either *U. prasinus* or *U. schoolcraftensis* is incorrect. Types in ANSP 1128 lost *fide* Johnson & Baker (1973: 167)].

Unio schoolcraftensis Lea, 1834: 37, pl. 3, fig. 9 [Named for "Mr. Schoolcraft" but the original spelling must stand. Type locality: "Fox River of Green Bay" [WI]. Types in ANSP 1128 lost *fide* Johnson & Baker (1973: 169)].

Margarita (Unio) pustulosus Lea, 1836: 5, 15.

Margarita (Unio) schoolcraftensis (Lea, 1834). Lea 1836: 5, 15.

Unio dorfeuillianus Lea, 1838a: 73, pl. 17, fig. 54 [Type locality: "Ohio River." Lectotype USNM 84215].

Margarita (Unio) dorfeuillianus Lea, 1838b: 15 [*nomen nudum*].

Unio pernodosus Lea, 1845: 71, pl. 3, fig. 8 [Type locality: "North Carolina." Lectotype USNM 84298].

Quadrula bullatus (Rafinesque, 1820). Agassiz, 1852: 48.

Margaron (Unio) pustulosus (Lea, 1831). Lea , 1852b: 22.

Margaron (Unio) schoolcraftensis (Lea, 1834). Lea, 1852b: 22.

Margaron (Unio) dorfeuillianus (Lea, 1838). Lea 1852b: 22.

Margaron (Unio) schoolcraftii (Lea, 1834). Lea, 1870: 33 [emendation for *Unio schoolcraftensis* Lea, 1834, named for "Mr. Schoolcraft"].

Margaron (Unio) pernodosus (Lea, 1845). Lea, 1870: 34.

Unio schoolcraftii (Lea, 1870). Wright, 1888: no pagination.

Unio bullatus var. *prasinus* (Conrad, 1834). Paetel, 1890: 146.

Unio bullatus var. *schoolcraftensis* (Lea, 1834). Paetel, 1890: 146.

Quadrula pustulosa (Lea, 1831). Baker, 1898b: 86–88, pl. 24, fig. 2, pl. 25, pl. 27, fig. 13) [who acknowledges his source of the combination as Simpson's unpublished manuscript which was printed two years later (1900a)].

Quadrula (Quadrula) pustulosa (Lea, 1831). Simpson, 1900a: 779–780.

Quadrula (Quadrula) pustulosa var. *pernodosa* (Lea, 1845). Simpson, 1900a: 780.

Quadrula pustulosa var. *schoolcraftensis* (Lea, 1834). Simpson, 1914: 850–851.

Quadrula pustulosa schoolcraftensis (Lea, 1834). Utterback, 1915: 132.

Quadrula pustulosa asperatus (Lea, 1861). Utterback, 1915: 134, pl. 17, figs. 42a, b [misidentification].

Quadrula pustulosa pernodosa (Lea, 1845). Vanatta, 1915: 557.

Quadrula pustulosa prasina (Conrad, 1834). Ortmann & Walker, 1922: 16.

Quadrula (Bullata) pustulosa (Lea, 1831). Frierson, 1927: 48.

Quadrula (Bullata) pustulosa prasina (Conrad, 1834). Frierson, 1927: 48.

Quadrula (Bullata) pustulosa asperata (Lea, 1861). Frierson, 1927: 48 [in part, misidentification].

Quadrula bullata (Rafinesque, 1820). Morrison, 1969: 24.

Quadrula (Pustulosa) pustulosa pustulosa (Lea, 1831). Haas, 1969: 292.

Quadrula (Pustulosa) pustulosa prasina (Conrad, 1834). Haas, 1969: 292.

Quadrula (Pustulosa) pustulosa pernodosa (Lea, 1845). Haas, 1969: 293.

Status. Not listed by either Ohio Division of Wildlife or U.S. Fish & Wildlife Service as of 2009.

ADULT SHELL

Shell. Medium size (to ca. 80 mm), moderately heavy, outline subrotund with occasional subtriangulate or subquadrate specimens.

Anterior margin. Noticeably flattened dorsally, becoming strongly curved ventrally.

Ventral margin. Rounded anteriorly, becoming straight to emarginate postventrally, sharply curved dorsally where the posterior ridge intersects the margin.

Posterior margin. Very gently curved, straight or emarginate.

Dorsal margin. Straight in the region of the ligament, curving beneath the umbos, becoming straight to scarcely curved passing anteriorly from the umbos into a distinctly curved anterior margin.

Umbo. Anterior of center, prominent; beak sculpture of 2–4 irregular congruent ridges becoming strong on posterior ridge, disappearing anteriorly; beak sculpture not modified into disc sculpture.

Ligament. Brown to tan, heavy, exposed.

Disc sculpture. Surface smooth from umbonal sculpture through one or several years of growth of the origin of pustules; anterior slope without sculpture; posterior slope occasionally with small tubercles which tend to form oblique rows or costae; some specimens with apustulose flattened or sulcoid zone between posterior and central ridges; pustules round, transversely elongate or lachrymal, typically restricted to central disc, becoming sparse to absent with age, some individuals and populations entirely apustulose.

Periostracum. Yellow in juveniles, passing through shades of tan to dark brown with age; central disc commonly with a single broad green or gray-green ray or blotch, present at least on the earliest portion of the shell; several narrower rays frequently on posterior slope; rays continuous or interrupted but disappearing with age.

Nacre. White, iridescent posteriorly; Tulberg layers common in some populations.

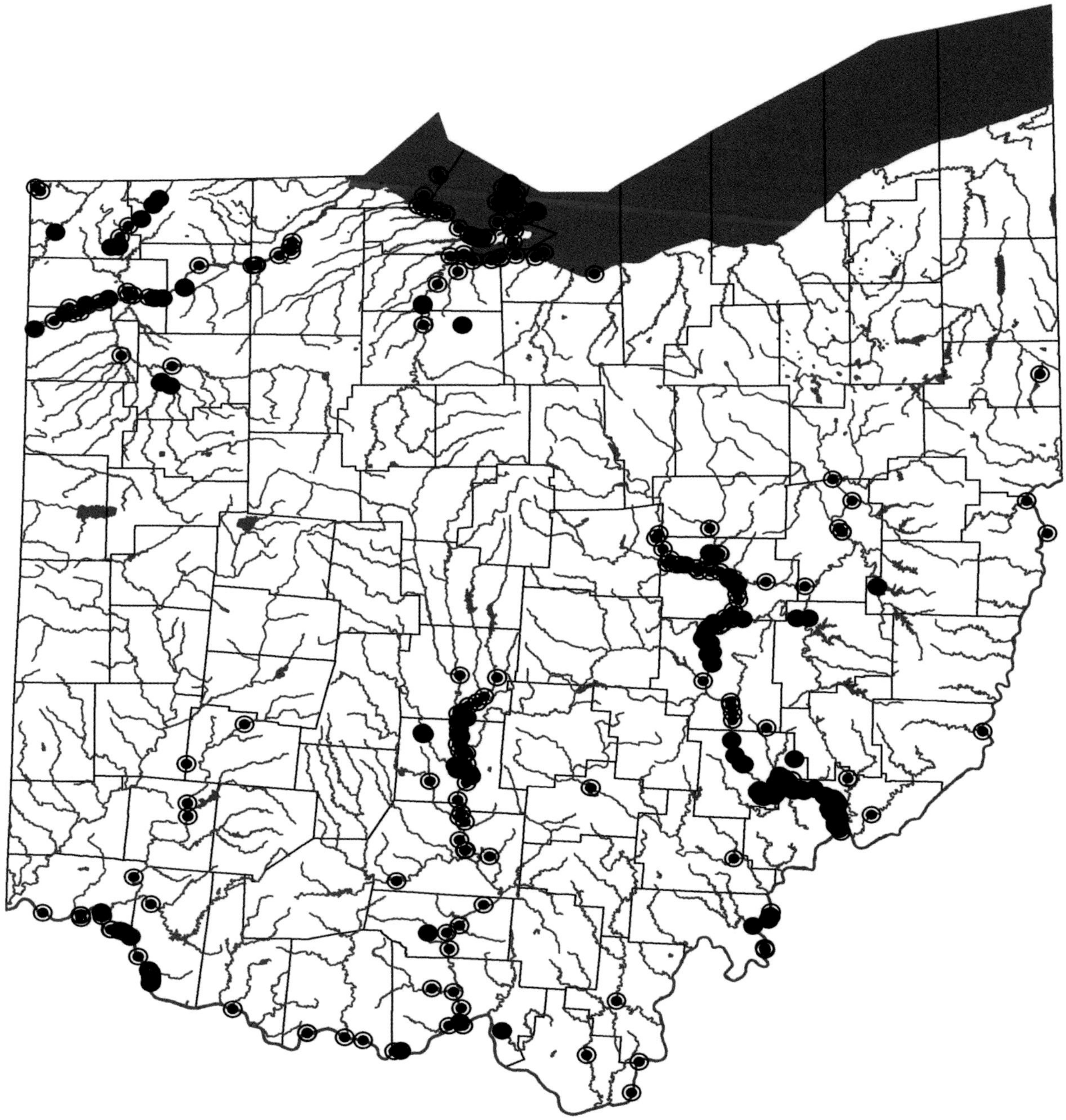

Fig. 184. ***Quadrula pustulosa pustulosa*** **(Lea, 1831)**
Solid dots—live or freshdead since 1980
Bullseye—all records, any condition

Hinge plate. Heavy, interdentum broad and flat from anterior terminus of laterals to posterior cardinal; two cardinals in left valve, one in right valve, with a tendency to form denticles at dorsal margins of cardinal fossae; laterals straight or scarcely curved, short, two in left valve, one in right, with a tendency to form a second at its inner base.
Muscle scars. Anterior adductor deep, rough, confluent with anterior retractor, distinct from anterior pedal protractor; posterior adductor typically distinct from posterior pedal retractor; dorsal suspensors multiple, with umbonal cavity on undersurface of posterior cardinal fossa and interdentum; pallial line complete, accessory present, frequently passing from anterior adductor postventrally to a point below the posterior adductor near shell margin.
Umbonal cavity. Deep and capacious in medium to large rivers; less deep and moderately compressed in small rivers and creeks.

Glochidia. Subelliptical, hinge margin straight, ventral margin curved without spines (Coker *et al.*, 1921: 147, pl. 11, figs. o, p; Howard, 1914a: 9, pl. 5, fig. 36). Literature records (Lefevre & Curtis, 1910a: 97; Surber, 1912: 9; Howard, 1914a: 9) indicate glochidial dimensions for length (0.22 to 0.23 mm) and height (0.28 to 0.32 mm). Howard (1914a: 9) noted, "The only glochidia here at all resembling it are those of *Lampsilis luteola* and *Plagiola securis*. *Lampsilis luteola* can be distinguished by its relatively greater width (0.25 mm) and hinge line (0.118 mm) [Howard gave a hinge length of 0.08 mm for *Q. pustulosa*], and *Plagiola securis* by its greater length (0.31 mm to 0.33 mm) and different proportions." In fact, neither *L. r. luteola*, with a much more elliptical glochidium, or *P. securis* [= *Ellipsaria lineolata*], with a glochidum approaching that of *Potamilus*, has glochidia similar to those of *Q. pustulosa*.

Reproductive biology. Tachytictic. All gills are marsupial (Ortmann, 1912a). Egg-bearing females appear in May, and glochidia develop from May to July (Ortmann, 1919; Bingham, 1968; Heath *et al.*, 1998). Glochidia are released by late August (Howard, 1913). Glochidia are released in fragile, white, leaf-shaped conglutinates (Utterback, 1915) that are of the amorphous mucus type (Watters, 2008). Metamorphosis has been observed only from catfishes.

Reported potential hosts

Ameiurus melas (Black Bullhead)—Howard, 1913 (NI); Howard, 1914a (LT); Coker *et al.*, 1921 (NI, LT).
Ameiurus nebulosus (Brown Bullhead)—Howard, 1914a (LT); Coker *et al.*, 1921 (LT).
Ictalurus punctatus (Channel Catfish)—Howard, 1913 (NI); Howard, 1914a (LT); Coker *et al.*, 1921 (NI, LT); Howard & Anson, 1923 (NI); Weiss & Layzer, 1995 (NI).
Pomoxis annularis (White Crappie)—Surber, 1913 (NI); Wilson, 1916 (NI).
Pylodictis olivaris (Flathead Catfish)—Howard, 1913 (NI); Howard, 1914a (LT); Wilson, 1916 (NI); Coker *et al.*, 1921 (NI).
Scaphirhynchus platorynchus (Shovelnose Sturgeon)—Coker *et al.*, 1921 (NI).

Habitat. This widespread species lives in moving water in muddy sand and cobble in large creeks and rivers, and Lake Erie. It is rarely found as a stray in small creeks (e.g., Fish Creek).

Range. Mississippi River system, east to New York and Pennsylvania, west to the Dakotas. Great Lakes drainages. Other subspecies occur in the Gulf drainages.

Range in Ohio. Widespread, but typically in rivers, where it may be abundant. Western basin of Lake Erie and its tributaries. Apparently absent from the Great Miami River.

Similar species. The green ray or blotch on the umbo is the hallmark of this species that sets it apart from the similar *Plethobasus cooperianus* and *Quadrula nodulata*. It also lacks the purple nacre of *Cyclonaias tuberculata*.

Comments. This species varies considerably in its growth rate. Specimens ca. 64 mm in length were between 16 and 46 years old.

This is an important commercial species.

Quadrula quadrula (Rafinesque, 1820)

Figures 185, 186

Etymology. L. *quadrula*, a little square; from L. *quadra*, a square + L. *-ula* (diminutive). This species appears square in lateral outline as much as or more so than any of our other native species. The length given by Rafinesque, 2 to 3 inches, is medium-sized for this species. *Quadrula* is a noun in apposition.
Vernacular name(s). *Mapleleaf*, Tear Shell.
Original description
Obliquaria (Quadrula) quadrula Rafinesque, 1820: 307, no figures.
Type locality. Rafinesque stated that the species was "assez commune dans l'Ohio" ["rather common in the Ohio (River)"]. However, the type locality of the neotype is "Salt River, Kentucky."
Type material. Johnson & Baker (1973: 167) noted, "The specimen identified by Rafinesque, mentioned by Vanatta,

Fig. 185. ***Quadrula quadrula*** **(Rafinesque, 1820). A.** OSUM 46774, Ohio and Erie Canal, Pike Co., OH. 145 mm. **B.** OSUM 19767.1, Lake Erie, Ottawa Co., OH. 96 mm. **C.** OSUM 58217.2, Maumee River, Defiance Co., OH. 83 mm. **D.** OSUM 40278.5, Muskingum River, Washington Co., OH. 81 mm. **E.** OSUM 2941, Lake Erie, [Co.?], OH. 21 mm. **F, Hinge detail:** OSUM 64389.6, Maumee River, Defiance Co., OH. 76 mm. **G.** OSUM 61508.2, Scioto River, Pickaway Co., OH. 91 mm. **H.** OSUM 64482.7, Ohio River, Scioto Co., OH. 65 mm. **Beak sculpture detail:** OSUM 64389.1, Maumee River, Defiance Co., OH. 36 mm total length.

1915, PANSP 67:556, is here selected neotype ANSP 20224 from the Salt River, Kentucky." The neotype, originally in the Poulson Collection, has a label in Rafinesque's hand noting "Rare Salt River" "Raf. sp. 35."

Synonymy

Obliquaria (*Quadrula*) *quadrula* Rafinesque, 1820: 307.

Unio lacrymosus (Lea, 1828): 272, pl. 6, fig. 8 [Type locality: "Ohio." Lectotype USNM 84165].

Unio asperrimus Lea, 1831: 71, pl. 5, fig. 3 [*asperimus* on plate] [Type locality: "Ohio River." Lectotype USNM 84174].

Unio quadrulus Say, 1834: no pagination.

Margarita (*Unio*) *lacrymosus* (Lea, 1828). Lea 1836: 14.

Margarita (*Unio*) *asperrimus* (Lea, 1831). Lea 1836: 14.

Margaron (*Unio*) *lacrymosus* (Lea, 1828). Lea 1852b: 21.

Margaron (*Unio*) *asperrimus* (Lea, 1831). Lea 1852b: 21.

Unio quadratus [*sic*] Reeve, 1864: pl. 6, fig. 24.

Unio unul atus [*sic*] Pratt, 1876: 167–168, pl. 31, figs. 1a, 1b [conditionally proposed, *lapsus calami pro "lunulatus."* Type locality: "opposite Rock Island Arsenal, Mississippi River, Iowa." Type not located].

Quadrula lachrymosa [*sic*] (Lea, 1828). Baker, 1898b: 83, pl. 25, fig. 1, pl. 12, fig. 2.

Quadrula (*Theliderma*) *lachrymosa* [*sic*] (Lea, 1828). Simpson, 1900a: 776–777.

Quadrula (*Theliderma*) *lachrymosa* [*sic*] var. *lunulatus* (Pratt, 1876). Simpson, 1900a: 777.

Quadrula quadrula (Rafinesque, 1820). Utterback, 1915: 135.

Quadrula quadrula contraryensis Utterback, 1915: 138, pl. 18, figs. 47a, b [Type locality: "Lake Contrary, at St. Joseph, Buchanan County, Missouri." Neotype CMNH 61.6892 by designation of Parodiz & Tripp, 1988].

Quadrula (*Quadrula*) *quadrula* (Rafinesque, 1820). Frierson, 1927: 47.

Quadrula (*Quadrula*) *quadrula* var. *bullocki* Baker, 1928b: 87–88, pl. 46, figs. 1–3 [Type locality: "Fox River near DePere, Wisconsin." Holotype UMMZ 209203].

Quadrula biangulata Morrison, 1942: 356.

Quadrula (*Quadrula*) *quadrula quadrula* (Rafinesque, 1820). Haas, 1969: 289.

Quadrula quadrula contrayensis [*sic*] Utterback, 1915. Johnson, 1969: 132.

Status. Not listed by either Ohio Division of Wildlife or U.S. Fish & Wildlife Service as of 2009.

ADULT SHELL

Shell. Large size (to 140 mm), thick, rectangular to nearly circular, compressed to quite inflated.

Anterior margin. Short and straight, then broadly rounded to the ventral border.

Ventral margin. Rounded to the sulcus, then quite concave to dorsal ridge.

Posterior margin. The posterior slope is extended into a blunt, broad flange; concave dorsally to the dorsal margin.

Dorsal margin. Very long, forming a moderately well developed wing, straight or arched, joining posterior margin with slightly obtuse or right angle.

Umbo. Rather narrow and projecting well above the dorsal margin; umbonal sculpture with approximately 5 chevron ridges, usually nodulose, particularly posteriorly, chevrons broken into individual pustules anteriorly, with well-defined sulcus.

Ligament. Short and prominent, tan to dark brown, not concealed by shells.

Disc sculpture. Sulcus bounded anteriorly and posteriorly by heavy, nodulose ribs; sulcus lacks pustules or other sculpture; anterior disc of shell covered with dense, broken, pustulose chevrons, particularly as juveniles; dorsal slope with nodulose, widely separated cords, often broken into a series of radiating pustules.

Periostracum. Green or yellow with lighter pustules (dark pustules rare), often with vague green stripes, particularly down the anterior-most row of knobs; some individuals may be nearly black, particularly when old.

Nacre. Porcelain white, often iridescent posteriorly; often with gold or brown Tulberg layers.

Hinge plate. Well-developed; right cardinal a single, massive, triangular denticle with a much smaller anterior (and occasionally posterior) ridge; left cardinal bifid, the teeth often far apart; laterals well-developed, straight or barely arched, short, one in right valve, two in left; interdentum flattened, broad, rather long.

Muscle scars. Anterior adductor impressed; posterior adductor flush; anterior adductor distinct from anterior pedal protractor scar, large posterior adductor attached but rarely confluent with posterior pedal retractor scar; dorsal suspensors located behind anterior adductor and in a line inside of interdentum, anterior- and posterior-most suspensor scars the largest. Pallial line entire, located far from shell margin. Accessory strong, reaching 75% the length of the shell.

Umbonal cavity. Very wide, somewhat compressed, and deep.

Glochidia. Subelliptical, dorsal margin slightly curved, lateral margins equally and gently curved, depressed, ventral margin rounded, without hooks (Surber, 1915: 8, pl. 1, fig. 11).

Literature records (Howard, 1914a: 16, pl. 5, fig. 29; Surber, 1915: 8, pl. 1, fig. 11) indicate glochidial dimensions for length (0.078 to 0.085 mm) and height (0.085 to

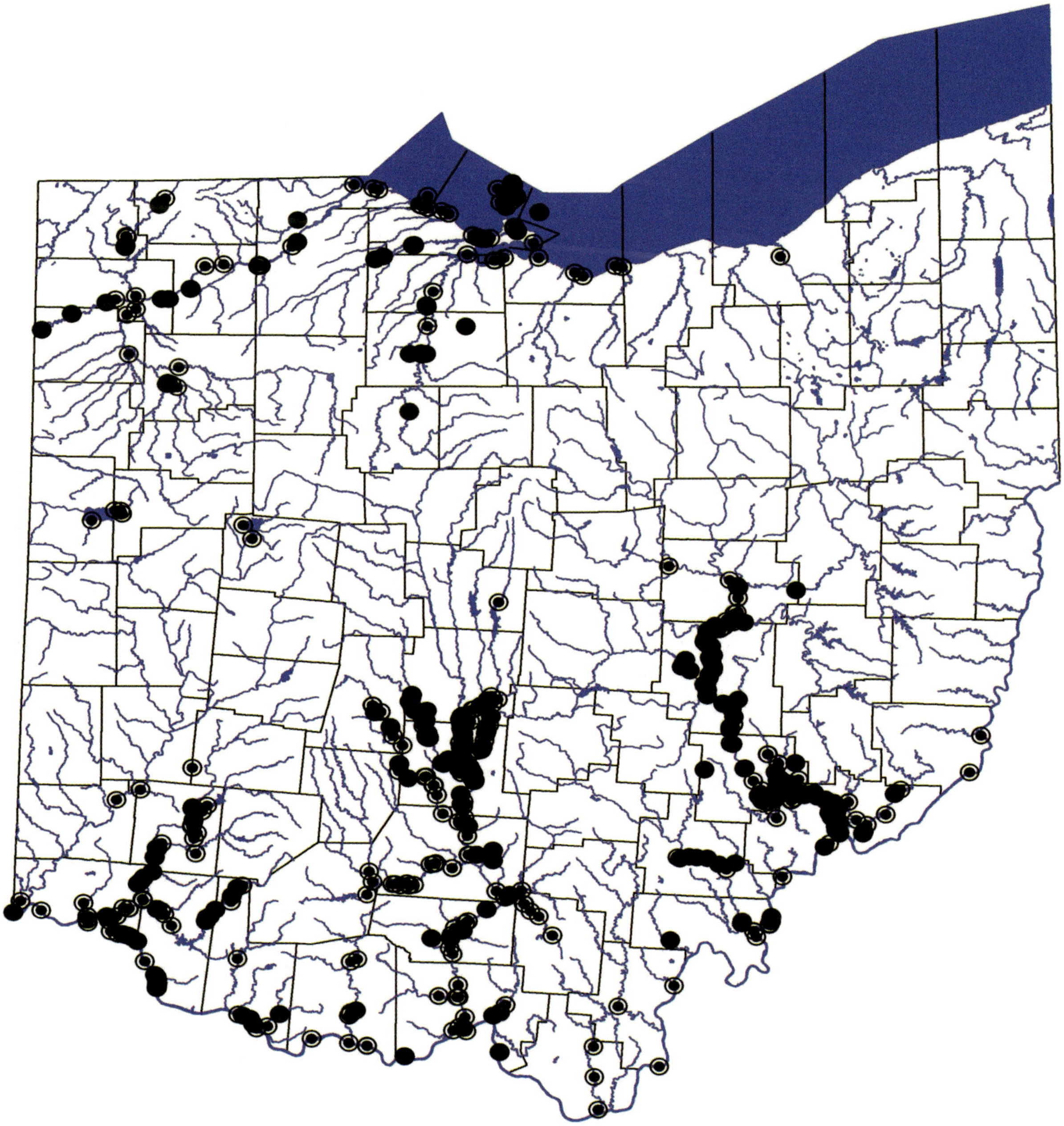

Fig. 186. ***Quadrula quadrula*** **(Rafinesque, 1820)**
Solid dots—live or freshdead since 1980
Bullseye—all records, any condition

0.090 mm). Surber (1915: 8) noted, "The shape of this glochidium is suggestive of *T. tuberculata* [*T. verrucosa*], but has a shorter hinge line and less depth . . . ." Howard (1914a: 16) added: "This is the smallest glochidium of the *Quadrula* group that I have seen."

Reproductive biology. Tachytictic. All gills are marsupial (Ortmann, 1912a). Eggs are present from May through July (Bingham, 1968), with glochidia present in July and August (Baker, 1928a). In Ohio, developing glochidia were found in aborted conglutinates in late June. Females release amorphous ribbon-like conglutinates composed entirely of glochidia (Watters, 2008). The females have a crenulated flap on the posterior-dorsal portion of the foot that may be analogous to mantle flaps used in luring hosts (Utterback, 1915). Metamorphosis has been observed only on Channel Catfish.

Reported potential hosts

Ictaluris punctatus (Channel Catfish)—Schwebach *et al.*, 2002 (LT).

Pylodictis olivaris (Flathead Catfish)—Howard & Anson, 1923 (NI).

Habitat. The Mapleleaf is found in moving water in muddy sand and cobble in large creeks and rivers, Lake Erie, and some large impoundments.

Range. Entire Mississippi River system, from the Red River in the north, east to Pennsylvania, in many western Mississippi River systems, south to Louisiana. Great Lakes drainages.

Range in Ohio. Widespread in larger rivers and the western basin of Lake Erie. Rare in northeastern Ohio.

Similar species. See under *Q. fragosa* and *Q. metanevra*.

Comments. Individuals gradually slow in growth during their entire life. Specimens may reach ca. 140 mm and 20 years old. The coarseness of the sculpture varies between populations; some may be quite nodulose, whereas others are nearly smooth. However, all individuals are sculptured as juveniles.

Simpsonaias Frierson, 1914

Etymology. This name was a patronym in honor of Charles Torrey Simpson (1846–1932), one of America's most productive students of unionoid molluscs. Also from Gr. *naias, naiados*, an aquatic nymph of freshwaters, especially of rivers, springs, brooks, and fountains. The translation is "Simpson's Naiad."

Original description

Simpsonaias Frierson, 1914: 7.

Type species. *Simpsonaias ambigua* (Say, 1825) by monotypy.

Synonymy

Simpsonaias Frierson, 1914a: 7 [*Hemilastena* Agassiz, 1852: 50, for which Agassiz gave *Alasmidonta ambigua* Say, 1825, as type, is a *lapsus calami* for *Hemistena* Rafinesque, 1820, and does not predate *Simpsonaias*.].

Simpsoniconcha Frierson, 1914b: 40 [In a later number (4) of the same volume of *The Nautilus*, Frierson proposes the name *Simpsoniconcha* for *Simpsonaias* in the belief that the earlier name was preoccupied. It apparently was not (Neave, 1940: 199), and the earlier name was eventually restored (Stansbery, 1974; Clarke, 1985).].

Geographic range. Upper Mississippi River and Great Lakes drainages.

Simpsonaias ambigua (Say, 1825)

Figures 187, 188, 189

Etymology. L. *ambiguus*, uncertain, doubtful; from L. *ambigo*, wander about, be undecided; from L. *ambi-*, around + L. *ago*, go. Say (1825: 13) noted in the original description that this species occupied a doubtful position as "a link between the genera *Alasmodonta* [*sic*] and *Anodonta*. When young the primary teeth are obvious, but when the shell arrives at the full growth the teeth are obsolete and in some instances, not at all visible." It was therefore "ambiguous" as to which of these two genera should receive this species.

Vernacular name(s). *Salamander Mussel*, Salamander Shell, Simpson's Shell.

Original description

Alasmodonta ambigua Say, 1825: 131, no figure given.

Type locality. ". . . obtained in the north-western territory, when traveling with major Long's party."

Type material. The original unionoid material of Say has apparently been lost (Johnson & Baker, 1973: 146).

Synonymy

Alasmodonta ambigua Say, 1825: 131.

Unio hildrethianus Lea, 1834: 36, pl. 3, fig. 8 [Type locality: "Ohio, near Marietta." Syntypes ANSP 41826].

Alasmodonta dubia Férussac, 1835: 26 [No type locality given. *Nomen nudum*. See Clarke, 1985: 61.]

Anodonta dubia "Say" Férussac, 1835: 31 [see above].

Margarita (*Unio*) *hildrethianus* (Lea, 1834). Lea, 1836: 28.

Hemilastena ambigua (Say, 1825). Agassiz, 1852: 50.

Margaron (*Margaritana*) *hildrethianus* (Lea, 1834). Lea, 1852b: 43.

Strophitus hildrethiana (Lea, 1834). Conrad, 1853: 263.

Fig. 187. ***Simpsonaias ambigua*** **(Say, 1825). A.** OSUM 17335.1, Muskingum River, Washington Co., OH. 38 mm. **B.** OSUM 4502, Big Walnut Creek, Franklin Co., OH. 28 mm. **C, H, Hinge detail:** OSUM 32520, East Fork Little Miami River, Clermont Co., OH. 37 mm. **D.** OSUM 60000.2, Grand River, Ashtabula Co., OH. 24 mm. **E.** OSUM 59695, St. Joseph River, Williams Co., OH. 45 mm. **F.** OSUM 34313, East Fork Little Miami River, Clermont Co., OH. 41 mm. **G.** OSUM 29802, West Branch St. Joseph River, Williams Co., OH. 48 mm. **Beak sculpture detail:** OSUM 39036.2, Wabash River, IN or IL. 23 mm total length.

Baphia hildrethiana (Lea, 1834). Adams & Adams, 1857: 499.
Margaritana ambigua (Say, 1825). Küster, 1862: 300, pl. 99, fig. 7.
Margaritana hildrethiana (Lea, 1834). Wright, 1888: no pagination.
Hemilastena ambigua (Say, 1825). Simpson, 1900a: 673.
Simpsonaias ambigua (Say, 1825). Frierson, 1914a: 17.
Simpsoniconcha ambigua (Say, 1825). Frierson, 1914b: 40.
Status. Listed as a Species of Concern by Ohio Division of Wildlife; not listed by U.S. Fish & Wildlife Service as of 2009.

ADULT SHELL

Shell. Small (to 50 mm), thin-shelled, somewhat inflated posteriorly, elongate.
Anterior margin. Elongated anteriorly, roundly arched into ventral margin.
Ventral margin. Nearly straight to broadly rounded, often indented posteriorly.
Posterior margin. Broadly rounded from ventral to dorsal margins.
Dorsal margin. Long, more or less straight, imperceptibly joining posterior margin.
Umbo. Low, pointed, narrow, located about 33% back of anterior margin; umbonal sculpture variable, ranging from absent to numerous fine threads forming wide inverted "V"s or several "W"s.
Ligament. Long, tan, very narrow, rarely preserved on dead shells.
Disc sculpture. Sculptureless; the posterior slope is usually broadly inflated.
Periostracum. Yellow or tan or olive as juveniles, becoming dark brown or gray with age. Never rayed; annular growth lines are darker; periostracum shiny, polished.
Nacre. Porcelain white and thickened in anterior half of shell, becoming thinner (often markedly so) and iridescent posteriorly; nearly all specimens have an extensive salmon, bronze, or purple flush under the umbos; brown Tulberg layers rare.
Hinge plate. Extremely thin; right cardinal a minute, triangular blade, left cardinal represented by only a small swelling in the interdentum; laterals absent at all stages of life.
Muscle scars. Anterior and posterior adductors small, flush; anterior adductor barely connected to anterior pedal protractor scar, small posterior adductor distinct or barely connected with posterior pedal retractor scar; multiple dorsal suspensor linear and located behind anterior adductor and deep within the umbonal cavity. Pallial line entire, located very far from shell margin. Accessory line weak, short.
Umbonal cavity. Very wide, shallow, and open.

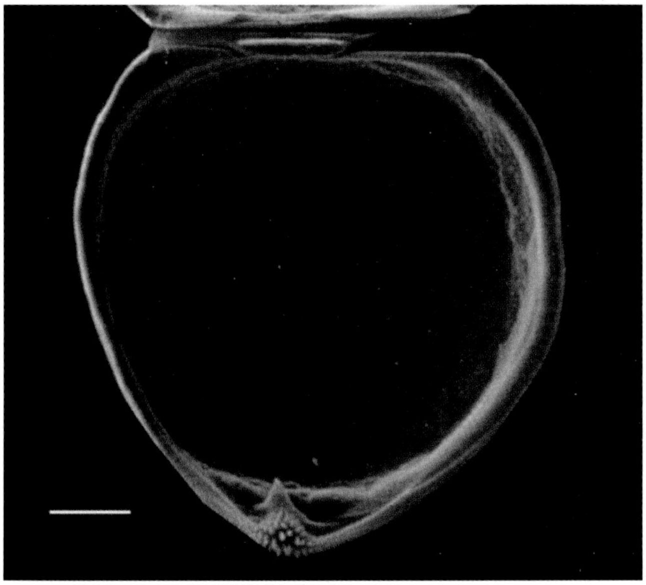

Fig. 188. ***Simpsonaias ambigua*** **(Say, 1825).** Glochidium. UWZY 22662, Wisconsin River, Richland Co., WI. Bar = 40 µm. (Hoggarth, 1999).

Glochidia. Ovate-subtriangular, slightly asymmetric, anterior margin slightly more rounded than the posterior margin, ventral terminus bluntly rounded, dorsal margin straight and fairly long. Exterior surface sculpturing beaded. Hooks styliform arising as a broad biconcave triangular plate from the ventral terminus and tapering to a sharp point (Lea, 1858c: 49, pl. 5, fig. 31; Howard, 1915: 8; Clarke, 1985: 64, fig. 20d; Hoggarth, 1999: figs. 15a–f).

Literature records (Howard, 1915: 8; Howard, 1951: 4, figs. 4a, b; Hoggarth, 1999: figs. 15a–f) indicate glochidial dimensions for length (0.25 to 0.26 mm), height (0.26 to 0.27 mm), and hinge (0.16 to 0.17 mm). Howard (1951: 4) noted, "The glochidia are clear white in color, of the triangular type, with well-developed hooks contrary to the description given by Lea who suggested, however, the possibility of hooks in more mature specimens. The dimensions are as follows: height 0.265 to 0.274 mm, length 0.247 to 0.255 mm." This glochidium can be distinguished from most others by its shape, and from that of the other Anodontinae by its size, hook structure, and exterior valve sculpturing.

Reproductive biology. The entire outer gills form the marsupia, becoming very swollen when gravid (Utterback, 1916a). Howard (1915) found infected mudpuppies in October. The glochidia did not metamorphose until the following May, overwintering on their hosts. Barnhart *et al.* (1998a) recovered glochidia and infected mudpuppies in April, with metamorphosis 19–28 days later. No conglutinates are formed (Watters, 2008).

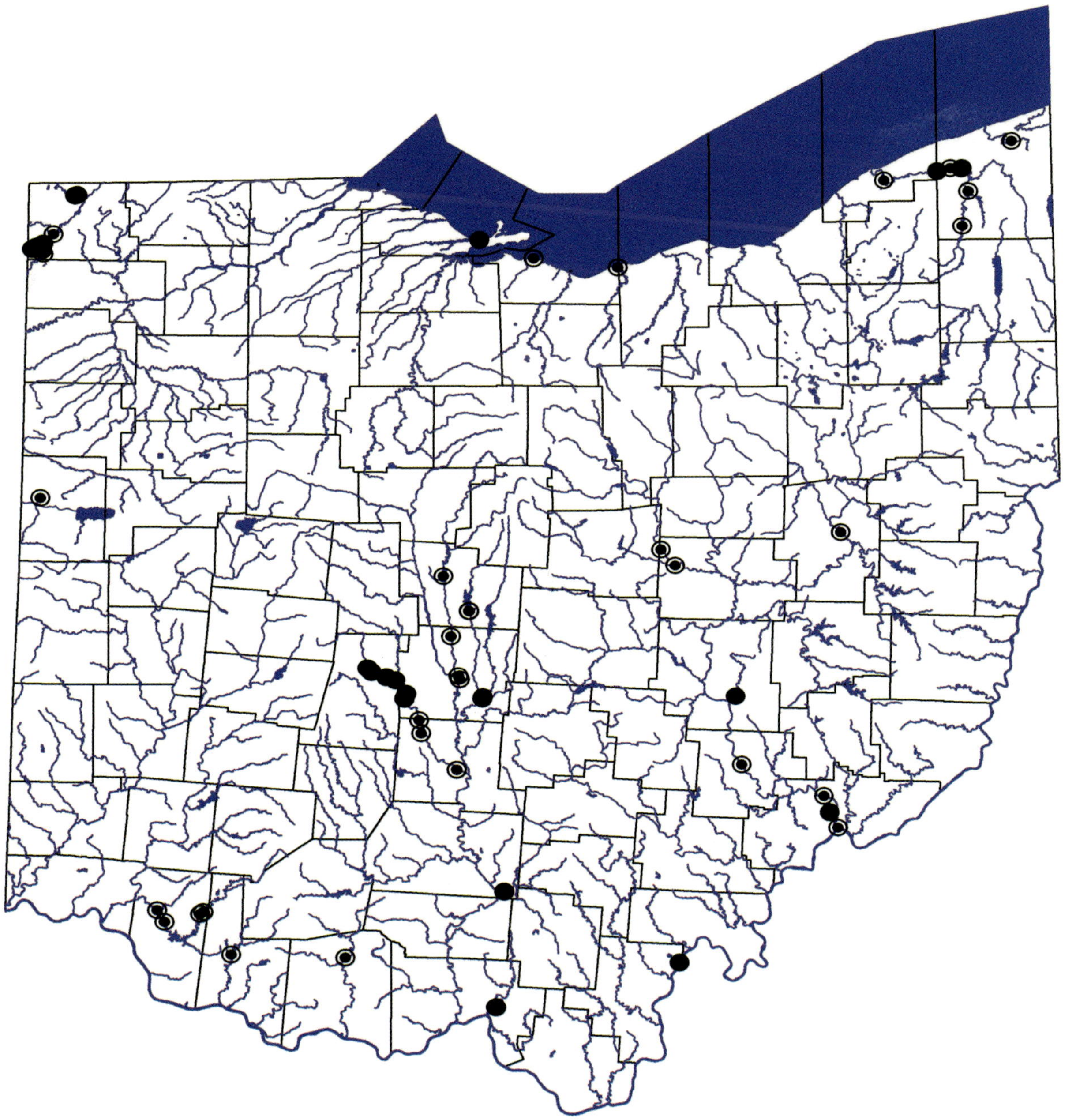

Fig. 189. ***Simpsonaias ambigua*** **(Say, 1825)**
Solid dots—live or freshdead since 1980
Bullseye—all records, any condition

Reported potential hosts

Necturus maculosus (Mudpuppy)—Howard, 1915 (NI, LT); Howard & Anson, 1923 (NS); Arey, 1932a (LT), 1932b (LT); Howard, 1951 (NI, LT); Barnhart *et al.*, 1998a (LT).

Habitat. Historically this species has been reported as abundant beneath large rocks where it occurred with its host, the Mudpuppy. Once distributed in creeks and rivers of all sizes, today it is most often found in creeks. It is rarely found buried in the substrate with other mussels.

Range. Upper Mississippi River, including the Cumberland River, the Green River of Kentucky, the Wabash and Kanawha rivers, west of the Mississippi River in the Meramec and Arkansas rivers. Lake Erie drainages.

Range in Ohio. Sporadic and rare. St. Joseph River system, Big Darby Creek, Salt Creek [of Scioto River], Little Miami River, Ohio Brush Creek, and Little Scioto River. A weathered shell was recently found in the Muskingum River at Marietta [the type locality of *hildrethianaus* Lea, 1834]. Historically recorded from the Ohio Canal at New Philadelphia, Tuscarawas River, Scioto River at Columbus, and Sandusky Bay (CMNH); the upper Scioto River, Licking River. Apparently most common in the Grand River.

Similar species. The near lack of characteristics ironically sets this species apart from all others. Unlike *Anodontoides*, *Pyganodon*, and *Strophitus*, it possesses rudimentary cardinal teeth. The lack of rays and lateral teeth, and the double-looped beak sculpture, differentiate it from *Villosa* species.

Comments. This peculiar species grows quickly for 2–3 years before abruptly slowing. It rarely reaches 50 mm in length (most are ca. 40 mm) and lives for 10 years (but most are only 4–5 years old).

Strophitus Rafinesque, 1820

Etymology. Gr. *strophos*, twisted + L. *-itus*, having the nature of, pertaining to. The root "twisted" may refer to the peculiar swelling of the hinge line near the umbos in species of this genus. Translation is "having a twisted hinge line."

Original description

Anodonta (*Strophitus*) Rafinesque, 1820: 316, no figures.

Type species. *Anodonta undulata* Say, 1817, by monotypy.

Synonymy

Anodonta (*Strophitus*) Rafinesque, 1820: 316.

Strophites Deshayes, 1832: 998 [unjustified emendation of *Strophitus* Rafinesque, 1820].

Strophilus Gray, 1847: 196 [unjustified emendation of *Strophitus* Rafinesque, 1820].

Uniopsis Agassiz, 1852: 49.

Geographic range. The range of the species of *Strophitus* appears to be limited to North America east of the Rocky Mountains.

Comments. This genus has a unique conglutinate, the mesoconglutinate, which is not found in any other genus (Watters, 2008).

Strophitus undulatus (Say, 1817)

Figures 190, 191, 192, 193, 194

Etymology. L. *unda*, wave + L. *atus*, having the nature of. Translation is "having an undulated hinge line."

Vernacular name(s). *Creeper*, Squawfoot, Sloughfoot, Common Creeper, Strange Floater.

Original description

Anodonta undulata Say, 1817: no pagination, under heading "Conchology," species no. 3, under Genus *Anodonta*, pl. 3, fig. 5 [Frierson (1927: 22) called Say's figure "wretched."].

Type locality. Of the 13 species described by Say in his 1817 article, *A. undulata* is one of two for which he failed to cite a source or "habitat." Haas (1969: 37) stated under this species that many of Say's species came from the Schuylkill River near Philadelphia, and Johnson (1970: 367) accordingly restricted the type locality to "Schuylkill River near Philadelphia, Philadelphia Co., Pennsylvania."

Type material. The original type of Say is apparently lost (Johnson, 1970: 367; Johnson & Baker, 1973: 146).

Synonymy

Anodonta undulata Say, 1817: no pagination, pl. 3, fig. 5.

Anodonta pensylvanica Lamarck, 1819: 86 [Type locality: "la rivière de Schuglkill, près de Philadelphie" ("Schuylkill River near Philadelphia"). Syntype in Geneva Museum. Lamarck's spelling of *pensylvanica* is not considered an incorrect original spelling and should not be emended. Linnaeus also had spelled this locality *pensylvanica*—i.e., *Mya pensylvanica* Linnaeus, 1758.].

Anodon rugosus Swainson, 1822: 2, pl. 96 [Type locality: "United States." Types in Museum of University of Manchester, merged with general collection (Kohn, 1992)].

Anodonta undulatus Say, 1817. Hildreth, 1828: 290.

Anodonta pennsylvanica [*sic*] Lamarck, 1819. Stark, 1828: 89.

Anodon areolatus Swainson, 1829: 1, pl. 18 [Type locality: "North America." Types in Museum of University of Manchester, merged with general collection (Kohn, 1992)].

Fig. 190. ***Strophitus undulatus*** **(Say, 1817). A.** OSUM 9094.3, Hocking River, [Co.?], OH. 90 mm. **B.** OSUM 32.3, St. Joseph River, Williams Co., OH. 97 mm. **C.** OSUM 29901.2, St. Joseph River, Williams Co., OH. 94 mm. **D.** OSUM 9974, East Fork Black River, Lorain Co., OH. 68 mm. **E.** OSUM 707, Lake Erie, Ottawa Co., OH. 38 mm. **F.** OSUM 4039.3, Big Walnut Creek, Franklin Co., OH. 85 mm. **G, H, Hinge detail:** OSUM 29606.2, Fish Creek, Williams Co., OH. 72 mm. **Beak sculpture detail:** OSUM 29606.1, Fish Creek, Williams Co., OH. 54 mm total length.

Alasmodonta edentula Say, 1829: 340 [Type locality: "Wabash River." Type ANSP lost *fide* Johnson, 1970].
Anodonta edentula (Say, 1829). Férussac, 1835: 25.
Anodonta areolatus (Swainson, 1829). Férussac, 1835: 25.
Anodon marginata (Say, 1818). Férussac, 1835: 25 [misidentification].
Anodonta ferussaciana (Lea, 1834). Férussac, 1835: 25 [misidentification].
Margarita (Anodonta) undulata (Say, 1817). Lea, 1836: 50.
Margarita (Anodonta) edentula (Say, 1829). Lea, 1836: 50.
Anodonta virgata Conrad, 1836: back cover [Type locality: "Buck Creek, Clarke Co., Ohio." Type ANSP lost *fide* Johnson, 1970].
Anodonta wardiana Lea, 1836: 50 [*nomen nudum*].
Anodonta pavonia Lea, 1838a: 78, pl. 21, fig. 65 [Type locality: "headwaters of the Little Beaver, Ohio." Lectotype USNM 86514].
Anodonta wardiana Lea, 1838a: 46, pl. 14, fig. 42 [Type locality: "near Chillicothe, Ohio." Figured lectotype USNM 86488].
Margarita (Anodonta) pavonia (Lea, 1836) Lea, 1838b: 30.
Anodon undulata (Say, 1817). Gould, 1841: 120, fig. 79.
Alasmodonta undulatus (Say, 1817). Adams, 1842: 165.
Anodon unadilla DeKay, 1843: 199, pl. 15, fig. 228 [Type locality: "Unadilla River, a tributary of the Susquehanna, Otsego Co., New York." Holotype in New York Lyceum of Natural History, destroyed by fire *fide* Johnson, 1970].
Anodon pavonia (Lea, 1836). DeKay, 1843: 203, pl. 40, fig. 358.
Anodonta rugosus (Swainson, 1822). Hanley, 1843: 217.
Anodonta tetragona Lea, 1845: 165 [Type locality: "Alexandria, Louisiana." Figured lectotype USNM 86682].
Anodon edentula (Say, 1829). Catlow & Reeve, 1845: 66.
Anodon undulata (Say, 1817). Catlow & Reeve, 1845: 68.
Strophitus undulatus (Say, 1817). Stimpson, 1851: 15.
Uniopsis edentula (Say, 1829). Agassiz, 1852: 49.
Anodonta shaefferiana Lea, 1852a: 288, pl. 26, fig. 50 [Type locality: "Horn Lake Creek, Tennessee." Figured lectotype USNM 86685].
Anodonta arkansasensis Lea, 1852a: 293, pl. 29, fig. 56 [Type locality: "Little Arkansas River, where the road to Santa Fé crosses it." Figured lectotype USNM 86603].
Margaron (Anodonta) wardiana (Lea, 1838). Lea, 1852b: 49.
Margaron (Anodonta) undulata (Say, 1817). Lea, 1852b: 49.
Margaron (Anodonta) edentula (Say, 1829). Lea, 1852b: 49.
Margaron (Anodonta) arkansasensis (Lea, 1852). Lea, 1852b: 50.
Margaron (Anodonta) pavonia (Lea, 1836). Lea, 1852b: 51.
Margaron (Anodonta) shaefferiana (Lea, 1852). Lea, 1852b: 51.
Margaron (Anodonta) tetragona (Lea, 1845). Lea, 1852b: 51.
Anodonta shefferiana [*sic*] Lea, 1852. Clessin, 1873: 243, pl. 17, figs. 5–7.
Unio undulata (Say, 1817). Deshayes, 1853: 217, pl. 30, figs. 8, 9.
Strophitus edentulus (Say, 1829). Conrad, 1853: 263.
Strophitus arkansasensis (Lea, 1852). Conrad, 1853: 262.
Strophitus virgatus (Conrad, 1836). Conrad, 1853: 263.
Strophitus unadilla (DeKay, 1843). Conrad, 1853: 263.
Strophitus tetragona (Lea, 1845). Conrad, 1853: 263.
Strophitus schaefferiana [*sic*] (Lea, 1852). Conrad, 1853: 263.
Alasmidon rhombica Anthony, 1865: 158, pl. 12, fig. 5 [Type locality: "Michigan." Figured lectotype MCZ 50296].
Anodon papyracea Anthony, 1865: 161, pl. 15, fig. 2 [Type locality: unknown, but Johnson (1970: 367) gave "Potomac River, Virginia." Figured lectotype MCZ 150656].
Anodon annulatus Sowerby, 1867: pl. 18, fig. 67 [Type locality: unknown. Type BMNH, lost *fide* Johnson, 1970].
Anodon quadriplicatus Sowerby, 1867: pl. 28, fig. 110 [Type locality: "Potomac River." Type in Walpole collection, assumed lost *fide* Johnson, 1970].
Anodon wardiana (Lea, 1838). Sowerby, 1867: pl. 28, fig. 114.
Anodonta shafferiana [*sic*] Lea, 1852. Currier, 1868: 12.
Anodon shaefferianus (Lea, 1852). Sowerby, 1870: pl. 35, fig. 143.
Margaron (Anodonta) papyracea (Anthony, 1865). Lea, 1870: 81.
Anodonta salmonia Clessin, 1873: 91, pl. 24, figs. 1, 2 [*non* Lea, 1838. Type locality: "Nordamerika, im Ohio-und Wabash-Fluss" ("North America in Ohio and Wabash Rivers"). Type in Stuttgart Museum, destroyed during WW II].
Anodonta shefferiana [*sic*] Lea, 1852. Clessin, 1873: 243, pl. 17, figs. 5–7.
Anodonta shafferiana [*sic*] Lea, 1852. Wright, 1888: no pagination.
Anodonta papyracea (Anthony, 1865). Wright, 1888: no pagination.
Margaritana rhombica (Anthony, 1865). Paetel, 1890: 173.
Strophitus pavonius (Lea, 1836). Bakerb, 1898: 68–69, pl. 3, fig. 5, pl. 5, fig. 3.
Strophitus edentulus (Say, 1829). Simpson, 1900a: 616–617.
Strophitus edentulus var. *pavonius* (Lea, 1836). Simpson, 1900a: 617–618.

Anodonta pennsylvanica [*sic*] Lamarck, 1819. Simpson, 1900a: 618.
Strophitus undulatus var. *quadriplicatus* (Sowerby, 1867). Simpson, 1914: 350.
Anodonta foliopsis De Gregorio, 1914: 63–64, pl. 11, fig. 4c [Type locality: "Sciota [Scioto] R. (Ohio)." Type Palermo Museum?].
Strophitus rugosus (Swainson, 1822). Ortmann & Walker, 1922: 40.
Strophitus undulatus pennsylvanicus [*sic*] (Lamarck, 1817). Frierson, 1927: 22.
Strophitus undulatus rugosus (Swainson, 1822). Frierson, 1927: 22.
Strophitus undulatus virgatus (Conrad, 1836). Frierson, 1927: 22.
Strophitus undulatus tennesseensis Frierson, 1927: 22 [Type locality: "Headwaters of the Tennessee River." Holotype UMMZ 74890?].
Strophitus undulatus ovatus Frierson, 1927: 22 [Type locality: "Lyon Creek, Edwards Co., IL." Lectotype UMMZ 87584].
Strophitus rugosus pavonius (Lea, 1836). Baker, 1928b: 202, pl. 73, figs. 1–5, pl. 74, figs. 7, 9, 10, pl. 67, figs. 1, 2.
Strophitus rugosus pepinensis Baker, 1928b: 204, pl. 74, fig. 8 [Type locality: "Lake Pepin near Lake City, Minnesota." Holotype UMMZ 209137].
Strophitus rugosus winnebagoensis Baker, 1928b: 205, pl. 74, figs. 1–6 [Type locality: "Lake Winnebago, Long Point Island, Wisconsin." Holotype UMMZ].
Strophitus rugosus lacustris Baker, 1928b: 207, pl. 75, figs. 6–8 [Type locality: "Oconomowoc Lake, Waukesha Co., Wisconsin." Holotype University of Illinois Museum of Natural History Z-22073].
Strophitus (*Strophitus*) *undulatus undulatus* (Say, 1817). Haas, 1969: 372.
Strophitus (*Strophitus*) *undulatus pennsylvanicus* [*sic*] (Say, 1817). Haas, 1969: 372.
Strophitus (*Strophitus*) *undulatus rugosus* (Swainson, 1822). Haas, 1969: 373.
Strophitus (*Strophitus*) *undulatus virgatus* (Conrad, 1836). Haas, 1969: 373.
Strophitus (*Strophitus*) *undulatus edentulus* (Say, 1829). Haas, 1969: 374.
Strophitus undulatus undulatus (Say, 1817). Oesch, 1984: 49.

Status. Not listed by either Ohio Division of Wildlife or U.S. Fish & Wildlife Service as of 2009.

ADULT SHELL

Shell. Small to medium-sized (rarely to 120 mm), thin-shelled as juveniles, becoming thicker with age, somewhat inflated, especially as adults, oval to rather elongate in profile.

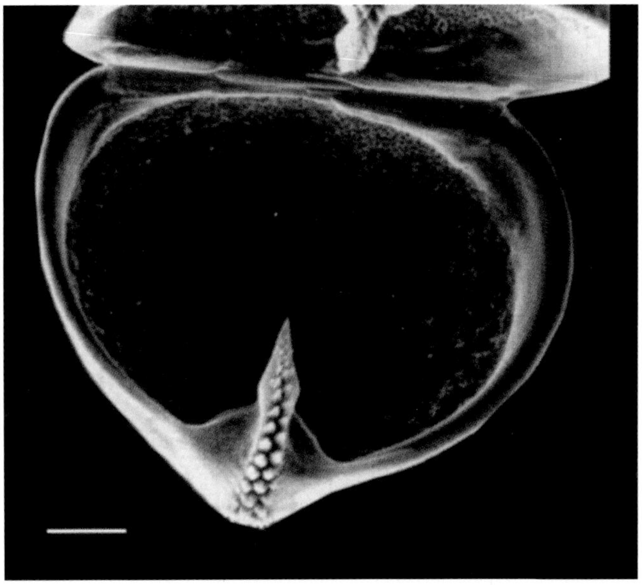

Fig. 191. *Strophitus undulatus* (Say, 1817). Glochidium. OSUM 52458.4, Ashuelot River, Cheshire Co., NH. Bar = 50 µm. (Hoggarth, 1999).

Anterior margin. Roundly arched into ventral margin, occasionally angled at juncture with dorsal margin.
Ventral margin. Broadly rounded, straight, or slightly concave.
Posterior margin. Ventral and posterior margins form wide, blunt acute angle at or below midline; posterior margin oblique and rounded, straight or rarely slightly concave to dorsal margin.
Dorsal margin. Fairly long, sinuate beneath umbo, joining posterior margin with very obtuse angle.
Umbo. Low and very wide, located about 33% to nearly 50% back from anterior margin; umbonal sculpture of 2–4 coarse, angular, "V"-shaped ridges, the "V"s tilted anterior and the posterior arm the shorter of the two coarse undulating or nearly linear ridges; some specimens nearly double-looped, occasionally somewhat nodulose on the descending angles.
Ligament. Fairly long, tan to dark brown, projecting both anterior and posterior of the umbo, usually not concealed by shells.
Disc sculpture. Smooth except for dorsal ridge.
Periostracum. Tan, brown, or greenish, becoming dark brown to black with age. Often with numerous green rays of various widths.
Nacre. White, yellowish, or silverish, iridescent posteriorly, often flushed with color in the umbonal cavity; many specimens seen have gold, white, or brown Tulberg layers.
Hinge plate. Thin to thickened, cord-like, sinuous. Very weakly produced into blade-like cardinal teeth in some specimens.

Muscle scars. Anterior adductor very slightly impressed; posterior adductor flush or shallowly impressed; anterior adductor distinct or weakly confluent with anterior pedal protractor scar, protractor elongate and parallel to adductor; large posterior adductor confluent and nearly inseparable from posterior pedal retractor scar; minute, multiple dorsal suspensors located within the umbonal cavity except for one or two larger anterior ones. Pallial line entire, located far from shell margin. Accessory line absent.
Umbonal cavity. Very wide, shallow, and open.

Glochidia. Depressed pyriform, asymmetric, dorsal margin straight, posterior margin broadly arcuate, anterior margin almost round, meeting the posterior margin at a slightly rounded and ventrally produced ventral terminus. Exterior surface sculpturing coarse and loose-looped, hooks styliform and covered with about 30 microstylets, arranged in two distal rows (Lea, 1858c: 50, pl. 5, figs. 37, 38; Hoggarth, 1999: figs. 16a–f).

Literature records (Ortmann, 1912a: 299–300; Surber, 1912: 8, pl. 1, fig. 3; Utterback, 1916: 312; Hoggarth, 1999: figs. 16a–f) indicate glochidial dimensions for length (0.35 to 0.37 mm), height (0.29 to 0.30 mm), and hinge (0.27 to 0.28 mm). The glochidium of *Strophitus* is morphologically depressed and pyriform and possesses a styliform hook at the ventral terminus of each valve. Lea (1858) described the glochidia of *Strophitus edentulus* (= *S. undulatus fide* Johnson, 1970) and *S. undulata* as subtriangular, with a long straight dorsal margin, inflated side margins (Lea erroneously viewed all glochidia as symmetrical about the dorso-ventral axis), and a large hook with four rows of "granules" proximally, reduced to two rows distally. Other than shape, Lea's description is surprisingly accurate. The glochidium of *S. undulatus* is unlike any so far described. Its shape, hook structure, and coarse-looped sculpture will distinguish it from other species examined.

Reproductive biology. Despite claims that this species can metamorphose without a host (Lefevre & Curtis, 1911, 1912; Howard & Anson, 1923), this has never been substantiated by recent experiments. Indeed, 37 potential hosts have been identified.

Bradytictic. The anterior portion of the outer gill is marsupial (Watters, unpubl.). Baker (1928a) gave the "breeding season" as July to following May. In central Ohio, this species releases numerous conglutinates in April. When released, each conglutinate consists of a white, solid column with a few (< 20) very large glochidia contained inside. Each glochidium is associated with a hole or pore in the side of the conglutinate. Within minutes after release by the female, the glochidia emerge through the pore and move to the outside of the conglutinate, where they open and await a host (Watters, 2002). The conglutinates are of the mesoconglutinate type (Watters, 2008). Conglutinates are attacked by shiners, dace, and other fish feeding in the water column, but the fish vigorously attempt to spit them out when taken into the mouth. Bottom-dwelling fishes such as darters are parasitized when they brush over them. Glochidia may attach anywhere on the fish. Newly parasitized fishes act in a very agitated and frantic manner, more so than for any other species of glochidia observed. This species is a generalist in host use.

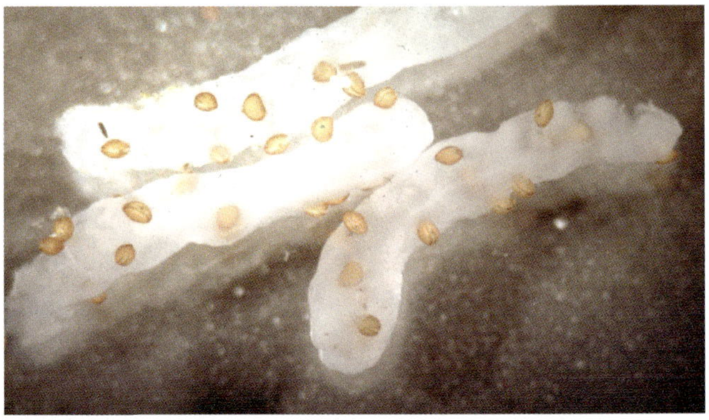

Fig. 192. *Strophitus undulatus* (Say, 1817). Mesoconglutinates with tethered glochidia.

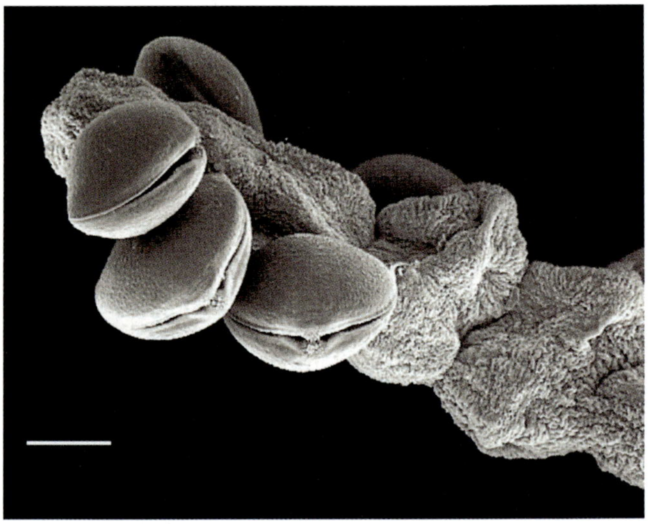

Fig. 193. *Strophitus undulatus* (Say, 1817). Portion of a mesoconglutinate with tethered glochidia. Bar = 200 µm. (Watters, 2008).

Reported potential hosts

Ambloplites rupestris (Rock Bass)—Gray *et al.*, 1999 (LT).
Ameiurus melas (Black Bullhead)—Hove, 1995a (LT); Hillegass & Hove, 1997 (LT); Hove *et al.*, 1997 (LT); Hove & Haas, 2001 (LT).
Ameiurus natalis (Yellow Bullhead)—Hove, 1995a (LT); Hove *et al.*, 1997 (LT); Hove & Haas, 2001 (LT).
Campsotoma anomalum (Central Stoneroller)—Hove & Haas, 2001 (LT).

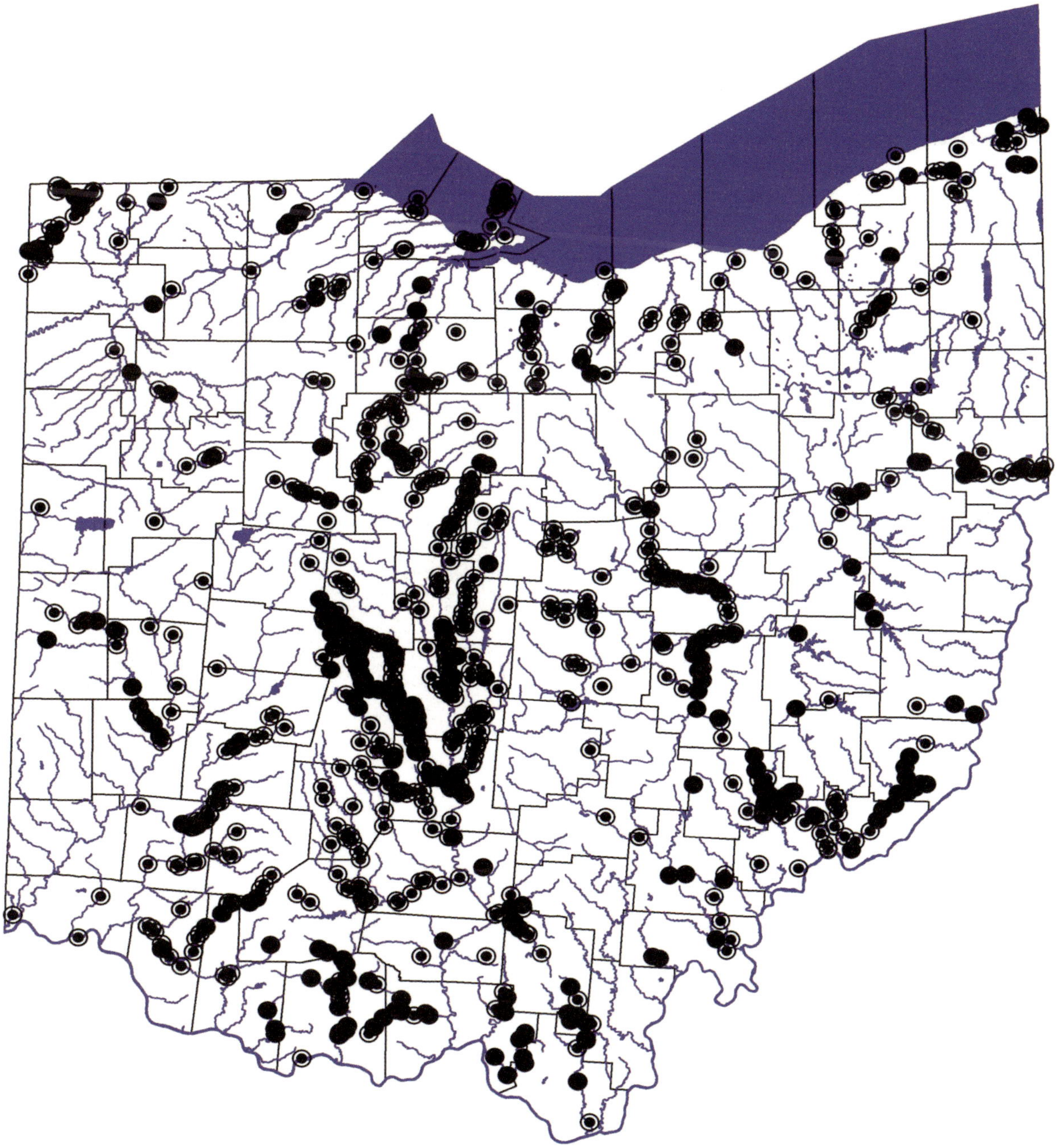

Fig. 194. ***Strophitus undulatus*** **(Say, 1817)**
Solid dots—live or freshdead since 1980
Bullseye—all records, any condition

Culaea inconstans (Brook Stickleback)—Hove & Haas, 2001 (LT).
Cyprinella spiloptera (Spotfin Shiner)—Hove, 1995a (LT); Hove *et al.*, 1997 (LT); Hove & Haas, 2001 (LT).
Etheostoma caeruleum (Rainbow Darter)—Hove & Haas, 2001 (LT); Watters *et al.*, 2005 (LT).
Etheostoma exile (Iowa Darter)—Hove & Haas, 2001 (LT).
Etheostoma flabellare (Fantail Darter)—Watters *et al.*, 1998a (LT), 1998d (LT); Hove & Haas, 2001 (LT).
Etheostoma nigrum (Johnny Darter)—Hove & Haas, 2001 (LT).
Etheostoma phoxocephala (Slenderhead Darter)—Hove & Haas, 2001 (LT).
Etheostoma zonale (Banded Darter)—Watters *et al.*, 1998a (LT), 1988d (LT).
Fundulus zebrinus (Plains Killifish)—Ellis & Keim, 1918 (LI).
Ictalurus punctatus (Channel Catfish)—Hove & Haas, 2001 (LT).
Lepomis cyanellus (Green Sunfish)—Ellis & Keim, 1918 (LI); Hillegass & Hove, 1997 (LT); Hove & Haas, 2001 (LT).
Lepomis gibbosus (Pumpkinseed)—Hove & Haas, 2001 (LT).
Lepomis macrochirus (Bluegill)—Hove, 1995a (LT); Hillegass & Hove, 1997 (LT); Hove *et al.*, 1997 (LT); Hove & Haas, 2001 (LT).
Lepomis megalotis (Longear Sunfish)—Watters *et al.*, 2005 (LT).
Lota lota (Burbot)—Hove & Haas, 2001 (LT).
Luxilus cornutus (Common Shiner)—Gray *et al.*, 1999 (LT); Hove & Haas, 2001 (LT).
Micropteris dolomieu (Smallmouth Bass)—Hove & Haas, 2001 (LT).
Micropterus salmoides (Largemouth Bass)—Howard *in* Baker, 1928a (LT); Hove, 1995a (LT); Hillegass & Hove, 1997 (LT); Hove *et al.*, 1997 (LT); Gray *et al.*, 1999 (LT); Hove & Haas, 2001 (LT).
Nocomis micropogon (River Chub)—Gray *et al.*, 1999 (LT).
Notropis ludibundus (Sand Shiner)—Watters *et al.*, 1998a (LT), 1998d (LT).
Perca flavescens (Yellow Perch)—Hillegass & Hove, 1997 (LT); Gray *et al.*, 1999 (LT); Hove & Haas, 2001 (LT).
Percina caprodes (Logperch)—Hove & Haas, 2001 (LT).
Percina maculata (Blackside Darter)—Hove & Haas, 2001 (LT).
Phoxinus eos (Northern Redbelly Dace)—Hove & Haas, 2001 (LT).
Pimephales notatus (Bluntnose Minnow)—Watters *et al.*, 1998a (LT), 1998d (LT); Hove & Haas, 2001 (LT).
Pimephales promelas (Fathead Minnow)—Hove, 1995a (LT); Hove *et al.*, 1997 (LT); Hove & Haas, 2001 (LT).
Pomoxis annularis (White Crappie)—Watters *et al.*, 1998a (LT), 1998d (LT).
Pomoxis nigromaculatus (Black Crappie)—Hove & Haas, 2001 (LT).
Rhinichthys atratulus (Blacknose Dace)—Hove & Haas, 2001 (LT).
Rhinichthys cataractae (Longnose Dace)—Watters *et al.*, 1998a (LT), 1998d (LT); Gray *et al.*, 1999 (LT); Hove & Haas, 2001 (LT).
Semotilus atromaculatus (Creek Chub)—Howard *in* Baker, 1928 (LT); Hove & Haas, 2001 (LT).
Sander vitreus (Walleye)—Hove, 1995a (LT); Hove *et al.*, 1997 (LT).
Umbra limi (Central Mudminnow)—Hove & Haas, 2001 (LT).

Habitat. This widespread species occurs in many habitats and water flows. It ranges from intermittent creeks to large rivers, and may occur wedged in the crevices of bedrock-bottomed streams.

Range. If this is indeed a single species, which is very doubtful, then it is one of the most widely distributed unionoids in the world. It ranges from the Red River of Canada, St. Lawrence River system, Mississippi River system, and Atlantic Slope from Nova Scotia to South Carolina.

Range in Ohio. Widespread and common, from the Ohio River to intermittent streams. Lake Erie and its tributaries.

Similar species. See under *Anodontoides ferussacianus*. The single-looped beak sculpture separates it from most *Pyganodon* and *Anodonta*.

Comments. Individuals grow very quickly for 2–3 years and then gradually slow. They rarely grow to 120 mm and 15 years old; most are half this size.

Toxolasma Rafinesque, 1831

Etymology. Gr. *toxon*, bow, arch + Gr. *elasma*, metal beaten out, metal plate. As in several other of his generic names, Rafinesque used the word *elasma* to refer to lateral teeth, which are often thin and plate-like. In the case of *Toxolasma*, he explained (1831: 2) that it differed from several other genera by having a "lamellar tooth not obliqual but arched . . . ."

Original description

Toxolasma Rafinesque, 1831: 2.

Type species. *Unio lividus (Toxolasma)* Rafinesque, 1831, by subsequent designation of Frierson (1914a: 7). Rafinesque included four species within his new genus: *T. cyclips* Rafinesque, 1831, *T. cinerescens* Rafinesque, 1831, *T. lividus* Rafinesque, 1831, and *T. flexus* Rafinesque, 1831. Ortmann & Walker (1922) and Johnson (1970) rejected the genus *Toxolasma* in favor of *Carunculina* Baker, 1898, on the grounds that the type species, *U. lividus*, was unidentifiable.

Synonymy

Toxolasma Rafinesque, 1831: 2.

Corunculina [*sic*] "Simpson" Baker, 1898b: 109 [*Corunculina* was corrected in index and errata to *Carunculina*. Type species: *Unio parvus* Barnes, 1823, by monotypy. Simpson's subsequent (1900a) designation of *Unio texasensis* Lea, 1857, as type is invalid, as that species was not among the species originally included in the genus, and the genus already had a type by monotypy.].

Geographic range. Mississippi River system from Minnesota to Louisiana. Gulf drainages from Choctawhatchee River to Texas. Atlantic Slope from the Neuse of North Carolina to the Altamaha River of Georgia.

Comments. Lee (2006) demonstrated that *Toxolasma* was neuter, requiring a change in agreement for most species names.

Toxolasma lividum (Rafinesque, 1831)

Figures 195, 196

Etymology. L. *lividum*, of bluish color. Rafinesque was apparently referring to the typical nacre color of this species when he noted "inside livid purplish" (1831: 2).

Vernacular name(s). *Purple Lilliput*, Little Purple.

Original description

Unio (*Toxolasma*) *lividus* Rafinesque,1831: 2, species no. 79, no figure.

Type locality. "In Rockcastle river [KY], exceedingly rare."

Type material. Neither Vanatta (1915: 549) nor Johnson & Baker (1973: 145) in reviewing the Rafinesque-Poulson "types" deposited at the Academy of Natural Sciences of Philadelphia mentioned locating material of *Unio lividum* Rafinesque, 1820. It appears that the primary type material of this species has been lost.

Synonymy

Unio (*Toxolasma*) *lividus* Rafinesque,1831: 2.

Unio glans Lea, 1831: 82, pl. 18, fig. 12 [Type locality: "Ohio River." Lectotype USNM 85312].

Margarita (*Unio*) *glans* (Lea, 1831). Lea, 1836: 28.

Unio moestus Lea, 1841: 82 [Type locality: "French Broad River, Tennessee." Lectotype USNM 85320].

Lampsilis glans (Lea, 1831). Agassiz, 1852: 44.

Margaron (*Unio*) *glans* (Lea, 1831). Lea, 1852b: 31.

Margaron (*Unio*) *moestus* (Lea, 1841). Lea, 1852b: 31.

Lampsilis (*Carunculina*) *glans* (Lea, 1831). Simpson, 1900a: 565.

Lampsilis (*Carunculina*) *moestus* (Lea, 1841). Simpson, 1900a: 565.

Carunculina glans (Lea, 1831). Ortmann, 1910: 119.

Eurynia (*Carunculina*) *glans* (Lea, 1831). Ortmann, 1912a: 339.

Lampsilis moesta (Lea, 1841). Simpson, 1914: 565.

Unio castus mirus De Gregorio, 1914: 27 [Type in Palermo Museum?].

Toxolasma lividum (Rafinesque, 1831). Ortmann, 1918: 573.

Carunculina moesta (Lea, 1841). Ortmann, 1921: 89.

Toxolasma livida (Rafinesque, 1831). Morrison, 1969: 24.

Carunculina glans glans (Lea, 1831). Stansbery, 1970: 18.

Carunculina glans moesta (Lea, 1841). Stansbery, 1970: 18.

Toxolasma glans (Lea, 1831). Valentine & Stansbery, 1971: 29.

Toxolasma glans glans (Lea, 1831). Stansbery, 1971: 14.

Toxolasma lividum lividum (Rafinesque, 1831). Stansbery, 1972: 46.

Toxolasma lividus lividus (Rafinesque, 1831). Stansbery, 1976: 48.

Toxolasma lividus glans (Lea, 1831). Stansbery, 1976: 48.

Status. Listed as Endangered by Ohio Division of Wildlife; not listed by U.S. Fish & Wildlife Service as of 2009.

ADULT SHELL

Shell. Small (to 40 mm), chunky, solid, oval to squarish in outline; moderately inflated. Females expanded posteriorly, males drawn to a blunt point.

Anterior margin. Very short, straight to broadly rounded to the ventral border.

Ventral margin. Very broadly rounded.

Posterior margin. Long, rather straight to the dorsal margin.

Dorsal margin. Long, straight, joining posterior margin in an obtuse angle.

Umbo. Narrow, very low, placed forward of midline; umbos face slightly forward; umbonal sculpture of irregular concentric lines, somewhat rugose posteriorly.

Ligament. Short, tan to dark brown, partially concealed by shell.

Disc sculpture. Disc of shell smooth except for low angulations on posterior slope.

Periostracum. Surface velvety or with a matte finish; juvenile shell yellowish-green, rarely with vague rays, adult shell dark brown to black.

Nacre. Deep purple, often paler around the margins, iridescent posteriorly; rarely with Tulberg layers.

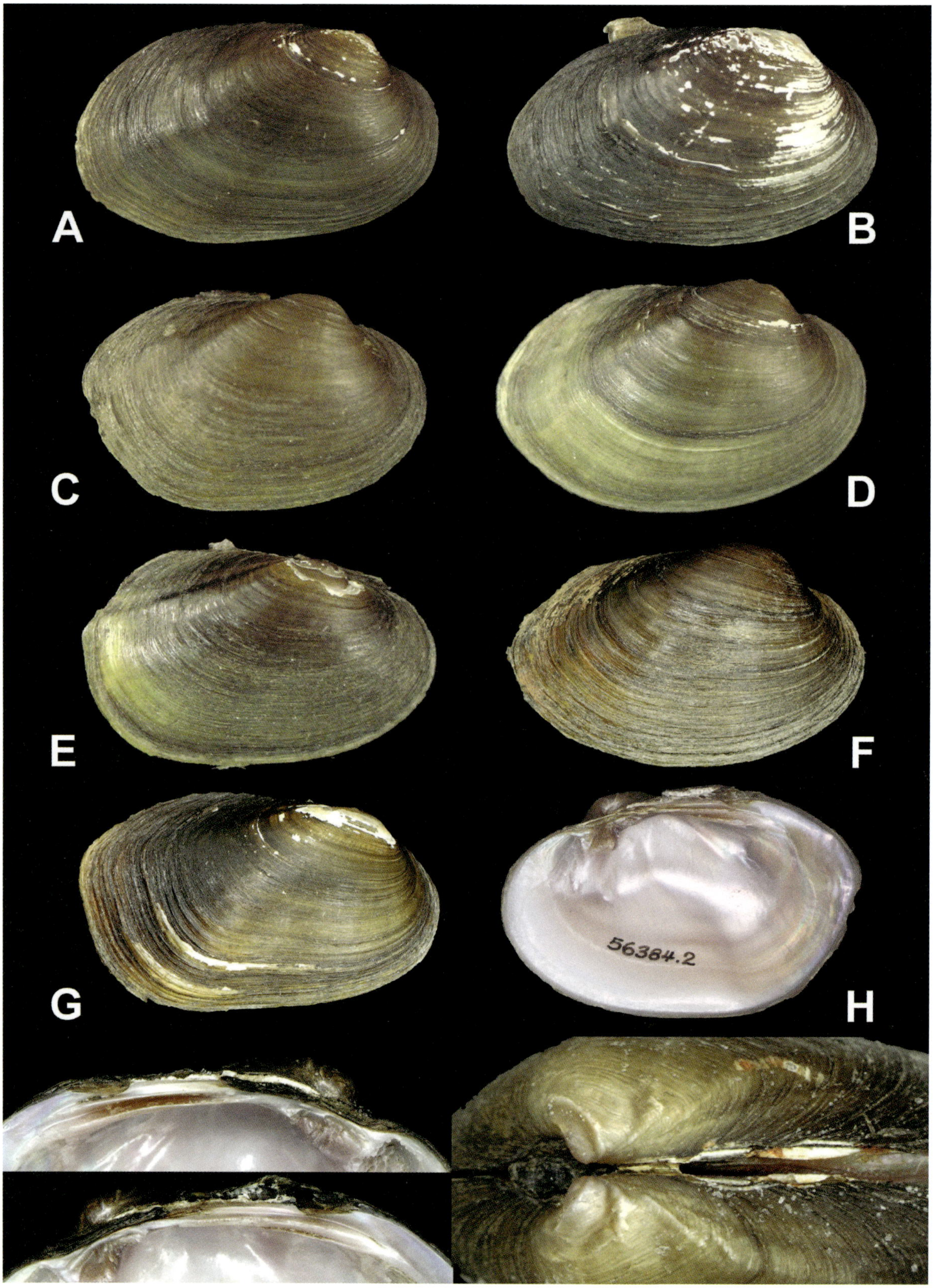

Fig. 195. ***Toxolasma lividum*** **(Rafinesque, 1831). A.** OSUM 42820, Little Miami River, Warren Co., OH. 35 mm. **B.** OSUM 56736.2, Blanchard River, Hancock Co., OH. 44 mm. **C, H.** OSUM 56384.2, Tippecanoe River, Tippecanoe Co., IN. 31 mm. **D.** OSUM 57320.2, Wabash River, Posey Co., IN. 31 mm. **E.** OSUM 12806.8, Mulberry Creek, Crawford Co., AR. 27 mm. **F.** OSUM 56488.3, Tippecanoe River, Carroll Co., IN. 44 mm. **G.** OSUM 68445.1, Lost Lake, Marshall Co., IN. 37 mm. **Hinge detail:** OSUM 56384.4, Tippecanoe River, Tippecanoe Co., IN. 34 mm total length. **Beak sculpture detail:** OSUM 56384.1, Tippecanoe River, Tippecanoe Co., IN. 19 mm total length.

Fig. 196. ***Toxolasma lividum*** **(Rafinesque, 1831)**
Solid dots—live or freshdead since 1980
Bullseye—all records, any condition

Hinge plate. Well-developed, short; right cardinal a single, lamellar, triangular denticle, often with small anterior accessory tooth; left cardinal bifid, the teeth well-separated; laterals well-developed, short, straight, one in right valve, two in left; interdentum very short.
Muscle scars. Anterior adductor small, oval, impressed; posterior adductor small, flush; tiny anterior adductor distinct from anterior pedal protractor scar, tiny posterior adductor also distinct; large dorsal suspensors located under umbo in a line or semicircle. Pallial line entire, located far from shell margin. Accessory line absent or weakly impressed.
Umbonal cavity. Very wide, shallow, open.

Glochidia. Subelliptical, dorsal margin short and straight, anterior and posterior margins curved, ventral margin rounded, hooks absent (Lea, 1863: 41).

There are no literature records for length and height for this glochidium. Lea (1863) stated, "Embryonic shell pouch shape, very near to that of *parvus*, which is near to *rectus*, but is more rounded." However, the glochidium of *L. rectus* is much larger than that of *T. parvum*, so this comparison is probably not accurate.

Reproductive biology. Bradytictic. Gravid females have been found from May to July (Ortmann, 1921). Only sunfish have been implicated as hosts.

Reported potential hosts
Lepomis cyanellus (Green Sunfish)—Jenkinson, 1982 (LT); Gooch *in* Hill, 1986 (LT).
Lepomis megalotis (Longear Sunfish)—Jenkinson, 1982 (LT); Gooch *in* Hill, 1986 (LT).

Habitat. Although once rather widespread, today *Toxolasma lividum* is most often found in small creeks. In the St. Joseph River drainage, it is found in very small creeks in packed sand and gravel.

Range. Ohio River system, including Cumberland and Tennessee rivers. Arkansas River. Lake Erie drainage.

Range in Ohio. Very rare if not extirpated. Previously in the Maumee River system, particularly Fish Creek, and St. Joseph and Blanchard rivers. The only living population in that system today is in Michigan. A single record for the Little Miami River exists.

Similar species. The small size, plump shape, and purple nacre differentiate this species. The related *Toxolasma parvum* lacks the purple nacre. *Villosa lienosa* is larger and more elongate.

Comments. A small but fairly long-lived species. Individuals may attain 40 mm in length and live 12 years.

Toxolasma parvum (Barnes, 1823)
Figures 197, 198

Etymology. L. *parvum*, little, small. In 1823, Barnes (p. 275) wrote, "The smallest and most beautiful of all the genus [*Unio*] yet discovered in America." *Toxolasma parvum* is certainly one of the smallest unionoid species, but what "most beautiful" characteristics Barnes saw are unknown.
Vernacular name(s). *Lilliput Shell.*
Original description
Unio parvus Barnes, 1823: "174" [274]–275, fig. 18 (outline only).
Type locality. "Inhabits the Fox River [WI]."
Type material. The types of Barnes, 1823, are apparently lost.
Synonymy
Unio parvus Barnes, 1823: "174" [274]–275, fig. 18.
Mya parva (Barnes, 1823). Eaton, 1826: 222.
Margarita (*Unio*) *parvus* (Barnes, 1823). Lea, 1836: 28.
Lampsilis parvus (Barnes, 1823). Agassiz, 1852: 44.
Margaron (*Unio*) *parvus* (Barnes, 1823). Lea, 1852b: 31.
Lampsilis (*Corunculina*) [*sic*] *parvus* (Barnes, 1823). Baker, 1898b: 109, pl. 13, fig. 3.
Lampsilis (*Carunculina*) *parvus* (Barnes, 1823). Simpson, 1900a: 564.
Eurynia (*Carunculina*) *parva* (Barnes, 1823). Ortmann, 1912a: 338.
Carunculina parva (Barnes, 1823). Wheeler, 1914: 75.
Unio pertenuis De Gregorio, 1914: 61, pl. 6, figs. 4a–f [*non* Lea, 1863. Type locality: Not stated, line apparently cut off by typesetter. Type in Palermo Museum?].
Toxolasma parvum (Barnes, 1823). Ortmann, 1919: 260.
Carunculina parva cahni Baker, 1927b: 222 [Type locality: "Neosho mill pond, Dodge Co., Wisconsin." Holotype University of Illinois Museum of Natural History Z-17341].
Toxolasma parva (Barnes, 1823). Valentine & Stansbery, 1971: 29.
Status. Not listed by either Ohio Division of Wildlife or U.S. Fish & Wildlife Service as of 2009.

ADULT SHELL
Shell. Small (to 50 mm), inflated, oblong.
Anterior margin. Straight, becoming evenly rounded towards the ventral margin.
Ventral margin. Straight or very gently rounded.
Posterior margin. Broadly rounded, reaching a slight angulation 25% from the dorsal margin, and another angulation at the dorsal margin.

Fig. 197. ***Toxolasma parvum*** **(Barnes, 1823). A.** OSUM 18683.2, Cowan Lake, Clinton Co., OH. 41 mm. **B.** OSUM 17071.8, Muskingum River, Washington Co., OH. 32 mm. **C, H.** OSUM 67992.3, Ohio and Erie Canal, Franklin Co., OH. 34 mm. **D, Beak sculpture detail:** OSUM 41595, Lake Erie, Erie Co., OH. 30 mm. **E.** OSUM 43963.4, Tuscarawas River, Tuscarawas Co., OH. 30 mm. **F.** OSUM 27636.21, Lake Erie, Erie Co., OH. 25 mm. **G.** OSUM 58658.2, Buckeye Lake, Fairfield Co., OH. 18 mm. **Hinge detail:** OSUM 18683.3, Cowan Lake, Clinton Co., OH. 41 mm total length.

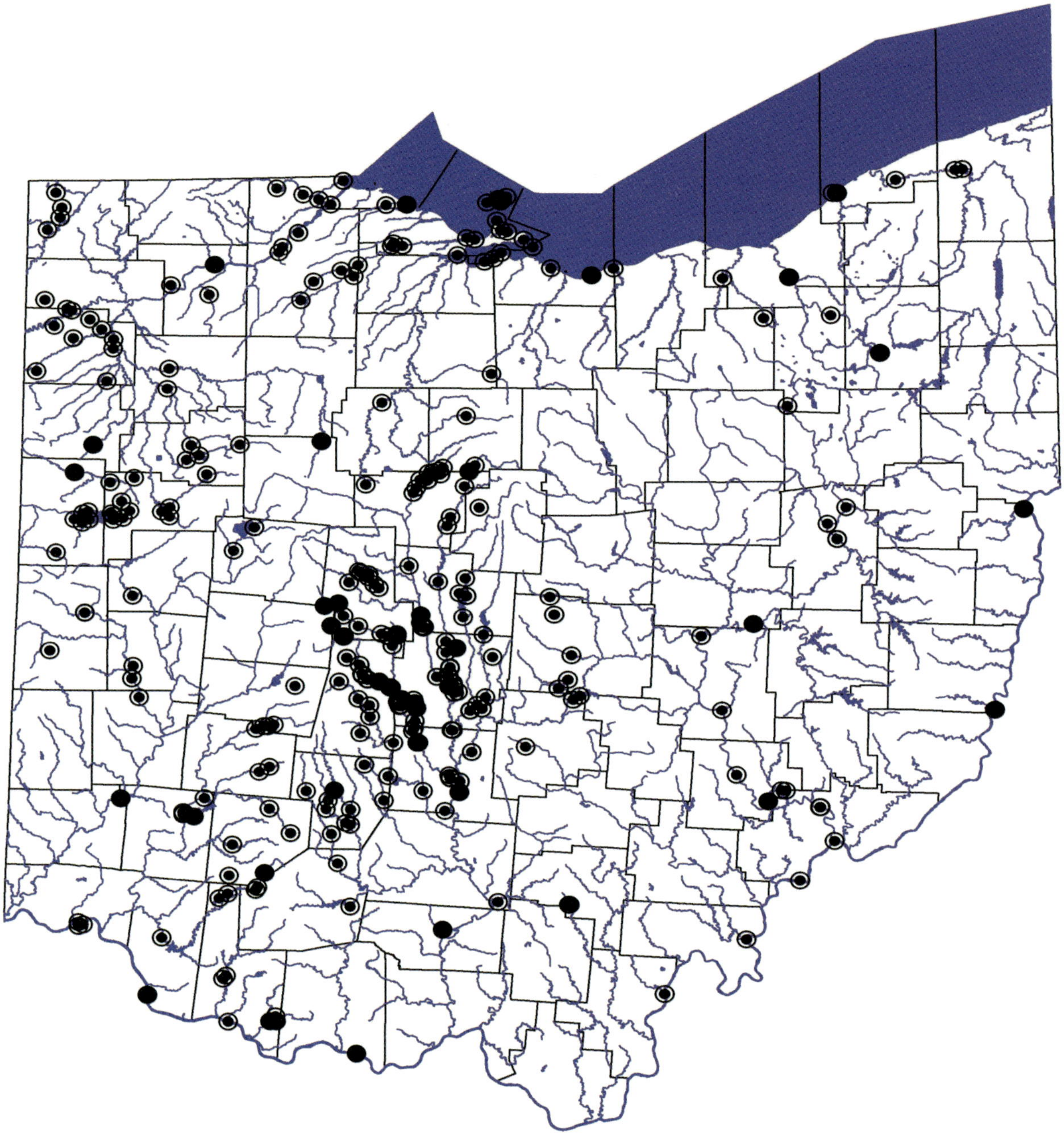

Fig. 198. ***Toxolasma parvum*** **(Barnes, 1823)**
Solid dots—live or freshdead since 1980
Bullseye—all records, any condition

Dorsal margin. Gently convex.
Umbo. Low, wide, prominent; beak sculpture consists of 4–6 coarse concentric ridges, aligned slightly obliquely so that the ridges open anterior to center; distal ridges may become wavy and broken; beak sculpture abruptly disappears at the first annular ring.
Ligament. Very small and inconspicuous, occupying 66% to 75% of the dorsal margin.
Disc sculpture. Sculptureless or with very low rib radiating to the posterior angulations.
Periostracum. Pale yellow, green, or gray, with a satiny sheen; old individuals darker, becoming uniformly black in very large specimens; the posterior half of the shell is often more darkly covered. Indistinct green rays may be present; these are most developed on the dorsal slope, which may have alternating light and dark green rays; the shell is overlaid with light and dark rings; annular rings tend to be dark, while interannular rings are light.
Nacre. White, iridescent posteriorly; copper-colored Tulberg layers are rare.
Hinge plate. Short, bowed, rather chunky for the size; interdentum very short and narrow, often absent, sometimes having a small lamellar tooth; one lamellar cardinal in right valve, often with a minute anterior accessory tooth; two cardinals in left valve; the posterior of the two is long and narrow, projecting out and above the hinge line; laterals straight and long, one in the right valve, two in the left.
Muscle scars. Anterior adductor moderately deep, distinct from anterior pedal protractor; posterior adductor flush with shell, partially confluent with posterior pedal retractor; suspensors multiple, elongate, located within the umbonal cavity; pallial line complete; accessory line present in small individuals, lost in larger ones.
Umbonal cavity. Moderately deep, but wide and open.

Glochidia. Subelliptical, dorsal margin short and straight, anterior and posterior margins gently curved, ventral margin rounded, hooks absent (Lea, 1860a: 39; Lea, 1874a: 68, pl. 21, fig. 3; Surber, 1915: 5, pl. 1, fig. 3).

Literature records (Surber, 1915: 5, pl. 1, fig. 3; Utterback, 1916a: 397; Utterback, 1916: 20; Ortmann, 1919: 259) indicate glochidial dimensions for length (0.17 to 0.18 mm) and height (0.20 mm). Surber (1915) noted, "In general outline this glochidium strongly suggests that of *L. ligamentina*, from which it differs in being much smaller and in having a differently shaped adductor; in size it approaches nearest *Q. metanevra*, which has a short hinge line, and *L. anodontoides*, in which the hinge line is slightly depressed."

Reproductive biology. Bradytictic. Marsupium formed by the posterior half of the outer gills (Ortmann, 1912a). This species is possibly hermaphroditic (Utterback, 1916). Eggs were found in June and August (Ortmann, 1919), and glochidia persist to the following July (Baker, 1928a). Glochidia are released in white, club-shaped conglutinates (Utterback, 1916). These are of the composite A type having the glochidia untethered and embedded in a core of unfertilized eggs (Watters, 2008). Females possess a tubercle (caruncle) on the posterior mantle margin in most non–Lake Erie populations; other populations may have a mantle flap instead (Watters, unpubl.). These move and undoubtedly act as lures.

Reported potential hosts
Etheostoma nigrum (Johnny Darter)—Watters *et al.*, 2005 (LT).
Lepomis cyanellus (Green Sunfish)—Mermilliod *in* Fuller, 1978 (NS); Hove, 1995c (LT).
Lepomis gulosus (Warmouth)—Wilson, 1916 (NI); Mermilliod *in* Fuller, 1978 (NS).
Lepomis humilis (Orangespotted Sunfish)—Mermilliod *in* Fuller, 1978 (NS).
Lepomis macrochirus (Bluegill)—Mermilliod *in* Fuller, 1978 (NS).
Pomoxis annularis (White Crappie)—Mermilliod *in* Fuller, 1978 (NS).

Habitat. Although found in variety of habitats, the Lilliput is most common in muddy sand or clay in creeks and impoundments.

Range. Mississippi River system from Minnesota to southern Louisiana, including the Ohio, Cumberland, and Tennessee rivers. East to New York. In many western drainages of the Mississippi River. Lake Erie drainages.

Range in Ohio. Widespread but usually uncommon, very rare in unglaciated and northeast Ohio. Found in all river reaches and in lakes, and locally common in impoundments.

Similar species. The small rectangular shape and coarse beak sculpture are diagnostic. It lacks the purple nacre of *T. lividum* and the rays and fine "V"-shaped beak sculpture of *Villosa iris*. *Villosa fabalis* also is rayed and has a thick hinge line.

Comments. This is a small, short-lived species. Most specimens examined were less than 5 years old. Exceptional specimens may reach 50 mm in length and 12 years in age. Individuals grow very rapidly for the first 2 years. Unlike many other members of the genus, this species shows no sexually dimorphic shell characteristics.

Tritogonia Agassiz, 1852

Etymology. Gr. *tritos,* third + Gr. *gonia,* corner, angle. Agassiz did not give his reason(s) for selecting the name *Tritogonia.* The roots seem clear, but their reference to features of this genus remain obscure. It may perhaps refer to the posterior third of the shell which may be deflected or "hooked."

Original description

Tritogonia Agassiz, 1852: 48.

Geographic range. Mississippi and Mobile river systems, eastern Texas.

Comments. Serb *et al.* (2003) synonymized this genus with *Quadrula* based on genetic evidence. See "Comments" under the genus *Quadrula* above.

Tritogonia verrucosa (Rafinesque, 1820)

Figures 199, 200, 201

Etymology. L. *verrucosus,* full of warts; from L. *verruca,* a wart. The shell of this species is typically covered with pustules.

Vernacular name(s). *Pistolgrip,* Deer Horn, Buck Horn, Turkey Wing.

Original description

Obliquaria (*Ellipsaria*) *verrucosa* Rafinesque, 1820: 304, pl. 81, figs. 10–12.

Type locality. "Dans l'Ohio [River]."

Type material. Johnson & Baker (1973: 174, pl. 1, fig. 7): "From the types mentioned by Vanatta, 1915, PANSP 67:554 [?] is here selected a lectotype ANSP 20235a; paralectotype ANSP 20235."

Synonymy

Obliquaria (*Ellipsaria*) *verrucosa* Rafinesque, 1820: 304, p. 81, figs. 10–12.

Unio tuberculatus Barnes, 1823: 125–126, fig. 8 [Type locality: "Ouisconsin [*sic*]." Types lost].

Mya tuberculata (Barnes, 1823). Eaton, 1826: 217.

Unio verrucosus (Rafinesque, 1820). Say, 1834: no pagination.

Margarita (*Unio*) *tuberculatus* (Barnes, 1823). Lea, 1836: 17.

Unio (*Theliderma*) *pustulata* Swainson, 1840: 271, fig. 54d [*non* Lea, 1834. Type locality: none given. Type in Museum of University of Manchester, merged with general collection *fide* Kohn (1992)].

Margaron (*Unio*) *tuberculatus* (Barnes, 1823). Lea, 1852b: 23.

Tritogonia verrucosus (Rafinesque, 1820). Agassiz, 1852: 48.

Unio gigas (Swainson, 1840). Sowerby, 1867: pl. 56, fig. 287 [misidentification].

Tritigonia [*sic*] *verrucosa* (Rafinesque, 1820). Smith, 1899: 291, pl. 81.

Unio conjugans Wright, 1899d: 89 [Type locality: "Hiawassa River, Tennessee." Holotype USNM 159501].

Tritogonia tuberculata (Barnes, 1823). Simpson, 1900a: 608.

Tritogonia tuberculata var. *obesa* Simpson, 1900a: 609 [Type locality: "Big Eddy, Nechez River, Texas." Lectotype USNM 125669].

Tritogonia conjugans (Walker, 1899). Simpson, 1900b: 79, pl. 4, fig. 1.

Quadrula tritogonia Ortmann, 1909a: 101 [unnecessary replacement for *Unio tuberculatus* Barnes, 1823].

Quadrula obesa (Simpson, 1900). Vanatta, 1910: 102.

Quadrula parkeri Geiser, 1911: 15 [unnecessary replacement for *Unio tuberculatus* Barnes, 1823].

Quadrula tuberculata (Barnes, 1823). Ortmann, 1912a: 254.

Quadrula verrucosa (Rafinesque, 1820). Ortmann & Walker, 1922: 18.

Quadrula (*Quadrula*) *quadrula conjugans* (Wright, 1899). Frierson, 1927: 48.

Quadrula (*Tritogonia*) *verrucosa* (Rafinesque, 1820). Frierson, 1927: 48.

Quadrula verrucosa (Rafinesque, 1820). Serb *et al.,* 2003: 8.

Status. Not listed by either Ohio Division of Wildlife or U.S. Fish & Wildlife Service as of 2009.

ADULT SHELL

Shell. Large size (to ca. 180 mm), thick, rectangular to quite elongate in profile, moderately inflated.

Anterior margin. Short and straight, then broadly rounded to the ventral border.

Ventral margin. Rounded to nearly straight as juveniles, adults decidedly concave at the juncture of the posterior slope.

Posterior margin. The posterior ridge is extended into a blunt point at the basal margin in males, or into a broad flange in females; concave dorsally to the dorsal margin.

Dorsal margin. Very long, straight or arched, joining posterior margin with very obtuse angle.

Umbo. Very wide and low; umbonal sculpture ca. 5 chevron ridges, usually nodulose, particularly posteriorly, chevrons broken into individual pustules anteriorly.

Ligament. Long, tan to dark brown, partially concealed by shells.

Disc sculpture. Disc of shell covered with dense, broken, pustulose chevrons; dorsal ridge with a prominent anterior border of large knobs or knobs fused into a single ridge; dorsal slope with massive corrugations.

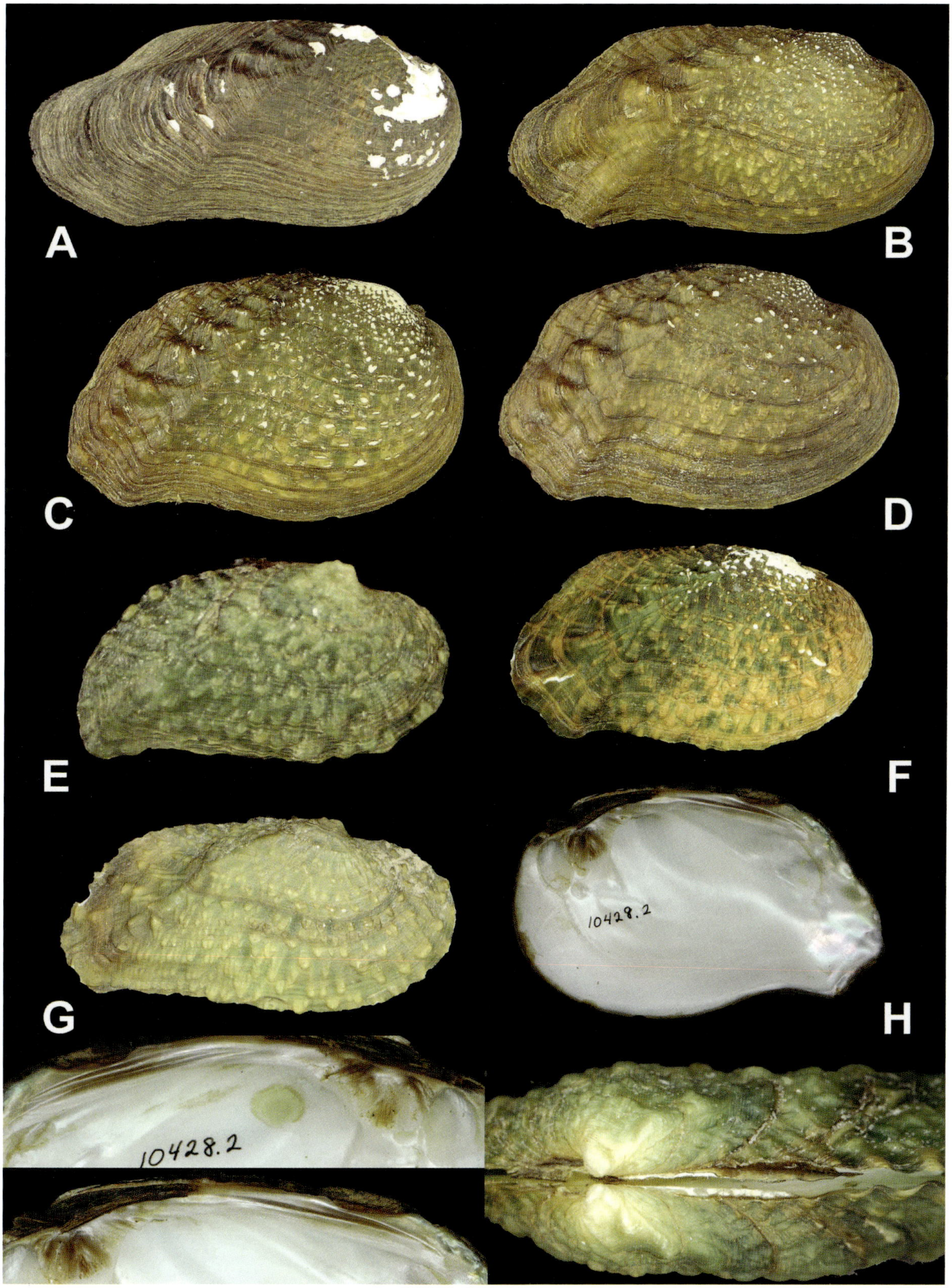

Fig. 199. ***Tritogonia verrucosa*** **(Rafinesque, 1820). A.** OSUM 59314.2, Symmes Creek, Gallia Co., OH. 178 mm. **B.** OSUM 1144.3, Big Walnut Creek, Franklin Co., OH. 137 mm. **C, H, Hinge detail:** OSUM 10428.2, Ohio Brush Creek, Adams Co., OH. 105 mm. **D.** OSUM 11328, Walhonding River, Coshocton Co., OH. 120 mm. **E, Beak sculpture detail:** OSUM 15565.1, Stones River, Davidson Co., TN. 26 mm. **F.** OSUM 68424.2, Green River, Green Co., KY. 55 mm. **G.** OSUM 76154.2, Big Darby Creek, Franklin Co., OH. 54 mm.

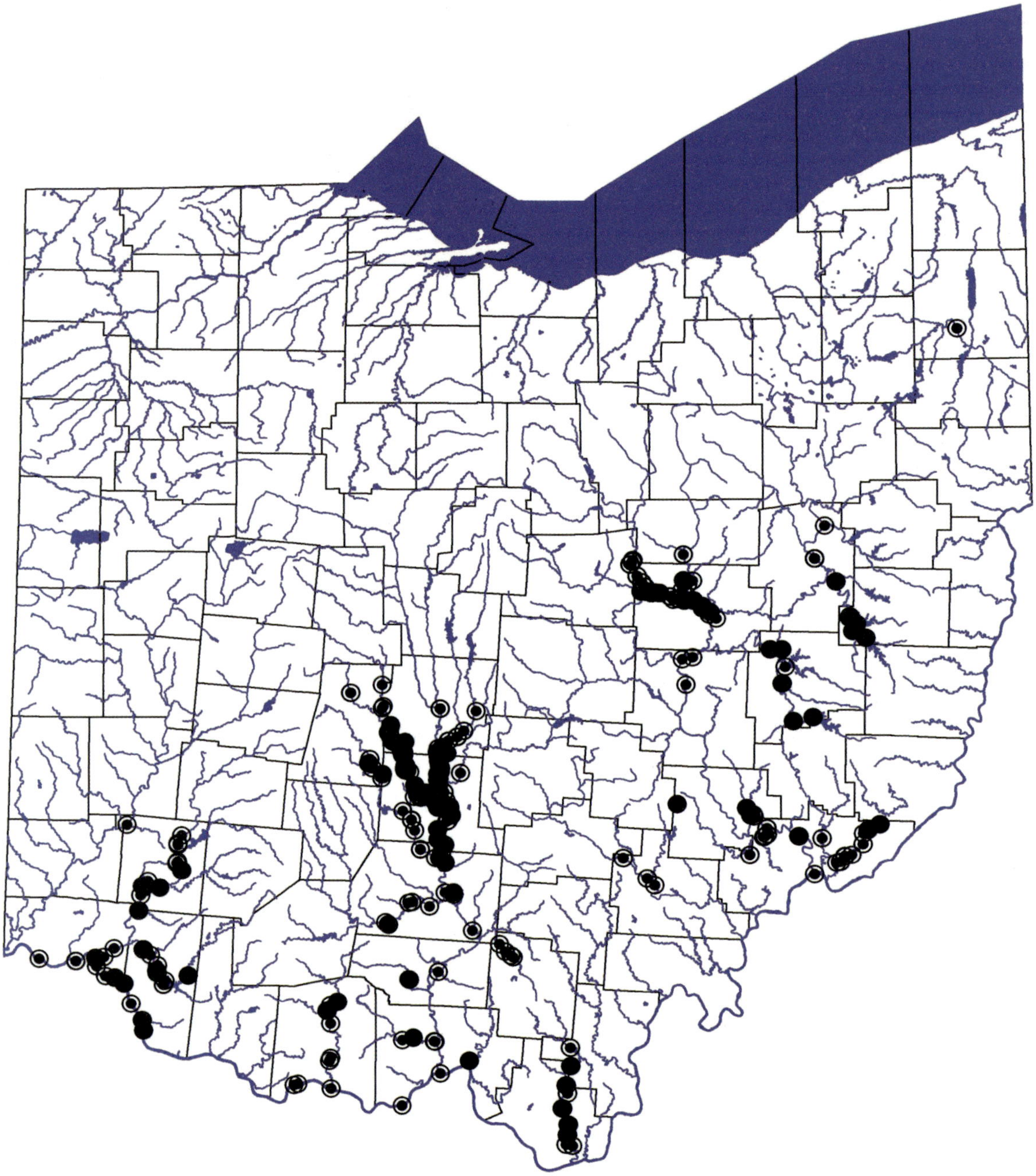

Fig. 200. ***Tritogonia verrucosa*** **(Rafinesque, 1820)**
Solid dots—live or freshdead since 1980
Bullseye—all records, any condition

Periostracum. Greenish with yellow or tan pustules, often with vague green stripes; some individuals may be nearly black, particularly when old.
Nacre. Porcelain white, often iridescent posteriorly; often with gold or brown Tulberg layers; some southern North American specimens may have purple nacre.
Hinge plate. Well-developed; right cardinal a single, massive, triangular denticle with a much smaller anterior ridge; left cardinal bifid, the teeth often far apart; laterals well-developed, straight, one in right valve, two in left; interdentum flattened, rather long.
Muscle scars. Anterior adductor impressed; posterior adductor flush; anterior adductor distinct from anterior pedal protractor scar, large posterior adductor confluent with posterior pedal retractor scar; dorsal suspensors located behind anterior adductor and in a line inside of interdentum. Pallial line entire, located far from shell margin. Accessory strong, reaching 33% the length of the shell.
Umbonal cavity. Very wide, compressed, and deep.

Glochidia. Subelliptical to subrotund, dorsal margin slightly curved, lateral margins gently and equally curved throughout their lengths, ventral margin semicircular. Exterior surface sculpturing rough, hooks absent, micropoints lanceolate and few in number (Hoggarth, 1999: figs. 32a–e).

Literature records (Surber, 1912: 9, pl. 2, fig. 31; Hoggarth, 1999: figs. 32a–e) indicate glochidial dimensions for length (0.085 to 0.094 mm), height (0.085 to 0.101 mm), and hinge (0.043 to 0.046 mm). Ortmann (1919: 44) noted that the glochidia are, ". . . unusually small." The glochidium of this species can be distinguished from all other glochidia by its very small size and the outline of its valve.

Reproductive biology. Tachytictic. Individuals spawn from March to July (Jirka & Neves, 1992). In Ohio, sperm were released in free-swimming, bowl-shaped aggregates in late April (Watters, unpubl.). Ortmann (1919) related that females develop (fertilized?) eggs by May. Glochidia form from May to July (Ortmann, 1919; Heath *et al.*, 1998). Glochidia are released in white or yellow, lanceolate conglutinates (Ortmann, 1919). When ready to release glochidia, the females display a large, swollen mantle margin around the posterior of the shell (Pepi & Hove, 1997). Individuals reach maturity from 4 to 6 years of age (Jirka & Neves, 1992). Catfishes have been reported as hosts.

Reported potential hosts
Ameiurus natalis (Yellow Bullhead)—Hove *et al.*, 1998b (LT).
Ameiurus nebulosa (Brown Bullhead)—Hove *et al.*, 1998b (LT).
Ictalurus punctatus (Channel Catfish)—Howells, 1997 (LT).
Pylodictis olivaris (Flathead Catfish)—Howells, 1996 (LT); Pepi & Hove, 1997 (LT); Howells, 1997 (LT); Hove *et al.*, 2004 (LT).

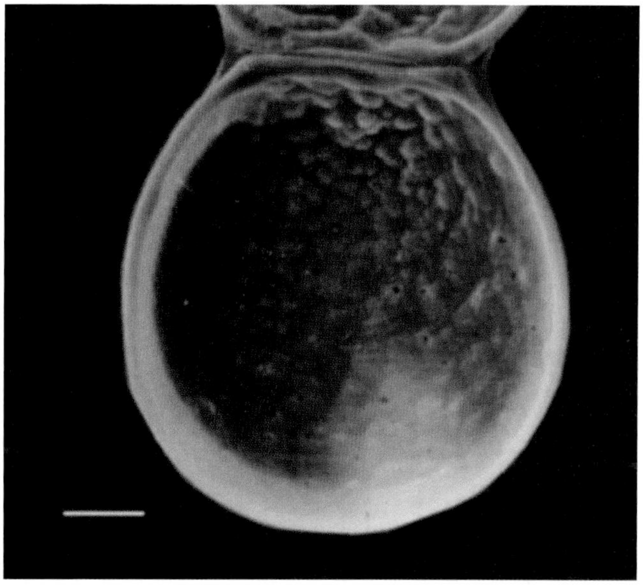

Fig. 201. *Tritogonia verrucosa* (Rafinesque, 1820). MAH 654.1, Big Darby Creek, Pickaway Co., OH. Bar = 15 µm. (Hoggarth, 1999).

Habitat. This species lives in large creeks with high water quality. It often is found along the water's edge, adjacent to moving water. Like the Rabbitsfoot, the Buck Horn is commonly found unburied, particularly during the summer.

Range. Mississippi River drainage including the Ohio, Cumberland, and Tennessee rivers. East to Pennsylvania. West to Texas and Oklahoma. Mobile River system, although those populations often have a purple nacre and may represent a separate species.

Range in Ohio. Historically in the Little Miami and much of the Scioto and Muskingum river drainages. A single record from the Great Miami River. Sporadic today: Big Darby and Killbuck creeks; Symmes, Salt, and White Oak creeks; Little Scioto River and Scioto Brush Creek. Absent from the Lake Erie drainages.

Similar species. This species may be confused with *Quadrula cylindrica*. That species is much smaller and usually more cylindrical and has a tent-like color pattern. The dense pustulose sculpture does not occur in *Amblema* and *Megalonaias*, which are rarely as elongated as *Tritogonia*.

Comments. Individuals gradually slow in growth during their entire life. Specimens may reach over 180 mm and 10 years old. The shells display sexual dimorphism, with the females having a much wider, inflated posterior ridge, often projecting well past the posterior margin.

Truncilla Rafinesque, 1819

Etymology. L. *truncus*, mutilated, deprived of some of its parts, cut off + L. *-illa* (diminutive). Rafinesque (1820: 300) explained the coining of the name *Truncilla:* "The name is derived from the remarkable oblique truncation, . . ." (Poulson translation, 1832). Members of the genus *Truncilla* are indeed characterized by a sharp posterior ridge and a flat (or nearly so) posterior slope, giving the shell a truncated appearance.

Original description

Truncilla Rafinesque, 1819: 427 [*Truncilla* dates from 1819 rather than 1820, despite the fact that no available names were included under that genus in 1819.].

Type species. *Truncilla truncata* Rafinesque, 1820, by subsequent designation of Herrmannsen (1849: 628).

Synonymy

Truncilla Rafinesque, 1819: 427.

Amygdalonaias Crosse & Fischer *in* Fischer & Crosse, 1893: 557 [Type species: *Unio cognata* Lea, 1860, by original designation].

Geographic range. Mississippi, Mobile, and Great Lakes drainages. Rio Grande, Colorado, Trinity, and Brazos rivers.

Comments. *Truncilla* was widely used for both *Epioblasma* and *Truncilla* prior to Frierson (1927), who recognized the differences between the two groups.

Truncilla donaciformis (Lea, 1828)

Figures 202, 203

Etymology. *Donax*, a genus of small marine bivalves, the coquinas (Donacidae) (from Gr. *donax*, reed or cane, shellfish) + L. *formis*, in the form or shape of, resembling. Lea so named this species because its shape reminded him of these marine bivalves, noting (1828: 267), "The characteristics of this little shell are its angulated posterior slope giving it in some measure the form of a donax, ..."

Vernacular name(s). *Fawnsfoot.*

Original description

Unio donaciformis Lea, 1828: 267, pl. 4, fig. 3.

Type locality. "Ohio [River]." Since the holotype is listed as from "T. G. Lea" whose residence was Cincinnati, Ohio, which is on the Ohio River, we herein restrict the type locality to the Ohio River at Cincinnati, Ohio.

Type material. Johnson & Baker (1973: 153) recognized Lea's figured specimen (USNM 84457) as the lectotype.

Synonymy

Unio donaciformis Lea, 1828: 267, pl. 4, fig. 3.

Unio zigzag Lea, 1829: 444, pl. 12, fig. 19 [Type locality: "Ohio." Types lost *fide* Johnson & Baker (1973: 175)].

Unio nervosa Rafinesque, 1820. Conrad, 1834: 70 [misidentification].

Unio nervosus Rafinesque, 1820. Say, 1834: no pagination [misidentification].

Margarita (*Unio*) *donaciformis* (Lea, 1828). Lea, 1836: 18.

Margarita (*Unio*) *zigzag* (Lea, 1829). Lea, 1836: 18.

Plagiola nervosa (Rafinesque, 1820). Agassiz, 1852: 48 [misidentification].

Margaron (*Unio*) *donaciformis* (Lea, 1828). Lea, 1852b: 24.

Margaron (*Unio*) *zigzag* (Lea, 1829). Lea, 1852b: 24.

Plagiola donaciformis (Lea, 1828). Baker, 1898b: 91, pl. 13, fig. 4.

Plagiola (*Amygdalonaias*) *donaciformis* (Lea, 1828). Simpson, 1900a: 605.

Amygdalonaias donaciformis (Lea, 1828). Ortmann, 1914: 67.

Unio zig-zag var. *illius* De Gregorio, 1914: 41, pl. 4, fig. 3 [Type locality: "Sciota [Scioto] R. (Ohio)." Type in Palermo Museum?].

Truncilla donaciformis (Lea, 1828). Ortmann & Walker, 1922: 50.

Status. Listed as Threatened by Ohio Division of Wildlife; not listed by U.S. Fish & Wildlife Service as of 2009.

ADULT SHELL

Shell. Small-sized (to ca. 50 mm); thick and heavy for its size, outline elongated, pointed posteriorly, umbo just anterior of center.

Anterior margin. Broadly curved.

Ventral margin. Very broadly curved, slightly angled in females; forming an acute point at the posterior margin at the base.

Posterior margin. Gently rounded or straight.

Dorsal margin. Short, straight; dorsal ridge sharp, forming ca. 90° angle, concave or straight.

Umbo. Very wide, low; beak sculpture microscopic, of ca. 7 double-looped ridges, the anterior loops wide, the posterior loops narrow and confined to the dorsal ridge, which is rib-like on the umbo; beak sculpture disappears early in the first year's growth.

Ligament. Tan to brown, short, not concealed by the valves.

Disc sculpture. Shell sculptureless except for posterior ridge.

Periostracum. Pale yellow, green, tan, or gray, becoming somewhat darker with age. Shells have a characteristic pattern of numerous dark green rays of all widths across the shell. These rays may be continuous or consist of

Fig. 202. ***Truncilla donaciformis*** **(Lea, 1828).** **A.** OSUM 22803.8, Muskingum River, Washington Co., OH. 54 mm. **B.** OSUM 22803.11, Muskingum River, Washington Co., OH. 49 mm. **C.** OSUM 67314.2, Scioto River, Franklin Co., OH. 42 mm. **D.** OSUM 13106.10, Lake Erie, Ottawa Co., OH. 28 mm. **E.** OSUM 67163.2, Scioto River, Pickaway Co., OH. 35 mm. **F.** OSUM 10249, Ohio River, Meigs Co., OH. 47 mm. **G.** OSUM 11149.1, Lake Erie, Ottawa Co., OH. 15 mm. **H.** OSUM 67314.4, Scioto River, Franklin Co., OH. 51 mm. **Hinge detail:** OSUM 67314.3, Scioto River, Franklin Co., OH. 48 mm total length. **Beak sculpture detail:** OSUM 11149.3, Lake Erie, Ottawa Co., OH. 17 mm total length.

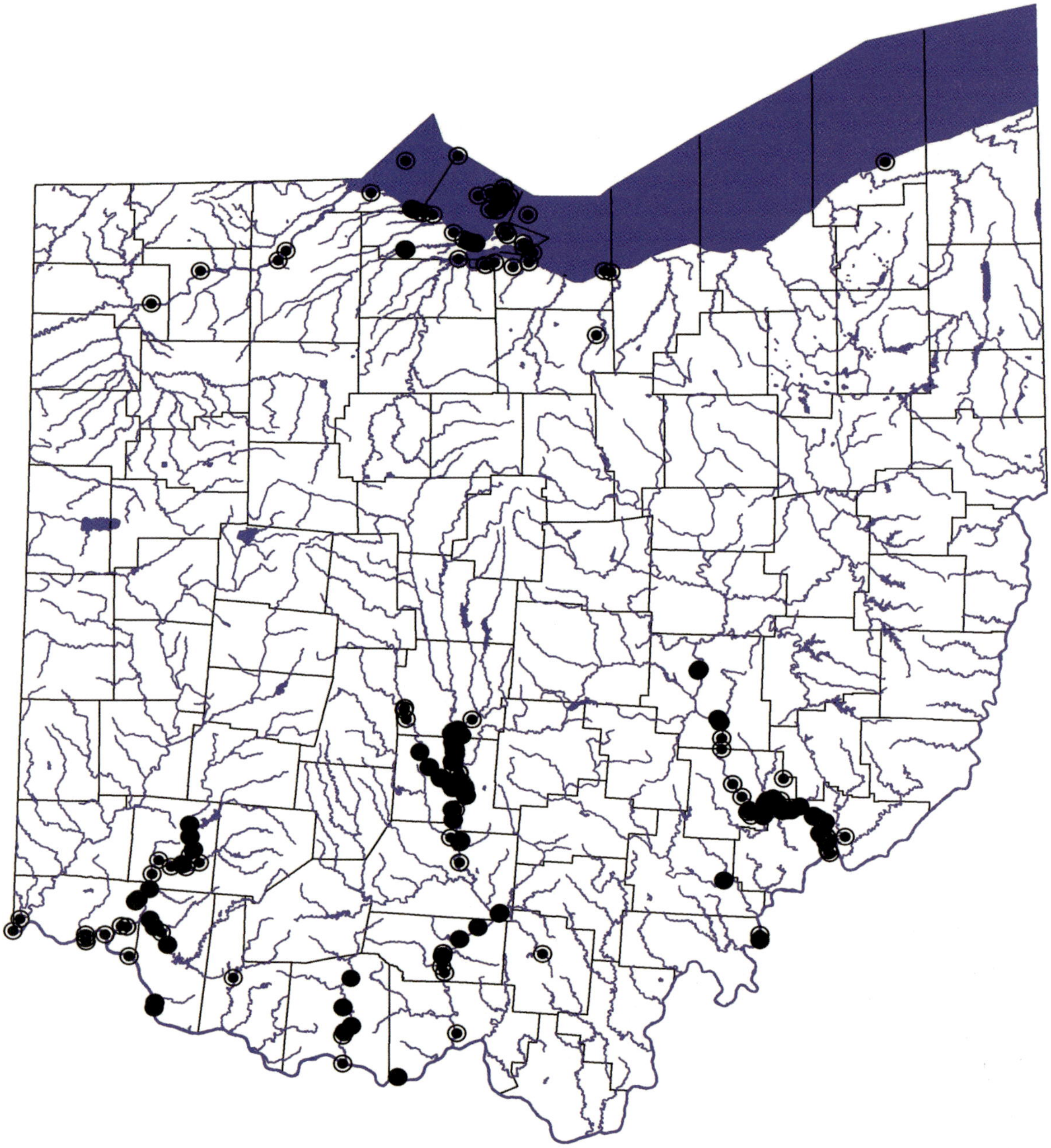

Fig. 203. ***Truncilla donaciformis*** **(Lea, 1828)**
Solid dots—live or freshdead since 1980
Bullseye—all records, any condition

distinct zig-zag or chevron markings; they may be distinct or smudged, and all combinations may exist on a single shell. Patterns often change between growth annuli. Young specimens tend to produce zig-zag markings, whereas older ones form rays.

Nacre. White, iridescent posteriorly; rarely with gold or green Tulberg layers.

Hinge plate. 60% of shell, straight or weakly angled, narrow; interdentum very narrow or absent; cardinal teeth lamellar or narrowly triangular, oriented longitudinally, two in left valve, one in right; lateral teeth short, straight, narrow, two in left, one in right.

Muscle scars. Anterior adductor muscle scar impressed, posterior adductor scar flush with shell surface, both distinct from protractor and retractor scars; dorsal suspensor scars a series of linear pits oriented obliquely beneath the interdentum; pallial line entire, accessory line short, extending to the shell margin.

Umbonal cavity. Wide, open, shallow.

Glochidia. Semicircular, dorsal margin slightly undulating and short, lateral margins equally and gently curved, ventral margin rounded, hooks absent (Surber, 1913: 116, pl. 30, fig. 8; Utterback, 1916a: 347).

Literature records (Surber, 1912: 9, pl. 2, fig. 29; Ortmann, 1914a: 67; Utterback, 1916a: 347) indicate glochidial dimensions for length (0.05 to 0.06 mm) and height (0.060 to 0.063 mm). Utterback (1916a: 347) noted, "Glochidia smallest on record, semicircular, spineless, hinge line short, very slightly undulated." He gave the length as 0.600 mm instead of 0.060.

Reproductive biology. Probably bradytictic. The distal edge of the posterior outer gills is marsupial. Ortmann (1919) reported gravid females from July and August. Fragile, white conglutinates are discharged (Utterback, 1916). Metamorphosis has not been observed in any host. In summer individuals may lie on the surface connected to long byssal threads; possibly these threads may entangle the host (Watters, unpubl.).

Reported potential hosts

Aplodinotus grunniens (Freshwater Drum)—Howard, 1913 (NI); Surber, 1913 (NI); Howard, 1914b (NI); Wilson, 1916 (NI); Howard & Anson, 1923 (NS).

Sander canadensis (Sauger)—Surber, 1913 (NI); Wilson, 1916 (NI).

Habitat. A species of rivers and lakes, rarely straying into smaller streams. It may be locally abundant in packed sand and gravel.

Range. Mississippi River system from Minnesota to southern Louisiana. Mobile River system to eastern Texas. Great Lakes drainages.

Range in Ohio. Little Miami, Scioto, Muskingum, and Hocking rivers, and Ohio Brush Creek of the Ohio River. Western basin of Lake Erie, Maumee, Portage, and Vermilion rivers.

Similar species. *Truncilla donaciformis* resembles *Truncilla truncata* but is more elongate and lacks the sharp dorsal ridge.

Comments. This species is fast growing but fairly short-lived. Growth is accelerated for the first 2 years, and the animal reaches 75% or more of its adult length in that time. Sexual dimorphism in the shell is not at all pronounced and is most noticeable from the less elongate female profile and slightly angled ventral margin. This small species rarely exceeds 50 mm and 7 years in age. A single long byssal thread is retained for several years.

Unlike its congener, *Truncilla truncata*, this species rarely if ever has non-white nacre.

Truncilla truncata Rafinesque, 1820

Figures 204, 205

Etymology. L. *truncatus*, maimed, cut off, shortened by having been cut off. Rafinesque (1820: 301) noted "posterior surface truncated" (Poulson translation, 1832), which is the typical appearance of this species.

Vernacular name(s). *Deertoe*, Deerfoot.

Original description

Truncilla truncata Rafinesque, 1820: 301, species no. 19, no figure.

Type locality. Rafinesque did not give a type locality; however, the title of his paper was "Monographie des Coquilles Bivalves Fluviatiles de la Rivière [System] Ohio, ..." The lectotype selected by Johnson & Baker (1973: 173) is from the Falls of the Ohio at Louisville, Kentucky. This becomes the type locality.

Type material. The type mentioned by Vanatta (1915: 553) was selected as lectotype (ANSP 20217 from "[Falls of the] Ohio River") by Johnson & Baker (1973: 173).

Synonymy

Truncilla truncata Rafinesque, 1819: 426 [*nomen nudum*].

Truncilla truncata Rafinesque, 1820: 301.

Truncilla truncata var. *fusca* Rafinesque, 1820: 301 [Type locality: none given. Type not located].

Truncilla truncata var. *vermiculata* Rafinesque, 1820: 301 [Type locality: none given. Type not located].

Fig. 204. ***Truncilla truncata*** **Rafinesque, 1820. A.** OSUM 58226.14, Maumee River, Defiance Co., OH. 67 mm. **B.** OSUM 3795.4, Lake Erie, Ottawa Co., OH. 43 mm. **C.** OSUM 11767.6, Mississippi River, Henderson Co., IL. 36 mm. **D, F, Hinge detail:** OSUM 58608.2, Miami River, [Co.?], OH. 64 mm. **E.** OSUM 56484.20, Tippecanoe River, Carroll Co., IN. 43 mm. **G.** OSUM 22813.1, Scioto River, Ross Co., OH. 88 mm. **H.** OSUM 58191.2, Tiffin River, Williams Co., OH. 54 mm. **Beak sculpture detail:** OSUM 20628, Paint Rock River, Jackson Co., AL. 17 mm total length.

Unio elegans Lea, 1831: 83, pl. 9, fig. 13 [Type locality: "Ohio River." Figured lectotype USNM 84449, paralectotype ANSP lost *fide* Johnson & Baker (1973: 154)].
Unio truncatus (Rafinesque, 1820). Say, 1834: no pagination.
Margarita (*Unio*) *elegans* (Lea, 1831). Lea, 1836: 18.
Plagiola truncata (Rafinesque, 1820). Agassiz, 1852: 48.
Margaron (*Unio*) *elegans* (Lea, 1831). Lea, 1852b: 23.
Unio truncatus var. *elegans* (Lea, 1831). Paetel, 1890: 170.
Plagiola elegans (Lea, 1831). Baker, 1898b: 91, pl. 21, fig. 1.
Plagiola (*Amygdalonaias*) *elegans* (Lea, 1831). Simpson, 1900a: 604.
Amygdalonajas [*sic*] *elegans* (Lea, 1831). Ortmann, 1912a: 328.
Unio elegans var. *elegantopsis* De Gregorio, 1914: 41, pl. 4, fig. 6 [Type locality: "Sciota" (Scioto River, Ohio). Type in Palermo Museum?].
Unio elegans var. *magnelegans* De Gregorio, 1914: 41, pl. 5, figs. 1a–c [Type locality: "Sciota" (Scioto River, Ohio). Type in Palermo Museum?].
Amygdalonaias truncata (Rafinesque, 1820). Utterback, 1916a: 348.
Truncilla truncata Rafinesque, 1820. Ortmann & Walker, 1922: 49.
Truncilla truncata var. Baker, 1924: 132 [described in Baker, 1928, below].
Truncilla truncata vermiculata Rafinesque, 1820. Frierson, 1927: 89.
Truncilla truncata var. *lacustris* Baker, 1928a: 227–228, pl. 78, figs. 1, 2 [Type locality "Long Point Island, Lake Winnebago [WI]." Holotype UMMZ 209214].

Status. Listed as a Species of Concern by Ohio Division of Wildlife; not listed by U.S. Fish & Wildlife Service as of 2009.

ADULT SHELL

Shell. Small to medium-sized (to 90 mm); heavy for its size, outline triangular, pointed posteriorly, umbo just anterior of center.
Anterior margin. Broadly curved.
Ventral margin. Very broadly curved anteriorly, then straight or concave to acute point at the posterior margin.
Posterior margin. Short, straight, forming distinct obtuse angle with dorsal slope.
Dorsal margin. Short, straight; dorsal ridge sharp, forming ca. 90° angle or less, concave or straight.
Umbo. Very wide, low; beak sculpture microscopic, of ca. 3 double-looped, nearly indistinct ridges, the anterior loops wide, the posterior loops narrow and confined to the dorsal ridge, which is rib-like on the umbo; beak sculpture disappears early in the first year's growth.
Ligament. Brown to black, short, partially concealed by the valves.
Disc sculpture. Shell sculptureless except for posterior ridge and occasionally 1–3 fine threads on posterior slope.
Periostracum. Pale yellow, orangish, green, tan, or gray, becoming somewhat darker with age. Shells have a characteristic pattern of numerous dark green rays of all widths across the shell. As in *T. donaciformis*, these rays may be continuous or consist of distinct zig-zag or chevron markings; they may be distinct or smudged. But unlike that species, *T. truncata* does not tend to produce all combinations on a single shell or change them between growth annuli.
Nacre. White, occasionally pink or salmon, iridescent posteriorly; rarely with gold or green Tulberg layers.
Hinge plate. 60% of shell, strongly curved or angled, fairly thick; interdentum narrow but thick; cardinal teeth chunky, triangular, oriented vertically, two in left valve, one in right, often with weak anterior accessory tooth; lateral teeth short, straight or gently curved, narrow, two in left, one in right.
Muscle scars. Anterior adductor muscle scar impressed, posterior adductor scar flush with shell surface, anterior pedal protractor scar distinct, posterior pedal retractor scar narrowly confluent or separate; dorsal suspensor scars a series of linear pits oriented obliquely beneath the interdentum; pallial line entire, accessory line very long, extending to the shell margin.
Umbonal cavity. Wide, open, shallow.

Glochidia. Semicircular, dorsal margin straight, anterior and posterior margins gently curved to a broadly rounded ventral margin, hooks absent (Ortmann, 1912a: 328; Utterback, 1916a: 347).

Literature records (Lefevre & Curtis, 1910a: 146, fig. g; Surber, 1912: 9, pl. 2, fig. 30; Ortmann, 1912a: 328; Utterback, 1916a: 347) indicate glochidial dimensions for length (0.06 to 0.08 mm) and height (0.07 to 0.09 mm). Utterback (1916a) noted, "Glochidia a little larger—otherwise these structures are identical with those of *donaciformis*."

Reproductive biology. The distal edge of the posterior outer gills is marsupial (Utterback, 1916). Surber (1912) reported gravid females from May to July. In Ohio, gravid females were found in April (Watters, unpubl.). No durable conglutinate is formed (Watters, 2008). Metamorphosis has not been observed in any host.

Reported potential hosts

Aplodinotus grunniens (Freshwater Drum)—Wilson, 1916 (NI); Howard & Anson, 1923 (NS).

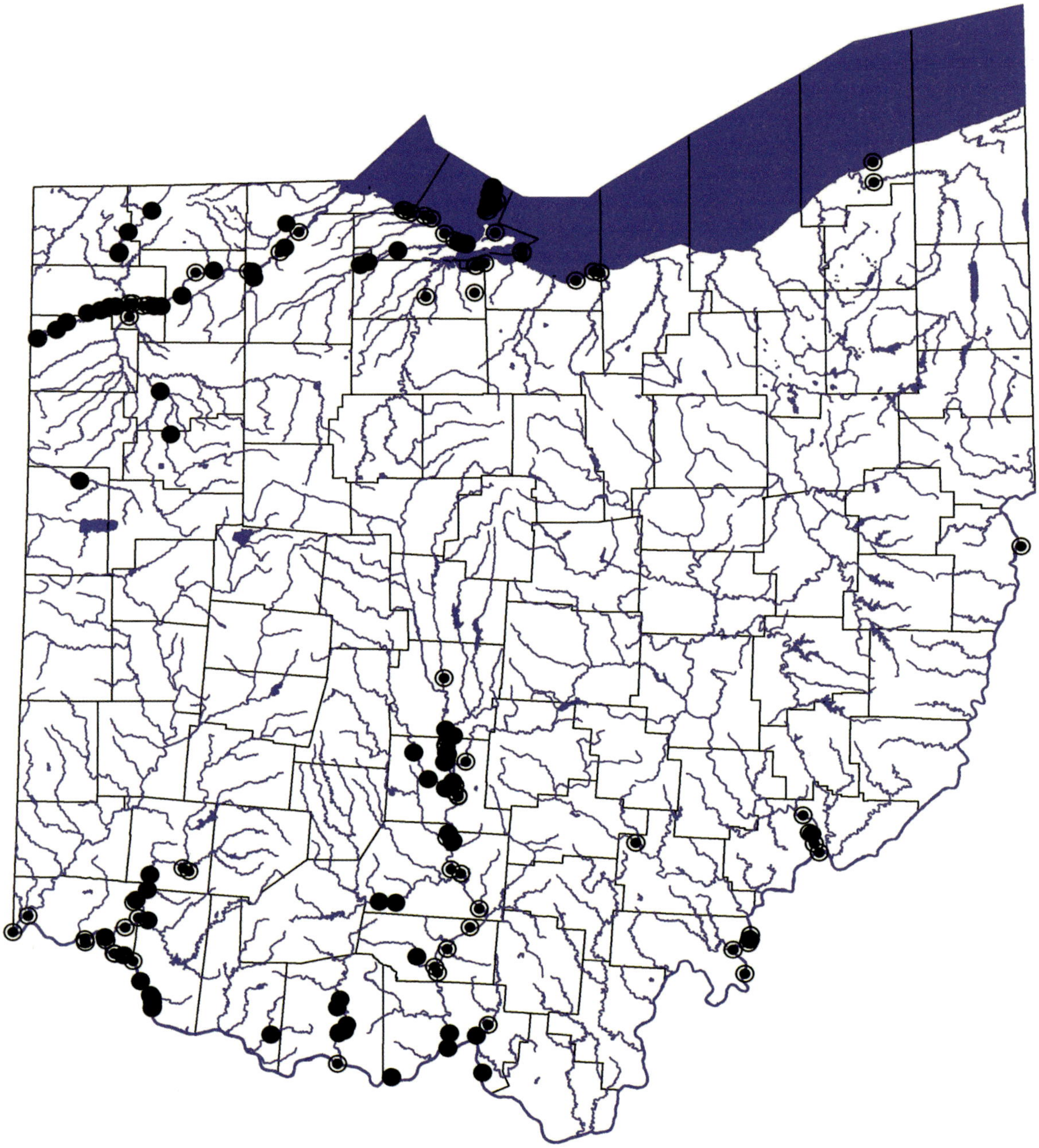

Fig. 205. ***Truncilla truncata*** **Rafinesque, 1820**
Solid dots—live or freshdead since 1980
Bullseye—all records, any condition

Sander canadensis (Sauger)—Wilson, 1916 (NI).

Habitat. A species of rivers and lakes, rarely straying into smaller streams. It may be locally abundant in packed sand and gravel.

Range. Mississippi River system from Minnesota to southern Louisiana. Mobile River system to eastern Texas. Great Lakes drainages.

Range in Ohio. In Lake Erie and rivers and large streams. Scioto River, Big Darby Creek, Ohio Brush Creek, Little Scioto, Portage, Grand, Maumee, and Muskingum rivers. Locally abundant (Maumee River in particular).

Similar species. See under *Truncilla donaciformis*.

Comments. Like *T. donaciformis*, this species is fast growing but fairly short-lived. Growth is accelerated for the first 2 years and then gradually slows. Sexual dimorphism in the shell is not pronounced and is most noticeable from the higher female profile and more angled ventral margin. This species reaches 90 mm, but generally averages ca. 50 mm. Specimens from the Maumee River are particularly large. It grows to 11 years in age.

Uniomerus Conrad, 1853

Etymology. L. *unio*, + L. *merus*, pure, genuine.
Original description
Uniomerus Conrad, 1853: 268.
Type species. *Unio excultus* Conrad, 1838, by subsequent designation of Clench & Turner (1956). [Say included eight species under *Uniomerus*, but did not designate a type. None of these eight were *Unio tetralasmus* Say, 1831. Simpson (1900a) and Haas (1969) designated *U. tetralasmus* Say, 1831, as type, but these are invalid actions, as *U. tetralasmus* was not one of the original listed species (ICZN Art. 69.2). Clench & Turner (1956) validly designated *Unio excultus* Conrad, 1838, one of the original eight species included by Conrad, as type.]
Geographic range. Eastern North America except for the Atlantic Gulf Slope; rare in the Great Lakes.

Uniomerus tetralasmus (Say, 1831)

Figures 206, 207

Etymology. Gr. *tetra-*, four + Gr. *elasma*, metal beaten out, metal plate. Say (1831) noted in his original description of *Unio tetralasmus* that the species is characterized by ". . . lateral teeth two in each valve." The internal right valve of the holotype clearly has two lateral teeth—hence the name *tetralasmus*. However, over 300 specimens of *U. tetralasmus* in the OSUM collection were examined relative to the presence of double dentition in the right valve, and none were found to match the figure in *American Conchology*. One must conclude that the holotype of this species is atypical. As in most dentate shells, there occasionally develops in the right valve a low shelf at the ventral base of the single well-developed lateral tooth that could be construed as a poorly developed second tooth. Specimens with this feature, however, do not come close to the figured holotype with two distinct laterals in the right valve. It is unfortunate that the character selected by the describer of this species to coin its name is rarely ever found in specimens. *Tetralasmus* is a noun in apposition.
Vernacular name(s). *Pondhorn*, Prairie Horn Shell.
Original description
Unio tetralasmus Say, 1831b: no pagination, pl. 23.
Type locality. ". . . I am indebted to Mr. Lesueur, who obtained it in Bayou St. John, near New-Orleans" (Say, 1831b: no pagination). Three nominal taxa of *Uniomerus* all have "New Orleans" as their type locality: *U. tetralasmus* Say, 1831, *U. camptodon* Say, 1832, and *U. excultus* Conrad, 1838. Recent workers have generally considered these as synonymous with *U. tetralasmus*, but their exact relationships to each other have not been determined. The types for all three species have been lost.
Type material. The original type material of Say has apparently been lost (Johnson & Baker, 1973: 146).
Synonymy
Unio tetralasmus Say, 1831b: no pagination, pl. 23.
Unio camptodon Say, 1832: 42 [Type locality: "New Orleans, Louisiana." Types lost].
Unio geometricus Lea, 1834: 38, pl. 4, fig. 10 [Type locality: "Bayou Teche, Louisiana." Type not located].
Margarita (*Unio*) *tetralasmus* (Say, 1831). Lea, 1836: 30.
Margarita (*Unio*) *camptodon* (Say, 1832). Lea, 1836: 30.
Unio declivis Say, 1831. Conrad, 1836: 45, pl. 23, fig. 1 [misidentification].
Unio excultus Conrad, 1838: 99, pl. 54, fig. 1 [Type locality: "New Orleans, Louisiana." Figured lectotype ANSP 20427 lost *fide* Johnson & Baker, 1973].
Unio sayi Ward *in* Tappan, 1839: 268, pl. 3, fig. 1 [Type locality: "Walnut Creek, Circleville, Ohio." Type not located].
Unio parallelus Conrad, 1841: 20 [Type locality: "Jackson, Mississippi." Figured lectotype ANSP 42847].
Margaron (*Unio*) *tetralasmus* (Say, 1831). Lea, 1852b: 32.
Margaron (*Unio*) *camptodon* (Say, 1831). Lea, 1852b: 32.
Unio rivularis Conrad, 1853: 257 [Misidentification. Replacement for *Unio declivis* "Say" Conrad, 1836: 45, pl. 23, fig. 1].

Fig. 206. *Uniomerus tetralasmus* (Say, 1831). A. OSUM 1472.1, Hellbranch Run, Franklin Co., OH. 110 mm. **B, H, Hinge detail:** OSUM 57147.6, Blue Creek, Union Co., OH. 109 mm. **C.** OSUM 9101.1, Little Scioto River, Marion Co., OH. 115 mm. **D.** OSUM 68827.2, Miami-Erie Canal, [Co.?], OH. 82 mm. **E, Beak sculpture detail:** OSUM 39851.1, "Ohio River." 33 mm. **F.** OSUM 68829.2, Ohio River, Hamilton Co., OH. 70 mm. **G.** OSUM 3718, Lake Erie, Ottawa Co., OH. 79 mm.

Unio porrectus Conrad, 1854: 296, pl. 26, fig. 7 [unnecessary new name for *Unio parallelus* Conrad, 1841, for presumed homonymy with *Unio parallellus* Sowerby, 1840].
Unio subcroceus Conrad, 1854: 297, pl. 27, fig. 1 [Type locality: "one of the tributaries to Canadian River, Arkansas." Figured lectotype ANSP 41406a].
Unio manubius Gould, 1855: 229 [Type locality: "Chihuahua, sixty miles from Camp Ringgold, Mexico." Holotype MCZ 169447].
Unio jamesianus Lea, 1857b: 84 [Type locality: "Jackson, Mississippi." Holotype USNM 85365].
Unio electrinus Reeve, 1865: pl. 25, fig. 121 [Type locality: none given. Type in BMNH lost].
Margaron (*Unio*) *jamesianus* (Lea, 1857). Lea, 1870: 50.
Margaron (*Unio*) *manubius* (Gould, 1855). Lea, 1870: 54.
Unio manubrius [*sic*] Paetel, 1890: 158.
Unio (*Uniomerus*) *tetralasmus* Say, 1831. Simpson, 1900a: 739.
Unio (*Uniomerus*) *tetralasmus* var. *camptodon* Say, 1832. Simpson, 1900a: 740.
Unio (*Uniomerus*) *tetralasmus* var. *manubius* Gould, 1855. Simpson, 1900a: 740.
Unio (*Uniomerus*) *tetralasmus* var. *sayi* Ward, 1839. Simpson, 1900a: 741.
Unio declivis geometricus Lea, 1834. Frierson, 1903: 49, pl. 3, middle fig.
Unio tetralasmus sayi Ward, 1839. Sterki, 1907: 393.
Uniomerus tetralasmus (Say, 1834). Ortmann, 1912a: 272.
Unio (*Uniomerus*) *tetralasmus* var. *manubus* [*sic*] Gould, 1855. Simpson, 1914: 708.
Uniomerus tetralasmus comptodon [*sic*] (Say, 1832). Utterback, 1915b: 48.
Elliptio (*Uniomerus*) *tetralasmus* (Say, 1831). Frierson, 1927: 34.
Elliptio (*Uniomerus*) *tetralasmus camptodon* (Say, 1832). Frierson, 1927: 34.
Elliptio (*Uniomerus*) *tetralasmus* var. *sayi* (Ward, 1839). Frierson, 1927: 34.
Elliptio (*Uniomerus*) *tetralasmus manubius* (Gould, 1855). Frierson, 1927: 34.
Uniomerus tetralasmus sayi (Ward, 1839). Baker, 1928a: 137.
Uniomerus tatralasmus [*sic*] (Say, 1834). Baker, 1928a: 494.
Elliptio tetralasmus manubius (Gould, 1855). Strecker, 1931: 16.
Uniomerus tetralasmus camptodon (Say, 1832). Murray & Leonard, 1968: 29.

Status. Listed as Threatened by Ohio Division of Wildlife; not listed by U.S. Fish & Wildlife Service as of 2009.

ADULT SHELL

Shell. Medium to fairly large sized (to 120 mm); thin, outline elongated, sharply pointed posteriorly, umbos ca. 33% back from anterior margin.
Anterior margin. Broadly curved.
Ventral margin. Very broadly curved or straight, forming an acute point at the posterior margin at the base.
Posterior margin. Straight to concave.
Dorsal margin. Very long, straight or slightly convex.
Umbo. Wide, low, barely rising above the hinge line; beak sculpture of 5–8 coarse concentric ribs, oriented obliquely towards the antero-dorsal margin; the anterior portions of the more distal ribs may be lost, leaving only the posterior arms; beak sculpture disappears early in the first year's growth.
Ligament. Tan to brown, very long, largely concealed by the valves.
Disc sculpture. Shell sculptureless except for 2 low, broad ribs on the dorsal slope, divided by shallow grooves; this sculpture may be nearly lost in some specimens. Some individuals have a wrinkled appearance along the posterior-ventral portion of the shell.
Periostracum. Pale yellow, becoming somewhat darker with age; shells rarely with faint green rays. Growth annuli dark brown.
Nacre. White, iridescent posteriorly; most specimens have gold Tulberg layers.
Hinge plate. Long, straight, very thin; interdentum long but narrow; cardinal teeth lamellar or narrowly triangular, two in left valve, one in right with an accessory tooth anterior to cardinal; lateral teeth short, straight or slightly bowed, very thin, two in left, one in right.
Muscle scars. Both adductor muscle scars flush with shell surface, anterior adductor separate from anterior pedal protractor; posterior adductor confluent with posterior pedal retractor scar; dorsal suspensor scars a cluster of irregular pits in the umbonal cavity; pallial line entire, accessory line absent.
Umbonal cavity. Wide, open, shallow, nearly nonexistent.

Glochidia. Subelliptical, dorsal margin straight, lateral margins evenly and gently curved, ventral margin broadly rounded, hooks absent (Utterback, 1915:2 45, fig. 5; Utterback, 1916b: 20).

Literature records (Utterback, 1915: 245, fig. 5; Utterback, 1916b: 20) indicate glochidial dimensions for length (0.16 mm) and height (0.21 mm). Utterback (1915: 245) stated, "Glochidia spatulate, very regular in outline, hinge line short and straight, medium in size."

Reproductive biology. Probably bradytictic. The outer gills are marsupial (Ortmann, 1912a). The brooding period is unknown. Conglutinates are white and sole-shaped, with transparent stripes across their width (Utterback,

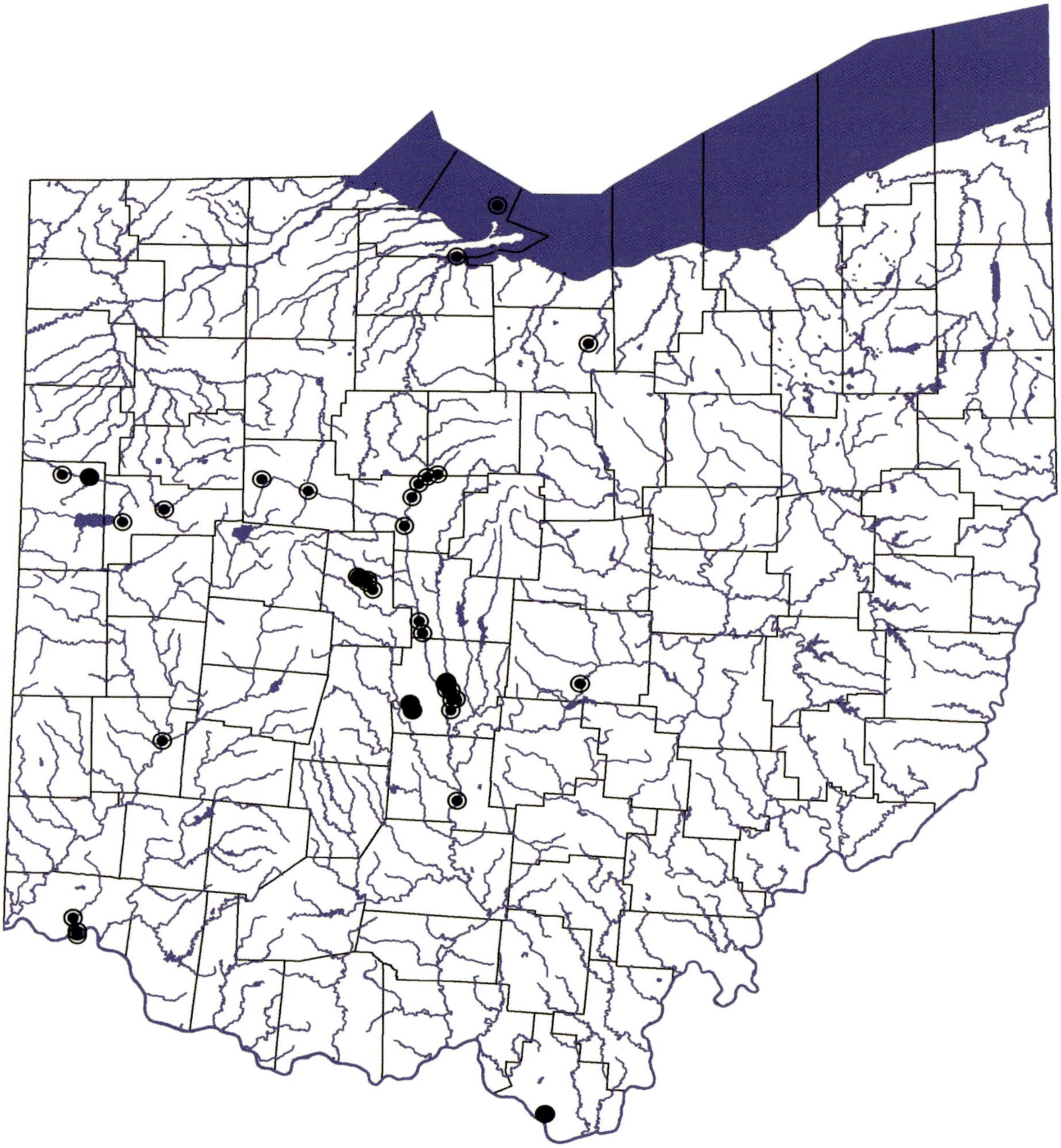

Fig. 207. ***Uniomerus tetralasmus*** **(Say, 1831)**
Solid dots—live or freshdead since 1980
Bullseye—all records, any condition

1915; Fuller, 1971). Metamorphosis has not been observed in any host.

Reported potential hosts

Notemigonus crysoleucas (Golden Shiner)—Stern & Felder, 1978 (NI).

Habitat. Most specimens are found in muddy sand.

Range. One of the most widely distributed unionoids in North America. Mississippi River system from Kansas to Louisiana, west to eastern Texas. Lower Ohio River system. Lake Erie. If Gould's *Unio manubius* is this species, then as far south as Chihuahua State, Mexico.

Range in Ohio. This species has a very sporadic and unusual distribution in Ohio. It has been recorded from small creeks (Hellbranch Run), small rivers (Olentangy River), and Lake Erie embayments (Metzger Marsh) and tributaries (Grand River). *Unio sayi* Ward *in* Tappan, 1839, a synonym of this species, was described from Walnut Creek, near Circleville. It is not known if this population still exists there. Historically this species also occurred in the Ohio Canal at Columbus, Cedar Creek of Lake Erie, and St. Mary's River (CMNH); rare in the Auglaize River, Sandusky River, Licking River. Ohio is on the eastern extent of this species' range. It seems to occur in prairie areas in the state.

Similar species. See under *Uniomerus declivis*. *Uniomerus tetralasmus* may be confused with *Pyganodon grandis* at first glance, but that species lacks any dentition and has double-looped beak sculpture.

Comments. This species is very fast growing, but very short-lived. Growth gradually slows over the life of the animal, most of its size being obtained in the first 2 years of life. No sexual dimorphism in the shell is evident. Individuals may exceed 120 mm in length, but rarely grow over 10 years old. Most specimens are less than 5 years old, although an individual from the Auglaize River was at least 15 years old.

Utterbackia Baker, 1927

Etymology. L. *Utterbackia*. Utterback is a patronym in honor of Professor William Irvin Utterback, author of *The Naiades of Missouri* (1915–1916) and many other studies on freshwater mussels.

Original description

Utterbackia Baker, 1927b: 220–223, no plates or figures.

Type species. *Anodonta imbecillis* Say, 1829, by original designation.

Synonymy

Utterbackia Baker, 1927b: 220–223, no plates or figures.

Utterbachia [*sic*] Baker, 1927b: 221.

Geographic range. This genus has been found from the Rocky Mountains to the Eastern Seaboard and from Gulf of Mexico streams north to Lakes Erie and Ontario as well as Michigan and Minnesota.

Comments. Hoeh (1990) studied the North American Anodontas and concluded that *Utterbackia* had at least two species: *U. imbecillis* (Say, 1829) and *U. peggyae* (Johnson, 1965). Bogan & Hoeh (1995) added their new species *U. peninsularis*. *Utterbackia peggyae* has a range restricted to streams tributary to the Gulf of Mexico from Alabama to the Florida Panhandle. *Utterbackia peninsularis* occurs in western peninsular Florida. In contrast, *U. imbecillis* has a distribution covering most of the eastern United States east of the Rocky Mountains.

Utterbackia imbecillis (Say, 1829)

Figures 208, 209, 210

Etymology. From L. *imbecillus*, feeble, weak. The shell of this species is unusually thin and fragile.

Vernacular name(s). *Paper Pondshell*, Papershell, Paper Floater.

Original description

Anodonta imbecillis Say, 1829: 355–356, no plates or figures.

Type locality. "Inhabits Wabash River." Clench & Turner (1956) restricted the type locality to "Wabash River, New Harmony, Indiana." But the neotype (SMF 4301) selected by Haas (1930) is from "St. Ohne, Fundortaugabe."

Type material. It appears that the primary type material(s) of all or nearly all of Say's species have been lost. Hass (1930: 326) has, however, selected SMF No. 4301 to serve as the neotype of *A. imbecillis* Say, 1829.

Synonymy

Anodonta imbecillis Say, 1829: 355–356.

Anodonta incerta Lea, 1834: 46, pl. 6, fig. 16 [Type locality: "Ohio River near Cincinnati [OH]." Types lost *fide* Johnson, 1970].

Margarita (*Anodonta*) *incerta* Lea, 1834. Lea, 1836: 51.

Anodon incerta Say, 1829. Catlow & Reeve, 1845: 67.

Margaron (*Anodonta*) *imbecillis* Say, 1829. Lea, 1852b: 50.

Anodonta imbecilis [*sic*] Say, 1829. Lapham, 1852: 370.

Anodon horda Gould, 1855: 229 [Type locality: "Comanche Creek, Texas" (tributary of the Llano River, close to the present site of Mason, Mason Co.) (Johnson, 1970). Lea (1870: 80) noted, "I have several times had specimens from Texas and have no doubt of its being Say's *imbecillis*, which has such an extraordinary wide geographical distribution in

Fig. 208. ***Utterbackia imbecillis*** **(Say, 1829).** **A.** OSUM 7708, Little Darby Creek, Madison Co., OH. 80 mm. **B.** OSUM 61952.14, Scioto River, Pickaway Co., OH. 64 mm. **C.** OSUM 45525.4, Ohio River, Hamilton Co., OH. 67 mm. **D.** OSUM 53528.2, Scioto River, Delaware Co., OH. 88 mm. **E.** OSUM 33564.2, Olentangy River, Delaware Co., OH. 75 mm. **F.** OSUM 10966.1, Lake Erie, Erie Co., OH. 39 mm. **G H, Hinge detail:** OSUM 22724.4, Kankakee River, Newton Co., IN. 72 mm. **Beak sculpture detail:** OSUM 61952.1, Scioto River, Pickaway Co., OH. 28 mm total length.

the Southern and Western States." Holotype USNM 678301]

Anodonta henryiana Lea, 1857d: 102 [Type locality: "Matamoras, Tamaulipas [State], Mexico." Simpson (1914: 397) observed that *A. henryiana* Lea, 1857, was "very close to *A. imbecillis* and probably only a southwestern variety of it, as there are intermediate specimens, which are very hard to place." Lectotype USNM 86692].

Anodon incertus Say, 1829. Sowerby, 1867: pl. 17, fig. 59.

Anodon imbecillis Say, 1829. Sowerby, 1870: pl. 27, fig. 102.

Anodonta hordeum Gould, 1855. Paetel, 1890: 180.

Anodonta phalena De Gregorio, 1914: 64, pl. 11, figs. 3a–c [Type locality: "Sciota [Scioto] R. (Ohio)." Type in Palermo Museum?].

Lastena ohiensis (Rafinesque, 1820). Utterback, 1915: 260, pl. 23, figs. 74a, b [misidentification].

Anodonta ohiensis Rafinesque, 1820. Ortmann, 1919: 162, pl. 11, fig. 4 [misidentification].

Utterbackia imbecillis fusca Baker, 1927b: 222 [Type locality: "Sturgeon Bay, Door Co., Wisconsin." Holotype UMMZ 209141].

Anodonta (*Lastena*) *ohiensis* Rafinesque, 1820. Frierson, 1927: 17 [misidentification].

Anodonta (*Lastena*) *ohiensis horda* (Gould, 1855). Frierson, 1927: 17.

Utterbackia imbecillis (Say, 1829). Baker, 1928a: 172, pl. 68, figs. 6–8.

Anodonta (*Utterbackia*) *imbecilis* [*sic*] Say, 1829. Clarke & Berg, 1959: 40, fig. 42.

Status. Not listed by either Ohio Division of Wildlife or U.S. Fish & Wildlife Service as of 2009.

ADULT SHELL

Shell. Medium size (to 115 mm), very thin shelled throughout life, elongated, usually very inflated.

Anterior margin. Broadly rounded, often somewhat angular towards the ventral margin.

Ventral margin. Nearly straight to broadly rounded.

Posterior margin. Ventral and posterior margins form wide, blunt acute angle below midline, often somewhat expanded ventrally; posterior margin oblique and straight to dorsal margin.

Dorsal margin. Long, more or less straight, joining posterior margin with very obtuse angle, forming low wing in some specimens (particularly in small individuals).

Umbo. Flush with dorsal margin, wide and flattened, umbonal sculpture of 2–7 weakly double-looped ridges, becoming oblique distally from the umbo if more than 4 ridges are present.

Ligament. Very long, tan to dark brown, completely concealed by shells.

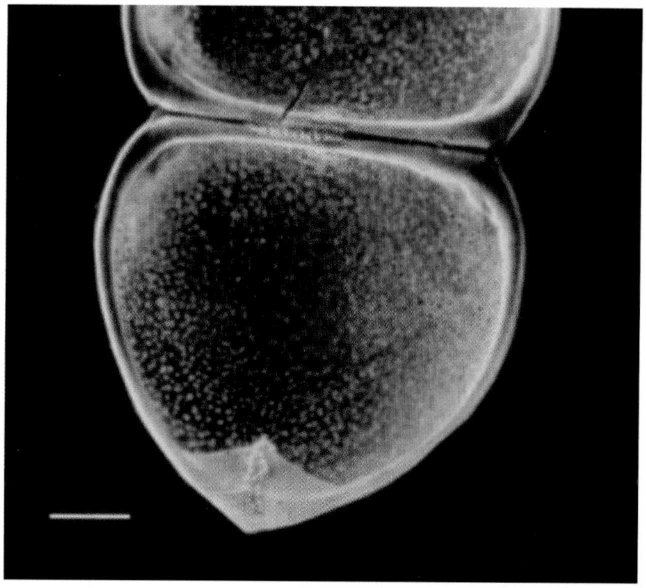

Fig. 209. *Utterbackia imbecillis* (Say, 1829). Glochidium. UWZY 24971.1, Baraboo River, Sauk Co., WI. Bar = 60 µm. (Hoggarth, 1999).

Disc sculpture. Smooth except for two weak posterior ridges. Some populations have a peculiar set of vertical corrugations on the shell disc as adults.

Periostracum. Green, yellow, or tan, usually with vague rays, particularly posteriorly.

Nacre. Uniformly silver-white, highly iridescent posteriorly. Tan Tulberg layers very rare.

Hinge plate. Very thin, straight, edentulous.

Muscle scars. Adductors not impressed, anterior protractor scar oval and separate from anterior pedal adductor; posterior protractor fused along base with top of posterior pedal retractor; dorsal suspensors variable, from a line of several small scars to irregular patches under the umbo; pallial line entire, located far from shell margin. Accessory line absent.

Umbonal cavity. Very wide, shallow, and open.

Glochidia. Subtriangular, dorsal margin straight, anterior and posterior margins slightly and evenly curved, ventral margin broadly pointed. Exterior surface sculpturing tight-looped, hooks styliform, microstylets lanceolate near the distal margin of the hook and multifaceted near the proximal border (Lea, 1858c: 50, pl. 5, fig. 36; Ortmann, 1911: 346, pl. 89, fig. 13; Howard, 1915: 9; Hoggarth, 1999: figs. 13a–f).

Literature records (Ortmann, 1912a: 291; Surber, 1912: 8, pl. 1, fig. 2; Utterback, 1915: 260; Tucker, 1927: 287–288, figs. 1, 2; Hoggarth, 1999: figs. 13a–f) indicate glochidial dimensions for length (0.29 to 0.31 mm), height (0.29 to 0.31 mm), and hinge (0.24 to 0.26 mm). This glochidium can be distinguished from others in the subfamily by its exterior valve sculpture, its rather small size, and its relatively long central ligament. It resembles *A. suborbiculata* in these last two characters but differs in valve sculpture and hook structure. The outline of this glochidium is like that of *A. cygnea,* while the hook allies this species to *P. grandis.*

Reproductive biology. Howard (1914b, 1915), Howard & Anson (1923), and Allen (1924) all reported that this species did not require a host to complete metamorphosis. Baker (1928a: 173) stated that "the presence of juveniles in the marsupium during November in a majority of the specimens examined indicates that the metamorphosis may be completed in the fall [without a host]." Heard (1975) concluded that there was no evidence for this claim. However, Dickinson & Sietman (2008) confirmed that this species may indeed bypass the host requirement. The extent of this phenomenon is not known.

Bradytictic. Baker (1928a) reported gravid females year round. Ortmann (1912) believed this species was hermaphroditic. Watters (unpubl.) found gravid females from April through September in Ohio. Females release strands of mucus containing glochidia, which become entangled on objects and form net-like structures. Fishes are parasitized on the fins when they swim through these nets. A similar method of parasitizing hosts occurs in *A. suborbiculata* and *A. ferussacianus.* This species is a host generalist and may use exotic aquaria fish (*) and amphibians (**) (see following list).

Reported potential hosts

Ambloplites rupestris (Rock Bass)—Trdan & Hoeh, 1982 (LT).

Ambystoma tigrinum (Tiger Salamander)—Watters & O'Dee, 1997a (LT); Watters & O'Dee, 1998a (LT)**.

Aplocheilus lineatus (Panchax Killifish)—Watters & O'Dee, 1997a (LT); Watters & O'Dee, 1998a (LT)*.

Aulanacara stuartgranti (Peacock Fish)—Watters & O'Dee, 1997a (LT); Watters & O'Dee, 1998a (LT)*.

Barbus semifasciolatus (Golden Barb)—Watters & O'Dee, 1997a (LT); Watters & O'Dee, 1998a (LT)*.

Betta splendens (Siamese Fighting Fish)—Watters & O'Dee, 1997a (LT); Watters & O'Dee, 1998a (LT)*.

Brachydanio kerri (Blue Danio)—Watters & O'Dee, 1997a (LT); Watters & O'Dee, 1998a (LT)*.

Carassius auratus (Goldfish)—Watters & O'Dee, 1997a (LT); Watters & O'Dee, 1998a (LT)*.

Colisa lalia (Flame Gourami)—Watters & O'Dee, 1997a (LT); Watters & O'Dee, 1998a (LT)*.

Cyprinella spiloptera (Spotfin Shiner)—Hove *et al.*, 1995 (LT); Watters & O'Dee, 1997 (LT)*.

Danio malabaricus (Giant Danio)—Watters & O'Dee, 1997a (LT); Watters & O'Dee, 1998a (LT)*.

Eigenmannia virescens (Glass Knifefish)—Watters & O'Dee, 1997a (LT); Watters & O'Dee, 1998a (LT)*.

Etheostoma lepidum (Greenthroat Darter)—Howells, 1997 (LT).

Fundulus diaphanus (Banded Killifish)—Trdan & Hoeh, 1982 (LT).

Gambusia affinis (Mosquitofish)—Stern & Felder, 1978 (NI).

Gyrinocheilus aymonieri (Chinese Algae-eater)—Watters & O'Dee, 1997a (LT); Watters & O'Dee, 1998a (LT)*.

Haplochromis venustus (Marbled Peacock Fish)—Watters & O'Dee, 1997a (LT); Watters & O'Dee, 1998a (LT)*.

Hemigrammus nanus (Silver Tip Tetra)—Watters & O'Dee, 1997a (LT); Watters & O'Dee, 1998a (LT)*.

Hyphessobrycon erythrozonus (Glowlight Tetra)—Watters & O'Dee, 1997a (LT); Watters & O'Dee, 1998a (LT)*.

Hyphessobrycon flammeus (Von Rio Tetra)—Watters & O'Dee, 1997a (LT); Watters & O'Dee, 1998a (LT)*.

Ictalurus punctatus (Channel Catfish)—Keller & Ruessler, 1997 (LT).

Lepomis cyanellus (Green Sunfish)—Tucker, 1927 (LT); Tucker, 1928 (LT); Trdan & Hoeh, 1982 (LT).

Lepomis gibbosus (Pumpkinseed)—Trdan & Hoeh, 1982 (LT).

Lepomis gulosus (Warmouth)—Stern & Felder, 1978 (NI).

Lepomis macrochirus (Bluegill)—Stern & Felder, 1978 (NI); Keller & Ruessler, 1997 (LT); Trdan & Hoeh, 1982 (LT).

Lepomis marginatus (Dollar Sunfish)—Stern & Felder, 1978 (NI).

Lepomis megalotis (Longear Sunfish)—Parker *et al.*, 1980 (LT).

Melanotaenia maccullochi (Rainbow Fish)—Watters & O'Dee, 1997a (LT); Watters & O'Dee, 1998a (LT)*.

Metynnis argenteus (Silver Dollar)—Watters & O'Dee, 1997a (LT); Watters & O'Dee, 1998a (LT)*.

Micropterus salmoides (Largemouth Bass)—Trdan & Hoeh, 1982 (LT); Watters & O'Dee, 1997a (LT); Watters & O'Dee, 1997*; Watters & O'Dee, 1998a (LT).

Moenkhausia oligolepis (Redeye Tetra)—Watters & O'Dee, 1997a (LT); Watters & O'Dee, 1998a (LT)*.

Notemigonus crysoleucas (Golden Shiner)—Keller & Ruessler, 1997 (LT).

Pangasius sutchii (Pangasius Catfish)—Watters & O'Dee, 1997a (LT); Watters & O'Dee, 1998a (LT)*.

Pangio kuhlii myersi (Black Loach)—Watters & O'Dee, 1997a (LT); Watters & O'Dee, 1998a (LT)*.

Perca flavescens (Yellow Perch)—Trdan & Hoeh, 1982 (LT).

Poecilia reticulata (Guppy)—Watters & O'Dee, 1997a (LT); Watters & O'Dee, 1998a (LT)*.

Pomoxis nigromaculatus (Black Crappie)—Hove *et al.*, 1995a (LT).

Pseudotropheus zebra (Zebra Malawi Cichlid)—Watters & O'Dee, 1997a (LT); Watters & O'Dee, 1998a (LT)*.

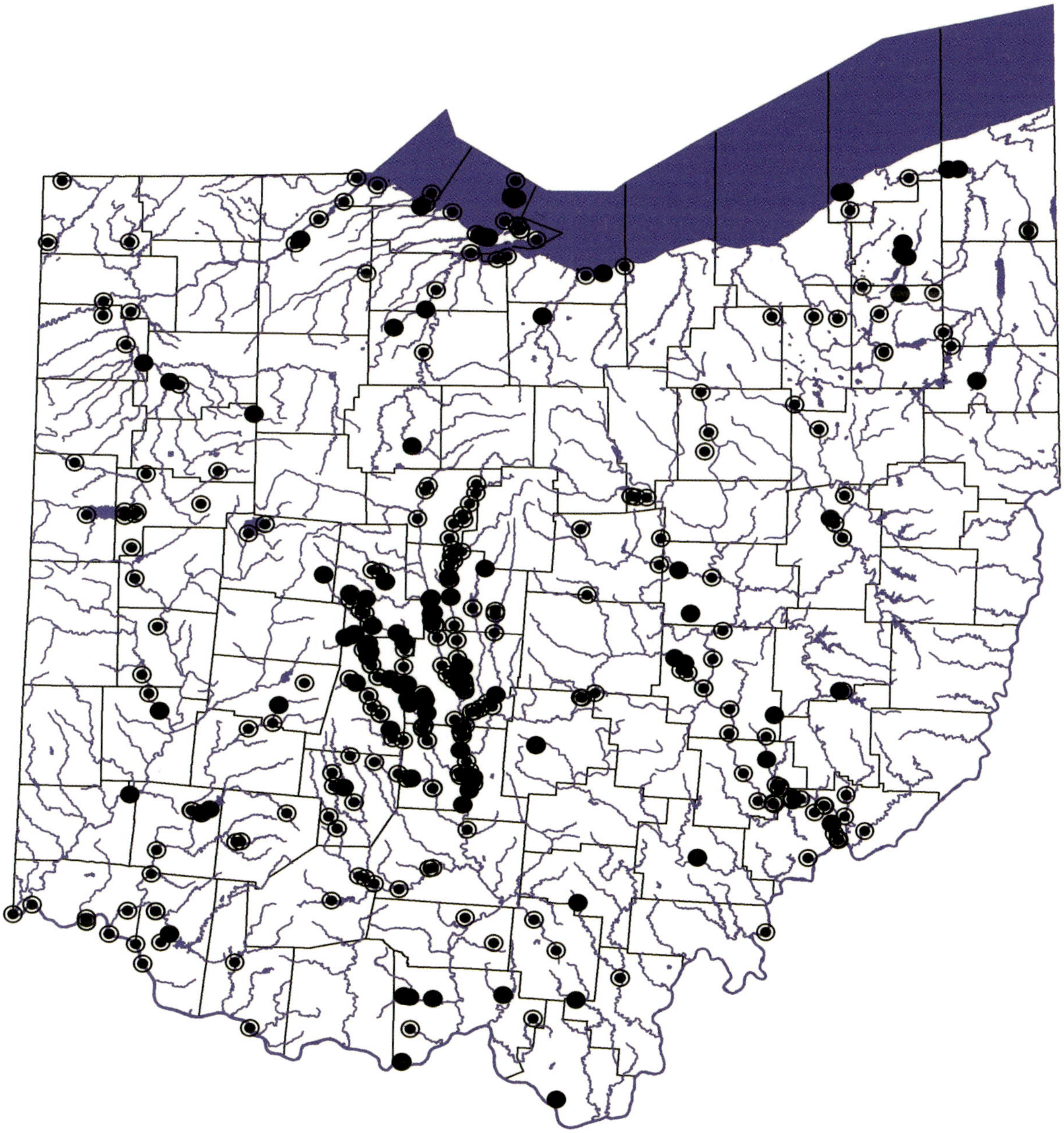

Fig. 210. ***Utterbackia imbecillis*** **(Say, 1829)**
Solid dots—live or freshdead since 1980
Bullseye—all records, any condition

Pterophyllum scalare (Angelfish)—Watters & O'Dee, 1997a (LT); Watters & O'Dee, 1998a (LT)*.
Rana catesbeiana (Bullfrog Tadpole)—Watters & O'Dee, 1997a (LT); Watters & O'Dee, 1998a (LT)**.
Rana pipiens (Northern Leopard Frog tadpole)—Watters & O'Dee, 1997a (LT); Watters & O'Dee, 1998a (LT)**.
Rasbora einthoveni (Brilliant Rasbora)—Watters & O'Dee, 1997a (LT); Watters & O'Dee, 1998a (LT)*.
Semotilus atromaculatus (Creek Chub)—Clarke & Berg, 1959 (NI).
Trichogaster leeri (Pearl Gourami)—Watters & O'Dee, 1997a (LT); Watters & O'Dee, 1998a (LT)*.
Trichogaster trichopterus (Lavender Gourami)—Watters & O'Dee, 1997a (LT); Watters & O'Dee, 1998a (LT)*.
Xenopus laevis (African Clawed Frog adult)—Watters & O'Dee, 1997a (LT); Watters & O'Dee, 1998a (LT)**.
Xiphophorus helleri (Painted Sword)—Watters & O'Dee, 1997a (LT); Watters & O'Dee, 1998a (LT)*.

Habitat. Usually found in soft substrates in lakes, ponds, and impoundments.

Range. Mississippi River system from Minnesota to Louisiana. Atlantic Slope and Gulf drainages from Pennsylvania and Maryland, throughout Florida, to Texas. South to Tamaulipas State, Mexico.

Range in Ohio. Widespread but seldom common outside of impoundments.

Similar species. The thin, elongate shell, the lack of dentition, and the flattened umbo flush with the dorsal margin are key characteristics. *Anodonta suborbiculata* also has a flat umbo, but that species is nearly circular in outline.

Comments. Individuals grow quickly for 2–3 years and then gradually slow. They grow to ca. 115 mm and rarely live for more than 5 years.

Villosa Frierson, 1927

Etymology. L. *villosa*, hairy, shaggy, rough. The name *Villosa* was first used for a subgenus by Frierson and originally used in reference to a southern species (*Unio villosus* Wright, 1898). Frierson mentioned its "deeply striate" and "rough" periostracum. Most species of *Villosa*, including Ohio species, however, typically have smooth exteriors.
Original description
Lampsilis (*Villosa*) Frierson, 1927: 80.
Type species. *Unio villosus* Wright, 1898, by tautonomy. Frierson (1927) listed four species under his *Lampsilis* (*Villosa*): *L. villosa* (Wright, 1898), *L. amygdalum* (Lea, 1843), *L. vesicularis* (Lea, 1870), and *L. pellucida* Frierson, 1927. He did not designate a type species as stated by Johnson (1970: 372), but the type becomes *Unio villosus* by tautonomy.
Synonymy
Micromya Agassiz, 1852: 47 [*non* Randani, 1840 (Insecta: Diptera). Agassiz listed three species but gave no type. Type species: *Unio lapillus* Say, 1832 (= *Unio fabalis* Lea, 1831) by subsequent designation of Herrmannsen, 1852: 83].
Lampsilis (*Villosa*) Frierson, 1927: 80.
Geographic range. Mississippi River drainage from Minnesota to Louisiana. Great Lakes drainages. Atlantic Slope from the Rappahanock and James rivers of Virginia through peninsular Florida, Gulf drainages from Florida to eastern Texas.

Villosa fabalis (Lea, 1831)

Figures 211, 212

Etymology. L. *fabalis*, of or pertaining to beans; from L. *faba*, a bean. Although Lea didn't mention his reason for using the name *fabalis*, this species is small, solid, and roughly bean-shaped and bean-size.
Vernacular name(s). *Rayed Bean.*
Original description
Unio fabalis Lea, 1831: 86–87, pl. 10, fig. 16.
Type locality. "Hab. Ohio river, T. G. Lea" (Lea 1831: 86). This locality is very likely the Ohio River at Cincinnati, Ohio, since Thomas G. Lea made his home there and very frequently sent specimens to his brother Isaac Lea from the Ohio River at that city. We herein restrict the type locality to the Ohio River at Cincinnati, Ohio.
Type material. Figured lectotype USNM 85270a (Johnson, 1974: 55).
Synonymy
Unio fabalis Lea, 1831: 86.
Unio capillus Say, 1831c: 528 [*nomen nudum*].
Unio lapillus Say, 1832: pl. 41 [unjustified emendation for *Unio capillus* Say, 1831. Types lost].
Margarita (*Unio*) *fabalis* (Lea, 1831). Lea, 1836: 28.
Micromya lapillus (Say, 1832). Agassiz, 1852: 47.
Margaron (*Unio*) *fabalis* (Lea, 1831). Lea, 1852b: 31.
Micromya fabalis (Lea, 1831). Call, 1900: 458.
Eurynia (*Micromya*) *fabalis* (Lea, 1831). Ortmann, 1912a: 339.
Micromya fabale (Lea, 1838). Wilson & Clark, 1912: 51.
Unio donacopsis De Gregorio, 1914: 60–61, pl. 10, figs. 5a–d [Type locality "Sciota river" [Scioto River, OH]. Type in Palermo Museum?].
Lemiox fabalis (Lea, 1831). Frierson, 1927: 93.
Villosa fabalis (Lea, 1831). Stein, 1963: 19.

Fig. 211. ***Villosa fabalis*** **(Lea, 1831)** **A.** OSUM 65330.40, Blanchard River, Hancock Co., OH. 33 mm. **B.** OSUM 3780.1, Lake Erie, Ottawa Co., OH. 22 mm. **C.** OSUM 4646.4, Olentangy River, Delaware Co., OH. 38 mm. **D.** OSUM 68538, Auglaize River, Defiance Co., OH. 33 mm. **E, H, Hinge detail:** OSUM 32795.25, French Creek, Venango Co., PA. 39 mm. **F.** OSUM 21372, Alum Creek, Delaware Co., OH. 25 mm. **G, Beak sculpture detail:** OSUM 25243, "Ohio River." 10 mm total length.

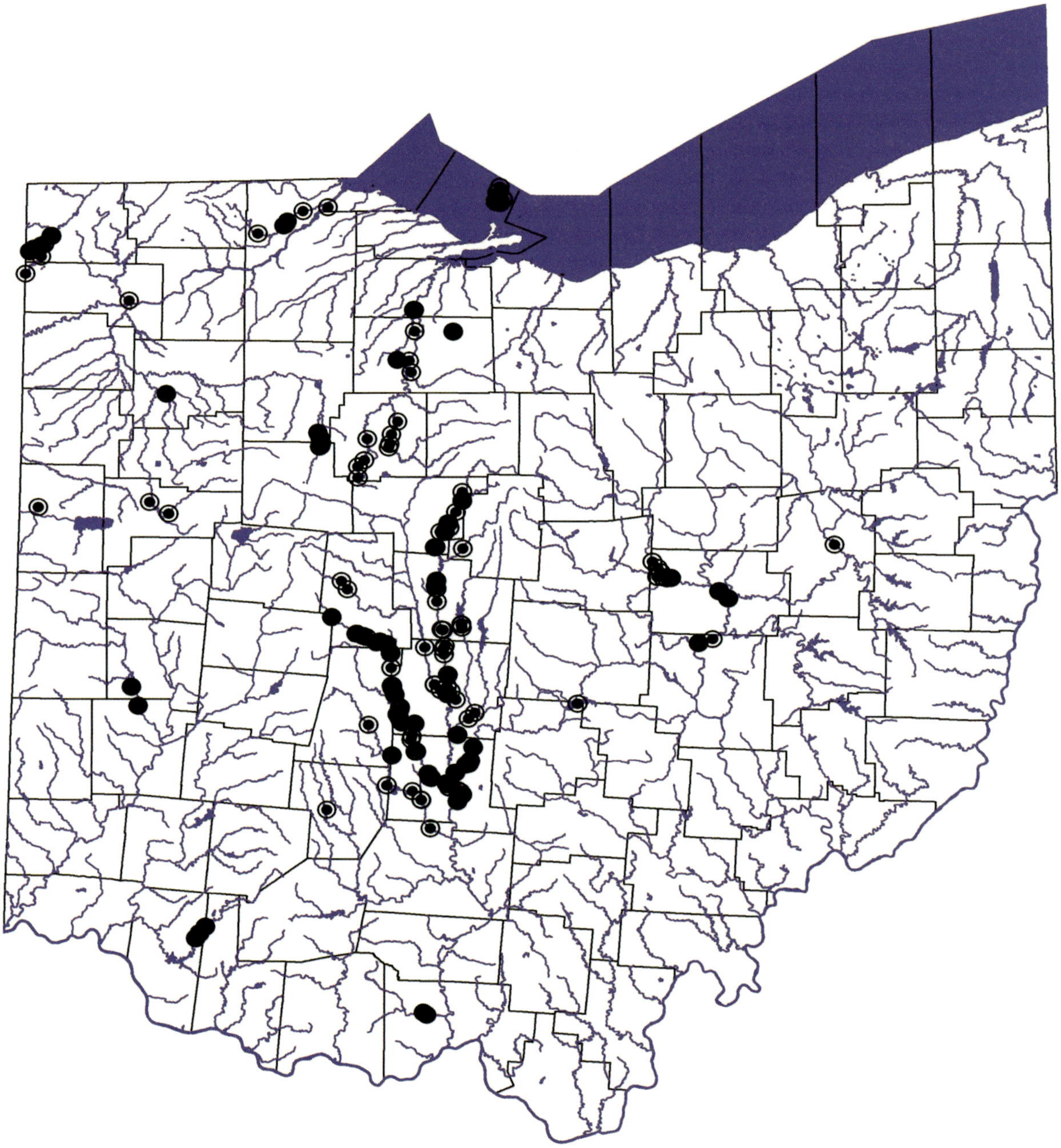

Fig. 212. ***Villosa fabalis*** **(Lea, 1831)**
Solid dots—live or freshdead since 1980
Bullseye—all records, any condition

Status. Listed as Endangered by Ohio Division of Wildlife; not listed by U.S. Fish & Wildlife Service as of 2009.

ADULT SHELL

Shell. Small (to 40 mm), solid, rounded, tapering posteriorly to a posterior point in males, more rounded in females.

Anterior margin. Straight or convex, becoming evenly rounded.

Ventral margin. Broadly rounded in males, straight or slightly concave to the beginning of the marsupial region in females.

Posterior margin. Males come to an acute point near the ventral margin; females have an abrupt angulation at the base of the marsupial region, then straight to a second angulation between the base and halfway up to the dorsal margin.

Dorsal margin. Convex in both sexes, more so in females.

Umbo. Wide, very low, but pointed; umbos have only the slightest indications of sculpture, composed of 2–3 very fine concentric ridges; this sculpture disappears within the first year of growth.

Ligament. Occupies most of the dorsal margin, low in males, more massive in females, partially sunken between the valves.

Disc sculpture. Sculptureless except for one or two weak angulations on the dorsal slope.

Periostracum. Yellow, tan, or green with numerous dark green rays; these rays may nearly completely cover the shell, often arranged in groups of three or more, of various thickness; rays are continuous but often wavy; rays persist throughout the life of the mussel.

Nacre. White, iridescent posteriorly; extensive gold Tulberg layers common.

Hinge plate. Very short, massive, often deformed in older individuals; interdentum very short, thick, barely distinct from rest of shell, occasionally with a short tooth or lamellum; single irregular cardinal in right valve usually accompanied by an anterior and posterior accessory tooth, which may be as strong as the cardinal; two chunky cardinals in the left valve; laterals short, bowed, thick, one in right valve, two in left.

Muscle scars. Adductors small, subcircular, deep, smooth; anterior protractor partially confluent or distinct from anterior pedal adductor; posterior retractor distinct, minute; dorsal suspensors multiple, attached beneath interdentum; pallial line complete; accessory rarely absent, short, terminating anteriorly of the anterior adductor.

Umbonal cavity. Wide, very shallow, open.

Glochidia. Subspatulate, dorsal margin straight, lateral margins equally and gently curved, ventral margin broadly rounded, hooks absent (Ortmann, 1919: 263).

Literature records (Ortmann, 1919: 263) indicate glochidial dimensions for length (0.17 mm) and height (0.20 mm). Ortmann (1919: 263) noted the glochidia are "of the usual shape, subspatulate, higher than long, but of rather small size."

Reproductive biology. Bradytictic. The posterior part of the outer gills is marsupial (Ortmann, 1912). Gravid females were found in May (Ortmann, 1919). Females display a flickering of white marsupia behind dark-rimmed mantle tissue (Watters, unpubl.).

Reported potential hosts

Etheostoma tippecanoe (Tippecanoe Darter)—White *et al.*, 1996 (LT) [based on genetic comparison of glochidia].

Habitat. The Rayed Bean lives in sand and cobble in high quality creeks and small rivers. Ortmann (1919: 264) found this species in and near riffles, generally in water weeds, and deeply buried in sand and gravel bound together by roots.

Range. Upper Mississippi River system, including the upper Tennessee River. Great Lakes drainages. It has been extirpated from the Thames and Detroit rivers in Canada, but a reproducing population still exists in the Sydenham River in Ontario (West *et al.*, 2000). Probably the largest population is in French Creek in Pennsylvania.

Range in Ohio. This species is perhaps more widely distributed than its current known range would suggest. It is a small, secretive species that is difficult to locate. Recent records include the Stillwater River, Fish Creek, Big and Little Darby creeks, Scioto Brush Creek, the Little Miami River, the East Fork of the Little Miami River, the Walhonding River, Blanchard River, and Swan Creek. Historically widely distributed in the Scioto and Sandusky rivers. Bass Islands of Lake Erie.

Similar species. This species is most like *Villosa iris*, which is much thinner and larger and has prominent umbo sculpture, and *Toxolasma parvum*, which is often darker, lacks the thin wavy rays, and is much thinner.

Comments. Although small, this is a fairly long-lived species. Individuals may live for 11 years or more, but only reach 40 mm in length. They grow quickly during the first three years, then abruptly slow. Growth rates are fairly uniform across populations. Female characteristics become apparent in the shells as early as the second year.

Villosa iris (Lea, 1829)

Figures 213, 214, 215

Etymology. L. *iris*, goddess of the rainbow, a rainbow; from Gr. *iris*, rainbow. Lea (1829: 439) noted that the nacre was "iridescent posteriorly," a characteristic especially prominent in this species. *Iris* is a noun in apposition.

Vernacular name(s). *Rainbow*, Rainbow Shell.

Original description

Unio iris Lea, 1829: 439, pl. 11, fig. 18.

Type locality. "Ohio."

Type material. Johnson (1974: 76) noted "Type [lost]."

Synonymy. We have refrained from including those nominal taxa from the Cumberland, Tennessee, and Mobile River systems that probably are not conspecific with *V. iris*.

Unio iris Lea, 1829: 439, pl. 11, fig. 18.

Margarita (*Unio*) *iris* (Lea, 1829). Lea, 1836: 37.

Unio novi-eboraci Lea, 1838a: 104, pl. 24, fig. 114 [Type locality: "Oak Orchard Creek, New York." Lectotype USNM 86042].

Margarita (*Unio*) *novi-eboraci*. Lea, 1838b: 19.

Margaron (*Unio*) *novi-eboraci* (Lea, 1838). Lea, 1852b: 27.

Margaron (*Unio*) *iris* (Lea, 1829). Lea, 1852b: 38.

Unio subrostratus Say, 1831. Küster, 1861: 203, pl. 67, fig. 3 [misidentification].

Unio opalina Anthony, 1866: 146, pl. 7, fig. 2 [Type locality: "Shears Lake, Michigan." Holotype MCZ 161893].

Lampsilis iris (Lea, 1829). Baker, 1898b: 105, pl. 13, fig. 1, pl. 14, fig. 2.

Lampsilis novi-eboraci (Lea, 1838). Sterki, 1907: 389.

Eurynia (*Micromya*) *iris* (Lea, 1829). Ortmann, 1912a: 341, fig. 23.

Unio fatuus var. De Gregorio, 1914: 53 [No name given. Type locality: "Sciota" [Scioto River, OH]. Type in Palermo Museum].

Eurynia iris (Lea, 1929). Ortmann, 1919: 234.

Eurynia (*Micromya*) *iris novi-eboraci* (Lea, 1838). Ortmann, 1919: 268, pl. 16, figs. 8, 9.

Micromya iris (Lea, 1829). Ortmann, 1924: 46.

Micromya iris novi-eboraci (Lea, 1838). Ortmann, 1924: 115.

Lampsilis fasciata opalina (Anthony, 1866). Frierson, 1927: 72.

Lampsilis nervosa (Rafinesque, 1820). Frierson, 1927: 77 [misidentification].

Ligumia iris novi-eboraci (Lea, 1838). Baker, 1928a: 260, pl. 86, figs. 8–12.

Villosa iris (Lea, 1829). Clarke & Berg, 1959: 53, fig. 45.

Ligumia nervosa (Rafinesque, 1820). Haas, 1969: 445–446 [misidentification].

Villosa iris iris (Lea, 1829). Starrett, 1971: 333.

Status. Not listed by either Ohio Division of Wildlife or U.S. Fish & Wildlife Service as of 2009.

ADULT SHELL

Shell. Medium-sized (to 80 mm), thin, becoming thick only in gerontic specimens, elongated.

Anterior margin. Short, straight, convex or concave, then evenly rounded to ventral margin.

Ventral margin. Males are gently rounded, nearly straight in elongated specimens; females are moderately inflated in the marsupial region, with a short convex area anteriorly.

Posterior margin. Males terminate in one or two blunt points 33% to 50% the distance from the ventral margin; females more rounded, without noticeable angulations.

Dorsal margin. Forming a wide angle with the posterior margin, straight or somewhat convex.

Umbo. Placed well anterior of center, low, broad, not prominent; beak sculpture of 4–10 irregular ribs; these begin as "V"-shaped ribs pointing toward the beaks, but tend to become a series of wavy or broken ribs, often obliquely situated; beak sculpture abruptly disappears at an annular ring.

Ligament. Occupies entire posterior slope but is very narrow, barely perceptible, largely obscured by the valves.

Disc sculpture. Sculptureless except for 1–2 weak angulations that delimit the posterior slope.

Periostracum. Yellow, green, or tan, becoming darker with age; patterned across entire shell with radiating dark green or brown rays; these may be continuous or broken by annular rings. Populations with broken rays tend to have only a few bold and thick rays; those with continuous rays have much more subtle, often extremely fine and numerous rays; rays continue throughout the life of the individual; posterior slope uniformly dark colored in some populations.

Nacre. White, iridescent posteriorly; extensive golden Tulberg layers common in some populations.

Hinge plate. Narrow in all but the oldest individuals; interdentum long, cord-like; two small cardinal teeth in left valve, occasionally obscure; a single cardinal in right valve, chunky and triangular; a very weak accessory tooth may be found anterior, nearly flush with the shell margin; laterals long, low, concave, one in right valve, two in left.

Muscle scars. Adductors small, smooth; anterior adductor fairly deep, distinct from anterior pedal protractor; posterior adductor flush with shell surface, mostly confluent with posterior pedal retractor; pallial line complete, accessory present but very fine, short.

Umbonal cavity. Shallow, wide, and open.

Glochidia. Subspatulate, dorsal margin straight, anterior and posterior margins subequal, dorsally divergent and ventrally parallel, ventral margin gently rounded. Dorsal alae long; loose-looped exterior sculpturing covers the valves. Hooks absent, micropoints lanceolate and arranged in broken vertical rows (Lea, 1858c: 47, pl. 5, fig.

Fig. 213. ***Villosa iris*** **(Lea, 1829). A.** OSUM 29462.2, Fish Creek, Williams Co., OH. 61 mm. **B.** OSUM 19207.3, Sydenham River, Lambton Co., Ontario. 60 mm. **C.** OSUM 25085.10, Scioto Brush Creek, Scioto Co., OH. 47 mm. **D, H, Hinge detail:** OSUM 10287, Scioto River, Franklin Co., OH. 69 mm. **E.** OSUM 37824.4, Grand River, Kent Co., MI. 46 mm. **F.** OSUM 61601.3, Honey Creek, Seneca Co., OH. 64 mm. **G.** OSUM 8016, Little Darby Creek, Madison Co., OH. 37 mm. **Beak sculpture detail:** OSUM 32943.3, Big Darby Creek, Franklin Co., OH. 28 mm total length.

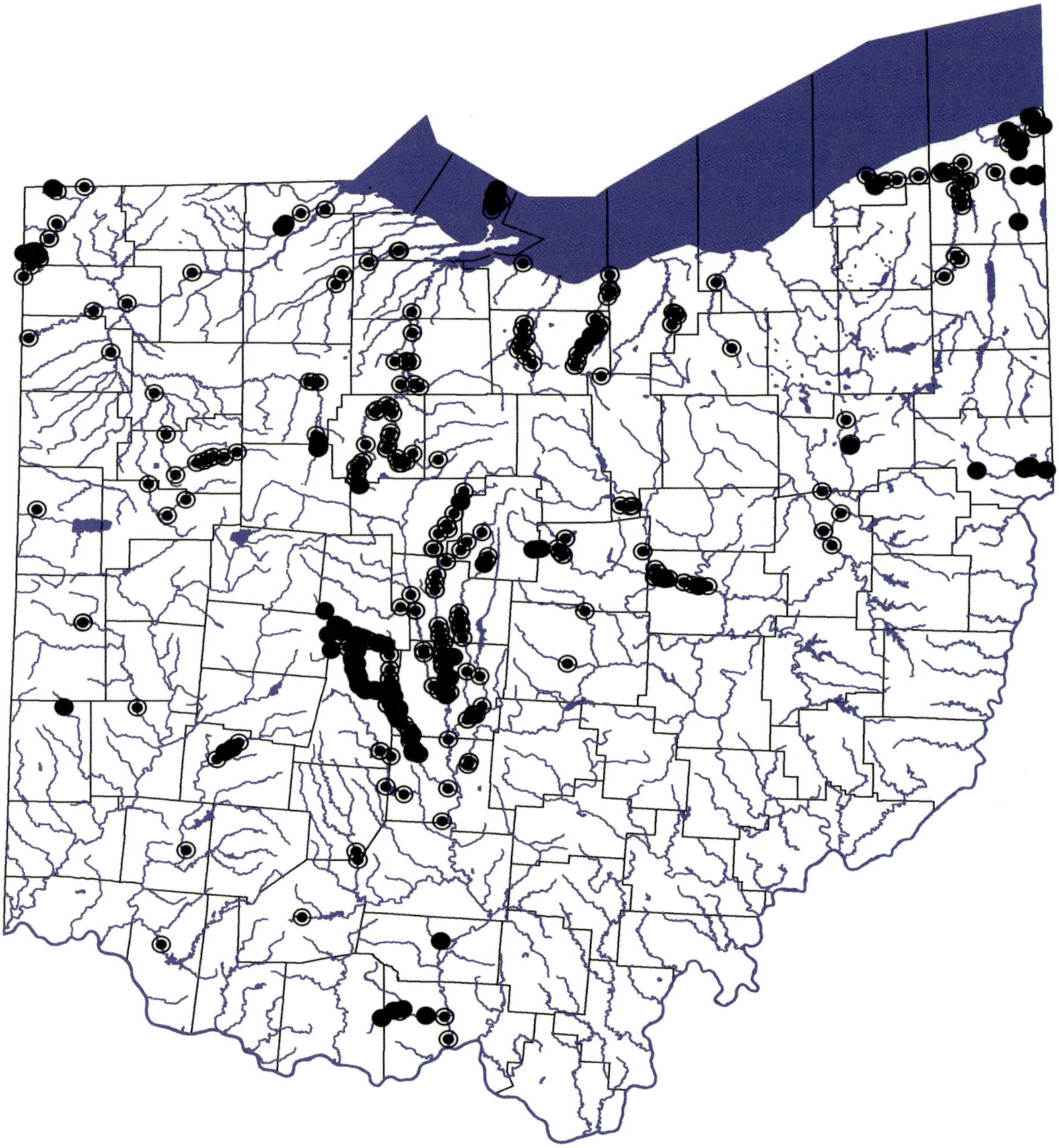

Fig. 214. ***Villosa iris*** **(Lea, 1829)**
Solid dots—live or freshdead since 1980
Bullseye—all records, any condition

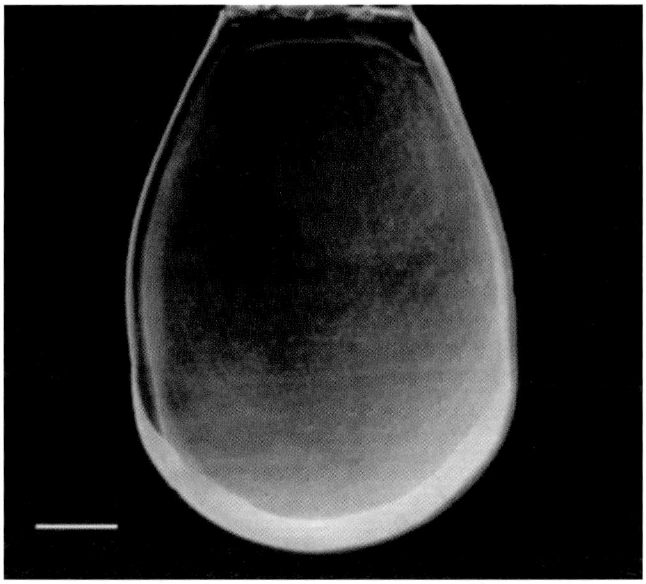

Fig. 215. *Villosa iris* (Lea, 1829). Glochidium. MAH 641.1, Little Darby Creek, Madison Co., OH. Bar = 40 µm. (Hoggarth, 1999).

14; Ortmann, 1911: 347, pl. 89, fig. 20; Hoggarth, 1999: figs. 57a–f).

Literature records (Surber, 1912: 9, pl. 3, fig. 46; Ortmann, 1912a: 342; Utterback, 1916a: 433; Hoggarth, 1999: figs. 57a–f) indicate glochidial dimensions for length (0.22 to 0.24 mm), height (0.28 to 0.31 mm), and hinge (0.11 to 0.12 mm). This glochidium can be distinguished from all other subspatulate glochidia examined by its loose-looped exterior valve sculpture, its wide unsculptured distal flange border, and smaller dorsal alae.

Reproductive biology. Bradytictic. Marsupium formed by the posterior half to two-thirds of the outer gills (Ortmann, 1912a). Ortmann (1919) reported gravid females from September to the following May. Females display long black or brown tentacles. This species appears to be a host generalist.

Reported potential hosts

Ambloplites rupestris (Rock Bass)—Zale & Neves, 1982b (LT), 1982c (NS); Neves, 1983 (NS).
Betta splendens (Siamese Fighting Fish)—Watters & O'Dee, 1997c (LT)—exotic.
Cottus bairdi (Mottled Sculpin)—Watters *et al.*, 2005 (LT).
Erimystax dissimilis (Streamline Chub)—Watters & O'Dee, 1997c (LT).
Ethiostoma blennioides (Greenside Darter)—Watters & O'Dee, 1997c (LT).
Ethiostoma caeruleum (Rainbow Darter)—Watters & O'Dee, 1997c (LT).
Ethiostoma camurum (Bluebreast Darter)—Watters & O'Dee, 1997c (LT).
Gambusia affinis (Mosquitofish)—Neves *et al.*, 1985 (LT).
Lepomis cyanellus (Green Sunfish)—Watters & O'Dee, 1997c (LT).
Luxilus chrysocephalus (Striped Shiner)—Watters & O'Dee, 1997c (LT).
Micropterus dolomieu (Smallmouth Bass)—Zale & Neves, 1982b (LT), 1982c (NS); Neves, 1983 (NS); Watters & O'Dee, 1997c (LT); Watters *et al.*, 2005 (LT).
Micropterus notius (Suwannee Bass)—Neves *et al.*, 1985 (LT).
Micropterus punctulatus (Spotted Bass)—Neves *et al.*, 1985 (LT).
Micropterus salmoides (Largemouth Bass)—Neves *et al.*, 1985 (LT); Watters & O'Dee, 1997c (LT).
Perca flavescens (Yellow Perch)—Watters & O'Dee, 1997c (LT).

Habitat. The rainbow is most common in small creeks in sandy cobble, usually in moving water. It is rare in larger rivers and occurs in portions of Lake Erie.

Range. In its many forms, this species occurs throughout most of the upper Mississippi River and Great Lakes drainages.

Range in Ohio. Widespread in rivers, creeks, and lakes, but prefers small flowing creeks. Seldom common. Western basin of Lake Erie. Virtually absent from unglaciated Ohio.

Similar species. This species differs from *Venustaconcha* in its longer, finer hinge and different beak sculpture. *Villosa fabalis* is smaller but has a relatively thicker shell and a short, thick hinge. *Villosa lienosa* generally lacks the bold rays and has double-looped beak sculpture. *Ptychobranchus fasciolaris* has a similar color pattern but differs in its very short, massive hinge.

Comments. Different populations grow at different rates, with a 50-mm specimen varying from 5 to 11 years old. Large individuals may reach 80 mm. Specimens over 15 years old are rare. Individuals grow quickly for the first 2 to 4 years. Females develop marsupial shell characteristics by age 3. This suggests that this species may reach sexual maturity at an early age.

This common species has received many names, largely by Isaac Lea. Frierson (1927: 78) mentioned that "Lea gave a name to every trivial difference he could discern." *Villosa iris* may well represent a species complex.

Villosa lienosa (Conrad, 1834)

Figures 216, 217

Etymology. Possible origins of the Latin(ized) term *lienosa* include L. *linea*, line; French *lien*, band, bond, tie; L. *lienosus*, or pertaining + L. *-osa*, full of. Conrad also may have used the Latin word *lienosus*, meaning splenetic or melancholy, to refer to the somber colors of the outside of the shell in same way that other authors have used *lugubris* (mournful) to describe dark-colored shells. In his original description of this species, Conrad (1834: 339) mentioned "concentric lines coarse and prominent." He may have been using the French word *lien* to form *lienosus* ("full of bands"). There is also the possibility that *lienosus* was a printer error for *lineosus* (L. *linea*, line + L. *osus*, full of) that Conrad chose not to correct. He did mention "lines" in his original description and in fact later refers once to this species as *U. lineosus* (1838: 91).

Vernacular name(s). *Little Spectaclecase*, Black Creek Shell.

Original description

Unio lienosus Conrad, 1834: 339, pl. 1, fig. 4.

Type locality. "Inhabits small streams in South Alabama."

Type material. Conrad (1834: 339) listed his specimen as having a length of 2.8 inches and being in the "Cabinet of the Academy of Natural Sciences of Philadelphia." He later (Conrad 1838: 61, pl. 34, fig. 2) mentioned "Cab. A. N. S. No. 20407" in his monography. But Clench & Turner (1956) and Johnson & Baker (1973: 160) noted "Syntypes ANSP 9747 from Conrad mentioned as in the ANSP."

Synonymy

Unio lienosus Conrad, 1834: 339, pl. 1, fig. 4.

Margarita (Unio) lienosus (Conrad, 1834). Lea, 1836: 26.

Unio saxeus Conrad, 1838: Species no. 11, back cover [*nomen nudum*].

Unio saxeus Conrad, 1840: 109, pl. 60, fig. 1 [Type locality: "Alabama River, Claiborne, Alabama." Syntype ANSP 9747].

Unio caliginosus Lea, 1845: 165 [Type locality: "Red River, Alexandria, Louisiana." Lectotype USNM 85108].

Unio gouldii Lea, 1845: 165 [Type locality: "Tuscaloosa, Alabama." Lectotype USNM 85160].

Unio nigerrimus Lea, 1852a: 268, pl. 18, fig. 23 [Type locality: "Alexandria, Louisiana." Lectotype USNM 85307].

Margaron (Unio) saxeus (Conrad, 1840). Lea, 1852b: 27.

Margaron (Unio) lienosus (Conrad, 1834). Lea, 1852b: 29.

Margaron (Unio) caliginosus (Lea, 1845). Lea, 1852b: 29.

Margaron (Unio) gouldii (Lea, 1845). Lea, 1852b: 29.

Margaron (Unio) nigerrimus (Lea, 1852). Lea, 1852b: 31.

Unio fuligo Reeve, 1856: pl. 30, fig. 159 [Types in BMNH lost].

Unio concestator Lea, 1857a: 31 [Type locality: "creeks near Columbus, Georgia." Lectotype USNM 85102].

Unio apicinus Lea, 1857a: 32 [Type locality: "Oothkalooga Creek, Georgia." Lectotype ANSP A14418 lost *fide* Johnson, 1974].

Unio fallax Lea, 1857a: 32 [Type locality: "creeks near Columbus, Georgia" and "French Broad River, Tennessee." Type not located].

Unio intercedens Lea, 1857a: 32 [Type locality: "streams near Columbus, Georgia." Lectotype USNM 85122].

Unio prattii Lea, 1858c: 166 [Type locality: "Chattahoochee River, Roswell, Georgia" and "Tobesaufke Creek, near Macon, Georgia." Possible lectotype USNM 85145 *fide* Johnson, 1974].

Unio porphyreus Lea, 1861c: 60 [Type locality: "Coosa River, Alabama." Lectotype USNM 84851].

Unio contiguus Lea, 1861d: 392 [Type locality: "Stewarts Mill Dam, North Carolina." Lectotype USNM 85120].

Unio bi-caelatus Reeve, 1865: pl. 26, fig. 130 [Types in BMNH lost].

Unio fontanus Conrad, 1866: 279, pl. 15, fig. 13 [Type locality: "rivulet near Vicksburg, Mississippi." Figured Lectotype ANSP 56458].

Margaron (Unio) apicinus (Lea, 1857). Lea, 1870: 44.

Margaron (Unio) concestator (Lea, 1857). Lea, 1870: 45.

Margaron (Unio) fallax (Lea, 1857). Lea, 1870: 45.

Margaron (Unio) intercedens (Lea, 1857). Lea, 1870: 45.

Margaron (Unio) prattii (Lea, 1858). Lea, 1870: 45.

Margaron (Unio) contiguus (Lea, 1861). Lea, 1870: 45.

Unio unicostatus Wright, 1899c: 69 [Type locality: "Spring Creek, Georgia." Lectotype USNM 159966].

Lampsilis lienosus (Conrad, 1834). Simpson, 1900a: 547.

Lampsilis lienosus var. *unicostata* (Wright, 1899). Simpson, 1900a: 547.

Lampsilis concestator (Lea, 1857). Simpson, 1900a: 548.

Lampsilis prattii (Lea, 1858). Simpson, 1900a: 550.

Lampsilis nigerrimus (Lea, 1852). Simpson, 1900a: 551.

Lampsilis apicinus (Lea, 1857). Simpson, 1900a: 551.

Lampsilis gouldii (Lea, 1845). Simpson, 1900a: 569.

Eurynia (Micromya) lienosa (Conrad, 1834). Ortmann, 1912a: 340–341.

Lampsilis nigerrima (Lea, 1852). Simpson, 1914: 100-101.

Lampsilis lienosa var. *unicostata* (Wright, 1899). Simpson, 1914: 101–102.

Lampsilis apicinus (Lea, 1857). Simpson, 1914: 112.

Lampsilis lienosa (Conrad, 1834). Simpson, 1914: 113–114.

Unio niggerimus [*sic*] Lea, 1852. Frierson, 1927: 74.

Unio pratti [*sic*] Lea, 1858. Frierson, 1927: 75.

Villosa (Micromya) lienosa (Conrad, 1834). Parmalee, 1967: 76.

Ligumia lienosa (Conrad, 1834). Haas, 1969: 438.

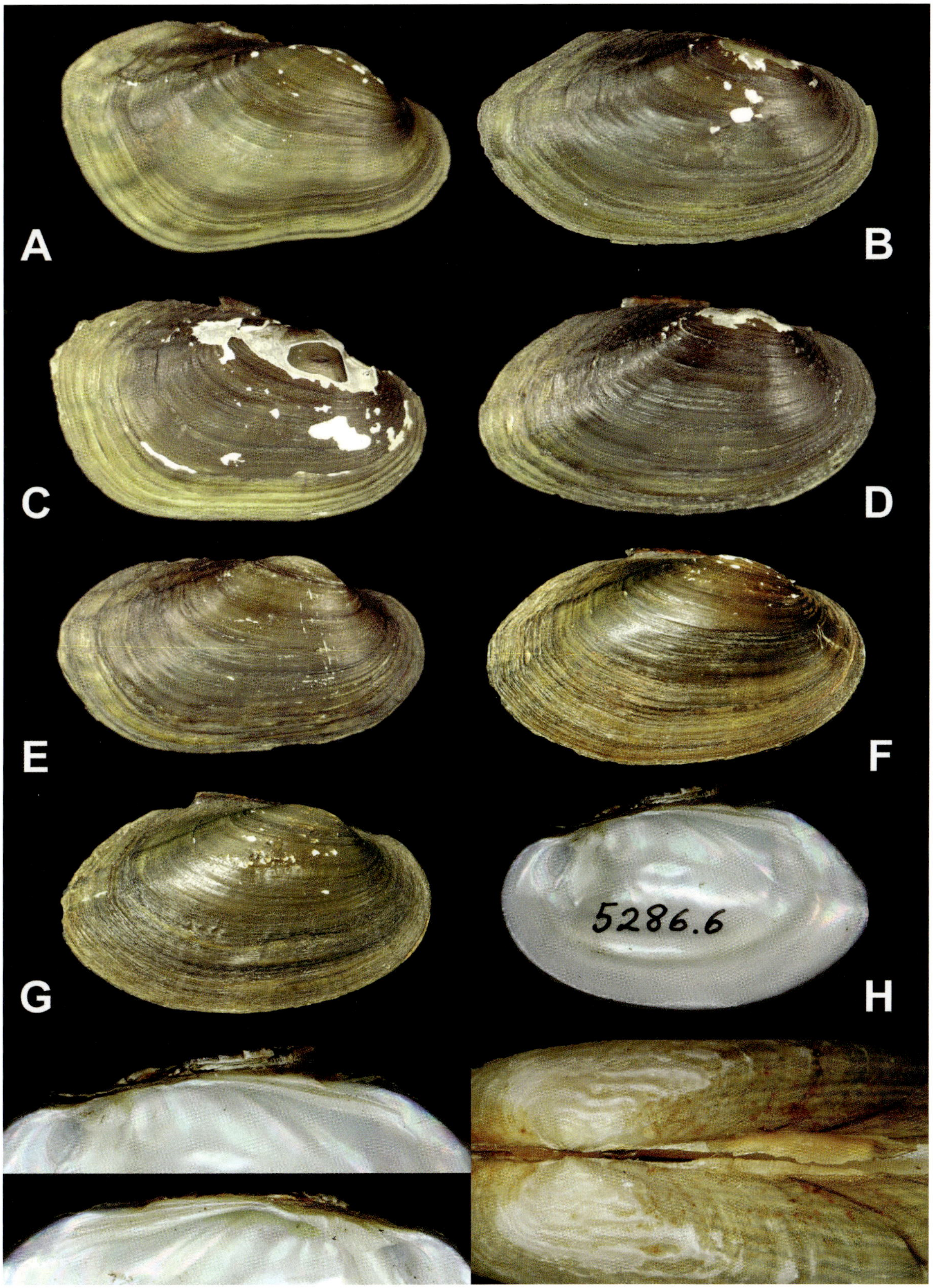

Fig. 216. *Villosa lienosa* **(Conrad, 1834). A.** OSUM 42819.3, East Fork Little Miami River, Clermont Co., OH. 50 mm. **B.** OSUM 26470.3, Buffalo Creek, Lawrence Co., OH. 48 mm. **C.** OSUM 65527.2, Symmes Creek, Jackson Co., OH. 46 mm. **D.** OSUM 30043.2, Symmes Creek, Gallia Co., OH. 56 mm. **E.** OSUM 24197.4, Flat Rock River, Shelby Co., IN. 6 mm. **F.** OSUM 17511.1, Green River, Hart Co., KY. 49 mm. **G, H, Hinge detail:** OSUM 5286.6, Ohio River, Clark Co., IN. 44 mm total length. **Beak sculpture detail:** OSUM 5286.1, Ohio River, Clark Co., IN. 25 mm total length.

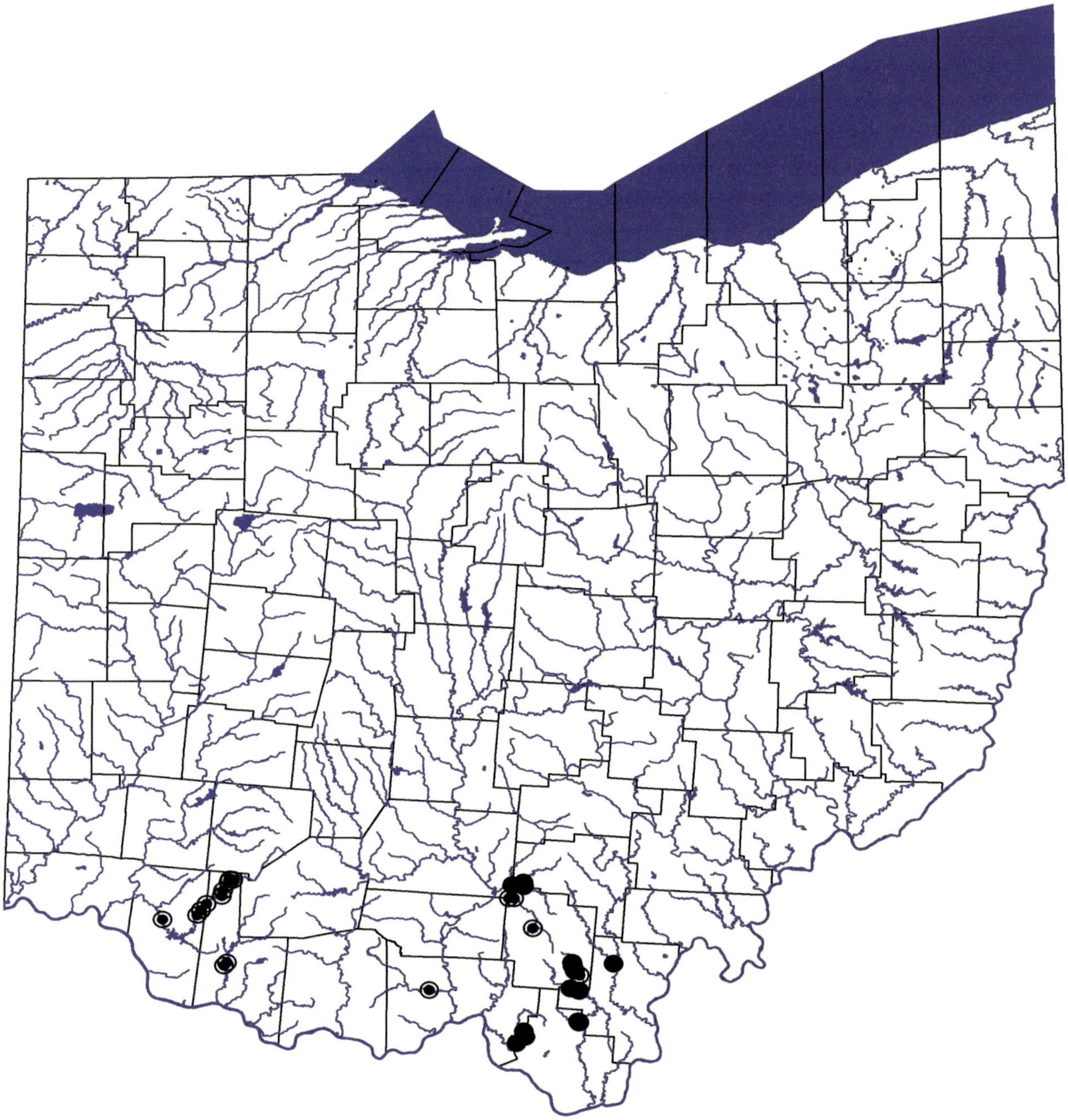

Fig. 217. ***Villosa lienosa*** **(Conrad, 1834)**
Solid dots—live or freshdead since 1980
Bullseye—all records, any condition

Villosa lienosa lienosa (Conrad, 1834). Oesch, 1984: 204–206.

Status. Listed as Endangered by Ohio Division of Wildlife; not listed by U.S. Fish & Wildlife Service as of 2009.

ADULT SHELL

Shell. Medium-sized (to 70 mm); lightweight but sturdy, outline elongate oval, truncate in females.

Anterior margin. Straight, then narrowly curved.

Ventral margin. Straight, slightly concave in females; gently curving to posterior margin in males, forming obtuse or nearly right angle on postero-ventral margin in females.

Posterior margin. Forming blunt acute angle in males about at halfway mark; forming nearly right angle in females distinctly above halfway point.

Dorsal margin. Long, straight.

Umbo. Wide, low, barely rising above the hinge line; in Ohio specimens the beak sculpture consists of 4–6 double "V"-shaped ridges; anterior "V" may become lost distally; beak sculpture disappears early in the first year's growth. Throughout its wide range this sculpture may be reduced to 2 "V"-shaped ridges or expanded to numerous "V"-shaped ridges that break up into tiny pustules across the disc of the first year's shell.

Ligament. Tan to brown, long, not concealed by the valves.

Disc sculpture. Shell sculptureless except for 2 low, broad ribs on the dorsal slope.

Periostracum. In Ohio specimens, pale yellowish-green to tan, the early portions of the shell darker; most shells have fine, faint green rays, at least on the posterior slope. These may coalesce there to form broader rays. Outside of Ohio, specimens may be nearly uniformly dark brown or black.

Nacre. White, iridescent posteriorly; often with gold or green Tulberg layers. Outside of Ohio, specimens may have white, rose, or purple nacre, but only white-nacred individuals have been found in Ohio.

Hinge plate. Long, only slightly angled, very delicate; interdentum very long, cord-like, nearly obsolete; cardinal teeth chunky, triangular, two in left valve, one in right with an accessory tooth anterior to cardinal; lateral teeth very long, straight or slightly bowed, thin, two in left, one in right.

Muscle scars. Anterior adductor muscle scars impressed, posterior flush with shell interior, anterior distinct from anterior pedal protractor scar, posterior confluent with retractor scar; dorsal suspensor scars in a tight group under the interdentum; pallial line entire, accessory line very weak to absent.

Umbonal cavity. Wide, open, shallow, nearly nonexistent.

Glochidia. Subspatulate, dorsal margin long and slightly curved, anterior and posterior margins equally divergent dorsally and nearly parallel ventrally, the posterior margin being slightly more rounded; ventral margin gently curved, hooks absent (Lea, 1874a: 68, pl. 21, figs. 4, 8; Surber, 1915: 5, fig. 15).

Literature records (Ortmann, 1912a: 341, pl. 20, fig. 5; Surber, 1915: 5, fig. 15; Utterback, 1916a: 432) indicate glochidial dimensions for length (0.19 to 0.20 mm) and height (0.24 to 0.27 mm). Surber (1915: 5) noted, "In general outline this glochidium is of the type represented by *Lampsilis trabalis* and *Obovaria circulus* [*O. subrotunda*], being intermediate in size between these two species, but probably nearer *trabalis* in form, particularly in the rather obliquely rounded posterior margin."

Reproductive biology. Probably bradytictic. Marsupium formed by the posterior half or more of the outer gills (Ortmann, 1912a). Females discharge large, club-shaped conglutinates (Utterback, 1916a).

Reported potential hosts

Ameiuris nebulosus (Brown Bullhead)—Keller & Ruessler, 1997 (LT).

Ictaluris punctatus (Channel Catfish)—Keller & Ruessler, 1997 (LT).

Lepomis macrochirus (Bluegill)—Keller & Ruessler, 1997 (LT).

Micropterus salmoides (Largemouth Bass)—Keller & Ruessler, 1997 (LT).

Habitat. In Ohio, this species usually is found in cool-water creeks, often quite small in size, in silty sand and clay.

Range. Mississippi River system, most Gulf drainages from peninsular Florida to Texas. North of the Ohio River, it now occurs only in the southern-most streams of Ohio, Indiana, and Illinois, although it once was distributed much further north, to at least latitude 41°, based on recent surveys (Ecological Specialists, Inc., 1992).

Range in Ohio. Small, southern creeks and their tributaries: Little Miami River, Ohio Brush Creek, Pine Creek, Symmes Creek, Salt Creek.

Similar species. *Villosa iris* does not have double-looped beak sculpture and is more boldly rayed. *Toxolasma lividum* is shorter and has concentric beak sculpture.

Comments. This species is very slow growing in Ohio, probably as a result of the cold-water habitat it prefers there. Individuals from the Little Miami River grow somewhat faster. In the southern part of its range it is a fast-growing species. For instance, a 45-mm specimen from Symmes Creek in Ohio may be over 11 years old, whereas a similarly sized individual from the Tombigbee River of Alabama may be only 6 years old. Ohio specimens reach 70 mm and live for 13 years. Growth is accelerated for the first 3 years. Sexual dimorphism in the shell is very pronounced, becoming evident by the third year.

Specimens from Ohio, and specimens from the upper Ohio River in general, have white nacre, whereas many southern specimens typically have dark purple nacre. It may be that the northern forms represent a distinct subspecies. However, none of the many names applied to this species were given to northern forms.

Questionable and Potential Ohio Occurrences

Species included in this section fall into two groups: those that have been recorded from Ohio, apparently in error, and those that may turn up in the future. The latter group includes species that occur close to Ohio but have not as yet been found in the state.

Margaritifera Schumacher, 1816

Etymology. Gr. *margaron* = L. *margarita*, pearl = L. *-ferre*, to bear or to carry. These are the pearl-producing mussels of the Old World.

Original description

Margaritana Schumacher, 1816: 14, 123–124.

Type species. *Margaritana fluviatilis* Schumacher, 1816 (= *Margaritifera margaritifera* (Linnaeus, 1758)), by original designation based on Chemnitz, 1782: pl. 1, fig. 5.

Synonymy

Baphiae Meuschen, 1787: 472 [*non binominal*].

Margaritana Schumacher, 1816: 14, 123–124.

Damaris Swainson, 1823: 3, caption to pl. 3.

Damalis [*sic*] "Leach" Gray, 1847: 196 [*non* Fabricius, 1805 (Insecta: Diptera), *nec* Huebner, 1819 (Insecta: Lepidoptera), *nec* Smith, 1827 (Mammalia: Bovidae), *nec* Gray, 1846 (Mammalia: Bovidae)].

Geographic range. Nearly circum-temperate and circumboreal, except for North America, where this genus is found in the northeast and northwest, with isolated species in Alabama and Louisiana. Numerous nominal species have been recorded from the former U.S.S.R—the status of these taxa requires additional study (see Zhadin, 1952). *Margaritifera laosensis* (Lea, 1863) occurs in southeast Asia, and *Margaritifera moroccana* Haas, 1940, occurs in Morocco.

Margaritifera margaritifera (Linnaeus, 1758)

Figure 218

Etymology. Gr. *margaron* = L. *margarita*, pearl = L. *-ferre*, to bear or to carry.

Vernacular name(s). *Eastern Pearlshell*, Old World Pearlshell, Holarctic Pearlshell.

Original description

Mya margaritifera Linnaeus, 1758: 671, species no. 20 under the genus *Mya*.

Type locality. "Habitat in totius orbis arctici cataractis" ["Dwelling in all of the waterfalls of the arctic world"].

Type material: ?

Synonymy

Mya margaritifera Linnaeus, 1758: 671, species no. 20.

Unio margaritiferus (Linnaeus, 1758). Retzius, 1788: 16.

Unio margaritifera (Linnaeus, 1758). Draparnaud, 1801: 107.

Margaritana margaritifera (Linnaeus, 1758). Schumacher, 1817: 124, pl. 10, fig. 4.

Unio elongata Lamarck, 1819: 70 [Type locality: "dans les rivières de l'Angleterre, et probablement du nord de l'Europe" ("in the rivers of England, and probably in northern Europe"). Syntypes in MNHN].

Unio elongatus (Lamarck, 1819). Nilsson, 1822: 106.

Alasmodonta arcuata Barnes, 1823: 277–278, pl. 12, fig. 20 [Type locality: "West Canada Creek" and "a small stream in Tappan." Type lost].

Mya arcuata (Barnes, 1823). Eaton, 1826: 222.

Alasmodon margaritiferum (Linnaeus, 1758). Fleming, 1828: 417.

Unio roissyi Michaud, 1831: 112, pl. 16, fig. 28 [Type not located].

Alasmodonta margaritifera (Linnaeus, 1758). Conrad, 1834: 72.

Unio margaritifer [*sic*] var. *minor* Rossmassler, 1835: 19, pl. 9, fig. 129 [Types in SMF?].

Margarita (*Unio*) *crassissimus* "Klein" Lea, 1836: 40 [Type locality: none given. Klein's 1753 name is pre-Linnean, and authorship must be attributed to Lea.].

Margarita (*Magaritana*) *margaritifera* (Linnaeus, 1758). Lea, 1836: 45.

Alasmodon margaritiferus (Linnaeus, 1758). Brown, 1836: 112, pl. 21, fig. 13, pl. 22, figs. 1–3.

Alasmodon arcuata (Barnes, 1823). Gould, 1841: 113, fig. 75.

Unio tristis Morelet, 1845: 107, pl. 31, fig. 2 [Holotype BMNH 93.2.4.1961].

Margaritana arcuata (Barnes, 1823). Stimpson, 1851: 15.

Margaron (*Margaritana*) *margaritifera* (Linnaeus, 1758). Lea, 1852b: 43.

Fig. 218. ***Margaritifera margaritifera*** **(Linnaeus, 1758). A.** OSUM 58026.1, Trout Brook, Washington Co., ME. 106 mm. **B.** OSUM 25273.4, Salmon Brook, Hartford Co., CT. 98 mm. **C.** OSUM 24353.3, Stewiacke River, Colchester Co., Nova Scotia. 107 mm. **D.** OSUM 56455.1, Old Stream, Washington Co., ME. 50 mm. **E.** OSUM 24441.1, Shunock Brook, New London Co., CT. 36 mm. **F.** OSUM 33145.1, Pearl Hill Brook, Middlesex Co., MA. 75 mm. **G, H, Hinge detail:** OSUM 24590.7, Hackensack River, Rockland Co., NY. 101 mm. **Beak sculpture detail:** no specimens available.

Damaris elongata (Lamarck, 1819). Leach, 1852: 322.
Unio margaritifer [*sic*] (Linnaeus, 1758). Küster, 1856: 130, pls. 138, 139.
Baphia margaritifera (Linnaeus, 1758). Adams & Adams, 1857: 499, pl. 117, figs. 2, 2a.
Unio (*Margaritana*) *margaritifer* [*sic*] (Linnaeus, 1758). Schrenck, 1867: 700.
Margaritana margaritiferus (Linnaeus, 1758). Westerlund, 1890: 184.
Margaritana elongata (Lamarck, 1819). Westerlund, 1890: 185.
Margaritana roissyi (Michaud, 1831). Westerlund, 1890: 186.

Status. Not listed by Ohio Division of Wildlife; not listed by U.S. Fish & Wildlife Service as of 2009**.**

ADULT SHELL

Shell. Medium size (to 140 mm), thin, lanceolate, not inflated.
Anterior margin. Long and straight or concave before the umbo, broadly rounded to the ventral border.
Ventral margin. Very broadly rounded as juveniles, becoming increasingly concave with age in some populations.
Posterior margin. Forming a broad, somewhat expanded point near the ventral margin, then obliquely ascending to the dorsal margin.
Dorsal margin. Long, gently arched, joining posterior margin with very obtuse angle; juncture may be imperceptible.
Umbo. Very wide and low; umbonal sculpture eroded away on even the smallest specimens examined.
Ligament. Long, tan to dark brown, largely concealed by shells.
Disc sculpture. Sculptureless.
Periostracum. Waxy yellow or tan as smallest juveniles, becoming dark brown to black with age; rayless.
Nacre. White, gray, or vaguely purplish, often iridescent posteriorly; nearly all specimens seen have extensive gold or brown Tulberg layers.
Hinge plate. Narrow; right cardinal a single, chunky, triangular denticle; left cardinal bifid, the posterior portion the larger of the two; laterals weak but present in juveniles, usually absent when adult; interdentum very long and narrow.
Muscle scars. Anterior adductor impressed; posterior adductor less so; anterior adductor distinct but abutting small pedal protractor scar, large posterior adductor confluent or not with posterior pedal retractor scar; linear dorsal suspensor located behind interdentum. Center of shell with characteristic "punctures"(microfossae)—numerous small scars arranged obliquely towards the posterior-ventral line. These scars are mantle attachment sites and resemble small comets; they may be difficult to discern in large adults. Pallial line entire, located far from shell margin. Accessory line absent.
Umbonal cavity. Very wide and shallow.

Glochidia. Triangulate subelliptical, with lateral margins equally curved to an attenuate ventral terminus, dorsal margin straight. Exterior surface granulate, without pits, and ventral margin without hooks. Micropoints located on a wide ventral flange (Pekkarinen & Valovirta, 1996: 413–417, fig. 1a–f).

Literature records (Harms, 1909; Bykhovskaya-Pavlovskaya *et al.*, 1964; Smith, 1976; Young & Williams, 1984; Bauer, 1994; Nezlin *et al.*, 1994; Pekkarinen and Valovirta, 1996) indicate glochidial dimensions for length (0.05–0.07 mm) and height (0.07–0.08 mm).

The size, shape, and structure of this glochidium will distinguish it from all others that may be found in Ohio.

Reproductive biology. Once mature at 12–13 years (Young & Williams, 1984) females reproduce for the remainder of their life, never reaching senescence (Bauer, 1987c). Individuals in excess of 200 years of age have been reported based on studies of shell microgrowth patterns (Mutvei & Westermark, 2001). At low population densities, females may become hermaphrodites capable of self-fertilization (Bauer, 1987c). Glochidia are shed from July to October (Smith, 1976; Young & Williams, 1984) and may overwinter on the host until the following May (Young *et al.*, 1987) or longer. Once metamorphosed, juveniles may live 10 or more years buried within the substrate (Bauer, 1986). Typically, salmonids are hosts for margaritiferids. The host associations reported below are for North American fish.

Reported potential hosts

Catostomus tahoensis (Tahoe Sucker)—Murphy, 1942 (LT).
Oncorhynchus clarki (Cutthroat Trout)—Meyers & Millemann, 1977 (LI); Karna & Millemann, 1978 (NT).
Oncorhynchus kisutch (Coho Salmon)—Meyers & Millemann, 1977 (LI); Karna & Millemann, 1978 (NT).
Oncorhynchus mykiss (Rainbow Trout)—Davis, 1934 (NT, LT); Meyers & Millemann, 1977 (LT); Karna & Millemann, 1978 (NT).
Oncorhynchus nerka (Sockeye Salmon)—Meyers & Millemann, 1977 (LT).
Oncorhynchus tshawytscha (Chinook Salmon)—Murphy, 1942 (NT, LT); Meyers & Millemann, 1977 (LT); Fustish & Millemann, 1978 (NI); Karna & Millemann, 1978 (NT).
Rhinichthys osculus (Speckled Dace)—Murphy, 1942 (NT).
Richardsonius egregius (Lahantan Redside)—Murphy, 1942 (NT).

Salmo salar (Atlantic Salmon)—Meyers & Millemann, 1977 (LI); Cunjak & McGladdery, 1991 (NT); Nezlin *et al.*, 1994 (LI).
Salmo trutta (Brown Trout)—Murphy, 1942 (NT, LT); Clarke & Berg, 1959 (LI).
Salvelinus fontinalis (Brook Trout)—Murphy, 1942 (NT); Clarke & Berg, 1959 (LI); Smith, 1976 (LT).

Habitat. This species occurs in cold streams, rarely associated with more than a very few other mussel species. Bauer *et al.* (1991) hypothesized that it has a lower metabolism than other mussels and could utilize food-poor streams not available to others.

Range. In North America, this species occurs from eastern Pennsylvania to Newfoundland, Labrador, and Nova Scotia. It is found in Europe and Asia from Spain, the British Isles, and Scandinavia, through central Europe to northern Asia and Japan. Some of these populations have been separated as distinct species but the systematics have not been adequately studied.

Range in Ohio. Goodrich & van der Schalie (1932: 12) noted, "*The Jay Catalogue of Recent Shells*, edition of 1852, mentions *Margaritana margaritifera* (Linnaeus) as from Lake Erie. This is an error." We would agree since the nearest known vouchered record (Ortmann, 1919: 3) is in a different drainage system (e g., Susquehanna River) from any in Ohio, and we are not aware of any stream captures, Pleistocene or Recent, that could explain this occurrence. Furthermore, a search was made of all four editions of Jay's *Catalogue* (1835, 1836, 1839, 1852) and its supplement (1852), but the citation listed above was not found. In Jay's (1852: 69) fourth edition he did, however, list under *Margaritana* Schumacher 1816 (= *Margaritifera* Schumacher, 1816) the following entry: "1951 *marginata* . . . Lake Erie." This may have been mistakenly cited for *M. margaritifera* since it lies only three lines below listings in Jay's *Catalogue* for this latter species. We believe that this is very likely an error. We are not aware of any valid Ohio records for this species.

Similar species. The large, black, elongate shell differs from all other Ohio species except *Cumberlandia monodonta* (Say, 1829) by possessing small, puncture-like muscle scars on the central portion of the nacre.

Comments. Individuals grow at a fairly constant rate for most of their life, only very gradually slowing. North American specimens may grow to ca. 140 mm and live for more than 50 years. European specimens have been found in excess of 200 years old (Mutvei & Westermark, 2001).

Arcidens Simpson, 1900

Etymology. L. *arcus*, bow + L. *dens, dentis*, tooth. Simpson (1914: 476) noted of this genus, ". . . left valve with an arched, somewhat elongated tooth under the beak, which often curves upward; . . . ." This character is typical of *Arcidens*, but this arched tooth or fusion of teeth is also found in *Arkansia*, and some *Alasmidonta* and *Lasmigona*.

Original description

Arcidens Simpson, 1900a: 661, no figure.

Type species. *Arcidens confragosus* Say, 1829, by monotypy.

Synonymy

Arcidens Simpson, 1900a: 661.
Arcidens (*Arcidens*) Simpson, 1900. Clarke, 1981: 84–85.

Geographic range. Mississippi River from Minnesota to Louisiana, but only in the lower Ohio and Tennessee rivers. Gulf drainages from Alabama to eastern Texas.

Arcidens confragosus (Say, 1829)

Figures 219, 220

Etymology. L. *con-*, together, with + L. *fragosus*, broken, rough, uneven.

Vernacular name(s). *Rock Pocketbook*, Rock Shell, Black Rock Shell, Bastard, Queen, Rough Rock Shell.

Original description

Alasmodonta confragosa Say, 1829: 339–340, no figures. Say, however, notes that this is "no other than the doubtful species figured in the Encyl. Meth., pl. 248, fig. 9 [Bruguière, 1789], and referred to as the *nodosus*, Linn., but recently on comparison with that figure, not being able to make it correspond satisfactorily, I have concluded to give it a distinct name." We have examined Bruguière's figure and find the umbonal sculpture and cardinal teeth of the left valve similar to but not within the range of the *A. confragosa* material we have studied. This species was redescribed in Say, 1831b: no pagination, pl. 21.

Type locality. Say gave two localities. "Inhabits Wabash River" (Say, 1829: 339) was emended in 1831 to "This species was first obtained by Mr. O Evans in a side stream of the Wabash, called Fox river" [Fox River, Illinois] (Say, 1831b: no pagination). Say (1829) included a second locality as well: "New Orleans," emended by him in 1832 to "Bayou Teche in the Parish of St. Mary, Louisiana." Johnson (1980: 115) incorrectly gave the type locality as Little Wabash River, Richland Co., IL.

Type material. The original unionoid type material of Say has apparently been lost (Johnson & Baker, 1973: 146).

Synonymy

Alasmodonta confragosa Say, 1829: 339–340.

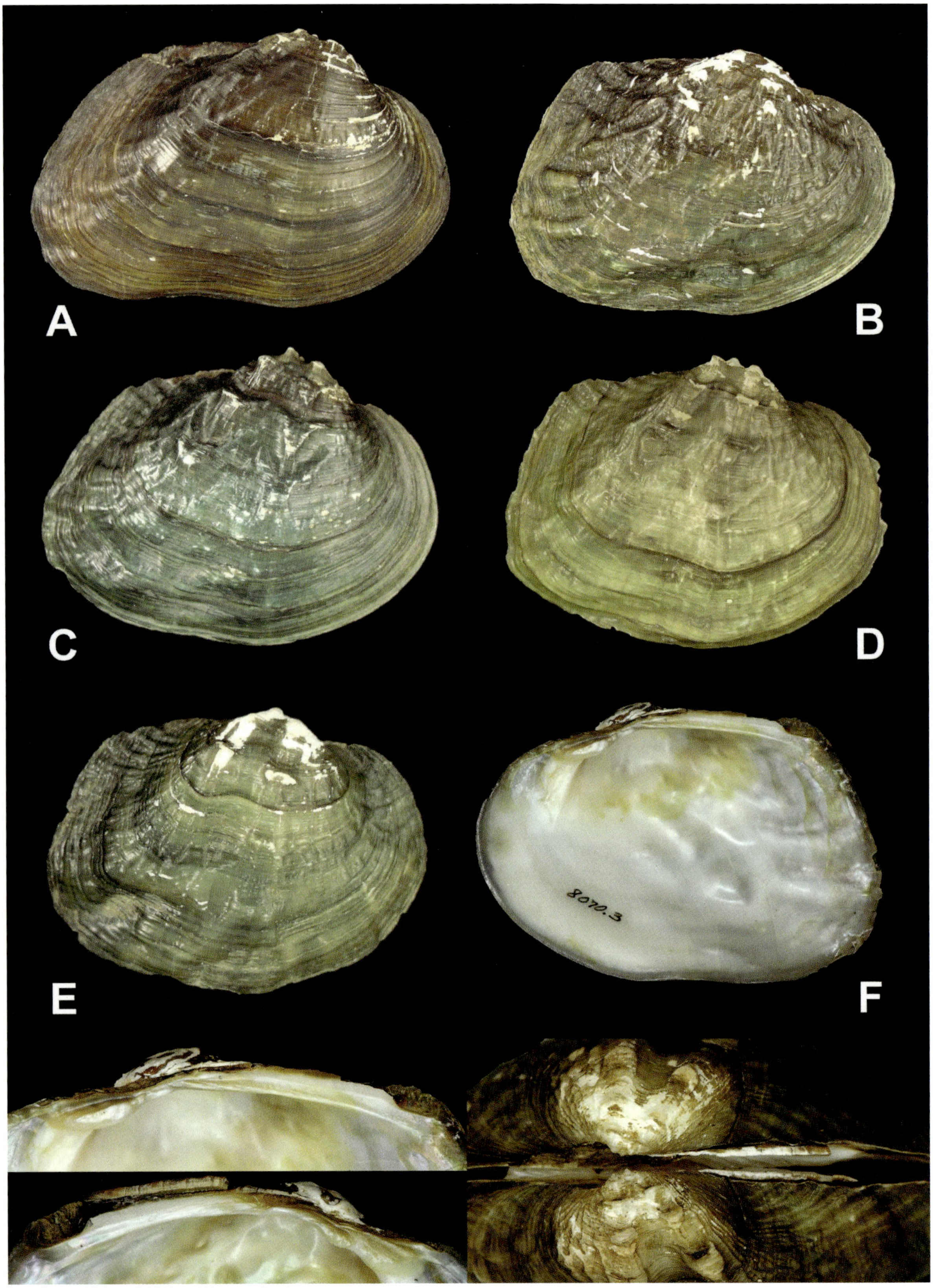

Fig. 219. *Arcidens confragosus* (Say, 1829). A. OSUM 51167, Mississippi River, Crawford Co., WI. 115 mm. **B, F, Hinge detail:** OSUM 8070.3, Mud River, Logan Co., KY. 85 mm. **C.** OSUM 17717, Illinois River, Cass Co., IL. 78 mm. **D, Beak sculpture detail:** OSUM 43866, Sangamon River, [Co.?], IL. 57 mm. **E.** OSUM 50614, Ohio River, Pulaski Co., IL. 38 mm.

Unio confragosa (Say, 1829). Deshayes, 1835: 552.
Margarita (Margaritana) confragosa (Say, 1829). Lea, 1836: 43.
Unio confragosus (Say, 1829). Catlow & Reeve, 1845: 552.
Complanaria confragosus (Say, 1829). Agassiz, 1852: 49.
Margaron (Margaritana) confragosa (Say, 1829). Lea, 1852b: 42.
Baphia confragosa (Say, 1829). Adams & Adams, 1857: 500.
Margaritana confragosa (Say, 1829). Calkins, 1874: 46.
Arcidens confragosus (Say, 1829). Simpson, 1914: 475.
Arcidens confragosa jacintoensis Strecker, 1931: 13 [Type locality: "San Jacinto River, [San Jacinto Co.] Texas." Types in Baylor University Museum, Waco, Texas].
Arcidens (Arcidens) confragosus (Say, 1829). Clarke, 1981a: 89–97, figs. 30–32.

Status. Listed as Extirpated by Ohio Division of Wildlife; not listed by U.S. Fish & Wildlife Service as of 2009.

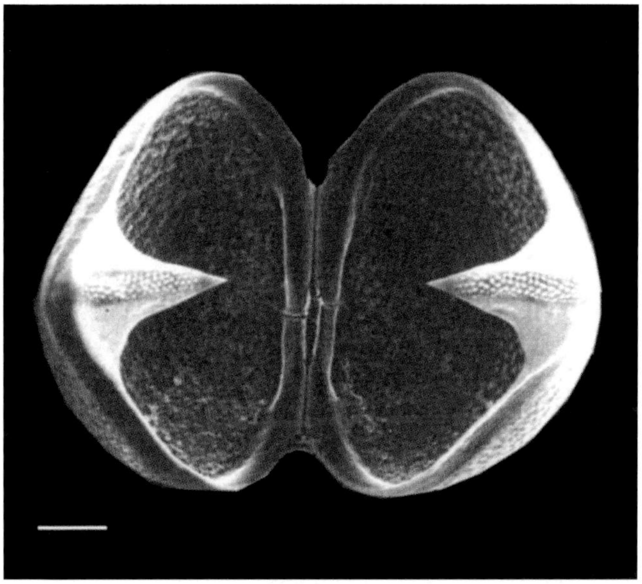

Fig. 220. *Arcidens confragosus* (Say, 1829). Glochidium. OSUM 52015, Green River, Warren Co., KY. Bar = 65 µm. (Hoggarth, 1999).

ADULT SHELL

Shell. Medium to fairly large (to 150 mm), thin-shelled for its size, particularly as juveniles, moderately compressed, rather rectangular in profile.
Anterior margin. Roundly arched into ventral margin.
Ventral margin. Nearly straight to broadly rounded, often indented near the midline.
Posterior margin. Ventral and posterior margins form wide, blunt acute angle; posterior margin oblique and straight to dorsal margin.
Dorsal margin. Long, more or less straight, joining posterior margin with very obtuse, sharp angle.
Umbo. Broad, prominent, located just anterior of the mid-dorsal line; umbonal sculpture of 3–6 very coarse "W"-shaped ridges, the bottoms of the "W" forming erect elongated ridges. The beak sculpture persists onto the adult shell (see below).
Ligament. Long, tan to dark brown, barely concealed by shells.
Disc sculpture. Pronounced and complex. The "W" sculpture of the beaks persists on the adult, the bottoms of the "W" forming pronounced concave ridges, with the center of the "W" becoming smooth. On the posterior half of the shell are superimposed oblique undulating ribs that turn towards the posterior margin, resulting in a corrugated posterior slope; these ribs may bifurcate. On the anterior portion of their shell are superimposed numerous radiating corrugations—these may become lost on some old individuals.
Periostracum. Green, yellow, or tan as juveniles, becoming dark brown to black with age. Regions adjacent to annular rings often darker. Rarely with very faint rays.
Nacre. Porcelain white, often iridescent posteriorly; many specimens seen have extensive gold or brown Tulberg layers; the outer rim of the shell interiorly is dark where the outer colors show through.
Hinge plate. Rather weak for the size of the shell; right cardinal a single, chunky, triangular denticle, often accompanied by an anterior accessory denticle; left cardinal with an anterior, weakly developed, lamellate tooth and a large, concave tooth posterior in the interdentum; laterals weakly developed, usually only one in each valve.
Muscle scars. Anterior adductor only slightly or not at all impressed; posterior adductor flush; anterior adductor distinct from anterior pedal protractor scar in juveniles, becoming confluent with age, posterior adductor confluent with posterior pedal retractor scar; multiple dorsal suspensor muscle scars located deep within umbonal cavity. Pallial line entire, located very far from shell margin. Accessory line weak, short.
Umbonal cavity. Very wide, open, and fairly deep.

Glochidia. Pyriform, about as long as high, asymmetric, dorsal margin straight, posterior margin produced dorsally and incurved ventrally with its maximum inflation occurring between 30%–40% from dorsal to ventral. Anterior margin broadly rounded dorsally and incurved just before the ventral terminus so that its maximum inflation occurs at about 50% from the dorsal margin. Exterior surface densely covered with rosette sculpturing. The styliform hook extends from the ventral terminus as a very strongly biconcave triangular plate. It has about 80 pyramidal microstylets arranged in about five rows. The number of rows of microstylets remains the same from proximal to distal ends of the hook. Micropoints occur on the rim of the valve, at the ventral terminus, and along the microstylet border, leaving a wide unsculptured distal hook margin (Hoggarth, 1999: 44–45, figs. 23a–g).

Literature records (Surber, 1912: 8, pl. 1, fig. 5; Utterback, 1915: 254; Clarke, 1981a: 92–94, fig. 31; Hoggarth, 1999: 44–45, figs. 23a–g) indicate glochidia dimensions for length (0.35–0.36 mm), height (0.35–0.36 mm), and hinge length (0.23–0.25 mm). This glochidium is similar to that of some members of the genera *Alasmidonta* and *Lasmigona* but can be distinguished by its equal height and length, its exterior valve sculpture, and the arrangement of microstylets on the hook.

Reproductive biology. Bradytictic. Marsupia occupy the outer gills (Utterback, 1915). Gravid females are found from September to June (Baker, 1928a). No conglutinates are formed (Watters, 2008). Actual transformation has been recorded only from Channel Catfish.

Reported potential hosts

Ambloplites rupestris (Rock Bass)—Surber, 1913 (NI); Wilson, 1916 (NI).

Anguilla rostrata (American Eel)—Wilson, 1916 (NI).

Aplodinotus grunniens (Freshwater Drum)—Wilson, 1916 (NI).

Dorosoma cepedianum (Gizzard Shad)—Surber, 1913 (NI); Wilson, 1916 (NI).

Ictalurus punctatus (Channel Catfish)—Howells, 1994a (LT); Howells, 1997 (LT).

Pomoxis annularis (White Crappie)—Surber, 1913 (NI); Wilson, 1916 (NI).

Habitat. Muddy sand and silt in rivers.

Range. Mississippi River from Minnesota to Louisiana, but only in the lower Ohio River to Indiana and lower Tennessee River. Gulf drainages from Alabama to eastern Texas.

Range in Ohio. Sterki (1907: 373) recorded this species from the "Ohio River at Cincinnati." We have not seen this Ohio record or any other for this species vouchered by a specimen, and so understandably we have doubts. Clarke's (1981a) distribution map for this species included this Cincinnati record, but it was not recorded in his list of sites. Cincinnati would seem a likely place for it to have been or to appear in Ohio, but valid records appear to be absent for the present.

Similar species. The highly sculptured shell is reminiscent of *Amblema*, *Megalonaias*, and *Quadrula*. All of those genera have well-developed, even massive, hinges compared to the weak hinge of *Arcidens*.

Comments. This species grows very quickly for 3–4 years, before slowing. Despite its size (to 150 mm), most individuals are less than 12 years old.

Glebula Conrad, 1853

Etymology. L. *Glaeba*, a small lump of earth, a clod + L. *-ula* (diminutive). Conrad did not mention the reason for choosing the name *Glebula* for this monotypic genus. However, the form of *G. rotundata* (Lamarck, 1819) is typically roundish, solid, and not unlike a lump of clay.

Original description

Glebula Conrad, 1853: 268.

Type species. *Unio rotundata* Lamarck, 1819, by monotypy.

Geographic range. Lower Mississippi River and Gulf drainages from Texas to Florida, very rarely in the Ohio River.

Glebula rotundata (Lamarck, 1819)

Figure 221

Etymology. L. *rotundus*, circular, round, spherical. This shell is often nearly round in profile and inflated to the point of being spherical.

Vernacular name(s). *Round Pearlshell.*

Original description

Unio rotundata Lamarck, 1819: 75, unfigured.

Type locality. None given. Restricted by Clench & Turner (1956) to "Bayo [*sic*] Teche, St. Mary Parish, Louisiana."

Type material. Holotype USNM 85760.

Synonymy

Unio rotundata Lamarck, 1819: 75.

Unio suborbiculata Lamarck, 1819: 81 [Type locality restricted to "Bayou Teche [LA]" by Clench & Turner (1956: 192). Holotype USNM 85760, based on same specimen as *Unio rotundata* Lamarck, 1819, *fide* Johnson, 1969].

Unio glebulus Say, 1831c: 526 [Type locality: "Bayou Teche, Louisiana." Type lost].

Unio subglobosus Lea, 1834: 30, pl. 2, fig. 3 [Type locality: "Bayou Teche, Louisiana." Figured lectotype USNM 159969, ANSP paralectotypes lost *fide* Johnson & Baker, 1973].

Unio suborbiculatus Lamarck, 1819. Férussac, 1835: 26.

Margarita (*Unio*) *rotundatus* (Lamarck, 1819). Lea, 1836: 33.

Unio rotundatus Lamarck, 1819. Hanley, 1842: 201.

Margaron (*Unio*) *rotundatus* (Lamarck, 1819). Lea, 1852b: 34.

Glebula rotundata (Lamarck, 1819). Conrad, 1853: 268.

Unio grandensis Conrad, 1855: 256 [Type locality: "Rio Grande, Texas," probably in error (Howells *et al.*,

Fig. 221. ***Glebula rotundata*** **(Lamarck, 1819). A, F, Hinge detail:** OSUM 56566.1, Sunflower River, Washington Co., MS. 79 mm. **B.** OSUM 15704.27, Bayou Teche, St. Mary Parish, LA. 92 mm. **C.** OSUM 5088, Ohio River, Jefferson Co., KY. 85 mm. **D, Beak sculpture detail:** OSUM 6081, a bayou, East Baton Rouge Parish, LA. 34 mm. **E.** OSUM 5090, "Ohio." 75 mm.

1996, reported no records of this species from that river). ANSP type lost *fide* Johnson & Baker, 1973].
Unio granadensis [*sic*] Conrad, 1855. Simpson, 1900a: 598.
Glebula suborbiculatus (Lamarck, 1819). Frierson, 1927: 91.

Status. Not listed by either Ohio Division of Wildlife or U.S. Fish & Wildlife Service as of 2009.

ADULT SHELL

Shell. Medium-sized (to ca. 110 mm), solid, oblong to nearly round, generally very inflated.

Anterior margin. Rounded or slightly indented.

Ventral margin. Broadly rounded, often forming a wide, obtuse angle posteriorly, then straight or concave before a rounded acute angulation; females are slightly more rounded posteriorly.

Posterior margin. Short, convex in small specimens, becoming straight or concave in older ones.

Dorsal margin. Very short, straight or convex.

Umbo. Very broad and low in small individuals, becoming pronounced in older ones; no beak sculpture could be discerned in any specimens examined, and Vidrine (1993) states that this species has no beak sculpture.

Ligament. Very large, prominent, brown to black, occupying nearly entire dorsal margin.

Disc sculpture. Shell sculptureless except for 1–3 radiating ridges that define the posterior slope; these often curve ventrally, sometimes giving a "hooked" appearance to the shell.

Periostracum. Pale green or yellow as juveniles, becoming darker with age; some old individuals are nearly black, but most develop wide, alternating, concentric rings of brown, tan, and gray; shell rayless except for darkly colored dorsal ridges.

Nacre. White with a metallic sheen as juveniles, iridescent posteriorly; adults may be flushed with light purple or rose on the hinge plate and posteriorly; gold or pink Tulberg layers are commonly seen.

Hinge plate. Rather delicate as juveniles, becoming massive with age, distinctly bowed; interdentum long, somewhat flattened laterally; cardinal teeth lamellar in juveniles, two in left valve with posterior one reduced, one in right valve with well-developed accessory anterior to cardinal. In adults the cardinal teeth become chunky and triangular, often subdivided into numerous serrated teeth; lateral teeth short, delicate, and bowed in juveniles, thick and straight in adults; in juveniles there are two laterals in the left valve and one in right; in adults the right valve develops a ventral accessory tooth that may be nearly as well-developed as the lateral.

Muscle scars. Anterior adductor impressed, particularly in old individuals, separated from anterior pedal protractor; posterior adductor flush with shell, entirely confluent with posterior pedal retractor; dorsal suspensors numerous, arranged in a line from the interdentum along the inside of the anterior adductor fossa; pallial line entire, accessory line very long but indistinct.

Umbonal cavity. Open and wide, becoming deep with age.

Glochidia. Not reported.

Reproductive biology. Parker *et al.* (1984) studied the reproductive biology of this species in Louisiana. They found it capable of as many as three broods per year. Females had developing embryos in March and glochidia from April through October, although this pattern was not synchronized among females. Each female carried an estimated 531,000 glochidia per brood. No durable conglutinate is formed (Watters, 2008). Metamorphosis occurred after 16 days at 22°C, or five days at 30°C. Metamorphosis has been reported only from sunfishes.

Reported potential hosts

Anchoa mitchilli (Bay Anchovy)—Parker *et al.,* 1984 (NI).
Cyprinus carpio (Common Carp)—Parker *et al.,* 1984 (NI).
Lepisosteus oculatus (Spotted Gar)—Parker *et al.,* 1984 (NI).
Lepomis cyanellus (Green Sunfish)—Parker *et al.,* 1984 (LT).
Lepomis macrochirus (Bluegill)—Parker *et al.,* 1984 (LT).
Morone chrysops (White Bass)—Parker *et al.,* 1984 (NI).
Trinectes maculatus (Hogchoker)—Parker *et al.,* 1984 (NI).

Habitat. This is a species of bayous, marshes, backwaters, and impoundments in quiet water in a muddy substrate.

Range. Lower Mississippi River and Gulf drainages from Texas to Florida, very rarely in the Ohio River.

Range in Ohio. Not yet recorded from the state, but may occur in the Ohio River.

Similar species. The combination of a greatly inflated, nearly egg-shaped, thin shell, ragged periostracum, thin hinge, and purplish nacre distinguish this species from all others.

Comments. This is a long-lived species. It grows very quickly for 2–3 years, but abruptly slows to a very low growth rate. Growth rates vary considerably: a 95-mm individual may be from 11 to 29 years old. Specimens may reach over 110 mm in length and live for over 30 years. Sexual dimorphism in the shell is not pronounced and may be indistinguishable from normal variation in the roundness of the shell.

Lasmigona Rafinesque, 1831

(see p. 162 for generic discussion)

Lasmigona subviridis (Conrad, 1835)

Figures 222, 223

Etymology. L. *sub-*, under, below, somewhat less than. L. *viridis*, green. Specimens of the species are nearly always greenish and sometimes bright green.

Vernacular name(s). *Green Floater*, Eastern Creekshell, Mountain Creekshell, Appalachian Creekshell.

Original description

Unio viridis (*subviridis*) Conrad, 1835: Appendix: 4, pl. 9, fig. 1. [Conrad (1835: 4) noted that the species described here may be "the *viridis* of Rafinesque . . . found in Kentucky." If not "I propose . . . to call the present shell *subviridis*." *Unio viridis* Rafinesque, 1820, is indeed a quite different species (which see herein). Although proposed as a conditional name, because *U. subviridis* was introduced prior to 1960 it is an available name (ICZN Art. 15.1).].

Type locality. "Inhabits Schuykill river; Juniata river; creeks in Lancaster co., Penn."

Type material. Figured lectotype ANSP 2105 lost *fide* Johnson, 1970.

Synonymy

Unio subviridis Conrad, 1835: Appendix: 4, pl. 9, fig. 1.

Margarita (*Unio*) *tappanianus* Lea, 1836: 39; 1838: 25 [*nomen nudum*].

Unio viridis Rafinesque, 1820. Conrad, 1836: 35, pl. 17, fig. 1 [misidentification].

Unio tappanianus Lea, 1838a: 62, pl. 17, fig. 55 [unnecessary replacement name for Conrad's 1836 misidentification. Same type locality as *Unio subviridis* Conrad, 1835. Figured lectotype USNM 85240 *fide* Johnson, 1970, but corrected to 86133 in Johnson, 1974. However, the type of the replacement name must be the same as that of *Unio subviridis* Conrad, 1835.].

Unio hyalinus Lea, 1845: 164, pl. 2, fig. 4 [Type locality: [James River] "Richmond, Virginia." Lectotype USNM 86131 *fide* Johnson, 1970].

Unio tappianus [*sic*] Lea, 1838. Catlow & Reeve, 1845: 64.

Margaron (*Unio*) *tappanianus* (Lea, 1838). Lea, 1852b: 39.

Margaron (*Unio*) *hyalinus* (Lea, 1845). Lea, 1852b: 39.

Margaritana quadrata Lea, 1861a: 41 [Type locality: "East Tennessee"; probably in error. Possible lectotype USNM 86264].

Unio pertenuis Lea, 1863: 193 [Type locality: "Neuse River, near Raleigh, N. Carolina." Possible lectotype USNM 86139].

Margaron (*Unio*) *pertenuis* (Lea, 1863). Lea, 1870: 62.

Margaron (*Margaritana*) *quadrata* (Lea, 1861). Lea, 1870: 68.

Margaritana (*Alasmodonta*) *quadrata* Lea, 1861. Clessin, 1876: 273, pl. 83, figs. 5, 6.

Symphynota (*Symphynota*) *viridis* "Conrad." Simpson, 1900a: 663–664.

Symphynota (*Symphynota*) *quadrata* (Lea, 1861). Simpson, 1900a: 664.

Lasmigona (*Platynaias*) *subviridis* (Conrad, 1835). Clarke, 1985: 51–57, figs. 15–17.

Unio pertenius [*sic*] Lea, 1863. Clarke, 1985: 51.

Status. Not listed by either Ohio Division of Wildlife or U.S. Fish & Wildlife Service as of 2009.

ADULT SHELL

Shell. Small to medium-sized (to 70 mm), thin, oval to somewhat rectangular, compressed to moderately inflated.

Anterior margin. Short, straight to broadly rounded to the ventral border.

Ventral margin. Very broadly rounded to straight.

Posterior margin. Long, straight, forming two angles basally.

Dorsal margin. Long, straight, joining posterior margin with obtuse angle.

Umbo. Low, small; umbos directed slightly forward; umbonal sculpture of 2–4 irregular heavy ribs, often in double loops but occasionally not.

Ligament. Long but inconspicuous, tan to dark brown, partially concealed by shells.

Disc sculpture. Shell smooth except for weak angulations forming the angles at the ventro-posterior margin.

Periostracum. Shiny, yellow, yellow-green, or tan with fine green lines or rays; some rare populations are predominately brown.

Nacre. Porcelain white anteriorly, iridescent posteriorly; often with a flush of salmon in the umbonal cavity; often with extensive gold or brown Tulberg layers.

Hinge plate. Well-developed but narrow; right cardinal a single, flattened, triangular denticle; left cardinal bifid, not well-developed, the teeth widely separated; laterals poorly developed, long, curved, one in right valve, two in left; interdentum flat, narrow, long.

Muscle scars. Anterior adductor moderate, oval, and strongly impressed; posterior adductor flush; large anterior adductor abuts longitudinal anterior pedal protractor scar; large posterior adductor distinct, joined with small posterior pedal retractor scar; cluster of small dorsal suspensors located behind cardinal teeth and interdentum, nearly hidden from view. Pallial line entire, located far from shell margin. Accessory line inconspicuous and very short.

Umbonal cavity. Very wide, shallow.

Fig. 222. ***Lasmigona subviridis*** **(Conrad, 1835). A.** OSUM 55234.17, Greenbrier River, Pocahontas Co., WV. 45 mm. **B.** OSUM 55234.12, Greenbrier River, Pocahontas Co., WV. 35 mm. **C.** OSUM 25653.1, Potomac River, Jefferson Co., WV. 42 mm. **D.** OSUM 37612.10, Erie Canal, Herkimer Co., NY. 35 mm. **E, F, Hinge detail:** OSUM 22868.29, Kanawha River, Fayette Co., WV. 50 mm. **Beak sculpture detail:** OSUM 34550.1, Potomac River, Fairfax Co., VA. 29 mm total length.

Glochidia. Depressed pyriform, posterior margin, strongly curved; anterior margin broadly rounded and dorsal margin straight. Exterior valve surface, coarsely malleated and uniformly pitted except within a narrow marginal band. Ventral terminus, broadly rounded, not produced or nipple-like as in the other members of the genus, and located about 40% from posterior to anterior. Exterior surface sculpturing, loose-looped. Styliform hook armed with about 25 microstylets and numerous micropoints. Microstylets, lanceolate, about five to six abreast proximally and arranged in a double offset row distally. Micropoints located on the ventral margin of the valve and upon the lateral surfaces of the hook but not over the edge of the valve at the ventral terminus, leaving a wide, unsculptured distal hook margin (Hoggarth, 1999: 47–48, figs. 25a–c).

Literature records (Lea, 1874b: 69, pl. 12, fig. 14; Ortmann, 1912a: 282; Ortmann, 1919: 122; Clarke, 1985: 54–55, fig. 16; Hoggarth, 1999: 47–48, figs. 25a–c) indicate glochidia dimensions for length (0.36–0.38 mm), height (0.28–0.32 mm), and hinge length (0.25–0.26 mm). This glochidium can be distinguished by its more broadly curved lateral margins and the absence of micropoints extending onto the valve at the ventral terminus.

Reproductive biology. This is one of two confirmed North American species to metamorphose without a host, *Utterbackia imbecillis* being the other species. Fully formed juveniles were found in the marsupia in April from Pennsylvania and North Carolina populations (Barfield & Watters, 1998; Lellis & King, 1998).

Reported potential hosts. No host required (Barfield & Watters, 1998; Lellis & King, 1998) although Neves (pers. comm., 2000) reported transformation on unspecified hosts.

Habitat. This species occurs in creeks and small rivers, usually in sandy substrate.

Range. Atlantic Slope from the Hudson River to North Carolina. Across the divide to the Kanawha River system in West Virginia and Virginia, and Tygarts Creek in Kentucky.

Range in Ohio. While there are no vouchered Ohio records of this species in the Ohio material examined, Zeto (1980: 45) has collected this species from Tygarts Creek, Kentucky. The collection site is in Greenup County which borders the Ohio River opposite Scioto County, Ohio. OSUM has specimens of *L. subviridis* from the Kanawha River below Kanawha Falls, Fayette County, West Virginia, where this river downstream to the Ohio has few individuals of any species of unionoids in the present day.

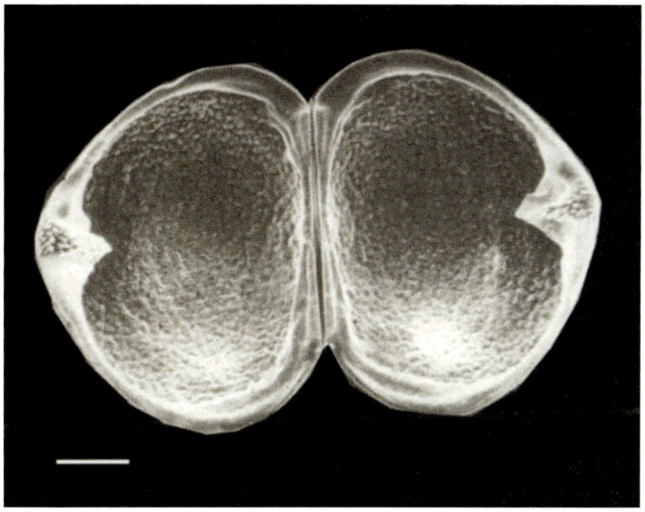

Fig. 223. ***Lasmigona subviridis* (Conrad, 1835).** Glochidium. OSUM 27131.68, Little River, Floyd Co., VA. Bar = 80 µm. (Hoggarth, 1999).

The presence of this species in the mountain streams of North Carolina, Virginia, and West Virginia in the headwaters of the Kanawha River suggests that its range may once have extended into Ohio, and perhaps to the west beyond by way of the pre-Pleistocene Teays River. Relict populations may yet be found living in unglaciated Ohio (especially south-central and southeastern counties from Monroe to Adams County) having relatively undamaged tributaries of this ancient master stream. *Lasmigona subviridis* is not typically a large river species, and it should be sought in medium to small streams in good condition.

Similar species. This species outwardly resembles *Alasmidonta viridis* but possesses lateral teeth and a finer hinge. *Lasmigona compressa* is larger and more elongate and possesses a tooth-like extension under the hinge of the left valve.

Comments. Exceptional individuals may grow to 70 mm, but most are from 30–40 mm in length. This is a very short-lived, fast-growing species, rarely attaining 7 years, usually 3–4.

Ligumia Swainson, 1840

(see p. 182 for generic discussion)

Ligumia subrostrata (Say, 1831)

Figure 224

Etymology. L. *sub.*, under, less than, almost, near, somewhat + L. *rostratus*, beaked, having a snout or bill; "somewhat beaked."

Vernacular name(s). *Pondmussel,* Western Pondshell, Long Pondshell, Narrow Pondshell, Prairie Shell.

Original description

Unio subrostratus Say, 1831b: no pagination, no figures.

Type locality. "Wabash." Say (1831b: no pagination) noted, "This may be said to be the analogue of the *U. nasutus* Nob. of the Western waters. . . . The rostrum however is not so definite, and it is a more convex shell. The aged shell is dark brown, but near the beaks reddish brown."

Type material. Neotype SMF No. 4348 by Haas (1930: 328).

Synonymy

Unio subrostratus Say, 1831b: 26.

Unio nashvillianus Lea, 1834: 100, pl. 14, fig. 43 [Type locality: "Cumberland River; Ohio [River] at Louisville," but possible lectotype label reads "Nashville." Possible lectotype USNM 85114, two possible paralectotypes MCZ 16380].

Margarita (*Unio*) *nashvillianus* (Lea, 1834). Lea, 1836: 26.

Unio mississippiensis Conrad, 1850: 277, pl. 38, fig. 11 [Type locality: "one of the tributaries of the lower Mississippi [River]." Probable figured lectotype MCZ 178777].

Lampsilis subrostratus (Say, 1831). Agassiz, 1852: 44.

Margaron (*Unio*) *nashvillianus* (Lea, 1834). Lea, 1852b: 29.

Margaron (*Unio*) *mississippiensis* (Conrad, 1850). Lea, 1852b: 29.

Unio rutersvillensis Lea, 1859b: 155 [Type locality: "Rutersville, Texas," but lectotype label reads "Clear Creek." Lectotype USNM 84976].

Unio cocoduensis Reeve, 1865: pl. 24, fig. 117 [Type locality: "Hab?" Holotype BMNH 1965187].

Unio topekaensis Lea, 1868: 144 [Type locality: "Topeka, Kansas," but holotype label reads "Falls City, Nebraska." Lectotype USNM 84978].

Margaron (*Unio*) *rutersvillensis* (Lea, 1859). Lea, 1870: 43.

Margaron (*Unio*) *topekaensis* (Lea, 1868). Lea, 1870: 43.

Margaron (*Unio*) *nashvilliensis* (Lea, 1834). Lea, 1870: 45 [unjustified emendation].

Lampsilis subrostratus (Say, 1831). Simpson, 1900a: 546.

Lampsilis (*Ligumia*) *subrostrata* (Say, 1831). Frierson, 1927: 77.

Eurynia (*Eurynia*) *subrostrata* (Say, 1831). Ortmann, 1912a: 344.

Lampsilis subrostrata var. *furva* Simpson, 1914: 100 [Type locality: "North Indiana, Maumee River." Lectotype USNM 126921].

Eurynia subrostrata (Say, 1831). Utterback, 1916b: 15.

Ligumia subrostrata (Say, 1831). Grier & Mueller, 1922: 100.

Status. Not listed by Ohio Division of Wildlife or by U.S. Fish & Wildlife Service as of 2009.

ADULT SHELL

Shell. Medium-sized (to 80 mm), sturdy but not heavy, elongated.

Anterior margin. Straight, then evenly rounded to the ventral margin.

Ventral margin. Broadly rounded in males, meeting the posterior margin in an acute angulation (or two) about halfway between the dorsal and ventral margins; females with conspicuous marsupial region, ventral margin straight or concave anterior to marsupia, forming an obtuse angle with the posterior margin at the base.

Posterior margin. Long, straight or concave in males; females straight from the ventral angulation, then forming an acute angle approximately 33% down from the dorsal margin, then straight to the dorsal margin.

Dorsal margin. Long, straight or somewhat convex.

Umbo. Low, wide, barely rising above the hinge line; beak sculpture of 5–7 coarse "V"-shaped ridges, pointed towards the umbo, the peaks aligned obliquely toward the ventral-posterior border; this sculpture abruptly disappears by the end of the first year's growth; in some specimens the last "V"s become nearly linear and broken into two parts.

Ligament. Brown, long, occupying most of the dorsal margin, partially enclosed by the valves.

Disc sculpture. Shell smooth except for 1–3 ridges defining the posterior slope.

Periostracum. Base color yellow or greenish; with age, the older portions of the shell become dark brown, leaving the newer annuli lighter; shell usually with well-defined, dark green rays; these cover the entire shell and are darkest on the posterior slope; rays continuous or broken across annuli, rarely wavy.

Nacre. Pearly white, very iridescent posteriorly; green or gold Tulberg layers common and extensive.

Hinge plate. Very light, interdentum long, almost imperceptible in many examples; cardinals lamellate; two in left valve, but posterior one may be very weak; one in right valve, but with well-developed accessory tooth anterior to cardinal nearly confluent with shell margin; laterals very long, straight or slightly bowed, two in left, one in right.

Muscle scars. Anterior adductor barely indented, smooth, well-separated from anterior pedal protractor; posterior adductor flush with shell, partially confluent with posterior pedal retractor; dorsal suspensors few, inside umbonal cavity; pallial line entire, accessory very long but nearly imperceptible.

Umbonal cavity. Wide, open, moderately deep.

Glochidia. Subelliptical, higher than long with narrowly arched lateral margins to a broadly rounded ventral margin; hookless (Lefevre and Curtis, 1912: pl. 8, fig. 13).

Fig. 224. ***Ligumia subrostrata*** **(Say, 1831). A.** OSUM 36842.1, Straight River, Dubois Co., IN. 85 mm. **B.** OSUM 56267.1, Tippecanoe River, Kosciusko Co., IN. 62 mm. **C, H. Hinge detail:** OSUM 16800, Mississippi River, Henderson Co., IL. 80 mm. **D.** OSUM 3714.28, Schreiber Farm Creek, Macoupin Co., IL. 36 mm. **E.** OSUM 69544.2, Trim Cane Creek, Oktibbeha Co., MS. 43 mm. **F.** OSUM 57246.1, Eagle Lake, Noble Co., IN. 40 mm. **G.** OSUM 57246.5, Eagle Lake, Noble Co., IN. 65 mm. **Beak sculpture detail:** OSUM 3714.32. Schreiber Farm Creek, Macoupin Co., IL. 32 mm total length.

Lefevre and Curtis stated that measurements for this glochidium are found in their text figure I, but *L. subrostrata* is not found in that figure.

Reproductive biology. Probably bradytictic. Utterback (1916b) reported gravid females in June. Conglutinates are club-shaped (Utterback, 1916b). Centrarchids are the only reported hosts.

Reported potential hosts

Lepomis cyanellus (Green Sunfish)—Lefevre & Curtis, 1912 (LT); Stern & Felder, 1978 (NI).

Lepomis gulosus (Warmouth)—Evermann & Clark, 1920 (LT); Stern & Felder, 1978 (NI).

Lepomis macrochirus (Bluegill)—Stern & Felder, 1978 (NI).

Micropterus salmoides (Largemouth Bass)—Lefevre & Curtis, 1912 (LT).

Habitat. Sandy mud, often in ponds or slackwater areas.

Range. Upper Mississippi River to the lower Ohio River systems (as far as the Wabash River), lower Cumberland River. Gulf drainages from Louisiana to Texas.

Range in Ohio. *Ligumia subrostrata* is one of a few species of unionoids that, because of both its range out into the western grasslands and the extension of its habitat well upstream into intermittent creeks, brooks, and temporary ponds, can be called a prairie species. It was listed for Ohio by Sterki (1907: 389), La Rocque (1967: 267–269), Cummings & Mayer (1992: 138), and Parmalee & Bogan (1998: 154). We strongly suspect that the later records are based, directly or indirectly, on Sterki's speculation of 1907. He noted, "I have seen no specimens (although it is doubtless at least in the north-western part of the state)."

The above considered in view of the superficial similarity of *L. nasuta*, a common species in Lake Erie and a few of its tributaries until recently, renders the presence of *L. subrostrata* in Ohio as very doubtful. There are no OSUM voucher specimens from Ohio. Our four documented lots from Indiana are from the Wabash River system rather than the Maumee River basin. We agree with La Rocque that those headwaters of the Wabash in Ohio would be the most likely area in Ohio to find this western species, but it has not been recorded from there as yet. Nevertheless, Simpson's *Lampsilis subrostrata* var. *furva* Simpson, 1914, was described from the Maumee River.

Similar species. See under *Ligumia nasuta*.

Comments. This is a short-lived species. It grows rapidly for the first three years, and most specimens seen are only 4–5 years old, although its rarely attains 10 years. Growth rates are variable between populations, and specimens 80 mm long may be from 5 to 10 years old. It rarely reaches 90 mm in length. Female shell characteristics are apparent in the second year, suggesting a very early sexual maturity.

Potamilus Rafinesque, 1818

(see p. 241 for generic discussion)

Potamilus capax (Green, 1832)

Figure 225

Etymology. L. *capex,* roomy, spacious. Green was obviously referring to the highly inflated, "roomy" shell.

Vernacular name(s). *Fat Pocketbook.*

Original description

Unio capax Green, 1832: 290, no figure.

Type locality. "Falls of St. Anthony [Minneapolis, Hennepin Co., MN] and Bayou Teche" [latter in error; see Vidrine, 1993].

Type material. Lost *fide* Johnson, 1980.

Synonymy

Unio capax Green, 1832: 290.

Symphynota globosa Lea, 1832: 41, pl. 4, fig. 12 [Type locality: "River Ohio, 150 miles below Louisville [KY]." Syntypes ANSP 56609].

Lampsilis capax (Green, 1832). Smith, 1899: pl. 84.

Lampsilis (*Lampsilis*) *capax* (Green, 1832). Simpson, 1900a: 529–530.

Proptera capax (Green, 1832). Ortmann, 1914: 67.

Potamilus capax (Say, 1817). Morrison, 1969: 24.

Status. Listed as Extirpated by the Ohio Division of Wildlife; listed as Endangered by the U.S. Fish & Wildlife Service as of 2009. A Federal recovery plan has been completed (USFWS, 1984c).

ADULT SHELL

Shell. Large (to at least 130 mm), oval, extremely inflated, often polished.

Anterior margin. Concave, forming a slight wing, then abruptly and evenly curving ventrally.

Ventral margin. Broadly curved in males, straight or concave before marsupial region in females.

Posterior margin. Broadly rounded in both sexes, but female marsupial region projecting slightly.

Dorsal margin. Very short, bowed.

Umbo. Swollen, prominent, wide; beak sculpture consists of 2–4 nearly indistinguishable "V"-shaped ridges; beak sculpture disappears well before the end of the first year.

Ligament. Brown, large, occupying most of dorsal margin, partially concealed by valves.

Disc sculpture. Sculptureless except for 1–3 low ridges on the posterior slope.

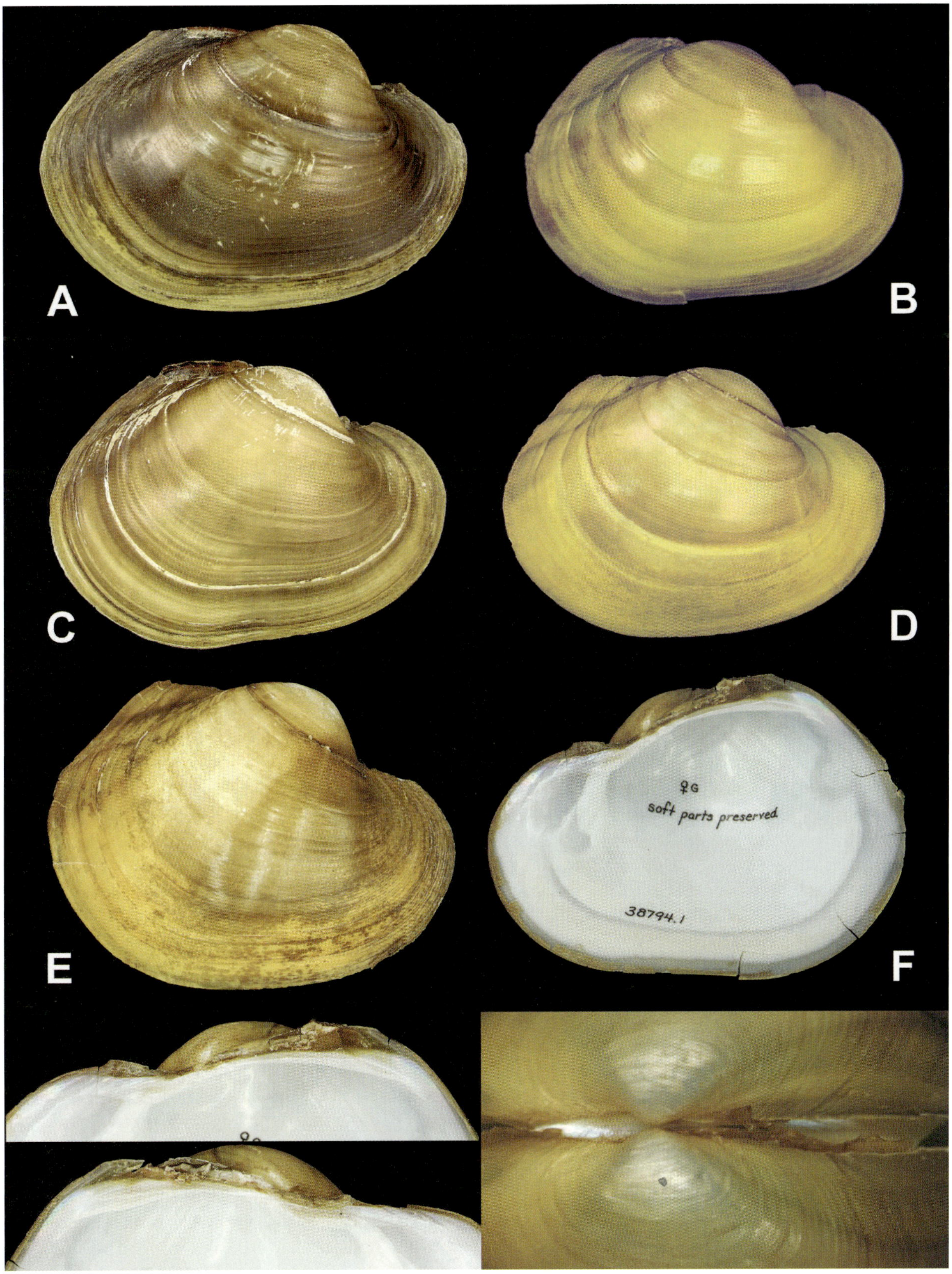

Fig. 225. *Potamilus capax* (Green, 1832). A. OSUM 34861, St. Francis River, Poinsett Co., AR. 128 mm. **B, F, Hinge detail:** OSUM 38794.1, White River, Knox Co., IN. 75 mm. **C.** OSUM 59714, Wabash River, Posey Co., IN. 113 mm. **D.** OSUM 50096, Ohio River, Vanderburgh Co., IN. 79 mm. **E.** OSUM 30984.1, Ohio River, Livingston Co., KY. 93 mm. **Beak sculpture detail:** OSUM 40200.4, no data. 44 mm total length.

Periostracum. Yellow, tan, or greenish, darker in individuals living in mud; suffused with irregular gray blotches that may cover much of the shell disc; very faint light green rays may be present; posterior slope rayed or darkly colored; annular rings wide and dark.
Nacre. White with an iridescent sheen, rarely flushed with pink or salmon.
Hinge plate. Light, distinctly "S"-shaped, bowed down anteriorly, up posteriorly; interdentum short, bowed, flattened; cardinals lamellate; anterior cardinal in left valve with two teeth, one anterior to the other, and often on the same plane so as to appear nearly confluent; a single cardinal in the right valve, but a well-formed accessory tooth occurs dorsally, confluent with the shell margin, so that two cardinal teeth may appear to be present; laterals moderately long, distinctly bowed, two in left valve, one in right.
Muscle scars. Anterior adductor only slightly impressed, smooth, removed from large anterior pedal protractor; posterior adductor completely flush with shell, posterior pedal retractor confluent only at ventral point; dorsal suspensors multiple, in an arc within the umbonal cavity; pallial line complete, accessory line present but weakly formed.
Umbonal cavity. Very deep but wide and capacious.

Glochidia. Ligulate with lanceolate hooks at the lateral margins of the ventral flange. Dorsal margin straight with incurved lateral margins. Vermiculate sculpturing covers the exterior surface of the valve, and micropoints are extremely reduced and appear unorganized (Cummings *et al.*, 1990: 16–17, figs. 6a–d, 9c).

Literature records (Coker & Surber, 1911: 179, pl. 1, figs. 4, 4a; Surber, 1912: 9, pl. 1, fig. 9; Utterback, 1916a: 394; Cummings *et al.*, 1990: 16–17, figs. 6a–d, 9c) indicate glochidia dimensions for length (0.10–11 mm), height (0.15–0.19 mm), and hinge length (0.05–0.06 mm).

Reproductive biology. Bradytictic. The posterior part of the outer gills is marsupial (Ortmann, 1912a). Gravid females have been found from June to December (Cummings & Mayer, 1993). Fragile, white conglutinates are discharged (Utterback, 1916). Freshwater Drum is the only known host.

Reported potential hosts

Aplodinotus grunniens (Freshwater Drum)—Cummings & Mayer, 1993 (NI); Barnhart, 1996 (LT); Barnhart & Roberts, 1997 (LT); Barnhart & Riusech, 1997 (LT).

Habitat. This species occurs in sand and sandy silt, often in backwaters or slow-moving rivers.

Range. Lower Ohio River system, Mississippi River drainages in Arkansas and Nebraska.

Range in Ohio. Not positively recorded from the state but occurs in adjacent Indiana.

Similar species. *Potamilus capax* does not resemble any other species of *Potamilus* but is very similar to *Lampsilis cardium*. That species is often rayed (not so in *P. capax*), has concentric-looped beak sculpture rather than double-looped, and lacks the distinctly "S"-shaped hinge line of *P. capax*.

Comments. Despite its large size, this is a short-lived species. It is an extremely rapid grower, reaching 90% of its total length in the first 3 years. Even the largest individuals, ca. 130 mm long, are only 4–5 years old.

Pyganodon Fischer & Crosse, 1893

(see p. 254 for generic discussion)

Pyganodon lacustris (Lea, 1857)

Figure 226

Etymology. L. *lacus*, a large body of water, a lake. L. *lacustris*, pertaining to a lake. This species is based upon specimens collected by Dr. James Lewis from Crooked and Little lakes in New York state (Lea, 1857b: 84).
Vernacular name(s). *Lake Floater*, Long Floater, Delicate Floater.
Original description
Anodonta lacustris Lea, 1857b: 84. [Redescribed in Lea, 1860c: 363–364, pl. 62, fig. 188].
Type locality. ". . . Hab. Crooked Lake and Little Lakes, New York. James Lewis, M.D."
Type material. Lectotype USNM 86597.
Synonymy
Anodonta lacustris Lea, 1857b: 84.
Anodon lacustris (Lea, 1857). Sowerby, 1867: pl. 19, fig. 73.
Margaron (*Anodonta*) *lacustris* (Lea, 1857). Lea, 1870: 80.
Pyganodon lacustris (Lea, 1857). Hoeh & Burch, 1989: 263–265.
Status. Not listed by either Ohio Division of Wildlife or U.S. Fish & Wildlife Service as of 2009.

ADULT SHELL
Shell. Medium size (to 100 mm), very thin, elongate, bluntly pointed posteriorly; somewhat compressed to inflated.

Fig. 226. ***Pyganodon lacustris*** **(Lea, 1857). A, Beak sculpture detail:** OSUM 11424, Cedar Lake, Herkimer Co., NY. 95 mm. **B.** OSUM 45526.2, Little Lakes, Herkimer Co., NY. 64 mm. **C.** OSUM 24399.2, Little Lakes, Herkimer Co., NY. 81 mm. **D, F, Hinge detail:** OSUM 43552.2, Black Lake, Jefferson Co., NY. 77 mm. **E.** OSUM 11422, Little Lakes, Herkimer Co., NY. 87 mm.

Anterior margin. Very short, straight to broadly rounded to the ventral border.
Ventral margin. Very broadly rounded.
Posterior margin. Long, rather straight to the dorsal margin.
Dorsal margin. Long, straight, joining posterior margin with very obtuse angle.
Umbo. Narrow, very low, placed far forward; umbos nearly face each other; umbonal sculpture of 7–9 fine ridges, the earliest double-looped but becoming concentric or irregular further from the umbo.
Ligament. Long, poorly developed, tan to dark brown, not concealed by shells.
Disc sculpture. Disc of shell smooth except for low angulations on posterior slope.
Periostracum. Shiny even as adults; base tan or pale straw, usually without rays except on the posterior slope ridges, annular rings dark.
Nacre. Iridescent white, particularly posteriorly, occasionally flushed in the umbonal region with salmon or light purple; often with extensive gold or brown Tulberg layers; commonly the nacre shows signs of disease, parasites, or damage.
Hinge plate. Edentulous.
Muscle scars. Anterior adductor large, "D"-shaped, flush with shell; posterior adductor large, flush; large anterior adductor distinct from anterior pedal protractor scar, small posterior adductor confluent with small posterior pedal retractor scar; dorsal suspensors located under umbo in cluster, line, or semicircle. Pallial line entire, located far from shell margin. Accessory line absent.
Umbonal cavity. Very wide, shallow, open.

Glochidia. Described by Lea (1863: 46) as subtriangular with equal height and length; styliform hook present, very much like *Pyganodon grandis*.

Reproductive biology. The reproductive biology is presumably like that of other *Pyganodon*.

Reported potential hosts. None reported. This species has been confused with other eastern anodontines, and hosts attributed to those species actually may belong here.

Habitat. Lakes and slow-moving streams and rivers.

Range. Hoeh & Burch (1989) gave the range of this species as the St. Lawrence, upper Susquehanna and Hudson, upper Mississippi, and southern Hudson Bay drainages.

Range in Ohio. Specimens having the general form of *P. lacustris* have been found at Deer Creek but are lacking umbonal sculpture. The fact that well-preserved umbonal sculpture has not been found on these Ohio specimens suggests that their habitat was acidic ponds, lakes, or streams. Hoeh & Burch (1989: 272) found both *P. grandis* and *P. lacustris* living in Four Mile Lake, Washtenaw County, Michigan, without intermediate individuals. Four Mile Lake is in the Lake Erie drainage basin and less than 80 km from the Ohio-Michigan state boundary. It would not be surprising to find *P. lacustris* living within Ohio in view of the proximity of its known range and the known presence of populations having its shell form. The striking similarity of this form to some individuals of *P. c. marginata, P. c. cataracta,* and *P. grandis* indicates the possibility of sibling species.

Similar species. See under *Pyganodon cataracta marginata*.

Comments. Individuals grow quickly for 2–4 years before slowing. Individuals grow to ca. 100 mm and 15 years old.

Uniomerus Conrad, 1853

(see p. 317 for generic discussion)

Uniomerus declivis (Say, 1831)

Figure 227

Etymology. L. *declivus*, downhill, sloping. The shell of this species has a long, sloping posterior margin.
Vernacular name(s). *Tapered Pondhorn.*
Original description
Unio declivus Say, 1831c: 527, no figures.
Type locality. "Bayou Teche, Louisiana."
Type material. Johnson (1970) gave ANSP 41698 as three syntypes, none of which was the specimen figured by Say in 1832 (pl. 35).
Synonymy
Unio declivus Say, 1831c: 527.
Margarita (*Unio*) *declivis* (Say, 1831). Lea, 1836: 32.
Unio sayanus Conrad, 1838: 102, pl. 56, fig. 2 [Type locality: "small stream Greene Co., Alabama." Type not found *fide* Johnson & Baker, 1973].
Unio symmetricus Lea, 1845: 164 [Type locality: "Red River, at Alexandria, Louisiana." Holotype USNM 85604].
Margaron (*Unio*) *symmetricus* (Lea, 1845). Lea, 1852b: 32.
Margaron (*Unio*) *declivis* (Say, 1831). Lea, 1852b: 33.
Unio (*Uniomerus*) *tetralasmus* var. *declivis* Say, 1831. Simpson, 1900a: 740.
Uniomerus tetralasmus Say, 1831. Utterback, 1915a: 244, pl. 21, figs. 69a, b [in part, misidentification].
Elliptio (*Uniomerus*) *declivis* (Say, 1831). Frierson, 1927: 34.

Fig. 227. ***Uniomerus declivis*** **(Say, 1831). A.** OSUM 58111.2, Hogan Creek, Dallas Co., AL. 74 mm. **B.** OSUM 44304.2, Little Ohoopee River, Johnson Co., GA. 67 mm. **C.** OSUM 32053, Panther Creek, Butler Co., AL. 79 mm. **D.** OSUM 66334.2, Yellow Creek, Lamar Co., AL. 92 mm. **E, F, Hinge detail:** OSUM 59050.6, Childers Creek, Dallas Co., AL. 60 mm. **Beak sculpture detail:** OSUM 58111.1, Hogan Creek, Dallas Co., AL. 53 mm total length.

Elliptio declivis (Say, 1831). Strecker, 1931: 16.
Uniomerus tetralasmus declivis (Say, 1831). Murray & Roy, 1968: 29.
Uniomerus declivis (Say, 1831). Morrison, 1976: 10.
Status. Not listed by either Ohio Division of Wildlife or U.S. Fish & Wildlife Service as of 2009.

ADULT SHELL
Shell. Medium to fairly large sized (to ca. 120 mm); thin, outline elongated, sharply pointed posteriorly, umbos ca. 30% back from anterior margin.
Anterior margin. Broadly curved.
Ventral margin. Very broadly curved, straight, or slightly concave, forming an acute point at the posterior margin at the base.
Posterior margin. Straight to concave.
Dorsal margin. Very long, straight or slightly convex.
Umbo. Wide, low, barely rising above the hinge line; beak sculpture of 4–6 coarse concentric ribs, oriented obliquely towards the antero-dorsal margin; the anterior portions of the more distal ribs may be lost, leaving only the posterior arms; beak sculpture disappears early in the first year's growth.
Ligament. Tan to brown, very long, largely concealed by the valves.
Disc sculpture. Shell sculptureless except for 2 low, broad ribs on the dorsal slope, divided by shallow grooves; this sculpture may be nearly lost in some specimens.
Periostracum. Yellowish-green when juvenile, becoming black with age; shells rarely with faint green rays. Growth annuli dark brown in juveniles.
Nacre. White to faint purplish, iridescent posteriorly; most specimens have gold Tulberg layers.
Hinge plate. Long, straight or curved, fairly thick; interdentum long but very narrow; cardinal teeth chunky and triangular, two in left valve, one in right with an accessory tooth anterior to cardinal; lateral teeth short, straight or slightly bowed, thick, two in left, one in right.
Muscle scars. Anterior adductor muscle scar somewhat impressed, posterior adductor scar flush with shell surface, anterior adductor separate from anterior pedal protractor; posterior adductor confluent with posterior pedal retractor scar; dorsal suspensor scars a cluster of irregular pits in the umbonal cavity, the posterior-most one being quite large; pallial line entire, accessory line absent.
Umbonal cavity. Wide, open, shallow, nearly nonexistent.

Glochidia. Not reported.

Reproductive biology. In Texas, females were gravid in May and June (Howells *et al.*, 1996).

Reported potential hosts. None reported.

Habitat. Howells *et al.* (1996) reported this species from "lentic waters with clay bottoms; it may be adapted to intermittent streams and withstands desiccation."

Range. Gulf systems from Florida to Texas and Mexico. Morrison (1976: 11) stated that it was found in Lake Erie. This probably is in error, although *U. tetralasmus* occurs there.

Range in Ohio. Not yet recorded from Ohio. We include it because of Morrison's (1976) record for Lake Erie.

Similar species. *Uniomerus tetralasmus* is yellow or tan with a polished shell. *Uniomerus declivus* is much darker, even black, with a coarse periostracum. This species somewhat resembles *Elliptio dilatata* but has different beak sculpture and a lighter purple nacre.

Comments. This species is much slower growing than *U. tetralasmus*. Growth is fastest during the first two years. No sexual dimorphism in the shell is evident. Individuals may approach 120 mm in length and grow to 15 years old.

Venustaconcha Thiele, 1934

Etymology. L. *venusta*, lovely, charming, beautiful, graceful + L. *concha*, shell.
Original description
Venustaconcha Thiele, 1934: 837.
Type species. *Unio venustus* Lea, 1838 [= *Unio ellipsiformis* Conrad, 1836], by original designation.
Synonymy
Lampsilis (*Venusta*) Frierson, 1927 [*non* Boettger, 1877 (Mollusca: Clausiliidae). Type species: *Unio venustus* Lea, 1838 (= *Unio ellipsiformis* Conrad, 1836), by original designation].
Venustaconcha Thiele, 1934: 837 [replacement name for *Venusta* Frierson, 1927].
Geographic range. Upper Mississippi River drainage, including lower Ohio River drainage but not in the Cumberland or Tennessee rivers. Lake Michigan drainage, but not Lake Erie. Gordon (1995) placed *Unio simus* (Lea, 1838) from Tennessee in this genus, but Parmalee & Bogan synonymized *U. simus* with *Villosa iris* (Lea, 1829).

Venustaconcha ellipsiformis ellipsiformis (Conrad, 1836)

Figures 228, 229

Etymology. From L. *ellipsis*, from G. *elleipsis*, a defective circle, ellipse + L. *forma*, shape. Like many freshwater

mussels, this species is elliptical in outline.
Vernacular name(s). *Ellipse.*
Original description
Unio ellipsiformis Conrad, 1836: 60, pl. 34, fig. 2.
Type locality. "Michigan."
Type material. Figured lectotype ANSP 42114a.
Synonymy
Unio ellipsiformis Conrad, 1836: 60, pl. 34, fig. 2.
Margarita (Unio) venustus Lea, 1836: 24 [*nomen nudum*].
Unio venustus Lea, 1838a: 4, pl. 2, fig. 4 [Type locality: "Potosi, Missouri" and "Cincinnati, Ohio," the latter probably in error. Lectotype USNM 84971].
Margarita (Unio) venustus Lea, 1838b: 24 [*nomen nudum*].
Unio spatulatus Lea, 1845: 164 [Type locality: "Rock River, Wisconsin." Lectotype USNM 84982].
Margaron (Unio) venustus Lea, 1870: 43.
Lampsilis spatulatus (Lea, 1845). Baker, 1898b: 106, pl. 10, fig. 5, pl. 13, fig. 2.
Lampsilis (Venusta) ellipsiformis (Conrad, 1836). Frierson, 1927: 81.
Ligumia ellipsiformis (Conrad, 1836). Baker, 1928a: 263–265, pl. 85, figs. 1–4.
Venustaconcha ellipsiformis ellipsiformis (Conrad, 1836). Oesch, 1984: 149–151.
Status. Not listed by either Ohio Division of Wildlife or U.S. Fish & Wildlife Service as of 2009.

ADULT SHELL
Shell. Medium size (to ca. 80 mm), heavy and thick for size, compressed, oval.
Anterior margin. Rounded or broadly pointed.
Ventral margin. Broadly rounded in males; females straight or nearly so for first 66%, then with a very obtuse angulation, then straight or convex to 1–2 angulations approximately 50% of the way between the dorsal and ventral margins.
Posterior margin. Forming an acute angle with the ventral margin, straight to the dorsal margin.
Dorsal margin. Long, straight.
Umbo. Low, wide, pointed; beak sculpture apparently absent on some well-preserved specimens, present on others as 2–4 minute "V"-shaped ridges; beak sculpture quickly disappears before the end of the first year's growth.
Ligament. Brown to black, prominent, occupying nearly entire dorsal margin, barely contained between valves.
Disc sculpture. Sculptureless, occasionally with a poorly developed angulation delimiting the posterior slope.
Periostracum. Base color of yellow, older individuals becoming darker; shells with dark green rays, often arranged in groups of numerous, very fine rays, particularly on the posterior half of the shell, but dorsal slope often rayless; rays continuous across annuli, often wavy; periostracum is easily abraded, and many specimens have a distinctly "rough" appearance.

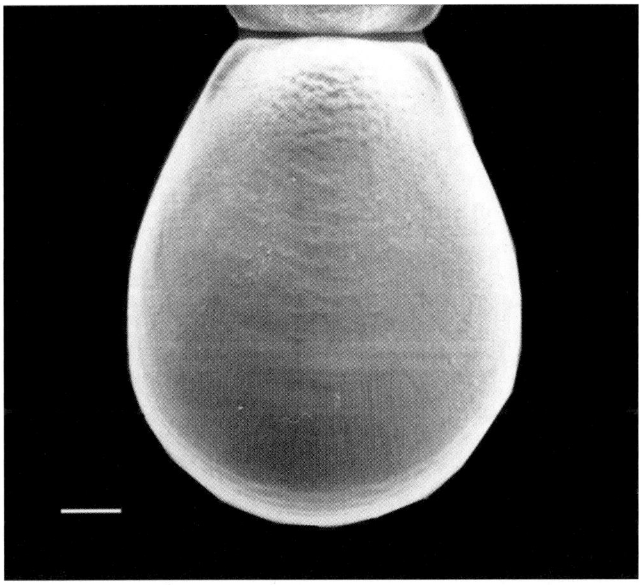

Fig. 228. ***Venustaconcha ellipsiformis ellipsiformis* (Conrad, 1836).** Glochidium. MAH 947.2, Kalamazoo River, Jackson Co., MI. Bar = 35 µm. (Hoggarth, 1999).

Nacre. White, iridescent posteriorly; in the subspecies *ellipsiformis*, the hinge and occasionally more of the shell are flushed with orange-brown; in the subspecies *pleasi*, the hinge is white; green Tulberg layers common in some populations.
Hinge plate. Short, thick, bowed; interdentum short, angled, with a peculiar "V"-shaped indentation caused by one of the dorsal suspensors; cardinals short, triangular, two in left valve, one in right with well-developed accessory tooth anterior to cardinal; laterals long, straight, two in left valve, one in right.
Muscle scars. Anterior adductor deep, rough, well-separated from elongated anterior pedal protractor; posterior adductor slightly indented, smooth, confluent with posterior pedal retractor; dorsal suspensors numerous, arranged in a "Y"-configuration, with the largest scar deeply impressed on the interdentum; pallial line entire, accessory absent or weak.
Umbonal cavity. Wide, open, shallow.

Glochidia. Subelliptical, symmetrical, dorsal margin straight, and anterior and posterior margins equally and gently curved to their maximum inflation at about 70% from dorsal to ventral. Exterior surface of the valve finely malleated near the umbo with fine, loose-looped exterior valve sculpture. Micropoints lanceolate and located on the ventral rim of the valve and on a moderately wide ventral flange. Micropoints arranged in broken vertical rows that extend about half the distance of the flange, leaving a wide unsculptured distal flange margin.

Literature records (Lea, 1857e: 47, pl. 5, fig. 9; Lea, 1863: 60; van der Schalie & van der Schalie, 1963: 7–8,

Fig. 229. ***Venustaconcha ellipsiformis ellipsiformis*** **(Conrad, 1836). A, F, Hinge detail:** OSUM 32733.3, Grand River, Kent Co., MI. 61 mm. **B.** OSUM 32728.2, Buck Creek, Kent Co., MI. 50 mm. **C.** OSUM 35184.4, Kankakee River, Kankakee Co., IL. 82 mm. **D.** OSUM 10243, Fox River, [Co.?], IL. 44 mm. **E, Beak sculpture detail:** OSUM 60524, Kankakee River, Kankakee Co., IL. 44 mm.

pl. 1, fig. center-left; Hoggarth, 1999: 86–88, figs. 53a–g) indicate glochidia dimensions for length (0.22–0.24 mm), height (0.28–0.29 mm), and hinge length (0.10–0.11 mm). Van der Schalie & van der Schalie (163: 7) noted, "The glochidium of *Actinonaias ellipsiformis* . . . compares very favorably in outline and general appearance with that of *Actinonaias carinata*. . . ."

Reproductive biology. Bradytictic. Marsupia are formed from the posterior half of the outer gills (Utterback, 1916). Individuals become sexually mature at three years (van der Schalie & van der Schalie, 1963). Baker (1928a) reported the "breeding season" in July and August. Sculpins and darters may be the main hosts.

Reported potential hosts (including *Venustaconcha ellipsiformis pleasi* Marsh, 1891)**.**

Cottus carolinae (Mottled Sculpin)—Hove & Kurth, 1998 (LT).
Cottus cognatus (Slimy Sculpin)—Hove & Anderson, 1997 (LT).
Culaea inconstans (Brook Stickleback)—Hove & Kurth, 1998 (LT); Hove *et al.*, 1998a (LT).
Etheostoma blennioides (Greenside Darter)—Riusech & Barnhart, 1998 (LT); Riusech & Barnhart, 2000 (LT).
Etheostoma caeruleum (Rainbow Darter)—Riusech & Barnhart, 1998 (LT); Riusech & Barnhart, 2000 (LT).
Etheostoma exile (Iowa Darter)—Hove & Kurth, 1998 (LT).
Etheostoma flabellare (Fantail Darter)—Hove & Kurth, 1998 (LT).
Etheostoma juliae (Yoke Darter)—Riusech & Barnhart, 1998 (LT); Riusech & Barnhart, 2000 (LT).
Etheostoma nigra (Johnny Darter)—Hove & Anderson, 1997 (LT).
Etheostoma spectabile (Orangethroat Darter)—Riusech & Barnhart, 2000 (LT).
Etheostoma whipplei (Redfin Darter)—Riusech & Barnhart, 2000 (LT).
Luxilus cardinalis (Cardinal Shiner)—Riusech & Barnhart, 2000 (LT).
Percina caprodes (Logperch)—Hove & Kurth, 1998 (LT).
Percina maculata (Blackside Darter)—Hillegass & Hove, 1997 (LT); Hove & Kurth, 1998 (LT).

Habitat. This species lives in sand and gravel in small rivers and streams in moving water.

Range. Upper Mississippi River north of the Missouri River. Lake Michigan drainage.

Range in Ohio. Lea (1938) listed his *Unio venustus* from Cincinnati, but this may be an error. This species comes very close to Ohio but apparently has not been recently found there. It occurs in the St. Joseph River system of Lake Michigan only 30 km from the St. Joseph River system of Lake Erie.

Similar species. This species is similar to *Villosa iris* but has a thicker, shorter hinge and double-looped beak sculpture.

Comments. This species grows quickly for the first 2–3 years, then abruptly slows down. It rarely attains 80 mm in length, and few individuals live for over 10 years.

Key to Species in This Book

1a Hinge without teeth, but hinge may be thickened 2
1b Hinge with distinct teeth 12

2a Umbos flush with dorsal margin or nearly so 3
2b Umbos projecting above dorsal margin 6

3a Shell quite circular in profile, nearly as high as long
Anodonta suborbiculata (p. 54)
3b Shell noticeably elongate, much longer than high 4

4a Umbos barely rising above dorsal margin, with green rays, rudimentary teeth 5
4b Umbos flush with dorsal margin, with vague green or yellowish unbroken rays, no teeth
Utterbackia imbecillis (p. 321)

5a Rays bold, broken; umbo more than 25% of the total length from the anterior margin
Hemistena lata (p. 132)
5b Rays distinct but not bold; umbo in anterior 25% of shell
Leptodea leptodon (p. 179)

6a Double-looped beak sculpture, at least initially 7
6b Concentric beak sculpture 10

7a Beak sculpture initially double-looped, becoming a series of zig-zag lines 8
7b Beak sculpture entirely double-looped 9

8a Beak sculpture fine, ribs crowded together and separated by less than their width
Pyganodon lacustris (p. 354)
8b Beak sculpture coarse, ribs separated by more than their width
Pyganodon cataracta marginata (p. 254)

9a Each rib of beak sculpture uniformly strong along its length, not nodulose; umbo in 25% anterior of shell
Pyganodon cataracta marginata (p. 254)
9b Each rib of beak sculpture most developed at bottom (distal) of each loop, often nodulose, umbo in 25% to 33% anterior of shell
Pyganodon grandis s.l. (p. 258)

10a Shell brownish or gray, rayless, with small, indistinct teeth
Simpsonaias ambigua (p. 288)
10b Shell brown, yellow, or greenish, often with rays; hinge may be gently "S"-shaped with no teeth present 11

11a Beak sculpture of fine, concentric loops; shell often elongate and cylindrical; common in extreme headwaters
Anodontoides ferussacianus (p. 59)
11b Beak sculpture of coarse, concentric loops; shell often rather compressed; common throughout most stream reaches
Strophitus undulatus (p. 292)

12a Nacre at least partially purple or pinkish purple in anterior 50% of shell (does not include iridescent reflections; shell material itself must be purple) 13
12b Nacre not purple 29

13a Shell nearly circular in profile, heavily sculptured with pustules; hinge very thick, short
Cyclonaias tuberculata (p. 63)
13b Shell not as above 14

14a Shell with a distinct wing posterior to the umbo, flattened; OR if wing is absent, dark dorsal slope contrasts with yellow shell disc 15
14b Shell without wing, dorsal slope not contrasting with disc of shell 17

15a Exterior of shell yellowish, darker toward the umbos; nacre may be purple toward the beak cavities but white or yellow elsewhere
Leptodea fragilis (p. 175)
15b Shell not as above 16

16a Exterior of shell dark brown, black, or green with distinct rays
Potamilus alatus (p. 241)
16b Exterior of shell greenish, tan, or gray, shiny; may have a wing anterior to umbo
Potamilus ohiensis (p. 245)

17a Nacre of shell with minute pits in center portion of shell 18
17b Nacre not pitted 19

18a Shell very elongate with broad, prominent dorsal ridge giving shell a twisted appearance
Cumberlandia monodonta (p. 27)
18b Shell 2 to 3 times as long as high, dorsal ridge low and scarcely perceptible
Margaritifera margaritifera (p. 338)

19a Shell less than 2 times as long as high, usually not pointed posteriorly 20

19b Shell elongate, 2 to 4 times as long as high, pointed posteriorly 27

20a Shell highest posterior to umbo, hinge teeth small 21
20b Shell highest at umbo or equally high with posterior of umbo, often with a "humped" appearance; hinge teeth large or small 22

21a Shell laterally compressed, periostracum silky and rough
Elliptio complanata (p. 77)
21b Shell not laterally compressed, periostracum with matte finish or smooth
Villosa lienosa (p. 334)

22a Ventral posterior margin with serrated, wide, raised rib
Epioblasma obliquata obliquata (p. 95)
22b Not as above 23

23a Hinge teeth light, often lamellar, hinge long and broadly arched, interdentum absent
Glebula rotundata (p. 344)
23b Hinge teeth heavy, not lamellar, hinge short and angled, interdentum present or absent 24

24a Umbos prominent, shell as high as wide or higher 25
24b Umbos low, shell obviously longer than high 26

25a Shell as high as long, umbos dorsal
Obovaria retusa (p. 204)
25b Shell noticeably longer than high, umbo at anterior margin
Epioblasma obliquata obliquata (p. 95)

26a Interdentum wide, thick; shell large as adult
Elliptio crassidens (p. 80)
26b Interdentum narrow or absent; shell small as adult
Toxolasma lividum (p. 299)

27a Shell with indistinct rays or rayless, periostracum dull
Elliptio dilatata (p. 84)
27b Shell, at least when small, with distinct broad rays; periostracum shiny 28

28a Hinge very weakly developed, laterals absent
Hemistena lata (p. 132)
28b Hinge well-developed, laterals present
Ligumia recta (p. 186)

29a Shell without lateral teeth 30
29b Shell with lateral teeth 37

30a Shell about 1.5 or more times longer than high 31
30b Shell nearly as high as long, quite flattened
Lasmigona complanata (p. 162)

31a Shell sculptureless and rayless, usually brownish, gray, or black 32
31b Shell colored with rays and/or possessing corrugations on posterior slope 34

32a Nacre with minute pits in central portion of shell 33
32b Nacre without pits
Simpsonaias ambigua (p. 288)

33a Shell very elongate with broad, prominent dorsal ridge giving shell a twisted appearance
Cumberlandia monodonta (p. 27)
33b Shell 2 to 3 times as long as high, dorsal ridge low and scarcely perceptible
Margaritifera margaritifera (p. 338)

34a Shell less than twice as long as high, chunky, no corrugations on posterior slope
Alasmidonta viridis (p. 45)
34b Shell about 1.5 to 2 times as long as high, corrugations present 35

35a Shell with coarse ribs and corrugations over disc of shell, beak sculpture of 2 rows of heavy knobs
Arcidens confragosus (p. 341)
35b Not as above 36

36a Corrugations fine, beak sculpture of heavy ribs
Alasmidonta marginata (p. 41)
36b Corrugations heavy, beak sculpture of fine ribs
Lasmigona costata (p. 170)

37a Disc of shell sculptureless or not sculptured as above 38
37b Disc of shell predominately sculptured with distinct pustules rather than ribs; largest pustules not arranged in a single row 39

38a Shell with coarse ribs and corrugations over disc of shell, beak sculpture of 2 rows of heavy knobs
Arcidens confragosus (p. 341)
38b Not as above 48

39a Shell nearly round in profile; sulcus absent or narrow and indistinct, sometimes present only on small specimens 40
39b Shell not as above; sulcus present or replaced by a massive, nodulose rib 44

40a Shell with broad, green ray or blotch radiating from umbo, which may disappear on mature shell
Quadrula pustulosa pustulosa (p. 280)
40b Shell without ray or with numerous very fine lines 41

41a Disc of shell sculptured with numerous pustules scattered over shell 42
41b Disc of shell sculptured with a few pustules arranged in two indistinct, radiating rows
Quadrula nodulata (p. 276)

42a Beak cavity deep, compressed; shell with or without vague rays 43
42b Beak cavity shallow, open; shell with distinct, dotted, numerous rays
Cyprogenia stegaria (p. 68)

43a Posterior slope flattened into a wing with narrow ribs on posterior slope radiating out to shell margin
Cyclonaias tuberculata (p. 63)

43b Posterior slope without wing or with wing very poorly developed; ribs on posterior slope arranged parallel to posterior ridge
Plethobasus cooperianus (p. 215)

44a Shell elongate, at least twice as long as high 45
44b Shell not elongate 46

45a Shell cylindrical, about as wide as high
Quadrula cylindrica cylindrica (p. 266)
45b Shell compressed, obviously higher than wide
Tritogonia verrucosa (p. 306)

46a Shell with a massive, nodulose rib radiating from umbo to posterior ventral margin, patterned with tent-like markings
Quadrula metanevra (p. 273)
46b Shell with a sulcus, not patterned as above 47

47a Posterior wing with narrow, well-defined granulose ribs
Quadrula quadrula (p. 284)
47b Posterior wing with coarse, irregular ribs and large pustules
Quadrula fragosa (p. 269)

48a Sculpture predominately of a single row of knobs radiating from the umbo to the ventral margin 49
48b Sculpture of broad, oblique ribs; pustules may also be present; OR shell is sculptureless 51

49a Knobs large, alternating in position from left to right valves, usually with a corrugated posterior ridge, usually rayed
Obliquaria reflexa (p. 196)
49b Knobs low, not obviously alternating in position; posterior ridge not corrugated; usually rayless or weakly rayed 50

50a Umbos at or before anterior margin, facing forward
Plethobasus cicatricosus (p. 211)
50b Umbos 25% to 33% of the total length behind the anterior margin; umbos face each other
Plethobasus cyphyus (p. 218)

51a Shell sculptured with broad, oblique ribs; pustules may also be present 52
51b Shell sculptureless except for growth lines or low rib bounding the sulcus 53

52a Pustules absent anterior of umbo, hinge massive
Amblema plicata plicata (p. 49)
52b Pustules occur anterior of umbo, hinge not massive
Megalonaias nervosa (p. 191)

53a Shell oval or nearly circular, umbo central, without a posterior or anterior wing, patternless or with a broad green ray on the umbo 54
53b Shell not as above 56

54a Shell with a broad, green ray or blotch on umbo
Quadrula pustulosa pustulosa (p. 280)
54b Shell without green ray 55

55a Hinge heavy; interdentum present
Obovaria subrotunda (p. 207)
55b Hinge light; interdentum absent
Glebula rotundata (p. 344)

56a Shell quite flattened, rounded, plate-like, winged, with heavy "V"-shaped beak sculpture; hinge heavy
Lasmigona complanata (p. 162)
56b Shell not as above 57

57a Shell elongate, about 2 to 4 times longer than high 58
57b Shell shorter 70

58a Shell yellowish or tan with distinct green rays 59
58b Shell not as above 63

59a Shell thinner, hinge delicate, interdentum narrow or absent 60
59b Shell thick, hinge thick, interdentum wide 62

60a Teeth rudimentary, no beak sculpture
Hemistena lata (p. 132)
60b Teeth small but well-developed, wavy or "V"-shaped beak sculpture 61

61a Shell highest at umbo, posterior point in ventral half of posterior margin, beak sculpture wavy
Villosa iris (p. 330)
61b Shell highest behind umbo, posterior point in dorsal half of posterior margin, beak sculpture "V"-shaped
Ligumia subrostrata (p. 349)

62a Rays interrupted, often checkerboard patterned
Ptychobranchus fasciolaris (p. 249)
62b Rays continuous, wavy, narrow
Venustaconcha ellipsiformis ellipsiformis (p. 358)

63a Shell inflated, pointed posteriorly, with coarse, concentric beak sculpture that opens forward
Uniomerus tetralasmus (p. 317)
63b Shell not as above 64

64a Shell dark and/or rayed 65
64b Shell waxy yellowish, rayless
Lampsilis teres (p. 157)

65a Beak sculpture of heavy, trapezoidal ribs, indistinctly rayed if at all, laterals heavy, hinge short
Elliptio dilatata (p. 84)
65b Shell not as above 66

66a Nacre with minute pits in central portion of shell 67
66b Nacre without pits 68

67a Shell very elongate with broad, prominent dorsal ridge giving shell a twisted appearance
Cumberlandia monodonta (p. 27)
67b Shell 2 to 3 times as long as high, dorsal ridge low and scarcely perceptible
Margaritifera margaritifera (p. 338)

68a Shell flattened, thin, drawn to pronounced point posteriorly that ends at or above the halfway point between the ventral and dorsal margins; OR posterior point twisted to end below halfway point 69
68b Shell may be cylindrical, thick, posterior more evenly rounded and ending about halfway between ventral and dorsal margins
Ligumia recta (p. 186)

69a Shell length less than 2 times the height; posterior point forms an angle greater than 80°
Ligumia subrostrata (p. 349)
69b Shell length more than 2 times the height; posterior point forms an angle less than 80°
Ligumia nasuta (p. 183)

70a Shell wedge-shaped, triangular, umbo near anterior margin, with wide, green, broken rays 71
70b Shell not as above 72

71a Shell as high as or higher than long, highly polished and waxy
Pleurobema bournianum (see *Pleurobema clava*, p. 222)
71b Shell longer than high, rarely waxy and polished
Pleurobema clava (p. 222)

72a Umbos either face each other OR are low and turned forward 73
72b Umbos prominent and turned forward 84

73a Shell obviously sulcate 74
73b Shell not sulcate 82

74a Sulcus bounded posteriorly by low rib or swelling that protrudes past the margin of the shell, causing the margin anterior to this to be concave 75
74b Not as above 77

75a Shell oval or round in profile, concave margin slight; ribs bounding sulcus barely protrude past margin
Epioblasma personata (p. 102)
75b Shell not oval or round, concave margin deep; ribs bounding sulcus extend well beyond the margin to form digitations 76

76a Rib anterior to sulcus colored dark green distally
Epioblasma lewisi (p. 92)
76b Rib anterior to sulcus with fine rays but not darkly colored
Epioblasma flexuosa (p. 89)

77a Shell yellowish, elongate; may have broad, green, broken rays
Ptychobranchus fasciolaris (p. 249)
77b Shell not as above 78

78a Shell drawn to a blunt point posteriorly, the point about halfway between dorsal and ventral margins 79
78b Shell rounded posteriorly; OR if drawn to a point, point is near ventral margin
Fusconaia flava (p. 124)

79a Sulcus bordered anteriorly by low swellings or prominent knobs delimited by growth lines; sulcus causes a shallow indentation in the shell outline as early as the third year 80
79b Sulcus smoothly joins disc of shell, not causing an indentation in the shell outline except in older individuals; outline smoothly rounded
Epioblasma obliquata perobliqua (p. 99)

80a Region anterior of sulcus with low swellings
Epioblasma torulosa rangiana (p. 108)
80b Region anterior of sulcus with prominent knobs 81

81a Knobs small, numerous, crowded
Epioblasma philipsii (p. 105)
81b Knobs large, few, distantly spaced
Epioblasma torulosa torulosa (p. 112)

82a Posterior ridge sharp, concave or straight, shell about as high as long, drawn to a distinct point posteriorly about halfway between the dorsal and ventral margins; umbos facing each other; beak cavities shallow, open
Truncilla truncata (p. 313)
82b Shell not as above 83

83a Posterior ridge prominent, straight or concave; shell noticeably longer than high, inflated; periostracum satiny if present; posterior slope with numerous fine ridges, chevron markings
Epioblasma triquetra (p. 116)
83b Shell not as above 97

84a Beak cavity deep, compressed 85
84b Beak cavity shallow, open 93

85a Sulcus absent 86
85b Sulcus present 89

86a Hinge not massive, cardinal teeth aligned vertically 87
86b Hinge massive, cardinal teeth aligned horizontally 88

87a Umbos prominent, near anterior margin
Fusconaia subrotunda subrotunda (p. 128)
87b Umbos low, more centrally located
Fusconaia flava (p. 124)

88a Juvenile shell rayed
Fusconaia subrotunda subrotunda (p. 128)
88b Juvenile shell unrayed except for a yellow ray on the dorsal slope
Fusconaia ebenus (p. 120)

89a Cardinal teeth aligned horizontally
Fusconaia ebenus (p. 120)
89b Cardinal teeth aligned vertically 90

90a Sulcus weak or absent, umbos at or behind anterior margin 91
90b Sulcus well-developed, umbos at or before margin 92

91a Shaped like isosceles triangle in profile, umbos prominent
Pleurobema plenum (p. 230)
91b Not shaped like isosceles triangle, umbos wide and low
Pleurobema sintoxia (p. 236)

92a Shaped like scalene triangle; umbos before anterior margin; nacre rose or white
Pleurobema rubrum (p. 233)
92b Not shaped like a scalene triangle; umbos slightly before anterior margin; nacre white
Pleurobema cordatum (p. 226)

93a Cardinal teeth aligned horizontally
Obovaria olivaria (p. 200)
93b Cardinal teeth aligned vertically 94

94a Sulcus distinct, periostracum satiny
Pleurobema cordatum (p. 226)
94b Sulcus absent; OR if present, periostracum not satiny 95

95a Umbos prominent, shell pointed posteriorly
Pleurobema sintoxia (p. 236)
95b Shell not as above 96

96a Shell quadrate; sulcus is usually a flattened area; nacre white or salmon
Fusconaia flava (p. 124)
96b Shell rounded; sulcus is usually absent; nacre white, pink, or salmon
Pleurobema sintoxia (p. 236)

97a Posterior of shell with a distinct outward swelling beyond margin of shell 98
97b Posterior of shell without swelling, although posterior of entire shell may be noticeably inflated 103

98a Posterior swelling expansive, not ridged 99
98b Posterior swelling not greatly expanded beyond the margin, swelling radially ridged 101

99a Region anterior of sulcus with low swellings
Epioblasma torulosa rangiana (p. 108)
99b Region anterior of sulcus with prominent knobs 100

100a Knobs small, numerous, crowded
Epioblasma philipsii (p. 105)
100b Knobs large, few, distantly spaced
Epioblasma torulosa torulosa (p. 112)

101a Shell nearly round, posterior swelling low and not well-developed
Epioblasma personata (p. 102)
101b Shell oval or elongate, posterior swelling well-developed 102

102a Shell oval, not obviously inflated; beaks not prominent
Epioblasma obliquata perobliqua (p. 99)
102b Shell more elongate, very inflated; beaks prominent
Epioblasma triquetra (p. 116)

103a Hinge on left valve with a tongue-like tooth under the beak in the interdentum posterior to the cardinal teeth and a small, wing-like extension on the posterior dorsum
Lasmigona compressa (p. 166)
103b Shell not as above 104

104a Shell thin, inflated, pointed posteriorly, with sharp concentric beak sculpture, the loops facing forward
Uniomerus tetralasmus (p. 317)
104b Shell not as above 105

105a Shell yellowish, compressed, thick; beaks low; hinge massive, not oval or round in outline 106
105b Shell not as above 107

106a Shell elongate, with interrupted green rays
Ptychobranchus fasciolaris (p. 249)
106b Shell almost as high as long, with narrow rays containing darker, chevron markings
Ellipsaria lineolata (p. 73)

107a Hinge delicate, with small, chunky teeth; nacre may be iridescent; dorsal ridge smoothly rounded 108
107b Hinge not delicate, teeth more massive; nacre white or flushed with pink, not iridescent or iridescent only posteriorly; dorsal ridge may be sharp 112

108a Shell compressed and thin, yellowish with a dark dorsal slope, rayless or with vague green rays
Leptodea fragilis (p. 175)
108b Shell not as above 109

109a Shell small (about 1 in.), without rays, quite cylindrical
Toxolasma parvum (p. 302)
109b Shell up to 2 in., with rays, compressed to cylindrical 110

110a Cardinal teeth chunky, triangular, beak sculpture wavy
Villosa iris (p. 330)
110b Cardinal teeth lamellar or rudimentary, beak sculpture double-looped or lacking 111

111a Shell elongate, no beak sculpture
Leptodea leptodon (p. 179)
111b Shell oval, beak sculpture double-looped
Lasmigona subviridis (p. 347)

112a Shell oval, heavy, yellow or greenish often with broad rays; hinge massive; umbo very low 113
112b Shell not as above 115

113a Shell abruptly truncated posteriorly
Lampsilis abrupta (p. 136)
113b Shell gently rounded posteriorly 114

114a Shell yellowish, rayless or with vague green rays; umbos in anterior 25% of shell; umbos wide and massive
Lampsilis abrupta (p. 129)
114b Shell yellowish, greenish, or brown, usually with well-developed green rays; umbos located 25% or more back from anterior margin, umbos wide but low
Actinonaias ligamentina s.l. (p. 34)

115a Shell small (less than 2 in.), somewhat elongate, pointed posteriorly 116
115b Shell may be larger, oval or cylindrical, usually not pointed posteriorly 117

116a Shell very thick for its size, with greenish rays; hinge short and thick
Villosa fabalis (p. 326)
116b Shell not thick, hinge not as above, shell with rayed or checkerboard pattern of various colors
Truncilla donaciformis (p. 310)

117a Shell noticeably highest at umbo 118
117b Shell highest posterior to umbo; OR umbo flush with posterior margin 121

118a Shell greatly inflated, hinge line gently "S"-shaped
Potamilus capax (p. 352)
118b Shell inflated or not; hinge line straight, curved, or angled 119

119a Shell oval, chunky, very thick; hinge massive; umbos large
Ellipsaria lineolata (p. 73)
119b Shell not as above 120

120a Dorsal ridge concave, sharp
Lampsillis ovata (p. 149)
120b Dorsal ridge straight or convex, rounded
Lampsilis cardium (p. 140)

121a Shell oval, yellowish, with numerous fine, wavy, dark green to brown rays
Lampsilis fasciola (p. 145)
121b Shell not as above 122

122a Shell pointed posteriorly, inflated, thick, with interrupted rays
Epioblasma triquetra (p. 116)
122b Shell not as above 123

123a Shell blackish, not obviously rayed
Villosa lienosa (p. 334)
123b Shell yellowish or greenish, usually rayed 124

124a Shell small (about 1 in.), thick, greenish with darker rays; hinge short, thick
Villosa fabalis (p. 326)
124b Shell may be larger, not noticeably thick, yellowish; may have widely spaced green or brown rays; hinge not as above
Lampsilis radiata luteola (p. 153)

Glossary

acuminate. Sharply pointed; terminating acutely.
acute. Sharp, pointed.
adductor muscle (scar). Muscles connecting the two shells anteriorly and posteriorly, used to close the shells. The muscles leave characteristic scars on the shell where they attach.
alate. Having a wing-like projection of the two valves of the shell, extending dorsally beyond the hinge line. See **wing**.
allotype. Type material of a species of the opposite sex to the holotype. See **cotype, holotype, lectotype, neotype, paralectotype, paratype, syntype, topotype.**
amphidetic. Having the ligament present on both sides of the umbo.
analogous character. One of two or more characters that are functionally similar but structurally and developmentally different.
annulus (pl. **annuli**). Major growth lines on the outside of the shell, often considered annular in nature.
anterior. The front part of something.
anterior adductor muscle (scar). See **adductor muscle.**
anterior protractor muscle (scar). A muscle used to move the foot; it has a characteristic scar.
apomorphic character. In a pair of homologous characters, the inherited change of a previously existing character into a new character. See **pleisiomorphic character.**
apposed. Set opposite to one another.
arcuate. Curved like a bow, arched.
beak. See **umbo.**
beak cavity. See **umbonal cavity.**
bivalvia. A class of the phylum Mollusca characterized by an external, two-piece, calcareous shell, a muscular foot, and the lack of a head or radula.
bradytictic. Reproduction in which the glochidia are overwintered in the marsupial demibranch of the female unionid. Also called winter breeders.
branchiae. Gills; those organs of an aquatic organism that maximize the area of contact of living permeable tissue with the environment; comparable to the lungs of terrestrial forms; the site of exchange of those fluids used in metabolism with those produced by metabolism.
branchial aperture. An opening, formed by the mantle (pallial) margins, through which water passes into the branchial cavity. Same as **incurrent aperture.**
byssus (byssal thread). A filament or very few filaments of material by means of which the juvenile of some individuals of some species are attached to other, usually larger or heavier, objects; as a result, the juvenile remain tethered to a site on or in the stream substrate.
cardinal teeth. The anterior-most hinge teeth (when present). Often short and triangular. Also known as pseudocardinal teeth in the Unionacea.
claustra. Elongate ridges of the mesial-posterior hinge plate of anodontines which are neither parallel to nor homologous with the lateral or lamellar teeth.

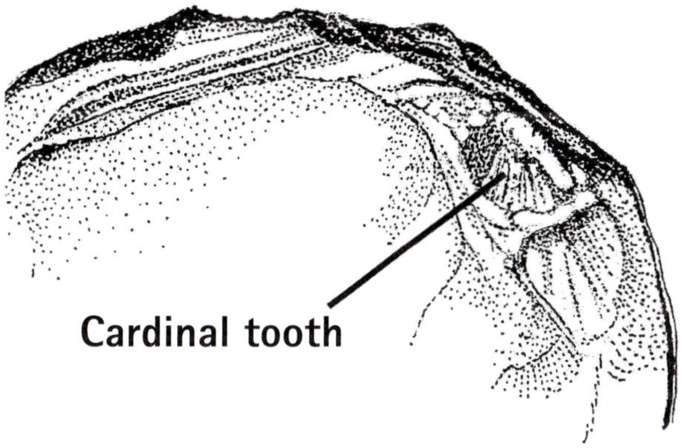

clavate. Club-shaped.
compressed. Flattened from side to side; the opposite of **inflated.**
conglutinate. The packages of glochidia formed by the female; often resemble host prey items. See Watters (2008).
corrugated. Sculptured with undulating ribs or folds.
cotype. A specimen identified by the author of a species as an example of that species. May be part of the type series. See **allotype, holotype, lectotype, neotype, paralectotype, paratype, syntype, topotype.**
depressed. Low, flattened.
diaphragm. In unionoid anatomy, a horizontal, transverse, perforate dividing wall of tissue between the branchial and suprabranchial cavities, frequently attached to the body mass mesially and to the mantle lobes laterally; consisting of the basal proximal parts of all four demibranches.
dimyarian. Having two adductor muscles.
disc. The central portion of the shell.
divaricate. Spreading out widely, "V"-shaped, forked.
dorsal. Pertaining to, or situated on or near, the back.
dorsal suspensor muscles (scars). Minute muscles arranged in bundles beneath the umbo.
ecophenotype. A form of a species caused by environmental rather than genetic factors.
edentulous. Without teeth.
elliptical. Having the shape of an ellipse.
emarginate. Having a margin or an edge something less than a straight line; slightly concave.
equivalve. Having valves equal or very nearly so in size.
excurrent aperture. The opening through which water is expelled from a bivalve mollusc. Same as **suprabranchial aperture.**
extinct. A species that has been globally lost.
extirpated. A species that has been lost from a particular area (such as a state) but still exists elsewhere.
gills. See **branchiae.**

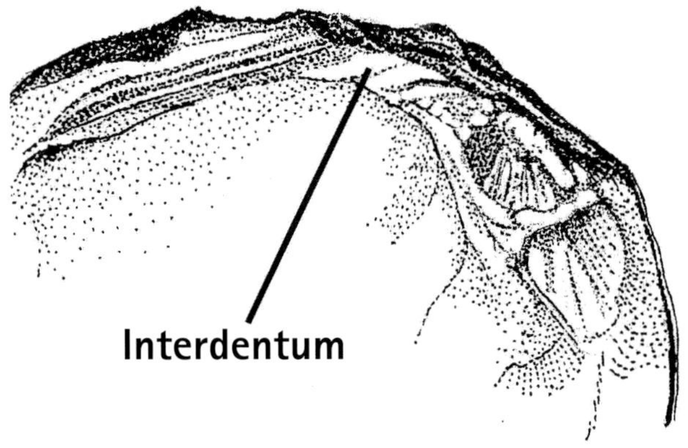

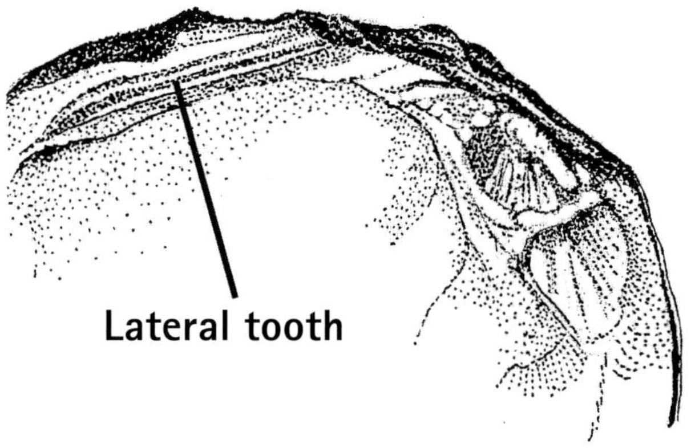

glochidium (pl. **glochidia**). The parasitic larva of freshwater mussels.

growth lines. See **annulus.**

hinge plate. The dorsal portion of the shell that may have interdigitating teeth; may be absent in some species.

holotype. The original specimen or illustration upon which the author of a species bases his description. See **allotype, cotype, lectotype, neotype, paralectotype, paratype, syntype, topotype.**

homologous character. One of a pair of characters in which one is derived directly from the other. Three or more characters are homologues if each gave rise to the next in linear sequence. Contrast with **analogous character.**

homonym. Taxonomic names that are spelled the same but were coined by different authors or at different times. The oldest name is the senior homonym and the valid name. Other names are junior homonyms. For example, *Unio childreni* Hanley, 1843 is a junior homonym of *Unio childreni* Gray, 1834. See **synonym.**

ICZN. International Code of Zoological Nomenclature, a collection of "rules" drafted by the International Commission of Zoological Nomenclature (1999), a paralegislative body that, in addition to writing the Code, rules on special cases.

impoundment. A portion of a river or stream that has been pooled by the presence of a dam.

incurrent aperture. The opening through which water is drawn into a bivalve mollusc by ciliary action. Same as **branchial aperture.**

inflated. Swollen from side to side; the opposite of **compressed.**

interdentum. The portion of the hinge plate positioned between the cardinal and lateral teeth.

iridescent. Rainbow-like colors produced from crystalline interactions in the shell.

lamellate. In reference to hinge teeth, blade-like.

lateral teeth. The posterior-most hinge teeth (when present). Often long and bladelike. Also known as pseudolateral teeth.

lectotype. One of several syntypes, designated by an author after the original publication of a species name as the type specimen; designated only where there was no original holotype. See **allotype, cotype, holotype, neotype, paralectotype, paratype, syntype, topotype.**

ligament. The non-living connection between the shells on the dorsum; acts as a spring to open the shells.

marsupium. The portion of the female gills or brachiae that are used to contain the glochidia until release.

microfossae. Minute muscle scars within the pallial line; lateral attachments of the mantle to the nacre.

mollusca. A phylum of soft-bodied animals typically producing a shell of calcium carbonate and/or conchiolin by means of a mantle and moving by means of a muscular foot.

monotypy. A species that became the type of a genus if it was the only species an author originally listed for the genus. Later additions by other workers, or the original author himself, are invalid. See **original designation, subsequent designation, subsequent monotypy, tautonomy.**

nacre. The prismatic inner lining of the shell; also called mother-of-pearl.

neotype. The designation made if the type material is lost, destroyed, or deemed inadequate for identification. See **allotype, cotype, holotype, lectotype, paralectotype, paratype, syntype, topotype.**

nomen dubium. A taxonomic name that cannot be confidently matched with a species because of deficiencies in the original description and/or illustration. Although it may act as a synonym or a homonym, it does not become the valid name for any taxon.

nomen nudum. A taxonomic name that was not validly introduced according to the ICZN rules. It has no taxonomic value and cannot be a synonym or a homonym.

nomen oblitum. A "forgotten" taxonomic name. If such a name is discovered to be a senior synonym or homonym, but it has not been used in the literature for the past 50 years, it may be disregarded under the ICZN Law of Priority.

oblique. Slanting or inclined; neither perpendicular nor horizontal.

obtuse. Angles greater than 90° and less than 180°.

orbicular. Having the form of an orb; round, spherical.

original designation. A designation made if a genus was described and the original author specified a species to be the type of the genus. See **monotypy, subsequent designation, subsequent monotypy, tautonomy.**

ovate. Having the shape of a chicken egg or of a longitudinal section of one; ellipsoidal but not elliptical (which see).

pallial line. The linear muscle scar that marks where the mantle attaches to the shell.

pallial retractor. A muscle that withdraws the distal mantle margin within the shell.

paralectotype. A designation made if a species is described from a series of specimens or illustrations and the original author

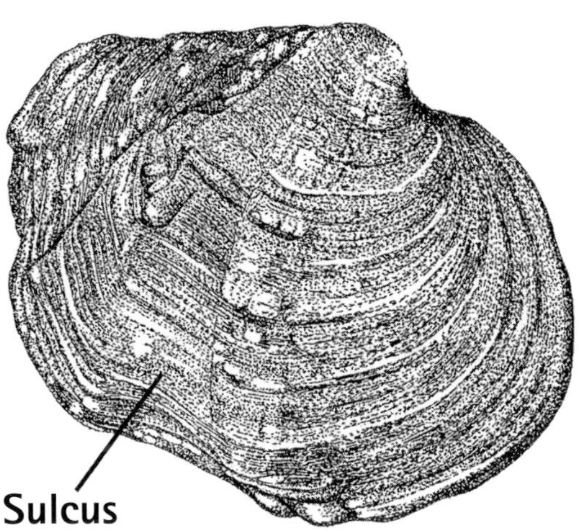

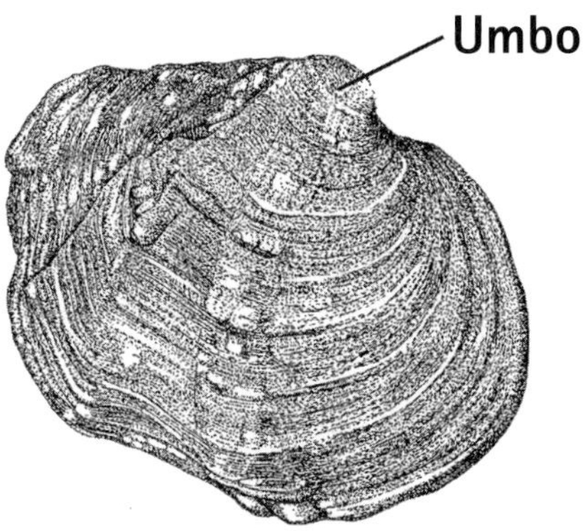

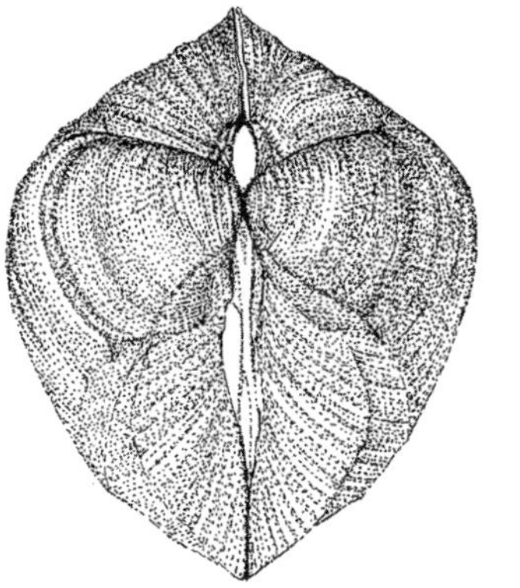

Facing each other

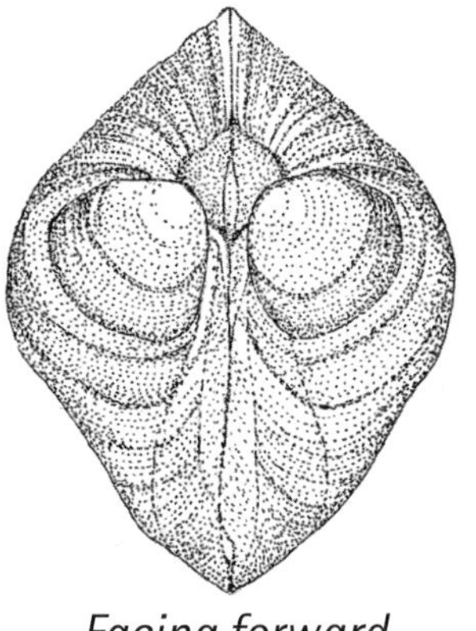

Facing forward

of the species did not choose one to represent the holotype; a subsequent worker chooses one as the lectotype. The remaining specimens from the type series are paralectotypes. See **allotype, cotype, holotype, lectotype, neotype, paratype, syntype, topotype.**

paratype. A designation made from remaining specimens of a type series if a species is described from a series of specimens or illustrations and the original author of the species chose one to represent the holotype. See **allotype, cotype, holotype, lectolectotype, neotype, syntype, topotype.**

patronym. A name formed by the addition of a prefix or suffix indicating relationship to the name of one's father or paternal ancestor; a name coined in honor of a particular person.

pedal levitor. A muscle that functions in the lifting of the foot. The proximal terminus is in or near the umbonal cavity of the unionoid.

pedal retractor. A muscle that withdraws the foot within the shell.

periostracum. The non-living outer layer of the shell; may be eroded away in old individuals or completely lost in weathered or sub-fossil dead shells.

plesiomorphic character. The original, preexisting character in a pair of homologous characters. Contrast with **apomorphic character.**

plications. Parallel ridges or folds on the surface of the shell.

posterior. The back end of something.

posterior adductor muscle (scar). See **adductor muscle.**

posterior retractor muscle (scar). A muscle used to move the foot; it has a characteristic scar.

posterior ridge. A ridge extending from the umbo posteriorly across the dorsum to the posterior margin; may be prominent or barely perceptible.

proximal. Nearest to the point of origin or attachment.

pustule. A bead-like sculpture on the shell.

pyriform. Shaped like a pear.

quadrate. Square or approximately square.

rotund. Round or rounded out, spherical.

subsequent designation. A designation made if a genus was described with more than one species of that genus, and the original author did not specify which one was to be the type of the genus; the genus remains without a type until someone subsequently designates one. See **monotypy, original designation, subsequent monotypy, tautonomy.**

subsequent monotypy. A designation made if a genus was described with more than one species, but all but one are unidentifiable (*nomen dubia*); the remaining species becomes the type of the genus. See **monotypy, original designation, subsequent designation, tautonomy.**

sulcus. On the shell, a depression or groove extending from the umbo to the ventral margin.

suprabranchial aperture. Same as excurrent aperture (which see).

symplesiomorphic character. A homologous character shared by a group of taxa and having originated in an ancestor earlier than the common ancestor of the group.

synonym. Taxonomic names that apply to the same biological

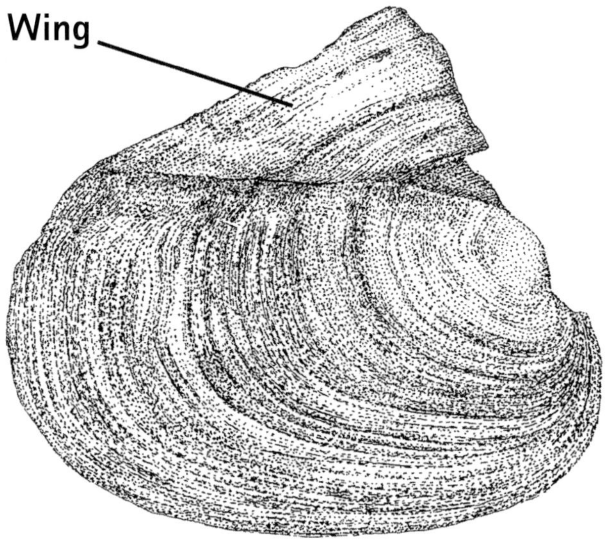

entity. The oldest name is the senior synonym and usually is the valid name for the taxon. Other names are junior synonyms. For example, *Unio foliatus* Hildreth, 1828 is a junior synonym of *Obliquaria flexuosa* Rafinesque, 1820 because both names represent the same species. See **homonym.**

syntype. A specimen from the type series from which no holotype or lectotype has been chosen. See **allotype, cotype, holotype, paralectotype, lectotype, neotype, topotype.**

tachytictic. Reproduction in which the glochidia are not overwintered in the marsupial demibranch of the female unionoid. Fertilization, larval development, and parasitic period all occur within the same calendar year. Also called summer breeders.

tautonomy (tautonymy). A name applied if a list of species accompanies the original description of a genus, and one of those species names is an obvious variant of the genus name; that species becomes the type of the genus. For example, *Margaritifera margaritifera* and *Villosa villosa* are tautonomous names. See **monotypy, original designation, subsequent designation, subsequent monotypy.**

tooth. Interdigitating portions of the hinge of the two shells. Usually of two types: cardinal and lateral.

topotype. A specimen collected from the original type locality. See **allotype, cotype, holotype, paralectotype, lectotype, neotype, syntype.**

triangulate. Having the shape of a triangle.

Tulberg layers. Blemishes in the nacre, often brown, gold, or gray.

type locality. The source of a species as stated by the original author. If no type locality was given, one may be designated by a subsequent worker. Type localities also may be amplified or corrected.

umbo. A projecting point or knob on a surface; specifically, the swollen dorsal region of a shell valve; plural **umbos.** Also referred to as beaks. Umbos may face each other or face forward.

umbonal cavity. The depression on the dorsal aspect of the inside of the shell that is the inner aspect of the umbo; may be deep or shallow.

valve. A single piece, actually half, of a bivalve shell.

ventral. Pertaining to, or situated on or near, the underside.

vernacular name. A name in the common native speech, as distinguished from scientific nomenclature.

wing. A projection of the valves on the dorsum forming a triangular process; may be either anterior or posterior to the umbo, or both.

References

Adams, C.B. 1842. Fresh water and land shells of Vermont. Pp. 151–169. [*In:*] Z. Thompson, *History of Vermont,* Part 1. Privately printed. 169 pp.

Adams, H. & A. Adams. 1853–1858. *The genera of recent Mollusca; arranged according to their organization.* London: John van Vorst, Paternoster Row. Vol. 1: 1–256, pls. 1–32 [1853], 257–484, pls. 33–60 [1854]; Vol. 2: 1–92, pls. 61–72 [1854], 93–284, pls. 73–96 [1855], 285–381, pls. 97–112 [1856], 413–540, pls. 113–128 [1857], 541–661 + i–xl, pls. 129–138 [1858].

Agassiz, L. 1846. *Nomenclatoris Zoologici.* Soloduri: Jent & Gassmann. i–viii. 393 pp.

Agassiz, L. 1852. Über die Gattungen unter den nord amerikanischen Najaden. *Archiv für Naturgeschichte* 18: 41–50.

Allen, E. 1924. The existence of a short reproductive cycle in *Anodonta imbecillis. Biological Bulletin* 46: 88–94.

Allen, W.R. 1914. The food and feeding habits of freshwater mussels. *Biological Bulletin* 27: 127–147.

Allen, W.R. 1921. Studies of the biology of freshwater mussels. *Biological Bulletin* 40: 210–241.

Amyot, J.P. & J.A. Downing. 1998. Locomotion in *Elliptio complanata* (Mollusca: Unionidae): A reproductive function? *Freshwater Biology* 39: 351–358.

Anders, K. & V. Wiese. 1993. Glochidia of the freshwater mussel, *Anodonta anatina,* affecting the anadromous European smelt (*Osmerus eperlanus*) from the Eider estuary, Germany. *Journal of Fish Biology* 42: 411–419.

Anderson, K.B., R.E. Sparks, & A.A. Paparo. 1978. Rapid assessment of water quality using the fingernail clam, *Musculium transversum. University of Illinois, Water Resources Center, Urbana, UILU WRC 78 0133, Research Report* (133). 115 pp.

Anthony, J.G. 1865. Descriptions of new species of North American Unionidae. *American Journal of Conchology* 1: 155–164, 5 pls.

Arey, L.B. 1921. An experimental study on glochidia and the factors underlying encystment. *Journal of Experimental Zoölogy* 33: 463–499.

Arey, L.B. 1924a. Observations on an acquired immunity to a metazoan parasite. *Journal of Experimental Zoölogy* 38: 377–381.

Arey, L.B. 1924b. Glochidial cuticulae, teeth, and the mechanics of attachment. *Journal of Morphology and Physiology* 39: 323–335.

Arey, L.B. 1932a. The formation and structure of the glochidial cyst. *Biological Bulletin* 62: 212–221.

Arey, L.B. 1932b. The nutrition of glochidia during metamorphosis. *Journal of Morphology* 53: 201–221.

Arey, L.B. 1932c. A microscopical study of glochidial immunity. *Journal of Morphology* 53: 367–379.

Baird, M.S. 2000. Life history of the Spectaclecase, *Cumberlandia monodonta* (Say 1829) (Bivalvia, Unionoidea, Margaritiferidae). Master's Thesis, Southwest Missouri State University, Springfield.

Frank Collins Baker
(Courtesy of Illinois Natural History Survey)

Baker, F.C. 1898a. The molluscan fauna of western New York. *Transactions of the Academy of Science of St. Louis* 8(5): 71–94, 1 pl.

Baker, F.C. 1898b. The Mollusca of the Chicago area, Part I: The Pelecypoda. *Bulletin of the Chicago Academy of Science* 3(1): 1–130, 27 pls.

Baker, F.C. 1905. The molluscan fauna of McGregor, Iowa. *Transactions of the Academy of Science of St. Louis* 15(3): 249–258.

Baker, F.C. 1906. A catalogue of the Mollusca of Illinois. *Bulletin of the Illinois State Laboratory of Natural History* 7(6): 53–136, 1 map.

Baker, F.C. 1916. The relation of mollusks to fish in Oneida Lake. *Technical Publication No. 4. New York State College of Forestry and Syracuse University* 16(21): 1–366.

Baker, F.C. 1918. The relation of shellfish to fish in Oneida Lake, New York. *New York State College of Forestry at Syracuse University, Circular* (21): 11–34, 16 figs.

Baker, F.C. 1920. The life of the Pleistocene or glacial period as recorded in the deposits laid down by the great ice sheets. *University of Illinois Bulletin* 17(41): xiv–475, 57 pls.

Baker, F.C. 1922. New species and varieties of Mollusca from Lake Winnebago, Wisconsin, with new records from the state. *The Nautilus* 35: 130–133; 36: 19–21.

Baker, F.C. 1923. A new *Anodontoides* from Wisconsin. *The Nautilus* 36: 123–125.

Baker, F.C. 1924. The fauna of Lake Winnebago region. A quantitative and qualitative survey, with reference to the Mollusca. *Transactions of the Wisconsin Academy of Sciences, Arts and Letters* 21: 109–146.

Baker, F.C. 1926. The naiad fauna of the Rock River system: A study of the law of stream distribution. *Transactions of the Illinois State Academy of Science* 19: 103–112.

Horace Burrington Baker
(Courtesy of MCZ)

Baker, F.C. 1927a. Molluscan associations of White Lake, Michigan: A study of a small inland lake from an ecological and systematic viewpoint. *Ecology* 8: 353–370.

Baker, F.C. 1927b. On the division of the Sphaeriidae into two subfamilies: And the descriptions of a new genus of Unionidae, with descriptions of new varieties. *American Midland Naturalist* 10: 220–223.

Baker, F.C. 1928a. The fresh water Mollusca of Wisconsin. Part II. Pelecypoda. *Bulletin of the Wisconsin Geological and Natural History Survey* 70(2): i–vi + 1–495.

Baker, F.C. 1928b. The Mollusca of Chautaugua Lake, New York, with descriptions of a new variety of *Ptychobranchus* and of *Helisoma*. *The Nautilus* 42: 48–60.

Baker, H.B. 1964. Some of Rafinesque's unionid names. *The Nautilus* 77: 140–142.

Bangham, R.V. 1940. Parasites of fish of Algonquin Park lakes. *Transactions of the American Fisheries Society for 1940:* 161–171.

Bangham, R.V. 1955. Studies on fish parasites of Lake Huron and Manitoulin Island. *American Midland Naturalist* 53: 184–194.

Barfield, M. & G.T. Watters. 1998. Non-parasitic life cycle in the Green Floater, *Lasmigona subviridis* (Conrad, 1835). *Triannual Unionid Report* (16): 22.

Barnes, D.H. 1823. On the genera *Unio* and *Alasmodonta;* with introductory remarks. *American Journal of Science and Arts* 6(1): 107–127; 6(2): 258–280, 13 pls.

Barney, R.L. 1922. An indication of the value of artificial propagation of pearly mussels. *The Nautilus* 36: 53–58.

Barnhart, M.C. 1996. Fat pocketbook pearlymussel (*Potamilus capax*). *Endangered Species Bulletin* 21(6): 23.

Barnhart, M.C. 1997. Sterile eggs in unionid mussels and their roles in conglutinate function. *Bulletin of the North American Benthological Society* 14(1): 56 [abstract].

Barnhart, M.C. & A.D. Roberts. 1997. Reproduction and fish hosts of unionids from the Ozark uplifts. Pp. 16–20. [*In:*] K.S. Cummings, A.C. Buchanan, C.A. Mayer, & T.J. Naimo (eds.), *Conservation and management of freshwater mussels II: Initiatives for the future.* Proceedings of a UMRCC symposium, St. Louis, MO: Upper Mississippi River Conservation Committee, Rock Island, IL.

Barnhart, M.C., F.A. Riusech, & A.D. Roberts. 1997. Fish hosts of the federally endangered pink mucket, *Lampsilis abrupta. Triannual Unionid Report* (13): 35.

Barnhart, M.C., F.A. Riusech, & A.D. Roberts. 1998a. Host of salamander mussel (*Simpsonaias ambigua*) and snuffbox (*Epioblasma triquetra*) from the Meramec River system, Missouri. *Triannual Unionid Report* (16): 34.

Barnhart, M.C., F.A. Riusech, & A.D. Roberts. 1998b. Drum is host of the scaleshell, *Leptodea leptodon. Triannual Unionid Report* (16): 35.

Barnhart, M.C., A.D. Roberts, & A.P. Farnsworth. 1995. Fish hosts of four unionids from Missouri and Kansas. *Triannual Unionid Report* (7): 22.

Bates, J.M. 1962. The impact of impoundment on the mussel fauna of Kentucky Reservoir, Tennessee River. *American Midland Naturalist* 68: 232–236.

Bates, J.M. & S.D. Dennis. 1985. Mussel resource survey, State of Tennessee. *Tennessee Wildlife Resource Agency Technical Report* 85–83, Nashville, Tennessee. 125 pp.

Bauer, G. 1986. The status of the freshwater pearl mussel *Margaritifera margaritifera* L. in the south of its European range. *Biological Conservation* 38: 1–9.

Bauer, G. 1987a. The parasitic stage of the freshwater pearl mussel (*Margaritifera margaritifera* L.). II. Susceptibility of brown trout. *Archiv für Hydrobiologie, Supplement* 76: 403–412.

Bauer, G. 1987b. The parasitic stage of the freshwater pearl mussel (*Margaritifera margaritifera* L.). III. Host relationships. *Archiv für Hydrobiologie, Supplement* 76: 413–423.

Bauer, G. 1987c. Reproductive strategy of the freshwater pearl mussel *Margaritifera margaritifera. Journal of Animal Ecology* 56: 691–704.

Bauer, G. 1994. The adaptive value of offspring size among freshwater mussels (Bivalvia: Unionidea). *Journal of Animal Ecology* 63: 933–944.

Bauer, G. & C. Vogel. 1987. The parasitic stage of the freshwater pearl mussel (*Margaritifera margaritifera* L.). I. Host response to glochidiosis. *Archiv für Hydrobiologie, Supplement* 76: 393–402.

Bauer, G. & K. Wächtler. 2001. Ecology and evolution of the naiads. Pp. 83–388. [*In:*] G. Bauer & K. Wächtler (eds.), *Ecology and evolution of the freshwater mussels Unionoida. Ecological Studies,* Vol. 145. Berlin: Springer-Verlag.

Bauer, G., S. Hochwald, & W. Silkenat. 1991. Spatial distribution of freshwater mussels: The role of host fish and metabolic rate. *Freshwater Biology* 26: 377–386.

Bauer, G., E. Schrimpff, W. Thomas, & R. Herrmann. 1980. Zusammenhänge zwischen dem Bestandsrückgang der Flußperlmuschel (*Margaritifera margaritifera*) im Fichtelgebirge und der Gewässerbelastung. *Archiv für Hydrobiologie* 88: 505–513.

Belt, C.B. 1975. The 1973 flood and man's constriction of the Mississippi River. *Science* 189: 681–684.

Berrow, S.D. 1991. Predation by the hooded crow *Corvus corone cornix* on freshwater pearl mussels *Margaritifera margaritifera. Irish Naturalists' Journal* 23: 492–493.

Bickel, D. 1968. Checklist of the Mollusca of Tennessee. *Sterkiana* 31: 15–39.

Bielz, E.A. 1869. *Verzeichniss der Mollusken-und Conchyliensammlung.* 5th ed. Hermannstadt. 47 pp.

Bingham, R.L. 1968. Reproductive seasons of eight freshwater mussels from the Wabash, White, and East Fork of the White Rivers of Indiana. *Survey of the Commercial Mussel Fauna of the Wabash and White Rivers, Sub Project No. 4 10 R.* Indiana Department of Natural Resources. 101 pp.

Bisbee, G.D. 1984. Ingestion of phytoplankton by two species of freshwater mussels, the black sandshell, *Ligumia recta,* and the three ridge, *Amblema plicata,* from the Wisconsin River in Oneida County, Wisconsin. *Bios* 55: 219–225.

Blainville, H.M.D. de. 1825. *Manuel de malacologie et de conchyliologie.* Paris & Strasbourg: F.G. Levrault, 2 vols. viii + 667, 2 pls., atlas with 109 pls.

Blalock, H.N. & J.B. Sickel. 1996. Changes in mussel (Bivalvia: Unionidae) fauna within the Kentucky portion of Lake Barkley since impoundment of the lower Cumberland River. *American Malacological Bulletin* 13: 111–116.

Blystad, C.N. 1924. Significance of larval mantle of fresh water mussels during parasitism, with notes on a new mantle condition exhibited by *Lampsilis luteola. Bulletin of the United States Bureau of Fisheries* 39: 203–219.

Bogan, A.E. 1993. *Workshop on freshwater bivalves of Pennsylvania.* Freshwater Molluscan Research, Sewell, NJ. 80 pp.

Bogan, A.E. 1998. A resolution of the nomenclatural confusion surrounding *Plagiola* Rafinesque, *Epioblasma* Rafinesque, and *Dysnomia* Agassiz (Mollusca: Bivalvia: Unionidae). *Malacological Review* 30: 77–86.

Bogan, A.E. & W.R. Hoeh. 1995. *Utterbackia peninsularis,* a newly recognized freshwater mussel (Bivalvia: Unionidae: Anodontinae) from peninsular Florida, U.S.A. *Walkerana* 7: 275–287.

Bogan, A.E. & P.W. Parmalee. 1983. *Tennessee's rare wildlife, 2: The mollusks.* Tennessee Wildlife Resource Agency, Ellington Agricultural Center, Nashville, Tennessee 37204, or Tennessee Heritage Program, Department of Conservation, 701 Broadway, Nashville, Tennessee 37203. 123 pp.

Bogan, A.E. & J.D. Williams. 1986. *Megalonaias* Utterback, 1915 (Mollusca, Bivalvia): Proposed conservation by the suppression of *Magnonaias* Utterback, 1915. Z.N.(S.)2512. *Bulletin of Zoological Nomenclature* 43: 273–276.

Bogan, A.E., J.D. Williams, & J.T. Garner. 2006. Case 3353. *Obovaria* Rafinesque, 1819 (Mollusca, Bivalvia): Proposed conservation of usage by designation of *Unio retusa* Lamarck, 1819 as the type species of *Obovaria* conservation. *Bulletin of Zoologial Nomenclature* 63: 226–230.

Boss, K.J. & W.J. Clench. 1967. Notes on *Pleurobema collina* (Conrad) from the James River, Virginia. *Occasional Papers on Mollusks, Museum of Comparative Zoology* 3(37): 45–52.

Brady, T., M. Hove, C. Nelson, R. Gordon, D. Hornbach, & A. Kapuscinski. 2004. Suitable host fish determined for hickorynut and pink heelsplitter. *Ellipsaria* 6: 14–15.

Brim Box, J. & J.D. Williams. 2000. Unionid molluscs of the Apalachicola Basin in Alabama, Florida, and Georgia. *Bulletin of the Alabama Museum of Natural History* 21: 143 pp.

Bruguière, J. G. 1789, 1792. *Encyclopédie méthodique. Histoire naturelle des vers.* Paris. i–xviii, 1–344 (1789); 345–758 (1792).

Buchanan, A.C. 1980. *Mussels (Naiades) of the Meramec River Basin.* Missouri Department of Conservation, Aquatic Series (17). 68 pp.

Buddensiek, V., H. Engel, S. Fleischauer Rössing, & K. Wächtler. 1993. Studies on the chemistry of interstitial water taken from defined horizons in the fine sediments of bivalve habitats in several northern German lowland waters. II: Microhabitats of *Margaritifera margaritifera* L., *Unio crassus* (Philipsson) and *Unio tumidus* Philipsson. *Archiv für Hydrobiologia* 127: 151–166.

Buhay, J.E., J.M. Serb, C.R. Dean, Q. Parham, & C. Lydeard. 2002. Conservation genetics of two endangered unionid bivalve species, *Epioblasma florentina walkeri* and *E. capsaeformis* (Unionidae: Lampsilini). *Journal of Molluscan Studies* 68: 385–391.

Burch, J.B. 1973. Freshwater Unionacean clams (Mollusca: Pelecypoda) of North America. *Biota of Freshwater Ecosystems. Identification Manual 11.* U.S. Environmental Protection Agency. 176 pp.

Henri Marie Ducrotay de Blainville

Burch, J.B. 1975. *Freshwater Unionacean clams (Mollusca: Pelecypoda) of North America.* Rev. ed. Hamburg, MI: Malacological Publications. 204 pp.

Burdick, R.C. & M.M. White. 2007. Phylogeography of the Wabash pigtoe, *Fusconaia flava* (Rafinesque, 1820) (Bivalvia: Unionidae). *Journal of Molluscan Studies* 73: 367–375.

Burkhead, N., J. Williams, & B.J. Freeman. 1992. A river under siege. *Georgia Wildlife* 2(2): 10–17.

Bykhovskaya-Pavlovskaya, I.E., A.V. Gusev, M.N. Dubinina, N.A. Izyumova, T.S. Smirnova, I.L Sokolovskaya, G.A. Shtein, S.S. Shul'man, & V.M. Epshtein. 1964. *Key to parasites of freshwater fish of the U.S.S.R.* Israel Program for Scientific Translation, Jerusalem. 919 pp.

Calkins, W.W. 1874. The land and fresh water shells, of LaSalle County, Ills. *Proceedings of the Ottawa Academy of Science for 1874.* 48 pp., 1 pl.

Call, R.E. 1880. Polymorphous *Anodonta*—variability of the American forms. *American Naturalist* 14: 529–530.

Call, R.E. 1900. A descriptive illustrated catalogue of the Mollusca of Indiana. *Indiana Department of Geology and Natural Resources, 24th Annual Report for 1899.* 335–535, 78 pls.

Calloway, C.B. & R.D. Turner. 1979. New techniques for preparing shells of bivalve larvae for examination with the scanning electron microscope. *Bulletin of the American Malacological Union, Inc. for 1978:* 17–24.

Campbell, D.C., J.M. Serb, J.E. Buhay, K.J. Roe, R.L. Minton, & C. Lydeard. 2005. Phylogeny of North American amblemines (Bivalvia, Unionoida): Prodigious polyphyly proves pervasive across genera. *Invertebrate Biology* 124: 131–164.

Catlow, A. & L. Reeve. 1845. *The conchologist's nomenclator. A catalogue of all the Recent species of shells, included under the subkingdom "Mollusca," with their authorities, synonymes, and references to works where figured or described.* London: Reeve Brothers. viii + 326 pp.

Chadwick, G.H. 1906. Notes on Wisconsin Mollusca. *Bulletin of the Wisconsin Natural History Society* 4(3): 67–99.

Chamberlain, T.K. 1934. The glochidial conglutinates of the Arkansas fanshell, *Cyprogenia aberti* (Conrad). *Biological Bulletin* 66: 55–61.

Chemnitz, J.H. 1782. *Neues systematisches Conchylien-Cabinet geordnet und beschrieben von F.H.W. Martini, und unter dessen Aufsicht nach der Natur gezeichnet und mit lebenden Farben erleuchtet.* Nürnberg: Gabriel Nicolaus Raspe. 6, pls. 1–36.

Chenu, J.C. 1859, 1862. *Manuel de conchyliologie et de paléontologie conchyliologique.* Paris: Librairie Victor Masson, Place de L'École de Médecine. Vol. 1: 1–508 [1859]; Vol. 2: 1–327, 1,236 figs. [1862].

Churchill, E.P. & S.I. Lewis. 1924. Food and feeding in fresh water mussels. *Bulletin of the United States Bureau of Fisheries* [Document 963] 39: 439–471.

Claassen, C. 1994. Washboards, pigtoes, and muckets: Historic musseling industries in the Mississippi watershed. *Historic Archaeology* 28(2): 1–145.

Clark, H.W. & G.H. Gillette. 1911. Some observations made on Little River, near Wichita, Kansas, with reference to the unionids. *Proceedings of the Biological Society of Washington* 24: 63–68.

Clark, H.W. & S. Stein. 1921. Glochidia in surface towings. *The Nautilus* 35: 16–20.

Clarke, A.H. 1973. The freshwater molluscs of the Canadian Interior Basin. *Malacologia* 13: 1–509.

Clarke, A.H. 1981a. The tribe Alasmidontini (Unionidae: Anodontinae), Part I: *Pegias, Alasmidonta,* and *Arcidens. Smithsonian Contribution to Zoology* 326: 1–101.

Clarke, A.H. 1981b. *The freshwater molluscs of Canada.* National Museums of Canada, Ottawa. 446 pp.

Clarke, A.H. 1985. The tribe Alasmidontini (Unionidae: Anodontinae), Part II: *Lasmigona* and *Simpsonaias. Smithsonian Contributions to Zoology* 399: 1–75.

Clarke, A.H. 1986. The mesoconch: A record of juvenile life in Unionidae. *Malacology Data Net* 1(2): 21–36.

Clarke, A.H. & C.O. Berg. 1959. The freshwater mussels of central New York. *Memoir Cornell University Agricultural Experiment Station, N.Y. State College of Agriculture, Ithaca, N.Y.* 367: 1–79.

Clarke, A.H. & W.J. Clench. 1965. *Amblema* Rafinesque, 1820 (Lamellibranchiata): Proposed addition to the official list and proposed suppression of *Amblema* Rafinesque 1819. Z.N.(S.)1699. *Bulletin of Zoological Nomenclature* 22: 196–197.

Clench, W.J. & R.D. Turner. 1956. Freshwater mollusks of Alabama, Georgia and Florida from the Escambia to the Suwannee River. *Bulletin of the Florida State Museum* 1: 97–239.

Clessin, S. 1873. *Beitrage zur Molluskenfauna der oberbayrischen Seen.* [not seen].

Clessin, S. 1876. *Deutsche Excursions—Molluskenfaunen.* Nürnberg. 581 pp.

Coker, R.E. 1914. The protection of fresh water mussels. *U.S. Bureau of Fisheries Document* 793: 1–23.

Coker, R.E. 1919. Fresh water mussels and mussel industry of the United States. *U.S. Bureau of Fisheries Bulletin* 36: 13–89.

Coker, R.E. & T. Surber. 1911. A note on the metamorphosis of the mussel *Lampsilis laevissimus. Biological Bulletin (Woods Hole)* 20: 179–182.

Coker, R.E., A.F. Shira, H.W. Clark, & A.D. Howard. 1921. Natural history and propagation of fresh water mussels. *Bulletin of the Bureau of Fisheries* [issued separately as *U.S. Bureau of Fisheries Document* 893] 37(1919–1920): 77–181, 17 pls.

Conner, C.H. 1905. Glochidia of *Unio* on fishes. *The Nautilus* 18: 142–143.

Conner, C.H. 1907. The gravid period of unios. *The Nautilus* 22: 87–89.

Conrad, T.A. 1834. *New freshwater shells of the United States, with coloured illustrations; and a monograph of the genus* Anculotus *of Say; also a synopsis of the American naiades.* Philadelphia: J. Dobson. 1–76, 8 pls.

Timothy Abbot Conrad

Conrad, T.A. 1835–1840? *Monography of the family Unionidae, or naiades of Lamarck, (fresh water bivalve shells,) of North America, illustrated by figures drawn on stone from nature.* Vol. 1: 1–12, pls. 1–5 [Dec. 1835]; 17–24, pls. 6–10 [Jan. 1836]; 25–32, pls. 11–15 [Feb. 1836]; 33–40, pls. 16–20 [Mar. 1836]; 41–48, pls. 21–25 [June 1836]; 49–56, pls. 26–30 [July 1836]; 57–64, pls. 31–35 [Dec. 1836]; 65–72, pls. 36–40 [Feb. 1837]; 73–80, pls. 41–45 [Mar. 1837]; 81–94 + 2, pls. 46–51 [May 1838]. Vol. 2: 95–102, pls. 52–56 [Nov. 1838]; 103–110, pls. 57–60 [no date, 1840?]; 111–118, pls. 61–65 [no date, 1840?].

Conrad, T.A. 1850. Descriptions of new fresh water and marine shells. *Journal of the Academy of Natural Sciences of Philadelphia* (n.s.)1: 275–280, pls. 37–39.

Conrad, T.A. 1853. A synopsis of the family of Naïades of North America, with notes, and a table of some of the genera and sub-genera of the family, according to their geographical distribution, and descriptions of genera and sub-genera. *Proceedings of the Academy of Natural Sciences of Philadelphia* 6: 243–269.

Conrad, T.A. 1854. Descriptions of new species of *Unio. Journal of the Academy of Natural Sciences of Philadelphia* (n.s.)2: 295–298, pls. 26, 27.

Conrad, T.A. 1855. Descriptions of three new species of *Unio. Proceedings of the Academy of Natural Sciences of Philadelphia* 7: 256.

Convey, L.E., J.M. Hanson, & W.C. MacKay. 1989. Size selective predation on unionid clams by muskrats. *Journal of Wildlife Management* 53: 654–657.

Cooper, W. 1834. List of shells collected by Mr. Schoolcraft in western and northwestern territory. *Appendix to Narrative of Expedition through Upper Missouri to Itasca Lake, etc., under the direction of Henry B. Schoolcraft.* New York: Harper & Bros. 153–156.

Cope, W.G. & D.L. Waller. 1995. Evaluation of freshwater mussel relocation as a conservation and management strategy. *Regulated Rivers: Research & Management* 11: 147–155.

Corwin, R.S. 1920. Raising freshwater mussels in enclosures. *Transactions of the American Fisheries Society* 49: 81–84.

Georges Cuvier

James Ellsworth DeKay

Corwin, R.S. 1921. Further notes on raising freshwater mussels in enclosures. *Transactions of the American Fisheries Society* 50: 307–311.

Couch, K.J. 1997. *An illustrated guide to the unionid mussels of Kansas.* Privately printed. 123 pp.

Cragin, F.W. 1887. A new species of *Unio* from Indian territory. *Bulletin of the Washburn College Laboratory of Natural History* 1(8): 6.

Crownhart, A., B. Sietman, M. Hove, & N. Rudh. 2006. *Quadrula metanevra* glochidia metamorphose on select minnow species. *Ellipsaria* 8(3): 6.

Culp, J.J. A.C. Shepard, & M.A. McGregor. 2006. New host fish identification for the Pyramid Pigtoe, *Pleurobema rubrum. Ellipsaria* (8)3: 5–6.

Cummings, K.S. & J.M.K. Berlocher. 1990. The naiades or freshwater mussels (Bivalvia: Unionidae) of the Tippecanoe River, Indiana. *Malacological Review* 23: 83–98.

Cummings, K.S. & C.A. Mayer. 1992. Field guide to freshwater mussels of the Midwest. *Illinois Natural History Survey Manual* 5. 194 pp.

Cummings, K.S. & C.A. Mayer. 1993. Distribution and host species of the federally endangered freshwater mussel, *Potamilus capax* (Green, 1832) in the Lower Wabash River, Illinois and Indiana. *Illinois Natural History Survey, Center for Biodiversity Technical Report* 1993(1): 1–29.

Cummings, K.S., M.E. Retzer, C.A. Mayer, & L.M. Page. 1990. Life history aspects and status of the federally endangered fat pocketbook, *Potamilus capax* (Green, 1832) (Mollusca: Unionidae) in the lower Wabash River, Illinois and Indiana. *Illinois Natural History Survey Center for Biodiversity Technical Report for 1990* (1): 1–37.

Cunjak, R.A. & S.E. McGladdery. 1991. The parasite-host relationship of glochidia (Mollusca: Margaritiferidae) on the gills of young-of-the-year Atlantic salmon (*Salmo salar*). *Canadian Journal of Zoology* 69: 353–358.

Currier, A.O. 1868. List of the shell bearing Mollusca of Michigan, especially of Kent and adjoining counties. *Kent Scientific Institute, Miscellaneous Publications* 1: 1–12.

Cuvier, G. 1817. *Le règne animal distribué d'après son organisation, pour servir de base à l'histoire naturelle des animaux et d'introduction à l'anatomie comparée.* 3rd ed. Paris: Fortin, Masson et Cie, Libraires. xviii + 532 pp.

Dance, S.P. 1966. *Shell collecting. An illustrated history.* Berkeley: University of California Press. 344 pp., 35 pls.

Danglade, E. 1922. The Kentucky River and its mussel resources. *U.S. Bureau of Fisheries Document* 934: 1–8.

Daniels, L.E. 1903. A check list of Indiana Mollusca, with localities. *27th Annual Report of the Indiana Geological Survey for 1902:* 629–652.

Darrigran, G. & G. Pastorino. 1993. Bivalvos invasores en el Río de la Plata, Argentina. *Comunicaciones de la Sociedad Malacológica del Uruguay* 7: 309–313.

Dartnall, H.J.G. & M. Walkey. 1979. The distribution of glochidia of the swan mussel, *Anodonta cygnea* (Mollusca) on the three spined stickleback *Gasterosteus aculeatus* (Pisces). *Journal of Zoology, London* 189: 31–37.

Davids, C. 1973. The relations between mites of the genus *Unionicola* and the mussels *Anodonta* and *Unio*. *Hydrobiologia* 41: 37–44.

Davis, H.S. 1934. Care and diseases of trout. *U.S. Bureau of Fisheries Investigational Report* (22): 1–69.

Davis, G.M. & S.L.H. Fuller. 1981. Genetic relationships among Recent Unionacea (Bivalvia) of North America. *Malacologia* 20: 217–253, 2 app.

DeCamp, W.H. 1881. List of the shell bearing Mollusca of Michigan. *Kent Scientific Institute, Miscellaneous Publication* (5): 1–15, 1 pl.

De Gregorio, A. de Marchese. 1914. Su taluni molluschi di acqua dolce di America. *Il Naturalista Siciliano* (2)22(2–3): 31–72, pls. 3–12.

DeKay, J.E. 1843. *Zoology of New York, Part 5: Mollusca.* Albany: Carroll & Cook. 271 pp., 40 pls.

Delessert, B. 1841. *Recueil de coquilles décrites par Lamarck dans son* Histoire naturelle des animaux sans vertèbres, *et non encore figurées.* Paris: Fortin, Masson et Cie, Libraires. 100 pp., 40 pls.

Deshayes, G.P. 1830. *Encyclopédie methodique, histoire des Verspar Bruguière et Lamarck, complete par Deshayes. Mollusques,* Vol. 2. Pp. 471–553. Paris: Panckoucke.

Deshayes, G.P. 1835. Pp. 524–574. [*In:*] J.B.P.A. de M. de [C. de]

Lamarck, *Histoire naturelle des animaux sans vertèbres. Tome sixième. Histoire des mollusques.* Deuxième Edition (les Nayades). Paris: J.B. Bailliere, Libraire.

Deshayes, G.P. 1839–1858. *Traité élémentaire de conchyliologie, avec les applications de cette science à la géologie.* 2 vols. Paris. 824 pp.

Desmarest, 1859. P. 212. [*In:*] J.C. Chenu, *Manuel de conchyliologie et de paléontologie conchyliologique.* Paris: Librairie Victor Masson, Place de L'École de Médecine.

DeVillez, J. 1991. *The commercial mussel harvest in the mid-west and mid-south, a comparative/collective study.* Kuttawa, KY: Privately printed. 13 pp.

Dickinson, B.D. and B.E. Sietman. 2008. Recent observation of metamophosis without parasitism in *Utterbackia imbecillis. Ellipsaria* 10(1): 7–8.

Downing, J.A., Y. Rochon, & M. Pérusse. 1993. Spatial aggregation, body size, and reproductive success in the freshwater mussel *Elliptio complanata. Journal of the American Benthological Society* 12: 148–156.

Draparnaud, J.P.R. 1801. *Tableau des mollusques terrestres et fluviatiles de la France.* Montpellier & Paris. 116 pp.

Draxler, B., M. Hove, S. Schieffer, M. Berg, G. Widiker, B. Sietman, D. Allen, & D. Hornback. 2006. Suitable host fishes for fatmucket (*Lampsilis siliquoidea*) and pocketbook (*Lampsilis cardium*) evaluated by high school and university researchers. *Ellipsaria* 8(1): 12.

Dudgeon, D. & B. Morton. 1984. Site selection and attachment duration of *Anodonta woodiana* (Bivalvia: Unionacea) glochidia on fish hosts. *Journal of Zoology, London* 204: 355–362.

Eaton, A. 1826. *Zoological text book, containing Cuvier's four grand divisions of animals.* Albany, NY. 288 pp.

Ecological Specialists, Inc. 1992. Final report for the Tippecanoe River unionid fish and habitat study. *Final Report to the Indiana Department of Natural Resources & U.S. Fish and Wildlife Service.* 108 pp. + appendices.

Ecological Specialists, Inc. 1993. Unionid survey of the lower Muskingum River (RM 34.1–0). *Final Report to the Ohio Division of Wildlife.* 51 pp. + appendices.

Elderkin, C.L., A.D. Christian, C.C. Vaughn, J.L. Metcalfe-Smith, & D.J. Berg. 2007. Population genetics of the freshwater mussel, *Amblema plicata* (Say 1817) (Bivalvia: Unionidae): Evidence of high dispersal and post-glacial colonization. *Conservation Genetics* 8: 355–372.

Ellis, M.M. 1931. A survey of conditions affecting fisheries in the upper Mississippi River. *Bureau of Fisheries, Fishery Circular* (5): 1–18.

Ellis, M.M. 1936. Erosion silt as a factor in aquatic environments. *Ecology* 17: 29–42.

Ellis, M.M. & M.D. Ellis. 1926. Growth and transformation of parasitic glochidia in physiological nutrient solutions. *Science* 64: 579–580.

Ellis, M.M. & M. Keim. 1918. Notes on the glochidia of *Strophitus edentulus pavonis* (Lea) from Colorado. *The Nautilus* 32: 17–18.

Engler, R.M. 1979. Bioaccumulation of toxic substances from contaminated sediments by fish and benthic organisms. Pp. 325–354. [*In:*] S.A. Peterson & K.K. Randolph (eds.), Management of bottom sediments containing toxic substances. *Proceedings of the 4th United States Japan Experts Meeting, Tokyo. U.S. Environmental Protection Agency, Corvallis, OR. EPA* 600 3: 79–102.

Evermann, B.W. & H.W. Clark. 1918. The Unionidae of Lake Maxinkukee. *Proceedings of the Indiana Academy of Science for 1917:* 251–285.

Evermann, B.W. & H.W. Clark. 1920. *Lake Maxinkukee, a physical and biological survey.* Indiana Department of Conservation, Indianapolis. 512 pp.

Farris, J.L. & J.H. Van Hassel (eds.). 2007. *Freshwater bivalve ecotoxicology.* Boca Raton, FL: CRC Press. 375 pp.

Fassler, C. R. 1991. The return of the American pearl, three feisty farmers take on the Japanese. *Aquaculture Magazine* (November/December): 1–8.

Faussek, V. 1895. Über den Parasitismus der *Anodonta* Larven in der Fischhaut. *Biologisches Centralblatt* 15: 115–125.

Ferriss, J.H. 1900. A new *Lampsilis* from Arkansas. *The Nautilus* 14: 38–39.

Férussac, A.E. de [Baron]. 1835. Observations adressées en forme de lettre à M.M. Th. Say, C.S. Rafinesque, Is. Lea, S.P. Hildreth, T.A. Conrad, et C.A. Poulson, sur la synonomie de coquilles bivalves de l'Amerique Septentrionale, et essai d'une table de concordance à ce sujet. *Magasin de Zoologie (Paris)* 5(59–60): 1–36.

Fichtel, C. & D.G. Smith. 1995. The freshwater mussels of Vermont. *Nongame & Natural Heritage Program, Vermont Fish & Wildlife Department, Technical Report* (18): 1–54.

Fikes, M.H. 1972. Maintenance of the naiad *Amblema plicata* (Say, 1817) in an artificial system. *Bulletin of the American Malacological Union for 1972:* 35.

Fischer, P. & H. Crosse. 1880–1902. Études sur les mollusques terrestres et fluviatiles de Mexique et de Guatemala. *Mission Scientifique au Mexique et dans l'Amerique Centrale. Recherches Zoologiques* (2)2. Paris. 731 pp.

Fischer von Waldheim, G. 1807. *Muséum Démidoff mis en ordre systématique et décrit,* Vol. 3. Moscow. 330 pp.

Fischerstrom, I. 1761. De concharum margaritiferarum natura. *Commentarii de Rebus in Scientia Naturali et Medicina Gestis* 10: 204–205.

Fleming, J. 1828. *A history of British animals.* Edinburgh & London. xxiii + 565 pp.

Forel, F.A. 1866. Einige Beobachtungen über die Entwicklung des zelligen Muskelgewebes. *Beiträge zur Entwicklungsgeschichte der Najaden. Inaugural Abhandlung der medicinischen Facultät zu Würzburg.* 40 pp., 3 pls.

Freitag, T.M. 1984. Recent naiad molluscs of the Detroit River. *American Malacological Bulletin* 3: 105.

Frierson, L.S. 1903. The specific value of *Unio declivus,* Say. *The Nautilus* 17: 49–51, 1 pl.

Frierson, L.S. 1910. Description of a new species of *Anodonta. The Nautilus* 23: 113–114, 1 pl.

Frierson, L.S. 1911. Remarks on *Unio varicosus, cicatricosus* and *Unio compertus,* new species. *The Nautilus* 25: 51–54, 2 pls.

Frierson, L.S. 1914a. Remarks on classification of the Unionidae. *The Nautilus* 28: 6–8.

Frierson, L.S. 1914b. Observations of the genus *Symphynota* Lea. *The Nautilus* 28: 40.

Frierson, L.S. 1917. New genera and species of South American naiades. *The Nautilus* 31: 47–49.

Frierson, L.S. 1927. *A classification and annotated check list of the North American naiades.* Waco, TX: Baylor University Press. 111 pp.

Fuller, S.L.H. 1971. A brief field guide to the fresh water mussels (Mollusca: Bivalvia: Unionacea) of the Savannah River system. *ASB Bulletin* 18: 137–146.

Fuller, S.L.H. 1974. Clams and mussels (Mollusca: Bivalvia). Pp. 215–273. [*In:*] C.W. Hart & S.L.H. Fuller (eds.), *Pollution ecology of freshwater invertebrates.* New York: Academic Press.

Fuller, S.L.H. 1978. Fresh water mussels (Mollusca: Bivalvia:

Unionidae) of the upper Mississippi River: Observations at selected sites within the 9 foot channel navigation project on behalf of the U.S. Army Corps of Engineers. *Final Report to the U.S. Army Corp of Engineers,* No. 78-33. 401 pp.

Fustish, C.A. & R.E. Millemann. 1978. Glochidiosis of salmonid fishes. II. Comparison of tissue response of Coho and Chinook Salmon to experimental infection with *Margaritifera margaritifera* (L.) (Pelecypoda: Margaritanidae). *Journal of Parasitology* 64: 155–157.

Gardiner, D.B., H. Silverman, & T.H. Dietz. 1991. Musculature associated with the water canals in freshwater mussels and response to monamines *in vitro. Biological Bulletin* 180: 453–465.

Gatenby, C.M., R.J. Neves, & B.C. Parker. 1993. Preliminary observations from a study to culture recently metamorphosed mussels. *Bulletin of the North American Benthological Society* 10: 128 [abstract].

Geiser, S.W. 1911. The correct name for *Tritogonia tuberculata. The Academician* 1: 15.

Ghent, A.W., R. Singer, & L. Johnson Singer. 1978. Depth distributions determined with SCUBA, and associated studies of the freshwater unionid clams *Elliptio complanata* and *Anodonta grandis* in Lake Bernard, Ontario. *Canadian Journal of Zoology* 56: 1654–1663.

Gmelin, J.F. 1791. *Caroli a Linné systema naturae per regne tria naturae.* 13th ed. Lipsiae. 1(6) Vermes. 3021–3910.

Goodrich, C. 1932. The Mollusca of Michigan. *Michigan Handbook Series No. 5.* Ann Arbor: University of Michigan Press. 120 pp., 7 pls.

Goodrich, C. & H. van der Schalie. 1932. I. On an increase in the naiad fauna of Saginaw Bay, Michigan. II. The naiad species of the Great Lakes. *Occasional Papers of the Museum of Zoology, University of Michigan* (238). 14 pp.

Goodrich, C. & H. van der Schalie. 1939. Aquatic mollusks of the Upper Peninsula of Michigan. *Miscellaneous Publications, Museum of Zoology, University of Michigan* 43: 1–45.

Goodrich, C. & H. van der Schalie. 1944. A revision of the Mollusca of Indiana. *American Midland Naturalist* 32: 257–326.

Gordon, M.E. 1995. *Venustaconcha sima* (Lea), an overlooked freshwater mussel (Bivalvia: Unionoidea) from the Cumberland River Basin of central Tennessee. *The Nautilus* 108: 55–60.

Gordon, M.E. & D.G. Smith. 1990. Autumnal reproduction in *Cumberlandia monodonta* (Unionoidea: Margaritiferidae). *Transactions of the American Microscopical Society* 109: 407–411.

Gordon, M.J., B.K. Swan, & C.G. Patterson. 1978. *Baeoctenus bicolor* (Diptera: Chironomidae) parasitic in unionid bivalve molluscs, and notes on other chironomid bivalve associations. *Journal of the Fisheries Research Board of Canada* 35: 154–157.

Goudraeu, S.E., R.J. Neves, & R.J. Sheehan. 1993. Effects of wastewater treatment plant effluents on freshwater mollusks in the upper Clinch River, Virginia, USA. *Hydrobiologia* 252: 211–230.

Gould, A.A. 1841. *A report on the Invertebrata of Massachusetts, comprising the Mollusca, Annelida, and Radiata.* Cambridge, MA. 373 pp., 15 pls.

Gould, A.A. 1855. New species of land and fresh-water shells from western (N.) America. (Cont.). *Proceedings of the Boston Society of Natural History* 5: 228–229.

Gould, A.A. 1870. *Report on the Invertebrata of Massachusetts, second edition, comprising the Mollusca.* W.G. Binney (ed.). Boston: Wright & Potter, State Printers. 524 pp., 12 pls.

Grace, T.B. & A.C. Buchanan. 1981. *Naiades (mussels) of the lower Osage River, Tavern Creek, and Maries River, Missouri.* Jefferson City, MO: Missouri Department of Conservation. 147 pp.

A.A. Gould

Graf, D.L. 2000. The Etherioidea revisited: A phylogenetic analysis of hyriid relationships (Mollusca: Bivalvia: Paleoheterodonta: Unionoida). *Occasional Papers of the Museum of Zoology, University of Michigan* (729): 1–21.

Graf, D.L. & D. Ó Foighil. 2000. The evolution of brooding characters among the freshwater pearly mussels (Mollusca: Bivalvia: Unionoidea). *Journal of Molluscan Studies* 66: 157–170.

Gray, E. van S., W.A. Lellis, J.C. Cole, & C.S. Johnson. 1999. Hosts of *Pyganodon cataracta* (eastern floater) and *Strophitus undulatus* (squawfoot) from the Upper Susquehanna River basin, Pennsylvania. *Triannual Unionid Report* (18): 6.

Gray, J.E. 1847. A list of the genera of recent Mollusca, their synonyma and types. *Proceedings of the Zoological Society of London* 15: 129–219.

Green, J. 1827. Some remarks on the Unios of the United States, with a description of a new species. *Contributions to Maclurian Lyceum to the Arts and Sciences* 1: 41–47.

Green, J. 1832. Notes of a naturalist. *Cabinet of Natural History and American Rural Sports* 2: 290–291.

Grier, N.M. 1918. New varieties of naiads from Lake Erie. *The Nautilus* 32: 9–12.

Grier, N.M. 1922. Final report on the study and appraisal of mussel resources in selected areas of the Upper Mississippi River. *American Midland Naturalist* 8: 1–33, 1 pl.

Grier, N.M. & J.F. Mueller. 1922. Notes on the naiad fauna of the Upper Mississippi River. II. The naiades of the Upper Mississippi drainage. *The Nautilus* 35: 46–49.

Grier, N.M. & J.F. Mueller. 1926. Further studies in correlation of shape and station in fresh water mussels. *Bulletin of the Wagner Free Institute of Science* 1: 11–28.

Guérin-Méneville, F.E. 1830–1844. Mollusques. [*In:*] *Iconographie du règne animal du G. Cuvier: Ou representation d'après nature de l'une des espèces les plus remarquables, et souvent non encore figurées, de chacque genre d'animaux.* Vol. 2: 37 pls.; Vol. 3: 64 pls. Paris: J.B. Baiaere.

Haag, W.R., D.J. Berg, D.W. Garton, & J.L. Farris. 1993. Reduced survival and fitness in native bivalves in response to fouling by the introduced zebra mussel (*Dreissena polymorpha*) in western Lake Erie. *Canadian Journal of Fisheries and Aquatic Sciences* 50: 13–19.

J.E. Gray

Fritz Haas (courtesy MCZ)

Haag, W.R., J.T. Butler, & P.D. Hartfield. 1995. An extraordinary reproductive strategy in freshwater bivalves: Prey mimicry to facilitate larval dispersal. *Freshwater Biology* 34: 471–476.

Haas, F. 1930. Über nord und mittelamerikanische Najaden. *Senckenbergiana Biologica* 12: 317–330.

Haas, F. 1940. A tentative classification of the Palearctic unionids. *Zoological Series of Field Museum of Natural History* 24(11): 115–141.

Haas, F. 1969. Superfamilia Unionacea. *Das Tierreich* (Berlin) 88: x + 663 pp.

Haggerty, T.M., J.T. Garner, G.H. Patterson, & L.C. Jones. 1995. A quantitative assessment of the reproductive biology of *Cyclonaias tuberculata* (Bivalvia: Unionidae). *Canadian Journal of Zoology* 73: 83–88.

Haldeman, S.S. 1842. Description of five new species of North American fresh water shells. *Journal of the Academy of Natural Sciences of Philadelphia* 8: 200–202.

Hankinson, T.L. 1908. *A biological survey of Walnut Lake, Michigan.* Wynkoop Hullenbeck Crawford Co., State Printers, Lansing, Michigan. pp. 155–288.

Hanley, S. 1842–1856. *An illustrated and descriptive catalogue of Recent bivalve shells.* London: Williams & Norgate. i–xviii, 1–392, 1–24, suppl. pls. 9–24.

Hanson, J.M., W.C. MacKay, & E.E. Prepas. 1989. Effect of size selective predation by muskrats (*Ondatra zebithicus*) on a population of unionid clams (*Anodonta grandis simpsonianus*). *Journal of Animal Ecology* 58: 15–28.

Harman, W.N. 1974. The effects of reservoir construction and channelization on the mollusks of the upper Delaware watershed. *Bulletin of the American Malacological Union for May* (1974): 12–14.

Harms, W. 1909. Postembryonale Entwicklungsgeschichte der Unioniden. *Zoologische Jahrbücher, Abteilung für Anatomie und Ontogenie* 28: 325–386, pls. 13–16.

Hartfield, P. 1993. Headcuts and their effect on freshwater mussels. Pp. 131–141. [*In:*] K.S. Cummings, A.C. Buchanan, & L. M. Koch (eds.), *Conservation and management of freshwater mussels: Proceedings of a UMRCC symposium, St. Louis, MO.* Upper Mississippi River Conservation Committee, Rock Island, IL.

Hartfield, P. & R. Butler. 1997. Observations on the release of superconglutinates by *Lampsilis perovalis* (Conrad 1834). Pp. 11–14. [*In:*] K.S. Cummings, A.C. Buchanan, C.A. Mayer, & T.J. Naimo (eds.), *Conservation and management of freshwater mussels II: Initiatives for the future. Proceedings of a UMRCC symposium, St. Louis, MO.* Upper Mississippi River Conservation Committee, Rock Island, IL.

Haukioja, E. & T. Hakala. 1974. Vertical distribution of freshwater mussels (Pelecypoda, Unionidae) in southwestern Finland. *Annales Zoologici Fennici* 11: 127–130.

Heard, W.H. 1975. Sexuality and other aspects of reproduction in *Anodonta* (Pelecypoda: Unionidae). *Malacologia* 15: 81–103.

Heard, W. H. 1979. Hermaphroditism in *Elliptio* (Pelecypoda: Unionidae). *Malacological Review* 12: 21–28.

Heard, W.H. & R.H. Guckert. 1971. A re-evaluation of the recent Unionacea (Pelecypoda) of North America. *Malacologia* 10: 333–355.

Heath, D., M. Hove, R. Benjamin, M. Endris, R. Kenyon, & J. Kurth. 1998. *Quadrula fragosa* exhibit unusual reproductive behaviors. *Triannual Unionid Report* (16): 33.

Hebert, P.D.N., B.W. Muncaster, & G.L. Mackie. 1989. Ecological and genetic studies on *Dreissena polymorpha* (Pallas): A new mollusc in the Great Lakes. *Canadian Journal of Fisheries and Aquatic Sciences* 46: 1587–1591.

Heinricher, J.R. & J.B. Layzer. 1999. Reproduction by individuals of a nonreproducing population of *Megalonaias nervosa* (Mollusca: Unionidae) following translocation. *American Midland Naturalist* 141: 140–148.

Henderson, J. 1924. Mollusca of Colorado, Utah, Montana, Idaho and Wyoming. *University of Colorado Studies, Boulder, Colorado* 13: 65–223.

Henderson, J. 1928. *Margaritifera* vs. *Margaritana. The Nautilus* 41: 91.

Hendrix, S.S., M.F. Vidrine, & R.H. Hartenstine. 1985. A list of records of freshwater aspidogastrids (Trematoda) and their hosts in North America. *Proceedings of the Helminthological Society of Washington* 52: 289–296.

Henley, W.F. & R.J. Neves. 1997. Chemosensory abilities of female freshwater mussels and glochidia (Unionidae). P. 282. [*In:*] K.S. Cummings, A.C. Buchanan, C.A. Mayer, & T.J. Naimo (eds.), *Conservation and management of freshwater mussels II: Initiatives for the future. Proceedings of a UMRCC symposium, St. Louis, MO.* Upper Mississippi River Conservation Committee, Rock Island, IL.

Herrmannsen, A.N. 1846–1852. *Indicis generum malacozoorum primordia.* Cassel: T. Fischer. Vol. 1: i–xxvii + 1–232 [1846], 233–637 [1847]; Vol. 2: 1–352 [1847], 353–492 [1848], 493–717 + xxix–xlii [1849]; *Supplementa et corrigenda:* i–v + 1–140 [1852].

Samuel Hildreth

Higgins, E. 1930. Fresh water mussel investigations. *Report of the Commissioner of Fisheries for 1929, Appendix* 10: 670–673.

Higgins, F. 1858. A catalogue of the shell-bearing species of Mollusca, inhabiting the vicinity of Columbus, Ohio, with some remarks thereon. *Ohio State Board of Agriculture with an Abstract of the Proceedings of the County Agricultural Society's 12th Annual Report, for 1857:* 548–555.

Hildreth, S.P. 1828. Observations on, and descriptions of the shells found in the waters of the Muskingum River, Little Muskingum, and Duck Creek, in the vicinity of Marietta, Ohio. *American Journal of Science and Arts* 14(2): 276–291, 2 pls. [reprinted in *Sterkiana* 8(1962): 39–48].

Hildreth, S.P. 1837. Miscellaneous observations made during a tour in May, 1835, to the Falls of the Cayahoga [*sic*], near Lake Erie: extracted from the diary of a naturalist. *American Journal of Science and Arts* 31: 1–84.

Hildreth, S.P. 1830. *Catalogue of fluviatile shells.* Unpublished original at Dawes Memorial Library, Marietta College, Marietta, OH. 104–118.

Hill, D.M. 1986. *Cumberlandian mollusks conservation program, activity 3: Identification of fish hosts.* Office of Natural Resources and Economic Development, Tennessee Valley Authority, Knoxville. 57 pp.

Hillegass, K.R. & M.C. Hove. 1997. Suitable fish hosts for glochidia of three freshwater mussels: Strange floater, ellipse, and snuffbox. *Triannual Unionid Report* (13): 25.

Hoeh, W.R. 1990. Phylogenetic relationships among eastern North American *Anodonta* (Bivalvia: Unionidae). *Malacological Review* 23: 63–82.

Hoeh, W.R. & J.B. Burch. 1989. The taxonomic status of *Anodonta lacustris* Lea (Bivalvia: Unionidae). *Walkerana* 3: 263–276.

Hoeh, W.R., M.B. Black, R. Gustafson, A.E. Bogan, R.A. Lutz, & R.C. Vrijenhoek. 1998. Testing alternative hypotheses of *Neotrigonia* (Bivalvia: Trigonioda) phylogenetic relationships using cytochrome C oxidase subunit 1 DNA sequences. *Malacologia* 40: 267–278.

Hoggarth, M.A. 1987. Determination of anterior posterior orientation of glochidia by the examination of glochidial valves present within the umbos of juvenile unionid clams (Mollusca: Bivalvia). *Ohio Journal of Science* 87: 93–95.

Hoggarth, M.A. 1992. An examination of the glochidia-host relationships reported in the literature for North American species of Unionacea (Mollusca: Bivalvia). *Malacology Data Net* 3: 1–30.

Hoggarth, M. A. 1999. Description of some of the glochidia of the Unionidae (Mollusca: Bivalvia). *Malacologia* 41: 1–118.

Hoggarth, M.A., D.L. Rice, & D.M. Lee. 1995. Discovery of the federally endangered freshwater mussel, *Epioblasma obliquata obliquata* (Rafinesque, 1820) (Unionidae) in Ohio. *Ohio Journal of Science* 95: 298–299.

Holland, H.E. & M.L. Huston. 1985. Distribution and food habits of young of year fishes in a backwater lake of the upper Mississippi River USA. *Journal of Freshwater Ecology* 3: 81–92.

Holland-Bartels, L.E. 1990. Physical factors and their influence on the mussel fauna of a main channel border habitat of the upper Mississippi River. *Journal of the North American Benthological Society* 9: 327–335.

Holland-Bartels, L.E. & T.W. Kammer. 1989. Seasonal development of *Lampsilis cardium, Amblema plicata plicata,* and *Potamilus alatus* (Pelecypoda: Unionidae) in the upper Mississippi River. *Journal of Freshwater Ecology* 5: 87–92.

Houghton, W. 1862. On the parasitic nature of the fry of *Anodonta cygnea. Quarterly Journal of Microscopical Science* (n.s.)2: 162–168.

Houslet, B.S. & J.B. Layzer. 1997. Difference in growth between two populations of *Villosa taeniata* in Horse Lick Creek, Kentucky. Pp. 37–44. [*In:*] K.S. Cummings, A.C. Buchanan, C.A. Mayer, & T.J. Naimo (eds.), *Conservation and management of freshwater mussels II: Initiatives for the future. Proceedings of a UMRCC symposium, St. Louis, MO.* Upper Mississippi River Conservation Committee, Rock Island, IL.

Hove, M.C. 1995a. Early life history research on the squawfoot, *Strophitus undulatus. Triannual Unionid Report* (7): 28–29

Hove, M.C. 1995b. Host research on round pigtoe glochidia. *Triannual Unionid Report* (8): 8.

Hove, M.C. 1995c. Suitable fish hosts of the lilliput, *Toxolasma parvus. Triannual Unionid Report* (8): 9.

Hove, M.C. 1997. Ictalurids serve as suitable hosts for the purple wartyback. *Triannual Unionid Report* (11): 4.

Hove, M.C. & T.W. Anderson. 1997. Mantle waving behavior and suitable fish hosts of the ellipse. *Triannual Unionid Report* (11): 3.

Hove, M.C. & M. Haas. 2001. Creeper glochidia appear to be generalists. *Ellipsaria* 3(1): 19–20.

Hove, M. & J. Kurth. 1997. *Cyclonaias tuberculata* glochidia transform on catfish barbels. *Triannual Unionid Report* (13): 21

Hove, M. & J. Kurth. 1998. Darters, sculpins, and sticklebacks serve as suitable hosts for *Venustaconcha ellipsiformis* glochidia. *Triannual Unionid Report* (14): 8.

Hove, M. & R. Neves. 1991. Distribution and life history of the James River spinymussel. *Endangered Species Technical Bulletin* 16: 9.

Hove, M., R. Engelking, E. Evers, M. Peteler, & E. Peterson. 1994. *Cyclonaias tuberculata* host suitability tests. *Triannual Unionid Report* (5): 1 p.

Hove, M.C., R.A. Engelking, E.M. Long, M.E. Peteler, & E.M. Peterson. 1995a. *Anodontoides ferussacianus* and *Anodonta imbecillis* host suitability tests. *Triannual Unionid Report* (6): 1 p.

Hove, M.C., R.A. Engelking, E.M. Long, M.E. Peteler, & E.M. Peterson. 1995b. Life history research on the creek heelsplitter, *Lasmigona compressa. Triannual Unionid Report* (6): 1 p.

Hove, M.C., R.A. Engelking, M.E. Peteler, E.M. Peterson, A.R. Kapuscinski, L.A. Sovell, & E.R. Evers. 1997. Suitable fish hosts for glochidia of four freshwater mussels. Pp. 21–25. [*In:*] K.S.

Cummings, A.C. Buchanan, C.A. Mayer, & T.J. Naimo (eds.), *Conservation and management of freshwater mussels II: Initiatives for the future. Proceedings of a UMRCC symposium, St. Louis, MO.* Upper Mississippi River Conservation Committee, Rock Island, IL.

Hove, M.C., R.A. Engelking, M. Peteler, & L. Sovell. 1994. Life history research on *Ligumia recta* and *Lasmigona costata*. *Triannual Unionid Report* (4): 1 p.

Hove, M.C., J.E. Kurth, D.J. Heath, R.L. Benjamin, M.B. Endris, R.L. Kenyon, A.R. Kapuscinski, K.R. Hillegass, T.W. Anderson, V.E. Pepi, & C.J. Lee. 1998a. Hosts and host attracting behaviors of five upper Mississippi River mussels. *Abstracts, World Congress of Malacology, Washington, DC:* 159.

Hove, M.C., J.E. Kurth, & A.R. Kapuscinski.1998b. Brown bullhead suitable host for *Tritogonia verrucosa; Cumberlandia monodonta* host(s) remain elusive. *Triannual Unionid Report* (15): 13.

Hove, M.C., J. Medland, P. Cliff, M. Haas, B. Whaley, J. Woods, & A. Kapuscinski. 2001. Winged mapleleaf glochidial metamorphosis on channel catfish verified. *Ellipsaria* 3: 16–17.

Hove, M., M. Berg, K. Dietrich, C. Gonzalez, D. Hornbach, K. Juleen, M. Ledford, M. Marzec, M. McGill, C. Nelson, B.J. Ritger, J. Selander, & A. Kapuscinski. 2003. High school students participate in Snuffbox host suitability trials. *Ellipsaria* 5(3): 19.

Howard, A.D. 1913. The catfish as a host for fresh water mussels. *Transactions of the American Fisheries Society* 42: 65–70.

Howard, A.D. 1914a. Experiments in propagation of fresh water mussels of the *Quadrula* group. *Report of the U.S. Commissioner of Fisheries for 1913. Appendix* 4: 1–52, 6 pls. [issued separately as *U.S. Bureau of Fisheries Document* 801].

Howard, A.D. 1914b. Some cases of narrowly restricted parasitism among commercial species of fresh water mussels. *Transactions of the American Fisheries Society* 44: 41–44.

Howard, A.D. 1914c. A new record in rearing fresh water pearl mussels. *Transactions of the American Fisheries Society* 44: 45–47.

Howard, A.D. 1914d. A second case of metamorphosis without parasitism in the Unionidae. *Science* 40: 353–355.

Howard, A.D. 1915. Some exceptional cases of breeding among the Unionidae. *The Nautilus* 29: 4–11.

Howard, A.D. 1916. An artificial infection with glochidia on the river herring. *Transactions of the American Fisheries Society* 46: 93–100.

Howard, A.D. 1922. Experiments in the culture of fresh water mussels. *Bulletin of the Bureau of Fisheries* 38: 63–89 [issued separately as *U.S. Bureau of Fisheries Document* 916].

Howard, A.D. 1951. A river mussel parasitic on a salamander. *Natural History Miscellanea* 77: 1–6.

Howard, A.D. & B.J. Anson. 1923. Phases in the parasitism of the Unionidae. *Journal of Parasitology* 9: 68–82, 2 pls.

Howells, R.G. 1994a. Host fish determination. *Info Mussel Newsletter* 2(2): 3–4.

Howells, R.G. 1994b. Student wins science fair with mussel project. *Info Mussel Newsletter* 2(3): 5–6.

Howells, R.G. 1996. Pistolgrip and Gulf mapleleaf hosts. *Info Mussel Newsletter* 4(3): 3.

Howells, R.G., R.W. Neck, & H.D. Murray. 1996. *Freshwater Mussels of Texas.* Texas Parks and Wildlife Press, Austin, TX. iv + 218 pp.

Howells, R.G. 1997. New fish hosts for nine freshwater mussels (Bivalvia: Unionidae) in Texas. *Texas Journal of Science* 49: 255–258.

Hruska, J. 1992. The freshwater pearl mussel in South Bohemia: Evaluation of the effect of temperature on reproduction, growth and age structure of the population. *Archiv für Hydrobiologia* 126: 181–191.

Hudson, R.G. & B.G. Isom. 1984. Rearing juveniles of the freshwater mussels (Unionidae) in a laboratory setting. *The Nautilus* 98: 129–135.

Huebner, J.D. & K.S. Pynnönen. 1992. Viability of glochidia of *Anodonta* exposed to low pH and selected metals. *Canadian Journal of Zoology* 70: 2348–2355.

Humphrey, C. 1987a. Effects of mine waters on freshwater mussels. Pp. 100–103 [*In:*] *Alligator Rivers Region Research Institute, annual research summary for 1985–86.* Australian Government Publishing Service, Canberra. 142 pp.

Humphrey, C. 1987b. Freshwater mussels (*Velesunio angasi*). Pp. 107–108. [*In:*] *Alligator Rivers Region Research Institute, annual research summary for 1986–1987.* Australian Government Publishing Service, Canberra.

Humphrey, C.L. & R.D. Simpson. 1985. The biology and ecology of *Velesunio angasi* (Bivalvia: Hyriidae) in the Malaga Creek, Northern Territory. *Report to the Office of Supervising Scientist for the Alligator Rivers Region, Open File Record* (38): 476 pp., 2 appendices.

Imlay, M.J. & M.L. Paige. 1972. Laboratory growth of freshwater sponges, unionid mussels, and sphaeriid clams. *Progressive Fish Culturist* 34: 210–216.

International Commission on Zoological Nomenclature. 1999. *International code of zoological nomenclature.* 4th ed. International Trust for Zoological Nomenclature, London. 306 pp.

International Commission on Zoological Nomenclature. 2008. *Obovaria* Rafinesque, 1819 (Mollusca, Bivalvia): Usage conserved by designation of *Unio retusa* Lamarck, 1819 as the type species. Opinion 2188 (Case 3353). *Bulletin of Zoological Nomenclature* 65: 64–65.

Isely, F.B. 1914. Experimental study of the growth and migration of fresh water mussels. *Report of the Commissioner of Fisheries for the Fiscal Year 1913, Appendix* 3: 1–24, 3 pls.

Isom, B.G. 1969. The mussel resource of the Tennessee River. *Malacologia* 7: 397–425.

Isom, B.G. 1971. Mussel fauna found in Fort Loudoun Reservoir Tennessee River, Knox County, Tennessee, in December, 1970. *Malacological Review* 4: 127–130.

Isom, B.G. 1983. Potential uses of *in vitro* culture of freshwater mussel glochidia for conservation. Pp. 42–45. [*In:*] A.C. Miller (ed.), *Report of freshwater mollusks workshop (26–27 October 1982).* U.S. Army Engineer Waterways Experimental Station, Vicksburg, Mississippi.

Isom, B.G & R.G. Hudson. 1982. *In vitro* culture of parasitic freshwater mussel glochidia. *The Nautilus* 96: 147–151.

Jacobson, P.J., J.L. Farris, D.S. Cherry, & R.J. Neves. 1993. Juvenile freshwater mussel (Bivalvia: Unionidae) responses to acute toxicity testing with copper. *Environmental Toxicology and Chemistry* 12: 879–883.

Jansen, W. 1991. Seasonal prevalence, intensity of infestation, and distribution of glochidia of *Anodonta grandis simpsoniana* Lea on Yellow Perch, *Perca flavescens. Canadian Journal of Zoology* 69: 964–972.

Jansen, W. & J.M. Hanson. 1991. Estimates of the number of glochidia produced by clams (*Anodonta grandis simpsonianus* Lea), attaching to Yellow Perch (*Perca flavescens*), and surviving to various ages in Narrow Lake, Alberta. *Canadian Journal of Zoology* 69: 973–977.

Jay, J.C. 1835. *Catalogue of Recent shells in the cabinet of John C. Jay, M.D.* 1st ed. New York, NY. 55 pp.

Jay, J.C. 1836. *Catalogue of Recent shells, with descriptions of new or rare species, in the collection of John C. Jay, M.D.* 2nd ed. New York, NY. 87 pp.

Jay, J.C. 1839. *A catalogue of the shells, arranged according to the Lamarckian system; together with descriptions of new or rare species, contained in the collection of John C. Jay, M.D.* 3rd ed. New York: Wiley and Putnam. 125 pp., 10 pls.

Jay, J.C. 1850. *A catalogue of the shells, arranged according to the Lamarckian system, with their authorities, synonomies, and references to works where figured or described, contained in the collection of John C. Jay M.D.* 4th ed. New York: R. Craighead. 459 pp.

Jay, J.C. 1852. *A catalogue of the shells, arranged according to the Lamarckian system, with their authorities, synonomies, and references to the works where figured or described, contained in the collection of John C. Jay, M.D.* 4th ed., with supplement. New York, NY. 479 pp.

Jenkinson, J.J. 1982. Cumberlandian mollusk conservation program. Pp. 95–103. [*In:*] A.C. Miller (compiler), *Report of freshwater mollusks workshop (19–20 May 1981).* U.S. Army Engineer Waterways Experimental Station, Vicksburg, Mississippi. 184 pp.

Jenkinson, J.J. 1983. Status report on the Tennessee Valley Authority Cumberlandian Mollusk Conservation Program. Pp. 79–83. [*In:*] A.C. Miller (compiler), *Report of freshwater mollusks workshop (26–27 October 1982).* U.S. Army Engineer Waterways Experimental Station, Vicksburg, Mississippi. 196 pp.

Jirka, K.J. & R.J. Neves. 1992. Reproductive biology of four species of freshwater mussels (Mollusca: Unionidae) in the New River, Virginia and West Virginia. *Journal of Freshwater Ecology* 7: 35–44.

Johnson, R.I. 1953. A study of Lamarck's types of Unionidae and Mutelidae. (Cont.). *The Nautilus* 66: 90–95.

Johnson, R.I. 1956. The types of naiades (Mollusca: Unionidae) in the Museum of Comparative Zoology. *Bulletin of the Museum of Comparative Zoology* (115): 101–142, 2 pls.

Johnson, R.I. 1969. Illustrations of Lamarck's types of North American Unionidae mostly in the Paris Museum. *The Nautilus* 83: 52–61.

Johnson, R.I. 1970. The systematics and zoogeography of the Unionidae (Mollusca: Bivalvia) of the southern Atlantic Slope Region. *Bulletin of the Museum of Comparative Zoology* 140: 263–449.

Johnson, R.I. 1971. The types and figured specimens of Unionacea (Mollusca: Bivalvia) in the British Museum (Natural History). *Bulletin of the British Museum (Natural History)* 20: 75–108, 2 pls.

Johnson, R.I. 1974. Lea's unionid types or recent and fossil taxa of Unionacea and Mutelacea introduced by Isaac Lea, including the location of all the extant types. *Special Occasional Publication, Museum of Comparative Zoology, Harvard University* (2). 159 pp.

Johnson, R.I. 1975. Simpson's unionid types and miscellaneous unionid types in the National Museum of Natural History. *Special Occasional Papers, Museum of Comparative Zoology, Harvard University* 4: 1–56, 3 pls.

Johnson, R.I. 1977. Arnold Edward Ortmann, a bibliography of his work on mollusks, with a catalogue of his recent molluscan taxa. *Occasional Papers on Mollusks, Museum of Comparative Zoology, Harvard University* 4: 229–241.

Johnson, R.I. 1978. Systematics and zoogeography of *Plagiola* (= *Dysnomia* = *Epioblasma*), an almost extinct genus of freshwater mussels (Bivalvia: Unionidae) from Middle North America. *Bulletin of the Museum of Comparative Zoology* 148: 239–321.

Johnson, R.I. 1980. The types of Unionacea (Mollusca: Bivalvia) in the Academy of Natural Sciences of Philadelphia: Additions and corrections. *Proceedings of the Academy of Natural Sciences of Philadelphia* 132: 277–278.

Johnson, R.I. & H.B. Baker. 1973. The types of Unionacea (Mollusca: Bivalvia) in the Academy of Natural Sciences of Philadelphia. *Proceedings of the Academy of Natural Sciences of Philadelphia* 125: 145–186.

Jones, R.L. 1991. Population status of endangered mussels in the Buttahatchee River, Mississippi and Alabama, Segment 2, 1990. *Mississippi Museum of Natural Science, Museum Technical Report* (14): 1–36 + appendices.

Jones, R.O. 1950. Propagation of fresh water mussels. *Progressive Fish Culturist* 12: 13–25.

Jones, J.W. & R.J. Neves. 2000. Annual progress report for 1999: Life history and artificial culture of endangered mussels. Report submitted to Tennessee Wildlife Resources Agency, Nashville. 66 pp.

Jones, J.W. & R.J. Neves. 2002. Life history and propagation of the endangered Fanshell Pearlymussel, *Cyprogenia stegaria* Rafinesque (Bivalvia: Unionidae). *Journal of the North American Benthological Society* 21: 76–88.

Jones, J.W., R. Mair, & R.J. Neves. 2003. Annual progress report for 2002: Life history and articial culture of endangered mussels. *Report submitted to Tennessee Wildlife Resources Agency, Nashville.* 80 pp.

Jones, J.W., R.J. Neves, S.A. Ahlstedt, & E.M. Hallerman. 2006. Holistic approach to taxonomic evaluation of two closely related endangered freshwater mussel species, the Oyster Mussel *Epioblasma capsaeformis* and Tan Riffleshell *Epioblasma florentina walkeri* (Bivalvia: Unionidae). *Journal of Molluscan Studies* 72: 267–283.

Joseph, P. 1998. The battle of the dams. *Smithsonian* 29(8): 48–61.

Karna, D.W. & R.E. Millemann. 1978. Glochidiosis of salmonid fishes. III. Comparative susceptibility to natural infection with *Margaritifera margaritifera* (L.) (Pelecypoda: Margaritanidae) and associated histopathology. *Journal of Parasitology* 64: 528–537.

Kat, P.W. 1983. Sexual selection and simultaneous hermaphroditism among the Unionidae (Bivalvia: Mollusca). *Journal of Zoology, London* 201: 395–416.

Kay, E.A. 1965. The Reverend John Lightfoot, Daniel Solander, and the Portland Catalogue. *The Nautilus* 79: 10–19.

Kays, W.T., H. Silverman, & T.H. Dietz. 1990. Water channels and water canals in the gill of the freshwater mussel, *Ligumia subrostrata:* Ultrastructure and histochemistry. *Journal of Experimental Zoology* 254: 256–269.

Keller, A.E. 1993. Acute toxicity of several pesticides, organic compounds and a wastewater effluent to the freshwater mussel, *Anodonta imbecilis, Ceriodaphnia dubia* and *Pimephales promelas. Bulletin of Environmental Contamination and Toxicology* 51: 696–702.

Keller, A.E. & D.S. Ruessler. 1997. Determination or verification of host fish for nine species of unionid mussels. *American Midland Naturalist* 138: 402–407.

Keller, A.E. & S.G. Zam. 1991. The acute toxicity of selected metals to the freshwater mussel, *Anodonta imbecillis. Environmental Toxicology and Chemistry* 10: 539–546.

Kelley, H.M. 1899. A statistical study of the parasites of the Unionidae. *Bulletin of the Illinois State Laboratory of Natural History* 5: 399–418.

Khym, J.R. & J.B. Layzer. 2000. Host fish suitability for glochia of *Ligumia recta. American Midland Naturalist* 143: 178–184.

Jared Potter Kirtland

Jean-Baptiste Pierre Antoine de Monet, Chevalier de Lamarck

Kirtland, J.P. 1838. Report on the zoology of Ohio. A catalogue of the mammals, birds, reptiles, fishes, Testacea, and Crustacea in Ohio. *Second Annual Report, Geological Survey of Ohio* 2: 157–200.

Klippel, W.E. & P.W. Parmalee. 1979. The naiad fauna of Lake Springfield, Illinois: An assessment after two decades. *The Nautilus* 94: 189–197.

Knudsen, K.A. & M.C. Hove. 1997. Spectaclecase (*Cumberlandia monodonta*) conglutinates unique, host(s) elusive. *Triannual Unionid Report* (11): 2.

Kohn, A.J. 1992. *A chronological taxonomy of* Conus, *1758–1840.* Washington, DC: Smithsonian Institution Press. 315 pp., 26 pls.

Kondo, T. 1990. Reproductive biology of a small bivalve *Grandidieria burtoni* in Lake Tanganyika. *Venus* 49: 120–125.

Kraemer, L. 1970. The mantle flap in three species of *Lampsilis* (Pelecypoda: Unionidae). *Malacologia* 10: 225–282.

Kunz, G.F. 1898. A brief history of the gathering of fresh water pearls in the United States. *Bulletin of the U.S. Fish Commission* 17: 321–335.

Küster, H.C. 1839, 1848–1862. *Unio & Hyria. Systematisches Conchylien Cabinet von Martini und Chemnitz.* Küster edition. 9 (Part 2): 318 pp., pls. 1–101.

Lamarck, J.B.P.A. de M. de [C. de]. 1799. Prodrome d'une nouvelle classification des coquilles, comprenent une rédaction appropriée des caractères génériques, et l'établissement d'un grand nombre de genres nouveaux. *Memoirs de la Sociétié d'Histoire Naturelle, Paris,* Vol. 1: 63–91.

Lamarck, J.B.P.A. de M. de [C. de]. 1819. Les nayades. [*In:*] *Histoire naturelle des animaux sans vertèbres.* 1st ed. Paris, Vol. 5: 67–100.

Lapham, I.A. 1852. Catalogue of the Mollusca of Wisconsin. *Transactions of the Wisconsin State Agricultural Society* 2: 367–370.

La Rocque, A. 1953. Catalogue of the Recent Mollusca of Canada. *National Museum of Canada Bulletin* 129: 1–406.

La Rocque, A. 1960. Molluscan faunas of the Flagstaff Formation of central Utah. *Geological Society of America Memoir* 78. 100 pp.

La Rocque, A. 1967. Pleistocene Mollusca of Ohio. *Bulletin of the Geological Survey of Ohio* 62: vii–xiv + 113–356.

La Rocque, A. & J. Oughton. 1937. A preliminary account of the Unionidae of Ontario. *Canadian Journal of Research* 15: 147–155.

Latchford, F.R. 1882. Notes on the Unionidae found in the vicinity of Ottawa, Ont. *Transactions of the Ottawa Field-Naturalists' Club* 1(3): 48–57 [reprinted in *Sterkiana* 8(1962): 19–28].

Layzer, J.B., M.E. Gordon, & R.M. Anderson. 1993. Mussels: The forgotten fauna of regulated rivers. A case study of the Caney Fork River. *Regulated Rivers: Research & Management* 8: 63–71.

Lea, I. 1828. Description of six new species of the genus *Unio*, embracing the anatomy of the oviduct of one of them, together with some anatomical observations on the genus. *Transactions of the American Philosophical Society* 3(n.s.): 259–273, 4 pls.

Lea, I. 1829. Description of a new genus of the family of naïades, including eight new species, four of which are new; also the description of eleven new species of the genus *Unio* from the rivers of the United States: With observations on some of the characters of the naïades. *Transactions of the American Philosophical Society* 3(n.s.): 403–457, 8 pls.

Lea, I. 1831. Observations on the naïades, and descriptions of new species of that and other families. *Transactions of the American Philosophical Society* 4(n.s.): 63–121, 16 pls.

Lea, I. 1832–1834. Observations on the naiades; and descriptions of new species of that and other families. *Transactions of the American Philosophical Society* 5(n.s.): 23–119, 19 pls.

Lea, I. 1836. *A synopsis of the family of naiades.* Philadelphia, PA: Carey, Lea & Blanchard, i–viii + 1–59, frontispiece, addenda, errata [also printed 1836 by J. Miller, London].

Lea, I. 1838a. Descriptions of new freshwater and land shells. *Transactions of the American Philosophical Society* 6(n.s.): 1–154, pls. 1–24.

Lea, I. 1838b. *A synopsis of the family of naiades.* 2nd ed. Philadelphia, PA: Carey, Lea & Blanchard. i–xx + 17–88.

Lea, I. 1840. Descriptions of new fresh water and land shells. *Proceedings of the American Philosophical Society* 1(13): 284–289.

Lea, I. 1841. Continuation of Ms. Lea's paper on fresh water and land shells. *Proceedings of the American Philosophical Society* 2: 81–83.

Lea, I. 1842. Descriptions of new fresh water and land shells. *Transactions of the American Philosophical Society* 2(n.s.): 163–250, pls. 5–27.

Lea, I. 1845. Descriptions of new fresh water and land shells. *Proceedings of the Academy of Natural Sciences of Philadelphia* 4(33): 162–168.

Lea, I. 1848. Descriptions of new fresh water and land shells. *Transactions of the American Philosophical Society* 10(n.s.)(1): 67–101.

Lea, I. 1852a. Descriptions of new species of the family Unionidae. *Transactions of the American Philosophical Society* 10(n. s.)(2): 253–294, pls. 12–29.

Lea, I. 1852b. *A synopsis of the family of naiades.* 3rd ed. Philadelphia, PA: Blanchard & Lea. 1–44.

Isaac Lea

Lea, I. 1857a. Description of thirteen new species of uniones, from Georgia. *Proceedings of the Academy of Natural Sciences of Philadelphia* 9: 31–32.

Lea, I. 1857b. Description of eight new species of naïades from various parts of the United States. *Proceedings of the Academy of Natural Sciences of Philadelphia* 9: 84.

Lea, I. 1857c. Description of twelve new species of naïades from North Carolina. *Proceedings of the Academy of Natural Sciences of Philadelphia* 9: 85–86.

Lea, I. 1857d. Description of six new species of fresh water and land shells of Texas and Tamaulipas, from the collection of the Smithsonian Institution. *Proceedings of the Academy of Natural Sciences of Philadelphia* 9: 101–102.

Lea, I. 1857e. *Observations on the genus* Unio, *together with descriptions of new species in the family Unionidae.* Privately printed. Pp. 7–48, 13 pls.

Lea, I. 1858a. Descriptions of new species of *Unio,* from Tennessee, Alabama and North Carolina. *Proceedings of the Academy of Natural Sciences of Philadelphia* 10: 40–41.

Lea, I. 1858b. Descriptions of seven new species of Margaritanae, and four new species of Anodontinae. *Proceedings of the Academy of Natural Sciences of Philadelphia* 10: 138–139.

Lea, I. 1858c. Descriptions of the embryonic forms of thirty eight species of Unionidae. *Journal of the Academy of Natural Sciences of Philadelphia* (2)4: 43–50, 1 pl.

Lea, I. 1859a. Descriptions of eight new species of Unionidae, from Georgia, Mississippi and Texas. *Proceedings of the Academy of Natural Sciences of Philadelphia* 11: 112–113.

Lea, I. 1859b. Descriptions of seven new species of uniones, from Georgia. *Proceedings of the Academy of Natural Sciences of Philadelphia* 11: 154–155.

Lea, I. 1860a. *Observations on the genus* Unio, *together with descriptions of new species, their soft parts, and embryonic forms, in the family Unionidae.* Privately printed. 93 pp., 25 pls.

Lea, I. 1860b. Descriptions of seven new species of Unionidae from the United States. *Proceedings of the Academy of Natural Sciences of Philadelphia* 12: 306–307.

Lea, I. 1860c. New Unionidae of the United States and Northern Mexico. *Journal of the Academy of Natural Sciences of Philadelphia* 4(n.s.)(4): 327–374, pls. 51–66.

Lea, I. 1861a. Descriptions of twenty five new species of Unionidae from Georgia, Alabama, Mississippi, Tennessee and Florida. *Proceedings of the Academy of Natural Sciences of Philadelphia* 13: 38–41.

Antoni van Leeuwenhoek

Lea, I. 1861b. Descriptions of two new species of Anodontae, from Arctic America. *Proceedings of the Academy of Natural Sciences of Philadelphia* 13: 56.

Lea, I. 1861c. Descriptions of seven new species of uniones, from Alabama. *Proceedings of the Academy of Natural Sciences of Philadelphia* 13: 59–60.

Lea, I. 1861d. Descriptions of eleven new species of the genus *Unio* from the United States. *Proceedings of the Academy of Natural Sciences of Philadelphia* 13: 391–393.

Lea, I. 1862. Descriptions of ten new species of Unionidae from the United States. *Proceedings of the Academy of Natural Sciences of Philadelphia* 14: 168–69.

Lea, I. 1863. Descriptions of the soft parts of one hundred and forty three species and some embryonic forms of Unionidae of the United States. *Journal of the Academy of Natural Sciences of Philadelphia* 5(n.s.): 401–456.

Lea, I. 1868. Descriptions of sixteen new species of the genus *Unio* from the United States. *Proceedings of the Academy of Natural Sciences of Philadelphia* 20: 143–145.

Lea, I. 1870. *A synopsis of the family Unionidae.* 4th ed. Philadelphia: H.C. Lea. i–xxx + 23–184.

Lea, I. 1871. Descriptions of twenty new species of uniones from the United States. *Proceedings of the Academy of Natural Sciences of Philadelphia* 23(3): 189–193.

Lea, I. 1872. Descriptions of twenty nine new species of Unionidae from the United States. *Proceedings of the Academy of Natural Sciences of Philadelphia* 24: 155–161.

Lea, I. 1874a. Supplement to Isaac Lea's paper on Unionidae. *Journal of the Academy of Natural Sciences of Philadelphia* 7(n. s.): 55–69.

Lea, I. 1874b. *Observations on the genus* Unio, *together with descriptions of new species in the family Unionidae, and descriptions of embryonic forms and soft parts, also, new species of Strepomatidae, Limnaeidae,* Vol. 13. Privately printed. 75 pp., 22 pls.

Leach, W.E. 1852. *A synopsis of the Mollusca of Great Britain.*

Lee, H.G. 2006. Musings on a local specimen of *Toxolasma paulum* (I. Lea, 1840), the Iridescent Lilliput. *Jacksonville Shell Club Shell-o-Gram* 47(5): 3–6.

Leeuwenhoek, A. van. 1697. *Continuatio Arcanorum Naturæ Detectorum.* Henricum a Kroonevelt. 192 pp. + index.

Lefevre, G. & W.C. Curtis. 1910a. Reproduction and parasitism in the Unionidae. *Journal of Experimental Zoology* 9: 79–115, 5 pls.

Lefevre, G. & W.C. Curtis. 1910b. Experiments in the artificial propagation of fresh water mussels. *Bulletin of the United States Bureau of Fisheries* [*Document* 671] 28: 615–626.

Lefevre, G. & W.C. Curtis. 1911. Metamorphosis without parasitism in the Unionidae. *Science* 33: 863–865.

Carl Linnaeus

Martin Lister

Lefevre, G. & W.C. Curtis. 1912. Studies on the reproduction and artificial propagation of fresh water mussels. *Bulletin of the Bureau of Fisheries* 30(1910): 105–201, 12 pls. [issued separately as *U.S. Bureau of Fisheries Document* 756. Reprinted in *Sterkiana* 47, 48 (1972); 49, 51 (1973); 57 (1975); and 61, 63, 64 (1976)].

Lellis, W.A. & T.L. King. 1998. Release of metamorphosed juveniles by the Green Floater, *Lasmigona subviridis*. *Triannual Unionid Report* (16): 23.

Letson, E.J. 1905. Check list of the Mollusca of New York. *New York State Museum Bulletin* 88: 1–112.

Lewandowski, K. 1976. Unionidae as a substratum for *Dreissena polymorpha* Pall. *Polskie Archivum Hydrobiologii* 23(3): 409–420.

Lewandowski, K. & A. Stanczykowska. 1975. The occurrence and role of bivalves of the family Unionidae in Mikolajskie Lake. *Ekologia Polska* 23: 317–334.

Lewis, J. 1870. Catalogue of the shells of the Coosa River, Alabama. *American Journal of Conchology* 5: 166–169.

[Lightfoot, J.]. 1786. *A catalogue of the Portland Museum, lately the property of the Duchess Dowager of Portland, deceased*. Privately printed, London. 194 pp.

Ligon, F.K., W.E. Dietrich, & W.J. Trush. 1995. Downstream ecological effects of dams. *BioScience* 45:183–192.

Linnaeus, C. 1758. *Systema naturae per regna tria naturae. Editio decima, 1. Regnum Animale*. Stockholm: L. Salvii. 824 pp.

Liquori, V.M. & G.D. Insler. 1985. Gill parasites of the White Perch: Phenologies in the lower Hudson River. *New York Fish and Game Journal* 32: 71–76.

Lister, M. 1685. *Historiae sive synopsis methodicae conchyliorum quorum ommium picturae, ad vivum delineatae, exhibetur. Liber Primus, qui est de cochleis terrestribus*. Oxford. iv + 12 + 77 + 6 pp., 1,059 + 22 pls.

Loar, J.M., L.L. Dye, R.R. Turner, & S.G. Hildebrand. 1980. Analysis of environmental issues related to small-scale hydroelectric development, Vol. I. Dredging. *Oak Ridge National Laboratory, Environmental Sciences Division Publication 1565*. 134 pp.

Luo, M. 1993. Host fishes of four species of freshwater mussels and development of an immune response. Unpublished Master's Thesis in Biology, Tennessee Technological University, Cookville. 32 pp.

Lydeard, C. & R.L. Mayden. 1995. A diverse and endangered aquatic ecosystem of the southeast United States. *Conservation Biology* 9: 800–805.

Lydeard, C. & K.J. Roe. 1998. Phylogenetic systematics: The missing ingredient in the conservation of freshwater unionid bivalves. *Fisheries* 23: 16–17.

Lydeard, C., M. Mulvey, & G.M. Davis. 1996. Molecular systematics and evolution of reproductive traits of North American freshwater unionacean mussels (Mollusca: Bivalvia) as inferred from 16S rRNA gene sequences. *Philosophical Transactions of the Royal Society, London* B, 351: 1593–1603.

Lynn, J.W. 1987. Release of motile spermatophores from the freshwater mussel *Anodonta grandis*. *American Zoologist* 27: 90A [abstract].

Mäkelä, T.P. & A.O.J. Oikari. 1992. The effects of low water pH on the ionic balance in freshwater mussel *Anodonta anatina* L. *Annales Zoologici Fennici* 29: 169–175.

Mansur, M.C.D. & N.M.R. de Campos Velho. 1990. Técnicas para o estudo dos gloquídios de Hyriidae (Mollusca, Bivalvia, Unionoida). *Acta Biologica Leopoldensia* 12: 5–18.

Marsh, W.A. 1891. Description of two new species of *Unio*, from Arkansas. *The Nautilus* 5: 1–2.

Marsh, W.A. 1897. New American Unionidae. *The Nautilus* 10: 103–104, 1 pl.

Marsh, W.A. 1901. Description of a new *Unio* from Missouri. *The Nautilus* 15: 74–75.

Marshall, W.B. 1916. A new genus and species of naiad from the James River at Huron, South Dakota. *The Nautilus* 29: 133–135, pl. 4.

Marshall, W.B. 1929. New fossil land and fresh-water mollusks from the Reynosa Formation of Texas. *Proceedings of the United States National Museum* 76(2798): 1–6, 1 pl.

Martens, K.E. von. 1890–1901. Land and freshwater Mollusca. [*In:*] F.D. Godman & O. Salvin (eds.), *Biologia Centrali Americana*. London: R.H. Porter. Pp. 1–40, pl. 1 [1890]; 41–96, pls. 2–5 [1891]; 91–176, pls. 6–9 [1892]; 177–248, pls. 10–12 [1893]; pls. 13–15 [1894]; 249–288, pl. 16 [1897]; 289–368, pls. 17–20 [1898]; 369–472, pls. 21–28 [1899]; 473–608, pls. 29–41 [1900]; 609–706 + xxviii, pls. 42–44 [1901].

Marzolf, G.R. 1978. The potential effects of clearing and snagging on stream ecosystems. *U.S. Fish and Wildlife Service, FWS/OBS 78/14*. Washington, DC.

Matteson, M.R. 1948. Life history of *Elliptio complanatus* (Dillwyn, 1817). *American Midland Naturalist* 40: 690–723.

Matteson, M.R. 1955. Studies on the natural history of the Unionidae. *American Midland Naturalist* 53: 126–145.

McGill, M., M. Hove, T. Diedrich, C. Nelson, W. Taylor, & A. Kapuscinski. 2002. Several fishes are suitable hosts for Creek Heelsplitter glochidia. *Ellipsaria* 4(2): 18–19.

McGregor, M.A. & M.E. Gordon. 1992. *Commercial Musseling in Tennessee.* Department of Tennessee Wildlife Resources, Nashville, Tennessee. 7 pp.

Menke, C.T. 1828. *Synopsis methodica molluscorum generum omnium et specierum earum, quae in Museo Menkeano adservantur; cum synonymia critica et novarum specierum diagosibus.* Pyrmonti: Henrici Gelpke. 62–66, 90–91.

Meuschen, F.C. 1787. *Museum Geversianum, sive index rerum naturalium . . . comparavit A. Gevers publice distrahendam.* Rotterdam: J. van Karnebeek. 659 pp.

Meyers, T.R. & R.E. Millemann. 1977. Glochidiosis of salmonid fishes. I. Comparative susceptibility to experimental infection with *Margaritifera margaritifera* (L.) (Pelecypoda: Margaritanidae). *Journal of Parasitology* 63: 728–733.

Michaud, A.L.G. 1831. *Complément de l'histoire naturelle des mollusques de France terrestres et fluviatiles de J.P.R. Draparnaud.* Verdun. vi + 116 + 12 pp.

Miles, M. 1861. A catalogue of the mammals, birds, reptiles and mollusks, of Michigan. Pp. 219–241. [*In:*] A. Winchell. (ed.), *First biennial report of the progress of the Geological Survey of Michigan, embracing observations of the geology, zoölogy and botany of the Lower Peninsula.* Lansing, MI: Homer & Kerr.

Miller, A.C. & B.S. Payne. 1992. Characterization of a freshwater mussel (Unionidae) community immediately downriver of Kentucky Lock and Dam in the Tennessee River. *Transactions of the Kentucky Academy of Science* 53: 154–161.

Mills, E.L., J.H. Leach, J.T. Carlton, & C.L. Secor. 1993. Exotic species in the Great Lakes: A history of biotic crises and anthropogenic introductions. *Journal of Great Lakes Research* 19: 1–54.

Mitchell, R. & J.L. Wilson. 1965. New species of water mites (*Unionicola*) from Tennessee unionid mussels. *Journal of the Tennessee Academy of Sciences* 40: 104–107.

Mörch, O.A.L. 1852–1853. *Catalogus conchyliorum quae reliquit D. Alphonso D'Aguirra et Gadea Comes de Yoldi.* Copenhagen. Fasc. 1: 170 pp. [1852; Gastropoda]; Fasc. 2: 74 pp. [post 1 April 1853; Bivalvia].

Morelet, A. 1845. *Description des mollusques terrestres et fluviatiles de Portugal.* Paris: Bailliere. i–vii + 1–116 pp., 14 pls.

Morris, P.J. & P.G. Williamson. 1988. *Pleiodon* (Conrad) (Bivalvia: Mutelidae: Pleiodoninae) from the Late Cretaceous of Montana: A first North American record for the Mutelidae. *Journal of Paleontology* 62: 758–765.

Morris, T.J. & L.D. Corkum. 1996. Assemblage structure of freshwater mussels (Bivalvia: Unionidae) in rivers with grassy and forested riparian zones. *Journal of the North American Benthological Society* 15: 576–586.

Morrison, J.P.E. 1929. A preliminary list of the Mollusca of Dane County, Wisconsin. *Transactions of the Wisconsin Academy of Sciences, Arts and Letters* 24: 405–425.

Morrison, J.P.E. 1932. A report on the Mollusca of the northeastern Wisconsin lake district. *Transactions of the Wisconsin Academy of Sciences, Arts and Letters* 27: 359–396.

Morrison, J.P.E. 1942. Preliminary report on mollusks found in the shell mounds of the Pickwick Landing basin in the Tennessee River valley. *Bureau of American Ethnology Bulletin* 129: 337–392.

Morrison, J.P.E. 1969. The earliest names for North American naiads. *American Malacological Union, Inc. Annual Reports* 36: 22–24.

Morrison, J.P.E. 1976. Species of the genus *Uniomerus. Bulletin of the American Malacological Union, Inc. for 1976:* 10–11.

Mulvey, M., C. Lydeard, D.L. Pyer, K.M. Hicks, J. Brim-Box, J.D. Williams, & R.S. Butler. 1997. Conservation genetics of North American freshwater mussels *Amblema* and *Megalonaias. Conservation Biology* 11: 868–878.

Murphy, G. 1942. Relationship of the fresh water mussel to trout in the Truckee River. *California Fish and Game* 28: 89–102.

Murray, H.D. & A.B. Leonard. 1962. Handbook of unionid mussels in Kansas. *Miscellaneous Publication, Museum of Natural History, University of Kansas. Lawrence, KS* (28). 184 pp.

Murray, H.D. & E.C. Roy. 1968. Checklist of fresh water and land mollusks of Texas. *Sterkiana* 30: 25–42.

Mutvei, H. & T. Westermark. 2000. How environmental information can be obtained from unionid shells. Pp. 367–379. [*In:*] G. Bauer & K. Wächtler (eds.), *Ecology and evolution of the freshwater mussels Unionoida.* Berlin: Springer-Verlag.

Nagel, K.O. 1992. Das Schalenwachstum dreier Muschelarten (Bivalvia: Unionidae) in der Schwalm, einem nordhessischen Mittelgebirgsfluß. *Decheniana* 145: 165–176.

Neave, S.A. 1940. *Nomenclator Zoologicus,* Vol. 4. Zoological Society, London. 758 pp.

Neck, R.W. & R.G. Howells. 1994. Status survey of Texas Heelsplitter, *Potamilus amphichaenus* (Frierson, 1898). *Texas Parks and Wildlife Department, Resource Protection Division and Inland Fisheries Division, Special Report.*

Neel, J.K & W.R. Allen. 1964. The mussel fauna of the Upper Cumberland Basin before its impoundment. *Malacologia* 1: 427–459.

Nelson, K.L. 1993. Instream sand and gravel mining. Pp. 189–196. [*In:*] C.F. Bryan & D.A. Rutherford (eds.), *Impacts on warmwater streams: Guidelines for evaluation.* Southern Division, American Fisheries Society, Little Rock, AR.

Nelson, R.W., J.R. Dwyer, & W.E. Greenberg. 1987. Regulated flushing in a gravel bed river for channel habitat maintenance: A Trinity River case study. *Environmental Management* 11: 479–493.

Neves, R.J. 1983. The status of freshwater mussel research in Virginia. Pp. 155–168. [*In:*] A.C. Miller (compiler), *Report of freshwater mollusks workshop (26–27 October 1982).* U.S. Army Engineer Waterways Experimental Station, Vicksburg, Mississippi. 196 pp.

Neves, R.J. & M.C. Odom. 1989. Muskrat predation on endangered freshwater mussels in Virginia. *Journal of Wildlife Management* 53: 934–941.

Neves, R.J. & J.C. Widlak. 1988. Occurrence of glochidia in stream drift and on fishes of the Upper North Fork Holston River, Virginia. *American Midland Naturalist* 119: 111–120.

Neves, R.J., L.R. Weaver, & A.V. Zale. 1985. An evaluation of host fish suitability for glochidia of *Villosa vanuxemi* and *V. nebulosa* (Pelecypoda: Unionidae). *American Midland Naturalist* 113: 13–19.

Nezlin, L.P., R.A. Cunjak, A.A. Zotin, & V.V. Ziuganov. 1994. Glochidium morphology of the Freshwater Pearl Mussel (*Margaritifera margaritifera*) and glochidiosis of Atlantic Salmon (*Salmo salar*): A study by scanning electron microscopy. *Canadian Journal of Zoology* 72: 15–21.

Nichols, S.J. & D. Garling. 1998. Food web dynamics of Unionidae in a canopied river and a non-canopied lake. *Program & Abstracts, Freshwater Mussel Symposium, Columbus, OH, 6–8 March, 1998.* 28–29.

Alcide Charles Victor Marie Dessalines d'Orbigny

Arnold Edward Ortmann

Nilsson, S. 1822. *Historia Molluscorum Sueciae terrestrium et fluviatilium breviter delineata,* Vols. 1–20. Berlin: Lundae. 1–124.

Nitzsch, C.L. 1820. *Anodonta.* Pp. 189–190. [*In:*] J.S. Ersch & J. G. Gruber (eds.), *Allgemeine Encyclopädie der Wissenschaften und Künste,* Vol. 4. Leipzig. 1–475.

O'Dee, S.H. & G.T. Watters. 2000. New or confirmed host identification for ten freshwater mussels. Pp. 77–82. [*In:*] R.A. Tankersley, D.I. Warmolts, G.T. Watters, B.J. Armitage, P.D. Johnson, & R.S. Butler (eds.), *Proceedings of the Conservation, Captive Care and Propagation of Freshwater Mussels Symposium.* Ohio Biological Survey, Columbus.

Oesch, R.D. *Missouri naiades. A guide to the mussels of Missouri.* Missouri Department of Conservation. Jefferson City, Missouri. vii + 270 pp.

Ohio Division of Wildlife. 1997. *Ohio State Management Plan for Aquatic Nuisance Species.* Ohio Department of Natural Resources, Division of Wildlife, Columbus, Ohio. 10 pp.

Ohio Sea Grant. 1995. *Sea Grant Zebra Mussel Report: An Update of Research and Outreach, 1988–1994.* Ohio Sea Grant, The Ohio State University, Columbus, Ohio. 54 pp.

Oken, L. 1815–1816. *Lehrbuch der Naturgeschichte,* Vol. 3. Zoologie. Jena, Germany: August Schmidt & Co. 850 pp.

d'Orbigny, A.C.V.D. 1846. Voyage dans l'Amérique Méridionale 1835–1843. *Libraire de la Société Géologique de France, Paris,* Vol. 5, part 3, Mollusques. 594–622.

Ortmann, A.E. 1909a. The destruction of the fresh-water fauna in western Pennsylvania. *Proceedings of the American Philosophical Society* 48: 90–110.

Ortmann, A.E. 1909b. A preliminary list of the Unionidae of Western Pennsylvania, with new localities for species from Eastern Pennsylvania. *Annals of the Carnegie Museum* 5: 178–210.

Ortmann, A.E. 1910. A new system of the Unionidae. *The Nautilus* 23: 114–120.

Ortmann, A.E. 1911. A monograph of the najades of Pennsylvania. Parts I and II. *Memoirs of the Carnegie Museum* 4: 279–347, 4 pls.

Ortmann, A.E. 1912a. Notes upon the families and genera of the najades. *Annals of the Carnegie Museum* 8: 222–365, 3 pls.

Ortmann, A.E. 1912b. *Cumberlandia,* a new genus of naiades. *The Nautilus* 26: 13–14.

Ortmann, A.E. 1913a. The Alleghenian Divide, and its influence on the freshwater fauna. *Proceedings of the American Philosophical Society* 52: 287–390, pls. 12–14.

Ortmann, A.E. 1913b. Studies in najades. *The Nautilus* 27: 88–91.

Ortmann, A.E. 1914. Studies in najades. (Cont.). *The Nautilus* 28: 20–22, 28–34, 41–47, 65–69.

Ortmann, A.E. 1915. Studies in najades. (Cont.). *The Nautilus* 28: 106–108, 129–131, 141–143; 29: 63–67.

Ortmann, A.E. 1916. Studies in najades. (Cont.). *The Nautilus* 30: 54–57.

Ortmann, A.E. 1918. The nayades (freshwater mussels) of the upper Tennessee drainage, with notes on synonymy and distribution. *Proceedings of the American Philiosophical Society* 57: 521–626.

Ortmann, A.E. 1919. A monograph of the naiades of Pennsylvania. Part III. Systematic account of the genera and species. *Memoirs of the Carnegie Museum* 8: 1–385, 21 pls., 34 figs.

Ortmann, A.E. 1920. Correlation of shape and station in freshwater mussels (naiades). *Proceedings of the American Philiosophical Society* 59: 269–312.

Ortmann, A.E. 1921. The anatomy of certain mussels from the upper Tennessee. *The Nautilus* 34: 81–91.

Ortmann, A.E. 1924. The naiad fauna of Duck River in Tennessee. *American Midland Naturalist* 9: 18–62.

Ortmann, A.E. 1925. The naiad fauna of the Tennessee River system below Walden Gorge. *American Midland Naturalist* 9: 321–372.

Ortmann, A.E. 1926. The naiades of the Green River drainage in Kentucky. *Annals of the Carnegie Museum* 17: 167–188, pl. 8.

Ortmann, A.E. & B. Walker. 1922. On the nomenclature of certain North American naiades. *Occasional Papers of the Museum of Zoology, University of Michigan* (112): 2–75.

Paetel, F. 1890. *Catalog der Conchylien Sammlung von Fr. Paetel. Nebst Übersicht des angewandten Systems.* 4th ed. Vol. 3. Berlin. 1–256 + 23.

Parker, R.S., C.T. Hackney, & M.F. Vidrine. 1984. Ecology and reproductive strategy of a south Louisiana freshwater mussel, *Glebula rotundata* (Lamarck) (Unionidae: Lampsilini). *Freshwater Invertebrate Biology* 3: 53–58.

Parker, R.S., M.F. Vidrine, & C.T. Hackney. 1980. A new centrarchid host for the Paper Pond Shell, *Anodonta imbecillis* Say (Bivalvia: Unionidae). *ASB Bulletin* 27: 54–55.

Parmalee, P.W. 1967. The fresh water mussels of Illinois. *Illinois State Museum Popular Science Series* 8: 1–108.

Parmalee, P.W. & A.E. Bogan. 1998. *The freshwater mussels of Tennessee.* Knoxville: University of Tennessee Press. 328 pp.

Parmalee, P.W. & M.H. Hughes. 1993. Freshwater mussels (Mollusca: Pelecypoda: Unionidae) of Tellico Lake: Twelve years after impoundment of the Little Tennessee River. *Annals of the Carnegie Museum* 62: 81–93.

Parmalee, P.W., W.E. Klippel, & A.E. Bogan. 1982. Aboriginal and modern freshwater mussel assemblages (Pelecypoda: Unionidae) from the Chickamauga Reservoir, Tennessee. *Brimleyana* (8): 75–90.

Parodiz, J.J. 1967. Types of North American Unionidae in the collection of the Carnegie Museum. *Sterkiana* 28: 21–30.

Payne, B.S. & A.C. Miller. 1989. Growth and survival of recent recruits to a population of *Fusconaia ebena* (Bivalvia: Unionidae). *American Midland Naturalist* 121: 99–104.

Peacock, E. 2000. Assessing bias in archaeological shell assemblages. *Journal of Field Archaeology* 27: 183–196.

Peacock, E., W.R. Haag, and M.L. Warren, Jr. 2005. Prehistoric decline in freshwater mussels coincident with the advent of maize agriculture. *Conservation Biology* 19: 547–551.

Pearse, A.S. 1924. The parasites of lake fishes. *Transactions of the Wisconsin Academy of Sciences, Arts and Letters* 21: 161–194.

Pekkarinen, M. & I. Valovirta. 1996. Anatomy of the glochidia of the freshwater pearl mussel, *Margaritifera margaritifera* (L.). *Archiv für Hydrobiologie* 137: 411–423.

Penn, G.H. 1939. A study of the life cycle of the freshwater mussel, *Anodonta grandis*, in New Orleans. *The Nautilus* 52: 99–101.

Pepi, V.E. & M.C. Hove. 1997. Suitable fish hosts and mantle display behavior of *Tritogonia verrucosa*. *Triannual Unionid Report* (11): 5.

Petrbok, J. 1930. Verzeichniss der pleistozänen Mollusken von Bulgarien. *Zetschrift der Bulgarischen geologischen Gesellschaft, Sofia* 2(3): 9–16.

Petts, G.E. 1984. *Impounded rivers, perspectives for ecological management*. New York: John Wiley & Sons. 326 pp.

Philippi, R.A. 1853. *Handbuch der Conchyliologie und Malakozoologie*. Halle, Germany: Eduard Anton. 547 pp.

Philipsson, L.M. 1788. [*In:*] A.J. Retzius, *Dissertatio historico-naturalis sistems nova testaceorum genera*. University of Lund, Sweden. 23 pp.

Potiez, V.L.V. & A.L.G. Michaud. 1844. *Galerie des mollusques ou catalogue méthodique, descriptif et raisonné des Mollusques et Coquilles du Muséum de Douai*. Paris: J.B. Baillière.

Poulson, C.A. 1832. *A monograph of the fluviatile bivalve shells of the River Ohio containing twelve genera and sixty eight species* [English translation of Rafinesque, 1820]. J. Dobson, 108 Chestnut Street, Philadelphia, Pennsylvania. 72 pp.

Poupart. 1706. Remarques sur les coquillages à deux coquilles, & premièrement sur les moules. *Histoire de l'Académie Royale des Sciences, Paris pour 1706:* 51–61, 5 figs.

Pratt, W.H. 1876. Description of a *Unio* shell found on the south bank of the Mississippi River, opposite the Rock Island Arsenal, in 1870. *Proceedings of the Davenport Academy of Natural Sciences* 1: 167–168, 1 pl.

Prévost, I. 1826. De la génération chez la moule de peintres (*Unio Pictorum*). *Annales des Sciences Naturelles, Paris* 7: 447–454.

Price, S.F. 1900. The Mollusca of southern Kentucky. *The Nautilus* 14: 75–79.

Rafinesque, C.S. 1818. Discoveries in natural history, made during a journey through the western region of the United States. *American Monthly Magazine and Critical Review* 3: 354–355.

Rafinesque, C.S. 1819. Prodrome de 70 nouveaux genres d'animaux découverts dans l'intérieur des États Unis d'Amérique, durant l'année 1818. *Journal de Physique, de Chimie, d'Histoire Naturelle* 88: 417–429.

Contantine Samuel Rafinesque-Schmaltz

Rafinesque, C.S. 1820. Monographie des coquilles bivalves fluviatiles de la Rivière Ohio, contenant douze genres et soixante huit espèces. *Annales Generales des Sciences Physique, Bruxelles* 5: 287–322, 2 pls.

Rafinesque, C.S. 1831. *Continuation of a monograph of the bivalve shells of the river Ohio and other rivers of the western states*. Philadelphia, PA. 8 pp.

Rafinesque, C.S. 1832. *Odatelia* N.G. of N. America bivalve fluviatile shell. *Atlantic Journal and Friend of Knowledge* 4: 154.

Rand, T.G. & M. Wiles. 1982. Species differentiation of the glochidia of *Anodonta cataracta* Say, 1817 and *Anodonta implicata* Say, 1829 (Mollusca: Unionidae) by scanning electron microscopy. *Canadian Journal of Zoology* 60: 1722–1727.

Rathke, J. 1797. *Om Dammuslingen. Naturhistoire Selskabets Skrifter*, Vol. 4. Copenhagen. 139–179.

Read, L.B. & K.H. Oliver. 1953. Notes on the ecology of the fresh water mussels of Dallas County. *Field and Laboratory* 21: 75–80.

Reaka-Kudla, M.L., D.E. Wilson, & E.O. Wilson (eds.). 1997. *Biodiversity II, Understanding and Protecting our Biological Resources*. Washington, DC: Joseph Henry Press. 549 pp.

Reardon, L. 1929. A contribution to our knowledge of the anatomy of the freshwater mussels of the District of Columbia. *Proceedings of the U.S. National Museum* 75(11): 1–12, 5 pls.

Rees, W.J. 1965. The aerial dispersal of Mollusca. *Proceedings of the Malacological Society of London* 36: 269–282.

Reeve, L.A. 1843–1878. *Conchologica Iconica: Or illustrations of the shells of molluscous animals*. London: L. Reeve & Co. 20 vols., 2,727 pls.

Rehder, H.A. 1967. Valid zoological names of the Portland Catalogue. *Proceedings of the United States National Museum* 121(3579): 1–51.

Retzius, A.J. 1788. *Dissertatio historico-naturalis sistems nova testaceorum genera*. University of Lund, Sweden. 23 pp.

Reuling, F.H. 1919. Acquired immunity to an animal parasite. *Journal of Infectuous Diseases* 24: 337–346.

Rhoads, S.N. 1899. On a recent collection of Pennsylvanian mollusks from the Ohio River system below Pittsburgh. *The Nautilus* 12: 133–138.

Richard, P.E., T.H. Dietz, & H. Silverman. 1991. Structure of the gill during reproduction in the unionids *Anodonta grandis*,

Ligumia subrostrata, and *Carunculina parva texasensis. Canadian Journal of Zoology* 69: 1744–1754.

Richards, H.G. 1948. Fossil mollusks from the Triassic of Pennsylvania. *Notulae Naturae* (206): 1–4, figs. 1–4.

Richardson, R.E. 1928. The bottom fauna of the middle Illinois River, 1913–1925. Its distribution, abundance, valuation, and index value in the study of stream pollution. *Illinois State Natural History Survey Bulletin* 17: 387–475.

Riggs, C.D. & G.R. Webb. 1956. The mussel population of an area of loamy sand bottom of Lake Texoma. *American Midland Naturalist* 56: 197–203.

Riusech, F.A. & M.C. Barnhart. 1998. Host suitability differences among *Venustaconcha ellipsiformis* (Bivalvia: Unionidae) from different river drainages. *Conservation, Captive Care, & Propagation, Freshwater Mussel Symposium, Columbus, OH, Program & Abstracts:* 32 [abstract].

Riusech, F.A. & M.C. Barnhart. 2000. Host suitability and utilization in *Venustaconcha ellipsiformis* and *Venustaconcha pleasii* (Bivalvia: Unionidae) from the Ozark Plateaus. Pp. 83–91. [*In:*] R.A. Tankersley, D.I. Warmoltz, G.T. Watters, B.J. Armitage, P.D. Johnson, & R.S. Butler (eds.), *Freshwater mollusk symposium proceedings, Ohio Biological Survey, Columbus.*

Roberts, A.D. 1997. Reproductive biology of the flat floater mussel, *Anodonta suborbiculata,* Say, 1831 (Mollusca: Unionidae). Master's Thesis, Southwest Missouri State University. 74 pp.

Robertson, I.C.S. & C.L. Blakeslee. 1948. The Mollusca of the Niagara frontier region. *Bulletin of the Buffalo Society of Natural Sciences* 19(3): xi–191 pp.

Roe, K.J. & W.R. Hoeh. 2003. Systematics of freshwater mussels (Bivalvia: Unionoida). Pp. 93–122. [*In:*] C. Lydeard & D. Lindberg (eds.), *Molecular systematics and phylogeography of mollusks.* Washington, DC, and London: Smithsonian Books.

Roe, K.J. & C. Lydeard. 1998. Molecular systematics of the freshwater mussel genus *Potamilus* (Bivalvia: Unionidae). *Malacologia* 39: 195–205.

Roe, K.J., P.D. Hartfield, & C. Lydeard. 2001. Phylogeographic analysis of the threatened and endangered superconglutinate-producing mussels of the genus *Lampsilis* (Bivalvia: Unionidae). *Molecular Ecology* 10: 2225–2234.

Rosenberg, G. & M.L. Ludyanskiy. 1994. A nomenclatural review of *Dreissena* (Bivalvia: Dreissenidae), with identification of the Quagga Mussel as *Dreissena bugensis. Canadian Journal of Fisheries and Aquatic Sciences* 51: 1474–1484.

Rossmässler, E.A. 1835–1854. *Ikonographie der Land und Süßwasser Mollusken, mit vorzhglicher Berücksichtigung der europäischen noch nicht abgebildeten Arten.* Dresden & Wiesbaden. Vols. 1–3, 90 pls.

Savazzi, E. & Y. Peiyi. 1992. Some morphological adaptations in freshwater bivalves. *Lethaia* 25: 195–209.

Say, T. 1817. Conchology. [*In:*] W. Nicholson (ed.), *American edition of the British encyclopedia or dictionary of arts and sciences, comprising an accurate and popular view of the present improved state of human knowledge,* Vol. 2. 1st ed. Philadelphia: S.A. Mitchell & H. Ames. No pagination.

Say, T. 1818. Description of a new genus of fresh water bivalve shells. *Journal of the Academy of Natural Sciences of Philadelphia* 1: 459–460.

Say, T. 1819. Conchology. [*In:*] W. Nicholson (ed.). *American edition of the British encyclopedia or dictionary of arts and sciences, comprising an accurate and popular view of the present improved state of human knowledge.* 3rd ed. Philadelphia: Samuel A. Mitchel and Horace Ames. 19 pp. + 4 pls.

Say, T. 1825. Descriptions of some new species of fresh water and land shells of the United States. *Journal of the Academy of Natural Sciences of Philadelphia* 5: 119–131.

Thomas Say

Say, T. 1829. Descriptions of some new terrestrial and fluviatile shells of North America. *The disseminator of useful knowledge; containing hints to the youth of the U. States, from the School of Industry* 2: 275–277.

Say, T. 1830a–1834? *American conchology, or descriptions of the shells of North America. Illustrated by colored figures from original drawings executed from nature.* New Harmony, IN. Vol. 1: unpaginated + pls. 1–10 [1830]; Vols. 2–3: unpaginated + pls. 11–30 [Apr., Sept. 1831a]; Vols. 4–5: unpaginated + pls. 31–50 [Mar., Aug. 1832]; Vol. 6: unpaginated + pls. 51–60 [Apr. 1834]; Vol. 7: unpaginated + pls. 61–68 [undated, posthumously published and edited by T.A. Conrad].

Say, T. 1830b–1831b. New terrestrial and fluviatile shells of North America. (Cont.). *The disseminator of useful knowledge; containing hints to the youth of the U. States, from the School of Industry* (2)1(27): no pagination [28 Dec. 1830]; (2)1(29): no pagination [15 Jan. 1831]; (2)1(31): no pagination [29 Jan. 1831].

Say, T. 1831c. Descriptions of several new species of shells and of a new species of *Lumbricus. Transylvania Journal of Medicine* 4: 525–528.

Say, T. 1834. An attempt to exhibit the synonymy of the western North American species of *Unio* and *Alasmodonta.* Appearing with Part 6 of *American Conchology.* Unpaginated [2 pp.]

[Say, T.]. 1840. *Descriptions of some new terrestrial and fluviatile shells of North America* [compilation by Lucy Say]. New Harmony, IN. 26 pp.

Scammon, R.E. 1906. The Unionidae of Kansas, Part I. An illustrated catalogue of the Kansas Unionidae. *Kansas University Science Bulletin* 3: 279–373, 21 pls.

Schaufuss, L.W. 1869. *Molluscorum systema et catalogus. System und Aufzählung sämmtlicher Conchylien der Sammlung von Fr. Paetel.* Dresden: Oscar Weiske. 119 pp.

Schierholz, C. 1889. Über Entwicklung der Unioniden. *Denkschriften der Kaiserlichen Akademie der Wissenschaften, Wien. Mathematisch Naturwissenschafliche Classe* 55: 183–214; 4 pls.

Schlüter, F. 1838. *Kurzgefasztes systematisches Verzeichniss meiner Conchyliensammlung nebst Andeutung aller bis jetzt von mir bei Halle gefundenen Land-und Fluszconchylien.* Halle, Germany. i–vii + 40 pp.

Schmidt, J.E. 1986. Notes on the historic and present naiad fauna of the Caney Fork River, central Tennessee. *American Malacological Bulletin* (4): 117.

Schrenck, L. 1867. Mollusken des Amur Landes und des Nordjapanischen Meeres. *Reisen und Forschungen im Amur Lande in den Jahren 1854–1856,* Vol. 2. 259–974.

Schulz, C. & K. Marbain. 1998. Hosts species for rare freshwater mussels in Virginia. *Triannual Unionid Report* (15): 32–38.

Schumacher, C.F. 1816. Oversigt over det Konglige Danske. *Videnskabernes Selskaps Forhandlinger* 7: 7.

Schumacher, C.F. 1817. *Essai d'un nouveau système des habitations des vers testacés.* Copenhagen: Schultz. 287 pp., 22 pls.

Schuster, G.A., R.S. Butler, & D.H. Stansbery. 1989. A survey of the unionids (Bivalvia: Unionidae) of Buck Creek, Pulaski County, Kentucky. *Transactions of the Kentucky Academy of Science* 50: 79–85.

Schwebach, M., D. Schriever, N. Dillon, M. Hove, M. McGill, C. Nelson, F. Thomas, & A. Kapuscinski. 2002. Channel Catfish is a suitable host species for Mapleleaf glochidia. *Ellipsaria* 4(3): 12–13.

Scruggs, G.D. 1960. Status of fresh water mussel stocks in the Tennessee River. *U.S. Fish & Wildlife Service, Special Scientific Report, Fisheries* (370). 41 pp.

Semenova, M.N., L.A. Karpytcheva, K.K. Voloshenko, & B.F. Bugaev. 1992. Comparative study of shell growth rate in Freshwater Pearl Mussel *Margaritifera margaritifera* (Bivalvia, Margaritiferidae) from different rivers. *Zoologicheskii Zhurnal* 71(5): 19–27.

Serb, J.M. 2006. Discovery of genetically distinct sympatric lineages in the freshwater mussel *Cyprogenia aberti* (Bivalvia: Unionidae). *Journal of Molluscan Studies* 72: 425–434.

Serb, J.M., J.E. Buhay, & C. Lydeard. 2003. Molecular systematics of the North American freshwater bivalve genus *Quadrula* (Unionidae: Ambleminae) based on mitochondrial ND1 sequences. *Molecular Phylogenetics and Evolution* 28: 1–11.

Sherman, R.A. 1993. Glochidial release and reproduction of the Snuffbox Mussel, *Epioblasma triquetra;* timing in southern Michigan. *Bulletin of the North American Benthological Society* 10: 197.

Short, C.W. & H.H. Eaton. 1831. Notices on western botany and conchology. *Transylvania Journal of Medicine* 4: 69–82.

Silverman, H., W.L. Steffens, & T.H. Dietz. 1985. Calcium from extracellular concretions in the gills of freshwater unionid mussels is mobilized during reproduction. *Journal of Experimental Zoology* 236: 137–147.

Simmons, G.M. & J.R. Reed. 1973. Mussels as indicators of biological recovery zone. *Journal of the Water Pollution Control Federation* 45: 2480–2492.

Simpson, C.T. 1892. Notes on the Unionidae of Florida and the southeastern states. *Proceedings of the United States National Museum* 15(911): 405–436, 26 pls.

Simpson, C.T. 1896. Description of four new Triassic unios from the Staked Plains of Texas. *Proceedings of the U.S. National Museum* 18: 381–385.

Simpson, C.T. 1900a. Synopsis of the naiades, or pearly freshwater mussels. *Proceedings of the United States National Museum* 22(1205): 501–1044.

Simpson, C.T. 1900b. New and unfigured Unionidae. *Proceedings of the Academy of Natural Sciences of Philadelphia* 52: 74–86, pls. 1–5.

Simpson, C.T. 1914. *A descriptive catalogue of the naiades, or pearly fresh-water mussels.* Detroit, MI: Bryant Walker. 3 vols. xii + 1,540 pp.

Sinclair, R.M. & B.G. Isom. 1963. *Further studies on the introduced Asiatic clam* (Corbicula) *in Tennessee.* Tennessee Stream and Pollution Control Board, Tennessee Department of Public Health, Nashville. 75 pp.

Siripattrawan, S., J-K. Park, & D. O'Foighil. 2000. Two lineages of the introduced Asian freshwater clam *Corbicula* occur in North America. *Journal of Molluscan Studies* 66: 423–429.

Charles Torrey Simpson

Smith, D.G. 1976. Notes on biology of *Margaritifera margaritifera margaritifera* (Lin.) in central Massachusetts. *American Midland Naturalist* 96: 252–256.

Smith, D.G. 1979. Marsupial anatomy of the demibranch of *Margaritifera margaritifera* (Lin.) in northeastern North America (Pelecypoda: Unionacea). *Journal of Molluscan Studies* 45: 39–44.

Smith, D.G. 2000a. On the taxonomic placement of *Unio ochraceus* Say, 1817 in the genus *Ligumia* (Bivalvia: Unionidae). *The Nautilus* 114: 155–160.

Smith, D.G. 2000b. Investigations of the byssal gland in juvenile unionids. Pp. 103–107. [*In:*] R.A. Tankersley, D.I. Warmolts, G.T. Watters, B.J. Armitage, P.D. Johnson, & R.S. Butler (eds.), *Proceedings of the Conservation, Captive Care and Propagation of Freshwater Mussels Symposium.* Ohio Biological Survey, Columbus.

Smith, D.G. & W.P. Wall. 1984. The Margaritiferidae reinstated: A reply to Davis and Fuller (1981), "Genetic relationships among recent Unioniacea (Bivalvia) of North America." *Occasional Papers on Mollusks, Museum of Comparative Zoology, Harvard University* 4: 321–330.

Smith, H.M. 1899. The mussel fishery and pearl button industry of the Mississippi River. *Bulletin of the U.S. Fish Commission* 18: 289–314.

Snyder, F.L., M.B. Hilgendorf, & D.W. Garton. 1990. Zebra mussels in North America: The invasion and its implications. *Ohio Sea Grant College Program OHSU-FS-045.* 4 pp.

Sowerby, G.B. 1839. *A conchological manual.* London: G.B. Sowerby. v + 130 pp., 24 pls.

Sowerby, G.B. 1842. *A conchological manual.* 2nd ed. London: G.B. Sowerby. v + 130 pp., 24 pls.

Sowerby, G.B., II. 1864–1868. Monograph of the genus *Unio.* [*In:*] L. Reeve & G.B. Sowerby (eds.), *Conchologica Iconica* 16: 1–163 + 96 pls.

Sowerby, J. & G.B. Sowerby. 1821–1834. *The genera of Recent and fossil shells.* London. Vol. 1: unpaginated, pls. 1–126 (1821–1825); Vol. 2: unpaginated, pls. 127–262 (1825?–1834).

St. John, M.E. 1973. Shell growth and variation in the *Actinonaias ligamentina* (Lamarck, 1819) complex (Mollusca: Bivalvia: Unionidae). Ph.D. Dissertation, The Ohio State University, Columbus. 129 pp.

Stansbery, D.H. 1966a. Observations on the habitat distribution of the naiad *Cumberlandia monodonta* (Say, 1829). *American Malacological Union, Inc. Annual Reports* 33: 29–30.

Stansbery, D.H. 1966b. Utilization of naiads by prehistoric man in the Ohio Valley. *American Malacological Union Annual Reports for 1966:* 41–43.
Stansbery, D.H. 1970. Eastern freshwater mollusks. Vol. I: The Mississippi and St. Lawrence River systems. American Malacological Union Symposium on Rare and Endangered Mollusks. *Malacologia* 10: 9–22.
Stansbery, D.H. 1972. The mollusk fauna of the North Fork Holston River at Saltville, Virginia. *Bulletin of the American Malacological Union for 1972:* 45–46.
Stansbery, D.H. 1973a. Dams and the extinction of aquatic life. *Garden Club of America Bulletin* 61: 43–46.
Stansbery, D.H. 1973b. A preliminary report on the naiad fauna of the Clinch River in the southern Appalachian Mountains of Virginia and Tennessee (Mollusca: Bivalvia: Unionoida). *Bulletin of the American Malacological Union for 1972:* 20–22.
Stansbery, D.H. 1976. Naiad mollusks. Endangered and threatened plants and animals of Alabama. Pp. 42–52. [*In:*] H.T. Boschung (ed.)., *Endangered and threatened plants and animals of Alabama. Bulletin of the Alabama Museum of Natural History* 2: 1–92.
Stansbery, D.H., K.E. Newman, K.G. Borror, & C.B. Stein. 1985. Literature records of bivalve mollusks of the Muskingum River system, Ohio. *Ohio Biological Survey Technical Report* 8: 1–472.
Stark, J. 1828. *Elements of natural history, adapted to the present state of the science, containing the generic characters of nearly the whole animal kingdom, and the descriptions of the principal species.* 2 vols. Edinburgh: W. Blackwood. 505 pp, 8 pls.
Starrett, W.C. 1971. A survey of the mussels (Unionacea) of the Illinois River: A polluted stream. *Illinois Natural History Survey Bulletin* 30: 267–403.
Staton, S.K., J.L. Metcalfe-Smith, & E.L. West. 2000. Status of the Northern Riffleshell, *Epioblasma torulosa rangiana* (Bivalvia: Unionidae), in Ontario and Canada. *Canadian Field-Naturalist* 114: 224–235.
Steg, M.B. & R.J. Neves. 1997. Fish host identification for Virginia listed unionids in the upper Tennessee River drainage. *Triannual Unionid Report* (13): 34.
Stein, C.B. 1963. The Unionidae (Mollusca: Pelecypoda) of the Olentangy River in central Ohio. Master's Thesis, The Ohio State University, Columbus.
Stein, C.B. 1968. The life history of *Amblema plicata* (Say, 1817) the Three-Ridge naiad (Mollusca: Bivalvia). Unpublished Ph.D. Dissertation, The Ohio State University, Columbus.
Steingraeber, M., M. Hove, M. Bartsch, D. Hornbach, C. Nelson, T. Newton, J. Kalas, A. Kapuscinski, & E. Simonsen. 2004. Two fish identified as hosts for Winged Mapleleaf (*Quadrula fragosa*). *Ellipsaria* 6(1): 7–8.
Sterki, V. 1898. Some observations on the genital organs of Unionidae, with the reference to classification. *The Nautilus* 12: 18–21, 28–32.
Sterki, V. 1907. A preliminary catalogue of the land and fresh water Mollusca of Ohio. *Proceedings of the Ohio State Academy of Science* 4: 367–402.
Sterki, V. 1914. Ohio Mollusca. Additions and corrections. *Ohio Naturalist* 14: 270–272.
Stern, E.M. & D.L. Felder. 1978. Identification of host fishes for four species of freshwater mussels (Bivalvia: Unionidae). *American Midland Naturalist* 100: 233–236.
Stimpson, W. 1851. *Shells of New England; a revision of the synonymy of the testaceous mollusks of New England, with notes on their structure and their geographical and bathymetrical distribution, with figures of new species.* Boston, Massachusetts. 58 pp., 2 pls.

William Swainson

Strayer, D.L. 1999. Effects of alien species on freshwater mollusks in North America. *Journal of the North American Benthological Association* 18: 74–98.
Strayer, D.L. & K.J. Jirka. 1997. The pearly mussels of New York State. *New York State Museum Memoir* 26: 113 pp., 27 pls.
Strecker, J.K. 1931. The distribution of the naiades or pearly fresh-water mussels of Texas. *Baylor University Museum Special Bulletin* 2: 1–71.
Sullivant, J. 1838. *An alphabetical catalogue of the shells, fossils, minerals, and zoophites in the cabinet of Joseph Sullivant, curator of the Philosophical and Historical Society of Ohio.* Columbus. 38 pp.
Suloway, L., J.J. Suloway, & E.E. Herricks. 1981. Changes in the freshwater mussel (Mollusca: Pelecypoda: Unionidae) fauna of the Kaskaskia River, Illinois, with emphasis on the effects of impoundment. *Transactions of the Illinois State Academy of Science* 74: 79–90.
Surber, T. 1912. Identification of the glochidia of freshwater mussels. *Report of the Commissioner of Fisheries for the Fiscal Year 1912 and Special Papers:* 1–10, 3 pls.
Surber, T. 1913. Notes on the natural hosts of fresh-water mussels. *Bulletin of the Bureau of Fisheries* [issued separately as *U.S. Bureau of Fisheries Document* 778] 32(1912): 103–116, 3 pls.
Surber, T. 1915. Identification of the glochidia of fresh-water mussels. *Report of the U.S. Commissioner of Fisheries for 1914,* Appendix 5: 1–9, 1 pl. [issued separately as *U.S. Bureau of Fisheries Document* 813].
Swainson, W. 1820–1833. *Zoological illustrations, or original figures and descriptions of new, rare, or interesting animals, selected chiefly from the classes of ornithology, entomology, and conchology.* London. Ser. 1, Vols. 1–3, pls. 1–182 [1820–1823]; Ser. 2, Vols. 1–3, pls. 1–136 [1829–1833].
Swainson, W. 1840. *A treatise on malacology: Or the natural classification of shells and shell fish.* London: Longman, Orme, Brown, Green, & Longmans, & J. Taylor. viii + 419, 130 figs.
Tankersley, R.A. & R.V. Dimock. 1992. Quantitative analysis of the structure of the marsupial gills of the freshwater mussel *Anodonta cataracta. Biological Bulletin* 182: 145–154.
Tankersley, R.A., J.J. Hart, & M.G. Weiber. 1997. Developmental shifts in feeding biodynamics of juvenile *Utterbackia imbecillis*

Benjamin Tappan

(Mollusca: Bivalvia). Pp. 282–283. [*In:*] K.S. Cummings, A.C. Buchanan, C.A. Mayer, & T.J. Naimo (eds.), *Conservation and management of freshwater mussels II: Initiatives for the future. Proceedings of a UMRCC symposium, St. Louis, MO.* Upper Mississippi River Conservation Committee, Rock Island, IL.

Tappan, B. 1839. Description of some new shells. *American Journal of Science and Arts* 35: 268–270.

Tedla, S. & C.H. Fernando. 1969. Observations on the glochidia of *Lampsilis radiata* (Gmelin) infesting Yellow Perch, *Perca flavescens* (Mitchill) in the Bay of Quinte, Lake Ontario. *Canadian Journal of Zoology* 47: 705–712.

Theler, J.L. 1991. Aboriginal utilization of freshwater mussels at the Aztalan Site, Wisconsin. *Illinois State Museum Scientific Papers* 23: 315–322.

Thiele, J. 1934. *Handbuch der Systematischen Weichtierkunde.* Amsterdam: A. Ashur & Co. v + 779–1154, figs. 784–897.

Thomson, T. 1820. Historical sketch of improvements in physical science during the year 1819. *Annals of Philosophy* 16: 81–130.

Threlfall, W. 1986. Seasonal occurrence of *Anodonta cataracta* Say, 1817, glochidia on Three-Spined Sticklebacks, *Gasterosteus aculeatus* Linnaeus. *The Veliger* 29: 231–234.

Tompa, A.S. 1979. Life cycle completion of the freshwater clam *Lasmigona compressa* (Bivalvia: Unionidae) on an experimental host, *Lebistes reticulatus. The Veliger* 22: 188–190.

Trdan, R.J. 1981. Reproductive biology of *Lampsilis radiata siliquoidea* (Pelecypoda: Unionidae). *American Midland Naturalist* 106: 243–248.

Trdan, R.J. & W.R. Hoeh. 1982. Eurytopic host use by two congeneric species of freshwater mussel (Pelecypoda: Unionidae: *Anodonta*). *American Midland Naturalist* 108: 381–388.

Tucker, J.K. 1994. Colonization of unionid bivalves by the Zebra Mussel, *Dreissena polymorpha,* in Pool 26 of the Mississippi River. *Journal of Freshwater Ecology* 9(2): 129–134.

Tucker, M.E. 1927. Morphology of the glochidium and juvenile of the mussel *Anodonta imbecillis. Transactions of the American Microscopical Society* 46: 286–293.

Tucker, M.E. 1928. Studies on the life cycles of two species of fresh water mussels belonging to the genus *Anodonta. Biological Bulletin (Woods Hole)* 54: 117–127.

Turgeon, D.D., J.F. Quinn, A.E. Bogan, E.V. Coan, F.G. Hochberg, W.G. Lyons, P.M. Mikkelsen, R.J. Neves, C.F.E. Roper, G. Rosenberg, B. Roth, A. Scheltema, F.G. Thompson, M. Vecchione, & J.D. Williams. 1998. *Common and scientific names of aquatic invertebrates from the United States and Canada: Mollusks.* 2nd ed. American Fisheries Society, Special Publication 26: ix–526 pp.

U.S. Fish and Wildlife Service. 1984c. *Orange-footed Pearly Mussel Recovery Plan.* Atlanta, GA. 46 pp.

U.S. Fish and Wildlife Service. 1984b. *White Warty-back Pearly Mussel Recovery Plan.* Atlanta, GA. 43 pp.

U.S. Fish and Wildlife Service. 1984c. *Recovery Plan for the Fat Pocketbook Pearly Mussel* Potamilus (= Proptera) capax *(Green, 1832).* Jackson, MS. 44 pp.

U.S. Fish and Wildlife Service. 1985a. *Recovery Plan for the Tubercled-blossom Pearly Mussel* Epioblasma (Dysnomia) torulosa torulosa (Rafinesque, 1820); *Turgid-blossom Pearly Mussel* Epioblasma (Dysnomia) turgidula (Lea, 1858) *and Yellow-blossom Pearly Mussel* Epioblasma (Dysnomia) florentina florentina. Atlanta, GA. 42 pp.

U.S. Fish and Wildlife Service. 1985b. *Recovery Plan for the Pink Mucket Pearly Mussel* Lampsilis orbiculata *(Hildreth, 1828).* Atlanta, GA. 47 pp.

U.S. Fish and Wildlife Service. 1990a. *White Cat's Paw Pearly Mussel Recovery Plan.* Twin Cities, MN. 42 pp.

U.S. Fish and Wildlife Service. 1990b. *Cracking Pearlymussel* (Hemistena (= Lastena) lata) *Recovery Plan.* Atlanta, GA. 24 pp.

U.S. Fish and Wildlife Service. 1991a. *Fanshell* (Cyprogenia stegaria (= Cyprogenia irrorata)) *Recovery Plan.* Atlanta, GA. 37 pp.

U.S. Fish and Wildlife Service. 1991b. *Ring Pink Mussel Recovery Plan.* Atlanta, GA. 26 pp.

U.S. Fish and Wildlife Service. 1992. *Purple Cat's Paw Pearlymussel Recovery Plan.* Atlanta, GA. 26 pp.

U.S. Fish and Wildlife Service. 1993. *Clubshell* (Pleurobema clava) *and Northern Riffleshell* (Epioblasma torulosa rangiana) *Recovery* Plan. Hadley, MA. 55 pp.

Utterback, W.I. 1915. The naiades of Missouri. *American Midland Naturalist* 4: 41–53, 96–152, 181–204, 244–273.

Utterback, W.I. 1916a. The naiades of Missouri. *American Midland Naturalist* 4: 311–327, 339–354, 387–400, 432–464, pls. 1–28.

Utterback, W.I. 1931. Sex behavior among naiades. *Proceedings of the West Virginia Academy of Science* 5: 43–45.

Valenciennes, A. 1827. Coquilles marines bivalves de l'Amérique Équinoxiale, recueillies pendant le voyage de *MM de Humboldt et Bonpland.* [*In:*] F.H.A. von Humboldt & A.J.A. Bonpland, *Voyage aux régions équinoxiales du Nouveau Continent.* Paris. (2)2: 217–224, pls. 48–50.

Valentine, B.D. & D.H. Stansbery. 1971. An introduction to the naiades of the Lake Texoma region, Oklahoma, with notes on the Red River fauna (Mollusca: Unionidae). *Sterkiana* 42: 1–40.

Valovirta, I. 1990. Conservation of *Margaritifera margaritifera* in Finland. *Council of Europe Environmental Encounters Series* (10): 59–63.

van der Schalie, H. 1938. Contributing factors in the depletion of naiades in eastern United States. *Basteria* 3: 51–57.

van der Schalie, H. 1966. Hermaphroditism among North American freshwater mussels. *Malacologia* 5: 77–78 [abstract].

van der Schalie, H. 1970. Hermaphroditism among North American freshwater mussels. *Malacologia* 10: 93–112.

van der Schalie, H. & A. van der Schalie. 1963. The distribution, ecology and life history of the mussel, *Actinonaias ellipsiformis*

Conrad, in Michigan. *American Malacological Union, Annula Reports for 1963:* 18.

Van Hassel, J.H., J.J. Ney, & D.J. Garling. 1980. Heavy metals in a stream ecosystem at sites near highways. *Transactions of the American Fisheries Society* 109: 636–643.

Vanatta, E.G. 1910. Unionidae from southeastern Arkansas and N.E. Louisiana. *The Nautilus* 23: 102–104.

Vanatta, E.G. 1915. Rafinesque's types of *Unio. Proceedings of the Academy of Natural Sciences of Philadelphia* 67: 549–559.

Vaughn, C.C. and C.C. Hakenkamp. 2001. The functional role of borrowing bivalves in freshwater ecosystems. *Freshwater Biology* 46: 1431–1446.

Vaught, K.C. 1989. *A classification of the living Mollusca.* American Malacologists, Inc., Melbourne, FL. 195 pp.

Vidrine, M.F. 1989. A summary of the mollusk-mite associations of Louisiana and adjacent waters. *Louisiana Environmental Professional* 6: 30–63.

Vidrine, M.F. 1993. *The historical distributions of freshwater mussels in Louisiana.* Gail Q. Vidrine Collectables. Eunice, Louisiana. xii–225, 20 pls.

Villa, A. & G.B. Villa. 1841. *Dispositio systematica cochyliarum terrestrium et fluviatilum quæ adservantur in collectione fratrum Ant. et Jo. Bapt. Villa plurium Academiarum scientiarum sodalium conspectu abnormitatum novarumque specierum descriptionibus adjectis.* Mediolani: Borroni & Scotti. 64 pp.

Walker, B. 1910. Description of a new species of *Truncilla. The Nautilus* 24: 42–44, 1 pl.

Walker, B. 1913. The unione fauna of the Great Lakes. *The Nautilus* 27: 18–23.

Walker, B. 1918. Notes on North American naiades, Vol. 1. *Occasional Papers of the Museum of Zoology, University of Michigan* 49: 1–6.

Waller, D.L. & L.G. Mitchell. 1989. Gill tissue reactions in Walleye *Stizostedion vitreum vitreum* and Common Carp *Cyprinus carpio* to glochidia of the freshwater mussel *Lampsilis radiata siliquoidea. Diseases of Aquatic Organisms* 6: 81–87.

Waller, D.L., L.E. Holland-Bartels, & L.G. Mitchell. 1988. Morphology of glochidia of *Lampsilis higginsi* (Bivalvia: Unionidae) compared with three related species. *American Malacological Bulletin* 6: 39–43.

Waller, D.L., L.E. Holland-Bartels, L.G. Mitchell, & T.W. Kammer. 1985. Artificial infestation of Largemouth Bass and Walleye with glochidia of *Lampsilis ventricosa* (Pelecypoda: Unionidae). *Freshwater Invertebrate Biology* 4: 152–153.

Warren, R.E. 1975. Prehistoric Unionacean (freshwater mussel) utilization at the Widows Creek site (1JA305), north east Alabama. Unpublished Master's Thesis, University of Nebraska, Lincoln. 245 pp.

Warren, R.E. 1983. Late Holocene archeological variation in Ozark Highland caves and rockshelters: Environmental correlates and foraging behavior. Ph.D. Dissertation, University of Missouri. 252 pp.

Warren, R.E. 1991. Freshwater mussels as paleoenvironmental indicators: A quantitative approach to assemblage analysis. *Illinois State Museum Scientific Papers* 23: 23–66.

Warren, R.E. 1995. Prehistoric mussel faunas from the northern Ozark Highland of Missouri: Cultural and geological implications. *Missouri Archeologist* 53: 80–100.

Warren, R.E. In press. Molluscan fauna: Freshwater mussels and marine shell. [*In:*] C.W. Cowan (ed.), Recent research at the Madisonville Village site, Hamilton County, Ohio. *Occasional Papers in Anthropology, Cincinnati Museum of Natural History* (1).

Watters, G.T. 1993. *A guide to the freshwater mussels of Ohio.* Rev. ed. Ohio Department of Natural Resources, Division of Wildlife, Columbus, Ohio. 106 pp.

Watters, G.T. 1994. Form and function of unionoidean shell sculpture and shape (Bivalvia). *American Malacological Bulletin* 11: 1–20.

Watters, G.T. 1995a. New hosts for *Anodontoides ferussacianus* (Lea, 1834). *Triannual Unionid Report* (7): 7.

Watters, G.T. 1995b. Sampling freshwater mussel populations: The bias of muskrat middens. *Walkerana* 7: 63–69.

Watters, G.T. 1996a. New hosts for *Lampsilis cardium. Triannual Unionid Report* (9): 8.

Watters, G.T. 1996b. Hosts for the Northern Riffle Shell (*Epioblasma torulosa rangiana*). *Triannual Unionid Report* (10): 14.

Watters, G.T. 1996c. Small dams as barriers to freshwater mussels (Bivalvia, Unionoida) and their hosts. *Biological Conservation* 75: 79–85.

Watters, G.T. 1996d. ClamNation—The first year. *Triannual Unionid Report* (10): 15.

Watters, G.T. 1997a. Glochidial metamorphosis of the freshwater mussel *Lampsilis cardium* (Bivalvia: Unionidae) on larval Tiger Salamanders, *Ambystoma tigrinum* ssp. (Amphibia: Ambystomidae). *Canadian Journal of Zoology* 75: 505–508.

Watters, G.T. 1997b. Individual based models of mussel fish interactions: A cautionary study. Pp. 45–62. [*In:*] K.S. Cummings, A.C. Buchanan, C.A. Mayer, & T.J. Naimo (eds.), *Conservation and management of freshwater mussels II: Initiatives for the future. Proceedings of a UMRCC symposium, St. Louis, MO.* Upper Mississippi River Conservation Committee, Rock Island, IL.

Watters, G.T. 1988. A survey of the mussels of the St. Joseph River system, with emphasis on the federally endangered White Cat's Paw Pearly Mussel. *Final Report to the Indiana Department of Natural Resources and U.S. Fish & Wildlife Service.* 127 pp.

Watters, G.T. 1999. Morphology of the conglutinate of the Kidneyshell freshwater mussel, *Ptychobranchus fasciolaris. Invertebrate Zoology* 118: 289–295.

Watters, G.T. 2001. The evolution of the Unionacea in North America and its implications for the worldwide fauna. Pp. 281–307. [*In:*] G. Bauer & K. Wächtler (eds.), *Ecology and evolution of the freshwater mussels Unionoida.* Berlin: Springer-Verlag.

Watters, G.T. 2002. The kinetic conglutinate of the Creeper freshwater mussel, *Strophitus undulatus* (Say, 1817). *Journal of Molluscan Studies* 68: 155–158.

Watters, G.T. 2007. A brief look at freshwater mussel (Unionacae) biology. Pp. 51–64. [*In:*] J.L. Farris & J.H. Van Hassel (eds.), *Freshwater bivalve ecotoxicology.* Boca Raton, FL: CRC Press.

Watters, G.T. 2008. The morphology of conglutinates and conglutinate-like structures in North American freshwater mussels: A scanning-electron microscopy survey. *Novapex* 9: 1–20.

Watters, G.T. & K. Kuehnl. 2004. Ohio Pigtoe host suitability trials. *Ellipsaria* 6(2): 10.

Watters, G.T. & S.H. O'Dee. 1996. Shedding of untransformed glochidia by fishes parasitized by *Lampsilis fasciola* Rafinesque, 1820 (Mollusca: Bivalvia: Unionidae): Evidence of acquired immunity in the field? *Journal of Freshwater Ecology* 11: 383–389.

Watters, G.T. & S.H. O'Dee. 1997a. Surrogate hosts: Transformation on exotic and non-piscine hosts. *Triannual Unionid Report* (11): 35.

Watters, G.T. & S.H. O'Dee. 1997b. Overwintering of freshwater mussel glochidia on fish hosts. *Bulletin of the North American Benthological Society* 14: 56–57 [abstract].

Watters, G.T. & S. H. O'Dee. 1997c. Potential hosts for *Villosa iris* (Lea, 1829). *Triannual Unionid Report* (12): 7.

Watters, G.T. & S.H. O'Dee. 1997d. Potential hosts for *Lampsilis radiata luteola* (Lamarck, 1819). *Triannual Unionid Report* (12): 7.

Watters, G.T. & S.H. O'Dee, 1997e. Identification of potential hosts. *Triannual Unionid Report* (13): 38–39.

Watters, G.T. & S. H. O'Dee. 1997f. *Utterbackia imbecillis* goes fishing with nets. *Triannual Unionid Report* (13): 40.

Watters, G.T. & S.H. O'Dee. 1998a. Metamorphosis of freshwater mussel glochidia (Bivalvia: Unionidae) on amphibians and exotic fishes. *American Midland Naturalist* 139: 49–57.

Watters, G.T. & S.H. O'Dee. 1998b. Potential hosts for the Washboard, *Megalonaias nervosa. Triannual Unionid Report* (14): 9.

Watters, G.T. & S.H. O'Dee. 1998c. Glochidial release as a function of water temperature. *Triannual Unionid Report* (14): 40.

Watters, G.T. & S.H. O'Dee. 1999. Glochidia of the freshwater mussel *Lampsilis* overwintering on fish hosts. *Journal of Molluscan Studies, London* 65: 453–459.

Watters, G.T., S.H. O'Dee, & S. Chordas. 1998a. New potential hosts. *Triannual Unionid Report* (15): 27–28.

Watters, G.T., S.H. O'Dee, & S. Chordas. 1998b. Infective and non-infective glochidia in *Lasmigona costata. Triannual Unionid Report* (15): 29.

Watters, G.T., S.H. O'Dee, & S. Chordas. 2001. Patterns of vertical migration in freshwater mussels (Bivalvia: Unionoida). *Journal of Freshwater Ecology* 16: 541–549.

Watters, G.T., T. Gibson, & B. Kelly. 2009. Host identifications or confirmations. *Ellipsaria* 11(1): 19.

Watters, G.T., S.H. O'Dee, S. Chordas, & J. Reiger. 1998c. Potential hosts for *Lampsilis reeviana brevicula, Obliquaria reflexa. Triannual Unionid Report* (16): 21–22.

Watters, G.T., S.W. Chordas, S.H. O'Dee, & J. Reiger. 1998d. Host identification studies for six species of Unionidae. *First Symposium of the Freshwater Mollusk Conservation Society, Chattanooga, TN. Program Guide and Abstracts.*

Watters, G.T., S.H. O'Dee, S. Chordas, & D. Glover. 1999. Seven potential hosts for *Ligumia recta* (Lamarck, 1819). *Triannual Unionid Report* (18): 5.

Watters, G.T., T. Menker, S. Thomas, & K. Kuehnl. 2005. Host identifications or confirmations. *Ellipsaria* 7(2): 11–12.

Weir, G.P. 1977. An ecology of the Unionidae in Otsego Lake with special references to immature stages. *Occasional Paper SUNY Oneonta Biological Field Station at Copperstown, New York* 4: 1–108.

Weiss, J.L. & J.B. Layzer. 1995. Infestations of glochidia on fishes in the Barren River, Kentucky. *American Macalogical Bulletin* 11: 153–159.

West, E.L., J.L. Metcalfe-Smith, & S.K. Staton. 2000. Status of the Rayed Bean, *Villosa fabalis* (Bivalvia: Unionidae), in Ontario and Canada. *Canadian Field-Naturalist* 114: 248–258.

Westerlund, C.A. 1890. Fauna der in der paläarctischen Region (Europa, Kaukasien, Sibirien, Turan, Persien, Kurdistan, Armenien, Mesopotamien, Kleinasien, Syrien, Arabien, Egypten, Tripolis, Tunesien, Algerien und Marocco) lebenden Binnenconchylien, Vol. 7. *Malacozoa Acephala,* Vol. 7. Berlin: Friesländer & Sohn. 319 + 16 + 15 pp.

Wheeler, H.E. 1914. The unione fauna of Cache River, with description of a new *Fusconaia* from Arkansas. *The Nautilus* 28: 73–78, 1 pl.

Wheeler, H.E. 1918. The Mollusca of Clark County, Arkansas. *The Nautilus* 31: 109–125.

William Wood

White, C.A. 1878. Descriptions of new species of invertebrate fossils from the Laramie Group. *Bulletin of the U. S. Geological and Geographical Survey of the Territories* 4: 707–719.

White, L.R., B.A. McPheron, & J.R. Stauffer, Jr. 1996. Molecular genetic identification tools for the unionids of French Creek, Pennsylvania. *Malacologia* 38: 181–202.

Wiles, M. 1975a. The glochidia of certain Unionidae (Mollusca) in Nova Scotia and their fish hosts. *Canadian Journal of Zoology* 53: 33–41.

Wiles, M. 1975b. Parasites of *Fundulus diaphanus* (LeSueur) (Pisces: Cyprinodontidae) in certain Nova Scotian freshwaters. *Canadian Journal of Zoology* 53: 1578–1580.

Williams, J.D., A.E. Bogan, and J.T. Garner. 2008. *Freshwater mussels of Alabama and the Mobile Basin of Georgia, Mississippi, and Tennessee.* Tuscaloosa: University of Alalbama Press. 908 pp.

Williams, J.D., S.L.H. Fuller, & R. Grace. 1992. Effects of impoundments on freshwater mussels (Mollusca: Bivalvia: Unionidae) in the main channel of the Black Warrior and Tombigbee rivers in western Alabama. *Bulletin of the Alabama Museum of Natural History* (13): 1–10.

Williams, J.D., M.L. Warren, K.S. Cummings, J.L. Harris, & R.J. Neves. 1993. Conservation status of freshwater mussels of the United States and Canada. *Fisheries* 18(9): 6–22.

Wilson, C.B. 1916. Copepod parasites of fresh water fishes and their economic relations to mussel glochidia. *Bulletin of the Bureau of Fisheries* 34: 333–374, 15 pls. [issued separately as *U.S. Bureau of Fisheries Document* 824].

Wilson, C.B. & H.W. Clark. 1912. The mussel fauna of the Maumee River. *Report and Special Papers of the U.S. Fish Commission for 1911:* 1–72 [issued separately as *U.S. Bureau of Fisheries Document* 757].

Wilson, K.A. & K. Ronald. 1967. Parasite fauna of the Sea Lamprey (*Petromyzon marinus* von Linné) in the Great Lakes region. *Canadian Journal of Zoology* 45: 1083–1092.

Wood, W. 1828. *Index testaceologicus, or, a catalogue of shells, British and foreign, arranged according to the Linnean system: With Latin and English names, references to authors, and places where found, 2nd ed., corrected and revised.* London. Printed for W. Willam Wood. xxii + 212 + 34 pp., 38 pls.

Wood, W. 1856. [*In:*] S. Hanley (ed.), *Index testaceologicus, an illustrated catalogue of British and foreign shells, containing about 2,800 figures accurately coloured after nature, by W. Wood, a new*

Berlin Hart Wright
(courtesy Jacksonville Shell Club)

and entirely revised edition. London: Willis & Sotheran. xx + 234 pp.

Woody, C.A. & L. Holland-Bartels. 1993. Reproductive characteristics of a population of the Washboard mussel *Megalonaias nervosa* (Rafinesque 1820) in the Upper Mississippi River. *Journal of Freshwater Biology* 8: 57–66.

Wright, B.H. 1888. *Check list of North American Unionidae and other fresh water bivalves.* Portland, Oregon. 8 pp.

Wright, B.H. 1897. Contribution to a knowledge of United States Unionidae. *The Nautilus* 10: 136–139.

Wright, B.H. 1898. New varieties of Unionidae. *The Nautilus* 11: 123–124.

Wright, B.H. 1899a. New southern Unios. (Cont.). *The Nautilus* 13: 31.

Wright, B.H. 1899b. New southern Unios. (Cont.). *The Nautilus* 13: 42–43.

Wright, B.H. 1899c. New southern Unios. (Cont.). *The Nautilus* 13: 69.

Wright, B.H. 1899d. New southern Unios. (Cont.). *The Nautilus* 13: 89–90.

Wright, B.H. & B. Walker. 1902. *Checklist of North American naiades.* Detroit, MI. Printed for the Authors. 19 pp.

Yeager, B.L. 1993. Dams. Pp. 57–113. [*In:*] C.F. Bryan & D. A. Rutherford (eds.), *Impacts on warmwater streams: Guidelines for evaluation.* Southern Division, American Fisheries Society, Little Rock, AR.

Yeager, B.L. & R.J. Neves. 1986a. Reproductive cycle and fish hosts of the Rabbit's Foot mussel, *Quadrula cylindrica strigillata* (Mollusca: Unionidae) in the Upper Tennessee River drainage. *American Midland Naturalist* 116: 329–340.

Yeager, B.L. & R.J. Neves. 1986b. *Quadrula cylindrica.* Pp. 7–11, 21–35. [*In:*] D.M. Hill (ed.), *Activity 3: Identification of fish hosts.* Tennessee Valley Authority Cumberlandian Mollusk Conservation Program, Norris, Tennessee. 57 pp.

Yeager, B.L. & C.F. Saylor. 1995. Fish hosts for four species of freshwater mussels (Pelecypoda: Unionidae) in the upper Tennessee River drainage. *American Midland Naturalist* 133: 1–6.

Yeager, M.M., D.S. Cherry, & R. Neves. 1993. Interstitial feeding behavior of juvenile unionid mussels. *ASB Bulletin* 40: 113 [abstract].

Yokley, P. 1972. Life history of *Pleurobema cordatum* (Rafinesque, 1820) (Bivalvia: Unionacea). *Malacologia* 11: 351–364.

Yokley, P. & C.H. Gooch. 1976. The effect of gravel dredging on reservoir primary production, invertebrate production, and mussel production, July 1, 1974–June 30, 1976. *Tennessee Wildlife Resource Agency, Project No. 2-245-R.* 32 pp.

Young, D. 1911. The implantation of the glochidium on the fish. *University of Missouri Bulletin Science Series* 2: 1–20.

Young, M. & J. Williams. 1984. The reproductive biology of the Freshwater Pearl Mussel *Margaritifera margaritifera* (Linn.) in Scotland. I. Field studies. *Archiv für Hydrobiologie* 99: 405–422.

Young, M., G.J. Purser, & B. al Mousawi. 1987. Infection and successful reinfection of Brown Trout [*Salmo trutta* (L.)] with glochidia of *Margaritifera margaritifera* (L.). *American Malacological Bulletin* 5: 125–128.

Zale, A.V. & R.J. Neves. 1982a. Identification of a fish host for *Alasmidonta minor* (Mollusca: Unionidae). *American Midland Naturalist* 107: 386–388.

Zale, A.V. & R.J. Neves. 1982b. Fish hosts of four species of lampsiline mussels (Mollusca: Unionidae) in Big Moccasin Creek, Virginia. *Canadian Journal of Zoology* 60: 2535–2542.

Zale, A.V. & R.J. Neves. 1982c. Reproductive biology of four freshwater mussel species (Mollusca: Unionidae) in Virginia. *Freshwater Invertebrate Biology* 1: 17–28.

Zeto, M.A. 1980. First record of the freshwater mussel *Lasmigona subviridis* in Kentucky. *Transactions of the Kentucky Academy of Science* 41: 45.

Zhadin, V.I. 1952. Mollyuski presnykh i solonovatykh vod SSSR. *Akademiya Nauk Soyuza Sovetskikh Sotsialisticheskikh Respublik* (46) [English translation 1965, Mollusks of fresh and brackish waters of the U.S.S.R., Israel Program for Scientific Translations, Jerusalem. 368 pp.].

Zimmerman, J.A. 1960. Pleistocene molluscan faunas of the Newell Lake deposit, Logan County, Ohio. *Ohio Journal of Science* 60: 13–39.

Index

Page numbers and entries in **bold/*bold*** refer to primary references, chapter titles, major section titles, and recognized common names.

A

E

Q

R

S

U

W

Y

Z